The Molecular Targets and Therapeutic Uses of Curcumin in Health and Disease

ADVANCES IN EXPERIMENTAL MEDICINE AND BIOLOGY

Recent Volumes in this Series

Volume 587
NEW TRENDS IN CANCER FOR THE 21st CENTURY
Edited by Antonio Llombart-Bosch, Jose López-Guerrero and Vincenzo Felipe

Volume 588
HYPOXIA AND EXERCISE
Edited by Robert C. Roach, Peter D. Wagner, and Peter H. Hackett

Volume 589
NEURAL CREST INDUCTION AND DIFFERENTIATION
Edited by Jean-Pierre Saint-Jeannet

Volume 590
CROSSROADS BETWEEN INNATE AND ADAPTIVE IMMUNITY
Edited by Peter D. Katsikis, Bali Pulendran and Stephen P. Schoenberger

Volume 591
SOMATIC CELL NUCLEAR TRANSFER
Edited by Peter Sutovsky

Volume 592
REGULATORY MECHANISMS OF STRIATED MUSCLE CONTRACTION
Edited by Setsuro Ebashi and Iwao Ohtsuki

Volume 593
MICROARRAY TECHNOLOGY AND CANCER GENE PROFILING
Edited by Simone Mocellin

Volume 594
MOLECULAR ASPECTS OF THE STRESS RESPONSE
Edited by Peter Csermely and Laszlo Vigh

Volume 595
THE MOLECULAR TARGETS AND THERAPEUTIC USES OF CURCUMIN IN HEALTH AND DISEASE
Edited by Bharat B. Aggarwal, Young-Joon Surh and Shishir Shishodia

---

A Continuation Order Plan is available for this series. A continuation order will bring delivery of each new volume immediately upon publication. Volumes are billed only upon actual shipment. For further information please contact the publisher.

Bharat B. Aggarwal
Young-Joon Surh
Shishir Shishodia
Editors

The Molecular Targets and Therapeutic Uses of Curcumin in Health and Disease

With 77 Illustrations

Springer

Editors

Bharat B. Aggarwal, Ph.D.
Ransom Horne, Jr., Professor of Cancer Research
Professor of Cancer Medicine (Biochemistry)
and Chief, Cytokine Research Section
Department of Experimental Therapeutics
The University of Texas M. D. Anderson Cancer Center
1515 Holcombe Boulevard, BOX 143, Houston, TX 77030, USA
aggarwal@mdanderson.org

Young-Joon Surh, Ph.D.
Chief and Professor
National Research Laboratory of Molecular Carcinogenesis
and Chemoprevention
College of Pharmacy
Seoul National University
Shillim-dong, Kwanak-gu, Seoul 151-742, South Korea
surh@plaza.snu.ac.kr

Shishir Shishodia, Ph.D.
Assistant Professor
Department of Biology
Texas Southern University
3100 Cleburne Street, Houston, TX 77004, USA
shishodia@gmail.com

Library of Congress Control Number: 2006938892

ISBN-13:978-0-387-46400-8 e-ISBN-13:978-0-387-46401-5

Printed on acid-free paper.

9 8 7 6 5 4 3 2 1

springer.com

Dedicated to our gurus and parents whose guidance continues to inspire us!

Sarve bhavantu sukhinah sarve santu niramayah
Sarve bhadrani pasyantu ma kascid duhkhabhag bhavet

"May all be happy; may all be healthy;
may all enjoy prosperity; may none suffer."

PREFACE

The subject of this monograph, curcumin, which gives the yellow color to turmeric, best known as *Haridra* in Sanskrit (means dear to Hari or Lord Krishna). Turmeric is known by several synonyms related to its appearance or use, including *Pita* (yellow, leading to the name *Peethamber dhari* for Lord Krishna based on wearing only yellow clothes), *Gauri* (brilliant), *Kanchani* (looks like gold), *Nisha* (beautiful as a full moon night), *Krimighni* (antibacterial and antihelmenthic), *Mahaghni* (antidiabetic), and *Yoshit priya* (gynecological disorders). In Hindi, turmeric is known as *Haldi*, in Japanese as *Ukon* or *Gajyutsu*, and in Korean as *Ulgeum* or *Gangwhang*.

Turmeric is mentioned in the writings of the Italian explorer Marco Polo, who was introduced to it during his voyage to China and India around 1290 AD. Although he gets credit for bringing Far East spices to Europe, turmeric was actually introduced in Europe in the 13th century AD by Arab traders. The Portuguese explorer Vasco de Gama visited the Indian subcontinent during the 15th century and brought turmeric and other spices of the Orient to the West. It was only during the rule of the British in India that turmeric was combined with various other spices and renamed "curry powder," as it is called in the West.

Turmeric became of special importance to man with the discovery that when added to various food preparations, its dried and powdered rhizome preserved their freshness and nutritive value and improved the palatability and presentation of food. The brilliant yellow color of turmeric, which persists even at very high dilutions, found its way to commercial use as a coloring agent for various items, including cotton, silk, paper, wood, foodstuffs, and cosmetics. In Ayurveda (science of long life), turmeric has been used internally as a stomachic, tonic, and blood purifier and topically in the prevention and treatment of skin diseases. Turmeric concoctions have been traditionally used for the treatment of flatulence, dyspepsia, liver disorder, jaundice, urinary tract diseases, cold, chronic otorrhea, parasitic skin infection, bruises, sprain, wound, infected wound, and inflammation.

We are currently living in an era when 80% of the world's population cannot afford modern medicine. Even for those 20% who can, much of modern medicine is ineffective and has numerous side effects. It is a good time to revive the medicinal use of the ancient medicine curcumin. In this volume we bring together the contribution of modern science to one of the most ancient spices known to mankind. Curcumin's beneficial role in health and disease and its molecular targets are the focus of this monograph. This volume is directed at clinicians and scientists working in the areas of experimental and molecular therapeutics, molecular medicine, translational cancer research, Ayurveda, herbal medicine, naturopathy, and biomedical sciences in general and, most importantly, to the end users of curcumin. We hope that this book will "add spice to everybody's life."

We would like to thank all of the contributors for their valuable contributions to this work. We would also like to thank those who have contributed significantly to curcumin research but could not, because of limitations on space, be invited to contribute.

Bharat B. Aggarwal, Ph.D.
Young-Joon Surh, Ph.D.
Shishir Shishodia, Ph.D.

FOREWORD

It is indeed a matter of pride and privilege to write the Foreword; to this scholarly contribution on curcumin—the major constituent of turmeric. The volume has been successful in seamlessly connecting the traditional knowledge available on turmeric to the findings of systematic modern research on turmeric and, based on this effort, building the possibilities of developing novel drugs to treat diverse diseases. Turmeric (*Curcuma longa*)—a widely cultivated tropical plant—has been used since ancient times as a spice, as a beauty care agent, and as a traditional medicine.

The rhizome of turmeric is highly aromatic and antiseptic. The medicinal properties of turmeric have been expounded in Ayurvedic and traditional Chinese medicine (TCM) texts. Turmeric is traditionally known as a stomachic, blood purifier and is useful for the common cold, leprosy, intermittent fevers, afflictions of the liver, indolent ulcer, pyogenic (forming pus) afflictions, wound-healing, and inflammation.

In recent years, the medicinal properties of turmeric have increasingly been recognized. It is being researched systematically even in the Western world. I remember fighting the "turmeric battle" on the wrong patent on the wound-healing properties of turmeric that was given by the US Patent Office almost a decade ago.

As per the US National Library of Medicine, 256 research papers were published last year on curcumin. The researchers have found in curcumin a near-perfect starting material for drug discovery. Thus, a variety of curcumin analogues have been prepared and evaluated biologically. Curcumin exhibits a wide range of activities [e.g., antibacterial, anti-inflammatory, hypolipidemic, hepatoprotective, lipoxygenase (LOX), cyclooxygenase (COX), protease inhibitory effects, in addition to being effective as an active oxygen scavenger and lipid peroxidase (a class of oxidoreductase enzymes) inhibitor]. Curcumin and the curcuminoids also lower cholesterol, reduce platelet aggregation, inhibit the proliferation of cancer cells, and improve digestion by increasing the flow of bile from the gallbladder. The desirable preventive or putative therapeutic properties of curcumin have been considered to be associated with its antioxidant and anti-inflammatory properties.

Curcumin has been found to modulate the activity of several key transcription factors and, in turn, the cellular expression profiles. The effect of curcumin has been examined on most of the targets discovered within the last three decades. Curcumin modulates several different transcription factors, cytokines, growth factors, kinases and other enzymes. The research results have been elaborately covered in this book and explanations provided would add to knowledge pool.

The National Institutes of Health has four clinical trials in progress on curcumin treatment, namely for pancreatic cancer, multiple myeloma, Alzheimer's disease, and colorectal cancer. Curcumin has been found to possess potential chemopreventive activities. It shows cytotoxic potential against tumor cells both *in vitro* and *in vivo*. Thus, curcumin fits well in the effort of chemoprevention by edible phytochemicals, which is now considered to be an inexpensive, readily applicable, and accessible approach to cancer management. The optimization of intervention trials of diet-derived putative chemopreventive agents is currently under development in normal populations as well as in high- risk groups. Curcumin is also a good immunomodulator. These biological activities warrant further studies of curcumin in the treatment and prevention of human neoplasm.

Curcumin has enormous potential as an antiangiogenic drug. It has been elaborately explained in the chapter discussing this. The property has been attributed to curcumin's ability to downregulate certain transcription factors and proangiogenic factors. Curcumin also has the necessary characteristics of a neuroprotective drug. The activity has been proven in a variety of disease models. Thus, it has great potential for the prevention of multiple neurological conditions for which current therapeutics are less than optimal. The chapter entitled "Neuroprotective Effects of Curcumin" embodies the research carried out on the subject and the existing necessity for further efforts. The curcumin-mediated regulation of COX and LOX enzymes for obtaining their beneficial effects in preventing diverse inflammatory diseases has been dwelt upon in another chapter. Interestingly, curcumin has an edge over conventional nonsteroidal anti-inflammatory drugs and selective COX-2 inhibitors. This might pave the way for path-breaking research in the domain.

This volume in fact covers the length and breadth of research undertaken on curcumin and research results thus far obtained. The diversity ranges from molecular targets, cell growth regulation, antioxidant and anti-inflammatory properties, chemosensitivity, radio protection, and radio sensitivity to immunomodulation, anticancer effects, cardioprotective effects, nephroprotective to hepatoprotective effects, protection from acute and chronic lung diseases to pharmacokinetics and pharmacodynamics and clinical studies undertaken with curcumin. The canvas thus covered is indeed brilliant.

As research advances, it poses newer challenges as well. Several questions in the light of the new drug development effort thus remain to be answered in order to put curcumin in a higher orbit. These pertain to the solubility and stability of curcumin, its optimum dose, pharmacokinetics, mechanism of action of curcumin for a given disease, bioavailability profile, and intricacies of prevention and cure of an identified disease. Further research is thus necessary on these aspects. There is also a need to find out whether other components of turmeric than curcumin have beneficial effects, either alone or in combination with curcumin.

I am happy to see that the contributions in this book have proven beyond doubt that curcumin—an ingredient of the traditional Indian spice turmeric—has enormous potential against a variety of malignant and nonmalignant diseases. I am confident that the state-of-the-art on curcumin research so nicely compiled and analyzed throughout this volume would provide an insight and learning not only to professionals in the field but also to budding researchers. I hope that they would be inspired to answer the unanswered questions on curcumin based on new research endeavors. I congratulate the editors of the volume and the contributors of the various chapters for creating this unique and scholarly marvel.

R.A. Mashelkar, FRS
Director General
Council of Scientific & Industrial Research, New Delhi, India
October 19, 2006

CONTENTS

EDITORS

Dr. Bharat B. Aggarwal received his Ph.D. in Biochemistry from the University of California, Berkeley, did his postdoctoral fellowship in endocrinology at the University of California Medical Center San Francisco, and then worked for a biotechnology company (Genentech Inc.) where he discovered two different TNFs, both essential components of the immune system. In 1989, Dr. Aggarwal accepted a position as a Professor and Chief of the Cytokine Research Section at the University of Texas M. D. Anderson Cancer in Houston. He currently holds the Ransom Horne Jr. Professorship in Cancer Research. He has published over 400 original peer-reviewed articles and reviews, edited 9 books, and been granted almost 35 patents. Since 2001, Dr. Aggarwal has been listed as one of the world's most highly cited scientists by the Institute of Scientific Information. Dr. Aggarwal received an Outstanding Scientists Award from Ranbaxy in 2005 and from the American Association of Indians in Cancer Research in 2006.

Dr. Young-Joon Surh is a Professor of Biochemistry and Molecular Toxicology, College of Pharmacy, Seoul National University, South Korea and Chief of National Research Laboratory of Molecular Carcinogenesis and Chemoprevention. Professor Surh earned a Ph.D. degree at the McArdle Laboratory for Cancer Research, University of Wisconsin–Madison and completed postdoctoral training at the Massachusetts Institute of Technology. In 1992, he was appointed as a tenure-track assistant professor at Yale University School of Medicine. Since relocating to Seoul National University in 1996, he has been leading the Chemoprevention Working Group, investigating the molecular mechanisms of cancer prevention by edible phytochemicals, with emphasis on intracellular signaling molecules as prime targets. He is currently a member of the editorial boards of more than 10 international journals, including *Carcinogenesis, Molecular Carcinogenesis, Cancer Letters, Mutation Research, Food and Chemical Toxicology, and Biofactors.* He is also

co-editor of the book *Oxidative Stress, Inflammation and Health* published by CRC Press in 2005. Dr. Surh has published more than 120 papers in peer-reviewed international journals and about 50 invited editorials, reviews, and book chapters. Dr. Surh has recently published a seminal article, entitled "Chemoprevention with Dietary Phytochemicals," in *Nature Reviews Cancer*.

Dr. Shishir Shishodia earned his Ph.D. in Biotechnology from Banaras Hindu University, Varanasi, India and did his postdoctoral fellowship at the University of Texas M. D. Anderson Cancer Center. Before joining M. D. Anderson Cancer Center, he served as an Assistant Professor at Patna University, Patna, India. He is currently an Assistant Professor at the Texas Southern University, Houston. Dr. Shishodia's research interests include cytokine signaling, the role of transcription factors in tumorigenesis, and regulation of transcription by natural products. He has identified several natural compounds that exhibit anticancer properties. He has published over 50 peer-reviewed papers and co-edited the book *Resveratrol in Health and Disease*. Dr. Shishodia is a recipient of the BHU Gold Medal and the Theodore N. Law Odyssey Award for outstanding scientific achievements.

CONTRIBUTORS

Bharat B. Aggarwal
Ransom Horne, Jr., Professor of Cancer Research
Professor of Cancer Medicine (Biochemistry) and
Chief, Cytokine Research Section
Department of Experimental Therapeutics
The University of Texas M.D. Anderson Cancer Center
1515 Holcombe Boulevard
BOX 143
Houston, TX 77030, USA
aggarwal@mdanderson.org

Jack L. Arbiser
Associate Professor
Department of Dermatology, Emory University School of Medicine
Winship Cancer Institute
1639 Pierce Drive WMB 5309
Atlanta, GA 30322, USA
jarbise@emory.edu

Mary Babu
Deputy Director and Head
Bio-Materials Laboratory
CLRI, Chennai-600020, India
marybabu@hotmail.com

Sulochana S. Bhandarkar
Department of Dermatology
Emory University School of Medicine
Winship Cancer Institute
1639 Pierce Drive WMB 5309
Atlanta, GA 30322, USA
ssbhand@emory.edu

John J. Bright
Senior Investigator and Director of Neuroscience Research Laboratory
Methodist Research Institute
Clarian Health, 1800 N Capital Avenue
Noyes Building, Suite E504
Indianapolis, IN 46202, USA
jbright1@clarian.org

Madan M. Chaturvedi
Professor
Department of Zoology
Delhi University
Delhi 110007, India
mchaturvedi@zoology.du.ac.in

Ann-Lii Cheng
Professor
Department of Internal Medicine and Oncology, National Taiwan University Hospital
Cancer Research Center, National Taiwan University College of Medicine
7 Chung-Shan S. Rd., Taipei, Taiwan
andrew@ha.mc.ntu.edu.tw

Kyung-Soo Chun
Laboratory of Molecular Carcinogenesis
National Institute of Environmental Health Sciences
Research Triangle Park, NC 27709, USA
chunkyungsoo@yahoo.co.kr

Gregory M. Cole
Professor of Medicine and Neurology
Greater Los Angeles Veterans Affairs Healthcare System

Geriatric Research, Education, and Clinic Center 11E
16111 Plummer Street
Sepulveda, CA 91343, USA
gmcole@ucla.edu

Scott Dulchavsky
Department of Surgery
Henry Ford Health System
2977 West Grand Boulevard
Detroit, MI 48202, USA
sdulcha1@hfhs.org

Sally A. Frautschy
Associate Professor of Neurology
Department of Medicine and Neurology
University of California, Los Angeles
Greater Los Angeles Healthcare System
(Veteran's Affairs Medical Center)
16111 Plummer St. North Hills, CA 91343, USA
frautsch@ucl.edu

Xiaohua Gao
Department of Surgery
Henry Ford Health System
2977 West Grand Boulevard
Detroit, MI 48202, USA
tgao1@hfhs.org

Subhash C. Gautam
Department of Surgery
Henry Ford Health System
2977 West Grand Boulevard
Detroit, MI 48202, USA
sgautam1@hfhs.org

Andreas J. Gescher
Professor of Biochemical Toxicology
Cancer Biomarkers and Prevention Group
Department of Cancer Studies and Molecular Medicine
University of Leicester
Leicester LE2 7LX, UK
ag15@leicester.ac.uk.

Chandrasekharan Guruvayoorappan
Department of Immunology
Amala Cancer Research Centre
Amala Nagar, Thrissur Kerala 680555, India
guru_appa2003@yahoo.co.in

Chih-Hung Hsu
Department of Oncology
National Taiwan University Hospital
7, Chung-Shan S. Rd., Taipei, Taiwan
chih@ha.mc.ntu.edu.tw

Haruyo Ichikawa
International Research and Educational Institute for Integrated Medical Sciences
Tokyo Women's Medical School
8-1 Kawada-cho, Shinjuku
Tokyo 162-8666, Japan
haruyo.ichikawa@gmail.com

Ganesh C. Jagetia
Professor & Head
Department of Radiobiology
Kasturba Medical College
Manipal—576 104, India
gc.jagetia@gmail.com

Jeena Joseph
Department of Cancer Biology
Rajiv Gandhi Centre for Biotechnology
Thiruvananthapuram-695 014, India
josephjeena@gmail.com

Devarajan Karunagaran
Professor
Department of Biotechnology

Indian Institute of Technology Madras
Chennai - 600 036, India
karuna@iitm.ac.in

Kuzhuvelil B. Hari Kumar
Department of Biochemistry
Amala Cancer Research Centre
Amala Nagar, Thrissur Kerala, India
kbharikumar@gmail.com

Thankayyan R. Santhosh Kumar
Department of Cancer Biology
Rajiv Gandhi Centre for Biotechnology
Thiruvananthapuram-695 014, India
santhaltr@yahoo.com

Girija Kuttan
Professor
Department of Immunology
Amala Cancer Research Centre
Amala Nagar, Thrissur Kerala, India
amalaresearch@rediffmail.com

Ramadasan Kuttan
Research Director
Amala Cancer Research Centre
Amala Nagar, Thrissur
Kerala, India
amalaresearch@rediffmail.com

Pornngarm Limtrakul
Chairperson, Department of Biochemistry
Faculty of Medicine
Chiang Mai University
Chiang Mai, Thailand 50200
plimtrak@mail.med.cmu.ac.th

Jen-Kun Lin
Professor
Institute of Biochemistry and Molecular Biology
College of Medicine
National Taiwan University
No. 1, Section 1, Jen-ai Road
Taipei, Taiwan
jklin@ha.mc.ntu.edu.tw

Dennis C. Liotta
Professor of Organic Chemistry
Department of Chemistry
Emory University
Atlanta, GA, 30322, USA
dliotta@emory.com

Radha K. Maheshwari
Professor of Pathology
Director, Center for Combat Casualty and Life Sustainment Research
Coordinator Indo-US activities
Uniformed Services University of the Health Sciences
Bethesda, MD 20814, USA
rmaheshwari@usuhs.mil

Nikita Malani
Cytokine Research Laboratory
Department of Experimental Therapeutics - Box 143
University of Texas M. D. Anderson Cancer Center
1515 Holcombe Boulevard
Houston, Texas, USA
nmalani@sas.upenn.edu

Venugopal P. Menon
Professor and Chairman
Department of Biochemistry &
Center for Micronutrient Research
Annamalai University
Annamalai Nagar-608 002
Chidambaram, Tamilnadu, India
biocmr@sify.com

Sumitra Miriyala
Department of Medicine
Carolina Cardiovascular Biology Center
University of North Carolina

Chapel Hill, North Carolina-27599, USA
sumishree@hotmail.com

Cara A. Mosley
Department of Chemistry
Emory University
Atlanta, GA, 30322, USA
camosle@emory.edu

Toshihiko Osawa
Professor
Laboratory of Food and Biodynamics
Nagoya University Graduate School of Bioagricultural Sciences
Furo-cho, Chikusa-ku, Nagoya
464-8601, Japan
osawat@agr.nagoya-u.ac.jp

Manikandan Panchatcharam
Department of Medicine
Carolina Cardiovascular Biology Center
University of North Carolina
Chapel Hill, North Carolina 27599, USA
spmani5@gmail.com

Durairaj Punithavathi
Bio-materials Laboratory
CLRI, Adyar, Chennai-600020, India
punithavathid@yahoo.com

Chinthalapally V. Rao
Professor of Medicine
Kerley-Cade Endowed Chair in Cancer Research
Hematology-Oncology Section
Director, Chemoprevention Program
University of Oklahoma Cancer Institute
OUHSC, P.O. Box 26901
975 NE 10th Street, BRC 1203
Oklahoma City, OK 73104, USA
cv-rao@ouhsc.edu

Puvanakrishnan Rengarajulu
Deputy Director and Head
Department of Biotechnology
CLRI, Chennai-600 020
India
puvanakrishnan@yahoo.com

Adluri Ram Sudheer
Department of Biochemistry & Biotechnology
Faculty of Science
Annamalai University
Annamalai Nagar-608002, Tamilnadu, India
biosudheer99@yahoo.co.in

Anuj Sharma
Uniformed Services University of Health Sciences,
Bethesda, MD, USA
Birla Institute of Technology and Science
Pilani 333031, Rajasthan, India
asharma@usuhs.mil

Ricky A. Sharma
Senior Fellow & Honorary Consultant
Radiation Oncology & Biology,
University of Oxford, Churchill Hospital, Oxford OX3 7LJ, UK
ricky.sharma@rob.ox.ac.uk

Shishir Shishodia
Assistant Professor
Department of Biology
Texas Southern University
3100 Cleburne Street
Houston, TX 77004, USA
shishodia@gmail.com;
shishodias@tsu.edu

Tulika Singh
Department of Biology
Texas Southern University

3100 Cleburne Street
Houston, TX 77004, USA
tsshishodia@gmail.com

James P. Snyder
Director of Biostructural Research
Department of Chemistry
Emory University
1515 Dickey Drive
Atlanta, GA 30322, USA
jsnyder@emory.edu

William P. Steward
Professor of Oncology
Cancer Biomarkers and Prevention Group
Department of Cancer Studies and Molecular Medicine
University of Leicester
Leicester Royal Infirmary
Leicester LE1 SWW, UK
wps1@leicester.ac.uk

Chitra Sundaram
Cytokine Research Laboratory
Department of Experimental Therapeutics - Box 143
University of Texas M. D. Anderson Cancer Center
1515 Holcombe Boulevard
Houston, Texas, USA
chitra_1964@yahoo.com

Young-Joon Surh
Chief and Professor
National Research Laboratory of Molecular Carcinogenesis and Chemoprevention
College of Pharmacy
Seoul National University
Shillim-dong, Kwanak-gu
Seoul 151-742, South Korea
surh@plaza.snu.ac.kr

Bruce Teter
Department of Medicine and Neurology
University of California, Los Angeles
Greater Los Angeles Healthcare System
(Veteran's Affairs Medical Center)
16111 Plummer St. North Hills, CA 91343, USA
bteter@ucla.edu

Rajesh L. Thangapazham
Uniformed Services University of Health Sciences
Bethesda, MD, USA
Birla Institute of Technology and Science
Pilani 333031, Rajasthan, India
rlthangapazham@usuhs.mil

Narayanan Venkatesan
Associate Professor
Faculte de Medecine
UMR-7561 CNRS UHP; B.P. 184
54505 Vandoeuvre lès Nancy
France
vnar12@yahoo.com

CURCUMIN: THE INDIAN SOLID GOLD

Bharat B. Aggarwal, Chitra Sundaram, Nikita Malani, and Haruyo Ichikawa

Abstract: Turmeric, derived from the plant *Curcuma longa*, is a gold-colored spice commonly used in the Indian subcontinent, not only for health care but also for the preservation of food and as a yellow dye for textiles. Curcumin, which gives the yellow color to turmeric, was first isolated almost two centuries ago, and its structure as diferuloylmethane was determined in 1910. Since the time of Ayurveda (1900 BC) numerous therapeutic activities have been assigned to turmeric for a wide variety of diseases and conditions, including those of the skin, pulmonary, and gastrointestinal systems, aches, pains, wounds, sprains, and liver disorders. Extensive research within the last half century has proven that most of these activities, once associated with turmeric, are due to curcumin. Curcumin has been shown to exhibit antioxidant, anti-inflammatory, antiviral, antibacterial, antifungal, and anticancer activities and thus has a potential against various malignant diseases, diabetes, allergies, arthritis, Alzheimer's disease, and other chronic illnesses. These effects are mediated through the regulation of various transcription factors, growth factors, inflammatory cytokines, protein kinases, and other enzymes. Curcumin exhibits activities similar to recently discovered tumor necrosis factor blockers (e.g., HUMIRA, REMICADE, and ENBREL), a vascular endothelial cell growth factor blocker (e.g., AVASTIN), human epidermal growth factor receptor blockers (e.g., ERBITUX, ERLOTINIB, and GEFTINIB), and a HER2 blocker (e.g., HERCEPTIN). Considering the recent scientific bandwagon that multitargeted therapy is better than monotargeted therapy for most diseases, curcumin can be considered an ideal *"Spice for Life"*.

1. INTRODUCTION

The questions of whether medicines discovered today are safer, more efficacious, and more affordable than generic medicines (whose patents have expired) or medicines that are centuries old could be answered "no" for most of the modern medicines. If so, then it is logical to revisit and revive these age-old medicines for the welfare of mankind. Curcumin is one such medicine. Its history goes back over 5000 years, to the heyday of Ayurveda (which means the science of long life). Turmeric derived from the rhizome of the plant *Curcuma longa* has been used by the people of the Indian subcontinent for centuries with no known side effects, not only as a component of food but also to treat a wide variety of ailments.

Turmeric is a spice of golden color that is used in cooking in the Indian subcontinent. Because of its color and taste, turmeric was named "Indian saffron" in Europe. Today, India is the primary exporter of turmeric (known as haldi in India). Although its ability to preserve food through its antioxidant mechanism, to give color to food, and to add taste to the food is well known, its health-promoting effects are less well recognized or appreciated. It was once considered a cure for jaundice, an appetite suppressant, and a digestive. In Indian and Chinese medicines, turmeric was used as an anti-inflammatory agents to treat gas, colic, toothaches, chest pains, and menstrual difficulties. This spice was also used to help with stomach and liver problems, to heal wounds and lighten scars, and as a cosmetic.

Turmeric was mentioned in the writings of Marco Polo concerning his 1280 journey to China and India and it was first introduced to Europe in the 13th century by Arab traders. Although Vasco de Gama (a Portugeese sailor) during 15th century, after his visit to India, truly introduced spices to the West, it was during the rule of British in India that turmeric was combined with various other spices and renamed "curry powder," as it is called in the West. What is there in turmeric that has therapeutic potential, how does this substance mediates its effects, with what types of receptor does it interact, and for what type of diseases is it effective? All of these questions will be addressed in this review.

2. COMPOSITION OF TURMERIC

Turmeric contains a wide variety of phytochemicals, including curcumin, demethoxycurcumin, bisdemethoxycurcumin, zingiberene, curcumenol, curcumol, eugenol, tetrahydrocurcumin, triethylcurcumin, turmerin, turmerones, and turmeronols.[1] Curcumin, demethoxycurcumin, and bisdemethoxycurcumin have also been isolated from *Curcuma mangga*,[2] *Curcuma zedoaria*,[3] *Costus speciosus*,[4] *Curcuma xanthorrhiza*,[4] *Curcuma aromatica*,[5] *Curcuma phaeocaulis*,[5] *Etlingera elatior*,[6] and *Zingiber cassumunar*[7] (Figure 1; see Table 1). Curcumin is the phytochemical that gives a yellow color to turmeric and is now recognized as being responsible for most of the therapeutic effects. It is estimated that 2–5% of turmeric is curcumin. Curcumin was first isolated from turmeric in 1815, and the structure was delineated in 1910 as diferuloylmethane. Most currently available preparations of curcumin contain approximately 77% diferuloylmethane, 18% demethoxycurcumin, and 5% bisdemethoxycurcumin. Curcumin is hydrophobic in nature and frequently soluble in dimethylsulfoxide, acetone, ethanol, and oils. It has an absorption maxima around 420 nm. When exposed to acidic conditions, the color of turmeric/curcumin turns from yellow to deep red, the form in which it is used routinely for various religious ceremonies.

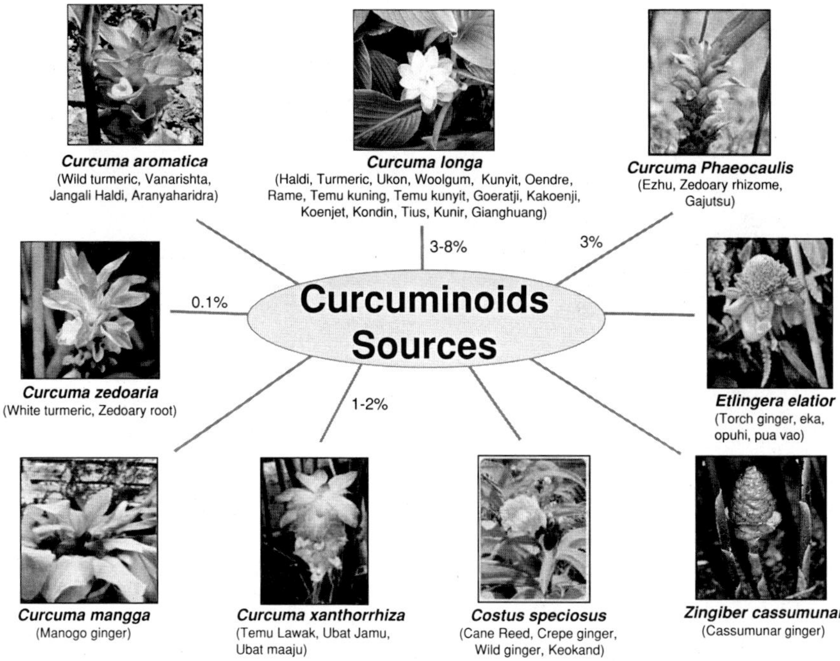

Figure 1. Sources of curcuminoids. (See also Plate 1 in the Color Plate Section.)

3. CURCUMIN ANALOGUES

As indicated earlier, turmeric contains three different analogues of curcumin (i.e., diferuloylmethane, also called curcumin, demethoxycurcumin, and bisdemethoxycurcumin (Figure 2). Whether all three analogues exhibit equal activity is not clear. Although in most systems curcumin was found to be most potent,[8,9] in some systems bisdemethoxycurcumin was found to exhibit higher activity.[3,10] There are also suggestions that the mixture of all three is more potent than either one alone.[11,12]

When administered orally, curcumin is metabolized into curcumin glucuronide and curcumin sulfonate.[13] However, when administered systemically or intraperitoneally, it is metabolized into tetrahydrocurcumin, hexahydrocurcumin, and hexahydrocurcuminol. Tetrahydrocurcumin has been shown to be active in some systems[14–18] and not in others.[13,19] Whether other metabolites of curcumin exhibit biological activity is not known.

4. USES OF CURCUMIN

The use of turmeric for health purposes is nothing new. As a folklore medicine, its use has been documented in both Indian and Chinese cultures. The long list of uses

Table 1. List of various species of curcuma.

C. aeruginosa	*C. coriacea*	*C. meraukensis*	*C. rubricaulis*
C. albicoma	*C. decipiens*	*C. montana*	*C. rubrobracteata*
C. albiflora	*C. domestica*	*C. musacea*	*C. sessilis*
C. alismatifolia	*C. ecalcarata*	*C. mutabilis*	*C. sichuanensis*
C. amada	*C. ecomata*	*C.neilgherrensis*	*C. singularis*
C. amarissima	*C. elata*	*C. nilamburensis*	*C. soloensis*
C. americana	*C. erubescens*	*C. ochrorhiza*	*C. sparganifolia*
C. angustifolia	*C. euchroma*	*C. officinalis*	*C. speciosa*
C. aromatica*	*C. exigua*	*C. oligantha*	*C. spicata*
C. attenuata	*C. ferruginea*	*C. ornata*	*C. stenochila*
C. aurantiaca	*C. flaviflora*	*C. pallida*	*C. strobilifera*
C. australasica	*C. glans*	*C. parviflora*	*C. sulcata*
C. bakeriana	*C. glaucophylla*	*C. parvula*	*C. sumatrana*
C. bicolor	*C. gracillima*	*C. peethapushpa*	*C. sylvatica*
C. brog	*C. grahamiana*	*C. petiolata*	*C. sylvestris*
C. burttii	*C. grandiflora*	***C. phaeocaulis****	*C. thalakaveriensis*
C. caesia	*C. haritha*	*C. pierreana*	*C. thorelii*
C. cannanorensis	*C. harmandii*	*C. plicata*	*C. trichosantha*
C. caulina	*C. heyneana*	*C. porphyrotaenia*	*C. vamana*
C. careyana	*C. inodora*	*C. prakasha*	*C. vellanikkarensis*
C. ceratotheca	*C. latiflora*	*C. pseudomontana*	*C. viridiflora*
C. chuanezhu	*C. latifolia*	*C. purpurascens*	*C. wenchowensis*
C. chuanhuangjiang	*C. leucorhiza*	*C. purpurea*	*C. wenyujin*
C. chuanyujin	*C. leucorrhiza*	*C. raktakanta*	***C. xanthorrhiza****
C. cochinchinensis	*C. loerzingii*	*C. ranadei*	*C. yunnanensis*
C. codonantha	***C. longa****	*C. reclinata*	*C. zanthorrhiza*
C. coerulea	*C. longiflora*	*C. rhabdota*	***C. zedoaria****
C. colorata	*C. longispica*	*C. rhomba*	*C. zerumbet*
C. comosa*	*C. lutea*	*C. roscoeana*	
C. cordata	*C. malabarica*	*C. rotunda*	
C. cordifolia	***C. mangga****	*C. rubescens*	

Note: Curcuma is indicated by C.
*Curcuminoids have been isolated from the plant indicated in bold.
Source: Modified from http://en.wikipedia.org/wiki/Curcuma.

include antiseptic, analgesic, anti-inflammatory, antioxidant, antimalarial, insect-repellant, and other activities associated to turmeric.[4,20–27] (Figure 3). Perhaps one of the most often prescribed uses is for wound-healing.[28] This activity is well known to people from the Indian subcontinent. Modern research has provided considerable evidence, and the mechanism by which turmeric/curcumin could accelerate wound-healing has been described.[29–36]

It is now well recognized that most chronic diseases are the result of disregulated inflammation,[37,38] Turmeric has been traditionally described as an anti-inflammatory agent. Recent scientific evidence has indeed demonstrated that turmeric, and curcumin in particular, exhibits potent anti-inflammatory activities as determined by a wide variety of systems.[39–49] Therefore, it is not too surprising that turmeric displays activities against a variety of diseases. Because curcumin also exhibits potent antioxidant activity, whether the anti-inflammatory activity of curcumin is mediated through its antioxidant mechanism is not clear. Since most

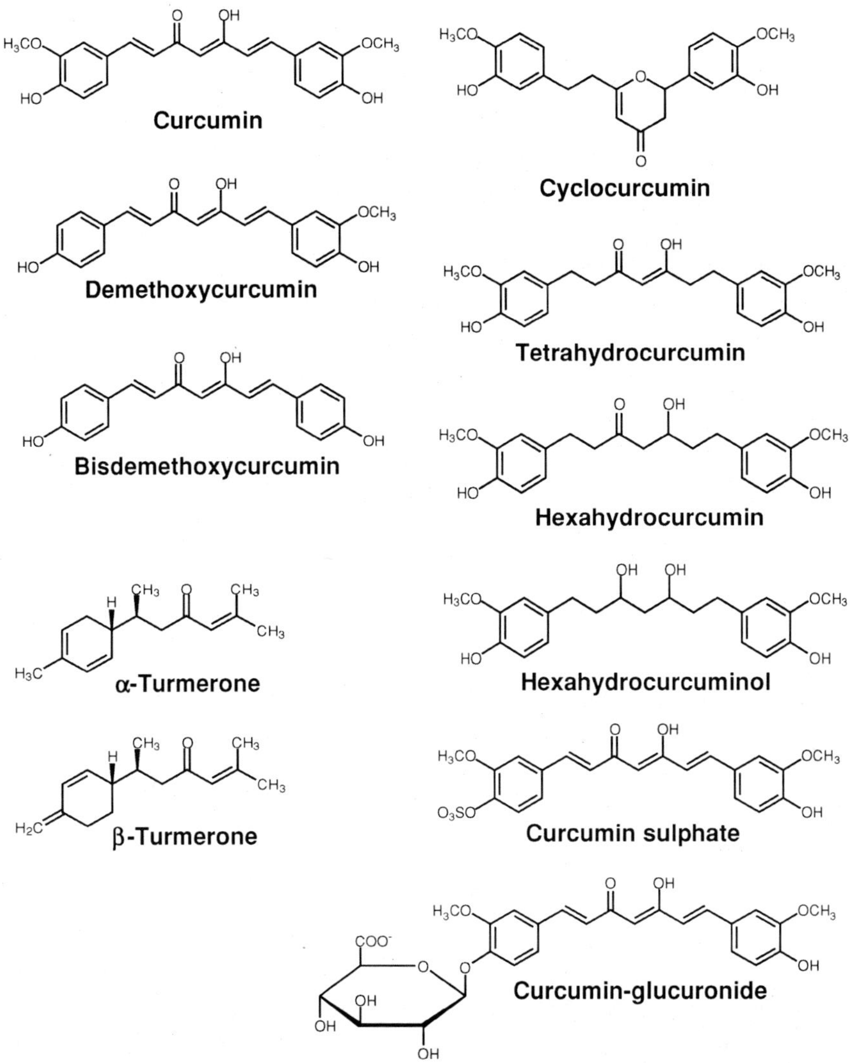

Figure 2. Chemical structures of curcumin and its analogues.

well-characterized antioxidants do not exhibit anti-inflammatory activity, it is unlikely that the anti-inflammatory activity of curcumin is due to its antioxidant activity.

5. MOLECULAR TARGETS OF TURMERIC/CURCUMIN

Most molecular targets established in modern biology were discovered within the last three decades. The effect of curcumin on most of these targets has

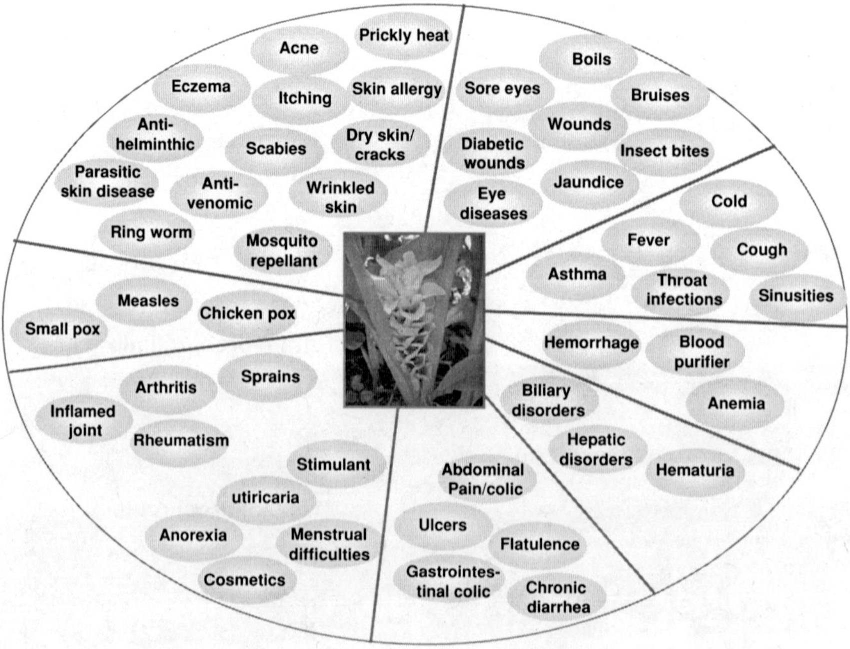

Figure 3. Traditional uses of curcmin. (See also Plate 2 in the Color Plate Section.)

been examined[10,12,45,50–201] (Figure 4). The results have revealed that curcumin can modulate several different transcription factors,[50–96,113,114] cytokines,[45,97–112] growth factors,[202–215] kinases,[115–128] and other enzymes.[91,129–159] Although most diseases are caused by dysregulated inflammation, to find a safe and efficacious anti-inflammatory agent is a real challenge in modern medicine. Steroids are perhaps the best known anti-inflammatory agents. However, there are numerous side effects associated with them. In addition to steroids, numerous nonsteroidal antiinflamatory drugs (NSAIDs) have been discovered within the last century, and these include salicylates, ibuprofen, sulindac, phenylbutazone, naproxen, diclofenac, indomethacin, and coxibs.[216] Experience over the years has indicated that most of these NSAIDs are associated with a constellation of side effects. Perhaps the best example is the cardiovascular system-related side effects recently identified with most coxibs.[217–219] Although the intake of such anti-inflammatory agents can be justified for chronic conditions, they are not appropriate as chemopreventive agents under normal conditions, because that purpose requires long periods of time. Thus, there is a great need for safer and efficacious anti-inflammatory agents.

Numerous lines of evidence suggest that curcumin is a potent anti-inflammatory agent (see Figure 5). First, curcumin suppresses the activation of the transcription factor NF–κB, which regulates the expression of pro-inflammatory gene products.[50–81] Second, curcumin downregulates the expression of COX-2, an

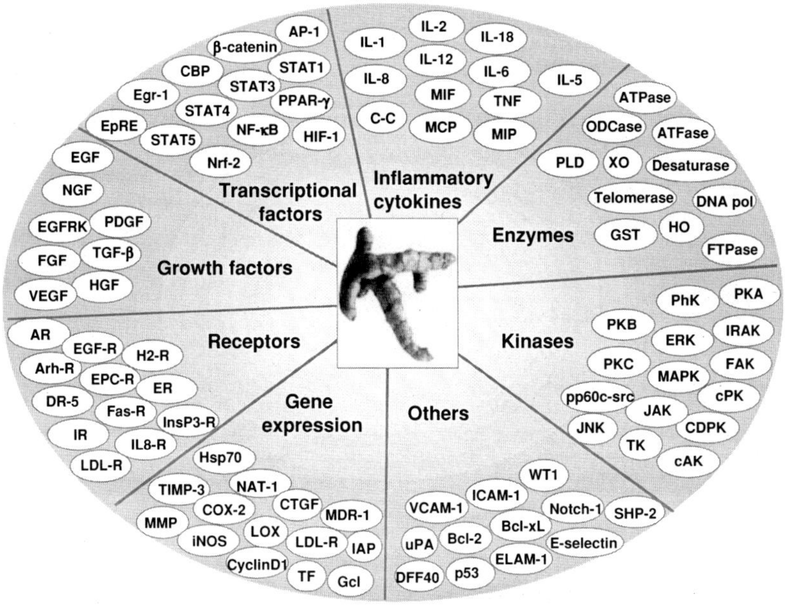

Figure 4. Molecular targets of curcumin. Abbreviations used: NF-κB, nuclear factor-κB; AP-1, activating protein-1; STAT, signal transducers and activators of transcription; Nrf-2, nuclear factor erythroid 2-related factor; Egr-1, early growth response gene-1; PPARγ, peroxisome preoliferator-activated receptor-γ; CBP, CREB-binding protein; EpRE, electrophile response element; CTGF, connective tissue growth factor; EGF, epidermal growth factor; EGFRK, EGF receptor-kinase; FGF, fibroblast growth factor; HGF, hepatocyte growth factor; NGF, nerve growth factor; PDGF, platelet-derived growth factor; TGF-β1, transforming growth factor-β1; VEGF, vascular endothelial growth factor; AR, androgen receptor; Arh-R, aryl hydrocarbon receptor; DR-5, death receptor-5; EGF-R, EGF-receptor; EPC-R, endothelial protein C-receptor; ER-α, estrogen receptor-α; Fas-R, Fas receptor; H2-R, histamine (2)-receptor; InsP3-R, inositol 1,4,5-triphosphate receptor; IR, integrin receptor; IL-8-R, interleukin-8-receptor; LDL-R, low-density lipoprotein-receptor; MMP, matrix metalloproteinase; TIMP, tissue inhibitor of metalloproteinase-3; iNOS, inducible nitric oxide synthase; COX-2, cyclooxygenase-2; LOX, lipoxygenase; Gcl, glutamate-cysteine ligase; NAT, arylamine *N*-acetyltransferases; IAP, inhibitory apoptosis protein; HSP-70, heat shock protein 70; MDR, multidrug resistance; TNF-α, tumor necrosis factor-α; IL, interleukin; MCP, monocyte chemoattractant protein; MIF, migration inhibition protein; MIP, macrophage inflammatory protein; cAK, autophosphorylation-activated protein kinase; CDPK, Ca^{2+}-dependent protein kinase; cPK, protamine kinase; ERK, extracellular receptor kinase; FAK, focal adhesion kinase; IARK, IL-1 receptor-associated kinase; JAK, janus kinase; JNK, c-jun N-terminal kinase; MAPK, mitogen-activated protein kinase; PhK, phosphorylase kinase; PKA, protein kinase A; PKB, protein kinase B; PKC, protein kinase C; pp60c-src, a nonreceptor protein tyrosine kinase c-Src, cellular src kinase; TK, protein tyrosine kinase; FPTase, farnesyl protein transferase; GST, gluthathione-*S*-transferase; HO, hemeoxygenase; ICAM-1, intracellular adhesion molecule-1; VCAM-1, vascular cell adhesion molecule-1; ELAM-1, endothelial leukocyte adhesion molecule-1; Bcl-2, B-cell lymphoma protein 2; SHP-2, Src homology 2 domain-containing tyrosine phosphatase 2, uPA, urokinase-type plasminogen activator, DFF40; DNA fragmentation factor, 40-kd subunit. (See also Plate 3 in the Color Plate Section.)

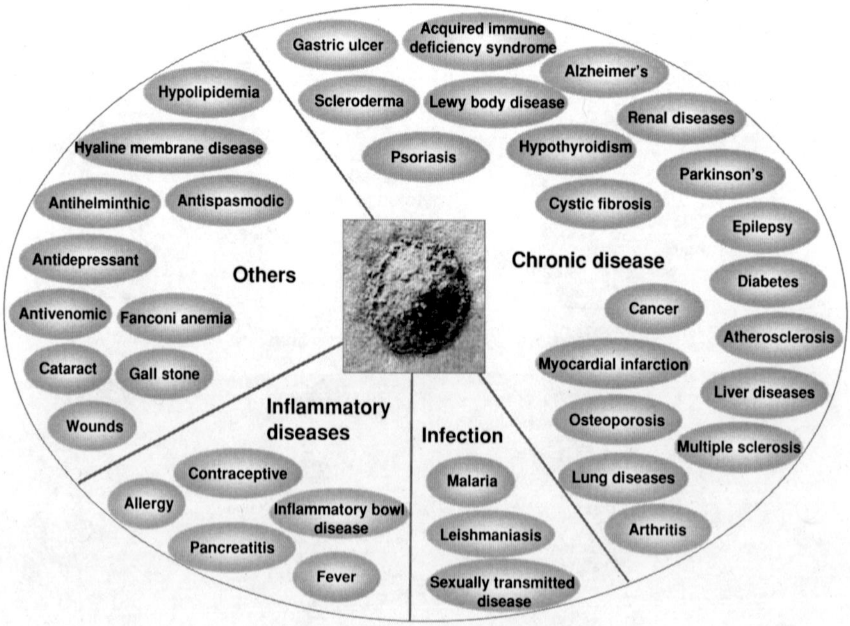

Figure 5. Potential uses of curcumin based on modern technology. (See also Plate 4 in the Color Plate Section.)

enzyme linked with most types of inflammations.[75,177–181,183] Third, curcumin inhibits the expression of another pro-inflammatory enzyme, 5-LOX.[177,182–184] Additionally, curcumin has been shown to bind to the active site of 5-LOX and inhibit its activity[183] Fourth, curcumin downregulates the expression of various cell surface adhesion molecules that have been linked with inflammation.[220–222] Fifth, curcumin downregulates the expression of various inflammatory cytokines, including TNF, IL-1, IL-6, IL-8, and chemokines.[45,97–112] Sixth, curcumin has been shown to inhibit the action of TNF, one of the most pro-inflammatory cytokines.[97–100] Seventh, curcumin is a potent antioxidant, which might contribute to its anti-inflammatory action.[16,19,31,159,223–279] All of this recent evidence confirms the anti-inflammatory action of curcumin, known for thousands of years. Its pharmacological safety combined with its anti-inflammatory action, makes it an ideal agent to explore for preventive and therapeutic situations.

Whereas pro-oxidants are considerd mediators of numerous diseases, antioxidants are generally believed to delay or halt the disease. However, this paradigm is not always valid, as most cytokines mediate their effects through pro-oxidant mechanisms. Reactive oxygen species (ROS) also play an important role in cell-mediated cytotoxicity (CMC) of the immune system. Numerous reports indicate that curcumin could mediate both pro-oxidant and antioxidant roles. First, curcumin could induce the expression of ROS,[8,280–282] which plays an important role in the antiproliferative effects of this molecule.[283] Second, curcumin binds

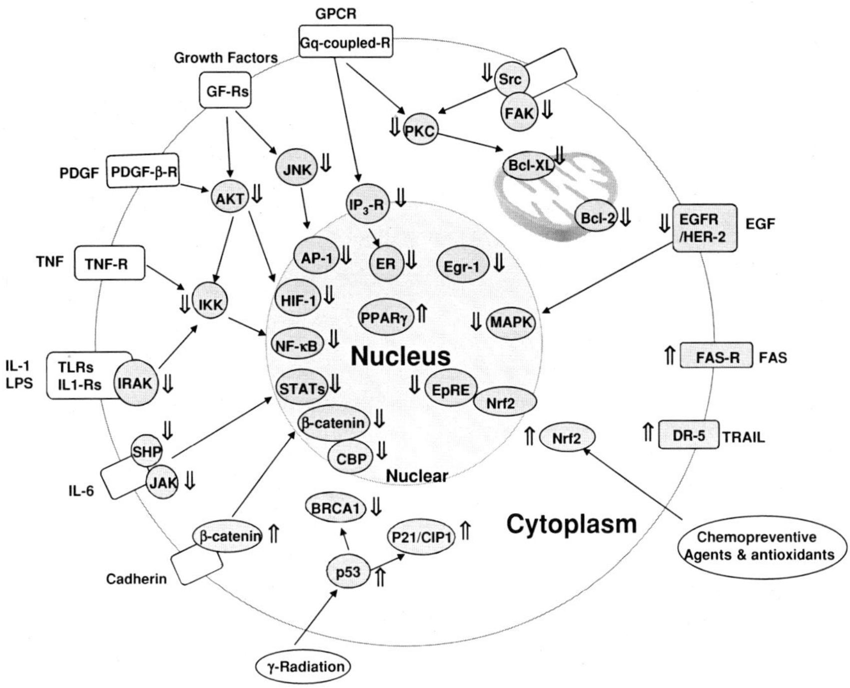

Figure 6. Signaling pathway modulated by curcumin. Intermediates upregulated by curcumin are indicated as ⇑ and those downregulated by curcumin are indicated as ⇓.

thioredoxin reductase (TR) and converts this enzyme to NADPH oxidase, thus leading to the production of ROS.[284] Because TR is overexpressed in tumor cells,[285–287] curcumin kills tumor cells through this mechanism. Third, curcumin suppresses lipid peroxidation.[224,226–228,232,234,238,252,256,264,265,268,288,289] Fourth, curcumin increases the expression of intracellular glutathione.[139,140,142,143,146,290–294] Fifth, curcumin could also play an antioxidant role through its ability to bind iron.[229] All of these reports combined suggest the ability of curcumin to modulate the redox status of the cells. That curcumin can modulate the cellular action of various growth factors and cytokines has also been demonstrated (Figure 6). First, curcumin has been shown to downregulate the effect of epidermal growth factor (EGF) through downregulation of expression and activity of EGF receptors (EGFR).[203,210–212] Second, curcumin has been shown to downregulate the activity of human EGFR-2 (called HER2/neu),[127] a growth factor receptor closely linked with cancer of the breast, lung, kidney, and prostate. Third, curcumin suppresses the action of interleukin (IL)-6 through the downregulation of STAT3 activation.[296] Fourth, curcumin modulates the action of TNF, a growth factor for tumor cells.[297] Fifth, curcumin negatively regulates the action of IL-2,[298] a growth factor for T cells. Thus, curcumin can affect the action of a wide variety of growth factors.[202–215]

Angiogenesis is a process of vascularization of the tissue, which is critical for the growth of solid tumors. Numerous molecules have been linked with angiogenesis. These include vascular endothelial growth factor (VEGF), COX-2, fibroblast growth factor (FGF), and TNF. Evidence suggests that curcumin could suppress angiogenesis.[113,205,208,299–303] Curcumin includes its ability to downregulate the expression of VEGF.[208] Likewise, it downregulates FGF-mediated angiogenesis.[205] Curcumin was found to negatively regulate the expression of COX-2[74,177–181] and suppresses both the expression and action of TNF.[97–100]

6. CURCUMIN RECEPTORS

Receptors are cellular proteins to which a molecule binds, leading to secondary cellular responses. Whether there are any authentic receptors for curcumin is not known. However, numerous molecules to which curcumin binds have been identified. These include serum albumin,[304,305,306] 5-LOX,[183,307] xanthine oxidase,[159] thioredoxin reductase,[284] iron,[295] COX-2,[308] IKK,[309] *p*-glycoprotein,[310,311] GST,[291] PKA,[115] PKC.[115] cPK,[115] PhK,[115] autophosphorylation-activated protein kinase,[115] pp60c-src tyrosine kinase,[115] Ca^{2+}-dependent protein kinase (CDPK),[116] Ca^{2+}-ATPase of sarcoplasmic reticulum,[131] aryl hydrocarbon receptor,[186] rat river cytochrome p450s,[291] Topo II isomerase,[312] inositol 1,4,5-triphosphate receptor,[313] and glutathione.[143]

7. DISEASE TARGETS OF CURCUMIN

Extensive research within the last half a decade has revealed that curcumin has potential against a wide variety of diseases, both malignant and nonmalignant (see Figure 5). The potential of curcumin, however, has not been systematically examined through the modern multicenter, randomized, double-blind, placebo-controlled clinical trials.[314–335] Its potential in humans is indicated either through preclinical studies, some pilot studies in humans, anecdotal studies in patients, or epidemiological studies. Curcumin has been shown to exhibit activity against numerous inflammatory diseases, including pancreatitis,[100,214,261,336,337] arthritis,[105,338–341] inflammatory bowel disease (IBD),[332] colitis,[342–344] gastritis,[345,346] allergy,[99,347,348] and fever,[349,350] possibly through the downregulation of inflammatory markers, as indicated earlier. The effect of curcumin against various autoimmune diseases has also been demonstrated; they include scleroderma,[351] psoriasis,[352]multiple sclerosis,[111,353] and diabetes.[354–362] Again, these effects of curcumin are through the regulation of pro-inflammatory signaling.

Although once thought to be distinct, the molecular targets for both the prevention and therapy of cancer are now considered the same,[363,364]. Numerous lines of evidence suggest the potential of curcumin against various types of cancer[11,56,76,83,95,145,153,155,273,283,298,309,365–462] (see Table 2). First,

Table 2. Chemopreventive and anticancer effects of curcumin.

Skin
External cancerous lesion[405]
Human basal cell carcinoma[469]
Human melanoma[412–414]
Human epidermal carcinoma[415]
Prevention from 7,12-dimethylbenz[a]anthracene[11,406]
Prevention from azoxymethanol[407]
Prevention from benz[a]pyrene and 12-O-tetradecanoylphorbol-13-acetate[408]
Prevention from 12-O-tetradecanoylphorbol-13-acetate[153,155,409]
Prevention from 12-O-tetradecanoylphorbol-13-acetate- and 7,12-dimethylbenz[a]anthracene[410]

Oral
Prevention from methyl-(acetoxymethyl)-nitrosamine[416]
Prevention from 4-nitroquinoline 1-oxide[417]
Prevention from 7,12-dimethylbenz[a]anthracene[418–420]

Esophageal
Prevention from N-nitrosomethylbenzylamine[421]

Forestomach
Prevention from benzo[a]pyrene[406,422,423]
Prevention from N-methyl-N'-nitro-N-nitrosoguanidine[424]

Intestine
Prevention from Min/+ mouse (a model of familial adenomatous polyposis)[425,426]

Colon
Colon adeno carcinoma[95,435–440]
Prevention from azoxymethane[427–433]
Prevention from 1,2-dimethylhydrazine dihydrochrolide[434]

Mammary gland
Prevention from 7,12-dimethylbenz[a]anthracene[11,427,441–443]
Prevention from diethylstilbestrol[444]
Prevention from radiation[365,455]

Liver
Human hepatoblastoma[371,462]
Prevention from diethylnitrosamine[366,367,369]
Prevention from N-nitrosodiethylamine and phenobarbital[370]

Prostate
Prevention from 3,2'-dimethyl-4-aminobiphenol (DMAB) and 2-amino-1-methylimidazo[4,5-b]pyridine (PhIP)[372]

Blood and Bone Marrow
Human leukemia[145,273,373–379]
T-lymphocyte[298,380,381]
Rat thymocytes[382]
Rat histhymocytoma[283]
B-cell lymphoma[56,383,384]
B-cell non-Hodgkin's lymphoma[385,386]
Burkitt's lymphoma[387]
Human multiple myeloma[83,309,388]
Primary effusion lymphoma[389]

Brain
Neuroblastoma[390,391]
Ehrlich's ascites carcinoma[456,480]
Astrocytoma[393]

Breast
Breast carcinoma [394–399]

Gatrointestinal
Gastric signet ring carcinoma[400]

Head and Neck
Head and neck squamous cell carcinoma[76,200,401]

Lung
Human lung[402,447]

Pancreas
Pancreatic carcinoma[403]

Ovarian
Human ovarian[404]

curcumin has been shown to suppress the proliferation of a wide variety of tumor cells through the downregulation of antiapoptotic gene products, activation of caspases, and induction of tumor suppressor genes such as *p53*.[95,145,283,298,313,351,373–384,389–393,396,397,399–403,411,412,415,435–440,463–499]

Second, curcumin has also been to shown to suppress the invasion of tumors through the downregulation of matrix metalloproteinases (MMPs) and cell surface adhesion molecules[134,208,220,301,302,340,346,500–507] Third, curcumin suppresses the angiogenesis of tumors through the suppression of angiogeneic cytokines.[508–512] Fourth, the anti-inflammatory effects of curcumin contribute to its antitumor activity as well.[39–49]

Curcumin has also been shown to play a role in diabetes mellitus type II, in which the patient develops a resistance to insulin.[354,356,359–361,513] Both NF–κB and TNF have been linked with the induction of resistance to insulin. Because curcumin can downregulate the activation of NF–κB and downregulate TNF expression and TNF signaling,[97–100] it can be exploited in diabetic patients. Several animal studies have demonstrated that curcumin can overcome insulin resistance.[514,515]

That curcumin prevents myocardial infarction and other cardiovascular diseases has also been demonstrated.[202,516–524] The effects of curcumin in cardiovascular diseases are linked to its ability to (1) inhibit platelet aggregation,[215,525–529] (2) inhibit inflammatory response,[90,202,530–532] (3) lower LDL and elevate HDL,[533–538] (4) inhibit fibrinogen synthesis,[539] and (5) inhibit oxidation of LDL.[288,531,540–542] All of these activities contribute to the cardiovascular effects of curcumin. Because curcumin can suppress amyloid-induced inflammation, curcumin has also been linked to the suppression of Alzheimer's disease.[150,297,327,543–554]

8. CONCLUSION

The above description and various other chapters in this volume prove that curcumin has enormous potential for a variety of diseases. There are, however, still several unanswered questions. First, phase I clinical trials have indicated that as high as 12 g of curcumin per day for over 3 months is well tolerated in humans.[334] What the optimum dose of curcumin is for the treatment of a given disease is not clear. Serum levels of curcumin tend to be low,[334] which might be responsible for its pharmacological safety, These data have led to the notion that curcumin has low bioavailability. Second, the tissue concentration of curcumin and how it compares to what is seen in cell culture conditions are not known. There are studies, however, that suggest that agents such as piperine (a component of black pepper) can enhance the bioavailability of curcumin through suppression of its glucuronidation occurring primarily in the liver and in the intestine.[317] Third, whether there are components of turmeric other than curcumin that have beneficial effects either alone or in combination with curcumin needs to be determined. For instance, numerous activities have been assigned to turmeric oil.[307,555–559] Fourth, what effect do other spices have on the pharmacology and the biology of curcumin needs to be determined. Fifth, structural analogues of curcumin that are more bioavailable and efficacious are needed. However, this might compromise the safety of curcumin. Sixth, well-controlled large clinical trials are required to determine the potential of curcumin both in the prevention and therapy of a disease. All of these studies should further add to the usefulness of curcumin. Overall, the biological safety, combined with its cost and efficacy, and thousands of years of experimentation justify calling curcumin "Indian Solid Gold."

ACKNOWLEDGMENTS

This research was supported by The Clayton Foundation for Research (to BBA) and by a P50 Head and Neck SPORE grant from the National Institutes of Health (to BBA). We would like to thank Walter Pagel for a careful review of the manuscript.

ABBREVIATIONS USED

EGF, epidermal growth factor; EGFR, EGF receptor; NF-κB, nuclear factor-κB; TNF, tumor necrosis factor; AP-1, activating protein-1; JNK, c-jun N-terminal kinase; MMP, matrix metalloprotease; COX-2, cyclooxygenase 2; iNOS, inducible nitric oxide synthase; HDL, high-density lipoprotein; LDL, low-density lipoprotein; TGF-β1, transforming growth factor beta 1; IL, interleukin; STAT, signal transducers and activators of transcription; proteins, low molecular weight proteins; NSAIDs, nonsteroidal anti-inflammatory drugs; amyloid precursor; GST, gluthathione-S–transferase; LOX, lipooxygenase; ROS, reactive oxygen species; VEGF, vascularendothelial growth factor; FGF, fibroblast growth factor; IKK, IκB kinase; PKC, protein kinase C; PKA, protein kinase A.

REFERENCES

1. I. Chattopadhyay, K. Biswas, U. Bandyopadhyay, and R. K. Banerjee, Turmeric and curcumin: Biological actions and medicinal applications. *Curr Sci* **87**, 44–50 (2004).
2. F. Abas, N. H. Lajis, K. Shaari, D. A. Israf, J. Stanslas, U. K. Yusuf, and S. M. Raof, A labdane diterpene glucoside from the rhizomes of *Curcuma mangga*. *J Nat Prod* **68**, 1090–1093 (2005).
3. W. J. Syu, C. C. Shen, M. J. Don, J. C. Ou, G. H. Lee, and C. M. Sun, Cytotoxicity of curcuminoids and some novel compounds from *Curcuma zedoaria*. *J Nat Prod* **61**, 1531–1534 (1998).
4. J. A. Duke, CRC Handbook of Medicinal Spices, 137–144 (2002). CRC Press.
5. C. Tohda, N. Nakayama, F. Hatanaka, and K. Komatsu, Comparison of anti-inflammatory activities of six curcuma rhizomes: A possible curcuminoid-independent

pathway mediated by *Curcuma phaeocaulis* extract. *Evid Based Complement Alternat Med* **3**, 255–260 (2006).

6. H. Mohamad, N. H. Lajis, F. Abas, A. M. Ali, M. A. Sukari, H. Kikuzaki, and N. Nakatani, Antioxidative constituents of *Etlingera elatior*. *J Nat Prod* **68**, 285–288 (2005).
7. T. Dechatowongse, Isolation of constituents from the rhizome of plai (*Zingiber cassumunar* Rpxb.). *Bull Dept Med Sci* **18**, 75 (1976).
8. H. Ahsan, N. Parveen, N. U. Khan, and S. M. Hadi, Pro-oxidant, anti-oxidant and cleavage activities on DNA of curcumin and its derivatives demethoxycurcumin and bisdemethoxycurcumin. *Chem Biol Interact* **121**, 161–175 (1999).
9. N. Sreejayan and M. N. Rao, Free radical scavenging activity of curcuminoids. *Arzneimittelforschung* **46**, 169–171 (1996).
10. R. Thapliyal and G. B. Maru, Inhibition of cytochrome P450 isozymes by curcumins in vitro and in vivo. *Food Chem Toxicol* **39**, 541–547 (2001).
11. M. T. Huang, Y. R. Lou, J. G. Xie, W. Ma, Y. P. Lu, P. Yen, B. T. Zhu, H. Newmark, and C. T. Ho, Effect of dietary curcumin and dibenzoylmethane on formation of 7,12-dimethylbenz[a]anthracene-induced mammary tumors and lymphomas/leukemias in Sencar mice. *Carcinogenesis* **19**, 1697–1700 (1998).
12. Sreejayan and M. N. Rao, Nitric oxide scavenging by curcuminoids. *J Pharm Pharmacol* **49**, 105–107 (1997).
13. C. Ireson, S. Orr, D. J. Jones, R. Verschoyle, C. K. Lim, J. L. Luo, L. Howells, S. Plummer, R. Jukes, M. Williams, W. P. Steward, and A. Gescher, Characterization of metabolites of the chemopreventive agent curcumin in human and rat hepatocytes and in the rat in vivo, and evaluation of their ability to inhibit phorbol ester-induced prostaglandin E2 production. *Cancer Res* **61**, 1058–1064 (2001).
14. Y. Sugiyama, S. Kawakishi, and T. Osawa, Involvement of the beta-diketone moiety in the antioxidative mechanism of tetrahydrocurcumin. *Biochem Pharmacol* **52**, 519–525 (1996).
15. S. M. Khopde, K. I. Priyadarsini, S. N. Guha, J. G. Satav, P. Venkatesan, and M. N. Rao, Inhibition of radiation-induced lipid peroxidation by tetrahydrocurcumin: possible mechanisms by pulse radiolysis. *Biosci Biotechnol Biochem* **64**, 503–509 (2000).
16. K. Okada, C. Wangpoengtrakul, T. Tanaka, S. Toyokuni, K. Uchida, and T. Osawa, Curcumin and especially tetrahydrocurcumin ameliorate oxidative stress-induced renal injury in mice. *J Nutr* **131**, 2090–2095 (2001).
17. L. Pari and P. Murugan, Protective role of tetrahydrocurcumin against erythromycin estolate-induced hepatotoxicity. *Pharmacol Res* **49**, 481–486 (2004).
18. L. Pari and D. R. Amali, Protective role of tetrahydrocurcumin (THC) an active principle of turmeric on chloroquine induced hepatotoxicity in rats. *J Pharm Pharm Sci* **8**, 115–123 (2005).
19. Y. Nakamura, Y. Ohto, A. Murakami, T. Osawa, and H. Ohigashi, Inhibitory effects of curcumin and tetrahydrocurcuminoids on the tumor promoter-induced reactive oxygen species generation in leukocytes in vitro and in vivo. *Jpn J Cancer Res* **89**, 361–370 (1998).
20. K. R. Chaudhri, *Turmeric, haldi or haridra, in eye diseases.*, Antiseptic. 1950 Jan; **47**(1), 67.
21. K. M. Nadkarni. Curcuma Longa in India Materia (1976) Popular Prakashan, 414–418 Mumbai.

22. C. Niederau and E. Gopfert, [The effect of chelidonium- and turmeric root extract on upper abdominal pain due to functional disorders of the biliary system. Results from a placebo-controlled double-blind study]. *Med Klin (Munich)* **94**, 425–430 (1999).
23. C. Li, L. Li, J. Luo, and N. Huang, [Effect of turmeric volatile oil on the respiratory tract]. *Zhongguo Zhong Yao Za Zhi* **23**, 624–625, inside back cover (1998).
24. Curcuma longa (turmeric). Monograph. *Altern Med Rev* **6(Suppl)**, S62–S66 (2001).
25. A. Tawatsin, S. D. Wratten, R. R. Scott, U. Thavara, and Y. Techadamrongsin, repellency of volatile oils from plants against three mosquito vectors. *J Vector Ecol* **26**, 76–82 (2001).
26. G. Bouvier, M. Hergenhahn, A. Polack, G. W. Bornkamm, and H. Bartsch, Validation of two test systems for detecting tumor promoters and EBV inducers: comparative responses of several agents in DR-CAT Raji cells and in human granulocytes. *Carcinogenesis* **14**, 1573–1578 (1993).
27. A. P. Saikia, V. K. Ryakala, P. Sharma, P. Goswami. and U. Bora, Ethnobotany of medicinal plants used by Assamese people for various skin ailments and cosmetics. *J Ethnopharmacol* **106**, 149–157 (2006).
28. T. K. Biswas and B. Mukherjee, Plant medicines of Indian origin for wound healing activity: A review. *Int J Low Extrem Wounds* **2**, 25–39 (2003).
29. G. S. Sidhu, A. K. Singh, D. Thaloor, K. K. Banaudha, G. K. Patnaik, R. C. Srimal, and R. K. Maheshwari, Enhancement of wound healing by curcumin in animals. *Wound Repair Regen* **6**, 167–177 (1998).
30. G. S. Sidhu, H. Mani, J. P. Gaddipati, A. K. Singh, P. Seth, K. K. Banaudha, G. K. Patnaik, and R. K. Maheshwari, Curcumin enhances wound healing in streptozotocin induced diabetic rats and genetically diabetic mice. *Wound Repair Regen* **7**, 362–374 (1999).
31. T. T. Phan, P. See, S. T. Lee, and S. Y. Chan, Protective effects of curcumin against oxidative damage on skin cells in vitro: its implication for wound healing. *J Trauma* **51**, 927–931 (2001).
32. T. R. Fray, A. L. Watson, J. M. Croft, C. D. Baker, J. Bailey, N. Sirel, A. Tobias, and P. J. Markwell, A combination of aloe vera, curcumin, vitamin C, and taurine increases canine fibroblast migration and decreases tritiated water diffusion across canine keratinocytes in vitro. *J Nutr* **134**, 2117S–2119S (2004).
33. D. Gopinath, M. R. Ahmed, K. Gomathi, K. Chitra, P. K. Sehgal, and R. Jayakumar, Dermal wound healing processes with curcumin incorporated collagen films. *Biomaterials* **25**, 1911–1917 (2004).
34. G. C. Jagetia and G. K. Rajanikant, Role of curcumin, a naturally occurring phenolic compound of turmeric in accelerating the repair of excision wound, in mice whole-body exposed to various doses of gamma-radiation. *J Surg Res* **120**, 127–138 (2004).
35. G. C. Jagetia and G. K. Rajanikant, Effect of curcumin on radiation-impaired healing of excisional wounds in mice. *J Wound Care* **13**, 107–109 (2004).
36. G. C. Jagetia and G. K. Rajanikant, Curcumin treatment enhances the repair and regeneration of wounds in mice exposed to hemibody gamma-irradiation. *Plast Reconstr Surg* **115**, 515–528 (2005).
37. A. Kumar, Y. Takada, A. M. Boriek, and B. B. Aggarwal, Nuclear factor-kappaB: Its role in health and disease. *J Mol Med* **82**, 434–448 (2004).
38. B. B. Aggarwal, Y. Takada, S. Shishodia, A. M. Gutierrez, O. V. Oommen, H. Ichikawa, Y. Baba, and A. Kumar, Nuclear transcription factor NF-kappa B: Role in biology and medicine. *Indian J Exp Biol* **42**, 341–353 (2004).

39. H. P. Ammon, M. I. Anazodo, H. Safayhi, B. N. Dhawan, and R. C. Srimal, Curcumin: A potent inhibitor of leukotriene B4 formation in rat peritoneal polymorphonuclear neutrophils (PMNL). *Planta Med* **58**, 226 (1992).
40. Y. Fujiyama-Fujiwara, R. Umeda, and O. Igarashi, Effects of sesamin and curcumin on delta 5-desaturation and chain elongation of polyunsaturated fatty acid metabolism in primary cultured rat hepatocytes. *J Nutr Sci Vitaminol (Tokyo)* **38**, 353–363 (1992).
41. R. Srivastava, Inhibition of neutrophil response by curcumin. *Agents Actions* **28**, 298–303 (1989).
42. H. P. Ammon, H. Safayhi, T. Mack, and J. Sabieraj, Mechanism of antiinflammatory actions of curcumine and boswellic acids. *J Ethnopharmacol* **38**, 113–119 (1993).
43. A. C. Reddy and B. R. Lokesh, Studies on anti-inflammatory activity of spice principles and dietary n-3 polyunsaturated fatty acids on carrageenan-induced inflammation in rats. *Ann Nutr Metab* **38**, 349–358 (1994).
44. B. Joe and B. R. Lokesh, Effect of curcumin and capsaicin on arachidonic acid metabolism and lysosomal enzyme secretion by rat peritoneal macrophages. *Lipids* **32**, 1173–1180 (1997).
45. Y. X. Xu, K. R. Pindolia, N. Janakiraman, C. J. Noth, R. A. Chapman, and S. C. Gautam, Curcumin, a compound with anti-inflammatory and anti-oxidant properties, down-regulates chemokine expression in bone marrow stromal cells. *Exp Hematol* **25**, 413–422 (1997).
46. B. Joe and B. R. Lokesh, Dietary n-3 fatty acids, curcumin and capsaicin lower the release of lysosomal enzymes and eicosanoids in rat peritoneal macrophages. *Mol Cell Biochem* **203**, 153–161 (2000).
47. E. A. Jones, A. Shahed, and D. A. Shoskes, Modulation of apoptotic and inflammatory genes by bioflavonoids and angiotensin II inhibition in ureteral obstruction. *Urology* **56**, 346–351 (2000).
48. M. Banerjee, L. M. Tripathi, V. M. Srivastava, A. Puri, and R. Shukla, Modulation of inflammatory mediators by ibuprofen and curcumin treatment during chronic inflammation in rat. *Immunopharmacol Immunotoxicol* **25**, 213–224 (2003).
49. R. C. Lantz, G. J. Chen, A. M. Solyom, S. D. Jolad, and B. N. Timmermann, The effect of turmeric extracts on inflammatory mediator production. *Phytomedicine* **12**, 445–452 (2005).
50. S. Singh and B. B. Aggarwal, Activation of transcription factor NF-kappa B is suppressed by curcumin (diferuloylmethane) [corrected]. *J Biol Chem* **270**, 24,995–25,000 (1995).
51. A. Bierhaus, Y. Zhang, P. Quehenberger, T. Luther, M. Haase, M. Muller, N. Mackman, R. Ziegler, and P. P. Nawroth, The dietary pigment curcumin reduces endothelial tissue factor gene expression by inhibiting binding of AP-1 to the DNA and activation of NF-kappa B. *Thromb Haemost* **77**, 772–782 (1997).
52. A. Munzenmaier, C. Lange, E. Glocker, A. Covacci, A. Moran, S. Bereswill, P. A. Baeuerle, M. Kist, and H. L. Pahl, A secreted/shed product of *Helicobacter pylori* activates transcription factor nuclear factor-kappa B. *J Immunol* **159**, 6140–6147 (1997).
53. U. R. Pendurthi, J. T. Williams, and L. V. Rao, Inhibition of tissue factor gene activation in cultured endothelial cells by curcumin. Suppression of activation of transcription factors Egr-1, AP-1, and NF-kappa B. *Arterioscler Thromb Vasc Biol* **17**, 3406–3413 (1997).
54. Y. X. Xu, K. R. Pindolia, N. Janakiraman, R. A. Chapman, and S. C. Gautam, Curcumin inhibits IL1 alpha and TNF-alpha induction of AP-1 and NF-kB DNA-binding activity in bone marrow stromal cells. *Hematopathol Mol Hematol* **11**, 49–62 (1997).

55. P. Brennan and L. A. O'Neill, Inhibition of nuclear factor kappaB by direct modification in whole cells: Mechanism of action of nordihydroguaiaritic acid, curcumin and thiol modifiers. *Biochem Pharmacol* **55**, 965–973 (1998).
56. S. S. Han, S. T. Chung, D. A. Robertson, D. Ranjan, and S. Bondada, Curcumin causes the growth arrest and apoptosis of B cell lymphoma by downregulation of egr-1, c-myc, bcl-XL, NF-kappa B, and p53. *Clin Immunol* **93**, 152–161 (1999).
57. C. Jobin, C. A. Bradham, M. P. Russo, B. Juma, A. S. Narula, D. A. Brenner, and R. B. Sartor, Curcumin blocks cytokine-mediated NF-kappa B activation and proinflammatory gene expression by inhibiting inhibitory factor I-kappa B kinase activity. *J Immunol* **163**, 3474–3483 (1999).
58. S. M. Plummer, K. A. Holloway, M. M. Manson, R. J. Munks, A. Kaptein, S. Farrow, and L. Howells, Inhibition of cyclo-oxygenase 2 expression in colon cells by the chemopreventive agent curcumin involves inhibition of NF-kappaB activation via the NIK/IKK signalling complex. *Oncogene* **18**, 6013–6020 (1999).
59. Y. J. Surh, S. S. Han, Y. S. Keum, H. J. Seo, and S. S. Lee, Inhibitory effects of curcumin and capsaicin on phorbol ester-induced activation of eukaryotic transcription factors, NF-kappaB and AP-1. *Biofactors* **12**, 107–112 (2000).
60. S. E. Chuang, P. Y. Yeh, Y. S. Lu, G. M. Lai, C. M. Liao, M. Gao, and A. L. Cheng, Basal levels and patterns of anticancer drug-induced activation of nuclear factor-kappaB (NF-kappaB), and its attenuation by tamoxifen, dexamethasone, and curcumin in carcinoma cells. *Biochem Pharmacol* **63**, 1709–1716 (2002).
61. A. Grandjean-Laquerriere, S. C. Gangloff, R. Le Naour, C. Trentesaux, W. Hornebeck, and M. Guenounou, Relative contribution of NF-kappaB and AP-1 in the modulation by curcumin and pyrrolidine dithiocarbamate of the UVB-induced cytokine expression by keratinocytes. *Cytokine* **18**, 168–177 (2002).
62. S. S. Han, Y. S. Keum, H. J. Seo, and Y. J. Surh, Curcumin suppresses activation of NF-kappaB and AP-1 induced by phorbol ester in cultured human promyelocytic leukemia cells. *J Biochem Mol Biol* **35**, 337–342 (2002).
63. T. C. Hour, J. Chen, C. Y. Huang, J. Y. Guan, S. H. Lu, and Y. S. Pu, Curcumin enhances cytotoxicity of chemotherapeutic agents in prostate cancer cells by inducing p21(WAF1/CIP1) and C/EBPbeta expressions and suppressing NF-kappaB activation. *Prostate* **51**, 211–218 (2002).
64. K. S. Chun, Y. S. Keum, S. S. Han, Y. S. Song, S. H. Kim, and Y. J. Surh, Curcumin inhibits phorbol ester-induced expression of cyclooxygenase-2 in mouse skin through suppression of extracellular signal-regulated kinase activity and NF-kappaB activation. *Carcinogenesis* **24**, 1515–1524 (2003).
65. S. Philip and G. C. Kundu, Osteopontin induces nuclear factor kappa B-mediated promatrix metalloproteinase-2 activation through I kappa B alpha /IKK signaling pathways, and curcumin (diferulolylmethane) down-regulates these pathways. *J Biol Chem* **278**, 14,487–14,497 (2003).
66. S. Shishodia, P. Potdar, C. G. Gairola, and B. B. Aggarwal, Curcumin (diferuloylmethane) down-regulates cigarette smoke-induced NF-kappaB activation through inhibition of IkappaBalpha kinase in human lung epithelial cells: correlation with suppression of COX-2, MMP-9 and cyclin D1. *Carcinogenesis* **24**, 1269–1279 (2003).
67. A. Foryst-Ludwig, M. Neumann, W. Schneider-Brachert and M. Naumann, Curcumin blocks NF-kappaB and the motogenic response in *Helicobacter pylori*-infected epithelial cells. *Biochem Biophys Res Commun* **316**, 1065–1072 (2004).

68. I. A. Leclercq, G. C. Farrell, C. Sempoux, A. dela Pena, and Y. Horsmans, Curcumin inhibits NF-kappaB activation and reduces the severity of experimental steatohepatitis in mice. *J Hepatol* **41**, 926–934 (2004).
69. B. van't Land, N. M. Blijlevens, J. Marteijn, S. Timal, J. P. Donnelly, T. J. de Witte and L. M'Rabet, Role of curcumin and the inhibition of NF-kappaB in the onset of chemotherapy-induced mucosal barrier injury. *Leukemia* **18**, 276–284 (2004).
70. B. B. Aggarwal, S. Shishodia, Y. Takada, S. Banerjee, R. A. Newman, C. E. Bueso-Ramos and J. E. Price, Curcumin suppresses the paclitaxel-induced nuclear factor-kappaB pathway in breast cancer cells and inhibits lung metastasis of human breast cancer in nude mice. *Clin Cancer Res* **11**, 7490–7498 (2005).
71. S. K. Biswas, D. McClure, L. A. Jimenez, I. L. Megson, and I. Rahman, Curcumin induces glutathione biosynthesis and inhibits NF-kappaB activation and interleukin-8 release in alveolar epithelial cells: mechanism of free radical scavenging activity. *Antioxid Redox Signal* **7**, 32–41 (2005).
72. M. Farid, M. B. Reid, Y. P. Li, E. Gerken, and W. J. Durham, Effects of dietary curcumin or N-acetylcysteine on NF-kappaB activity and contractile performance in ambulatory and unloaded murine soleus. *Nutr Metab (Lond)* **2**, 20 (2005).
73. G. Y. Kim, K. H. Kim, S. H. Lee, M. S. Yoon, H. J. Lee, D. O. Moon, C. M. Lee, S. C. Ahn, Y. C. Park, and Y. M. Park, Curcumin inhibits immunostimulatory function of dendritic cells: MAPKs and translocation of NF-kappa B as potential targets. *J Immunol* **174**, 8116–8124 (2005).
74. K. W. Lee, J. H. Kim, H. J. Lee, and Y. J. Surh, Curcumin inhibits phorbol ester-induced up-regulation of cyclooxygenase-2 and matrix metalloproteinase-9 by blocking ERK1/2 phosphorylation and NF-kappaB transcriptional activity in MCF10A human breast epithelial cells. *Antioxid Redox Signal* **7**, 1612–1620 (2005).
75. J. Lee, Y. H. Im, H. H. Jung, J. H. Kim, J. O. Park, K. Kim, W. S. Kim, J. S. Ahn, C. W. Jung, Y. S. Park, W. K. Kang, and K. Park, Curcumin inhibits interferon-alpha induced NF-kappaB and COX-2 in human A549 non-small cell lung cancer cells. *Biochem Biophys Res Commun* **334**, 313–318 (2005).
76. M. M. LoTempio, M. S. Veena, H. L. Steele, B. Ramamurthy, T. S. Ramalingam, A. N. Cohen, R. Chakrabarti, E. S. Srivatsan, and M. B. Wang, Curcumin suppresses growth of head and neck squamous cell carcinoma. *Clin Cancer Res* **11**, 6994–7002 (2005).
77. S. Shishodia, H. M. Amin, R. Lai, and B. B. Aggarwal, Curcumin (diferuloylmethane) inhibits constitutive NF-kappaB activation, induces G1/S arrest, suppresses proliferation, and induces apoptosis in mantle cell lymphoma. *Biochem Pharmacol* **70**, 700–713 (2005).
78. S. Wessler, P. Muenzner, T. F. Meyer, and M. Naumann, The anti-inflammatory compound curcumin inhibits Neisseria gonorrhoeae-induced NF-kappaB signaling, release of pro-inflammatory cytokines/chemokines and attenuates adhesion in late infection. *Biol Chem* **386**, 481–490 (2005).
79. S. Aggarwal, H. Ichikawa, Y. Takada, S. K. Sandur, S. Shishodia, and B. B. Aggarwal, Curcumin (diferuloylmethane) down-regulates expression of cell proliferation and antiapoptotic and metastatic gene products through suppression of IkappaBalpha kinase and Akt activation. *Mol Pharmacol* **69**, 195–206 (2006).
80. M. Tomita, H. Kawakami, J. N. Uchihara, T. Okudaira, M. Masuda, N. Takasu, T. Matsuda, T. Ohta, Y. Tanaka, K. Ohshiro, and N. Mori, Curcumin (diferuloylmethane) inhibits constitutive active NF-kappaB, leading to suppression of cell growth of human T-cell leukemia virus type I-infected T-cell lines and primary adult T-cell leukemia cells. *Int J Cancer* **118**, 765–772 (2006).

81. W. M. Weber, L. A. Hunsaker, C. N. Roybal, E. V. Bobrovnikova-Marjon, S. F. Abcouwer, R. E. Royer, L. M. Deck, and D. L. Vander Jagt, Activation of NFkappaB is inhibited by curcumin and related enones. *Bioorg Med Chem* **14**, 2450–2461 (2006).
82. W. Q. Li, F. Dehnade, and M. Zafarullah, Oncostatin M-induced matrix metalloproteinase and tissue inhibitor of metalloproteinase-3 genes expression in chondrocytes requires Janus kinase/STAT signaling pathway. *J Immunol* **166**, 3491–3498 (2001).
83. A. C. Bharti, Y. Takada, and B. B. Aggarwal, Curcumin (diferuloylmethane) inhibits receptor activator of NF-kappa B ligand-induced NF-kappa B activation in osteoclast precursors and suppresses osteoclastogenesis. *J Immunol* **172**, 5940–5947 (2004).
84. H. Y. Kim, E. J. Park, E. H. Joe, and I. Jou, Curcumin suppresses Janus kinase-STAT inflammatory signaling through activation of Src homology 2 domain-containing tyrosine phosphatase 2 in brain microglia. *J Immunol* **171**, 6072–6079 (2003).
85. J. Rajasingh, H. P. Raikwar, G. Muthian, C. Johnson, and J. J. Bright, Curcumin induces growth-arrest and apoptosis in association with the inhibition of constitutively active JAK-STAT pathway in T cell leukemia. *Biochem Biophys Res Commun* **340**, 359–368 (2006).
86. D. Chendil, R. S. Ranga, D. Meigooni, S. Sathishkumar, and M. M. Ahmed, Curcumin confers radiosensitizing effect in prostate cancer cell line PC-3. *Oncogene* **23**, 1599–1607 (2004).
87. D. A. Dickinson, K. E. Iles, H. Zhang, V. Blank, and H. J. Forman, Curcumin alters EpRE and AP-1 binding complexes and elevates glutamate-cysteine ligase gene expression. *FASEB J* **17**, 473–475 (2003).
88. B. K. Prusty and B. C. Das, Constitutive activation of transcription factor AP-1 in cervical cancer and suppression of human papillomavirus (HPV) transcription and AP-1 activity in HeLa cells by curcumin. *Int J Cancer* **113**, 951–960 (2005).
89. M. Tomita, H. Kawakami, J. N. Uchihara, T. Okudaira, M. Masuda, N. Takasu, T. Matsuda, T. Ohta, Y. Tanaka, and N. Mori, Curcumin suppresses constitutive activation of AP-1 by downregulation of JunD protein in HTLV-1-infected T-cell lines. *Leuk Res* **30**, 313–321 (2006).
90. U. R. Pendurthi and L. V. Rao, Suppression of transcription factor Egr-1 by curcumin. *Thromb Res* **97**, 179–189 (2000).
91. E. Balogun, M. Hoque, P. Gong, E. Killeen, C. J. Green, R. Foresti, J. Alam, and R. Motterlini, Curcumin activates the haem oxygenase-1 gene via regulation of Nrf2 and the antioxidant-responsive element. *Biochem J* **371**, 887–895 (2003).
92. J. Xu, Y. Fu, and A. Chen, Activation of peroxisome proliferator-activated receptor-gamma contributes to the inhibitory effects of curcumin on rat hepatic stellate cell growth. *Am J Physiol Gastrointest Liver Physiol* **285**, G20–G30 (2003).
93. S. Zheng and A. Chen, Activation of PPARγ is required for curcumin to induce apoptosis and to inhibit the expression of extracellular matrix genes in hepatic stellate cells in vitro. *Biochem J* **384**, 149–157 (2004).
94. A. Chen and J. Xu, Activation of PPARγ by curcumin inhibits Moser cell growth and mediates suppression of gene expression of cyclin D1 and EGFR. *Am J Physiol Gastrointest Liver Physiol* **288**, G447–G456 (2005).
95. A. S. Jaiswal, B. P. Marlow, N. Gupta, and S. Narayan, Beta-catenin-mediated transactivation and cell-cell adhesion pathways are important in curcumin (diferuylmethane)-induced growth arrest and apoptosis in colon cancer cells. *Oncogene* **21**, 8414–8427 (2002).

96. C. H. Park, E. R. Hahm, S. Park, H. K. Kim, and C. H. Yang, The inhibitory mechanism of curcumin and its derivative against beta-catenin/Tcf signaling. *FEBS Lett* **579**, 2965–2971 (2005).
97. M. M. Chan, Inhibition of tumor necrosis factor by curcumin, a phytochemical. *Biochem Pharmacol* **49**, 1551–1556 (1995).
98. M. K. Jang, D. H. Sohn, and J. H. Ryu, A curcuminoid and sesquiterpenes as inhibitors of macrophage TNF-alpha release from Curcuma zedoaria. *Planta Med* **67**, 550–552 (2001).
99. H. Matsuda, S. Tewtrakul, T. Morikawa, A. Nakamura, and M. Yoshikawa, Anti-allergic principles from Thai zedoary: structural requirements of curcuminoids for inhibition of degranulation and effect on the release of TNF-alpha and IL-4 in RBL-2H3 cells. *Bioorg Med Chem* **12**, 5891–5898 (2004).
100. A. Gulcubuk, K. Altunatmaz, K. Sonmez, D. Haktanir-Yatkin, H. Uzun, A. Gurel and S. Aydin, Effects of curcumin on tumour necrosis factor-alpha and interleukin-6 in the late phase of experimental acute pancreatitis. *J Vet Med A Physiol Pathol Clin Med* **53**, 49–54 (2006).
101. J. P. Gaddipati, S. V. Sundar, J. Calemine, P. Seth, G. S. Sidhu, and R. K. Maheshwari, Differential regulation of cytokines and transcription factors in liver by curcumin following hemorrhage/resuscitation. *Shock* **19**, 150–156 (2003).
102. C. J. Lee, J. H. Lee, J. H. Seok, G. M. Hur, Y. C. Park, I. C. Seol, and Y. H. Kim, Effects of baicalein, berberine, curcumin and hesperidin on mucin release from airway goblet cells. *Planta Med* **69**, 523–526 (2003).
103. N. Jurrmann, R. Brigelius-Flohe and G. F. Bol, Curcumin blocks interleukin-1 (IL-1) signaling by inhibiting the recruitment of the IL-1 receptor-associated kinase IRAK in murine thymoma EL-4 cells. *J Nutr* **135**, 1859–1864 (2005).
104. Y. Moon, W. C. Glasgow, and T. E. Eling, Curcumin suppresses interleukin 1beta-mediated microsomal prostaglandin E synthase 1 by altering early growth response gene 1 and other signaling pathways. *J Pharmacol Exp Ther* **315**, 788–795 (2005).
105. M. Shakibaei, G. Schulze-Tanzil, T. John and A. Mobasheri, Curcumin protects human chondrocytes from IL-l1beta-induced inhibition of collagen type II and beta1-integrin expression and activation of caspase-3: An immunomorphological study. *Ann Anat* **187**, 487–497 (2005).
106. T. Kobayashi, S. Hashimoto and T. Horie, Curcumin inhibition of Dermatophagoides farinea-induced interleukin-5 (IL-5) and granulocyte macrophage-colony stimulating factor (GM-CSF) production by lymphocytes from bronchial asthmatics. *Biochem Pharmacol* **54**, 819–824 (1997).
107. Y. Abe, S. Hashimoto, and T. Horie, Curcumin inhibition of inflammatory cytokine production by human peripheral blood monocytes and alveolar macrophages. *Pharmacol Res* **39**, 41–47 (1999).
108. H. Hidaka, T. Ishiko, T. Furuhashi, H. Kamohara, S. Suzuki, M. Miyazaki, O. Ikeda, S. Mita, T. Setoguchi, and M. Ogawa, Curcumin inhibits interleukin 8 production and enhances interleukin 8 receptor expression on the cell surface:impact on human pancreatic carcinoma cell growth by autocrine regulation. *Cancer* **95**, 1206–1214 (2002).
109. B. Y. Kang, S. W. Chung, W. Chung, S. Im, S. Y. Hwang, and T. S. Kim, Inhibition of interleukin-12 production in lipopolysaccharide-activated macrophages by curcumin. *Eur J Pharmacol* **384**, 191–195 (1999).
110. B. Y. Kang, Y. J. Song, K. M. Kim, Y. K. Choe, S. Y. Hwang and T. S. Kim, Curcumin inhibits Th1 cytokine profile in CD4+ T cells by suppressing interleukin-12 production in macrophages. *Br J Pharmacol* **128**, 380–384 (1999).

111. C. Natarajan and J. J. Bright, Curcumin inhibits experimental allergic encephalomyelitis by blocking IL-12 signaling through Janus kinase-STAT pathway in T lymphocytes. *J Immunol* **168**, 6506–6513 (2002).
112. M. Tomita, B. J. Holman, C. P. Santoro, and T. J. Santoro, Astrocyte production of the chemokine macrophage inflammatory protein-2 is inhibited by the spice principle curcumin at the level of gene transcription. *J Neuroinflammation* **2**, 8 (2005).
113. M. K. Bae, S. H. Kim, J. W. Jeong, Y. M. Lee, H. S. Kim, S. R. Kim, I. Yun, S. K. Bae, and K. W. Kim, Curcumin inhibits hypoxia-induced angiogenesis via down-regulation of HIF-1. *Oncol Rep* **15**, 1557–1562 (2006).
114. H. Choi, Y. S. Chun, S. W. Kim, M. S. Kim, and J. W. Park, Curcumin inhibits hypoxia-inducible factor-1 by degrading aryl hydrocarbon receptor nuclear translocator: A mechanism of tumor growth inhibition. *Mol Pharmacol* **70**(5), 1664–1671 (2006).
115. S. Reddy and B. B. Aggarwal, Curcumin is a non-competitive and selective inhibitor of phosphorylase kinase. *FEBS Lett* **341**, 19–22 (1994).
116. M. Hasmeda and G. M. Polya, Inhibition of cyclic AMP-dependent protein kinase by curcumin. *Phytochemistry* **42**, 599–605 (1996).
117. Y. R. Chen and T. H. Tan, Inhibition of the c-Jun N-terminal kinase (JNK) signaling pathway by curcumin. *Oncogene* **17**, 173–178 (1998).
118. M. S. Squires, E. A. Hudson, L. Howells, S. Sale, C. E. Houghton, J. L. Jones, L. H. Fox, M. Dickens, S. A. Prigent, and M. M. Manson, Relevance of mitogen activated protein kinase (MAPK) and phosphotidylinositol-3-kinase/protein kinase B (PI3K/PKB) pathways to induction of apoptosis by curcumin in breast cells. *Biochem Pharmacol* **65**, 361–376 (2003).
119. T. H. Leu, S. L. Su, Y. C. Chuang, and M. C. Maa, Direct inhibitory effect of curcumin on Src and focal adhesion kinase activity. *Biochem Pharmacol* **66**, 2323–2331 (2003).
120. M. Hu, Q. Du, I. Vancurova, X. Lin, E. J. Miller, H. H. Simms, and P. Wang, Proapoptotic effect of curcumin on human neutrophils: activation of the p38 mitogen-activated protein kinase pathway. *Crit Care Med* **33**, 2571–2578 (2005).
121. L. R. Chaudhary and K. A. Hruska, Inhibition of cell survival signal protein kinase B/Akt by curcumin in human prostate cancer cells. *J Cell Biochem* **89**, 1–5 (2003).
122. J. Y. Liu, S. J. Lin, and J. K. Lin, Inhibitory effects of curcumin on protein kinase C activity induced by 12-O-tetradecanoyl-phorbol-13-acetate in NIH 3T3 cells. *Carcinogenesis* **14**, 857–861 (1993).
123. K. J. Mistry, M. Krishna, and R. K. Bhattacharya, Modulation of aflatoxin B1 activated protein kinase C by phenolic compounds. *Cancer Lett* **121**, 99–104 (1997).
124. P. Varadkar, P. Dubey, M. Krishna, and N. Verma, Modulation of radiation-induced protein kinase C activity by phenolics. *J Radiol Prot* **21**, 361–370 (2001).
125. J. K. Lin, Suppression of protein kinase C and nuclear oncogene expression as possible action mechanisms of cancer chemoprevention by Curcumin. *Arch Pharm Res* **27**, 683–692 (2004).
126. S. A. Rushworth, R. M. Ogborne, C. A. Charalambos, and M. A. O'Connell, Role of protein kinase C delta in curcumin-induced antioxidant response element-mediated gene expression in human monocytes. *Biochem Biophys Res Commun* **341**, 1007–1016 (2006).
127. R. L. Hong, W. H. Spohn, and M. C. Hung, Curcumin inhibits tyrosine kinase activity of p185neu and also depletes p185neu. *Clin Cancer Res* **5**, 1884–1891 (1999).
128. T. Dorai, N. Gehani, and A. Katz, Therapeutic potential of curcumin in human prostate cancer. II. Curcumin inhibits tyrosine kinase activity of epidermal growth factor receptor and depletes the protein. *Mol Urol* **4**, 1–6 (2000).

129. S. Kaul and T. P. Krishnakanth, Effect of retinol deficiency and curcumin or turmeric feeding on brain Na(+)−K+ adenosine triphosphatase activity. *Mol Cell Biochem* **137**, 101–107 (1994).
130. J. G. Bilmen, S. Z. Khan, M. H. Javed, and F. Michelangeli, Inhibition of the SERCA Ca2+ pumps by curcumin. Curcumin putatively stabilizes the interaction between the nucleotide-binding and phosphorylation domains in the absence of ATP. *Eur J Biochem* **268**, 6318–6327 (2001).
131. M. J. Logan-Smith, P. J. Lockyer, J. M. East and A. G. Lee, Curcumin, a molecule that inhibits the Ca2+-ATPase of sarcoplasmic reticulum but increases the rate of accumulation of Ca^{2+}. *J Biol Chem* **276**, 46,905–46.911 (2001).
132. Y. A. Mahmmoud, Curcumin modulation of Na,K-ATPase: phosphoenzyme accumulation, decreased K+ occlusion, and inhibition of hydrolytic activity. *Br J Pharmacol* **145**, 236–245 (2005).
133. J. Kang, J. Chen, Y. Shi, J. Jia, and Y. Zhang, Curcumin-induced histone hypoacetylation: The role of reactive oxygen species. *Biochem Pharmacol* **69**, 1205–1213 (2005).
134. W. H. Liu, X. M. Chen, and B. Fu, Thrombin stimulates MMP-9 mRNA expression through AP-1 pathway in human mesangial cells. *Acta Pharmacol Sin* **21**, 641–645 (2000).
135. S. Shimizu, S. Jareonkitmongkol, H. Kawashima, K. Akimoto, and H. Yamada, Inhibitory effect of curcumin on fatty acid desaturation in *Mortierella alpina* 1S-4 and rat liver microsomes. *Lipids* **27**, 509–512 (1992).
136. H. Kawashima, K. Akimoto, S. Jareonkitmongkol, N. Shirasaka, and S. Shimizu, Inhibition of rat liver microsomal desaturases by curcumin and related compounds. *Biosci Biotechnol Biochem* **60**, 108–110 (1996).
137. X. Chen, T. Hasuma, Y. Yano, T. Yoshimata, Y. Morishima, Y. Wang, and S. Otani, Inhibition of farnesyl protein transferase by monoterpene, curcumin derivatives and gallotannin. *Anticancer Res* **17**, 2555–2564 (1997).
138. H. M. Kang, K. H. Son, D. C. Yang, D. C. Han, J. H. Kim, N. I. Baek, and B. M. Kwon, Inhibitory activity of diarylheptanoids on farnesyl protein transferase. *Nat Prod Res* **18**, 295–299 (2004).
139. M. Susan and M. N. Rao, Induction of glutathione S-transferase activity by curcumin in mice. *Arzneimittelforschung* **42**, 962–964 (1992).
140. M. L. Iersel, J. P. Ploemen, I. Struik, C. van Amersfoort, A. E. Keyzer, J. G. Schefferlie, and P. J. van Bladeren, Inhibition of glutathione S-transferase activity in human melanoma cells by alpha,beta-unsaturated carbonyl derivatives. Effects of acrolein, cinnamaldehyde, citral, crotonaldehyde, curcumin, ethacrynic acid, and trans-2-hexenal. *Chem Biol Interact* **102**, 117–132 (1996).
141. J. T. Piper, S. S. Singhal, M. S. Salameh, R. T. Torman, Y. C. Awasthi, and S. Awasthi, Mechanisms of anticarcinogenic properties of curcumin: The effect of curcumin on glutathione linked detoxification enzymes in rat liver. *Int J Biochem Cell Biol* **30**, 445–456 (1998).
142. S. S. Singhal, S. Awasthi, U. Pandya, J. T. Piper, M. K. Saini, J. Z. Cheng, and Y. C. Awasthi, The effect of curcumin on glutathione-linked enzymes in K562 human leukemia cells. *Toxicol Lett* **109**, 87–95 (1999).
143. S. Awasthi, U. Pandya, S. S. Singhal, J. T. Lin, V. Thiviyanathan, W. E. Seifert, Jr., Y. C. Awasthi, and G. A. Ansari, Curcumin-glutathione interactions and the role of human glutathione S-transferase P1-1. *Chem Biol Interact* **128**, 19–38 (2000).
144. R. A. Sharma, C. R. Ireson, R. D. Verschoyle, K. A. Hill, M. L. Williams, C. Leuratti, M. M. Manson, L. J. Marnett, W. P. Steward, and A. Gescher, Effects of dietary

curcumin on glutathione S-transferase and malondialdehyde-DNA adducts in rat liver and colon mucosa: Relationship with drug levels. *Clin Cancer Res* **7**, 1452–1458 (2001).
145. A. Duvoix, F. Morceau, S. Delhalle, M. Schmitz, M. Schnekenburger, M. M. Galteau, M. Dicato, and M. Diederich, Induction of apoptosis by curcumin: mediation by glutathione S-transferase P1-1 inhibition. *Biochem Pharmacol* **66**, 1475–1483 (2003).
146. R. Blasius, A. Duvoix, F. Morceau, M. Schnekenburger, S. Delhalle, E. Henry, M. Dicato, and M. Diederich, Curcumin stability and its effect on glutathione S-transferase P1-1 mRNA expression in K562 cells. *Ann N Y Acad Sci* **1030**, 442–448 (2004).
147. N. Hill-Kapturczak, V. Thamilselvan, F. Liu, H. S. Nick and A. Agarwal, Mechanism of heme oxygenase-1 gene induction by curcumin in human renal proximal tubule cells. *Am J Physiol Renal Physiol* **281**, F851–F859 (2001).
148. G. Scapagnini, R. Foresti, V. Calabrese, A. M. Giuffrida Stella, C. J. Green and R. Motterlini, Caffeic acid phenethyl ester and curcumin: a novel class of heme oxygenase-1 inducers. *Mol Pharmacol* **61**, 554–561 (2002).
149. E. Balogun, R. Foresti, C. J. Green, and R. Motterlini, Changes in temperature modulate heme oxygenase-1 induction by curcumin in renal epithelial cells. *Biochem Biophys Res Commun* **308**, 950–955 (2003).
150. V. Calabrese, D. A. Butterfield, and A. M. Stella, Nutritional antioxidants and the heme oxygenase pathway of stress tolerance: Novel targets for neuroprotection in Alzheimer's disease. *Ital J Biochem* **52**, 177–181 (2003).
151. J. Gaedeke, N. A. Noble, and W. A. Border, Curcumin blocks fibrosis in anti-Thy 1 glomerulonephritis through up-regulation of heme oxygenase 1. *Kidney Int* **68**, 2042–2049 (2005).
152. H. Yamamoto, K. Hanada, K. Kawasaki, and M. Nishijima, Inhibitory effect on curcumin on mammalian phospholipase D activity. *FEBS Lett* **417**, 196–198 (1997).
153. Y. P. Lu, R. L. Chang, M. T. Huang, and A. H. Conney, Inhibitory effect of curcumin on 12-O-tetradecanoylphorbol-13-acetate-induced increase in ornithine decarboxylase mRNA in mouse epidermis. *Carcinogenesis* **14**, 293–297 (1993).
154. C. V. Rao, B. Simi, and B. S. Reddy, Inhibition by dietary curcumin of azoxymethane-induced ornithine decarboxylase, tyrosine protein kinase, arachidonic acid metabolism and aberrant crypt foci formation in the rat colon. *Carcinogenesis* **14**, 2219–2225 (1993).
155. C. Ishizaki, T. Oguro, T. Yoshida, C. Q. Wen, H. Sueki, and M. Iijima, Enhancing effect of ultraviolet A on ornithine decarboxylase induction and dermatitis evoked by 12-O-tetradecanoylphorbol-13-acetate and its inhibition by curcumin in mouse skin. *Dermatology* **193**, 311–317 (1996).
156. Y. Okazaki, M. Iqba,l and S. Okada, Suppressive effects of dietary curcumin on the increased activity of renal ornithine decarboxylase in mice treated with a renal carcinogen, ferric nitrilotriacetate. *Biochim Biophys Acta* **1740**, 357–366 (2005).
157. C. Ramachandran, H. B. Fonseca, P. Jhabvala, E. A. Escalon, and S. J. Melnick, Curcumin inhibits telomerase activity through human telomerase reverse transcritpase in MCF-7 breast cancer cell line. *Cancer Lett* **184**, 1–6 (2002).
158. S. Chakraborty, U. Ghosh, N. P. Bhattacharyya, R. K. Bhattacharya, and M. Roy, Inhibition of telomerase activity and induction of apoptosis by curcumin in K-562 cells. *Mutat Res* **596**, 81–90 (2006).

159. J. K. Lin and C. A. Shih, Inhibitory effect of curcumin on xanthine dehydrogenase/oxidase induced by phorbol-12-myristate-13-acetate in NIH3T3 cells. *Carcinogenesis* **15**, 1717–2171 (1994).
160. I. Brouet and H. Ohshima, Curcumin, an anti-tumour promoter and anti–inflammatory agent, inhibits induction of nitric oxide synthase in activated macrophages. *Biochem Biophys Res Commun* **206**, 533–540 (1995).
161. M. M. Chan, H. I. Huang, M. R. Fenton, and D. Fong, In vivo inhibition of nitric oxide synthase gene expression by curcumin, a cancer preventive natural product with anti-inflammatory properties. *Biochem Pharmacol* **55**, 1955–1962 (1998).
162. K. F. Soliman and E. A. Mazzio, In vitro attenuation of nitric oxide production in C6 astrocyte cell culture by various dietary compounds. *Proc Soc Exp Biol Med* **218**, 390–397 (1998).
163. M. Onoda and H. Inano, Effect of curcumin on the production of nitric oxide by cultured rat mammary gland. *Nitric Oxide* **4**, 505–515 (2000).
164. M. H. Pan, S. Y. Lin-Shiau and J. K. Lin, Comparative studies on the suppression of nitric oxide synthase by curcumin and its hydrogenated metabolites through down-regulation of IkappaB kinase and NFkappaB activation in macrophages. *Biochem Pharmacol* **60**, 1665–1676 (2000).
165. H. Narang and M. Krishna, Inhibition of radiation induced nitration by curcumin and nicotinamide in mouse macrophages. *Mol Cell Biochem* **276**, 7–13 (2005).
166. A. Mukhopadhyay, S. Banerjee, L. J. Stafford, C. Xia, M. Liu, and B. B. Aggarwal, Curcumin-induced suppression of cell proliferation correlates with down-regulation of cyclin D1 expression and CDK4-mediated retinoblastoma protein phosphorylation. *Oncogene* **21**, 8852–8861 (2002).
167. T. Choudhuri, S. Pal, T. Das, and G. Sa, Curcumin selectively induces apoptosis in deregulated cyclin D1-expressed cells at G2 phase of cell cycle in a p53-dependent manner. *J Biol Chem* **280**, 20,059–20,068 (2005).
168. Y. K. Kwon, J. M. Jun, S. W. Shin, J. W. Cho, and S. I. Suh, Curcumin decreases cell proliferation rates through BTG2-mediated cyclin D1 down-regulation in U937 cells. *Int J Oncol* **26**, 1597–1603 (2005).
169. D. Bech-Otschir, R. Kraft, X. Huang, P. Henklein, B. Kapelari, C. Pollmann, and W. Dubiel, COP9 signalosome-specific phosphorylation targets p53 to degradation by the ubiquitin system. *Embo J* **20**, 1630–1639 (2001).
170. P. J. Moos, K. Edes, J. E. Mullally, and F. A. Fitzpatrick, Curcumin impairs tumor suppressor p53 function in colon cancer cells. *Carcinogenesis* **25**, 1611–1617 (2004).
171. P. Tsvetkov, G. Asher, V. Reiss, Y. Shaul, L. Sachs, and J. Lotem, Inhibition of NAD(P)H:quinone oxidoreductase 1 activity and induction of p53 degradation by the natural phenolic compound curcumin. *Proc Natl Acad Sci USA* **102**, 5535–5540 (2005).
172. S. S. Kakar and D. Roy, Curcumin inhibits TPA induced expression of c-fos, c-jun and c-myc proto-oncogenes messenger RNAs in mouse skin. *Cancer Lett* **87**, 85–59 (1994).
173. M. T. Huang, W. Ma, Y. P. Lu, R. L. Chang, C. Fisher, P. S. Manchand, H. L. Newmark, and A. H. Conney, Effects of curcumin, demethoxycurcumin, bisdemethoxycurcumin and tetrahydrocurcumin on 12-O-tetradecanoylphorbol-13-acetate-induced tumor promotion. *Carcinogenesis* **16**, 2493–2497 (1995).
174. Y. P. Lu, R. L. Chang, Y. R. Lou, M. T. Huang, H. L. Newmark, K. R. Reuhl, and A. H. Conney, Effect of curcumin on 12-O-tetradecanoylphorbol-13-acetate- and ultraviolet

B light-induced expression of c-Jun and c-Fos in JB6 cells and in mouse epidermis. *Carcinogenesis* **15**, 2363–2370 (1994).

175. P. Limtrakul, S. Anuchapreeda, S. Lipigorngoson and F. W. Dunn, Inhibition of carcinogen induced c-Ha-ras and c-fos proto-oncogenes expression by dietary curcumin. *BMC Cancer* **1**, 1 (2001).
176. K. Nakamura, Y. Yasunaga, T. Segawa, D. Ko, J. W. Moul, S. Srivastava, and J. S. Rhim, Curcumin down-regulates AR gene expression and activation in prostate cancer cell lines. *Int J Oncol* **21**, 825–830 (2002).
177. M. T. Huang, T. Lysz, T. Ferraro, T. F. Abidi, J. D. Laskin, and A. H. Conney, Inhibitory effects of curcumin on in vitro lipoxygenase and cyclooxygenase activities in mouse epidermis. *Cancer Res* **51**, 813–819 (1991).
178. F. Zhang, N. K. Altorki, J. R. Mestre, K. Subbaramaiah, and A. J. Dannenberg, Curcumin inhibits cyclooxygenase-2 transcription in bile acid- and phorbol ester-treated human gastrointestinal epithelial cells. *Carcinogenesis* **20**, 445–451 (1999).
179. A. Goel, C. R. Boland, and D. P. Chauhan, Specific inhibition of cyclooxygenase-2 (COX-2) expression by dietary curcumin in HT-29 human colon cancer cells. *Cancer Lett* **172**, 111–118 (2001).
180. J. W. Cho, K. Park, G. R. Kweon, B. C. Jang, W. K. Baek, M. H. Suh, C. W. Kim, K. S. Lee, and S. I. Suh, Curcumin inhibits the expression of COX-2 in UVB-irradiated human keratinocytes (HaCaT) by inhibiting activation of AP-1: p38 MAP kinase and JNK as potential upstream targets. *Exp Mol Med* **37**, 186–192 (2005).
181. R. G. Tunstall, R. A. Sharma, S. Perkins, S. Sale, R. Singh, P. B. Farmer, W. P. Steward, and A. J. Gescher, Cyclooxygenase-2 expression and oxidative DNA adducts in murine intestinal adenomas: Modification by dietary curcumin and implications for clinical trials. *Eur J Cancer* **42**, 415–421 (2006).
182. D. L. Flynn, M. F. Rafferty, and A. M. Boctor, Inhibition of 5-hydroxy-eicosatetraenoic acid (5-HETE) formation in intact human neutrophils by naturally-occurring diarylheptanoids: inhibitory activities of curcuminoids and yakuchinones. *Prostaglandins Leukot Med* **22**, 357–360 (1986).
183. J. Hong, M. Bose, J. Ju, J. H. Ryu, X. Chen, S. Sang, M. J. Lee, and C. S. Yang, Modulation of arachidonic acid metabolism by curcumin and related beta-diketone derivatives: effects on cytosolic phospholipase A(2), cyclooxygenases and 5-lipoxygenase. *Carcinogenesis* **25**, 1671–1679 (2004).
184. N. S. Prasad, R. Raghavendra, B. R. Lokesh, and K. A. Naidu, Spice phenolics inhibit human PMNL 5-lipoxygenase. *Prostaglandins Leukot Essent Fatty Acids* **70**, 521–528 (2004).
185. P. F. Firozi, V. S. Aboobaker, and R. K. Bhattacharya, Action of curcumin on the cytochrome P450-system catalyzing the activation of aflatoxin B1. *Chem Biol Interact* **100**, 41–51 (1996).
186. H. P. Ciolino, P. J. Daschner, T. T. Wang, and G. C. Yeh, Effect of curcumin on the aryl hydrocarbon receptor and cytochrome P450 1A1 in MCF-7 human breast carcinoma cells. *Biochem Pharmacol* **56**, 197–206 (1998).
187. R. Thapliyal, S. S. Deshpande, and G. B. Maru, Effects of turmeric on the activities of benzo(a)pyrene-induced cytochrome P-450 isozymes. *J Environ Pathol Toxicol Oncol* **20**, 59–63 (2001).
188. T. Sugiyama, J. Nagata, A. Yamagishi, K. Endoh, M. Saito, K. Yamada, S. Yamada, and K. Umegaki, Selective protection of curcumin against carbon tetrachloride-induced inactivation of hepatic cytochrome P450 isozymes in rats. *Life Sci* **78**, 2188–2193 (2006).

189. M. J. Van Erk, E. Teuling, Y. C. Staal, S. Huybers, P. J. Van Bladeren, J. M. Aarts, and B. Van Ommen, Time- and dose-dependent effects of curcumin on gene expression in human colon cancer cells. *J Carcinog* **3**, 8 (2004).
190. C. Yan, M. S. Jamaluddin, B. Aggarwal, J. Myers, and D. D. Boyd, Gene expression profiling identifies activating transcription factor 3 as a novel contributor to the proapoptotic effect of curcumin. *Mol Cancer Ther* **4**, 233–241 (2005).
191. H. M. Wortelboer, M. Usta, A. E. van der Velde, M. G. Boersma, B. Spenkelink, J. J. van Zanden, I. M. Rietjens, P. J. van Bladeren, and N. H. Cnubben, Interplay between MRP inhibition and metabolism of MRP inhibitors: the case of curcumin. *Chem Res Toxicol* **16**, 1642–1651 (2003).
192. W. Chearwae, S. Anuchapreeda, K. Nandigama, S. V. Ambudkar, and P. Limtrakul, Biochemical mechanism of modulation of human P-glycoprotein (ABCB1) by curcumin I, II, and III purified from Turmeric powder. *Biochem Pharmacol* **68**, 2043–2052 (2004).
193. P. Limtrakul, S. Anuchapreeda, and D. Buddhasukh, Modulation of human multidrug-resistance MDR-1 gene by natural curcuminoids. *BMC Cancer* **4**, 13 (2004).
194. W. Chearwae, C. P. Wu, H. Y. Chu, T. R. Lee, S. V. Ambudkar, and P. Limtrakul, Curcuminoids purified from turmeric powder modulate the function of human multidrug resistance protein 1 (ABCC1). *Cancer Chemother Pharmacol* **57**, 376–388 (2006).
195. X. Q. Tang, H. Bi, J. Q. Feng, and J. G. Cao, Effect of curcumin on multidrug resistance in resistant human gastric carcinoma cell line SGC7901/VCR. *Acta Pharmacol Sin* **26**, 1009–1016 (2005).
196. J. Lee, H. H. Jung, Y. H. Im, J. H. Kim, J. O. Park, K. Kim, W. S. Kim, J. S. Ahn, C. W. Jung, Y. S. Park, W. K. Kang, and K. Park, Interferon-alpha resistance can be reversed by inhibition of IFN-alpha-induced COX-2 expression potentially via STAT1 activation in A549 cells. *Oncol Rep* **15**, 1541–1549 (2006).
197. C. Park, G. Y. Kim, G. D. Kim, B. T. Choi, Y. M. Park, and Y. H. Choi, Induction of G2/M arrest and inhibition of cyclooxygenase-2 activity by curcumin in human bladder cancer T24 cells. *Oncol Rep* **15**, 1225–1231 (2006).
198. W. J. Durham, S. Arbogast, E. Gerken, Y. P. Li, and M. B. Reid, Progressive nuclear factor-kappaB activation resistant to inhibition by contraction and curcumin in mdx mice. *Muscle Nerve* **34**(3), 298–303 (2006).
199. C. C. Su, G. W. Chen, J. G. Lin, L. T. Wu, and J. G. Chung, Curcumin inhibits cell migration of human colon cancer colo 205 cells through the inhibition of nuclear factor kappa B /p65 and down-regulates cyclooxygenase-2 and matrix metalloproteinase-2 expressions. *Anticancer Res* **26**, 1281–1288 (2006).
200. N. Chakravarti, J. N. Myers, and B. B. Aggarwal, Targeting constitutive and interleukin-6-inducible signal transducers and activators of transcription 3 pathway in head and neck squamous cell carcinoma cells by curcumin (diferuloylmethane). *Int J Cancer* **119**(6), 1268–1275 (2006).
201. A. M. Siddiqui, X. Cui, R. Wu, W. Dong, M. Zhou, M. Hu, H. H. Simms, and P. Wang, The anti-inflammatory effect of curcumin in an experimental model of sepsis is mediated by up-regulation of peroxisome proliferator-activated receptor-gamma*. *Crit Care Med* **34**(7), 1874–1882 (2006).
202. H. C. Huang, T. R. Jan, and S. F. Yeh, Inhibitory effect of curcumin, an anti-inflammatory agent, on vascular smooth muscle cell proliferation. *Eur J Pharmacol* **221**, 381–384 (1992).
203. L. Korutla and R. Kumar, Inhibitory effect of curcumin on epidermal growth factor receptor kinase activity in A431 cells. *Biochim Biophys Acta* **1224**, 597–600 (1994).

204. J. F. Santibanez, M. Quintanilla, and J. Martinez, Genistein and curcumin block TGF-beta 1-induced u-PA expression and migratory and invasive phenotype in mouse epidermal keratinocytes. *Nutr Cancer* **37**, 49–54 (2000).
205. R. Mohan, J. Sivak, P. Ashton, L. A. Russo, B. Q. Pham, N. Kasahara, M. B. Raizman, and M. E. Fini, Curcuminoids inhibit the angiogenic response stimulated by fibroblast growth factor-2, including expression of matrix metalloproteinase gelatinase B. *J Biol Chem* **275**, 10,405–10,512 (2000).
206. S. C. Shih and K. P. Claffey, Role of AP-1 and HIF-1 transcription factors in TGF-beta activation of VEGF expression. *Growth Factors* **19**, 19–34 (2001).
207. H. Mani, G. S. Sidhu, R. Kumari, J. P. Gaddipati, P. Seth and R. K. Maheshwari, Curcumin differentially regulates TGF-beta1, its receptors and nitric oxide synthase during impaired wound healing. *Biofactors* **16**, 29–43 (2002).
208. A. E. Gururaj, M. Belakavadi, D. A. Venkatesh, D. Marme, and B. P. Salimath, Molecular mechanisms of anti-angiogenic effect of curcumin. *Biochem Biophys Res Commun* **297**, 934–942 (2002).
209. J. Gaedeke, N. A. Noble, and W. A. Border, Curcumin blocks multiple sites of the TGF-beta signaling cascade in renal cells. *Kidney Int* **66**, 112–120 (2004).
210. P. C. Smith, J. F. Santibanez, J. P. Morales, and J. Martinez, Epidermal growth factor stimulates urokinase-type plasminogen activator expression in human gingival fibroblasts. Possible modulation by genistein and curcumin. *J Periodontal Res* **39**, 380–387 (2004).
211. A. Chen, J. Xu, and A. C. Johnson, Curcumin inhibits human colon cancer cell growth by suppressing gene expression of epidermal growth factor receptor through reducing the activity of the transcription factor Egr-1. *Oncogene* **25**, 278–287 (2006).
212. J. H. Kim, C. Xu, Y. S. Keum, B. Reddy, A. Conney, and A. N. Kong, Inhibition of EGFR signaling in human prostate cancer PC-3 cells by combination treatment with beta-phenylethyl isothiocyanate and curcumin. *Carcinogenesis* **27**, 475–482 (2006).
213. S. Zheng and A. Chen, Curcumin suppresses the expression of extracellular matrix genes in activated hepatic stellate cells by inhibiting gene expression of connective tissue growth factor. *Am J Physiol Gastrointest Liver Physiol* **290**, G883–G893 (2006).
214. A. Masamune, N. Suzuki, K. Kikuta, M. Satoh, K. Satoh, and T. Shimosegawa, Curcumin blocks activation of pancreatic stellate cells. *J Cell Biochem* **97**, 1080–1093 (2006).
215. X. Yang, D. P. Thomas, X. Zhang, B. W. Culver, B. M. Alexander, W. J. Murdoch, M. N. Rao, D. A. Tulis, J. Ren, and N. Sreejayan, Curcumin inhibits platelet-derived growth factor-stimulated vascular smooth muscle cell function and injury-induced neointima formation. *Arterioscler Thromb Vasc Biol* **26**, 85–90 (2006).
216. Y. Takada, A. Bhardwaj, P. Potdar, and B. B. Aggarwal, Nonsteroidal anti-inflammatory agents differ in their ability to suppress NF-kappaB activation, inhibition of expression of cyclooxygenase-2 and cyclin D1, and abrogation of tumor cell proliferation. *Oncogene* **23**, 9247–9258 (2004).
217. J. M. Dogne, J. Hanson, C. Supuran, and D. Pratico, Coxibs and cardiovascular side-effects: from light to shadow. *Curr Pharm Des* **12**, 971–975 (2006).
218. A. T. Chan, J. E. Manson, C. M. Albert, C. U. Chae, K. M. Rexrode, G. C. Curhan, E. B. Rimm, W. C. Willett, and C. S. Fuchs, Nonsteroidal antiinflammatory drugs, acetaminophen, and the risk of cardiovascular events. *Circulation* **113**, 1578–1587 (2006).
219. M. Hermann and F. Ruschitzka, Coxibs, non-steroidal anti-inflammatory drugs and cardiovascular risk. *Intern Med J* **36**, 308–319 (2006).

220. B. Gupta and B. Ghosh, Curcuma longa inhibits TNF-alpha induced expression of adhesion molecules on human umbilical vein endothelial cells. *Int J Immunopharmacol* **21**, 745–757 (1999).
221. B. Madan and B. Ghosh, Diferuloylmethane inhibits neutrophil infiltration and improves survival of mice in high-dose endotoxin shock. *Shock* **19**, 91–96 (2003).
222. B. Fuller, S. Dijk, P. Butler, V. Hoang, and B. Davidson, Pro-inflammatory agents accumulate during donor liver cold preservation: A study on increased adhesion molecule expression and abrogation by curcumin in cultured endothelial cells. *Cryobiology* **46**, 284–288 (2003).
223. O. P. Sharma, Antioxidant activity of curcumin and related compounds. *Biochem Pharmacol* **25**, 1811–1812 (1976).
224. V. K. Shalini and L. Srinivas, Lipid peroxide induced DNA damage: protection by turmeric (*Curcuma longa*). *Mol Cell Biochem* **77**, 3–10 (1987).
225. M. Nagabhushan, U. J. Nair, A. J. Amonkar, A. V. D'Souza, and S. V. Bhide, Curcumins as inhibitors of nitrosation in vitro. *Mutat Res* **202**, 163–169 (1988).
226. I. A. Donatus, Sardjoko, and N. P. Vermeulen, Cytotoxic and cytoprotective activities of curcumin. Effects on paracetamol-induced cytotoxicity, lipid peroxidation and glutathione depletion in rat hepatocytes. *Biochem Pharmacol* **39**, 1869–1875 (1990).
227. S. C. Sahu and M. C. Washington, Effect of ascorbic acid and curcumin on quercetin-induced nuclear DNA damage, lipid peroxidation and protein degradation. *Cancer Lett* **63**, 237–241 (1992).
228. K. K. Soudamini, M. C. Unnikrishnan, K. B. Soni, and R. Kuttan, Inhibition of lipid peroxidation and cholesterol levels in mice by curcumin. *Indian J Physiol Pharmacol* **36**, 239–243 (1992).
229. M. K. Unnikrishnan and M. N. Rao, Curcumin inhibits nitrite-induced methemoglobin formation. *FEBS Lett* **301**, 195–196 (1992).
230. B. Joe and B. R. Lokesh, Role of capsaicin, curcumin and dietary n-3 fatty acids in lowering the generation of reactive oxygen species in rat peritoneal macrophages. *Biochim Biophys Acta* **1224**, 255–263 (1994).
231. D. V. Rajakumar and M. N. Rao, Antioxidant properties of dehydrozingerone and curcumin in rat brain homogenates. *Mol Cell Biochem* **140**, 73–79 (1994).
232. A. C. Reddy and B. R. Lokesh, Effect of dietary turmeric (*Curcuma longa*) on iron-induced lipid peroxidation in the rat liver. *Food Chem Toxicol* **32**, 279–283 (1994).
233. A. C. Reddy and B. R. Lokesh, Studies on the inhibitory effects of curcumin and eugenol on the formation of reactive oxygen species and the oxidation of ferrous iron. *Mol Cell Biochem* **137**, 1–8 (1994).
234. Sreejayan and M. N. Rao, Curcuminoids as potent inhibitors of lipid peroxidation. *J Pharm Pharmacol* **46**, 1013–1016 (1994).
235. R. Selvam, L. Subramanian, R. Gayathri, and N. Angayarkanni, The anti-oxidant activity of turmeric (*Curcuma longa*). *J Ethnopharmacol* **47**, 59–67 (1995).
236. M. K. Unnikrishnan and M. N. Rao, Inhibition of nitrite induced oxidation of hemoglobin by curcuminoids. *Pharmazie* **50**, 490–492 (1995).
237. M. K. Unnikrishnan and M. N. Rao, Curcumin inhibits nitrogen dioxide induced oxidation of hemoglobin. *Mol Cell Biochem* **146**, 35–37 (1995).
238. S. Kaul and T. P. Krishnakantha, Influence of retinol deficiency and curcumin/turmeric feeding on tissue microsomal membrane lipid peroxidation and fatty acids in rats. *Mol Cell Biochem* **175**, 43–48 (1997).
239. A. Nogaki, K. Satoh, K. Iwasaka, H. Takano, M. Takahama, Y. Ida, and H. Sakagami, Radical intensity and cytotoxic activity of curcumin and gallic acid. *Anticancer Res* **18**, 3487–3491 (1998).

240. S. Bhaumik, M. D. Jyothi, and A. Khar, Differential modulation of nitric oxide production by curcumin in host macrophages and NK cells. *FEBS Lett* **483**, 78–82 (2000).
241. S. Kapoor and K. I. Priyadarsini, Protection of radiation-induced protein damage by curcumin. *Biophys Chem* **92**, 119–126 (2001).
242. M. R. Kelly, J. Xu, K. E. Alexander, and G. Loo, Disparate effects of similar phenolic phytochemicals as inhibitors of oxidative damage to cellular DNA. *Mutat Res* **485**, 309–318 (2001).
243. T. Masuda, T. Maekawa, K. Hidaka, H. Bando, Y. Takeda, and H. Yamaguchi, Chemical studies on antioxidant mechanism of curcumin: analysis of oxidative coupling products from curcumin and linoleate. *J Agric Food Chem* **49**, 2539–2547 (2001).
244. K. C. Das and C. K. Das, Curcumin (diferuloylmethane), a singlet oxygen ((1)O(2)) quencher. *Biochem Biophys Res Commun* **295**, 62–66 (2002).
245. G. K. Jayaprakasha, B. S. Jena, P. S. Negi, and K. K. Sakariah, Evaluation of antioxidant activities and antimutagenicity of turmeric oil: a byproduct from curcumin production. *Z Naturforsch [C]* **57**, 828–835 (2002).
246. R. Toniolo, F. Di Narda, S. Susmel, M. Martelli, L. Martelli, and G. Bontempelli, Quenching of superoxide ions by curcumin. A mechanistic study in acetonitrile. *Ann Chim* **92**, 281–288 (2002).
247. M. Balasubramanyam, A. A. Koteswari, R. S. Kumar, S. F. Monickaraj, J. U. Maheswari, and V. Mohan, Curcumin-induced inhibition of cellular reactive oxygen species generation: novel therapeutic implications. *J Biosci* **28**, 715–721 (2003).
248. A. Betancor-Fernandez, A. Perez-Galvez, H. Sies, and W. Stahl, Screening pharmaceutical preparations containing extracts of turmeric rhizome, artichoke leaf, devil's claw root and garlic or salmon oil for antioxidant capacity. *J Pharm Pharmacol* **55**, 981–986 (2003).
249. S. M. Chauhan, A. S. Kandadai, N. Jain, and A. Kumar, Biomimetic oxidation of curcumin with hydrogen peroxide catalyzed by 5,10,15,20-tetraarylporphyrinatoiron(III) chlorides in dichloromethane. *Chem Pharm Bull (Tokyo)* **51**, 1345–1347 (2003).
250. M. Iqbal, Y. Okazaki and S. Okada, In vitro curcumin modulates ferric nitrilotriacetate (Fe-NTA) and hydrogen peroxide (H2O2)-induced peroxidation of microsomal membrane lipids and DNA damage. *Teratog Carcinog Mutagen* **1(Suppl)**, 151–160 (2003).
251. B. D. Johnston and E. G. DeMaster, Suppression of nitric oxide oxidation to nitrite by curcumin is due to the sequestration of the reaction intermediate nitrogen dioxide, not nitric oxide. *Nitric Oxide* **8**, 231–234 (2003).
252. R. Rukkumani, M. Sri Balasubashini, and V. P. Menon, Protective effects of curcumin and photo-irradiated curcumin on circulatory lipids and lipid peroxidation products in alcohol and polyunsaturated fatty acid-induced toxicity. *Phytother Res* **17**, 925–929 (2003).
253. V. Eybl, D. Kotyzova, and M. Bludovska, The effect of curcumin on cadmium-induced oxidative damage and trace elements level in the liver of rats and mice. *Toxicol Lett* **151**, 79–85 (2004).
254. S. Fujisawa, T. Atsumi, M. Ishihara, and Y. Kadoma, Cytotoxicity, ROS-generation activity and radical-scavenging activity of curcumin and related compounds. *Anticancer Res* **24**, 563–569 (2004).
255. M. O. Iwunze and D. McEwan, Peroxynitrite interaction with curcumin solubilized in ethanolic solution. *Cell Mol Biol (Noisy-le-grand)* **50**, 749–752 (2004).
256. C. Kalpana and V. P. Menon, Modulatory effects of curcumin on lipid peroxidation and antioxidant status during nicotine-induced toxicity. *Pol J Pharmacol* **56**, 581–586 (2004).

257. R. K. Kempaiah and K. Srinivasan, Influence of dietary curcumin, capsaicin and garlic on the antioxidant status of red blood cells and the liver in high-fat-fed rats. *Ann Nutr Metab* **48**, 314–320 (2004).
258. B. Mishra, K. I. Priyadarsini, M. K. Bhide, R. M. Kadam, and H. Mohan, Reactions of superoxide radicals with curcumin: Probable mechanisms by optical spectroscopy and EPR. *Free Radical Res* **38**, 355–362 (2004).
259. R. Barreto, S. Kawakita, J. Tsuchiya, E. Minelli, K. Pavasuthipaisit, A. Helmy, and F. Marotta, Metal-induced oxidative damage in cultured hepatocytes and hepatic lysosomal fraction: beneficial effect of a curcumin/absinthium compound. *Chin J Dig Dis* **6**, 31–36,(2005).
260. J. Chen, D. Wanming, D. Zhang, Q. Liu, and J. Kang, Water-soluble antioxidants improve the antioxidant and anticancer activity of low concentrations of curcumin in human leukemia cells. *Pharmazie* **60**, 57–61 (2005).
261. S. Durgaprasad, C. G. Pai, Vasanthkumar, J. F. Alvres, and S. Namitha, A pilot study of the antioxidant effect of curcumin in tropical pancreatitis. *Indian J Med Res* **122**, 315–318 (2005).
262. V. Eybl, D. Kotyzova, L. Leseticky, M. Bludovska, and J. Koutensky, The influence of curcumin and manganese complex of curcumin on cadmium-induced oxidative damage and trace elements status in tissues of mice. *J Appl Toxicol* **26**(3), 207–212 (2005).
263. W. M. Weber, L. A. Hunsaker, S. F. Abcouwer, L. M. Deck, and D. L. Vander Jagt, Anti-oxidant activities of curcumin and related enones. *Bioorg Med Chem* **13**, 3811–3820 (2005).
264. M. Sreepriya and G. Bali, Effects of administration of Embelin and Curcumin on lipid peroxidation, hepatic glutathione antioxidant defense and hematopoietic system during N-nitrosodiethylamine/Phenobarbital-induced hepatocarcinogenesis in Wistar rats. *Mol Cell Biochem*, 1–7 (2006).
265. Q. Y. Wei, W. F. Chen, B. Zhou, L. Yang, and Z. L. Liu, Inhibition of lipid peroxidation and protein oxidation in rat liver mitochondria by curcumin and its analogues. *Biochim Biophys Acta* **1760**, 70–77 (2006).
266. A. R. Shahed, E. Jones, and D. Shoskes, Quercetin and curcumin up-regulate antioxidant gene expression in rat kidney after ureteral obstruction or ischemia/reperfusion injury. *Transplant Proc* **33**, 2988 (2001).
267. F. Bonte, M. S. Noel-Hudson, J. Wepierre and A. Meybeck, Protective effect of curcuminoids on epidermal skin cells under free oxygen radical stress. *Planta Med* **63**, 265–266 (1997).
268. S. Watanabe and T. Fukui, Suppressive effect of curcumin on trichloroethylene-induced oxidative stress. *J Nutr Sci Vitaminol (Tokyo)* **46**, 230–234 (2000).
269. J. L. Quiles, M. D. Mesa, C. L. Ramirez-Tortosa, C. M. Aguilera, M. Battino, A. Gil and M. C. Ramirez-Tortosa, *Curcuma longa* extract supplementation reduces oxidative stress and attenuates aortic fatty streak development in rabbits. *Arterioscler Thromb Vasc Biol* **22**, 1225–1231 (2002).
270. W. H. Chan, C. C. Wu, and J. S. Yu, Curcumin inhibits UV irradiation-induced oxidative stress and apoptotic biochemical changes in human epidermoid carcinoma A431 cells. *J Cell Biochem* **90**, 327–338 (2003).
271. P. Mahakunakorn, M. Tohda, Y. Murakami, K. Matsumoto, H. Watanabe, and O. Vajaragupta, Cytoprotective and cytotoxic effects of curcumin: dual action on H_2O_2-induced oxidative cell damage in NG108-15 cells. *Biol Pharm Bull* **26**, 725–728 (2003).

272. R. Rukkumani, K. Aruna, P. S. Varma, K. N. Rajasekaran, and V. P. Menon, Comparative effects of curcumin and an analog of curcumin on alcohol and PUFA induced oxidative stress. *J Pharm Pharm Sci* **7**, 274–283 (2004).
273. R. Banjerdpongchai and P. Wilairat, Effects of water-soluble antioxidants and MAPKK/MEK inhibitor on curcumin-induced apoptosis in HL-60 human leukemic cells. *Asian Pac J Cancer Prev* **6**, 282–285 (2005).
274. Y. D. Hsuuw, C. K. Chang, W. H. Chan, and J. S. Yu, Curcumin prevents methylglyoxal-induced oxidative stress and apoptosis in mouse embryonic stem cells and blastocysts. *J Cell Physiol* **205**, 379–386 (2005).
275. T. Mahesh, M. S. Balasubashini, and V. P. Menon, Effect of photo-irradiated curcumin treatment against oxidative stress in streptozotocin-induced diabetic rats. *J Med Food* **8**, 251–255 (2005).
276. K. U. Schallreuter and H. Rokos, Turmeric (curcumin): A widely used curry ingredient, can contribute to oxidative stress in Asian patients with acute vitiligo. *Indian J Dermatol Venereol Leprol* **72**, 57–59 (2006).
277. I. Chattopadhyay, U. Bandyopadhyay, K. Biswas, P. Maity, and R. K. Banerjee, Indomethacin inactivates gastric peroxidase to induce reactive-oxygen-mediated gastric mucosal injury and curcumin protects it by preventing peroxidase inactivation and scavenging reactive oxygen. *Free Radical Biol Med* **40**, 1397–1408 (2006).
278. K. Cleary and R. F. McFeeters, Effects of oxygen and turmeric on the formation of oxidative aldehydes in fresh-pack dill pickles. *J Agric Food Chem* **54**, 3421–3427 (2006).
279. G. Scapagnini, C. Colombrita, M. Amadio, V. D'Agata, E. Arcelli, M. Sapienza, A. Quattrone, and V. Calabrese, Curcumin activates defensive genes and protects neurons against oxidative stress. *Antioxid Redox Signal* **8**, 395–403 (2006).
280. M. Yoshino, M. Haneda, M. Naruse, H. H. Htay, R. Tsubouchi, S. L. Qiao, W. H. Li, K. Murakami, and T. Yokochi, Prooxidant activity of curcumin: Copper-dependent formation of 8-hydroxy-2'-deoxyguanosine in DNA and induction of apoptotic cell death. *Toxicol In Vitro* **18**, 783–789 (2004).
281. T. Atsumi, S. Fujisawa, and K. Tonosaki, Relationship between intracellular ROS production and membrane mobility in curcumin- and tetrahydrocurcumin-treated human gingival fibroblasts and human submandibular gland carcinoma cells. *Oral Dis* **11**, 236–242 (2005).
282. S. Fujisawa and Y. Kadoma, Anti- and pro-oxidant effects of oxidized quercetin, curcumin or curcumin-related compounds with thiols or ascorbate as measured by the induction period method. *In Vivo* **20**, 39–44 (2006).
283. S. Bhaumik, R. Anjum, N. Rangaraj, B. V. Pardhasaradhi, and A. Khar, Curcumin mediated apoptosis in AK-5 tumor cells involves the production of reactive oxygen intermediates. *FEBS Lett* **456**, 311–314 (1999).
284. J. Fang, J. Lu, and A. Holmgren, Thioredoxin reductase is irreversibly modified by curcumin: A novel molecular mechanism for its anticancer activity. *J Biol Chem* **280**, 25,284–25,290 (2005).
285. E. C. Herrmann and E. C. Moore, Purification of thioredoxin from rat Novikoff ascites hepatoma. *J Biol Chem* **248**, 1219–1223 (1973).
286. E. C. Moore, A thioredoxin–thioredoxin reductase system from rat tumor. *Biochem Biophys Res Commun* **29**, 264–8, (1967).
287. K. U. Schallreuter and J. M. Wood, The activity and purification of membrane–associated thioredoxin reductase from human metastatic melanotic melanoma. *Biochim Biophys Acta* **967**, 103–109 (1988).

288. J. L. Quiles, C. Aguilera, M. D. Mesa, M. C. Ramirez-Tortosa, L. Baro, and A. Gil, An ethanolic-aqueous extract of *Curcuma longa* decreases the susceptibility of liver microsomes and mitochondria to lipid peroxidation in atherosclerotic rabbits. *Biofactors* **8**, 51–57 (1998).
289. L. M. Antunes, J. D. Darin, and L. Bianchi Nde, Effects of the antioxidants curcumin or selenium on cisplatin-induced nephrotoxicity and lipid peroxidation in rats. *Pharmacol Res* **43**, 145–150 (2001).
290. A. Singh, S. P. Singh, and R. Bamezai, Postnatal modulation of hepatic biotransformation system enzymes via translactational exposure of F1 mouse pups to turmeric and curcumin. *Cancer Lett* **96**, 87–93 (1995).
291. S. Oetari, M. Sudibyo, J. N. Commandeur, R. Samhoedi, and N. P. Vermeulen, Effects of curcumin on cytochrome P450 and glutathione S–transferase activities in rat liver. *Biochem Pharmacol* **51**, 39–45 (1996).
292. A. Singh, S. P. Singh, and R. Bamezai, Effect of arecoline on the curcumin-modulated hepatic biotransformation system enzymes in lactating mice and translactationally exposed F1 pups. *Nutr Cancer* **25**, 101–110 (1996).
293. M. L. van Iersel, J. P. Ploemen, M. Lo Bello, G. Federici, and P. J. van Bladeren, Interactions of alpha, beta-unsaturated aldehydes and ketones with human glutathione S-transferase P1-1. *Chem Biol Interact* **108**, 67–78 (1997).
294. M. Iqbal, S. D. Sharma, Y. Okazaki, M. Fujisawa, and S. Okada, Dietary supplementation of curcumin enhances antioxidant and phase II metabolizing enzymes in ddY male mice: possible role in protection against chemical carcinogenesis and toxicity. *Pharmacol Toxicol* **92**, 33–38 (2003).
295. Y. Jiao, J. t. Wilkinson, E. Christine Pietsch, J. L. Buss, W. Wang, R. Planalp, F. M. Torti, and S. V. Torti, Iron chelation in the biological activity of curcumin. *Free Radical Biol Med* **40**, 1152–1160 (2006).
296. A. C. Bharti, N. Donato, and B. B. Aggarwal, Curcumin (diferuloylmethane) inhibits constitutive and IL-6-inducible STAT3 phosphorylation in human multiple myeloma cells. *J Immunol* **171**, 3863–3871 (2003).
297. R. K. Giri, V. Rajagopal, and V. K. Kalra, Curcumin, the active constituent of turmeric, inhibits amyloid peptide-induced cytochemokine gene expression and CCR5-mediated chemotaxis of THP-1 monocytes by modulating early growth response-1 transcription factor. *J Neurochem* **91**, 1199–1210 (2004).
298. D. Ranjan, C. Chen, T. D. Johnston, H. Jeon, and M. Nagabhushan, Curcumin inhibits mitogen stimulated lymphocyte proliferation, NFkappaB activation, and IL-2 signaling. *J Surg Res* **121**, 171–177 (2004).
299. J. L. Arbiser, N. Klauber, R. Rohan, R. van Leeuwen, M. T. Huang, C. Fisher, E. Flynn, and H. R. Byers, Curcumin is an in vivo inhibitor of angiogenesis. *Mol Med* **4**, 376–383 (1998).
300. D. Thaloor, A. K. Singh, G. S. Sidhu, P. V. Prasad, H. K. Kleinman, and R. K. Maheshwari, Inhibition of angiogenic differentiation of human umbilical vein endothelial cells by curcumin. *Cell Growth Differ* **9**, 305–312 (1998).
301. J. S. Shim, J. H. Kim, H. Y. Cho, Y. N. Yum, S. H. Kim, H. J. Park, B. S. Shim, S. H. Choi, and H. J. Kwon, Irreversible inhibition of CD13/aminopeptidase N by the antiangiogenic agent curcumin. *Chem Biol* **10**, 695–704 (2003).
302. P. Yoysungnoen, P. Wirachwong, P. Bhattarakosol, H. Niimi, and S. Patumraj, Antiangiogenic activity of curcumin in hepatocellular carcinoma cells implanted nude mice. *Clin Hemorheol Microcirc* **33**, 127–135 (2005).

303. P. Yoysungnoen, P. Wirachwong, P. Bhattarakosol, H. Niimi, and S. Patumraj, Effects of curcumin on tumor angiogenesis and biomarkers, COX-2 and VEGF, in hepatocellular carcinoma cell-implanted nude mice. *Clin Hemorheol Microcirc* **34**, 109–115 (2006).
304. A. Barik, K. I. Priyadarsini, and H. Mohan, Photophysical studies on binding of curcumin to bovine serum albumins. *Photochem Photobiol* **77**, 597–603 (2003).
305. F. Zsila, Z. Bikadi, and M. Simonyi, Unique, pH-dependent biphasic band shape of the visible circular dichroism of curcumin-serum albumin complex. *Biochem Biophys Res Commun* **301**, 776–782 (2003).
306. A. Barik, B. Mishra, L. Shen, H. Mohan, R. M. Kadam, S. Dutta, H. Y. Zhang, and K. I. Priyadarsini, Evaluation of a new copper(II)-curcumin complex as superoxide dismutase mimic and its free radical reactions. *Free Radical Biol Med* **39**, 811–822 (2005).
307. E. Skrzypczak-Jankun, K. Zhou, N. P. McCabe, S. H. Selman, and J. Jankun, Structure of curcumin in complex with lipoxygenase and its significance in cancer. *Int J Mol Med* **12**, 17–24 (2003).
308. R. S. Ramsewak, D. L. DeWitt, and M. G. Nair, Cytotoxicity, antioxidant and anti-inflammatory activities of curcumins I-III from *Curcuma longa*. *Phytomedicine* **7**, 303–308 (2000).
309. A. C. Bharti, N. Donato, S. Singh, and B. B. Aggarwal, Curcumin (diferuloylmethane) down-regulates the constitutive activation of nuclear factor-kappa B and IkappaBalpha kinase in human multiple myeloma cells, leading to suppression of proliferation and induction of apoptosis. *Blood* **101**, 1053–1062 (2003).
310. N. Romiti, R. Tongiani, F. Cervelli, and E. Chieli, Effects of curcumin on P-glycoprotein in primary cultures of rat hepatocytes. *Life Sci* **62**, 2349–2358 (1998).
311. S. Anuchapreeda, P. Leechanachai, M. M. Smith, S. V. Ambudkar, and P. N. Limtrakul, Modulation of P-glycoprotein expression and function by curcumin in multidrug-resistant human KB cells. *Biochem Pharmacol* **64**, 573–582 (2002).
312. R. D. Snyder and M. R. Arnone, Putative identification of functional interactions between DNA intercalating agents and topoisomerase II using the V79 in vitro micronucleus assay. *Mutat Res* **503**, 21–35 (2002).
313. J. L. Dyer, S. Z. Khan, J. G. Bilmen, S. R. Hawtin, M. Wheatley, M. U. Javed, and F. Michelangeli, Curcumin: a new cell-permeant inhibitor of the inositol 1,4,5-trisphosphate receptor. *Cell Calcium* **31**, 45–52 (2002).
314. R. R. Satoskar, S. J. Shah, and S. G. Shenoy, Evaluation of anti-inflammatory property of curcumin (diferuloyl methane) in patients with postoperative inflammation. *Int J Clin Pharmacol Ther Toxicol* **24**, 651–654 (1986).
315. G. J. Kelloff, C. W. Boone, J. A. Crowell, V. E. Steele, R. Lubet, and C. C. Sigman, Chemopreventive drug development: perspectives and progress. *Cancer Epidemiol Biomarkers Prev* **3**, 85–98 (1994).
316. J. S. James, Curcumin: Clinical trial finds no antiviral effect. *AIDS Treat News* (no. 242), 1–2 (1996).
317. G. Shoba, D. Joy, T. Joseph, M. Majeed, R. Rajendran, and P. S. Srinivas, Influence of piperine on the pharmacokinetics of curcumin in animals and human volunteers. *Planta Med* **64**, 353–356,(1998).
318. B. Lal, A. K. Kapoor, O. P. Asthana, P. K. Agrawal, R. Prasad, P. Kumar, and R. C. Srimal, Efficacy of curcumin in the management of chronic anterior uveitis. *Phytother Res* **13**, 318–322 (1999).

319. A. Rasyid and A. Lelo, The effect of curcumin and placebo on human gall-bladder function: an ultrasound study. *Aliment Pharmacol Ther* **13**, 245–249,(1999).
320. M. C. Heng, M. K. Song, J. Harker, and M. K. Heng, Drug-induced suppression of phosphorylase kinase activity correlates with resolution of psoriasis as assessed by clinical, histological and immunohistochemical parameters. *Br J Dermatol* **143**, 937–049 (2000).
321. R. Lodha and A. Bagga, Traditional Indian systems of medicine. *Ann Acad Med Singapore* **29**, 37–41 (2000).
322. A. L. Cheng, C. H. Hsu, J. K. Lin, M. M. Hsu, Y. F. Ho, T. S. Shen, J. Y. Ko, J. T. Lin, B. R. Lin, W. Ming-Shiang, H. S. Yu, S. H. Jee, G. S. Chen, T. M. Chen, C. A. Chen, M. K. Lai, Y. S. Pu, M. H. Pan, Y. J. Wang, C. C. Tsai, and C. Y. Hsieh, Phase I clinical trial of curcumin, a chemopreventive agent, in patients with high-risk or pre-malignant lesions. *Anticancer Res* **21**, 2895–2900 (2001).
323. S. M. Plummer, K. A. Hill, M. F. Festing, W. P. Steward, A. J. Gescher, and R. A. Sharma, Clinical development of leukocyte cyclooxygenase 2 activity as a systemic biomarker for cancer chemopreventive agents. *Cancer Epidemiol Biomarkers Prev* **10**, 1295–1299 (2001).
324. R. A. Sharma, H. R. McLelland, K. A. Hill, C. R. Ireson, S. A. Euden, M. M. Manson, M. Pirmohamed, L. J. Marnett, A. J. Gescher, and W. P. Steward, Pharmacodynamic and pharmacokinetic study of oral Curcuma extract in patients with colorectal cancer. *Clin Cancer Res* **7**, 1894–1900 (2001).
325. A. Rasyid, A. R. Rahman, K. Jaalam, and A. Lelo, Effect of different curcumin dosages on human gall bladder. *Asia Pacific J Clin Nutr* **11**, 314–318 (2002).
326. M. Bayes, X. Rabasseda, and J. R. Prous, Gateways to clinical trials. *Methods Find Exp Clin Pharmacol* **26**, 723–753 (2004).
327. G. M. Cole, T. Morihara, G. P. Lim, F. Yang, A. Begum, and S. A. Frautschy, NSAID and antioxidant prevention of Alzheimer's disease: Lessons from in vitro and animal models. *Ann N Y Acad Sci* **1035**, 68–84 (2004).
328. G. Garcea, D. J. Jones, R. Singh, A. R. Dennison, P. B. Farmer, R. A. Sharma, W. P. Steward, A. J. Gescher, and D. P. Berry, Detection of curcumin and its metabolites in hepatic tissue and portal blood of patients following oral administration. *Br J Cancer* **90**, 1011–1015 (2004).
329. R. A. Sharma, S. A. Euden, S. L. Platton, D. N. Cooke, A. Shafayat, H. R. Hewitt, T. H. Marczylo, B. Morgan, D. Hemingway, S. M. Plummer, M. Pirmohamed, A. J. Gescher, and W. P. Steward, Phase I clinical trial of oral curcumin: biomarkers of systemic activity and compliance. *Clin Cancer Res* **10**, 6847–6854 (2004).
330. M. Bayes, X. Rabasseda, and J. R. Prous, Gateways to clinical trials. *Methods Find Exp Clin Pharmacol* **27**, 711–738 (2005).
331. G. Garcea, D. P. Berry, D. J. Jones, R. Singh, A. R. Dennison, P. B. Farmer, R. A. Sharma, W. P. Steward, and A. J. Gescher, Consumption of the putative chemopreventive agent curcumin by cancer patients: assessment of curcumin levels in the colorectum and their pharmacodynamic consequences. *Cancer Epidemiol Biomarkers Prev* **14**, 120–125 (2005).
332. P. R. Holt, S. Katz, and R. Kirshoff, Curcumin therapy in inflammatory bowel disease: A pilot study. *Dig Dis Sci* **50**, 2191–2193 (2005).
333. D. Shoskes, C. Lapierre, M. Cruz-Corerra, N. Muruve, R. Rosario, B. Fromkin, M. Braun, and J. Copley, Beneficial effects of the bioflavonoids curcumin and quercetin on early function in cadaveric renal transplantation: A randomized placebo controlled trial. *Transplantation* **80**, 1556–1559 (2005).

334. C. D. Lao, M. T. t. Ruffin, D. Normolle, D. D. Heath, S. I. Murray, J. M. Bailey, M. E. Boggs, J. Crowell, C. L. Rock, and D. E. Brenner, Dose escalation of a curcuminoid formulation. *BMC Complement Altern Med* **6**, 10 (2006).
335. M. Cruz-Correa, D. A. Shoskes, P. Sanchez, R. Zhao, L. M. Hylind, S. D. Wexner, and F. M. Giardiello, Combination treatment with curcumin and quercetin of adenomas in familial adenomatous polyposis. *Clin Gastroenterol Hepatol* **4**, 1035–1038 (2006).
336. I. Gukovsky, C. N. Reyes, E. C. Vaquero, A. S. Gukovskaya, and S. J. Pandol, Curcumin ameliorates ethanol and nonethanol experimental pancreatitis. *Am J Physiol Gastrointest Liver Physiol* **284**, G85–G9, (2003).
337. A. Gulcubuk, K. Sonmez, A. Gurel, K. Altunatmaz, N. Gurler, S. Aydin, L. Oksuz, H. Uzun, and O. Guzel, Pathologic alterations detected in acute pancreatitis induced by sodium taurocholate in rats and therapeutic effects of curcumin, ciprofloxacin and metronidazole combination. *Pancreatology* **5**, 345–353 (2005).
338. B. Joe, U. J. Rao, and B. R. Lokesh, Presence of an acidic glycoprotein in the serum of arthritic rats: modulation by capsaicin and curcumin. *Mol Cell Biochem* **169**, 125–134 (1997).
339. A. Liacini, J. Sylvester, W. Q. Li, and M. Zafarullah, Inhibition of interleukin-1-stimulated MAP kinases, activating protein-1 (AP-1) and nuclear factor kappa B (NF-kappa B) transcription factors down-regulates matrix metalloproteinase gene expression in articular chondrocytes. *Matrix Biol* **21**, 251–262 (2002).
340. A. Liacini, J. Sylvester, W. Q. Li, W. Huang, F. Dehnade, M. Ahmad, and M. Zafarullah, Induction of matrix metalloproteinase-13 gene expression by TNF-alpha is mediated by MAP kinases, AP-1, and NF-kappaB transcription factors in articular chondrocytes. *Exp Cell Res* **288**, 208–217 (2003).
341. J. Sylvester, A. Liacini, W. Q. Li, and M. Zafarullah, Interleukin-17 signal transduction pathways implicated in inducing matrix metalloproteinase-3, -13 and aggrecanase-1 genes in articular chondrocytes. *Cell Signal* **16**, 469–476 (2004).
342. K. Sugimoto, H. Hanai, K. Tozawa, T. Aoshi, M. Uchijima, T. Nagata, and Y. Koide, Curcumin prevents and ameliorates trinitrobenzene sulfonic acid-induced colitis in mice. *Gastroenterology* **123**, 1912–1922 (2002).
343. B. Salh, K. Assi, V. Templeman, K. Parhar, D. Owen, A. Gomez-Munoz, and K. Jacobson, Curcumin attenuates DNB-induced murine colitis. *Am J Physiol Gastrointest Liver Physiol* **285**, G235–G243 (2003).
344. Y. Jiang, Z. S. Li, F. S. Jiang, X. Deng, C. S. Yao, and G. Nie, Effects of different ingredients of zedoary on gene expression of HSC-T6 cells. *World J Gastroenterol* **11**, 6780–6786 (2005).
345. D. C. Kim, S. H. Kim, B. H. Choi, N. I. Baek, D. Kim, M. J. Kim, and K. T. Kim, *Curcuma longa* extract protects against gastric ulcers by blocking H2 histamine receptors. *Biol Pharm Bull* **28**, 2220–2224 (2005).
346. S. Swarnakar, K. Ganguly, P. Kundu, A. Banerjee, P. Maity, and A. V. Sharma, Curcumin regulates expression and activity of matrix metalloproteinases 9 and 2 during prevention and healing of indomethacin-induced gastric ulcer. *J Biol Chem* **280**, 9409–9415 (2005).
347. O. S. Baek, O. H. Kang, Y. A. Choi, S. C. Choi, T. H. Kim, Y. H. Nah, D. Y. Kwon, Y. K. Kim, Y. H. Kim, K. H. Bae, J. P. Lim, and Y. M. Lee, Curcumin inhibits protease-activated receptor-2 and -4-mediated mast cell activation. *Clin Chim Acta* **338**, 135–141 (2003).
348. A. Ram, M. Das, and B. Ghosh, Curcumin attenuates allergen-induced airway hyperresponsiveness in sensitized guinea pigs. *Biol Pharm Bull* **26**, 1021–1024 (2003).

349. J. J. Lee, W. T. Huang, D. Z. Shao, J. F. Liao, and M. T. Lin, Blocking NF-kappaB activation may be an effective strategy in the fever therapy. *Jpn J Physiol* **53**, 367–375 (2003).
350. D. Z. Shao, J. J. Lee, W. T. Huang, J. F. Liao, and M. T. Lin, Inhibition of nuclear factor-kappa B prevents staphylococcal enterotoxin A-induced fever. *Mol Cell Biochem* **262**, 177–185 (2004).
351. E. Tourkina, P. Gooz, J. C. Oates, A. Ludwicka-Bradley, R. M. Silver, and S. Hoffman, Curcumin-induced apoptosis in scleroderma lung fibroblasts: role of protein kinase cepsilon. *Am J Respir Cell Mol Biol* **31**, 28–35 (2004).
352. B. Bosman, Testing of lipoxygenase inhibitors, cyclooxygenase inhibitors, drugs with immunomodulating properties and some reference antipsoriatic drugs in the modified mouse tail test, an animal model of psoriasis. *Skin Pharmacol* **7**, 324–334 (1994).
353. R. Verbeek, E. A. van Tol, and J. M. van Noort, Oral flavonoids delay recovery from experimental autoimmune encephalomyelitis in SJL mice. *Biochem Pharmacol* **70**, 220–228 (2005).
354. P. S. Babu and K. Srinivasan, Influence of dietary curcumin and cholesterol on the progression of experimentally induced diabetes in albino rat. *Mol Cell Biochem* **152**, 13–21 (1995).
355. P. S. Babu and K. Srinivasan, Hypolipidemic action of curcumin, the active principle of turmeric (*Curcuma longa*) in streptozotocin induced diabetic rats. *Mol Cell Biochem* **166**, 169–175 (1997).
356. G. B. Sajithlal, P. Chithra, and G. Chandrakasan, Effect of curcumin on the advanced glycation and cross-linking of collagen in diabetic rats. *Biochem Pharmacol* **56**, 1607–1614 (1998).
357. N. Arun and N. Nalini, Efficacy of turmeric on blood sugar and polyol pathway in diabetic albino rats. *Plant Foods Hum Nutr* **57**, 41–52 (2002).
358. R. K. Kempaiah and K. Srinivasan, Antioxidant status of red blood cells and liver in hypercholesterolemic rats fed hypolipidemic spices. *Int J Vitam Nutr Res* **74**, 199–208 (2004).
359. T. Mahesh, M. M. Sri Balasubashini, and V. P. Menon, Photo-irradiated curcumin supplementation in streptozotocin-induced diabetic rats: effect on lipid peroxidation. *Therapie* **59**, 639–644 (2004).
360. M. Kuroda, Y. Mimaki, T. Nishiyama, T. Mae, H. Kishida, M. Tsukagawa, K. Takahashi, T. Kawada, K. Nakagawa, and M. Kitahara, Hypoglycemic effects of turmeric (*Curcuma longa* L. rhizomes) on genetically diabetic KK-Ay mice. *Biol Pharm Bull* **28**, 937–939 (2005).
361. J. B. Majithiya and R. Balaraman, Time-dependent changes in antioxidant enzymes and vascular reactivity of aorta in streptozotocin-induced diabetic rats treated with curcumin. *J Cardiovasc Pharmacol* **46**, 697–705 (2005).
362. T. Osawa and Y. Kato, Protective role of antioxidative food factors in oxidative stress caused by hyperglycemia. *Ann NY Acad Sci* **1043**, 440–451 (2005).
363. B. B. Aggarwal, Y. Takada, and O. V. Oommen, From chemoprevention to chemotherapy: Common targets and common goals. *Expert Opin Invest Drugs* **13**, 1327–1338 (2004).
364. J. L. Abbruzzese and S. M. Lippman, The convergence of cancer prevention and therapy in early-phase clinical drug development. *Cancer Cell* **6**, 321–326 (2004).
365. H. Inano, M. Onoda, N. Inafuku, M. Kubota, Y. Kamada, T. Osawa, H. Kobayashi, and K. Wakabayashi, Potent preventive action of curcumin on radiation-induced initiation of mammary tumorigenesis in rats. *Carcinogenesis* **21**, 1835–1841 (2000).

366. S. E. Chuang, M. L. Kuo, C. H. Hsu, C. R. Chen, J. K. Lin, G. M. Lai, C. Y. Hsieh, and A. L. Cheng, Curcumin-containing diet inhibits diethylnitrosamine-induced murine hepatocarcinogenesis. *Carcinogenesis* **21**, 331–335 (2000).
367. S. E. Chuang, A. L. Cheng, J. K. Lin, and M. L. Kuo, Inhibition by curcumin of diethylnitrosamine-induced hepatic hyperplasia, inflammation, cellular gene products and cell-cycle-related proteins in rats. *Food Chem Toxicol* **38**, 991–995 (2000).
368. C. C. Chua, R. C. Hamdy, and B. H. Chua, Mechanism of transforming growth factor-beta1-induced expression of vascular endothelial growth factor in murine osteoblastic MC3T3-E1 cells. *Biochim Biophys Acta* **1497**, 69–76 (2000).
369. Y. Shukla and A. Arora, Suppression of altered hepatic foci development by curcumin in wistar rats. *Nutr Cancer* **45**, 53–59 (2003).
370. M. Sreepriya and G. Bali, Chemopreventive effects of embelin and curcumin against N-nitrosodiethylamine/phenobarbital-induced hepatocarcinogenesis in Wistar rats. *Fitoterapia* **76**, 549–555 (2005).
371. M. C. Jiang, H. F. Yang-Yen, J. K. Lin, and J. J. Yen, Differential regulation of p53, c-Myc, Bcl-2 and Bax protein expression during apoptosis induced by widely divergent stimuli in human hepatoblastoma cells. *Oncogene* **13**, 609–616 (1996).
372. K. Imaida, S. Tamano, K. Kato, Y. Ikeda, M. Asamoto, S. Takahashi, Z. Nir, M. Murakoshi, H. Nishino, and T. Shirai, Lack of chemopreventive effects of lycopene and curcumin on experimental rat prostate carcinogenesis. *Carcinogenesis* **22**, 467–472 (2001).
373. M. L. Kuo, T. S. Huang, and J. K. Lin, Curcumin, an antioxidant and anti-tumor promoter, induces apoptosis in human leukemia cells. *Biochim Biophys Acta* **1317**, 95–100 (1996).
374. Y. Wu, Y. Chen, and W. Chen, Effects of concurrent use of rh-IFN-gamma and curcumin on the anti-proliferative capacity of HL-60 cells. *J Tongji Med Univ* **19**, 267–270 (1999).
375. A. Bielak-Zmijewska, M. Koronkiewicz, J. Skierski, K. Piwocka, E. Radziszewska, and E. Sikora, Effect of curcumin on the apoptosis of rodent and human nonproliferating and proliferating lymphoid cells. *Nutr Cancer* **38**, 131–138 (2000).
376. Y. Chen, Y. Wu, J. He, and W. Chen, The experimental and clinical study on the effect of curcumin on cell cycle proteins and regulating proteins of apoptosis in acute myelogenous leukemia. *J Huazhong Univ Sci Technol Med Sci* **22**, 295–298 (2002).
377. A. Duvoix, F. Morceau, M. Schnekenburger, S. Delhalle, M. M. Galteau, M. Dicato, and M. Diederich, Curcumin-induced cell death in two leukemia cell lines: K562 and Jurkat. *Ann NY Acad Sci* **1010**, 389–392 (2003).
378. L. X. Wu, J. H. Xu, G. H. Wu, and Y. Z. Chen, Inhibitory effect of curcumin on proliferation of K562 cells involves down-regulation of p210(bcr/abl) initiated Ras signal transduction pathway. *Acta Pharmacol Sin* **24**, 1155–1160 (2003).
379. A. Bielak-Mijewska, K. Piwocka, A. Magalska, and E. Sikora, P-glycoprotein expression does not change the apoptotic pathway induced by curcumin in HL-60 cells. *Cancer Chemother Pharmacol* **53**, 179–185 (2004).
380. E. Sikora, A. Bielak-Zmijewska, K. Piwocka, J. Skierski, and E. Radziszewska, Inhibition of proliferation and apoptosis of human and rat T lymphocytes by curcumin, a curry pigment. *Biochem Pharmacol* **54**, 899–907 (1997).
381. K. Piwocka, K. Zablocki, M. R. Wieckowski, J. Skierski, I. Feiga, J. Szopa, N. Drela, L. Wojtczak, and E. Sikora, A novel apoptosis-like pathway, independent of mitochondria and caspases, induced by curcumin in human lymphoblastoid T (Jurkat) cells. *Exp Cell Res* **249**, 299–307 (1999).

382. E. Jaruga, S. Salvioli, J. Dobrucki, S. Chrul, J. Bandorowicz-Pikula, E. Sikora, C. Franceschi, A. Cossarizza, and G. Bartosz, Apoptosis-like, reversible changes in plasma membrane asymmetry and permeability, and transient modifications in mitochondrial membrane potential induced by curcumin in rat thymocytes. *FEBS Lett* **433**, 287–293 (1998).
383. D. Ranjan, T. D. Johnston, K. S. Reddy, G. Wu, S. Bondada, and C. Chen, Enhanced apoptosis mediates inhibition of EBV-transformed lymphoblastoid cell line proliferation by curcumin. *J Surg Res* **87**, 1–5 (1999).
384. H. L. Liu, Y. Chen, G. H. Cui, and J. F. Zhou, Curcumin, a potent anti-tumor reagent, is a novel histone deacetylase inhibitor regulating B-NHL cell line Raji proliferation. *Acta Pharmacol Sin* **26**, 603–609 (2005).
385. S. Shishodia, G. Sethi, and B. B. Aggarwal, Curcumin: Getting back to the roots. *Ann NY Acad Sci* **1056**, 206–217 (2005).
386. C. Sun, X. Liu, Y. Chen, and F. Liu, Anticancer effect of curcumin on human B cell non-Hodgkin's lymphoma. *J Huazhong Univ Sci Technolog Med Sci* **25**, 404–407 (2005).
387. Y. Wu, Y. Chen, J. Xu, and L. Lu, Anticancer activities of curcumin on human Burkitt's lymphoma. *Zhonghua Zhong Liu Za Zhi* **24**, 348–352 (2002).
388. A. C. Bharti, S. Shishodia, J. M. Reuben, D. Weber, R. Alexanian, S. Raj-Vadhan, Z. Estrov, M. Talpaz, and B. B. Aggarwal, Nuclear factor-kappaB and STAT3 are constitutively active in CD138+ cells derived from multiple myeloma patients, and suppression of these transcription factors leads to apoptosis. *Blood* **103**, 3175–3184 (2004).
389. S. Uddin, A. R. Hussain, P. S. Manogaran, K. Al-Hussein, L. C. Platanias, M. I. Gutierrez, and K. G. Bhatia, Curcumin suppresses growth and induces apoptosis in primary effusion lymphoma. *Oncogene* **24**, 7022–7030 (2005).
390. N. R. Jana, P. Dikshit, A. Goswami, and N. Nukina, Inhibition of proteasomal function by curcumin induces apoptosis through mitochondrial pathway. *J Biol Chem* **279**, 11,680–11,685 (2004).
391. A. Liontas and H. Yeger, Curcumin and resveratrol induce apoptosis and nuclear translocation and activation of p53 in human neuroblastoma. *Anticancer Res* **24**, 987–998 (2004).
392. M. H. Pan, W. L. Chang, S. Y. Lin-Shiau, C. T. Ho, and J. K. Lin, Induction of apoptosis by garcinol and curcumin through cytochrome c release and activation of caspases in human leukemia HL-60 cells. *J Agric Food Chem* **49**, 1464–1474 (2001).
393. S. Nagai, M. Kurimoto, K. Washiyama, Y. Hirashima, T. Kumanishi, and S. Endo, Inhibition of cellular proliferation and induction of apoptosis by curcumin in human malignant astrocytoma cell lines. *J Neurooncol* **74**, 105–111 (2005).
394. K. Mehta, P. Pantazis, T. McQueen, and B. B. Aggarwal, Antiproliferative effect of curcumin (diferuloylmethane) against human breast tumor cell lines. *Anticancer Drugs* **8**, 470–481 (1997).
395. C. Ramachandran and W. You, Differential sensitivity of human mammary epithelial and breast carcinoma cell lines to curcumin. *Breast Cancer Res Treat* **54**, 269–278 (1999).
396. T. Choudhuri, S. Pal, M. L. Agwarwal, T. Das, and G. Sa, Curcumin induces apoptosis in human breast cancer cells through p53-dependent Bax induction. *FEBS Lett* **512**, 334–340 (2002).
397. J. M. Holy, Curcumin disrupts mitotic spindle structure and induces micronucleation in MCF-7 breast cancer cells. *Mutat Res* **518**, 71–84 (2002).

398. Z. M. Shao, Z. Z. Shen, C. H. Liu, M. R. Sartippour, V. L. Go, D. Heber, and M. Nguyen, Curcumin exerts multiple suppressive effects on human breast carcinoma cells. *Int J Cancer* **98**, 234–240 (2002).
399. C. Ramachandran, S. Rodriguez, R. Ramachandran, P. K. Raveendran Nair, H. Fonseca, Z. Khatib, E. Escalon, and S. J. Melnick, Expression profiles of apoptotic genes induced by curcumin in human breast cancer and mammary epithelial cell lines. *Anticancer Res* **25**, 3293–3302 (2005).
400. L. Moragoda, R. Jaszewski, and A. P. Majumdar, Curcumin induced modulation of cell cycle and apoptosis in gastric and colon cancer cells. *Anticancer Res* **21**, 873–878 (2001).
401. S. Aggarwal, Y. Takada, S. Singh, J. N. Myers, and B. B. Aggarwal, Inhibition of growth and survival of human head and neck squamous cell carcinoma cells by curcumin via modulation of nuclear factor-kappaB signaling. *Int J Cancer* **111**, 679–692 (2004).
402. G. Radhakrishna Pillai, A. S. Srivastava, T. I. Hassanein, D. P. Chauhan, and E. Carrier, Induction of apoptosis in human lung cancer cells by curcumin. *Cancer Lett* **208**, 163–170 (2004).
403. L. Li, B. B. Aggarwal, S. Shishodia, J. Abbruzzese, and R. Kurzrock, Nuclear factor-kappaB and IkappaB kinase are constitutively active in human pancreatic cells, and their down-regulation by curcumin (diferuloylmethane) is associated with the suppression of proliferation and the induction of apoptosis. *Cancer* **101**, 2351–2362 (2004).
404. M. Shi, Q. Cai, L. Yao, Y. Mao, Y. Ming, and G. Ouyang, Antiproliferation and apoptosis induced by curcumin in human ovarian cancer cells. *Cell Biol Int* **30**, 221–226 (2006).
405. R. Kuttan, P. C. Sudheeran, and C. D. Josph, Turmeric and curcumin as topical agents in cancer therapy. *Tumori* **73**, 29–31 (1987).
406. M. A. Azuine and S. V. Bhide, Chemopreventive effect of turmeric against stomach and skin tumors induced by chemical carcinogens in Swiss mice. *Nutr Cancer* **17**, 77–83 (1992).
407. M. T. Huang, E. E. Deschner, H. L. Newmark, Z. Y. Wang, T. A. Ferraro, and A. H. Conney, Effect of dietary curcumin and ascorbyl palmitate on azoxymethanol-induced colonic epithelial cell proliferation and focal areas of dysplasia. *Cancer Lett* **64**, 117–121 (1992).
408. M. T. Huang, Z. Y. Wang, C. A. Georgiadis, J. D. Laskin, and A. H. Conney, Inhibitory effects of curcumin on tumor initiation by benzo[a]pyrene and 7,12-dimethylbenz[a]anthracene. *Carcinogenesis* **13**, 2183–2186 (1992).
409. M. T. Huang, W. Ma, P. Yen, J. G. Xie, J. Han, K. Frenkel, D. Grunberger, and A. H. Conney, Inhibitory effects of topical application of low doses of curcumin on 12-O-tetradecanoylphorbol-13-acetate-induced tumor promotion and oxidized DNA bases in mouse epidermis. *Carcinogenesis* **18**, 83–88 (1997).
410. P. Limtrakul, S. Lipigorngoson, O. Namwong, A. Apisariyakul, and F. W. Dunn, Inhibitory effect of dietary curcumin on skin carcinogenesis in mice. *Cancer Lett* **116**, 197–203 (1997).
411. M. C. Jiang, H. F. Yang-Yen, J. J. Yen, and J. K. Lin, Curcumin induces apoptosis in immortalized NIH 3T3 and malignant cancer cell lines. *Nutr Cancer* **26**, 111–120 (1996).
412. J. A. Bush, K. J. Cheung, Jr., and G. Li, Curcumin induces apoptosis in human melanoma cells through a Fas receptor/caspase-8 pathway independent of p53. *Exp Cell Res* **271**, 305–314 (2001).

413. M. Zheng, S. Ekmekcioglu, E. T. Walch, C. H. Tang, and E. A. Grimm, Inhibition of nuclear factor-kappaB and nitric oxide by curcumin induces G2/M cell cycle arrest and apoptosis in human melanoma cells. *Melanoma Res* **14**, 165–171 (2004).
414. D. R. Siwak, S. Shishodia, B. B. Aggarwal, and R. Kurzrock, Curcumin-induced antiproliferative and proapoptotic effects in melanoma cells are associated with suppression of IkappaB kinase and nuclear factor kappaB activity and are independent of the B-Raf/mitogen-activated/extracellular signal-regulated protein kinase pathway and the Akt pathway. *Cancer* **104**, 879–890 (2005).
415. W. H. Chan and H. J. Wu, Anti-apoptotic effects of curcumin on photosensitized human epidermal carcinoma A431 cells. *J Cell Biochem* **92**, 200–212 (2004).
416. M. A. Azuine and S. V. Bhide, Adjuvant chemoprevention of experimental cancer: Catechin and dietary turmeric in forestomach and oral cancer models. *J Ethnopharmacol* **44**, 211–217 (1994).
417. T. Tanaka, H. Makita, M. Ohnishi, Y. Hirose, A. Wang, H. Mori, K. Satoh, A. Hara, and H. Ogawa, Chemoprevention of 4-nitroquinoline 1-oxide-induced oral carcinogenesis by dietary curcumin and hesperidin: comparison with the protective effect of beta-carotene. *Cancer Res* **54**, 4653–4659 (1994).
418. K. Krishnaswamy, V. K. Goud, B. Sesikeran, M. A. Mukundan, and T. P. Krishna, Retardation of experimental tumorigenesis and reduction in DNA adducts by turmeric and curcumin. *Nutr Cancer* **30**, 163–166 (1998).
419. N. Li, X. Chen, J. Liao, G. Yang, S. Wang, Y. Josephson, C. Han, J. Chen, M. T. Huang, and C. S. Yang, Inhibition of 7,12-dimethylbenz[a]anthracene (DMBA)-induced oral carcinogenesis in hamsters by tea and curcumin. *Carcinogenesis* **23**, 1307–1313 (2002).
420. N. Li, X. Chen, C. Han, and J. Chen, [Chemopreventive effect of tea and curcumin on DMBA-induced oral carcinogenesis in hamsters]. *Wei Sheng Yan Jiu* **31**, 354–357 (2002).
421. J. Ushida, S. Sugie, K. Kawabata, Q. V. Pham, T. Tanaka, K. Fujii, H. Takeuchi, Y. Ito, and H. Mori, Chemopreventive effect of curcumin on N-nitrosomethylbenzylamine-induced esophageal carcinogenesis in rats. *Jpn J Cancer Res* **91**, 893–898 (2000).
422. M. T. Huang, Y. R. Lou, W. Ma, H. L. Newmark, K. R. Reuhl, and A. H. Conney, Inhibitory effects of dietary curcumin on forestomach, duodenal, and colon carcinogenesis in mice. *Cancer Res* **54**, 5841–5847 (1994).
423. S. V. Singh, X. Hu, S. K. Srivastava, M. Singh, H. Xia, J. L. Orchard, and H. A. Zaren, Mechanism of inhibition of benzo[a]pyrene-induced forestomach cancer in mice by dietary curcumin. *Carcinogenesis* **19**, 1357–1360 (1998).
424. S. Ikezaki, A. Nishikawa, F. Furukawa, K. Kudo, H. Nakamura, K. Tamura, and H. Mori, Chemopreventive effects of curcumin on glandular stomach carcinogenesis induced by N-methyl-N'-nitro-N-nitrosoguanidine and sodium chloride in rats. *Anticancer Res* **21**, 3407–3411 (2001).
425. S. Perkins, R. D. Verschoyle, K. Hill, I. Parveen, M. D. Threadgill, R. A. Sharma, M. L. Williams, W. P. Steward, and A. J. Gescher, Chemopreventive efficacy and pharmacokinetics of curcumin in the min/+ mouse, a model of familial adenomatous polyposis. *Cancer Epidemiol Biomarkers Prev* **11**, 535–540 (2002).
426. S. Perkins, A. R. Clarke, W. Steward, and A. Gescher, Age-related difference in susceptibility of Apc(Min/+) mice towards the chemopreventive efficacy of dietary aspirin and curcumin. *Br J Cancer* **88**, 1480–1483 (2003).
427. M. A. Pereira, C. J. Grubbs, L. H. Barnes, H. Li, G. R. Olson, I. Eto, M. Juliana, L. M. Whitaker, G. J. Kelloff, V. E. Steele, and R. A. Lubet, Effects of the phyto-

chemicals, curcumin and quercetin, upon azoxymethane-induced colon cancer and 7,12-dimethylbenz[a]anthracene-induced mammary cancer in rats. *Carcinogenesis* **17**, 1305–1311 (1996).

428. M. J. Wargovich, C. D. Chen, A. Jimenez, V. E. Steele, M. Velasco, L. C. Stephens, R. Price, K. Gray, and G. J. Kelloff, Aberrant crypts as a biomarker for colon cancer: evaluation of potential chemopreventive agents in the rat. *Cancer Epidemiol Biomarkers Prev* **5**, 355–360 (1996).
429. H. S. Samaha, G. J. Kelloff, V. Steele, C. V. Rao, and B. S. Reddy, Modulation of apoptosis by sulindac, curcumin, phenylethyl-3-methylcaffeate, and 6-phenylhexyl isothiocyanate: apoptotic index as a biomarker in colon cancer chemoprevention and promotion. *Cancer Res* **57**, 1301–1305 (1997).
430. T. Kawamori, R. Lubet, V. E. Steele, G. J. Kelloff, R. B. Kaskey, C. V. Rao, and B. S. Reddy, Chemopreventive effect of curcumin, a naturally occurring anti-inflammatory agent, during the promotion/progression stages of colon cancer. *Cancer Res* **59**, 597–601 (1999).
431. C. V. Rao, T. Kawamori, R. Hamid, and B. S. Reddy, Chemoprevention of colonic aberrant crypt foci by an inducible nitric oxide synthase-selective inhibitor. *Carcinogenesis* **20**, 641–644 (1999).
432. Y. Kwon, M. Malik, and B. A. Magnuson, Inhibition of colonic aberrant crypt foci by curcumin in rats is affected by age. *Nutr Cancer* **48**, 37–43 (2004).
433. S. R. Volate, D. M. Davenport, S. J. Muga, and M. J. Wargovich, Modulation of aberrant crypt foci and apoptosis by dietary herbal supplements (quercetin, curcumin, silymarin, ginseng and rutin). *Carcinogenesis* **26**, 1450–1456 (2005).
434. J. M. Kim, S. Araki, D. J. Kim, C. B. Park, N. Takasuka, H. Baba-Toriyama, T. Ota, Z. Nir, F. Khachik, N. Shimidzu, Y. Tanaka, T. Osawa, T. Uraji, M. Murakoshi, H. Nishino, and H. Tsuda, Chemopreventive effects of carotenoids and curcumins on mouse colon carcinogenesis after 1,2-dimethylhydrazine initiation. *Carcinogenesis* **19**, 81–85 (1998).
435. R. Hanif, L. Qiao, S. J. Shiff, and B. Rigas, Curcumin, a natural plant phenolic food additive, inhibits cell proliferation and induces cell cycle changes in colon adenocarcinoma cell lines by a prostaglandin-independent pathway. *J Lab Clin Med* **130**, 576–584 (1997).
436. H. Chen, Z. S. Zhang, Y. L. Zhang, and D. Y. Zhou, Curcumin inhibits cell proliferation by interfering with the cell cycle and inducing apoptosis in colon carcinoma cells. *Anticancer Res* **19**, 3675–3680 (1999).
437. R. Rashmi, T. R. Santhosh Kumar, and D. Karunagaran, Human colon cancer cells differ in their sensitivity to curcumin-induced apoptosis and heat shock protects them by inhibiting the release of apoptosis-inducing factor and caspases. *FEBS Lett* **538**, 19–24 (2003).
438. G. P. Collett and F. C. Campbell, Curcumin induces c-jun N-terminal kinase-dependent apoptosis in HCT116 human colon cancer cells. *Carcinogenesis* **25**, 2183–2189 (2004).
439. S. C. Wei, Y. S. Lin, P. N. Tsao, J. J. Wu-Tsai, C. H. Wu, and J. M. Wong, Comparison of the anti-proliferation and apoptosis-induction activities of sulindac, celecoxib, curcumin, and nifedipine in mismatch repair-deficient cell lines. *J Formos Med Assoc* **103**, 599–606 (2004).
440. G. Song, Y. B. Mao, Q. F. Cai, L. M. Yao, G. L. Ouyang, and S. D. Bao, Curcumin induces human HT-29 colon adenocarcinoma cell apoptosis by activating p53 and regulating apoptosis-related protein expression. *Braz J Med Biol Res* **38**, 1791–1798 (2005).

441. K. Singletary, C. MacDonald, M. Wallig, and C. Fisher, Inhibition of 7,12-dimethylbenz[a]anthracene (DMBA)-induced mammary tumorigenesis and DMBA-DNA adduct formation by curcumin. *Cancer Lett* **103**, 137–141 (1996).
442. S. S. Deshpande, A. D. Ingle, and G. B. Maru, Chemopreventive efficacy of curcumin-free aqueous turmeric extract in 7,12-dimethylbenz[a]anthracene-induced rat mammary tumorigenesis. *Cancer Lett* **123**, 35–40 (1998).
443. K. Singletary, C. MacDonald, M. Iovinelli, C. Fisher, and M. Wallig, Effect of the beta-diketones diferuloylmethane (curcumin) and dibenzoylmethane on rat mammary DNA adducts and tumors induced by 7,12-dimethylbenz[a]anthracene. *Carcinogenesis* **19**, 1039–1043 (1998).
444. H. Inano, M. Onoda, N. Inafuku, M. Kubota, Y. Kamada, T. Osawa, H. Kobayashi, and K. Wakabayashi, Chemoprevention by curcumin during the promotion stage of tumorigenesis of mammary gland in rats irradiated with gamma-rays. *Carcinogenesis* **20**, 1011–1018 (1999).
445. R. Kuttan, P. Bhanumathy, K. Nirmala and M. C. George, Potential anticancer activity of turmeric (Curcuma longa). *Cancer Lett* **29**, 197–202, (1985).
446. M. Nagabhushan and S. V. Bhide, Curcumin as an inhibitor of cancer. *J Am Coll Nutr* **11**, 192–128 (1992).
447. S. S. Hecht, P. M. Kenney, M. Wang, N. Trushin, S. Agarwal, A. V. Rao, and P. Upadhyaya, Evaluation of butylated hydroxyanisole, myo-inositol, curcumin, esculetin, resveratrol and lycopene as inhibitors of benzo[a]pyrene plus 4-(methylnitrosamino)-1-(3-pyridyl)-1-butanone-induced lung tumorigenesis in A/J mice. *Cancer Lett* **137**, 123–130 (1999).
448. A. Khar, A. M. Ali, B. V. Pardhasaradhi, Z. Begum, and R. Anjum, Antitumor activity of curcumin is mediated through the induction of apoptosis in AK-5 tumor cells. *FEBS Lett* **445**, 165–168 (1999).
449. M. Churchill, A. Chadburn, R. T. Bilinski, and M. M. Bertagnolli, Inhibition of intestinal tumors by curcumin is associated with changes in the intestinal immune cell profile. *J Surg Res* **89**, 169–175 (2000).
450. B. Lal, A. K. Kapoor, P. K. Agrawal, O. P. Asthana, and R. C. Srimal, Role of curcumin in idiopathic inflammatory orbital pseudotumours. *Phytother Res* **14**, 443–447 (2000).
451. S. Busquets, N. Carbo, V. Almendro, M. T. Quiles, F. J. Lopez-Soriano, and J. M. Argiles, Curcumin, a natural product present in turmeric, decreases tumor growth but does not behave as an anticachectic compound in a rat model. *Cancer Lett* **167**, 33–38 (2001).
452. G. P. Collett, C. N. Robson, J. C. Mathers, and F. C. Campbell, Curcumin modifies Apc(min) apoptosis resistance and inhibits 2-amino 1-methyl-6-phenylimidazo[4,5-b]pyridine (PhIP) induced tumour formation in Apc(min) mice. *Carcinogenesis* **22**, 821–825 (2001).
453. P. Sindhwani, J. A. Hampton, M. M. Baig, R. Keck, and S. H. Selman, Curcumin prevents intravesical tumor implantation of the MBT-2 tumor cell line in C3H mice. *J Urol* **166**, 1498–1501 (2001).
454. H. Inano and M. Onoda, Radioprotective action of curcumin extracted from *Curcuma longa* LINN: Inhibitory effect on formation of urinary 8-hydroxy-2'-deoxyguanosine, tumorigenesis, but not mortality, induced by gamma-ray irradiation. *Int J Radiat Oncol Biol Phys* **53**, 735–743 (2002).
455. H. Inano and M. Onoda, Prevention of radiation-induced mammary tumors. *Int J Radiat Oncol Biol Phys* **52**, 212–223 (2002).

456. N. Ozen, E. Uslu, M. Ozen, S. Aydin, T. Altug, A. Belce, and E. Kokoglu, Curcumin's effects on sialic acid level and sialidase activity in Ehrlich ascites tumor bearing mice. *Tohoku J Exp Med* **197**, 221–227 (2002).
457. N. Frank, J. Knauft, F. Amelung, J. Nair, H. Wesch, and H. Bartsch, No prevention of liver and kidney tumors in Long-Evans Cinnamon rats by dietary curcumin, but inhibition at other sites and of metastases. *Mutat Res* **523–524**, 127–135 (2003).
458. J. Gertsch, M. Guttinger, J. Heilmann, and O. Sticher, Curcumin differentially modulates mRNA profiles in Jurkat T and human peripheral blood mononuclear cells. *Bioorg Med Chem* **11**, 1057–1063 (2003).
459. J. Odot, P. Albert, A. Carlier, M. Tarpin, J. Devy, and C. Madoulet, In vitro and in vivo anti-tumoral effect of curcumin against melanoma cells. *Int J Cancer* **111**, 381–387 (2004).
460. M. Belakavadi and B. P. Salimath, Mechanism of inhibition of ascites tumor growth in mice by curcumin is mediated by NF-kB and caspase activated DNase. *Mol Cell Biochem* **273**, 57–67 (2005).
461. A. Pal and A. K. Pal, Radioprotection of turmeric extracts in bacterial system. *Acta Biol Hung* **56**, 333–343 (2005).
462. M. Notarbartolo, P. Poma, D. Perri, L. Dusonchet, M. Cervello, and N. D'Alessandro, Antitumor effects of curcumin, alone or in combination with cisplatin or doxorubicin, on human hepatic cancer cells. Analysis of their possible relationship to changes in NF-kB activation levels and in IAP gene expression. *Cancer Lett* **224**, 53–65 (2005).
463. A. K. Singh, G. S. Sidhu, T. Deepa, and R. K. Maheshwari, Curcumin inhibits the proliferation and cell cycle progression of human umbilical vein endothelial cell. *Cancer Lett* **107**, 109–115 (1996).
464. R. G. Mehta and R. C. Moon, Characterization of effective chemopreventive agents in mammary gland in vitro using an initiation-promotion protocol. *Anticancer Res* **11**, 593–596 (1991).
465. J. A. Sokoloski, K. Shyam, and A. C. Sartorelli, Induction of the differentiation of HL-60 promyelocytic leukemia cells by curcumin in combination with low levels of vitamin D3. *Oncol Res* **9**, 31–39 (1997).
466. S. C. Gautam, Y. X. Xu, K. R. Pindolia, N. Janakiraman, and R. A. Chapman, Nonselective inhibition of proliferation of transformed and nontransformed cells by the anticancer agent curcumin (diferuloylmethane). *Biochem Pharmacol* **55**, 1333–1337 (1998).
467. E. Jaruga, A. Sokal, S. Chrul, and G. Bartosz, Apoptosis-independent alterations in membrane dynamics induced by curcumin. *Exp Cell Res* **245**, 303–312 (1998).
468. E. Jaruga, A. Bielak-Zmijewska, E. Sikora, J. Skierski, E. Radziszewska, K. Piwocka, and G. Bartosz, Glutathione-independent mechanism of apoptosis inhibition by curcumin in rat thymocytes. *Biochem Pharmacol* **56**, 961–965 (1998).
469. S. H. Jee, S. C. Shen, C. R. Tseng, H. C. Chiu, and M. L. Kuo, Curcumin induces a p53-dependent apoptosis in human basal cell carcinoma cells. *J Invest Dermatol* **111**, 656–661 (1998).
470. S. M. D'Ambrosio, R. Gibson-D'Ambrosio, G. E. Milo, B. Casto, G. J. Kelloff, and V. E. Steele, Differential response of normal, premalignant and malignant human oral epithelial cells to growth inhibition by chemopreventive agents. *Anticancer Res* **20**, 2273–2280 (2000).
471. T. Dorai, N. Gehani, and A. Katz, Therapeutic potential of curcumin in human prostate cancer-I. curcumin induces apoptosis in both androgen-dependent and androgen-independent prostate cancer cells. *Prostate Cancer Prostatic Dis* **3**, 84–93 (2000).

472. T. M. Elattar and A. S. Virji, The inhibitory effect of curcumin, genistein, quercetin and cisplatin on the growth of oral cancer cells in vitro. *Anticancer Res* **20**, 1733–1738 (2000).
473. Y. Wu, Y. Chen, and M. He, The influence of curcumin on the cell cycle of HL-60 cells and contrast study. *J Tongji Med Univ* **20**, 123–125 (2000).
474. B. K. Batth, R. Tripathi, and U. K. Srinivas, Curcumin-induced differentiation of mouse embryonal carcinoma PCC4 cells. *Differentiation* **68**, 133–140 (2001).
475. B. Cipriani, G. Borsellino, H. Knowles, D. Tramonti, F. Cavaliere, G. Bernardi, L. Battistini, and C. F. Brosnan, Curcumin inhibits activation of Vgamma9Vdelta2 T cells by phosphoantigens and induces apoptosis involving apoptosis-inducing factor and large scale DNA fragmentation. *J Immunol* **167**, 3454–3462 (2001).
476. T. Dorai, Y. C. Cao, B. Dorai, R. Buttyan, and A. E. Katz, Therapeutic potential of curcumin in human prostate cancer. III. Curcumin inhibits proliferation, induces apoptosis, and inhibits angiogenesis of LNCaP prostate cancer cells in vivo. *Prostate* **47**, 293–303 (2001).
477. H. Mori, K. Niwa, Q. Zheng, Y. Yamada, K. Sakata, and N. Yoshimi, Cell proliferation in cancer prevention; effects of preventive agents on estrogen-related endometrial carcinogenesis model and on an in vitro model in human colorectal cells. *Mutat Res* **480–481**, 201–207 (2001).
478. D. Morin, S. Barthelemy, R. Zini, S. Labidalle, and J. P. Tillement, Curcumin induces the mitochondrial permeability transition pore mediated by membrane protein thiol oxidation. *FEBS Lett* **495**, 131–136 (2001).
479. A. Mukhopadhyay, C. Bueso-Ramos, D. Chatterjee, P. Pantazis, and B. B. Aggarwal, Curcumin downregulates cell survival mechanisms in human prostate cancer cell lines. *Oncogene* **20**, 7597–7609 (2001).
480. S. Pal, T. Choudhuri, S. Chattopadhyay, A. Bhattacharya, G. K. Datta, T. Das, and G. Sa, Mechanisms of curcumin-induced apoptosis of Ehrlich's ascites carcinoma cells. *Biochem Biophys Res Commun* **288**, 658–665 (2001).
481. K. Piwocka, E. Jaruga, J. Skierski, I. Gradzka, and E. Sikora, Effect of glutathione depletion on caspase-3 independent apoptosis pathway induced by curcumin in Jurkat cells. *Free Radical Biol Med* **31**, 670–678 (2001).
482. R. J. Anto, A. Mukhopadhyay, K. Denning, and B. B. Aggarwal, Curcumin (diferuloylmethane) induces apoptosis through activation of caspase-8, BID cleavage and cytochrome c release: its suppression by ectopic expression of Bcl-2 and Bcl-xl. *Carcinogenesis* **23**, 143–150 (2002).
483. M. J. Park, E. H. Kim, I. C. Park, H. C. Lee, S. H. Woo, J. Y. Lee, Y. J. Hong, C. H. Rhee, S. H. Choi, B. S. Shim, S. H. Lee, and S. I. Hong, Curcumin inhibits cell cycle progression of immortalized human umbilical vein endothelial (ECV304) cells by up-regulating cyclin-dependent kinase inhibitor, p21WAF1/CIP1, p27KIP1 and p53. *Int J Oncol* **21**, 379–383 (2002).
484. K. Piwocka, A. Bielak-Mijewska and E. Sikora, Curcumin induces caspase-3-independent apoptosis in human multidrug-resistant cells. *Ann NY Acad Sci* **973**, 250–254 (2002).
485. J. H. Bae, J. W. Park, and T. K. Kwon, Ruthenium red, inhibitor of mitochondrial Ca^{2+}uniporter, inhibits curcumin-induced apoptosis via the prevention of intracellular Ca^{2+} depletion and cytochrome c release. *Biochem Biophys Res Commun* **303**, 1073–1079 (2003).
486. D. Deeb, Y. X. Xu, H. Jiang, X. Gao, N. Janakiraman, R. A. Chapman, and S. C. Gautam, Curcumin (diferuloyl-methane) enhances tumor necrosis factor-related

apoptosis-inducing ligand-induced apoptosis in LNCaP prostate cancer cells. *Mol Cancer Ther* **2**, 95–103 (2003).

487. A. Pol, M. Bergers, and J. Schalkwijk, Comparison of antiproliferative effects of experimental and established antipsoriatic drugs on human keratinocytes, using a simple 96-well-plate assay. *In Vitro Cell Dev Biol Anim* **39**, 36–42 (2003).
488. T. Dorai, J. P. Dutcher, D. W. Dempster, and P. H. Wiernik, Therapeutic potential of curcumin in prostate cancer–V: Interference with the osteomimetic properties of hormone refractory C4-2B prostate cancer cells. *Prostate* **60**, 1–17 (2004).
489. J. Holy, Curcumin inhibits cell motility and alters microfilament organization and function in prostate cancer cells. *Cell Motil Cytoskeleton* **58**, 253–268 (2004).
490. R. Rashmi, S. Kumar, and D. Karunagaran, Ectopic expression of Hsp70 confers resistance and silencing its expression sensitizes human colon cancer cells to curcumin-induced apoptosis. *Carcinogenesis* **25**, 179–187 (2004).
491. D. W. Scott and G. Loo, Curcumin-induced GADD153 gene up-regulation in human colon cancer cells. *Carcinogenesis* **25**, 2155–2164 (2004).
492. C. Syng-Ai, A. L. Kumari and A. Khar, Effect of curcumin on normal and tumor cells: role of glutathione and bcl-2. *Mol Cancer Ther* **3**, 1101–1108 (2004).
493. M. Fullbeck, X. Huang, R. Dumdey, C. Frommel, W. Dubiel, and R. Preissner, Novel curcumin- and emodin-related compounds identified by in silico 2D/3D conformer screening induce apoptosis in tumor cells. *BMC Cancer* **5**, 97 (2005).
494. X. Gao, D. Deeb, H. Jiang, Y. B. Liu, S. A. Dulchavsky, and S. C. Gautam, Curcumin differentially sensitizes malignant glioma cells to TRAIL/Apo2L-mediated apoptosis through activation of procaspases and release of cytochrome c from mitochondria. *J Exp Ther Oncol* **5**, 39–48 (2005).
495. E. M. Jung, J. H. Lim, T. J. Lee, J. W. Park, K. S. Choi, and T. K. Kwon, Curcumin sensitizes tumor necrosis factor-related apoptosis-inducing ligand (TRAIL)-induced apoptosis through reactive oxygen species-mediated upregulation of death receptor 5 (DR5). *Carcinogenesis* **26**, 1905–1913 (2005).
496. S. Mishra, N. Kapoor, A. Mubarak Ali, B. V. Pardhasaradhi, A. L. Kumari, A. Khar, and K. Misra, Differential apoptotic and redox regulatory activities of curcumin and its derivatives. *Free Radica Biol Med* **38**, 1353–1360 (2005).
497. S. D. Park, J. H. Jung, H. W. Lee, Y. M. Kwon, K. H. Chung, M. G. Kim, and C. H. Kim, Zedoariae rhizoma and curcumin inhibits platelet-derived growth factor-induced proliferation of human hepatic myofibroblasts. *Int Immunopharmacol* **5**, 555–569 (2005).
498. R. Rashmi, S. Kumar, and D. Karunagaran, Human colon cancer cells lacking Bax resist curcumin-induced apoptosis and Bax requirement is dispensable with ectopic expression of Smac or downregulation of Bcl-XL. *Carcinogenesis* **26**, 713–723 (2005).
499. Q. Wang, A. Y. Sun, A. Simonyi, M. D. Jensen, P. B. Shelat, G. E. Rottinghaus, R. S. MacDonald, D. K. Miller, D. E. Lubahn, G. A. Weisman, and G. Y. Sun, Neuroprotective mechanisms of curcumin against cerebral ischemia-induced neuronal apoptosis and behavioral deficits. *J Neurosci Res* **82**, 138–148 (2005).
500. C. W. Lee, W. N. Lin, C. C. Lin, S. F. Luo, J. S. Wang, J. Pouyssegur, and C. M. Yang, Transcriptional regulation of VCAM-1 expression by tumor necrosis factor-alpha in human tracheal smooth muscle cells: involvement of MAPKs, NF-kappaB, p300, and histone acetylation. *J Cell Physiol* **207**, 174–186 (2006).
501. L. I. Lin, Y. F. Ke, Y. C. Ko, and J. K. Lin, Curcumin inhibits SK-Hep-1 hepatocellular carcinoma cell invasion in vitro and suppresses matrix metalloproteinase-9 secretion. *Oncology* **55**, 349–353 (1998).

502. J. I. Fenton, M. S. Wolff, M. W. Orth, and N. G. Hord, Membrane-type matrix metalloproteinases mediate curcumin-induced cell migration in non-tumorigenic colon epithelial cells differing in Apc genotype. *Carcinogenesis* **23**, 1065–1070 (2002).
503. A. Banerji, J. Chakrabarti, A. Mitra, and A. Chatterjee, Effect of curcumin on gelatinase A (MMP-2) activity in B16F10 melanoma cells. *Cancer Lett* **211**, 235–242 (2004).
504. R. Rukkumani, K. Aruna, P. S. Varma, and V. P. Menon, Curcumin influences hepatic expression patterns of matrix metalloproteinases in liver toxicity. *Ital J Biochem* **53**, 61–66 (2004).
505. Q. H. Yao, D. Q. Wang, C. C. Cui, Z. Y. Yuan, S. B. Chen, X. W. Yao, J. K. Wang, and J. F. Lian, Curcumin ameliorates left ventricular function in rabbits with pressure overload: inhibition of the remodeling of the left ventricular collagen network associated with suppression of myocardial tumor necrosis factor-alpha and matrix metalloproteinase-2 expression. *Biol Pharm Bull* **27**, 198–202 (2004).
506. S. Y. Kim, S. H. Jung, and H. S. Kim, Curcumin is a potent broad spectrum inhibitor of matrix metalloproteinase gene expression in human astroglioma cells. *Biochem Biophys Res Commun* **337**, 510–516 (2005).
507. M. S. Woo, S. H. Jung, S. Y. Kim, J. W. Hyun, K. H. Ko, W. K. Kim, and H. S. Kim, Curcumin suppresses phorbol ester-induced matrix metalloproteinase-9 expression by inhibiting the PKC to MAPK signaling pathways in human astroglioma cells. *Biochem Biophys Res Commun* **335**, 1017–1025 (2005).
508. E. Y. Shin, S. Y. Kim, and E. G. Kim, c-Jun N-terminal kinase is involved in motility of endothelial cell. *Exp Mol Med* **33**, 276–283 (2001).
509. P. V. Leyon and G. Kuttan, Studies on the role of some synthetic curcuminoid derivatives in the inhibition of tumour specific angiogenesis. *J Exp Clin Cancer Res* **22**, 77–83 (2003).
510. E. V. Bobrovnikova-Marjon, P. L. Marjon, O. Barbash, D. L. Vander Jagt, and S. F. Abcouwer, Expression of angiogenic factors vascular endothelial growth factor and interleukin-8/CXCL8 is highly responsive to ambient glutamine availability: role of nuclear factor-kappaB and activating protein-1. *Cancer Res* **64**, 4858–4869 (2004).
511. W. G. Cao, M. Morin, V. Sengers, C. Metz, T. Roger, R. Maheux, and A. Akoum, Tumour necrosis factor-alpha up-regulates macrophage migration inhibitory factor expression in endometrial stromal cells via the nuclear transcription factor NF-kappaB. *Hum Reprod* **21**, 421–428 (2006).
512. M. L. Cho, Y. O. Jung, Y. M. Moon, S. Y. Min, C. H. Yoon, S. H. Lee, S. H. Park, C. S. Cho, D. M. Jue, and H. Y. Kim, Interleukin-18 induces the production of vascular endothelial growth factor (VEGF) in rheumatoid arthritis synovial fibroblasts via AP-1-dependent pathways. *Immunol Lett* **103**, 159–166 (2006).
513. B. H. Babu, B. S. Shylesh, and J. Padikkala, Antioxidant and hepatoprotective effect of *Acanthus ilicifolius*. *Fitoterapia* **72**, 272–277 (2001).
514. S. Nishizono, T. Hayami, I. Ikeda, and K. Imaizumi, Protection against the diabetogenic effect of feeding tert–butylhydroquinone to rats prior to the administration of streptozotocin. *Biosci Biotechnol Biochem* **64**, 1153–1158 (2000).
515. P. Suryanarayana, M. Saraswat, T. Mrudula, T. P. Krishna, K. Krishnaswamy, and G. B. Reddy, Curcumin and turmeric delay streptozotocin-induced diabetic cataract in rats. *Invest Ophthalmol Vis Sci* **46**, 2092–2099 (2005).
516. M. Dikshit, L. Rastogi, R. Shukla, and R. C. Srimal, Prevention of ischaemia-induced biochemical changes by curcumin & quinidine in the cat heart. *Indian J Med Res* **101**, 31–35 (1995).

517. C. Nirmala and R. Puvanakrishnan, Protective role of curcumin against isoproterenol induced myocardial infarction in rats. *Mol Cell Biochem* **159**, 85–93 (1996).
518. C. Nirmala and R. Puvanakrishnan, Effect of curcumin on certain lysosomal hydrolases in isoproterenol-induced myocardial infarction in rats. *Biochem Pharmacol* **51**, 47–51 (1996).
519. H. W. Chen and H. C. Huang, Effect of curcumin on cell cycle progression and apoptosis in vascular smooth muscle cells. *Br J Pharmacol* **124**, 1029–1040 (1998).
520. W. Zhang, D. Liu, X. Wo, Y. Zhang, M. Jin, and Z. Ding, Effects of *Curcuma longa* on proliferation of cultured bovine smooth muscle cells and on expression of low density lipoprotein receptor in cells. *Chin Med J (Engl)* **112**, 308–311 (1999).
521. M. Sato, G. A. Cordis, N. Maulik, and D. K. Das, SAPKs regulation of ischemic preconditioning. *Am J Physiol Heart Circ Physiol* **279**, H901–H907 (2000).
522. P. Manikandan, M. Sumitra, S. Aishwarya, B. M. Manohar, B. Lokanadam, and R. Puvanakrishnan, Curcumin modulates free radical quenching in myocardial ischaemia in rats. *Int J Biochem Cell Biol* **36**, 1967–1980 (2004).
523. G. Ramaswami, H. Chai, Q. Yao, P. H. Lin, A. B. Lumsden, and C. Chen, Curcumin blocks homocysteine-induced endothelial dysfunction in porcine coronary arteries. *J Vasc Surg* **40**, 1216–1222 (2004).
524. K. T. Nguyen, N. Shaikh, K. P. Shukla, S. H. Su, R. C. Eberhart, and L. Tang, Molecular responses of vascular smooth muscle cells and phagocytes to curcumin-eluting bioresorbable stent materials. *Biomaterials* **25**, 5333–5346 (2004).
525. R. Srivastava, V. Puri, R. C. Srimal, and B. N. Dhawan, Effect of curcumin on platelet aggregation and vascular prostacyclin synthesis. *Arzneimittelforschung* **36**, 715–717 (1986).
526. K. C. Srivastava, A. Bordia, and S. K. Verma, Curcumin, a major component of food spice turmeric (*Curcuma longa*) inhibits aggregation and alters eicosanoid metabolism in human blood platelets. *Prostaglandins Leukot Essent Fatty Acids* **52**, 223–227 (1995).
527. B. H. Shah, Z. Nawaz, S. A. Pertani, A. Roomi, H. Mahmood, S. A. Saeed, and A. H. Gilani, Inhibitory effect of curcumin, a food spice from turmeric, on platelet-activating factor- and arachidonic acid-mediated platelet aggregation through inhibition of thromboxane formation and Ca^{2+} signaling. *Biochem Pharmacol* **58**, 1167–1172 (1999).
528. C. Sumbilla, D. Lewis, T. Hammerschmidt, and G. Inesi, The slippage of the Ca^{2+} pump and its control by anions and curcumin in skeletal and cardiac sarcoplasmic reticulum. *J Biol Chem* **277**, 13,900–13,906 (2002).
529. Y. Sasaki, H. Goto, C. Tohda, F. Hatanaka, N. Shibahara, Y. Shimada, K. Terasawa, and K. Komatsu, Effects of curcuma drugs on vasomotion in isolated rat aorta. *Biol Pharm Bull* **26**, 1135–1143 (2003).
530. C. M. Terry, J. A. Clikeman, J. R. Hoidal, and K. S. Callahan, Effect of tumor necrosis factor-alpha and interleukin-1 alpha on heme oxygenase-1 expression in human endothelial cells. *Am J Physiol* **274**, H883–H891 (1998).
531. M. C. Ramirez-Tortosa, M. D. Mesa, M. C. Aguilera, J. L. Quiles, L. Baro, C. L. Ramirez-Tortosa, E. Martinez-Victoria, and A. Gil, Oral administration of a turmeric extract inhibits LDL oxidation and has hypocholesterolemic effects in rabbits with experimental atherosclerosis. *Atherosclerosis* **147**, 371–378 (1999).
532. K. H. Thompson, K. Bohmerle, E. Polishchuk, C. Martins, P. Toleikis, J. Tse, V. Yuen, J. H. McNeill, and C. Orvig, Complementary inhibition of synoviocyte, smooth

muscle cell or mouse lymphoma cell proliferation by a vanadyl curcumin complex compared to curcumin alone. *J Inorg Biochem* **98**, 2063–2070 (2004).

533. K. Keshavarz, The influence of turmeric and curcumin on cholesterol concentration of eggs and tissues. *Poult Sci* **55**, 1077–1083 (1976).
534. K. Srinivasan and K. Sambaiah, The effect of spices on cholesterol 7 alpha-hydroxylase activity and on serum and hepatic cholesterol levels in the rat. *Int J Vitam Nutr Res* **61**, 364–369 (1991).
535. K. B. Soni and R. Kuttan, Effect of oral curcumin administration on serum peroxides and cholesterol levels in human volunteers. *Indian J Physiol Pharmacol* **36**, 273–275 (1992).
536. H. M. Arafa, Curcumin attenuates diet–induced hypercholesterolemia in rats. *Med Sci Monit* **11**, BR228–234, (2005).
537. R. K. Kempaiah and K. Srinivasan, Beneficial influence of dietary curcumin, capsaicin and garlic on erythrocyte integrity in high-fat fed rats. *J Nutr Biochem* **17**(7), 471–478 (2005).
538. C. Fan, X. Wo, Y. Qian, J. Yin, and L. Gao, Effect of curcumin on the expression of LDL receptor in mouse macrophages. *J Ethnopharmacol* **105**, 251–254 (2006).
539. A. Ramirez Bosca, A. Soler, M. A. Carrion-Gutierrez, D. Pamies Mira, J. Pardo Zapata, J. Diaz-Alperi, A. Bernd, E. Quintanilla Almagro, and J. Miquel, An hydroalcoholic extract of *Curcuma longa* lowers the abnormally high values of human-plasma fibrinogen. *Mech Ageing Dev* **114**, 207–210 (2000).
540. K. A. Naidu and N. B. Thippeswamy, Inhibition of human low density lipoprotein oxidation by active principles from spices. *Mol Cell Biochem* **229**, 19–23 (2002).
541. R. Olszanecki, J. Jawien, M. Gajda, L. Mateuszuk, A. Gebska, M. Korabiowska, S. Chlopicki, and R. Korbut, Effect of curcumin on atherosclerosis in apoE/LDLR-double knockout mice. *J Physiol Pharmacol* **56**, 627–635 (2005).
542. W. F. Chen, S. L. Deng, B. Zhou, L. Yang, and Z. L. Liu, Curcumin and its analogues as potent inhibitors of low density lipoprotein oxidation: H-atom abstraction from the phenolic groups and possible involvement of the 4-hydroxy-3-methoxyphenyl groups. *Free Radical Biol Med* **40**, 526–535 (2006).
543. S. A. Frautschy, W. Hu, P. Kim, S. A. Miller, T. Chu, M. E. Harris-White, and G. M. Cole, Phenolic anti-inflammatory antioxidant reversal of Abeta-induced cognitive deficits and neuropathology. *Neurobiol Aging* **22**, 993–1005 (2001).
544. D. S. Kim, S. Y. Park, and J. K. Kim, Curcuminoids from *Curcuma longa* L. (Zingiberaceae) that protect PC12 rat pheochromocytoma and normal human umbilical vein endothelial cells from betaA(1-42) insult. *Neurosci Lett* **303**, 57–61 (2001).
545. G. P. Lim, T. Chu, F. Yang, W. Beech, S. A. Frautschy, and G. M. Cole, The curry spice curcumin reduces oxidative damage and amyloid pathology in an Alzheimer transgenic mouse. *J Neurosci* **21**, 8370–8377 (2001).
546. M. Grundman and P. Delaney, Antioxidant strategies for Alzheimer's disease. *Proc Nutr Soc* **61**, 191–202 (2002).
547. S. Y. Park and D. S. Kim, Discovery of natural products from *Curcuma longa* that protect cells from beta-amyloid insult: a drug discovery effort against Alzheimer's disease. *J Nat Prod* **65**, 1227–1231 (2002).
548. L. Adlerz, M. Beckman, S. Holback, R. Tehranian, V. Cortes Toro, and K. Iverfeldt, Accumulation of the amyloid precursor-like protein APLP2 and reduction of APLP1 in retinoic acid-differentiated human neuroblastoma cells upon curcumin-induced neurite retraction. *Brain Res Mol Brain Res* **119**, 62–72 (2003).

549. L. Baum and A. Ng, Curcumin interaction with copper and iron suggests one possible mechanism of action in Alzheimer'3s disease animal models. *J Alzheimers Dis* **6**, 367–77; discussion 443–449 (2004).
550. K. Ono, K. Hasegawa, H. Naiki, and M. Yamada, Curcumin has potent anti–amyloidogenic effects for Alzheimer's beta-amyloid fibrils in vitro. *J Neurosci Res* **75**, 742–750 (2004).
551. G. M. Cole, G. P. Lim, F. Yang, B. Teter, A. Begum, Q. Ma, M. E. Harris-White and S. A. Frautschy, Prevention of Alzheimer's disease: Omega-3 fatty acid and phenolic anti-oxidant interventions. *Neurobiol Aging* **26(Suppl 1)**, 133–136 (2005).
552. H. Kim, B. S. Park, K. G. Lee, C. Y. Choi, S. S. Jang, Y. H. Kim, and S. E. Lee, Effects of naturally occurring compounds on fibril formation and oxidative stress of beta-amyloid. *J Agric Food Chem* **53**, 8537–8541 (2005).
553. J. M. Ringman, S. A. Frautschy, G. M. Cole, D. L. Masterman, and J. L. Cummings, A potential role of the curry spice curcumin in Alzheimer's disease. *Curr Alzheimer Res* **2**, 131–136 (2005).
554. F. Yang, G. P. Lim, A. N. Begum, O. J. Ubeda, M. R. Simmons, S. S. Ambegaokar, P. P. Chen, R. Kayed, C. G. Glabe, S. A. Frautschy, and G. M. Cole, Curcumin inhibits formation of amyloid beta oligomers and fibrils, binds plaques, and reduces amyloid in vivo. *J Biol Chem* **280**, 5892–5901 (2005).
555. A. Apisariyakul, N. Vanittanakom, and D. Buddhasukh, Antifungal activity of turmeric oil extracted from *Curcuma longa* (Zingiberaceae). *J Ethnopharmacol* **49**, 163–169 (1995).
556. P. S. Negi, G. K. Jayaprakasha, L. Jagan Mohan Rao, and K. K. Sakariah, Antibacterial activity of turmeric oil: A byproduct from curcumin manufacture. *J Agric Food Chem* **47**, 4297–4300 (1999).
557. M. Wuthi-udomlert, W. Grisanapan, O. Luanratana, and W. Caichompoo, Antifungal activity of *Curcuma longa* grown in Thailand. *Southeast Asian J Trop Med Public Health* **31(Suppl 1)**, 178–182 (2000).
558. J. Jankun, A. M. Aleem, S. Malgorzewicz, M. Szkudlarek, M. I. Zavodszky, D. L. Dewitt, M. Feig, S. H. Selman, and E. Skrzypczak-Jankun, Synthetic curcuminoids modulate the arachidonic acid metabolism of human platelet 12-lipoxygenase and reduce sprout formation of human endothelial cells. *Mol Cancer Ther* **5**, 1371–1382 (2006).
559. E. Skrzypczak-Jankun, N. P. McCabe, S. H. Selman, and J. Jankun, Curcumin inhibits lipoxygenase by binding to its central cavity: theoretical and X-ray evidence. *Int J Mol Med* **6**, 521–526 (2000).

Chemistry

560. M. E. Braga, P. F. Leal, J. E. Carvalho, and M. A. Meireles, Comparison of yield, composition, and antioxidant activity of turmeric (*Curcuma longa* L.) extracts obtained using various techniques. *J Agric Food Chem* **51**, 6604–6611 (2003).
561. A. C. Manzan, F. S. Toniolo, E. Bredow, and N. P. Povh, Extraction of essential oil and pigments from *Curcuma longa* [L] by steam distillation and extraction with volatile solvents. *J Agric Food Chem* **51**, 6802–6807 (2003).
562. M. Backleh-Sohrt, P. Ekici, G. Leupold, and H. Parlar, Efficiency of foam fractionation for the enrichment of nonpolar compounds from aqueous extracts of plant materials. *J Nat Prod* **68**, 1386–1389 (2005).

563. V. M. Dirsch, H. Stuppner, and A. M. Vollmar, The Griess assay: suitable for a bio-guided fractionation of anti-inflammatory plant extracts? *Planta Med* **64**, 423–426 (1998).
564. G. K. Jayaprakasha, L. Jagan Mohan Rao, and K. K. Sakariah, Improved HPLC method for the determination of curcumin, demethoxycurcumin, and bisdemethoxycurcumin. *J Agric Food Chem* **50**, 3668–3672 (2002).
565. X. Sun, C. Gao, W. Cao, X. Yang, and E. Wang, Capillary electrophoresis with amperometric detection of curcumin in Chinese herbal medicine pretreated by solid-phase extraction. *J Chromatogr A* **962**, 117–125 (2002).
566. B. Tang, L. Ma, H. Y. Wang, and G. Y. Zhang, Study on the supramolecular interaction of curcumin and beta-cyclodextrin by spectrophotometry and its analytical application. *J Agric Food Chem* **50**, 1355–1361 (2002).
567. Y. Pak, R. Patek, and M. Mayersohn, Sensitive and rapid isocratic liquid chromatography method for the quantitation of curcumin in plasma. *J Chromatogr B Analyt Technol Biomed Life Sci* **796**, 339–346 (2003).
568. M. Bernabe-Pineda, M. T. Ramirez-Silva, M. Romero-Romo, E. Gonzalez-Vergara, and A. Rojas-Hernandez, Determination of acidity constants of curcumin in aqueous solution and apparent rate constant of its decomposition. *Spectrochim Acta A Mol Biomol Spectrosc* **60**, 1091–1097 (2004).
569. M. Lechtenberg, B. Quandt, and A. Nahrstedt, Quantitative determination of curcuminoids in Curcuma rhizomes and rapid differentiation of *Curcuma domestica* Val. and *Curcuma xanthorrhiza* Roxb. by capillary electrophoresis. *Phytochem Anal* **15**, 152–158 (2004).
570. M. J. Ansari, S. Ahmad, K. Kohli, J. Ali, and R. K. Khar, Stability-indicating HPTLC determination of curcumin in bulk drug and pharmaceutical formulations. *J Pharm Biomed Anal* **39**, 132–138 (2005).
571. L. A. May, E. Tourkina, S. R. Hoffman, and T. A. Dix, Detection and quantitation of curcumin in mouse lung cell cultures by matrix-assisted laser desorption ionization time of flight mass spectrometry. *Anal Biochem* **337**, 62–69 (2005).
572. F. Wang, X. Wu, S. Liu, Z. Jia, and J. Yang, The sensitive fluorimetric method for the determination of curcumin using the enhancement of mixed micelle. *J Fluoresc* **16**, 53–59 (2006).
573. H. H. Tonnesen and J. Karlsen, Studies on curcumin and curcuminoids. VI. Kinetics of curcumin degradation in aqueous solution. *Z Lebensm Unters Forsch* **180**, 402–404 (1985).
574. H. H. Tonnesen, J. Karlsen, and G. B. van Henegouwen, Studies on curcumin and curcuminoids. VIII. Photochemical stability of curcumin. *Z Lebensm Unters Forsch* **183**, 116–122 (1986).
575. H. H. Tonnesen, H. de Vries, J. Karlsen, and G. Beijersbergen van Henegouwen, Studies on curcumin and curcuminoids. IX: Investigation of the photobiological activity of curcumin using bacterial indicator systems. *J Pharm Sci* **76**, 371–373 (1987).
576. T. A. Dahl, W. M. McGowan, M. A. Shand, and V. S. Srinivasan, Photokilling of bacteria by the natural dye curcumin. *Arch Microbiol* **151**, 183–185 (1989).
577. C. F. Chignell, P. Bilski, K. J. Reszka, A. G. Motten, R. H. Sik, and T. A. Dahl, Spectral and photochemical properties of curcumin. *Photochem Photobiol* **59**, 295–302 (1994).
578. T. A. Dahl, P. Bilski, K. J. Reszka, and C. F. Chignell, Photocytotoxicity of curcumin. *Photochem Photobiol* **59**, 290–294 (1994).
579. Y. J. Wang, M. H. Pan, A. L. Cheng, L. I. Lin, Y. S. Ho, C. Y. Hsieh, and J. K. Lin, Stability of curcumin in buffer solutions and characterization of its degradation products. *J Pharm Biomed Anal* **15**, 1867–1876 (1997).

580. S. V. Jovanovic, C. W. Boone, S. Steenken, M. Trinoga, and R. B. Kaskey, How curcumin works preferentially with water soluble antioxidants. *J Am Chem Soc* **123**, 3064–3068 (2001).
581. H. H. Tonnesen, Solubility, chemical and photochemical stability of curcumin in surfactant solutions. Studies of curcumin and curcuminoids, XXVIII. *Pharmazie* **57**, 820–824 (2002).
582. H. H. Tonnesen, M. Masson, and T. Loftsson, Studies of curcumin and curcuminoids. XXVII. Cyclodextrin complexation: solubility, chemical and photochemical stability. *Int J Pharm* **244**, 127–135 (2002).
583. E. M. Bruzell, E. Morisbak, and H. H. Tonnesen, Studies on curcumin and curcuminoids. XXIX. Photoinduced cytotoxicity of curcumin in selected aqueous preparations. *Photochem Photobiol Sci* **4**, 523–530 (2005).
584. S. M. Khopde, K. I. Priyadarsini, D. K. Palit, and T. Mukherjee, Effect of solvent on the excited-state photophysical properties of curcumin. *Photochem Photobiol* **72**, 625–631 (2000).
585. F. Ortica and M. A. Rodgers, A laser flash photolysis study of curcumin in dioxane-water mixtures. *Photochem Photobiol* **74**, 745–751 (2001).
586. C. Parkanyi, M. R. Stem-Beren, O. R. Martinez, J. J. Aaron, M. Bulaceanu-MacNair and A. F. Arrieta, Solvatochromic correlations and ground- and excited-state dipole moments of curcuminoid dyes. *Spectrochim Acta A Mol Biomol Spectrosc* **60**, 1805–1810 (2004).
587. G. Began, E. Sudharshan, K. Udaya Sankar, and A. G. Appu Rao, Interaction of curcumin with phosphatidylcholine: A spectrofluorometric study. *J Agric Food Chem* **47**, 4992–4997 (1999).
588. A. C. Pulla Reddy, E. Sudharshan, A. G. Appu Rao, and B. R. Lokesh, Interaction of curcumin with human serum albumin: A spectroscopic study. *Lipids* **34**, 1025–1029 (1999).
589. F. Zsila, Z. Bikadi, and M. Simonyi, Unique, pH-dependent biphasic band shape of the visible circular dichroism of curcumin-serum albumin complex. *Biochem Biophys Res Commun* **301**, 776–782 (2003).
590. F. Zsila, Z. Bikadi, and M. Simonyi, Circular dichroism spectroscopic studies reveal pH dependent binding of curcumin in the minor groove of natural and synthetic nucleic acids. *Org Biomol Chem* **2**, 2902–2910 (2004).
591. F. Zsila, Z. Bikadi, and M. Simonyi, Induced circular dichroism spectra reveal binding of the antiinflammatory curcumin to human alpha1-acid glycoprotein. *Bioorg Med Chem* **12**, 3239–3245 (2004).
592. F. Wang, J. Yang, X. Wu and S. Liu, Study of the interaction of proteins with curcumin and SDS and its analytical application. *Spectrochim Acta A Mol Biomol Spectrosc* **61**, 2650–6, (2005).
593. H. Jiang, B. N. Timmermann, and D. R. Gang, Use of liquid chromatography-electrospray ionization tandem mass spectrometry to identify diarylheptanoids in turmeric (*Curcuma longa* L.) rhizome. *J Chromatogr A* **1111**, 21–31 (2006).

Curcumin Modulates DNA

594. K. C. Thresiamma, J. George, and R. Kuttan, Protective effect of curcumin, ellagic acid and bixin on radiation induced toxicity. *Indian J Exp Biol* **34**, 845–847 (1996).
595. K. C. Thresiamma, J. George, and R. Kuttan, Protective effect of curcumin, ellagic acid and bixin on radiation induced genotoxicity. *J Exp Clin Cancer Res* **17**, 431–434 (1998).

596. Shishu, A. K. Singla, and I. P. Kaur, Inhibitory effect of curcumin and its natural analogues on genotoxicity of heterocyclic amines from cooked food. *Indian J Exp Biol* **40**, 1365–1372 (2002).
597. K. Premkumar, S. Kavitha, S. T. Santhiya, A. R. Ramesh, and J. Suwanteerangkul, Interactive effects of saffron with garlic and curcumin against cyclophosphamide induced genotoxicity in mice. *Asia Pacific J Clin Nutr* **13**, 292–294 (2004).
598. Vijayalaxmi, Genetic effects of turmeric and curcumin in mice and rats. *Mutat Res* **79**, 125–132 (1980).
599. M. Nagabhushan and S. V. Bhide, Nonmutagenicity of curcumin and its antimutagenic action versus chili and capsaicin. *Nutr Cancer* **8**, 201–210 (1986).
600. M. Nagabhushan, A. J. Amonkar, and S. V. Bhide, In vitro antimutagenicity of curcumin against environmental mutagens. *Food Chem Toxicol* **25**, 545–547 (1987).
601. V. K. Shalini and L. Srinivas, Lipid peroxide induced DNA damage: protection by turmeric (*Curcuma longa*). *Mol Cell Biochem* **77**, 3–10 (1987).
602. K. Polasa, B. Sesikaran, T. P. Krishna, and K. Krishnaswamy, Turmeric (*Curcuma longa*)-induced reduction in urinary mutagens. *Food Chem Toxicol* **29**, 699–706 (1991).
603. M. A. Azuine, J. J. Kayal, and S. V. Bhide, Protective role of aqueous turmeric extract against mutagenicity of direct-acting carcinogens as well as benzo [alpha] pyrene-induced genotoxicity and carcinogenicity. *J Cancer Res Clin Oncol* **118**, 447–452 (1992).
604. K. Polasa, T. C. Raghuram, T. P. Krishna, and K. Krishnaswamy, Effect of turmeric on urinary mutagens in smokers. *Mutagenesis* **7**, 107–109 (1992).
605. V. K. Goud, K. Polasa, and K. Krishnaswamy, Effect of turmeric on xenobiotic metabolising enzymes. *Plant Foods Hum Nutr* **44**, 87–92 (1993).
606. P. Verger, M. Chambolle, P. Babayou, S. Le Breton, and J. L. Volatier, Estimation of the distribution of the maximum theoretical intake for ten additives in France. *Food Addit Contam* **15**, 759–766 (1998).
607. Y. Shukla, A. Arora, and P. Taneja, Antimutagenic potential of curcumin on chromosomal aberrations in Wistar rats. *Mutat Res* **515**, 197–202 (2002).
608. L. Srinivas and V. K. Shalini, DNA damage by smoke: protection by turmeric and other inhibitors of ROS. *Free Radical Biol Med* **11**, 277–283 (1991).
609. S. K. Abraham, L. Sarma, and P. C. Kesavan, Protective effects of chlorogenic acid, curcumin and beta-carotene against gamma-radiation-induced in vivo chromosomal damage. *Mutat Res* **303**, 109–112 (1993).
610. M. Subramanian, Sreejayan, M. N. Rao, T. P. Devasagayam, and B. B. Singh, Diminution of singlet oxygen-induced DNA damage by curcumin and related antioxidants. *Mutat Res* **311**, 249–255 (1994).
611. Y. Oda, Inhibitory effect of curcumin on SOS functions induced by UV irradiation. *Mutat Res* **348**, 67–73 (1995).
612. L. M. Antunes, M. C. Araujo, F. L. Dias, and C. S. Takahashi, Modulatory effects of curcumin on the chromosomal damage induced by doxorubicin in Chinese hamster ovary cells. *Teratog Carcinog Mutagen* **19**, 1–8 (1999).
613. M. C. Araujo, F. L. Dias, and C. S. Takahashi, Potentiation by turmeric and curcumin of gamma-radiation-induced chromosome aberrations in Chinese hamster ovary cells. *Teratog Carcinog Mutagen* **19**, 9–18 (1999).
614. K. Polasa, A. N. Naidu, I. Ravindranath, and K. Krishnaswamy, Inhibition of B(a)P induced strand breaks in presence of curcumin. *Mutat Res* **557**, 203–213 (2004).

615. A. Pal and A. K. Pal, Radioprotection of turmeric extracts in bacterial system. *Acta Biol Hung* **56**, 333–343 (2005).
616. A. K. Giri, S. K. Das, G. Talukder, and A. Sharma, Sister chromatid exchange and chromosome aberrations induced by curcumin and tartrazine on mammalian cells in vivo. *Cytobios* **62**, 111–117 (1990).
617. H. Ahsan and S. M. Hadi, Strand scission in DNA induced by curcumin in the presence of Cu(II). *Cancer Lett* **124**, 23–30 (1998).
618. J. Blasiak, A. Trzeciak, and J. Kowalik, Curcumin damages DNA in human gastric mucosa cells and lymphocytes. *J Environ Pathol Toxicol Oncol* **18**, 271–276, (1999).
619. J. Blasiak, A. Trzeciak, E. Malecka-Panas, J. Drzewoski, T. Iwanienko, I. Szumiel, and M. Wojewodzka, DNA damage and repair in human lymphocytes and gastric mucosa cells exposed to chromium and curcumin. *Teratog Carcinog Mutagen* **19**, 19–31 (1999).
620. M. C. Araujo, L. M. Antunes, and C. S. Takahashi, Protective effect of thiourea, a hydroxyl-radical scavenger, on curcumin-induced chromosomal aberrations in an in vitro mammalian cell system. *Teratog Carcinog Mutagen* **21**, 175–180 (2001).
621. K. Sakano and S. Kawanishi, Metal-mediated DNA damage induced by curcumin in the presence of human cytochrome P450 isozymes. *Arch Biochem Biophys* **405**, 223–230 (2002).
622. M. A. Mukundan, M. C. Chacko, V. V. Annapurna, and K. Krishnaswamy, Effect of turmeric and curcumin on BP-DNA adducts. *Carcinogenesis* **14**, 493–496 (1993).
623. S. S. Deshpande and G. B. Maru, Effects of curcumin on the formation of benzo[a]pyrene derived DNA adducts in vitro. *Cancer Lett* **96**, 71–80 (1995).
624. J. C. Chen, J. M. Hwang, G. W. Chen, M. F. Tsou, T. C. Hsia, and J. G. Chung, Curcumin decreases the DNA adduct formation, arylamines N-acetyltransferase activity and gene expression in human colon tumor cells (colo 205). *In Vivo* **17**, 301–309 (2003).
625. Y. S. Chen, C. C. Ho, K. C. Cheng, Y. S. Tyan, C. F. Hung, T. W. Tan, and J. G. Chung, Curcumin inhibited the arylamines N-acetyltransferase activity, gene expression and DNA adduct formation in human lung cancer cells (A549). *Toxicol In Vitro* **17**, 323–333 (2003).
626. J. Nair, S. Strand, N. Frank, J. Knauft, H. Wesch, P. R. Galle, and H. Bartsch, Apoptosis and age-dependant induction of nuclear and mitochondrial etheno-DNA adducts in Long-Evans Cinnamon (LEC) rats: enhanced DNA damage by dietary curcumin upon copper accumulation. *Carcinogenesis* **26**, 1307–1315 (2005).
627. Y. Chen, Y. Wu, W. Chen, and J. He, The effect of curcumin on mismatch repair (MMR) proteins hMSH2 and hMLH1 after ultraviolet (UV) irradiation on HL-60 cells. *J Huazhong Univ Sci Technolog Med Sci* **23**, 124–126 (2003).
628. A. P. Kulkarni, Y. T. Ghebremariam, and G. J. Kotwal, Curcumin inhibits the classical and the alternate pathways of complement activation. *Ann NY Acad Sci* **1056**, 100–112 (2005).
629. J. Cao, L. Jia, H. M. Zhou, Y. Liu, and L. F. Zhong, Mitochondrial and nuclear DNA damage induced by curcumin in human hepatoma G2 cells. *Toxicol Sci* **91**(2), 476–483 (2006).

Curcumin Downregulates *p*-Glycoprotein

630. A. Bielak–Mijewska, K. Piwocka, A. Magalska, and E. Sikora, P-Glycoprotein expression does not change the apoptotic pathway induced by curcumin in HL-60 cells. *Cancer Chemother Pharmacol* **53**, 179–185 (2004).

Curcumin in Chemosensitization

631. T. C. Hour, J. Chen, C. Y. Huang, J. Y. Guan, S. H. Lu, and Y. S. Pu, Curcumin enhances cytotoxicity of chemotherapeutic agents in prostate cancer cells by inducing p21(WAF1/CIP1) and C/EBPbeta expressions and suppressing NF-kappaB activation. *Prostate* **51**, 211–218 (2002).
632. S. V. Bava, V. T. Puliappadamba, A. Deepti, A. Nair, D. Karunagaran, and R. J. Anto, Sensitization of taxol-induced apoptosis by curcumin involves down-regulation of nuclear factor-kappaB and the serine/threonine kinase Akt and is independent of tubulin polymerization. *J Biol Chem* **280**, 6301–6308 (2005).
633. D. D. Deeb, H. Jiang, X. Gao, G. Divine, S. A. Dulchavsky, and S. C. Gautam, Chemosensitization of hormone-refractory prostate cancer cells by curcumin to TRAIL-induced apoptosis. *J Exp Ther Oncol* **5**, 81–91 (2005).
634. D. Chirnomas, T. Taniguchi, M. de la Vega, A. P. Vaidya, M. Vasserman, A. R. Hartman, R. Kennedy, R. Foster, J. Mahoney, M. V. Seiden and A. D. D'Andrea, Chemosensitization to cisplatin by inhibitors of the Fanconi anemia/BRCA pathway. *Mol Cancer Ther* **5**, 952–961 (2006).

Curcumin in Radiosensitization

635. D. Chendil, R. S. Ranga, D. Meigooni, S. Sathishkumar, and M. M. Ahmed, Curcumin confers radiosensitizing effect in prostate cancer cell line PC-3. *Oncogene* **23**, 1599–1607 (2004).
636. A. Khafif, R. Hurst, K. Kyker, D. M. Fliss, Z. Gil, and J. E. Medina, Curcumin: A new radio-sensitizer of squamous cell carcinoma cells. *Otolaryngol Head Neck Surg* **132**, 317–321 (2005).

Synergistic Effects of Curcumin

637. Y. Liu, R. L. Chang, X. X. Cui, H. L. Newmark, and A. H. Conney, Synergistic effects of curcumin on all-trans retinoic acid- and 1 alpha,25-dihydroxyvitamin D3-induced differentiation in human promyelocytic leukemia HL-60 cells. *Oncol Res* **9**, 19–29 (1997).
638. S. P. Verma, E. Salamone, and B. Goldin, Curcumin and genistein, plant natural products, show synergistic inhibitory effects on the growth of human breast cancer MCF-7 cells induced by estrogenic pesticides. *Biochem Biophys Res Commun* **233**, 692–696 (1997).
639. A. Khafif, S. P. Schantz, T. C. Chou, D. Edelstein, and P. G. Sacks, Quantitation of chemopreventive synergism between (–)-epigallocatechin-3-gallate and curcumin in normal, premalignant and malignant human oral epithelial cells. *Carcinogenesis* **19**, 419–424 (1998).
640. A. Spingarn, P. G. Sacks, D. Kelley, A. J. Dannenberg, and S. P. Schantz, Synergistic effects of 13-cis retinoic acid and arachidonic acid cascade inhibitors on growth of head and neck squamous cell carcinoma in vitro. *Otolaryngol Head Neck Surg* **118**, 159–164 (1998).
641. I. Navis, P. Sriganth, and B. Premalatha, Dietary curcumin with cisplatin administration modulates tumour marker indices in experimental fibrosarcoma. *Pharmacol Res* **39**, 175–179 (1999).
642. M. A. Indap and M. S. Barkume, Efficacies of plant phenolic compounds on sodium butyrate induced anti-tumour activity. *Indian J Exp Biol* **41**, 861–864 (2003).

643. J. Y. Koo, H. J. Kim, K. O. Jung, and K. Y. Park, Curcumin inhibits the growth of AGS human gastric carcinoma cells in vitro and shows synergism with 5-fluorouracil. *J Med Food* **7**, 117–121 (2004).

644. B. Du, L. Jiang, Q. Xia, and L. Zhong, Synergistic inhibitory effects of curcumin and 5-fluorouracil on the growth of the human colon cancer cell line HT-29. *Chemotherapy* **52**, 23–28 (2006).

645. S. Lev-Ari, L. Strier, D. Kazanov, L. Madar-Shapiro, H. Dvory-Sobol, I. Pinchuk, B. Marian, D. Lichtenberg, and N. Arber, Celecoxib and curcumin synergistically inhibit the growth of colorectal cancer cells. *Clin Cancer Res* **11**, 6738–6744 (2005).

646. S. Sen, H. Sharma, and N. Singh, Curcumin enhances Vinorelbine mediated apoptosis in NSCLC cells by the mitochondrial pathway. *Biochem Biophys Res Commun* **331**, 1245–1252 (2005).

647. R. L. Eckert, J. F. Crish, T. Efimova, and S. Balasubramanian, Opposing action of curcumin and green tea polyphenol in human keratinocytes. *Mol Nutr Food Res* **50**, 123–129 (2006).

648. T. O. Khor, Y. S. Keum, W. Lin, J. H. Kim, R. Hu, G. Shen, C. Xu, A. Gopalakrishnan, B. Reddy, X. Zheng, A. H. Conney, and A. N. Kong, Combined inhibitory effects of curcumin and phenethyl isothiocyanate on the growth of human PC-3 prostate xenografts in immunodeficient mice. *Cancer Res* **66**, 613–621 (2006).

649. S. Lev-Ari, L. Strier, D. Kazanov, O. Elkayam, D. Lichtenberg, D. Caspi, and N. Arber, Curcumin synergistically potentiates the growth-inhibitory and pro-apoptotic effects of celecoxib in osteoarthritis synovial adherent cells. *Rheumatology (Oxford)* **45**, 171–177 (2006).

Curcumin Modulates Immune System

650. S. Yasni, K. Yoshiie, H. Oda, M. Sugano, and K. Imaizumi, Dietary *Curcuma xanthorrhiza* Roxb. increases mitogenic responses of splenic lymphocytes in rats, and alters populations of the lymphocytes in mice. *J Nutr Sci Vitaminol (Tokyo)* **39**, 345–354 (1993).

651. E. H. South, J. H. Exon, and K. Hendrix, Dietary curcumin enhances antibody response in rats. *Immunopharmacol Immunotoxicol* **19**, 105–119 (1997).

652. S. Antony, R. Kuttan, and G. Kuttan, Immunomodulatory activity of curcumin. *Immunol Invest* **28**, 291–303 (1999).

653. X. Gao, J. Kuo, H. Jiang, D. Deeb, Y. Liu, G. Divine, R. A. Chapman, S. A. Dulchavsky, and S. C. Gautam, Immunomodulatory activity of curcumin: suppression of lymphocyte proliferation, development of cell-mediated cytotoxicity, and cytokine production in vitro. *Biochem Pharmacol* **68**, 51–61 (2004).

654. S. E. Ilsley, H. M. Miller, and C. Kamel, Effects of dietary quillaja saponin and curcumin on the performance and immune status of weaned piglets. *J Anim Sci* **83**, 82–88 (2005).

655. G. Y. Kim, K. H. Kim, S. H. Lee, M. S. Yoon, H. J. Lee, D. O. Moon, C. M. Lee, S. C. Ahn, Y. C. Park, and Y. M. Park, Curcumin inhibits immunostimulatory function of dendritic cells: MAPKs and translocation of NF-kappa B as potential targets. *J Immunol* **174**, 8116–8124 (2005).

656. X. Li and X. Liu, Effect of curcumin on immune function of mice. *J Huazhong Univ Sci Technolog Med Sci* **25**, 137–140 (2005).

657. V. S. Yadav, K. P. Mishra, D. P. Singh, S. Mehrotra, and V. K. Singh, Immunomodulatory effects of curcumin. *Immunopharmacol Immunotoxicol* **27**, 485–497 (2005).

658. D. Ranjan, T. D. Johnston, G. Wu, L. Elliott, S. Bondada, and M. Nagabhushan, Curcumin blocks cyclosporine A-resistant CD28 costimulatory pathway of human T-cell proliferation. *J Surg Res* **77**, 174–178 (1998).
659. D. Ranjan, A. Siquijor, T. D. Johnston, G. Wu, and M. Nagabhuskahn, The effect of curcumin on human B-cell immortalization by Epstein-Barr virus. *Am Surg* **64**, 47–51; discussion 51–52 (1998).
660. M. Deters, C. Siegers, P. Muhl, and W. Hansel, Choleretic effects of curcuminoids on an acute cyclosporin-induced cholestasis in the rat. *Planta Med* **65**, 610–613 (1999).
661. S. C. Chueh, M. K. Lai, I. S. Liu, F. C. Teng, and J. Chen, Curcumin enhances the immunosuppressive activity of cyclosporine in rat cardiac allografts and in mixed lymphocyte reactions. *Transplant Proc* **35**, 1603–1605 (2003).
662. M. Deters, T. Klabunde, H. Meyer, K. Resch, and V. Kaever, Effects of curcumin on cyclosporine-induced cholestasis and hypercholesterolemia and on cyclosporine metabolism in the rat. *Planta Med* **69**, 337–343 (2003).
663. N. Tirkey, G. Kaur, G. Vij, and K. Chopra, Curcumin, a diferuloylmethane, attenuates cyclosporine-induced renal dysfunction and oxidative stress in rat kidneys. *BMC Pharmacol* **5**, 15 (2005).
664. H. R. Ju, H. Y. Wu, S. Nishizono, M. Sakono, I. Ikeda, M. Sugano, and K. Imaizumi, Effects of dietary fats and curcumin on IgE-mediated degranulation of intestinal mast cells in brown Norway rats. *Biosci Biotechnol Biochem* **60**, 1856–1860 (1996).

Antiviral Effect of Curcumin

665. Z. Sui, R. Salto, J. Li, C. Craik, and P. R. Ortiz de Montellano, Inhibition of the HIV-1 and HIV-2 proteases by curcumin and curcumin boron complexes. *Bioorg Med Chem* **1**, 415–422 (1993).
666. A. Mazumder, K. Raghavan, J. Weinstein, K. W. Kohn, and Y. Pommier, Inhibition of human immunodeficiency virus type-1 integrase by curcumin. *Biochem Pharmacol* **49**, 1165–1170 (1995).
667. W. C. Jordan and C. R. Drew, Curcumin–a natural herb with anti-HIV activity. *J Natl Med Assoc* **88**, 333 (1996).
668. S. Barthelemy, L. Vergnes, M. Moynier, D. Guyot, S. Labidalle, and E. Bahraoui, Curcumin and curcumin derivatives inhibit Tat-mediated transactivation of type 1 human immunodeficiency virus long terminal repeat. *Res Virol* **149**, 43–52 (1998).
669. M. Hergenhahn, U. Soto, A. Weninger, A. Polack, C. H. Hsu, A. L. Cheng, and F. Rosl, The chemopreventive compound curcumin is an efficient inhibitor of Epstein-Barr virus BZLF1 transcription in Raji DR-LUC cells. *Mol Carcinog* **33**, 137–145 (2002).
670. M. M. Taher, G. Lammering, C. Hershey, and K. Valerie, Curcumin inhibits ultraviolet light induced human immunodeficiency virus gene expression. *Mol Cell Biochem* **254**, 289–297 (2003).
671. H. Chai, S. Yan, P. Lin, A. B. Lumsden, Q. Yao, and C. Chen, Curcumin blocks HIV protease inhibitor ritonavir-induced vascular dysfunction in porcine coronary arteries. *J Am Coll Surg* **200**, 820–830 (2005).
672. O. Vajragupta, P. Boonchoong, G. M. Morris, and A. J. Olson, Active site binding modes of curcumin in HIV-1 protease and integrase. *Bioorg Med Chem Lett* **15**, 3364–3368 (2005).

Antibacterial Effect of Curcumin

673. P. S. Negi, G. K. Jayaprakasha, L. Jagan Mohan Rao, and K. K. Sakariah, Antibacterial activity of turmeric oil: A byproduct from curcumin manufacture. *J Agric Food Chem* **47**, 4297–4300 (1999).
674. G. B. Mahady, S. L. Pendland, G. Yun, and Z. Z. Lu, Turmeric (*Curcuma longa*) and curcumin inhibit the growth of *Helicobacter pylori*, a group 1 carcinogen. *Anticancer Res* **22**, 4179–4181 (2002).

Anti-fungal Effect of Curcumin

675. A. Tantaoui-Elaraki and L. Beraoud, Inhibition of growth and aflatoxin production in *Aspergillus parasiticus* by essential oils of selected plant materials. *J Environ Pathol Toxicol Oncol* **13**, 67–72 (1994).
676. M. Wuthi-udomlert, W. Grisanapan, O. Luanratana, and W. Caichompoo, Antifungal activity of *Curcuma longa* grown in Thailand. *Southeast Asian J Trop Med Public Health* **31(Suppl 1)**, 178–182 (2000).
677. M. K. Kim, G. J. Choi, and H. S. Lee, Fungicidal property of *Curcuma longa* L. rhizome-derived curcumin against phytopathogenic fungi in a greenhouse. *J Agric Food Chem* **51**, 1578–1581 (2003).

Antimalarial Effect of Curcumin

678. G. N. Roth, A. Chandra, and M. G. Nair, Novel bioactivities of *Curcuma longa* constituents. *J Nat Prod* **61**, 542–545 (1998).
679. R. C. Reddy, P. G. Vatsala, V. G. Keshamouni, G. Padmanaban, and P. N. Rangarajan, Curcumin for malaria therapy. *Biochem Biophys Res Commun* **326**, 472–474 (2005).
680. D. N. Nandakumar, V. A. Nagaraj, P. G. Vathsala, P. Rangarajan, and G. Padmanaban, Curcumin-artemisinin combination therapy for malaria. *Antimicrob Agents Chemother* **50**, 1859–1860 (2006).

Antiparasitic Effect of Curcumin

681. F. Kiuchi, Y. Goto, N. Sugimoto, N. Akao, K. Kondo, and Y. Tsuda, Nematocidal activity of turmeric: Synergistic action of curcuminoids. *Chem Pharm Bull (Tokyo)* **41**, 1640–1643 (1993).
682. M. Nose, T. Koide, Y. Ogihara, Y. Yabu, and N. Ohta, Trypanocidal effects of curcumin in vitro. *Biol Pharm Bull* **21**, 643–645 (1998).
683. T. Koide, M. Nose, Y. Ogihara, Y. Yabu, and N. Ohta, Leishmanicidal effect of curcumin in vitro. *Biol Pharm Bull* **25**, 131–133 (2002).
684. D. Saleheen, S. A. Ali, K. Ashfaq, A. A. Siddiqui, A. Agha, and M. M. Yasinzai, Latent activity of curcumin against leishmaniasis in vitro. *Biol Pharm Bull* **25**, 386–389 (2002).
685. M. M. Chan, N. S. Adapala, and D. Fong, Curcumin overcomes the inhibitory effect of nitric oxide on Leishmania. *Parasitol Res* **96**, 49–56 (2005).
686. L. Perez-Arriaga, M. L. Mendoza-Magana, R. Cortes-Zarate, A. Corona-Rivera, L. Bobadilla-Morales, R. Troyo-Sanroman, and M. A. Ramirez-Herrera, Cytotoxic effect of curcumin on *Giardia lamblia* trophozoites. *Acta Trop* **98**(2), 152–161 (2006).

Curcumin and Its Analogs

687. A. Mukhopadhyay, N. Basu, N. Ghatak, and P. K. Gujral, Anti-inflammatory and irritant activities of curcumin analogues in rats. *Agents Actions* **12**, 508–515 (1982).
688. T. S. Rao, N. Basu, and H. H. Siddiqui, Anti-inflammatory activity of curcumin analogues. *Indian J Med Res* **75**, 574–578 (1982).
689. D. E. Douglas, 4,4′-Diacetyl curcumin–in-vitro histamine-blocking activity. *J Pharm Pharmacol* **45**, 766 (1993).
690. A. Mazumder, N. Neamati, S. Sunder, J. Schulz, H. Pertz, E. Eich, and Y. Pommier, Curcumin analogs with altered potencies against HIV-1 integrase as probes for biochemical mechanisms of drug action. *J Med Chem* **40**, 3057–3063 (1997).
691. Y. Oyama, T. Masuda, M. Nakata, L. Chikahisa, Y. Yamazaki, K. Miura, and M. Okagawa, Protective actions of 5′-n-alkylated curcumins on living cells suffering from oxidative stress. *Eur J Pharmacol* **360**, 65–71 (1998).
692. T. Devasena, K. N. Rajasekaran,and V. P. Menon, Bis-1,7-(2-hydroxyphenyl)-hepta-1,6-diene-3,5-dione (a curcumin analog) ameliorates DMH-induced hepatic oxidative stress during colon carcinogenesis. *Pharmacol Res* **46**, 39–45 (2002).
693. C. Gomes Dde, L. V. Alegrio, L. L. Leon, and M. E. de Lima, Total synthesis and anti-leishmanial activity of some curcumin analogues. *Arzneimittelforschung* **52**, 695–698 (2002).
694. J. Ishida, H. Ohtsu, Y. Tachibana, Y. Nakanishi, K. F. Bastow, M. Nagai, H. K. Wang, H. Itokawa, and K. H. Lee, Antitumor agents. Part 214: Synthesis and evaluation of curcumin analogues as cytotoxic agents. *Bioorg Med Chem* **10**, 3481–3487 (2002).
695. V. D. John, G. Kuttan, and K. Krishnankutty, Anti-tumour studies of metal chelates of synthetic curcuminoids. *J Exp Clin Cancer Res* **21**, 219–224 (2002).
696. H. Ohtsu, Z. Xiao, J. Ishida, M. Nagai, H. K. Wang, H. Itokawa, C. Y. Su, C. Shih, T. Chiang, E. Chang, Y. Lee, M. Y. Tsai, C. Chang, and K. H. Lee, Antitumor agents. 217. Curcumin analogues as novel androgen receptor antagonists with potential as anti-prostate cancer agents. *J Med Chem* **45**, 5037–5042 (2002).
697. T. Devasena, K. N. Rajasekaran, G. Gunasekaran, P. Viswanathan, and V. P. Menon, Anticarcinogenic effect of bis-1,7-(2-hydroxyphenyl)-hepta-1,6-diene-3,5-dione a curcumin analog on DMH-induced colon cancer model. *Pharmacol Res* **47**, 133–140 (2003).
698. P. V. Leyon and G. Kuttan, Studies on the role of some synthetic curcuminoid derivatives in the inhibition of tumour specific angiogenesis. *J Exp Clin Cancer Res* **22**, 77–83 (2003).
699. K. Mohri, Y. Watanabe, Y. Yoshida, M. Satoh, K. Isobe, N. Sugimoto, and Y. Tsuda, Synthesis of glycosylcurcuminoids. *Chem Pharm Bull (Tokyo)* **51**, 1268–1272 (2003).
700. T. P. Robinson, T. Ehlers, I. R. Hubbard, X. Bai, J. L. Arbiser, D. J. Goldsmith, and J. P. Bowen, Design, synthesis, and biological evaluation of angiogenesis inhibitors: Aromatic enone and dienone analogues of curcumin. *Bioorg Med Chem Lett* **13**, 115–117 (2003).
701. A. Sundaryono, A. Nourmamode, C. Gardrat, S. Grelier, G. Bravic, D. Chasseau, and A. Castellan, Studies on the photochemistry of 1,7-diphenyl-1,6-heptadiene-3,5-dione, a non-phenolic curcuminoid model. *Photochem Photobiol Sci* **2**, 914–920 (2003).
702. B. K. Adams, E. M. Ferstl, M. C. Davis, M. Herold, S. Kurtkaya, R. F. Camalier, M. G. Hollingshead, G. Kaur, E. A. Sausville, F. R. Rickles, J. P. Snyder, D. C. Liotta, and M.

Shoji, Synthesis and biological evaluation of novel curcumin analogs as anti-cancer and anti-angiogenesis agents. *Bioorg Med Chem* **12**, 3871–3883 (2004).

703. E. R. Hahm, Y. S. Gho, S. Park, C. Park, K. W. Kim, and C. H. Yang, Synthetic curcumin analogs inhibit activator protein-1 transcription and tumor-induced angiogenesis. *Biochem Biophys Res Commun* **321**, 337–344 (2004).
704. K. Nakano, T. Nakayachi, E. Yasumoto, S. R. Morshed, K. Hashimoto, H. Kikuchi, H. Nishikawa, K. Sugiyama, O. Amano, M. Kawase, and H. Sakagami, Induction of apoptosis by beta-diketones in human tumor cells. *Anticancer Res* **24**, 711–717 (2004).
705. K. M. Youssef, M. A. El-Sherbeny, F. S. El-Shafie, H. A. Farag, O. A. Al-Deeb, and S. A. Awadalla, Synthesis of curcumin analogues as potential antioxidant, cancer chemopreventive agents. *Arch Pharm (Weinheim)* **337**, 42–54 (2004).
706. B. K. Adams, J. Cai, J. Armstrong, M. Herold, Y. J. Lu, A. Sun, J. P. Snyder, D. C. Liotta, D. P. Jones, and M. Shoji, EF24, a novel synthetic curcumin analog, induces apoptosis in cancer cells via a redox-dependent mechanism. *Anticancer Drugs* **16**, 263–275 (2005).
707. M. A. Al-Omar, K. M. Youssef, M. A. El-Sherbeny, S. A. Awadalla, and H. I. El-Subbagh, Synthesis and in vitro antioxidant activity of some new fused pyridine analogs. *Arch Pharm (Weinheim)* **338**, 175–180 (2005).
708. J. Annaraj, S. Srinivasan, K. M. Ponvel, and P. Athappan, Mixed ligand copper(II) complexes of phenanthroline/bipyridyl and curcumin diketimines as DNA intercalators and their electrochemical behavior under Nafion and clay modified electrodes. *J Inorg Biochem* **99**, 669–676 (2005).
709. S. Dutta, S. Padhye, K. I. Priyadarsini, and C. Newton, Antioxidant and antiproliferative activity of curcumin semicarbazone. *Bioorg Med Chem Lett* **15**, 2738–2744 (2005).
710. M. S. Furness, T. P. Robinson, T. Ehlers, R. B. t. Hubbard, J. L. Arbiser, D. J. Goldsmith, and J. P. Bowen, Antiangiogenic agents: Studies on fumagillin and curcumin analogs. *Curr Pharm Des* **11**, 357–373 (2005).
711. T. P. Robinson, R. B. t. Hubbard, T. J. Ehlers, J. L. Arbiser, D. J. Goldsmith, and J. P. Bowen, Synthesis and biological evaluation of aromatic enones related to curcumin. *Bioorg Med Chem* **13**, 4007–4013 (2005).
712. R. Rukkumani, K. Aruna, P. S. Varma, K. N. Rajasekaran, and V. P. Menon, Comparative effects of curcumin and its analog on alcohol- and polyunsaturated fatty acid-induced alterations in circulatory lipid profiles. *J Med Food* **8**, 256–260 (2005).
713. C. Selvam, S. M. Jachak, R. Thilagavathi, and A. K. Chakraborti, Design, synthesis, biological evaluation and molecular docking of curcumin analogues as antioxidant, cyclooxygenase inhibitory and anti-inflammatory agents. *Bioorg Med Chem Lett* **15**, 1793–1797 (2005).
714. M. Suzuki, T. Nakamura, S. Iyoki, A. Fujiwara, Y. Watanabe, K. Mohri, K. Isobe, K. Ono, and S. Yano, Elucidation of anti-allergic activities of curcumin-related compounds with a special reference to their anti-oxidative activities. *Biol Pharm Bull* **28**, 1438–1443 (2005).
715. S. Venkateswarlu, M. S. Ramachandra, and G. V. Subbaraju, Synthesis and biological evaluation of polyhydroxycurcuminoids. *Bioorg Med Chem* **13**, 6374–6380 (2005).
716. H. B. Woo, W. S. Shin, S. Lee, and C. M. Ahn, Synthesis of novel curcumin mimics with asymmetrical units and their anti–angiogenic activity. *Bioorg Med Chem Lett* **15**, 3782–6, (2005).

717. K. M. Youssef and M. A. El–Sherbeny, Synthesis and antitumor activity of some curcumin analogs. *Arch Pharm (Weinheim)* **338**, 181–189 (2005).
718. L. Lin, Q. Shi, C. Y. Su, C. C. Shih, and K. H. Lee, Antitumor agents 247. New 4-ethoxycarbonylethyl curcumin analogs as potential antiandrogenic agents. *Bioorg Med Chem* **14**, 2527–2534 (2006).
719. T. Devasena, V. P. Menon, and K. N. Rajasekharan, Prevention of 1,2-dimethylhydrazine-induced circulatory oxidative stress by bis-1,7-(2-hydroxyphenyl)-hepta-1,6-diene-3,5-dione during colon carcinogenesis. *Pharmacol Rep* **58**, 229–235 (2006).
720. Z. Y. Du, Y. D. Bao, Z. Liu, W. Qiao, L. Ma, Z. S. Huang, L. Q. Gu, and A. S. Chan, Curcumin analogs as potent aldose reductase inhibitors. *Arch Pharm (Weinheim)* **339**, 123–128 (2006).
721. T. Takeuchi, T. Ishidoh, H. Iijima, I. Kuriyama, N. Shimazaki, O. Koiwai, K. Kuramochi, S. Kobayashi, F. Sugawara, K. Sakaguchi, H. Yoshida, and Y. Mizushina, Structural relationship of curcumin derivatives binding to the BRCT domain of human DNA polymerase lambda. *Genes Cells* **11**, 223–235 (2006).
722. A. J. Ruby, G. Kuttan, K. D. Babu, K. N. Rajasekharan, and R. Kuttan, Anti-tumour and antioxidant activity of natural curcuminoids. *Cancer Lett* **94**, 79–83 (1995).
723. R. J. Anto, J. George, K. V. Babu, K. N. Rajasekharan, and R. Kuttan, Antimutagenic and anticarcinogenic activity of natural and synthetic curcuminoids. *Mutat Res* **370**, 127–131 (1996).
724. T. Masuda, K. Hidaka, A. Shinohara, T. Maekawa, Y. Takeda, and H. Yamaguchi, Chemical studies on antioxidant mechanism of curcuminoid: analysis of radical reaction products from curcumin. *J Agric Food Chem* **47**, 71–77 (1999).
725. A. Asai and T. Miyazawa, Occurrence of orally administered curcuminoid as glucuronide and glucuronide/sulfate conjugates in rat plasma. *Life Sci* **67**, 2785–2793 (2000).
726. D. S. Kim and J. Y. Kim, Total synthesis of Calebin-A, preparation of its analogues, and their neuronal cell protectivity against beta-amyloid insult. *Bioorg Med Chem Lett* **11**, 2541–2543 (2001).
727. J. H. Kim, J. S. Shim, S. K. Lee, K. W. Kim, S. Y. Rha, H. C. Chung, and H. J. Kwon, Microarray-based analysis of anti-angiogenic activity of demethoxycurcumin on human umbilical vein endothelial cells: crucial involvement of the down-regulation of matrix metalloproteinase. *Jpn J Cancer Res* **93**, 1378–1385 (2002).
728. A. Srivivasan, V. P. Menon, V. Periaswamy, and K. N. Rajasekaran, Protection of pancreatic beta-cell by the potential antioxidant bis-o-hydroxycinnamoyl methane, analogue of natural curcuminoid in experimental diabetes. *J Pharm Pharm Sci* **6**, 327–333 (2003).
729. Z. Y. Du, R. R. Liu, W. Y. Shao, X. P. Mao, L. Ma, L. Q. Gu, Z. S. Huang, and A. S. Chan, alpha-Glucosidase inhibition of natural curcuminoids and curcumin analogs. *Eur J Med Chem* **41**, 213–218 (2006).
730. T. Nishiyama, T. Mae, H. Kishida, M. Tsukagawa, Y. Mimaki, M. Kuroda, Y. Sashida, K. Takahashi, T. Kawada, K. Nakagawa, and M. Kitahara, Curcuminoids and sesquiterpenoids in turmeric (*Curcuma longa* L.) suppress an increase in blood glucose level in type 2 diabetic KK-Ay mice. *J Agric Food Chem* **53**, 959–963 (2005).
731. S. I. Hoehle, E. Pfeiffer, A. M. Solyom, and M. Metzler, Metabolism of curcuminoids in tissue slices and subcellular fractions from rat liver. *J Agric Food Chem* **54**, 756–764 (2006).

732. H. Jiang, A. Somogyi, N. E. Jacobsen, B. N. Timmermann, and D. R. Gang, Analysis of curcuminoids by positive and negative electrospray ionization and tandem mass spectrometry. *Rapid Commun Mass Spectrom* **20**, 1001–1012 (2006).
733. T. Nhujak, W. Saisuwan, M. Srisa-art, and A. Petsom, Microemulsion electrokinetic chromatography for separation and analysis of curcuminoids in turmeric samples. *J Separ Sci* **29**, 666–676 (2006).
734. H. Kawashima, K. Akimoto, S. Jareonkitmongkol, N. Shirasaka, and S. Shimizu, Inhibition of rat liver microsomal desaturases by curcumin and related compounds. *Biosci Biotechnol Biochem* **60**, 108–110 (1996).
735. K. I. Priyadarsini, Free radical reactions of curcumin in membrane models. *Free Radical Biol Med* **23**, 838–843 (1997).
736. G. Began, E. Sudharshan, and A. G. Appu Rao, Inhibition of lipoxygenase 1 by phosphatidylcholine micelles-bound curcumin. *Lipids* **33**, 1223–1228 (1998).
737. A. Simon, D. P. Allais, J. L. Duroux, J. P. Basly, S. Durand-Fontanier and C. Delage, Inhibitory effect of curcuminoids on MCF-7 cell proliferation and structure-activity relationships. *Cancer Lett* **129**, 111–116 (1998).
738. A. T. Dinkova-Kostova and P. Talalay, Relation of structure of curcumin analogs to their potencies as inducers of Phase 2 detoxification enzymes. *Carcinogenesis* **20**, 911–914 (1999).
739. E. Skrzypczak-Jankun, N. P. McCabe, S. H. Selman, and J. Jankun, Curcumin inhibits lipoxygenase by binding to its central cavity: theoretical and X-ray evidence. *Int J Mol Med* **6**, 521–526 (2000).
740. P. Venkatesan and M. N. Rao, Structure-activity relationships for the inhibition of lipid peroxidation and the scavenging of free radicals by synthetic symmetrical curcumin analogues. *J Pharm Pharmacol* **52**, 1123–1128 (2000).
741. T. Masuda, Y. Toi, H. Bando, T. Maekawa, Y. Takeda, and H. Yamaguchi, Structural identification of new curcumin dimers and their contribution to the antioxidant mechanism of curcumin. *J Agric Food Chem* **50**, 2524–2530 (2002).
742. K. I. Priyadarsini, D. K. Maity, G. H. Naik, M. S. Kumar, M. K. Unnikrishnan, J. G. Satav, and H. Mohan, Role of phenolic O-H and methylene hydrogen on the free radical reactions and antioxidant activity of curcumin. *Free Radical Biol Med* **35**, 475–484 (2003).
743. Y. Sugiyama, S. Kawakishi, and T. Osawa, Involvement of the beta-diketone moiety in the antioxidative mechanism of tetrahydrocurcumin. *Biochem Pharmacol* **52**, 519–52, (1996).
744. G. Rucker, E. Breitmaier, D. Manns, W. Maier, A. Marek, B. Heinzmann, K. Heiden, and S. Seggewies, Antimalarial activity of 1,4-epidioxy-bisabola-2,12-diene derivatives. *Arch Pharm (Weinheim)* **330**, 12–16 (1997).
745. S. Nishizono, T. Hayami, I. Ikeda, and K. Imaizumi, Protection against the diabetogenic effect of feeding tert-butylhydroquinone to rats prior to the administration of streptozotocin. *Biosci Biotechnol Biochem* **64**, 1153–1158 (2000).
746. C. Gomes Dde, L. V. Alegrio, M. E. de Lima, L. L. Leon, and C. A. Araujo, Synthetic derivatives of curcumin and their activity against Leishmania amazonensis. *Arzneimittelforschung* **52**, 120–124 (2002).
747. E. R. Hahm, G. Cheon, J. Lee, B. Kim, C. Park, and C. H. Yang, New and known symmetrical curcumin derivatives inhibit the formation of Fos-Jun-DNA complex. *Cancer Lett* **184**, 89–96 (2002).

748. J. S. Shim, D. H. Kim, H. J. Jung, J. H. Kim, D. Lim, S. K. Lee, K. W. Kim, J. W. Ahn, J. S. Yoo, J. R. Rho, J. Shin, and H. J. Kwon, Hydrazinocurcumin, a novel synthetic curcumin derivative, is a potent inhibitor of endothelial cell proliferation. *Bioorg Med Chem* **10**, 2439–2444 (2002).

749. J. S. Shim, D. H. Kim, H. J. Jung, J. H. Kim, D. Lim, S. K. Lee, K. W. Kim, J. W. Ahn, J. S. Yoo, J. R. Rho, J. Shin, and H. J. Kwon, Hydrazinocurcumin, a novel synthetic curcumin derivative, is a potent inhibitor of endothelial cell proliferation. *Bioorg Med Chem* **10**, 2987–2992 (2002).

750. A. P. Kumar, G. E. Garcia, R. Ghosh, R. V. Rajnarayanan, W. L. Alworth, and T. J. Slaga, 4-Hydroxy-3-methoxybenzoic acid methyl ester: a curcumin derivative targets Akt/NF kappa B cell survival signaling pathway: potential for prostate cancer management. *Neoplasia* **5**, 255–266 (2003).

751. S. R. Lamb and S. M. Wilkinson, Contact allergy to tetrahydrocurcumin. *Contact Dermat* **48**, 227 (2003).

752. C. M. Ahn, W. S. Shin, H. Bum Woo, S. Lee, and H. W. Lee, Synthesis of symmetrical bis-alkynyl or alkyl pyridine and thiophene derivatives and their antiangiogenic activities. *Bioorg Med Chem Lett* **14**, 3893–3896 (2004).

753. S. Gafner, S. K. Lee, M. Cuendet, S. Barthelemy, L. Vergnes, S. Labidalle, R. G. Mehta, C. W. Boone, and J. M. Pezzuto, Biologic evaluation of curcumin and structural derivatives in cancer chemoprevention model systems. *Phytochemistry* **65**, 2849–2859 (2004).

754. C. R. Girija, N. S. Begum, A. A. Syed, and V. Thiruvenkatam, Hydrogen-bonding and C-H...pi interactions in 1,7-bis(4-hydroxy-3-methoxyphenyl)heptane-3,5-dione (tetrahydrocurcumin). *Acta Crystallogr C* **60**, o611–o613 (2004).

755. J. T. Mague, W. L. Alworth, and F. L. Payton, Curcumin and derivatives. *Acta Crystallogr C* **60**, o608–o610 (2004).

756. S. Park, S. Chung, K. M. Kim, K. C. Jung, C. Park, E. R. Hahm, and C. H. Yang, Determination of binding constant of transcription factor myc-max/max-max and E-box DNA: the effect of inhibitors on the binding. *Biochim Biophys Acta* **1670**, 217–128 (2004).

757. J. S. Shim, J. Lee, H. J. Park, S. J. Park, and H. J. Kwon, A new curcumin derivative, HBC, interferes with the cell cycle progression of colon cancer cells via antagonization of the Ca^{2+}/calmodulin function. *Chem Biol* **11**, 1455–1463 (2004).

758. Y. Sumanont, Y. Murakami, M. Tohda, O. Vajragupta, K. Matsumoto, and H. Watanabe, Evaluation of the nitric oxide radical scavenging activity of manganese complexes of curcumin and its derivative. *Biol Pharm Bull* **27**, 170–173 (2004).

759. D. D. Heath, M. A. Pruitt, D. E. Brenner, A. N. Begum, S. A. Frautschy, and C. L. Rock, Tetrahydrocurcumin in plasma and urine: Quantitation by high performance liquid chromatography. *J Chromatogr B Analyt Technol Biomed Life Sci* **824**, 206–212 (2005).

760. Y. Mizushina, T. Ishidoh, T. Takeuchi, N. Shimazaki, O. Koiwai, K. Kuramochi, S. Kobayashi, F. Sugawara, K. Sakaguchi, and H. Yoshida, Monoacetylcurcumin: A new inhibitor of eukaryotic DNA polymerase lambda and a new ligand for inhibitor-affinity chromatography. *Biochem Biophys Res Commun* **337**, 1288–1295 (2005).

761. C. H. Park, J. H. Lee, and C. H. Yang, Curcumin derivatives inhibit the formation of Jun-Fos-DNA complex independently of their conserved cysteine residues. *J Biochem Mol Biol* **38**, 474–480 (2005).

Pharmacology and Metabolism of Curcumin

762. G. M. Holder, J. L. Plummer, and A. J. Ryan, The metabolism and excretion of curcumin (1,7-bis-(4-hydroxy-3-methoxyphenyl)-1,6-heptadiene-3,5-dione) in the rat. *Xenobiotica* **8**, 761–768 (1978).
763. B. Wahlstrom and G. Blennow, A study on the fate of curcumin in the rat. *Acta Pharmacol Toxicol (Copenh)* **43**, 86–92 (1978).
764. V. Ravindranath and N. Chandrasekhara, Absorption and tissue distribution of curcumin in rats. *Toxicology* **16**, 259–265 (1980).
765. V. Ravindranath and N. Chandrasekhara, In vitro studies on the intestinal absorption of curcumin in rats. *Toxicology* **20**, 251–257 (1981).
766. V. Ravindranath and N. Chandrasekhara, Metabolism of curcumin–studies with [3H]curcumin. *Toxicology* **22**, 337–344 (1981).
767. M. H. Pan, T. M. Huang, and J. K. Lin, Biotransformation of curcumin through reduction and glucuronidation in mice. *Drug Metab Dispos* **27**, 486–494 (1999).
768. C. R. Ireson, D. J. Jones, S. Orr, M. W. Coughtrie, D. J. Boocock, M. L. Williams, P. B. Farmer, W. P. Steward, and A. J. Gescher, Metabolism of the cancer chemopreventive agent curcumin in human and rat intestine. *Cancer Epidemiol Biomarkers Prev* **11**, 105–111 (2002).
769. D. Heath, M. A. Pruitt, D. E. Brenner, and C. L. Rock, Curcumin in plasma and urine: Quantitation by high-performance liquid chromatography. *J Chromatogr B Analyt Technol Biomed Life Sci* **783**, 287–295 (2003).
770. G. Garcea, D. J. Jones, R. Singh, A. R. Dennison, P. B. Farmer, R. A. Sharma, W. P. Steward, A. J. Gescher, and D. P. Berry, Detection of curcumin and its metabolites in hepatic tissue and portal blood of patients following oral administration. *Br J Cancer* **90**, 1011–1105 (2004).
771. K. Yuan, Q. Weng, H. Zhang, J. Xiong, and G. Xu, Application of capillary zone electrophoresis in the separation and determination of the curcuminoids in urine. *J Pharm Biomed Anal* **38**, 133–138 (2005).
772. G. J. Kelloff, J. A. Crowell, E. T. Hawk, V. E. Steele, R. A. Lubet, C. W. Boone, J. M. Covey, L. A. Doody, G. S. Omenn, P. Greenwald, W. K. Hong, D. R. Parkinson, D. Bagheri, G. T. Baxter, M. Blunden, M. K. Doeltz, K. M. Eisenhauer, K. Johnson, G. G. Knapp, D. G. Longfellow, W. F. Malone, S. G. Nayfield, H. E. Seifried, L. M. Swall, and C. C. Sigman, Strategy and planning for chemopreventive drug development: Clinical development plans II. *J Cell Biochem* **26(Suppl)**, 54–71 (1996).
773. R. A. Sharma, H. R. McLelland, K. A. Hill, C. R. Ireson, S. A. Euden, M. M. Manson, M. Pirmohamed, L. J. Marnett, A. J. Gescher, and W. P. Steward, Pharmacodynamic and pharmacokinetic study of oral Curcuma extract in patients with colorectal cancer. *Clin Cancer Res* **7**, 1894–1900 (2001).
774. A. Liu, H. Lou, L. Zhao, and P. Fan, Validated LC/MS/MS assay for curcumin and tetrahydrocurcumin in rat plasma and application to pharmacokinetic study of phospholipid complex of curcumin. *J Pharm Biomed Anal* **40**, 720–727 (2006).

AIDS

775. C. J. Li, L. J. Zhang, B. J. Dezube, C. S. Crumpacker, and A. B. Pardee, Three inhibitors of type 1 human immunodeficiency virus long terminal repeat-directed gene expression and virus replication. *Proc Natl Acad Sci USA* **90**, 1839–1842 (1993).

776. A. Mazumder, A. Gazit, A. Levitzki, M. Nicklaus, J. Yung, G. Kohlhagen, and Y. Pommier, Effects of tyrphostins, protein kinase inhibitors, on human immunodeficiency virus type 1 integrase. *Biochemistry* **34**, 15,111–15,122 (1995).

777. A. Mazumder, S. Wang, N. Neamati, M. Nicklaus, S. Sunder, J. Chen, G. W. Milne, W. G. Rice, T. R. Burke, Jr., and Y. Pommier, Antiretroviral agents as inhibitors of both human immunodeficiency virus type 1 integrase and protease. *J Med Chem* **39**, 2472–2481 (1996)

Antidepressant

778. Y. Xu, B. S. Ku, H. Y. Yao, Y. H. Lin, X. Ma, Y. H. Zhang, and X. J. Li, The effects of curcumin on depressive-like behaviors in mice. *Eur J Pharmacol* **518**, 40–46 (2005).

779. Y. Xu, B. S. Ku, H. Y. Yao, Y. H. Lin, X. Ma, Y. H. Zhang, and X. J. Li, Antidepressant effects of curcumin in the forced swim test and olfactory bulbectomy models of depression in rats. *Pharmacol Biochem Behav* **82**, 200–206 (2005).

Anti-spasmodic

780. C. Itthipanichpong, N. Ruangrungsi, W. Kemsri, and A. Sawasdipanich, Antispasmodic effects of curcuminoids on isolated guinea-pig ileum and rat uterus. *J Med Assoc Thai* **86(Suppl 2)**, S299–S309 (2003).

Antivenomic

781. K. S. Girish and K. Kemparaju, Inhibition of Naja naja venom hyaluronidase by plant-derived bioactive components and polysaccharides. *Biochemistry (Mosc)* **70**, 948–952 (2005).

Atherosclerosis

782. R. Olszanecki, J. Jawien, M. Gajda, L. Mateuszuk, A. Gebska, M. Korabiowska, S. Chlopicki, and R. Korbut, Effect of curcumin on atherosclerosis in apoE/LDLR-double knockout mice. *J Physiol Pharmacol* **56**, 627–635 (2005).

Contraceptive

783. T. Rithaporn, M. Monga, and M. Rajasekaran, Curcumin: A potential vaginal contraceptive. *Contraception* **68**, 219–223 (2003).

Cataract

784. S. Awasthi, S. K. Srivatava, J. T. Piper, S. S. Singhal, M. Chaubey, and Y. C. Awasthi, Curcumin protects against 4-hydroxy-2-trans-nonenal-induced cataract formation in rat lenses. *Am J Clin Nutr* **64**, 761–766 (1996).

785. U. Pandya, M. K. Saini, G. F. Jin, S. Awasthi, B. F. Godley, and Y. C. Awasthi, Dietary curcumin prevents ocular toxicity of naphthalene in rats. *Toxicol Lett* **115**, 195–204 (2000).

786. P. Suryanarayana, K. Krishnaswamy, and G. B. Reddy, Effect of curcumin on galactose-induced cataractogenesis in rats. *Mol Vis* **9**, 223–230 (2003).

787. S. Padmaja and T. N. Raju, Antioxidant effect of curcumin in selenium induced cataract of Wistar rats. *Indian J Exp Biol* **42**, 601–603 (2004).

788. P. A. Kumar, P. Suryanarayana, P. Y. Reddy, and G. B. Reddy, Modulation of alpha-crystallin chaperone activity in diabetic rat lens by curcumin. *Mol Vis* **11**, 561–568 (2005).
789. A. Matteucci, C. Frank, M. R. Domenici, M. Balduzzi, S. Paradisi, G. Carnovale-Scalzo, G. Scorcia, and F. Malchiodi-Albedi, Curcumin treatment protects rat retinal neurons against excitotoxicity: effect on N-methyl-D: -aspartate-induced intracellular Ca(2+) increase. *Exp Brain Res* **167**, 641–648 (2005).
790. P. Suryanarayana, M. Saraswat, T. Mrudula, T. P. Krishna, K. Krishnaswamy, and G. B. Reddy, Curcumin and turmeric delay streptozotocin-induced diabetic cataract in rats. *Invest Ophthalmol Vis Sci* **46**, 2092–2099 (2005).

Cyctic Fibrosis

791. A. Dragomir, J. Bjorstad, L. Hjelte, and G. M. Roomans, Curcumin does not stimulate cAMP-mediated chloride transport in cystic fibrosis airway epithelial cells. *Biochem Biophys Res Commun* **322**, 447–451 (2004).
792. M. E. Egan, M. Pearson, S. A. Weiner, V. Rajendran, D. Rubin, J. Glockner-Pagel, S. Canny, K. Du, G. L. Lukacs, and M. J. Caplan, Curcumin, a major constituent of turmeric, corrects cystic fibrosis defects. *Science* **304**, 600–602 (2004).
793. Y. Song, N. D. Sonawane, D. Salinas, L. Qian, N. Pedemonte, L. J. Galietta, and A. S. Verkman, Evidence against the rescue of defective DeltaF508-CFTR cellular processing by curcumin in cell culture and mouse models. *J Biol Chem* **279**, 40,629–633 (2004).
794. A. L. Berger, C. O. Randak, L. S. Ostedgaard, P. H. Karp, D. W. Vermeer, and M. J. Welsh, Curcumin stimulates cystic fibrosis transmembrane conductance regulator Cl^- channel activity. *J Biol Chem* **280**, 5221–5226 (2005).
795. 249. J. Lipecka, C. Norez, N. Bensalem, M. Baudouin-Legros, G. Planelles, F. Becq, A. Edelman, and N. Davezac, Rescue of {delta}F508-CFTR (cystic fibrosis transmembrane conductance regulator) by curcumin: Involvement of the keratin 18 network. *J Pharmacol Exp Ther* **317**, 500–505 (2006).

Epilepsy

796. Y. Sumanont, Y. Murakami, M. Tohda, O. Vajragupta, H. Watanabe, and K. Matsumoto, Prevention of kainic acid-induced changes in nitric oxide level and neuronal cell damage in the rat hippocampus by manganese complexes of curcumin and diacetylcurcumin. *Life Sci* **78**, 1884–1891 (2006).

Hyaline Membrane Disease (HMD)

797. A. Literat, F. Su, M. Norwicki, M. Durand, R. Ramanathan, C. A. Jones, P. Minoo, and K. Y. Kwong, Regulation of pro-inflammatory cytokine expression by curcumin in hyaline membrane disease (HMD). *Life Sci* **70**, 253–267 (2001).

Hypolipidemia

798. P. S. Babu and K. Srinivasan, Hypolipidemic action of curcumin, the active principle of turmeric (*Curcuma longa*) in streptozotocin induced diabetic rats. *Mol Cell Biochem* **166**, 169–175 (1997).

Liver Diseases

799. K. B. Soni and R. Kuttan, Effect of oral curcumin administration on serum peroxides and cholesterol levels in human volunteers. *Indian J Physiol Pharmacol* **36**, 273–275 (1992).
800. A. C. Reddy and B. R. Lokesh, Effect of curcumin and eugenol on iron-induced hepatic toxicity in rats. *Toxicology* **107**, 39–45 (1996).
801. E. J. Park, C. H. Jeon, G. Ko, J. Kim, and D. H. Sohn, Protective effect of curcumin in rat liver injury induced by carbon tetrachloride. *J Pharm Pharmacol* **52**, 437–440 (2000).
802. V. Rajakrishnan, A. Jayadeep, O. S. Arun, P. R. Sudhakaran, and V. P. Menon, Changes in the prostaglandin levels in alcohol toxicity: Effect of curcumin and N-acetylcysteine. *J Nutr Biochem* **11**, 509–514 (2000).
803. R. Akrishnan and V. P. Menon, Potential role of antioxidants during ethanol-induced changes in the fatty acid composition and arachidonic acid metabolites in male Wistar rats. *Cell Biol Toxicol* **17**, 11–22 (2001).
804. A. Asai and T. Miyazawa, Dietary curcuminoids prevent high-fat diet-induced lipid accumulation in rat liver and epididymal adipose tissue. *J Nutr* **131**, 2932–2935 (2001).
805. H. C. Kang, J. X. Nan, P. H. Park, J. Y. Kim, S. H. Lee, S. W. Woo, Y. Z. Zhao, E. J. Park, and D. H. Sohn, Curcumin inhibits collagen synthesis and hepatic stellate cell activation in-vivo and in-vitro. *J Pharm Pharmacol* **54**, 119–126 (2002).
806. R. Rukkumani, M. Sri Balasubashini, P. Vishwanathan, and V. P. Menon, Comparative effects of curcumin and photo-irradiated curcumin on alcohol- and polyunsaturated fatty acid-induced hyperlipidemia. *Pharmacol Res* **46**, 257–264 (2002).
807. A. A. Nanji, K. Jokelainen, G. L. Tipoe, A. Rahemtulla, P. Thomas, and A. J. Dannenberg, Curcumin prevents alcohol-induced liver disease in rats by inhibiting the expression of NF-kappa B-dependent genes. *Am J Physiol Gastrointest Liver Physiol* **284**, G321–G327 (2003).
808. R. S. Naik, A. M. Mujumdar, and S. Ghaskadbi, Protection of liver cells from ethanol cytotoxicity by curcumin in liver slice culture in vitro. *J Ethnopharmacol* **95**, 31–37 (2004).
809. N. Kamalakkannan, R. Rukkumani, P. S. Varma, P. Viswanathan, K. N. Rajasekharan, and V. P. Menon, Comparative effects of curcumin and an analogue of curcumin in carbon tetrachloride-induced hepatotoxicity in rats. *Basic Clin Pharmacol Toxicol* **97**, 15–21 (2005).
810. S. Padmaja and T. N. Raju, Protective effect of curcumin during selenium induced toxicity on dehydrogenases in hepatic tissue. *Indian J Physiol Pharmacol* **49**, 111–114 (2005).

Lung Diseases

811. N. Venkatesan and G. Chandrakasan, Modulation of cyclophosphamide-induced early lung injury by curcumin, an anti-inflammatory antioxidant. *Mol Cell Biochem* **142**, 79–87 (1995).
812. N. Venkatesan, V. Punithavathi, and G. Chandrakasan, Curcumin protects bleomycin-induced lung injury in rats. *Life Sci* **61**, PL51–PL58 (1997).
813. N. Venkatesan, Pulmonary protective effects of curcumin against paraquat toxicity. *Life Sci* **66**, PL21–PL28 (2000).

814. C. D. Huang, O. Tliba, R. A. Panettieri, Jr., and Y. Amrani, Bradykinin induces interleukin-6 production in human airway smooth muscle cells: modulation by Th2 cytokines and dexamethasone. *Am J Respir Cell Mol Biol* **28**, 330–338 (2003).
815. C. Kalpana and V. P. Menon, Inhibition of nicotine-induced toxicity by curcumin and curcumin analog: a comparative study. *J Med Food* **7**, 467–471 (2004).
816. C. Kalpana and V. P. Menon, Curcumin ameliorates oxidative stress during nicotine-induced lung toxicity in Wistar rats. *Ital J Biochem* **53**, 82–86 (2004).
817. G. Deby-Dupont, A. Mouithys-Mickalad, D. Serteyn, M. Lamy, and C. Deby, Resveratrol and curcumin reduce the respiratory burst of Chlamydia-primed THP-1 cells. *Biochem Biophys Res Commun* **333**, 21–27 (2005).
818. A. H. Gilani, A. J. Shah, M. N. Ghayur, and K. Majeed, Pharmacological basis for the use of turmeric in gastrointestinal and respiratory disorders. *Life Sci* **76**, 3089–3105 (2005).
819. C. Kalpana, K. N. Rajasekharan, and V. P. Menon, Modulatory effects of curcumin and curcumin analog on circulatory lipid profiles during nicotine-induced toxicity in Wistar rats. *J Med Food* **8**, 246–250 (2005).

Osteoporosis

820. L. M. Antunes, M. C. Araujo, J. D. Darin, and M. L. Bianchi, Effects of the antioxidants curcumin and vitamin C on cisplatin-induced clastogenesis in Wistar rat bone marrow cells. *Mutat Res* **465**, 131–137 (2000).
821. K. Naganuma, S. Amano, H. Takeda, S. Kitano, and S. Hanazawa, Role of transcriptional factor activation protein-1 in endogenous expression of the interleukin-1 beta gene involved in Porphyromonas gingivalis fimbria-stimulated bone resorption in the mouse calvarial system. *Oral Microbiol Immunol* **15**, 53–57 (2000).
822. K. Ozaki, Y. Kawata, S. Amano, and S. Hanazawa, Stimulatory effect of curcumin on osteoclast apoptosis. *Biochem Pharmacol* **59**, 1577–1581 (2000).
823. M. Notoya, H. Nishimura, J. T. Woo, K. Nagai, Y. Ishihara, and H. Hagiwara, Curcumin inhibits the proliferation and mineralization of cultured osteoblasts. *Eur J Pharmacol* **534**, 55–62 (2006).

Parkinson's Disease

824. V. Zbarsky, K. P. Datla, S. Parkar, D. K. Rai, O. I. Aruoma, and D. T. Dexter, Neuroprotective properties of the natural phenolic antioxidants curcumin and naringenin but not quercetin and fisetin in a 6-OHDA model of Parkinson's disease. *Free Radical Res* **39**, 1119–1125 (2005).

Renal Diseases

825. H. H. Cohly, A. Taylor, M. F. Angel, and A. K. Salahudeen, Effect of turmeric, turmerin and curcumin on H_2O_2-induced renal epithelial (LLC-PK1) cell injury. *Free Radical Biol Med* **24**, 49–54 (1998).
826. D. A. Shoskes, Effect of bioflavonoids quercetin and curcumin on ischemic renal injury: A new class of renoprotective agents. *Transplantation* **66**, 147–152 (1998).
827. P. Suresh Babu and K. Srinivasan, Amelioration of renal lesions associated with diabetes by dietary curcumin in streptozotocin diabetic rats. *Mol Cell Biochem* **181**, 87–96 (1998).

828. E. A. Jones and D. A. Shoskes, The effect of mycophenolate mofetil and polyphenolic bioflavonoids on renal ischemia reperfusion injury and repair. *J Urol* **163**, 999–1004 (2000).

829. N. Venkatesan, D. Punithavathi, and V. Arumugam, Curcumin prevents adriamycin nephrotoxicity in rats. *Br J Pharmacol* **129**, 231–234 (2000).

830. K. Okada, C. Wangpoengtrakul, T. Tanaka, S. Toyokuni, K. Uchida, and T. Osawa, Curcumin and especially tetrahydrocurcumin ameliorate oxidative stress-induced renal injury in mice. *J Nutr* **131**, 2090–2095 (2001).

831. B. H. Ali, N. Al-Wabel, O. Mahmoud, H. M. Mousa, and M. Hashad, Curcumin has a palliative action on gentamicin-induced nephrotoxicity in rats. *Fundam Clin Pharmacol* **19**, 473–477 (2005).

832. Y. Okazaki, M. Iqbal, and S. Okada, Suppressive effects of dietary curcumin on the increased activity of renal ornithine decarboxylase in mice treated with a renal carcinogen, ferric nitrilotriacetate. *Biochim Biophys Acta* **1740**, 357–366 (2005).

833. N. Kuwabara, S. Tamada, T. Iwai, K. Teramoto, N. Kaneda, T. Yukimura, T. Nakatani, and K. Miura, Attenuation of renal fibrosis by curcumin in rat obstructive nephropathy. *Urology* **67**, 440–446 (2006).

Others

834. D. Thaloor, K. J. Miller, J. Gephart, P. O. Mitchell, and G. K. Pavlath, Systemic administration of the NF-kappaB inhibitor curcumin stimulates muscle regeneration after traumatic injury. *Am J Physiol* **277**, C320–C329 (1999).

835. S. Kumar, K. K. Dubey, S. Tripathi, M. Fujii, and K. Misra, Design and synthesis of curcumin-bioconjugates to improve systemic delivery. *Nucleic Acids Symp Ser*, 75–76 (2000).

836. S. Kumar, A. Misra, S. Tripathi, and K. Misra, Study on curcumin-oligonucleotide conjugate as a probable anticancer agent: its hybridisation with telomere target sequence 5′-GGGATTGGGATT-3′. *Nucleic Acids Res Suppl* **1**, 137–138 (2001).

837. S. Kumar, U. Narain, S. Tripathi, and K. Misra, Syntheses of curcumin bioconjugates and study of their antibacterial activities against beta-lactamase-producing microorganisms. *Bioconjug Chem* **12**, 464–469 (2001).

838. S. Mishra, S. Tripathi, and K. Misra, Synthesis of a novel anticancer prodrug designed to target telomerase sequence. *Nucleic Acids Res Suppl* **2**, 277–278 (2002).

839. R. V. G and S. Divakar, Synthesis of guaiacol-alpha-D: -glucoside and curcumin-bis-alpha-D: -glucoside by an amyloglucosidase from Rhizopus. *Biotechnol Lett* **27**, 1411–1415 (2005).

840. S. Mishra, U. Narain, R. Mishra, and K. Misra, Design, development and synthesis of mixed bioconjugates of piperic acid-glycine, curcumin-glycine/alanine and curcumin-glycine-piperic acid and their antibacterial and antifungal properties. *Bioorg Med Chem* **13**, 1477–1486 (2005).

841. K. Mohammadi, K. H. Thompson, B. O. Patrick, T. Storr, C. Martins, E. Polishchuk, V. G. Yuen, J. H. McNeill, and C. Orvig, Synthesis and characterization of dual function vanadyl, gallium and indium curcumin complexes for medicinal applications. *J Inorg Biochem* **99**, 2217–2225 (2005).

842. H. Suhaimi, F. B. Ahmad, and S. E. Friberg, Curcumin in a model skin lotion formulation. *J Pharm Sci* **84**, 376–380 (1995).

843. S. Liao, J. Lin, M. T. Dang, H. Zhang, Y. H. Kao, J. Fukuchi, and R. A. Hiipakka, Growth suppression of hamster flank organs by topical application of

catechins, alizarin, curcumin, and myristoleic acid. *Arch Dermatol Res* **293**, 200–205 (2001).

844. V. Kumar, S. A. Lewis, S. Mutalik, D. B. Shenoy, Venkatesh, and N. Udupa, Biodegradable microspheres of curcumin for treatment of inflammation. *Indian J Physiol Pharmacol* **46**, 209–217 (2002).
845. J. Y. Fang, C. F. Hung, H. C. Chiu, J. J. Wang, and T. F. Chan, Efficacy and irritancy of enhancers on the in-vitro and in-vivo percutaneous absorption of curcumin. *J Pharm Pharmacol* **55**, 593–601 (2003).
846. A. Paradkar, A. A. Ambike, B. K. Jadhav, and K. R. Mahadik, Characterization of curcumin-PVP solid dispersion obtained by spray drying. *Int J Pharm* **271**, 281–286 (2004).
847. L. Li, F. S. Braiteh, and R. Kurzrock, Liposome-encapsulated curcumin: in vitro and in vivo effects on proliferation, apoptosis, signaling, and angiogenesis. *Cancer* **104**, 1322–1331 (2005).
848. S. H. Su, K. T. Nguyen, P. Satasiya, P. E. Greilich, L. Tang, and R. C. Eberhart, Curcumin impregnation improves the mechanical properties and reduces the inflammatory response associated with poly(L-lactic acid) fiber. *J Biomater Sci Polym Ed* **16**, 353–370 (2005).
849. S. Jung, N. Otberg, G. Thiede, H. Richter, W. Sterry, S. Panzner, and J. Lademann, innovative liposomes as a transfollicular drug delivery system: penetration into porcine hair follicles. *J Invest Dermatol* **126**(8), 1728–1732 (2006).
850. C. D. Lao, M. T. t. Ruffin, D. Normolle, D. D. Heath, S. I. Murray, J. M. Bailey, M. E. Boggs, J. Crowell, C. L. Rock, and D. E. Brenner, Dose escalation of a curcuminoid formulation. *BMC Complement Altern Med* **6**, 10 (2006).
851. T. S. Rao, N. Basu, S. D. Seth, and H. H. Siddiqui, Some aspects of pharmacological profile of sodium curcuminate. *Indian J Physiol Pharmacol* **28**, 211–215 (1984).
852. O. Vajragupta, P. Boonchoong, H. Watanabe, M. Tohda, N. Kummasud, and Y. Sumanont, Manganese complexes of curcumin and its derivatives: Evaluation for the radical scavenging ability and neuroprotective activity. *Free Radical Biol Med* **35**, 1632–1644 (2003).
853. M. Bernabe-Pineda, M. T. Ramirez-Silva, M. A. Romero-Romo, E. Gonzalez-Vergara, and A. Rojas-Hernandez, Spectrophotometric and electrochemical determination of the formation constants of the complexes Curcumin-Fe(III)-water and Curcumin-Fe(II)-water. *Spectrochim Acta A: Mol Biomol Spectrosc* **60**, 1105–1113 (2004).
854. H. Ligeret, S. Barthelemy, G. Bouchard Doulakas, P. A. Carrupt, J. P. Tillement, S. Labidalle, and D. Morin, Fluoride curcumin derivatives: New mitochondrial uncoupling agents. *FEBS Lett* **569**, 37–42 (2004).
855. K. H. Thompson, K. Bohmerle, E. Polishchuk, C. Martins, P. Toleikis, J. Tse, V. Yuen, J. H. McNeill, and C. Orvig, Complementary inhibition of synoviocyte, smooth muscle cell or mouse lymphoma cell proliferation by a vanadyl curcumin complex compared to curcumin alone. *J Inorg Biochem* **98**, 2063–2070 (2004).
856. O. Vajragupta, P. Boonchoong, and L. J. Berliner, Manganese complexes of curcumin analogues: Evaluation of hydroxyl radical scavenging ability, superoxide dismutase activity and stability towards hydrolysis. *Free Radical Res* **38**, 303–314 (2004).
857. A. Barik, B. Mishra, L. Shen, H. Mohan, R. M. Kadam, S. Dutta, H. Y. Zhang, and K. I. Priyadarsini, Evaluation of a new copper(II)-curcumin complex as superoxide dismutase mimic and its free radical reactions. *Free Radical Biol Med* **39**, 811–822 (2005).

858. J. B. Majithiya, R. Balaraman, R. Giridhar, and M. R. Yadav, Effect of bis[curcumino]oxovanadium complex on non-diabetic and streptozotocin-induced diabetic rats. *J Trace Elem Med Biol* **18**, 211–217 (2005).

Review

859. H. P. Ammon and M. A. Wahl, Pharmacology of Curcuma longa. *Planta Med* **57**, 1–7 (1991).
860. A. H. Conney, T. Lysz, T. Ferraro, T. F. Abidi, P. S. Manchand, J. D. Laskin, and M. T. Huang, Inhibitory effect of curcumin and some related dietary compounds on tumor promotion and arachidonic acid metabolism in mouse skin. *Adv Enzyme Regul* **31**, 385–396 (1991).
861. C. W. Boone, V. E. Steele, and G. J. Kelloff, Screening for chemopreventive (anticarcinogenic) compounds in rodents. *Mutat Res* **267**, 251–255 (1992).
862. R. Han, Recent progress in the study of anticancer drugs originating from plants and traditional medicines in China. *Chin Med Sci J* **9**, 61–69 (1994).
863. R. Han, Highlight on the studies of anticancer drugs derived from plants in China. *Stem Cells* **12**, 53–63 (1994).
864. G. J. Kelloff, C. W. Boone, J. A. Crowell, V. E. Steele, R. Lubet, and C. C. Sigman, Chemopreventive drug development: perspectives and progress. *Cancer Epidemiol Biomarkers Prev* **3**, 85–98 (1994).
865. G. D. Stoner and H. Mukhtar, Polyphenols as cancer chemopreventive agents. *J Cell Biochem Suppl* **22**, 169–180 (1995).
866. J. N. Commandeur and N. P. Vermeulen, Cytotoxicity and cytoprotective activities of natural compounds. The case of curcumin. *Xenobiotica* **26**, 667–680 (1996).
867. G. J. Kelloff, C. W. Boone, J. A. Crowell, V. E. Steele, R. A. Lubet, L. A. Doody, W. F. Malone, E. T. Hawk, and C. C. Sigman, New agents for cancer chemoprevention. *J Cell Biochem* **26(Suppl)**, 1–28 (1996).
868. M. Berwick and S. Schantz, Chemoprevention of aerodigestive cancer. *Cancer Metastasis Rev* **16**, 329–347 (1997).
869. C. W. Boone and G. J. Kelloff, Biomarker end-points in cancer chemoprevention trials. *IARC Sci Publ*, 273–280 (1997).
870. A. H. Conney, Y. R. Lou, J. G. Xie, T. Osawa, H. L. Newmark, Y. Liu, R. L. Chang, and M. T. Huang, Some perspectives on dietary inhibition of carcinogenesis: Studies with curcumin and tea. *Proc Soc Exp Biol Med* **216**, 234–245 (1997).
871. M. T. Huang, H. L. Newmark, and K. Frenkel, Inhibitory effects of curcumin on tumorigenesis in mice. *J Cell Biochem* **27(Suppl)**, 26–34 (1997).
872. J. K. Lin, Y. C. Chen, Y. T. Huang, and S. Y. Lin-Shiau, Suppression of protein kinase C and nuclear oncogene expression as possible molecular mechanisms of cancer chemoprevention by apigenin and curcumin. *J Cell Biochem Suppl* **28–29**, 39–48 (1997).
873. S. Ren and E. J. Lien, Natural products and their derivatives as cancer chemopreventive agents. *Prog Drug Res* **48**, 147–171 (1997).
874. M. J. Wargovich, Experimental evidence for cancer preventive elements in foods. *Cancer Lett* **114**, 11–17 (1997).
875. A. Gescher, U. Pastorino, S. M. Plummer, and M. M. Manson, Suppression of tumour development by substances derived from the diet-: Mechanisms and clinical implications. *Br J Clin Pharmacol* **45**, 1–12 (1998).
876. K. Krishnaswamy and N. Raghuramulu, Bioactive phytochemicals with emphasis on dietary practices. *Indian J Med Res* **108**, 167–181 (1998).

877. A. J. Vlietinck, T. De Bruyne, S. Apers, and L. A. Pieters, Plant-derived leading compounds for chemotherapy of human immunodeficiency virus (HIV) infection. *Planta Med* **64**, 97–109 (1998).
878. H. L. Bradlow, N. T. Telang, D. W. Sepkovic, and M. P. Osborne, Phytochemicals as modulators of cancer risk. *Adv Exp Med Biol* **472**, 207–221 (1999).
879. D. Eigner and D. Scholz, Ferula asa-foetida and *Curcuma longa* in traditional medical treatment and diet in Nepal. *J Ethnopharmacol* **67**, 1–6 (1999).
880. W. Henke, K. Ferrell, D. Bech-Otschir, M. Seeger, R. Schade, P. Jungblut, M. Naumann, and W. Dubiel, Comparison of human COP9 signalsome and 26S proteasome lid'. *Mol Biol Rep* **26**, 29–34 (1999).
881. G. J. Kelloff, J. A. Crowell, V. E. Steele, R. A. Lubet, C. W. Boone, W. A. Malone, E. T. Hawk, R. Lieberman, J. A. Lawrence, L. Kopelovich, I. Ali, J. L. Viner, and C. C. Sigman, Progress in cancer chemoprevention. *Ann NY Acad Sci* **889**, 1–13 (1999).
882. K. M. Mohandas and D. C. Desai, Epidemiology of digestive tract cancers in India. V. Large and small bowel. *Indian J Gastroenterol* **18**, 118–121 (1999).
883. M. A. Pereira, Prevention of colon cancer and modulation of aberrant crypt foci, cell proliferation, and apoptosis by retinoids and NSAIDs. *Adv Exp Med Biol* **470**, 55–63 (1999).
884. Y. Surh, Molecular mechanisms of chemopreventive effects of selected dietary and medicinal phenolic substances. *Mutat Res* **428**, 305–327 (1999).
885. M. Cuendet and J. M. Pezzuto, The role of cyclooxygenase and lipoxygenase in cancer chemoprevention. *Drug Metabol Drug Interact* **17**, 109–157 (2000).
886. E. De Clercq, Current lead natural products for the chemotherapy of human immunodeficiency virus (HIV) infection. *Med Res Rev* **20**, 323–349 (2000).
887. R. Gopalakrishna and S. Jaken, Protein kinase C signaling and oxidative stress. *Free Radical Biol Med* **28**, 1349–1361 (2000).
888. K. L. Grant and C. D. Schneider, Turmeric. *Am J Health Syst Pharm* **57**, 1121–1122 (2000).
889. J. P. Groten, W. Butler, V. J. Feron, G. Kozianowski, A. G. Renwick, and R. Walker, An analysis of the possibility for health implications of joint actions and interactions between food additives. *Regul Toxicol Pharmacol* **31**, 77–91 (2000).
890. S. M. Hadi, S. F. Asad, S. Singh, and A. Ahmad, Putative mechanism for anticancer and apoptosis-inducing properties of plant-derived polyphenolic compounds. *IUBMB Life* **50**, 167–171 (2000).
891. G. J. Kelloff, J. A. Crowell, V. E. Steele, R. A. Lubet, W. A. Malone, C. W. Boone, L. Kopelovich, E. T. Hawk, R. Lieberman, J. A. Lawrence, I. Ali, J. L. Viner, and C. C. Sigman, Progress in cancer chemoprevention: development of diet-derived chemopreventive agents. *J Nutr* **130**, 467S–471S (2000).
892. J. K. Lin, M. H. Pan and S. Y. Lin-Shiau, Recent studies on the biofunctions and biotransformations of curcumin. *Biofactors* **13**, 153–158 (2000).
893. R. Lodha and A. Bagga, Traditional Indian systems of medicine. *Ann Acad Med Singapore* **29**, 37–41 (2000).
894. M. M. Manson, A. Gescher, E. A. Hudson, S. M. Plummer, M. S. Squires, and S. A. Prigent, Blocking and suppressing mechanisms of chemoprevention by dietary constituents. *Toxicol Lett* **112–113**, 499–505 (2000).
895. N. Ahmad, S. K. Katiyar and H. Mukhtar, Antioxidants in chemoprevention of skin cancer. *Curr Probl Dermatol* **29**, 128–139 (2001).
896. C. C. Araujo and L. L. Leon, Biological activities of *Curcuma longa* L. *Mem Inst Oswaldo Cruz* **96**, 723–728 (2001).

897. A. J. Gescher, R. A. Sharma, and W. P. Steward, Cancer chemoprevention by dietary constituents: a tale of failure and promise. *Lancet Oncol* **2**, 371–379 (2001).
898. K. Jaga and H. Duvvi, Risk reduction for DDT toxicity and carcinogenesis through dietary modification. *J R Soc Health* **121**, 107–113 (2001).
899. M. S. Levi, R. F. Borne, and J. S. Williamson, A review of cancer chemopreventive agents. *Curr Med Chem* **8**, 1349–1362 (2001).
900. J. K. Lin and S. Y. Lin-Shiau, Mechanisms of cancer chemoprevention by curcumin. *Proc Natl Sci Counc Repub China B* **25**, 59–66 (2001).
901. P. Talalay, The importance of using scientific principles in the development of medicinal agents from plants. *Acad Med* **76**, 238–247 (2001).
902. M. J. Wargovich, Colon cancer chemoprevention with ginseng and other botanicals. *J Korean Med Sci* **16(Suppl)**, S81–S86 (2001).
903. F. Afaq, V. M. Adhami, N. Ahmad, and H. Mukhtar, Botanical antioxidants for chemoprevention of photocarcinogenesis. *Front Biosci* **7**, d784–d792 (2002).
904. P. Bremner and M. Heinrich, Natural products as targeted modulators of the nuclear factor–kappaB pathway. *J Pharm Pharmacol* **54**, 453–572 (2002).
905. D. P. Chauhan, Chemotherapeutic potential of curcumin for colorectal cancer. *Curr Pharm Des* **8**, 1695–1706 (2002).
906. A. T. Dinkova-Kostova, Protection against cancer by plant phenylpropenoids: induction of mammalian anticarcinogenic enzymes. *Mini Rev Med Chem* **2**, 595–610 (2002).
907. P. Greenwald, J. A. Milner, D. E. Anderson, and S. S. McDonald, Micronutrients in cancer chemoprevention. *Cancer Metastasis Rev* **21**, 217–230 (2002).
908. T. H. Leu and M. C. Maa, The molecular mechanisms for the antitumorigenic effect of curcumin. *Curr Med Chem Anticancer Agents* **2**, 357–370 (2002).
909. J. Miquel, A. Bernd, J. M. Sempere, J. Diaz-Alperi, and A. Ramirez, The curcuma antioxidants: Pharmacological effects and prospects for future clinical use. A review. *Arch Gerontol Geriatr* **34**, 37–46 (2002).
910. B. S. Reddy and C. V. Rao, Novel approaches for colon cancer prevention by cyclooxygenase-2 inhibitors. *J Environ Pathol Toxicol Oncol* **21**, 155–164 (2002).
911. J. R. Sorenson, Cu, Fe, Mn, and Zn chelates offer a medicinal chemistry approach to overcoming radiation injury. *Curr Med Chem* **9**, 639–662 (2002).
912. R. Steriti, Nutritional support for chronic myelogenous and other leukemias: A review of the scientific literature. *Altern Med Rev* **7**, 404–409 (2002).
913. S. J. Stohs, S. Ohia, and D. Bagchi, Naphthalene toxicity and antioxidant nutrients. *Toxicology* **180**, 97–105 (2002).
914. Y. J. Surh, Anti-tumor promoting potential of selected spice ingredients with antioxidative and anti-inflammatory activities: A short review. *Food Chem Toxicol* **40**, 1091–1097 (2002).
915. J. M. Wallace, Nutritional and botanical modulation of the inflammatory cascade–eicosanoids, cyclooxygenases, and lipoxygenases–as an adjunct in cancer therapy. *Integr Cancer Ther* **1**, 7–37; discussion 37 (2002).
916. B. B. Aggarwal, A. Kumar, and A. C. Bharti, Anticancer potential of curcumin: preclinical and clinical studies. *Anticancer Res* **23**, 363–398 (2003).
917. V. Calabrese, D. A. Butterfield, and A. M. Stella, Nutritional antioxidants and the heme oxygenase pathway of stress tolerance: Novel targets for neuroprotection in Alzheimer's disease. *Ital J Biochem* **52**, 177–181 (2003).
918. V. Calabrese, G. Scapagnini, C. Colombrita, A. Ravagna, G. Pennisi, A. M. Giuffrida Stella, F. Galli, and D. A. Butterfield, Redox regulation of heat shock protein expression

in aging and neurodegenerative disorders associated with oxidative stress: A nutritional approach. *Amino Acids* **25**, 437–444 (2003).

919. Y. Carter, G. Liu, J. Yang, A. Fier, and C. Mendez, Sublethal hemorrhage induces tolerance in animals exposed to cecal ligation and puncture by altering p38, p44/42, and SAPK/JNK MAP kinase activation. *Surg Infect (Larchmt)* **4**, 17–27 (2003).
920. N. Chainani-Wu, Safety and anti-inflammatory activity of curcumin: A component of tumeric (*Curcuma longa*). *J Altern Complement Med* **9**, 161–168 (2003).
921. A. H. Conney, Enzyme induction and dietary chemicals as approaches to cancer chemoprevention: The Seventh DeWitt S. Goodman Lecture. *Cancer Res* **63**, 7005–7031 (2003).
922. D. E. Corpet and F. Pierre, Point: From animal models to prevention of colon cancer. Systematic review of chemoprevention in min mice and choice of the model system. *Cancer Epidemiol Biomarkers Prev* **12**, 391–400 (2003).
923. R. H. Dashwood, Use of transgenic and mutant animal models in the study of heterocyclic amine-induced mutagenesis and carcinogenesis. *J Biochem Mol Biol* **36**, 35–42 (2003).
924. D. A. Dickinson, D. R. Moellering, K. E. Iles, R. P. Patel, A. L. Levonen, A. Wigley, V. M. Darley-Usmar and H. J. Forman, Cytoprotection against oxidative stress and the regulation of glutathione synthesis. *Biol Chem* **384**, 527–537 (2003).
925. G. Garcea, A. R. Dennison, W. P. Steward, and D. P. Berry, Chemoprevention of gastrointestinal malignancies. *ANZ J Surg* **73**, 680–686 (2003).
926. S. P. Gupta and A. N. Nagappa, Design and development of integrase inhibitors as anti-HIV agents. *Curr Med Chem* **10**, 1779–1794 (2003).
927. S. Kadota, Y. Tezuka, J. K. Prasain, M. S. Ali, and A. H. Banskota, Novel diarylheptanoids of *Alpinia blepharocalyx*. *Curr Top Med Chem* **3**, 203–225 (2003).
928. R. Sinha, D. E. Anderson, S. S. McDonald, and P. Greenwald, Cancer risk and diet in India. *J Postgrad Med* **49**, 222–228 (2003).
929. B. B. Aggarwal and S. Shishodia, Suppression of the nuclear factor-kappaB activation pathway by spice-derived phytochemicals: Reasoning for seasoning. *Ann NY Acad Sci* **1030**, 434–441 (2004).
930. B. B. Aggarwal, Y. Takada, and O. V. Oommen, From chemoprevention to chemotherapy: Common targets and common goals. *Expert Opin Invest Drugs* **13**, 1327–1338 (2004).
931. S. M. Choi and B. M. Lee, An alternative mode of action of endocrine-disrupting chemicals and chemoprevention. *J Toxicol Environ Health B Crit Rev* **7**, 451–463 (2004).
932. P. B. Davis and M. L. Drumm, Some like it hot: curcumin and CFTR. *Trends Mol Med* **10**, 473–475 (2004).
933. T. Dorai and B. B. Aggarwal, Role of chemopreventive agents in cancer therapy. *Cancer Lett* **215**, 129–140 (2004).
934. D. A. Dickinson, K. E. Iles, A. F. Wigley, and H. J. Forman, Analysis of transcription factor remodeling in phase II gene expression with curcumin. *Methods Enzymol* **378**, 302–318 (2004).
935. C. K. Ferrari, Functional foods, herbs and nutraceuticals: Towards biochemical mechanisms of healthy aging. *Biogerontology* **5**, 275–289 (2004).
936. A. Gescher, Polyphenolic phytochemicals versus non-steroidal anti-inflammatory drugs: which are better cancer chemopreventive agents? *J Chemother* **16(Suppl 4)**, 3–6 (2004).
937. B. Joe, M. Vijaykumar, and B. R. Lokesh, Biological properties of curcumin-cellular and molecular mechanisms of action. *Crit Rev Food Sci Nutr* **44**, 97–111 (2004).

938. J. K. Lin, Suppression of protein kinase C and nuclear oncogene expression as possible action mechanisms of cancer chemoprevention by Curcumin. *Arch Pharm Res* **27**, 683–692 (2004).
939. S. Narayan, Curcumin, a multi-functional chemopreventive agent, blocks growth of colon cancer cells by targeting beta-catenin-mediated transactivation and cell-cell adhesion pathways. *J Mol Histol* **35**, 301–307 (2004).
940. B. S. Reddy, Studies with the azoxymethane-rat preclinical model for assessing colon tumor development and chemoprevention. *Environ Mol Mutagen* **44**, 26–35 (2004).
941. F. H. Sarkar and Y. Li, Cell signaling pathways altered by natural chemopreventive agents. *Mutat Res* **555**, 53–64 (2004).
942. N. S. Shenouda, C. Zhou, J. D. Browning, P. J. Ansell, M. S. Sakla, D. B. Lubahn, and R. S. Macdonald, Phytoestrogens in common herbs regulate prostate cancer cell growth in vitro. *Nutr Cancer* **49**, 200–208 (2004).
943. P. Srinivasan and B. Libbus, Mining MEDLINE for implicit links between dietary substances and diseases. *Bioinformatics* **20(Suppl 1)**, I290–I296 (2004).
944. S. Zhou, L. Y. Lim, and B. Chowbay, Herbal modulation of P-glycoprotein. *Drug Metab Rev* **36**, 57–104 (2004).
945. B. E. Bachmeier, C. M. Iancu, M. Jochum, and A. G. Nerlich, Matrix metalloproteinases in cancer: Comparison of known and novel aspects of their inhibition as a therapeutic approach. *Expert Rev Anticancer Ther* **5**, 149–163 (2005).
946. F. C. Campbell and G. P. Collett, Chemopreventive properties of curcumin. *Future Oncol* **1**, 405–414 (2005).
947. M. D'Incalci, W. P. Steward, and A. J. Gescher, Use of cancer chemopreventive phytochemicals as antineoplastic agents. *Lancet Oncol* **6**, 899–904 (2005).
948. R. Di Santo, R. Costi, M. Artico, R. Ragno, G. Greco, E. Novellino, C. Marchand, and Y. Pommier, Design, synthesis and biological evaluation of heteroaryl diketohexenoic and diketobutanoic acids as HIV-1 integrase inhibitors endowed with antiretroviral activity. *Farmaco* **60**, 409–417 (2005).
949. J. Dulak, Nutraceuticals as anti-angiogenic agents: hopes and reality. *J Physiol Pharmacol* **56(Suppl 1)**, 51–69 (2005).
950. A. Duvoix, R. Blasius, S. Delhalle, M. Schnekenburger, F. Morceau, E. Henry, M. Dicato, and M. Diederich, Chemopreventive and therapeutic effects of curcumin. *Cancer Lett* **223**, 181–190 (2005).
951. B. Frank and S. Gupta, A review of antioxidants and Alzheimer's disease. *Ann Clin Psychiatry* **17**, 269–286 (2005).
952. A. K. Garg, T. A. Buchholz, and B. B. Aggarwal, Chemosensitization and radiosensitization of tumors by plant polyphenols. *Antioxid Redox Signal* **7**, 1630–1647 (2005).
953. D. Karunagaran, R. Rashmi, and T. R. Kumar, Induction of apoptosis by curcumin and its implications for cancer therapy. *Curr Cancer Drug Targets* **5**, 117–129 (2005).
954. S. Kawanishi, S. Oikawa, and M. Murata, Evaluation for safety of antioxidant chemopreventive agents. *Antioxid Redox Signal* **7**, 1728–1739 (2005).
955. J. D. Lambert, J. Hong, G. Y. Yang, J. Liao, and C. S. Yang, Inhibition of carcinogenesis by polyphenols: evidence from laboratory investigations. *Am J Clin Nutr* **81**, 284S–291S (2005).
956. J. S. Lee and Y. J. Surh, Nrf2 as a novel molecular target for chemoprevention. *Cancer Lett* **224**, 171–184 (2005).
957. J. P. Liu, E. Manheimer, and M. Yang, Herbal medicines for treating HIV infection and AIDS. *Cochrane Database Syst Rev*, CD003937 (2005).

958. M. Mall and K. Kunzelmann, Correction of the CF defect by curcumin: Hypes and disappointments. *BioEssays* **27**, 9–13 (2005).
959. M. M. Manson, P. B. Farmer, A. Gescher, and W. P. Steward, Innovative agents in cancer prevention. *Recent Results Cancer Res* **166**, 257–275 (2005).
960. M. M. Manson, Inhibition of survival signalling by dietary polyphenols and indole-3-carbinol. *Eur J Cancer* **41**, 1842–1853 (2005).
961. A. Ray, Cancer preventive role of selected dietary factors. *Indian J Cancer* **42**, 15–24 (2005).
962. R. A. Sharma, A. J. Gescher, and W. P. Steward, Curcumin: The story so far. *Eur J Cancer* **41**, 1955–1968 (2005).
963. S. Shishodia, G. Sethi, and B. B. Aggarwal, Curcumin: Getting back to the roots. *Ann NY Acad Sci* **1056**, 206–217 (2005).
964. C. Thornfeldt, Cosmeceuticals containing herbs: Fact, fiction, and future. *Dermatol Surg* **31**, 873–880; discussion 880 (2005).
965. B. B. Aggarwal and S. Shishodia, Molecular targets of dietary agents for prevention and therapy of cancer. *Biochem Pharmacol* **71**, 1397–1421 (2006).
966. M. S. Baliga and S. K. Katiyar, Chemoprevention of photocarcinogenesis by selected dietary botanicals. *Photochem Photobiol Sci* **5**, 243–253 (2006).
967. I. Barta, P. Smerak, Z. Polivkova, H. Sestakova, M. Langova, B. Turek, and J. Bartova, Current trends and perspectives in nutrition and cancer prevention. *Neoplasma* **53**, 19–25 (2006).
968. S. Bengmark, Curcumin, an atoxic antioxidant and natural NFkappaB, cyclooxygenase-2, lipooxygenase, and inducible nitric oxide synthase inhibitor: A shield against acute and chronic diseases. *J Parenter Enteral Nutr* **30**, 45–51 (2006).
969. H. J. Kwon, Discovery of new small molecules and targets towards angiogenesis via chemical genomics approach. *Curr Drug Targets* **7**, 397–405 (2006).
970. R. K. Maheshwari, A. K. Singh, J. Gaddipati, and R. C. Srimal, Multiple biological activities of curcumin: A short review. *Life Sci* **78**, 2081–2087 (2006).
971. R. S. Rapaka and P. M. Coates, Dietary supplements and related products: A brief summary. *Life Sci* **78**, 2026–2032 (2006).
972. S. Singh and A. Khar, Biological effects of curcumin and its role in cancer chemoprevention and therapy. *Anticancer Agents Med Chem* **6**, 259–270 (2006).
973. H. Taniura, J. C. Sng, and Y. Yoneda, Histone modifications in status epilepticus induced by kainate. *Histol Histopathol* **21**, 785–791 (2006).
974. G. Yuan, M. L. Wahlqvist, G. He, M. Yang, and D. Li, Natural products and anti-inflammatory activity. *Asia Pacific J Clin Nutr* **15**, 143–152 (2006).

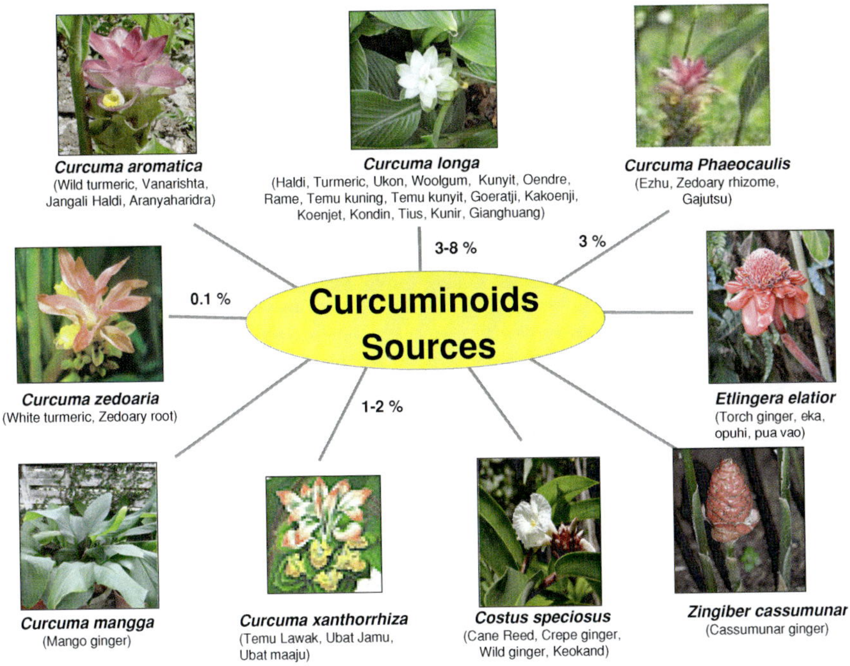

Plate 1.

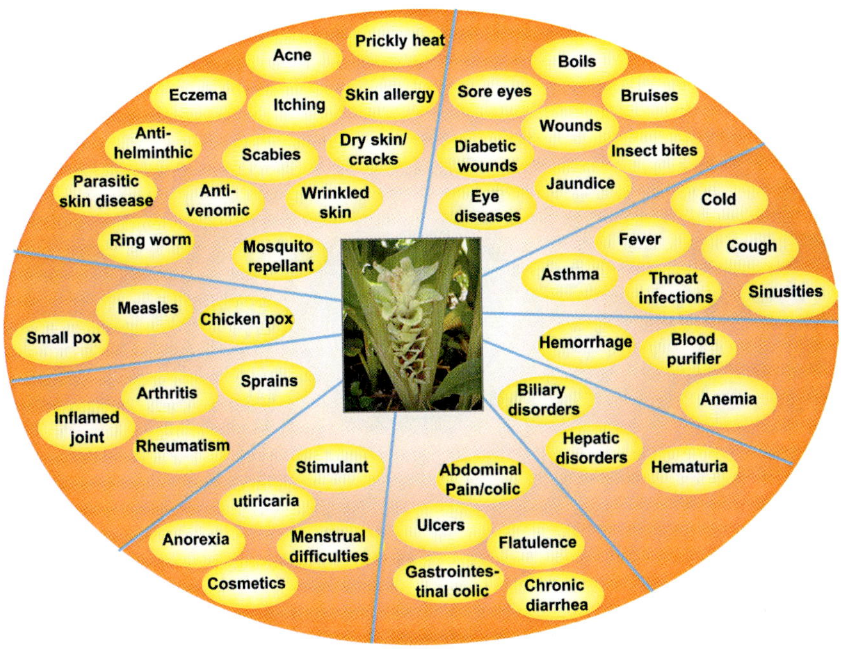

Plate 2.

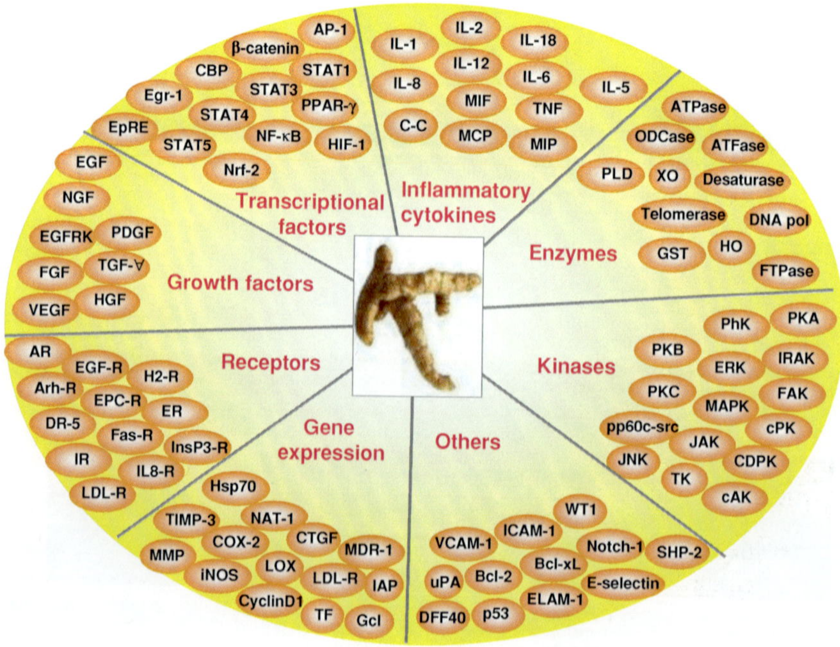

Plate 3.

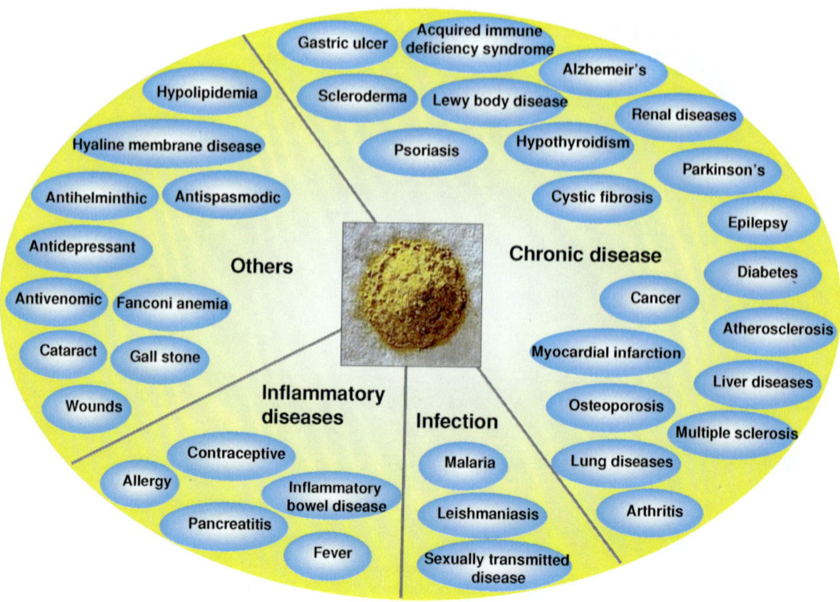

Plate 4.

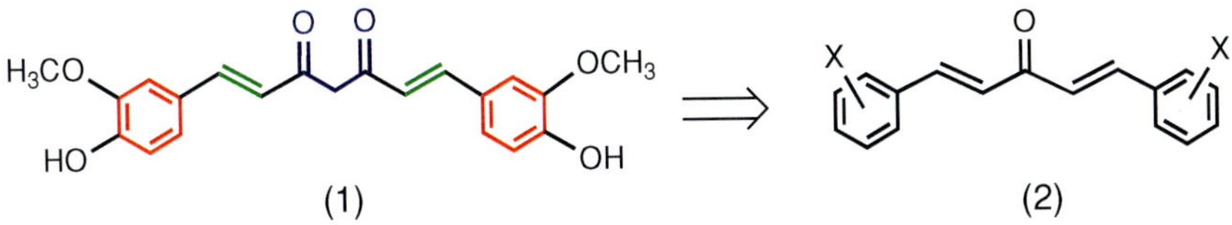

Plate 5.

HIGHLY ACTIVE ANTICANCER CURCUMIN ANALOGUES

Cara A. Mosley, Dennis C. Liotta, and James P. Snyder

Abstract: Curcumin, a compound in the human food supply, represents a near-perfect starting point for drug discovery. Consequently, a number of research groups have taken the natural product as a starting point to prepare and biologically evaluate a wide variety of curcumin analogues. One widely used structural modification truncates the central conjugated β-diketone in curcumin to the mono-carbonyl dienone. A diverse array of the latter compounds exhibit cytotoxicities against an equally diverse set of cancer-related cell lines. Importantly, these compounds still retain toxicity profiles in rodents comparable to the parent natural product, whereas some analogues (e.g., EF-24, **41**) exhibit good oral bioavailability and good pharmacokinetics in mice. Thiol conjugates of EF-24 analogues have been prepared that address stability and solubility issues while demonstrating cellular activities similar to the unmodified dienones. In parallel experiments, the factor VIIa–tissue factor complex (fVIIa-TF) has been exploited to develop a targeting strategy for the analogues. In particular, the EF24-FFRck-fVIIa protein conjugate is not only somewhat more effective relative to the drug alone against breast cancer and melanocyte cells. Both simple curcumin analogues and the protein conjugate evidence antiangiogenic activity in cell culture. The implication is that the fVIIa-TF targeting process, like the dienone drugs, permits a double-pronged attack with the potential to destroy a tumor directly by apoptosis.

1. INTRODUCTION

Many chemotherapeutic approaches to the range of diseases that fall in the cancer category have been explored. Perhaps the oldest of these involves the use of natural products. Because compounds made by plants and microorganisms often serve to attract allies or repel, disable, or kill competitors, it is not surprising that natural products have proved to be a rich source of potential anticancer therapies. However, a delicate balance must be struck between a compound's cancer-fighting capabilities and its toxicological profile for it to progress from a lead to a clinically useful agent. As a consequence, many promising natural product leads had to be structurally modified (natural product optimization) to produce compounds that exhibit more favorable pharmacologic profiles.

Today, the screening of natural products represents one of many approaches used to discover new drugs. Other methods include, *inter alia*, computer-assisted small-molecule drug design, combinatorial synthesis/high-throughput screening and the development of monoclonal antibodies. However, just as with natural products, drug candidates developed using these approaches can exhibit varying types of toxicity that manifest themselves as undesirable side effects. For example, Gleevec, a small-molecule drug used in treating chronic myeloid leukemia (CML), is generally well tolerated with only mild side effects.[1] Similarly, the antibody Herceptin exhibits drug-induced complications in only 1–4% of patients.[2] By contrast, many natural-product-derived chemotherapeutics are accompanied by significant toxicities. Taxol (paclitaxel, PTX), almost completely water insoluble, is delivered in a vehicle formulation of 50% ethanol and 50% polyethoxylated castor oil (Cremophor EL). The vehicle has been associated with various side effects, including hypersensitivity in 41–44% of all patients,[3,4] whereas PTX shows both neurotoxicity and cardiotoxicity in a subset of patients. Doxorubicin, a potent broad-spectrum inhibitor of human tumors, also exhibits severe adverse side effects. Among other things, the compound has been cited as the cause of irreversible degenerative cardiomyopathy and congestive heart failure.[5] Clearly, these serious side effects limit the overall clinical utility of these compounds.

In contrast to natural products that are prone to serious side effects, curcumin (**1**, Figure 1), the compound that imparts the color and spicy flavor to both turmeric and curry powder, is nontoxic. In the general population, it is consumed daily as a dietary spice at levels up to 100 mg/day.[6] In clinical trials, it has been administered at up to 8 g a day without showing untoward side effects.[7] This yellow spice has a long history in Eastern cultures as a treatment for a multitude of ailments, most commonly inflammation. Recently, curcumin has emerged as a key weapon in the fight against cancer. As a pleiotropic anticancer agent, the compound operates by a number of mechanisms as detailed in recent reviews.[6,8,10] Curcumin is cytotoxic to a variety of tumor cells, exhibits antimetastatic activity, inhibits the survival factor nuclear factor-κB (NF-κB), blocks angiogenesis, and is a potent antioxidant. Together, these findings imply that curcumin is a rare example of a substance that possesses both chemotherapeutic and chemopreventative properties without debilitating consequences for the patient.

Although the effects of curcumin on cellular pathways continue to be studied, there has been much research devoted to developing and understanding the structure–activity relationships (SARs) responsible for the drug's anticancer properties. By synthesizing families of analogues and subjecting them to biological

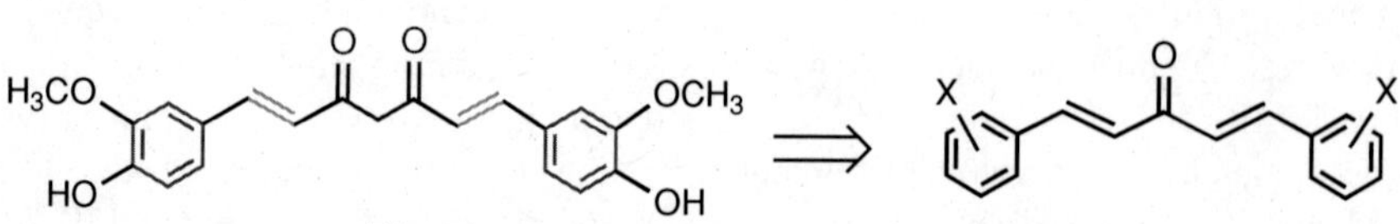

(1) (2)

Figure 1. Curcumin (**1**) and selected general structural permutations accomplished through analogue synthesis (**2**). (See also Plate 5 in the Color Plate Section.)

scrutiny, research workers hope to achieve an improvement in curcumin's natural anti-cancer and pharmacological profile while retaining its low toxicity.

2. BACKGROUND: CURCUMIN ANALOGUES IN CANCER TREATMENT

Structurally, there are three sectors in the molecular structure of curcumin that have been modified in the course of attempting to produce an "improved curcumin." These encompass the aromatic rings (red), the β-diketone moiety (blue), and the two flanking double bonds conjugated to the latter (green) (**1**, Figure 1). Successful synthesis of such analogues has resulted in the development of potential anticancer candidates that target various stages and/or processes in cancer cell growth.

Although there are exceptions, successful anticancer compounds based on curcuminoid structures ordinarily retain the conjugated α,β-unsaturated ketone moieties (**2**, Figure 1) Advances in antiangiogenic analogue SARs suggest that a variety of structural types are tolerated for potency. Inventive experiments using β-diketone curcumin analogues have been shown to be cytotoxic to prostate and breast cancer cell lines. Treatment of the corresponding cancers has traditionally involved hormonal therapy. Finally, SAR analysis has shown that *ortho*-substituted α,β-unsaturated ketones are the most potent antioxidants, an important finding for chemoprevention.[11]

2.1. Michael Acceptors Increase Anticarcinogenic Potency

The α,β-unsaturated β-diketone moiety of curcumin has received much attention in terms of its mechanistic role in promoting cytotoxicity. Generally, conjugated enones act as Michael acceptors (Figure 2) with thiols preferred over amino or hydroxy nucleophiles.[12]

Three recent articles describe attempts to exploit curcumin's potential as a Michael acceptor. Two of the studies imply the operation of such an interaction between curcumin and glutathione (GSH), but the products from the reactions are complex and ill-defined.[13,14] One investigation isolated two isomers from the combination of **1** and GSH separable by preparative high-performance liquid chromatography (HPLC). Both compounds by mass spectrometry furnished a molecu-

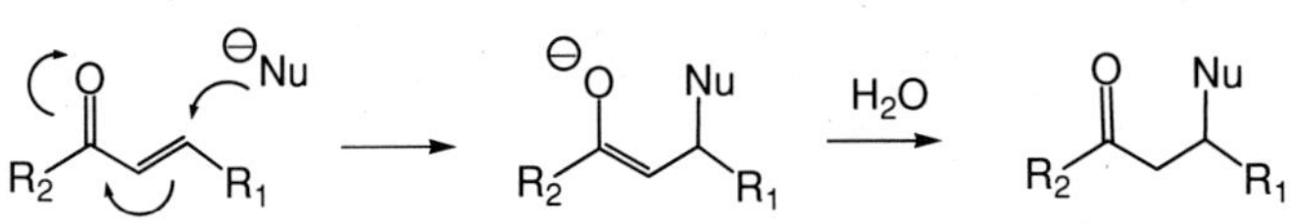

Figure 2. A Michael acceptor. The α,β-unsaturated ketone (**3**) undergoes attack at the β-position by the nucleophile (Nu) to generate an enolate intermediate (**4**). Aqueous quenching gives the product (**5**), a β-functionalized ketone.

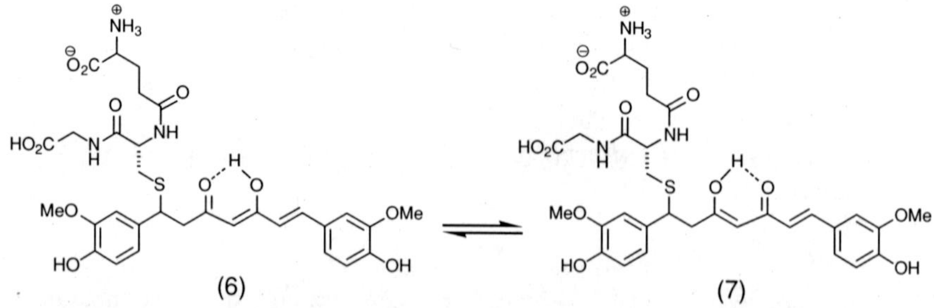

Figure 3. Hypothetical tautomeric isomers of curcumin–GSH (**1**-GSH).

lar weight (m/z) of 675.8, corresponding to a monoglutathionyl-curcumin adduct: **1**-GSH.[15] The two compounds are stable for up to 2 weeks when stored at $-20°C$, although no structures were reported. We surmise that the isomers, with HPLC retention times within 1 min of one another, are most likely the enol tautomeric Michael adducts **6** and **7**, as shown in Figure 3. If this is correct, it is not obvious why the compounds are unstable. As discussed below, certain monoketone analogues of curcumin, by contrast, form adducts with GSH that are isolable, stable, and water soluble. A likely scenario is that the addition is reversible. Subsequent exposure of curcumin to general acids and bases in air then promotes fragmentation and oxidation.[16,17]

The synthesis, cytotoxicity, and SAR of α, β-unsaturated monoketone systems have been examined. For example, a variety of 2,6-bis(arylidene)cyclohexanones (**37**, Scheme 1) are three to five times more potent than 5-fluorouracil (5-FU) as cytotoxins against L1210, Molt 4/C8, and CEM cell-based screens.[18,19] However, compounds with more than one conjugated carbonyl (e.g., **9**) are generally more active than those incorporating only a single conjugated C=O (e.g., **8**). This observation supports the "sequential cytotoxicity" hypothesis, which suggests that cancer cells are more susceptible to multiple chemical insults (in this case, two enones as opposed to one) by comparison with normal cells. For example, the *N*-acryloyl series **9** is reported to exhibit[20] remarkable cytotoxicity against murine P388, L1210,

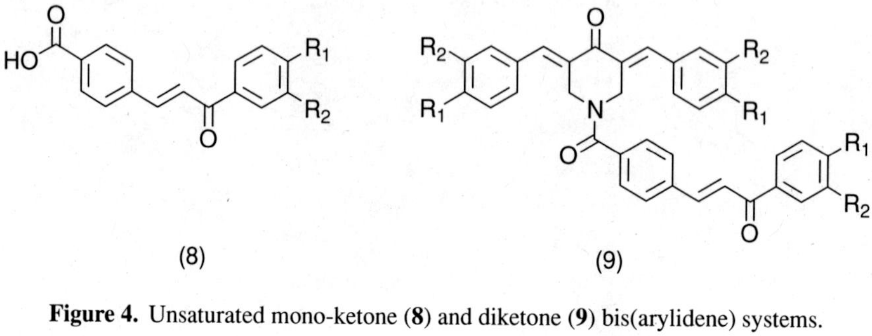

Figure 4. Unsaturated mono-ketone (**8**) and diketone (**9**) bis(arylidene) systems.

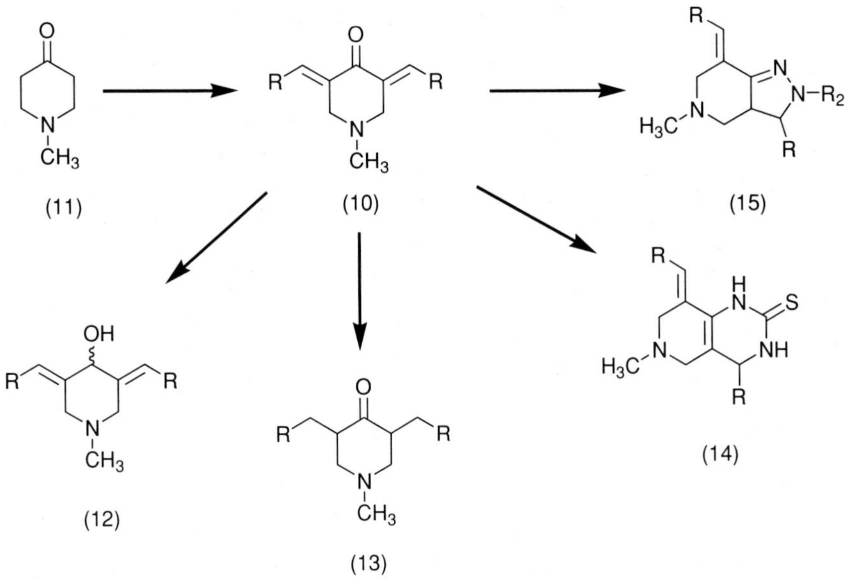

Figure 5. Classes of cytotoxic compounds derived from piperidone (**10**); R, aromatic moiety.

Molt 4/C8, and CEM T-lymphocytes with increases in potency of up to twofold. Although the specific protein targets of the compounds have not been identified, an alternative interpretation of the source of increased potency involves consideration of ligand binding. Series **9** incorporates both additional polar and hydrophobic functionality relative to **8**. This structural elaboration might contribute to the binding to protein targets both by enthalpic and entropic effects, thereby enhancing the overall binding free energy and associated K_a of the larger molecule.

Another class of conjugated α, β-unsaturated monoketone curcumin analogues based on piperidone (**10**, Figure 5) has been found[21] to be antineoplastic. These analogues show significant levels of cytotoxicity. For example, El-Suggabh and co-workers discovered that the α, β-unsaturated ketone moiety of the 3,5-(bisarylidine)-4-piperidones is vital to the series' antitumorigenic effects.[22] In an effort to determine the active pharmacophore of the molecular class in cancer efficacy assays, a series of chemical modifications on the unsaturated piperidone **10** were carried out (Figure 5). The ketone (**10**) was reduced to the racemic alcohol analogue (**12**), and the olefins were separately reduced to give the corresponding saturated derivatives (**13**). In addition, derivatives of **10** were subjected to cycloaddition reactions to yield various cycloadducts, including pyrido[4,3-*d*]pyrimidine derivatives (**14**) and pyrazolo[4,3-*c*] pyridine analogues (**15**). The racemic alcohol and the saturated series prove to be at least three times less potent against all National Cancer Institute (NCI)-tested cancer cell lines than the parent ketone structure, whereas the cycloaddition adducts gives both increased and decreased

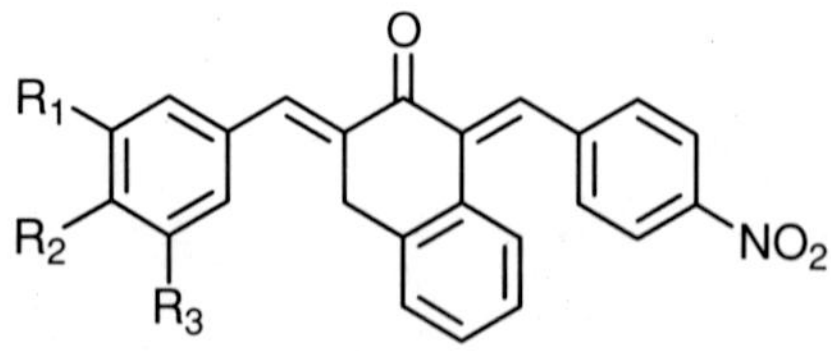

(16)

Figure 6. General structure of 3-arylidene-1-(4-nitrophenylmethylene)-3,4-dihydro-1H naphthalene-2-ones (**16**) capable of reversing MDR resistance.

potencies relative to the parent compound. Although no clear SAR was derived from the set of cycloadducts, those that showed potency are likewise selective for leukemia cell lines. This finding parallels Dimmock's earlier work, which demonstrated that *N*-acyl analogues of 3,5-(bisarylidine)-4-piperidones are specific for leukemia cell lines with IC_{50} values less than 10 μM.[21]

Most recently, a series of 3-arylidene-1-(4-nitrophenylmethylene)-3,4-dihydro-1H naphthalene-2-ones (**16**, Figure 6) and related structures were found to be selective against malignant cancer cells over normal cells.[23] Moreover, complementary data suggest these compounds reverse multidrug resistance (MDR), a defensive process that expels small molecules from the cell. Within this series, it was found that functionalized cyclohexanone ring-containing core structures are able to reverse MDR more than their cyclopentanone-containing counterparts. In addition, the presence of a third aromatic ring also proved to be beneficial for the reversal of MDR.

In contrast to their α,β-unsaturated counterparts, structures that lack conjugation between the carbonyl group and the terminal aromatic rings are inactive as cytotoxic agents. For example, in the case presented above, the racemic alcohol **12** and saturated ketone **13** are devoid of cytotoxic effects, whereas the conjugated dienone **10** is very potent in immortal cell line screening. If the dominant mechanism of action of such molecules is service as a Michael acceptor, then the absence of an electrophilic carbon in the β-position eliminates the active functional group, rendering the molecule impotent against cancer. This interpretation supports the notion that enones and dienones are alkylators capable of binding to intracellular components bearing sulfhydryl groups such as GSH and thioredoxin-1 (see discussion below).

2.2. Curcumin Analogues Target Angiogenesis

Recently, curcumin analogues have been found to be potent inhibitors of angiogenesis, the process by which cells acquire new or extended vascular networks.[24] Angiogenesis is required for most tumors to sustain growth via nourishment, waste disposal, and in many cases, metastasis. Curcumin and its naturally occurring analogues demethoxycurcumin (**17**) and bisdemethoxycurcumin (**18**) (Figure 7) have

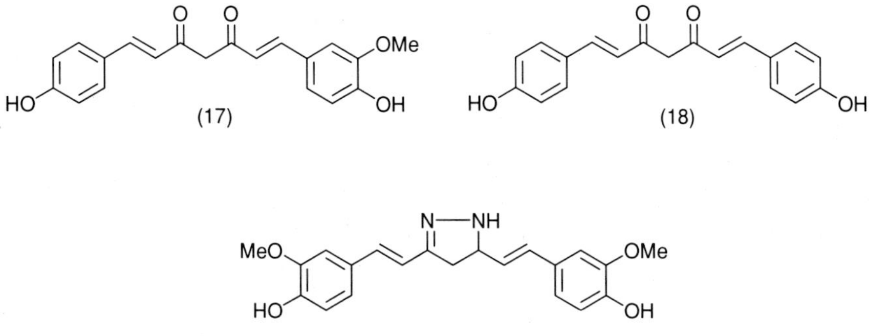

(17) (18)

(19)

Figure 7. Naurally occurring curcumin analogues demethoxycurcumin (**17**) and bis-demethoxycurcumin (**18**); hydrazinocurcumin (**19**).

been shown to inhibit basic fibroblast growth factor (bFGF) and induce proliferation of human umbilical vein endothelial cells (HUVECs) *in vitro* as well as angiogenesis in animal models.[25] Hydrazinocurcumin (**19**), a synthetic curcumin analogue, has also been used to study inhibition of angiogenesis.[26] The compound is 30-fold more potent than curcumin at preventing invasion of bovine aortic endothelial cells (bAECs; 0.52 μM) and the formation of capillary tubes in *vivo*, presumably because of the heteroaromatic pyrazole ring. Without knowing the specific targets of **19**, we presume that part of the increased activity for the compound might be due to improved solubility and biodistribution in comparison to **1**.

Diverse libraries of curcumin analogues have been used to study SARs and angiogenesis inhibition. Although chalcones are known to have anticancer activity, their effect on angiogenesis had not been fully probed.[11,27–31] Bowen and co-workers found that chalcones with 2,6-substitution in the terminal aromatic rings (**20**, Figure 8) are most efficacious. Further studies demonstrated that chalcones with electron-withdrawing groups in the 2-position and 6-position of the aromatic rings or an extra cyclohexane ring (tetralone, **21**) are the most potent inhibitors of endothelial cell proliferation.[32] The authors have suggested that the 2,6-substituents force the molecule out of a planar conformation and reduce or eliminate enone conjugation with the aromatic ring. Such a disconnection between enone and the terminal ring might be expected to cause increased susceptibility

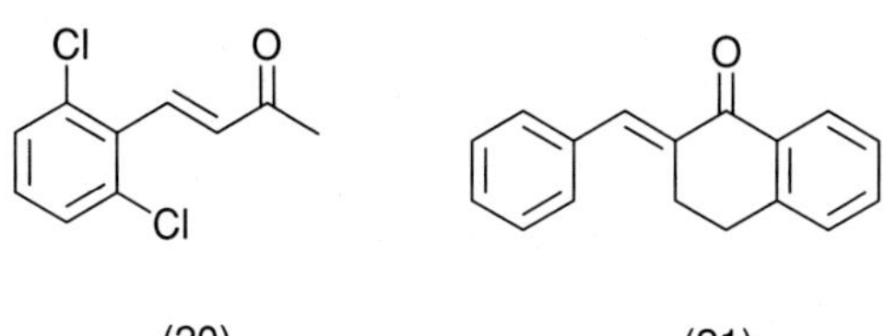

(20) (21)

Figure 8. Chalcone analogues possessing 2,6-substitutution (**20**) or a tetralone (**21**).

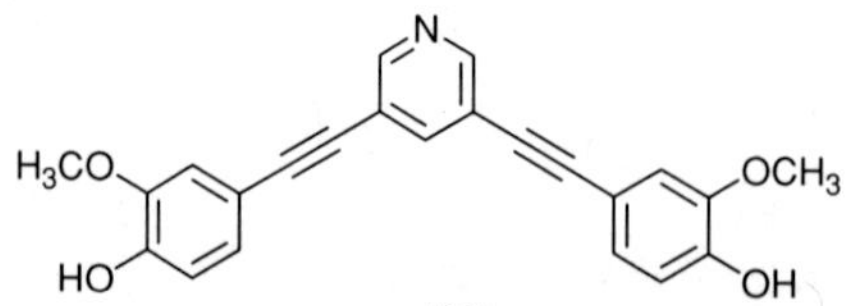
(22)

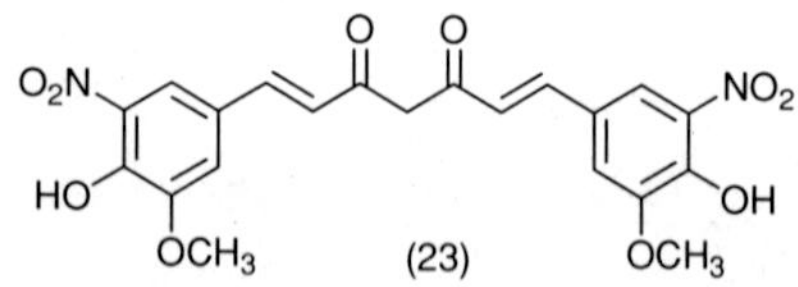
(23)

Figure 9. Antiangiogenic curcumin analogues. A variety of functional groups based on an enone structure lead to antiangiogenic compounds, including bis-alkynyl heteroaromatics (**22**) and BJC005 (**23**).

to nucleophilic attack at the enone fragment (i.e., Michael addition). This speculation assumes, of course, that the latter chemistry is intimately associated with the antiangiogenic effect. In the dienone series, both acetone and cyclohexanone linker regions (**28** and **29**, respectively Figure 12) are structural variations that are also reported to promote angiogenesis inhibition. Interestingly, 2,6-substitution does not increase potency in this series as it does for the chalcones.

Rigidity of the aromatic moieties of curcumin analogues has also been proposed to be important in enhancing antiangiogenic activity. Thus, instead of the traditional C=C double bonds serving as conjugating centers between the aromatic moieties and the central ketone, the terminal rings and a central heteroaromatic ring were linked by an alkyne spacer (Figure 9).[33] These interesting symmetrical bis-aromatic alkynyl pyridine (**22**) and thiophene derivatives exhibit potent inhibition of HUVECs, whereas the corresponding analogues with a flexible alkyl chain do not. This suggests that although rigidity of the central portion of the molecule might be important, active compounds do not require a semiflexible α,β-unsaturated β-diketone for antiantiogenic activity.

Another approach to inhibiting angiogenesis is to block matrix metalloproteinases (MMPs). In the presence of various angiogenic factors, endothelial cells are activated for transformation to malignant and metastatic tumor tissue by the disassociation of the cell from the basement membrane. This process is aided by the action of MMPs. Once a tumor cell detaches from the basement membrane, it will migrate into the blood or the lymph system and metastasize to another location. Shim and co-workers discovered[34] that curcumin is a potent inhibitor of the metalloproteinase CD13/Aminopeptidase N, whereas hydrazinocurcumin is not. Hahm and co-workers synthesized curcumin analogues (**23**, Figure 9) that inhibit the dimerization of the Fos-Jun heterodimer complex, which stimulates transcription of late-stage angiogenic genes.[35] Interestingly, the same nitro-containing analogue was found to inhibit the transcription of MMP-9 mRNA.

2.3. Curcumin Analogues for Traditional Hormone-Treated Cancers

Singletary and MacDonald tested a variety of curcumin analogues against mammary tumors induced by 7,12-dimethylbenz[*a*]anthracene (DMBA), an agent known to form an adduct with DNA. Dibenzoylmethane (**24**, Figure 10), a simple

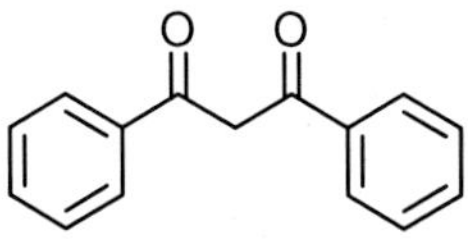

(24)

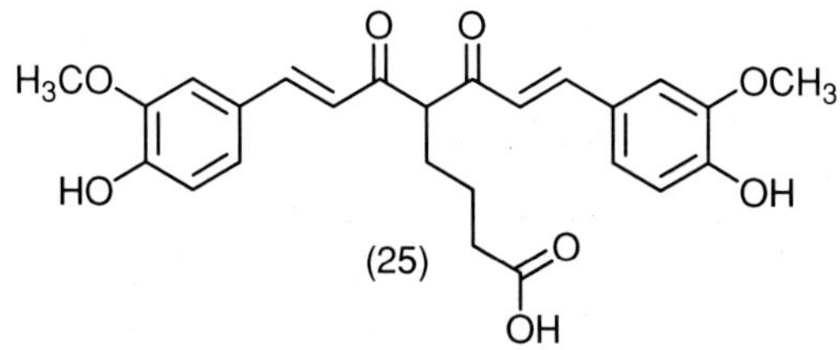

(25)

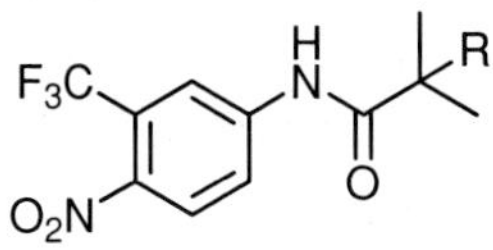

(26)

Figure 10. Dibenzoylmethane (**24**) is a curcumin-inspired analogue that inhibits breast cancer. Compound **25** is comparable in biological activity to hydroxyflutamide (**26**) in prostate cancer cell line inhibition studies.

β-diketone that lacks the allylic ketone moiety found in curcumin, proved to be effective as an antitumor drug. In addition, when included as 1.0% of a diet fed to female rats, the compound inhibited 82% of rat mammary tumorigenesis . By contrast, curcumin has no such effect. Furthermore, dibenzoylmethane inhibits the formation of DNA adducts in a non-neoplastic mammary cell line (MCF-10F) treated with the common chemical carcinogens benzo[*a*]pyrene (BP) and 1,6-dinitropyrene (1,6-DNP).[36] This suggests that β-diketone analogues of curcumin might serve as possible breast cancer chemopreventatives.

Recently, curcumin and a variety of closely related analogues (**25**, Figure 10) were discovered to be androgen receptor antagonists against PC-3 and DU-145 human prostate cancer cell lines in the presence of androgen receptor activator ARA70.[37,38] Antiandrogenic activity appears to be related to the presence of bis-(3,4-dimethoxyphenyl) groups and the α,β-unsaturated ketone(s). Extensive SAR studies suggest that the β-diketone and a hydrogen-bond donor on the aromatic ring must be coplanar. Surprisingly, five synthetic compounds similar to **25** exhibited comparable or increased antiandrogenic activity over the nonsteroidal antiandrogen hydroxyflutamide (**26**), which lacks structural similarity to curcumin. Further elucidation of the mechanisms responsible for this effect coupled to a more extensive exploration of the compound class could furnish a patient-friendly treatment alternative to the most common form of male cancer in the United States. The current regimen includes hydroxyflutamide combination therapy and possible castration.

2.4. Curcumin Analogues in Chemoprevention

Curcumin's action against cancer cells[8] has anticipated its introduction into the clinic. Presently, the compound is being evaluated against pancreatic cancer and multiple myeloma at M.D. Anderson.[39] If successful, certain analogues are sure to follow. Apart from directed therapy, however, the compounds can also be employed as chemopreventatives because they act as anti-inflammatory agents by inhibiting cyclooxygenase (COX) activity. COX enzymes are responsible for the conversion of arachadonic acid to prostaglandins and thromboxanes. Curcumin inhibits the COX-2 pathway, which is generally overexpressed in many malignant tissues.[40] The compound also suppresses two important enzymes, NF-κB-inducing kinase (NIK) and IκBa (IKK), in addition to downregulation of NF-κB.[41,42] Thus, curcumin analogues represent a possible alternative to side-effect-prone nonsteroidal anti-inflammatory agents (NSAIDS), which target COX-2 in addition to COX-1.[43] Related to the failure of the normal inflammatory response is the condition of sepsis, a bacterial infection that generates toxins causing the immune system to attack the body's own organs. In severe cases, this leads to organ failure and death. Recent studies show that curcumin offers protection against the endotoxins and predict the compound to offer a novel therapy against the effects of infection.[44,45]

In addition, it is well known that phenolic compounds and α,β-unsaturated ketones are highly effective antioxidants. Thus, it is not surprising that curcumin, a substituted bis-phenol derivative, is an excellent antioxidant.[46] Dinkova-Kostova et al. tested a multitude of conjugated monoketone curcumin analogues and found that bis-(benzylidene)cycloalkanones (**29**, Figure 12) are excellent inducers of phase 2 detoxification enzymes (e.g., glutathione-*S*-transferases). The latter are responsible for abating electrophilic toxicity and neoplasia, as well as being potent free-radical scavengers.[11,47,48] Interestingly, in the chalcone, phenylpropenoid, and bis-(benzylidene)cycloalkanone compound classes, introduction of an *ortho*-hydroxyl group significantly increases potency. The most pronounced example of this trend is exhibited by the bis-(benzylidene)cycloalkanone series in which monohydroxylation produces a 200-fold increase in antioxidant activity.

Curcumin analogues that lack phenolic groups but retain the β-diketone moiety still maintain[49] an antioxidant profile, presumably as a result of the ability to form a stable carbon radical on the central methylene carbon. Youssef and El-Sherbeny recently showed that locking the ketone into a rigid conformation to increase the degree of conjugation, namely in a piperidone ring, led to a very potent free-radical scavenger that displays broad-spectrum cytotoxicity (i.e., **27**, Figure 11).[50]

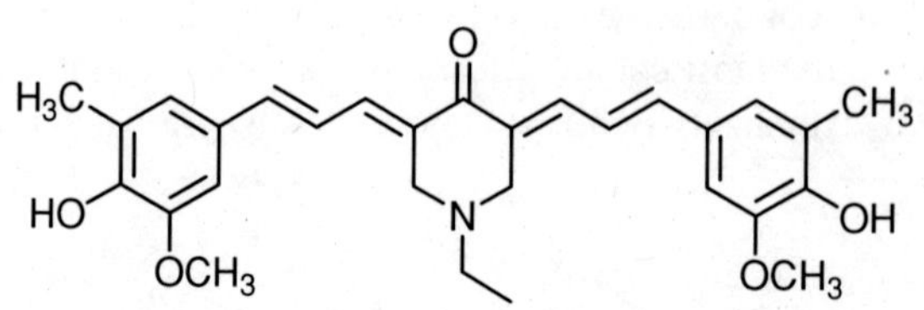

(27)

Figure 11. Youssef and El-Sherbeny's piperidone-curcumin analogue (**27**) with enforced enone conjugation.

A related piperidone shows minimal toxicity in a chemopreventive colon cancer model in mice and rats.[51] Because the latter, unlike curcumin, does not possess a methylene carbon sandwiched between carbonyl groups, radical-scavenging must operate by another mechanism. Most certainly, the terminal phenol groups leading to phenoxy radicals play this role.

3. EF-24 AND ANALOGUES: SYNTHESIS AND CYTOTOXICITY STUDIES

To identify a viable clinical candidate, it is generally necessary to synthesize large numbers of analogues. Our interest, similar to that of Dimmock, De Clerq, and El-Subbagh and coworkers,[19–23, 27] was centered on the monoketo diarylpentanoid class of molecules. These compounds are more compact than curcumin and are readily synthesized.

At Emory, prior to initiating compound synthesis, a topological similarity search of the available chemical database (ACD) was conducted with curcumin as the structural template. Two compounds were identified: 1,5-bis(3,4-dimethoxyphenyl)-1,4-pentadiene-3-one (BDMPP; **28,** Figure 12) and 2,6-bis((3-methoxy-4-hydroxyphenyl)-methylene)-cyclohexanone (BMHPC; **29**, Figure 12). Both compounds were known[52] to inhibit cell proliferation *in vitro* and *in vivo*, and in our hands, they proved to be twofold to threefold more potent than curcumin in melanoma and breast cancer cell lines (RPMI 7951 and MDA-MB-231, respectively).[53] The structural class was subsequently elaborated by synthesis of two related series: one involving the open α,β-unsaturated acetone linker between the two aryl groups (bis-benzylidene acetone analogues) and the other focusing on cyclic cyclohexanone, piperidone, or tetrahydropyranone linkers.

3.1. Chemistry

Chemical syntheses represented by Scheme 1 were typically carried out by combining a ketone with an aromatic aldehyde under basic aldol condensation conditions. The *o*-hydroxy (**30**) analogue, as well as the *m*-counterpart (**31**) and *p*-counterpart (**32**) were prepared to affirm the increased anticarcinogenic effect reported in the literature.[54] A variety of analogues were made, including the *o*-fluorine (**33**), *o*-methoxy (**34**), and *o*-acetoxy (**35**) variations. The racemic saturated alcohol **36**

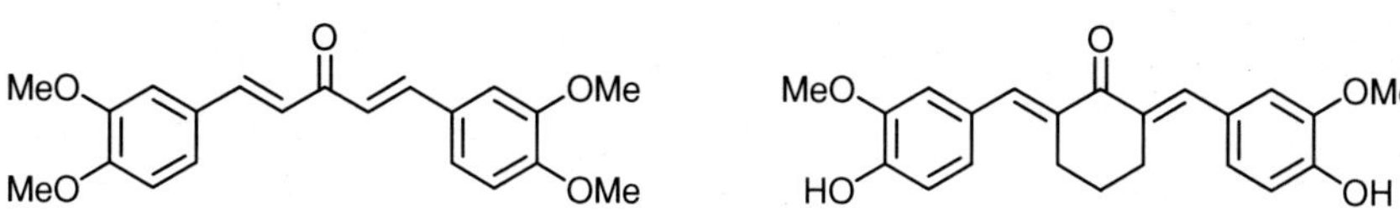

(28), BDMPP (29), BMHPC

Figure 12. Results of a Unity/Tripos 2D silimarity search. Two compounds, **28** and **29**, were identified that are similar to curcumin.

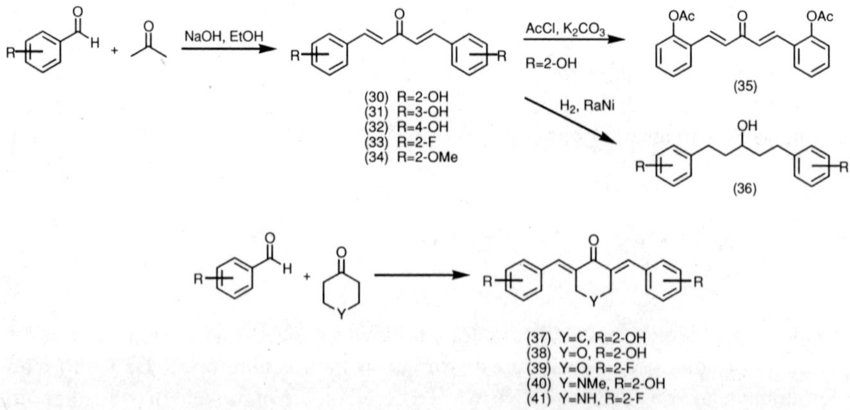

Scheme 1. Synthesis of monoketone α,β-unsaturated curcumin analogues. EF-24 is analogue **41.**

was also examined to confirm that conjugation across the linker region is essential for cytotoxicity. The expectation of lack of activity was fulfilled, although we note that both curcumin and tetrahydrocurcumin have been reported to exhibit cytotoxic and anti-angiogenic properties.[25] As pointed out earlier, curcumin operates as an antioxidant through both its central CH_2 group and its terminal phenolic moieties. The failure of the saturated dienones to exhibit cytotoxicity suggests that conjugation in the monoketones is the primary source of the observed bioactivity.

Annulation of the linker region to provide a cyclic dienone allowed us to probe the effect of a more rigidified structure on anticancer activity. The majority of these analogues (**37–41**) were also synthesized under basic aldol condensation conditions. However, analogue EF-24 (**41**)was obtained by means of glacial acetic acid-promoted aldol condensation.[55]

3.2. Curcumin Analogues Exhibit Anticancer and Antiangiogenic Properties *In Vitro* and Develop SAR

Among others, the unsaturated analogues derived from cyclic ketones, **37**, **38**, and EF-24 were screened in vitro against the NCI's 60 human tumor cell lines. This assay is designed to differentiate tumor- and subpanel-selective drugs from those exhibiting broad-spectrum activities. Each cell line was treated with 10-fold dilutions of drug at a minimum of five concentrations (0.01–100 μM) for 48 h. The compounds showed activity against all tested cell lines. However, EF-24, with an LC_{50} of 16.0 μM, responded with 4.2- and 5.7-fold improvement over curcumin and the common DNA-alkylating agent cisplatin, respectively. The substance exhibits high efficacy toward leukemia, colon, central nervous system, prostrate, and breast cancer cell lines with average 50% growth inhibition ($-GI_{50}$) values all less than 1.0 μM. In addition, EF-24 is at least 10-fold more potent against the latter cell types than both curcumin and cisplatin.

All synthetic analogues in Scheme 1 were tested at Emory against highly malignant melanoma (RPMI 7951) and breast (MDA-MB-321) cancer cell lines. These tests proved useful for development of an emerging SAR. As previously shown,[54] the *ortho*-hydroxy derivative 30 is more active ($GI_{50} = 1.0 \pm 0.5$ μM in melanoma and 2.3 ± 0.5 μM in breast) than the corresponding *meta* (**31**) and *para*-compounds (**32**) in the bis-benzylidene acetone analogue series. The specific moiety in the 2-position of the aromatic ring is crucial for cytotoxic activity, as the replacement of the 2-OH with fluorine (**33**) or methoxy (**34**) leads to decreased potency. As expected, the acetoxy (**35**) derivative shows only slightly decreased activity compared with **30**, undoubtedly due to the activity of esterases, which hydrolyze the compound to unmask the 2-OH functional group. In the unsaturated cyclic ketone series, all of the *ortho*-hydroxyl-containing compounds perform well. When Y is a heteroatom, such as oxygen (tetrahydropyran, **38** and **39**) or nitrogen (piperidone, **40** and **41**), the activity increases over the bis-benzylidene acetone analogue. EF-24 proved to be the most active compound in the NCI screen, with GI_{50} values of 0.7 μM and 0.8 μM in the melanoma and breast cancer cell lines, respectively. The anticancer potency of compound EF-24 is followed by oxygen analogue **38**, with the *N*-methyl-piperidone **40** being slightly less active. In contrast, the arylethyl alcohol analogue **36**, which lacks conjugation, shows no cytotoxicity, supporting the notion mentioned earlier that ketone conjugation is required for cell kill.

Curcumin analogues **37**, **38**, and EF-24 were also found to be potent antiangiogenesis agents in the NCI's *in vitro* antiangiogenesis screen. The analysis examined the inhibition of cell migration and cord formation.[53] The lead compound, EF-24, was comparable to the antiangiogenesis drug TNP-470 in the cell migration assay, with IC_{50} values of 0.8 μM and 0.6 μM, respectively. In addition, HUVEC cord formation was blocked significantly with an IC_{50} of 1.5 μM. Most importantly, EF-24 shows much higher activity in both of the assays compared to curcumin (>10 μM in the cord-formation assay). In terms of SAR, the same patterns of activity were seen in the antiangiogenesis screen as in the anticancer screens described earlier.

3.3. Curcumin Analogues Inhibit Tumorigeneis *In Vivo* with Minimal Toxicity

In addition to outperforming other analogues in the cytotoxicity and antiangiogenesis studies *in vitro*, EF-24 also induces breast tumor regression in athymic nude mice. Solid tumors derived from human breast cancer xenografts were grown on the flanks of female mice for 3 weeks, with subsequent subcutaneous drug administration for 2 weeks. At treatment levels from 2 to 100 mg/kg of EF-24, tumor weight shows a dose-dependent decrease. The average tumor weight in the 20-mg/kg group decreased by about 70% of the control.[53]

Most impressively, no toxicity was observed at up to 100 mg/kg of EF-24, which is well below the maximum tolerated dose (MTD) of 400 mg/kg. This illustrates that this curcuminoid derivative is much safer than cisplatin, which has an MTD of 10 mg/kg. The animals likewise gained weight during the experiment. In addition, sacrificed animals were examined to show no damage to the liver, kidney and spleen. As a result, EF-24 surfaced as an important lead compound, one that displays increased antitumor action *in vitro* and *in vivo* by comparison with curcumin, while evidencing little or no low toxicity in the preliminary evaluations.

3.4. Mechanism of Action: Redox-Dependent Induction of Apoptosis

The biological action of curcumin and its monoketone analogues is certainly pleiotropic.[4,8,9,10] Although the studies are in their early stages, we have examined aspects of the mechanism(s) of cell kill for EF-24.[56] In the context of apoptosis, the benchmark events in their sequence of occurrence are the following: depolarization of the mitochondrial membrane, caspase-3 activation, externalization of phosphatidylserine (PS), and DNA fragmentation (sub-G_1/G_0 accumulation of DNA). EF-24 perturbs all of the above processes. In addition, the compound reduces intracellular GSH and thioredoxin-1 (Trx-1) while increasing reactive oxygen species (ROS). Other studies with curcumin as the drug confirm the inverse concentration relationship between depletion of endogenous GSH and ROS, highlight the role of the antiapoptotic protein bcl-2, and review various indirect interactions between curcumin and bcl-2.[57] The complexity of the relationship among curcumin and analogues, GSH depletion, and levels of ROS should not be underestimated. Paradoxically, ROS can not only promote carcinogenesis but also induce apoptosis of tumor cells.[58] The precise mechanism of curcumnoids as antioxidants, which might have opposite effects on tumor growth,[58] has yet to be fully elucidated.

As mentioned earlier, EF-24 inhibits cell proliferation in the prostate (DU-145) and breast cancer (MDA-MB-231) cell lines. At a concentration of 10 μM, cell proliferation is blocked by 70–80% in the prostate and 100% in the breast cancer cell lines as monitored by incorporation of 5-bromo-2′deoxyuridine (BrdU) into actively dividing cells. This causes partial denaturation of double-stranded DNA, which can be detected immunochemically. When a cell incurs such DNA damage, the cell cycle is arrested. The cell might recover by means of DNA repair, but if the damage is not reparable, the apoptotic program is initiated. In the present case, flow-cytometric analysis was used to determine that EF-24 causes G_2/M cell cycle arrest after 48 h, followed by an increase in sub-G_1/G_0 cells after an additional 24 h. These biomarkers indicate that the cells are apoptotic.

Apoptotic cells acquire a "leaky" mitochondrial membrane. This causes the membrane to release cytochrome-*c* and various cations from the organelle into the cytosol.[59,60] Once discharged, cytochrome-*c* binds to apoptotic-inducing factors to form an apoptosome.[61] This complex activates cellular caspases, cysteine proteases, essential for protein degradation, chromatic condensation, and DNA fragmentation involved in the apoptotic pathway.[62] The apoptosome activates caspase-9, which subsequently activates caspase-3.[63] As curcumin is known to promote this cascade,[64–66] it was desirable to test EF-24's mitochondrial membrane depolarizing ability as well. By means of green/red fluorescence assays, EF-24 was indeed determined to depolarize the membrane in 80% of human breast cancer cells (MDA-MB-231) but only 50% in the human prostate cancer cell line (DU-145) 48 h after 20 μM drug treatment. In combination with the cell proliferation assay results, these experiments show that the malignant breast cancer cells are more susceptible to EF-24 treatment than the prostate cell line. The caspase-3 assay supported this increased sensitivity as well, with caspase activation being almost two times higher in MDA-MB-231 cells than DU-145 cells.

Early apoptosis is marked by the translocation of PS from inside the cellular plasma membrane to the external surface of the cell. This causes the cellular membrane to eventually become porous and break apart. EF-24 induces PS externalization after 72 h of treatment. Caspase activation of PS translocation is implicated, as a caspase inhibitor blocks PS externalization in both cell lines.

The major apoptotic events associated with EF-24 occur within 48 h after treatment, as indicated by G_2/M cell arrest. However, the detailed biochemistry behind curcumin analogue apoptosis is unknown. The study determined that within 48 h of drug treatment in both MDA-MB-231 and DU-145 cell lines, high levels of ROS are formed. ROS and thiol buffers GSH and Trx-1 are in exquisite balance in normal cells. However, cytotoxic events and chemical agents induce increased levels of ROS while simultaneously decreasing cellular levels of GSH.[67] Although the details are still unclear, it is believed that appropriate increases in ROS production leads to mitochondrial depolarization.[68] These phenomena are precisely paralleled in the EF-24 work.

Finally, we examined the mechanism of action for EF-24 that correlates directly with its α,β-unsaturated ketone structure. Whereas curcumin's reactions with GSH are characterized by unstable and incompletely characterized products (cf. Figure 3),[13, 14] EF-24 clearly acts as a Michael acceptor for both GSH and Trx-1 thiol functional groups (Scheme 2). Of two additional studies on EF-24 and analogues underway, one mirrors curcumin[41, 42] in blocking the action of NF-κB.[69] The second illustrates the pleiotropic nature of the analogues with respect to inhibition of a panel of Ser/Thr and Tyr kinases.[70]

3.5. Water-Soluble EF-24 Analogues: Glutathione Conjugates

EF-24 (**41**), like many potential drug candidates, is sparingly water insoluble in its free-base form. The compound is also light sensitive, leading to slow decomposition in solution under ambient conditions in room light. One approach to circumventing these problems takes advantage of EF-24's ability to serve as a Michael acceptor.

The synthesis of EF-24-GSH (**42**) conjugates proceeds as shown in Scheme 2. Treatment of EF-24 with an excess (at least 3.0 equivalents) of GSH in water produces the desired doubly-conjugated adduct, **42** which can be isolated as a white solid following HPLC purification.[71] Other similar GSH conjugates have also been prepared, and, similar to EF-24, all of them are both water soluble and light stable.

In addition to exhibiting improved physical properties, EF-24-GSH mimics EF-24 in cytotoxicity studies *in vitro*. When treated with either increasing EF-24 or EF-24-GSH concentrations, cytotoxicity against human breast cancer cells shows a near-identical dose-dependent increase (Figure 13). This finding strongly suggests that EF-24 and EF-24-GSH are in equilibrium (Figure 14) and that the conjugate

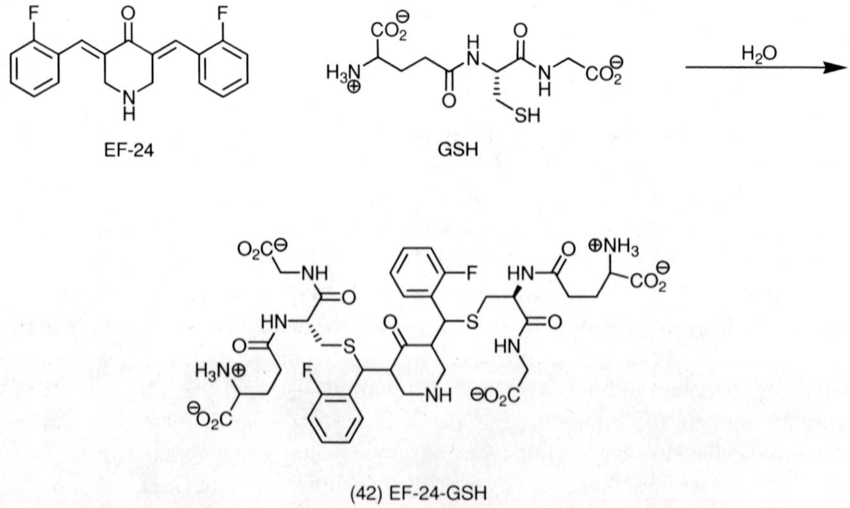

Scheme 2. Synthesis of water-soluble EF-24-GSH conjugate **42**.

can undergo a facile retro-Michael process to free the active EF-24 drug. Thus, GSH conjugates of curcumin analogues represent a means of addressing problems associated with poor water solubility and light stability while still retaining the parent drug's cytotoxic and anticarcinogenic effects.

Glutathione is not the only thiol capable of conjugation in the Michael sense with EF-24. Cysteine and a series of peptides containing cysteine have likewise been coupled to the latter. These compounds are also light-stable white crystalline solids that show cytotoxic behavior similar to the GSH adduct.[71]

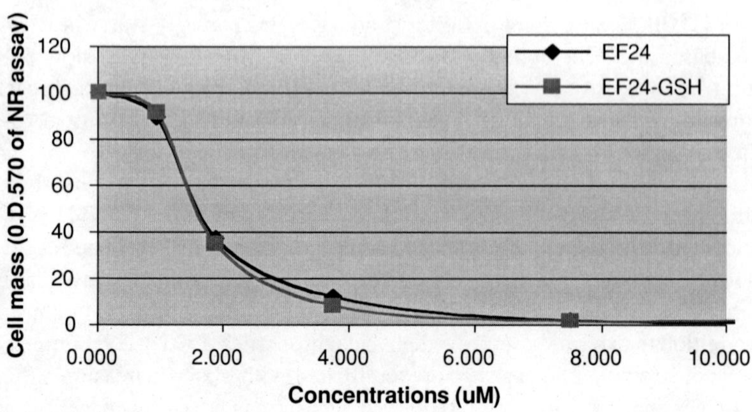

Figure 13. Cytotoxicity of EF-24 (**41**) and EF-24-GSH conjugate (**42**) against human breast cancer cells.

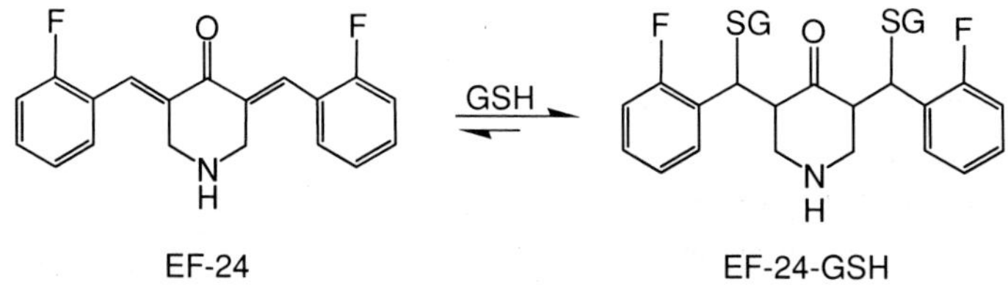

Figure 14. Suggested equilibrium between EF-24 (**41**) and EF-24-GSH (**42**).

4. TARGETING CANCER WITH A NOVEL DRUG DELIVERY SYSTEM: A PROTEIN CONJUGATE

Whereas many promising small-molecule chemotherapeutics have been developed in the fight against cancer, most possess debilitating side effects. The latter include nausea, chronic fatigue, depression, and hair and weight loss, a result of the non-specific nature of the drugs. Most cancer drugs operate on rapidly dividing cells. Their safety depends on differential rates of cell division for cancer cells relative to normal cells. Unfortunately, this implies that normal cells with higher rates of division or high concentrations of the drug target (e.g., tubulin) are also subject to chemotherapeutic cytotoxicity.

Various drug delivery systems have been developed in the attempt to counteract adverse side effects and increase patient compliance. For instance, doxorubicin is known to have cardiomyopathic side effects.[5] A synthetic cyclotriphosphazene-Gly-Phe-Leu-Gly-doxirubicin conjugate (**43**, Figure 15) is active against the leukemia L1210 cell line ($IC_{50} = 1.1\ \mu M$ versus doxorubicin $IC_{50} = 0.10\ \mu M$) with the possibility of being tumor selective.[5] The conjugate shows good water

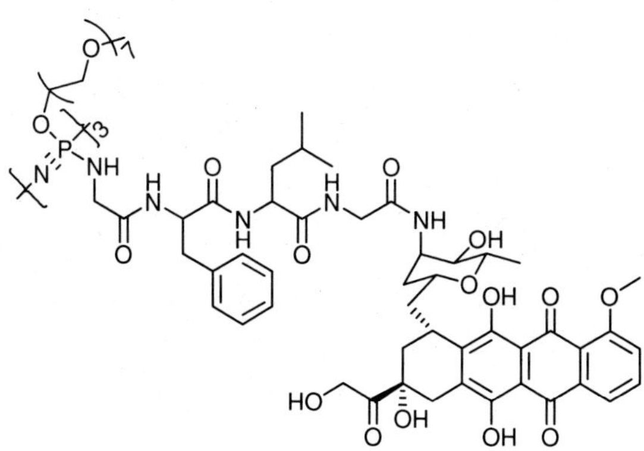

(43)

Figure 15. Cyclotriphosphazene-Gly-Phe-Leu-Gly-doxirubicin conjugate (**43**).

solubility, the drug has a high loading capacity (40%), and the peptide linker and phosphazene moieties are most likely biodegradable.

Ouchi and co-workers have recently developed a drug delivery system based on the high association constant for avidin–biotin binding, an interaction that involves four binding sites for the biotin ligand.[72] Synthetic biotin–drug and biotin–fluorescent dye complexes are bound to avidin such that one complex consists of one avidin, two biotin-linked drug molecules, and two biotin-linked dyes. The drug chosen was galactose, which is recognized by the asialoglycoprotein receptor. These hepatic receptors are specific for glycoproteins whose terminal sialic acid moiety has been removed. By means of confocal laser microscopy, the complex was observed to be internalized into HepG2 human hepatoma cells via receptor-mediated endocytosis.

Emory's approach, in spirit similar to that devised by Ouchi and Ohya, relies on the binding of the membrane-bound receptor tissue factor (TF) to the soluble protein factor VIIa (fVIIa), a process that takes place normally as part of the blood coagulation cascade. The fVIIa is an enzymatic serine protease. Shaw and co-workers showed that tripeptide chloromethyl ketones, such as phenylalanine–phenylalanine–arginine–chloromethyl ketone (FFRck), are irreversible inhibitors of fVIIa via covalent modification of a histidine residue in the catalytic triad that promotes endocytosis.[73] In addition, the FFRck-fVIIa conjugate has up to five times the affinity for TF than fVIIa alone. The basis for exploiting this targeting mechanism is that not only is TF overexpressed on the surface of tumor cells, but it is also present on vascular endothelial cells (VECs) of tumor vasculature due to overactive NF-κB.[74,75] Thus, covalently attaching antitumor and antiangiogenic agents to fVIIa represents a new drug delivery system with the capability of delivering a drug to the surface of TF-expressing cancer cells and to the angiogenic cells that carry nutrients to the growing solid tumor. Once in complex with TF, the protein–protein drug complex enters the target cells by endocytosis.[76] Inside the cell, the drug can be cleaved from its homing molecule by cellular esterases and subsequently perform its antitumor and antiangiogenic functions. We have illustrated the feasibility of this drug delivery mechanism with an EF24-FFRck-fVIIa conjugate, which is selectively cytotoxic to cancer cells *in vitro*[77] and *in vivo*.[78]

4.1. Chemistry

Prior to conjugation of drug to fVIIa, the synthesis of EF24-FFRck was undertaken. The optimum covalent linker between drug and tripeptide is succinate. Intermediate **44** (Scheme 3), a compound that shows two-thirds the activity of EF-24 in cytotoxicity studies, arises from treatment of EF-24 and succinic anhydride under basic conditions.

As a first step, the chloromethyl ketone tripeptide was prepared (Scheme 4). Boc-D-Arg(Mtr)-OH (**45**) was converted to the α-diazo ketone **46** via a modified Kettner method[79] which generates a mixed anhydride, and subsequent exposure

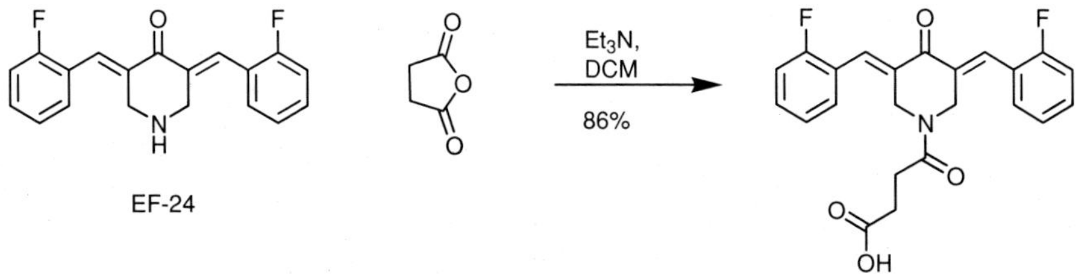

(44)

Scheme 3. Synthesis of EF24-linker **44**.

to diazomethane. Treatment of the diazo compound **46** with ethanolic HCl gives the corresponding deprotected chloromethyl ketone **47**, which was subjected to Kettner peptide-coupling conditions with *N*-Boc-Phe-Phe-OH to yield the desired tripeptide *N*-Boc-D-Arg(Mtr)-Phe-Phe chloromethyl ketone **48**. Deprotection of the Boc group was accomplished by treatment with ethanolic HCl to give **49**. The latter Mtr-protected FFRck was combined with an EF-24-succinate linker under dicyclohexylcarbodiimide (DCC)-coupling conditions to give EF-24-FFR(Mtr)ck. Deprotection of the arginine-protecting group was carried out with aqueous trifluoroacetic acid (TFA) in anisole to afford the final compound EF-24-FFRck (**50**). EF-24-FFRck was then conjugated to fVIIa through the catalytic histidine residue in a 1:1 ratio.

4.2. EF24-FFRck-fVIIa Inhibits Cell Growth *In Vitro* and *In Vivo*

The cytotoxicity of the EF24-FFRck-fVIIa conjugate (**51**) in TF-expressing cell lines was tested both *in vitro* and *in vivo*. For the former setting, the conjugate has a minimal effect on non-TF-expressing normal breast and melanocyte cells. In contrast, EF24-FFRck-fVIIa significantly inhibits cell viability in TF-expressing breast (MDA-MB-231) and melanocyte (RPM-7951) cells in a dose-dependent manner (Figure 16). In addition, EF24-FFRck lacking the fVIIa homing device shows no cytotoxic activity. This implies that the truncated conjugate does not interact with TF on the cell surface and, thus, does not gain cell entry.[77]

EF24-FFRck-fVIIa also shows inhibition of primary tumor xenograft growth as well as a reduction of tumor size *in vivo*.[78]

In addition to inhibiting tumor growth and size, the EF-24 conjugate inhibits angiogenesis in a rabbit corneal model (Figure 18). Treatment of a corneal pocket with vascular endothelial growth factor (VEGF) induces the formation of new blood vessels. The corneas were then exposed to the vehicle, EF-24 (5.0 μM), or EF24-FFRck-fVIIa (5.0 μM). Whereas treatment with either the vehicle or EF-24 had no effect on corneal angiogenesis, the EF-24 conjugate shows clear destruction of blood vessels. The results imply that VEGF induces VECs to express TF. The TF of the newly formed vessels is then targeted by fVIIa of the EF-24 conjugate.

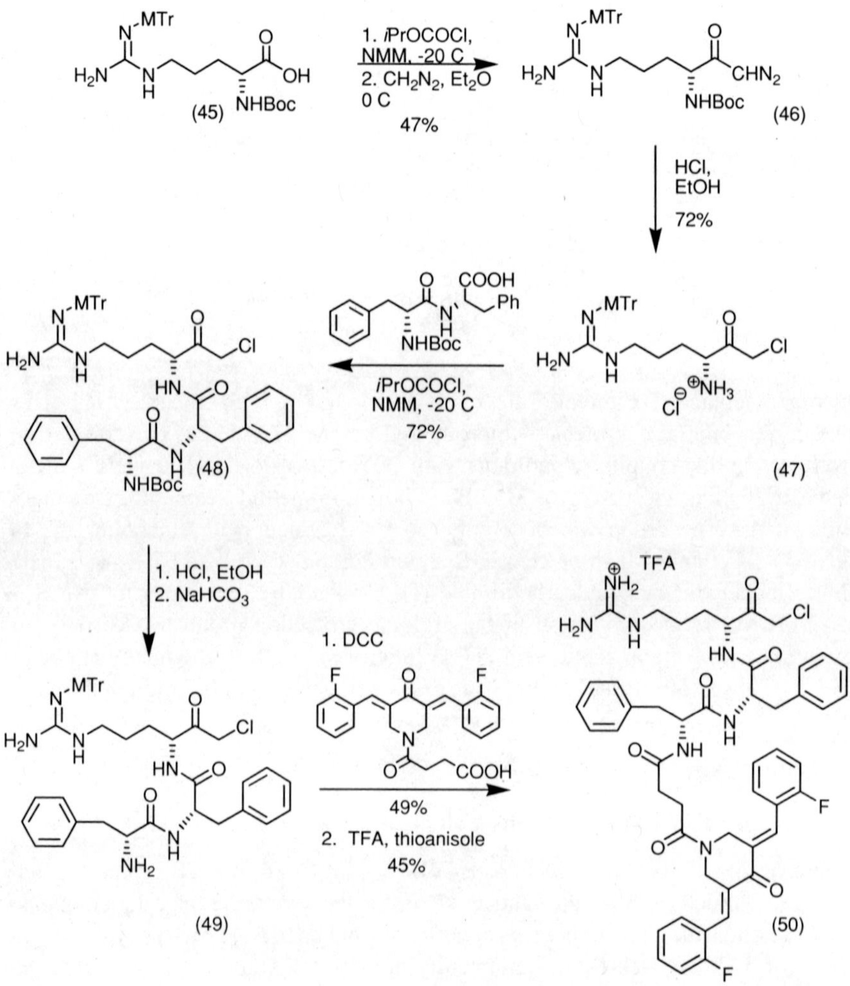

Scheme 4. *In vitro* effects of EF-24 (**41**), EF24-FFRck (**50**), and EF24-FFRck-fVIIa conjugate (**51**) in normal non-TF-expressing cells and their breast and melanocyte TF-expressing counterparts.

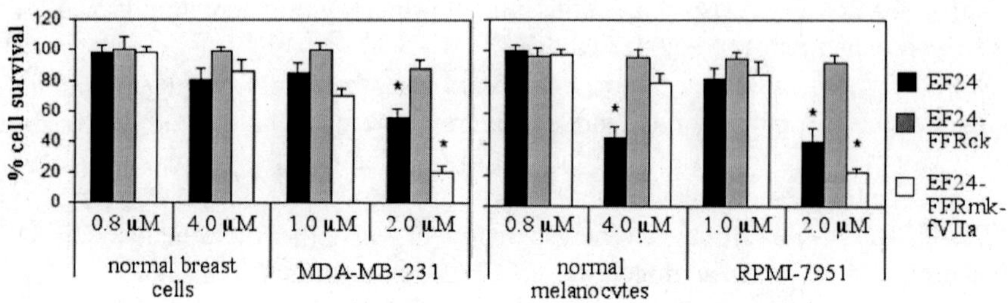

Figure 16. Synthesis of EF24-FFRck (**50**).

Once endocytosed by the TF-expressing VEC cells, EF-24 is freed from the carrier and subsequently blocks further angiogenesis by way of its cell-killing capacity.

5. CONCLUSIONS AND PROSPECTS

Curcumin represents a near-perfect starting point from a drug discovery and development perspective. Its mechanism of action involves multiple, synergistic pathways that have yet to exhibit any serious negative side effects in man or animals. Consequently, a number of research groups have taken the natural product as a starting point to develop a wide variety of curcumin analogues. Whereas some of the studies retained the three structural sectors indicated by the colors in structure (**1**), the majority of investigations truncated the central conjugated β-diketone to the monocarbonyl dienone represented by (**2**). The resulting diversity of compounds exhibit cytotoxicities against an equally diverse set of cancer-related cell lines. SARs have been developed and the interesting "sequential cytotoxicity" hypothesis exploited. Antiangiogenesis is among the properties of the analogues and promises to permit a double-pronged attack on target cells; namely, the compounds are not only cytotoxic to tumor cells but also carry the promise of being able to starve the tumors of nutrients.

Mechanisms of action of curcumin have been sufficiently well studied to suggest that the natural product is pleiotropic, namely acting by a series of mechanisms. Although the published investigations do not yet provide an integrated map of curcumin's actions in the cell or in tumors, it is clear that early studies focusing on analogues closely parallel those conducted for the parent molecule. An important pathway regulator that will most certainly surface as critical to the actions of curcumin and simpler analogues is the NF-κB complex. It is central to many integrated signaling pathways controlling cancer development, including antiapoptosis (e.g., bcl-xl, survivin, sodium dodecyl sulfate), proliferation (e.g., tumor necrosis factor (TNF), interleukins, cyclinD1) tumor promotion (e.g., COX2, INOS, MMP-9), metastasis (e.g., ICAM-1, VCAM-1), angiogenesis (e.g., VEGF, TNF), inflammation (e.g., TNF, chemokines), and immortality (e.g., telomerase). Not surprisingly, both curcumin and its ketone analogues operate directly on the proteins associated with NF-κB.

In our own laboratories, we have attempted to contribute to and take advantage of the still limited knowledge surrounding the molecular basis for curcumin-analogue modulation of cellular pathways. It has been possible to develop easily synthesized curcumin congeners that exhibit significantly improved cytotoxicity against a wide range of cell lines by comparison with curcumin, but still retain toxicity profiles in rodents comparable to the parent natural product. EF-24 (**41**) provides the added advantage of exhibiting good oral bioavailability and good pharmacokinetics in the same animals.[80] Thiol conjugates of EF-24 analogues have been prepared that address stability and solubility issues while demonstrating cellular activities similar to the unmodified dienones. In parallel experiments, the fVIIa-TF complex has been exploited to develop a targeting strategy for the analogues. In particular, the EF24-FFRck-fVIIa protein conjugate (**51**) is somewhat more effective relative to the drug alone against breast cancer and melanocyte cells (Figure 16).

In the face of these advances and those described elsewhere in this volume, we project that curcumin itself will become an important anticancer agent deployed clinically against a variety of cancers in the near future. Simultaneously, we expect that new analogues with improved antitumor properties will be developed in the near term and suggest themselves as second-generation curcuminoid clinical agents. This coincides with our goals, which is to identify analogues with the best overall pharmacologic profile, complete their preclinical assessment, and move the lead and a backup into clinical trials.

REFERENCES

1. B. J. Druker, C. L. Sawyers, H. Kantarjian, D. J. Resta, S. F. Reese, J. M. Ford, R. Capdeville, and M. Talpaz, Activity of a specific inhibitor of the BCR-ABL tyrosine kinase in the blast crisis of chronic myeloid leukemia and acute lymphoblastic leukemia with the Philadelphia chromosome. *N Engl J Med* **344**(14), 1038–1042 (2001).
2. M. A. Cobleigh, C. L. Vogel, D. Tripathy, N. J. Robert, S. Scholl, L. Fehrenbacher, J. M. Wolter, V. Paton, S. Shark, G. Lieberman, and D. J. Slamon, Multinational study of the efficacy and safety of humanized anti-HER2 monoclonal antibody in women who have HER2-overexpressing metastatic breast cancer that has progressed after chemotherapy for metastatic disease. *J Clin Oncol* **17**(9), 2639–2648 (1999).
3. R. B. Weiss, R. C. Donehower, P. H. Wiernik, T. Ohnuma, R. J. Gralla, D. L. Trump, J. R. Baker, Jr., D. A. Van Echo, D. D. Van Hoff, and B. Leyland-Jones, Hypersensitivity reactions from taxol. *J Clin Oncol* **8**(7), 1263–1268 (1990).
4. E. A. Eisenhauer, W. W. ten Bokkel Huinink, K. D. Swenerton, L. Gianni, J. Myles, M. E. van der Burg, I. Kerr, J. B. Vermorken, K. Buser, and N. Colombo, European–Canadian randomized trial of paclitaxel in relapsed ovarian cancer: high dose versus low-dose and long versus short infusion. *J Clin Oncol* **12**(12), 2654–2666 (1994).
5. D. Raghavan, B. Koczwara, and M. Javle, Evolving strategies of cytotoxic chemotherapy for advanced prostate cancer. *Eur J Cancer* **33**(4), 566–574 (1997).
6. S. Shishodia, G. Sethi, and B. B. Aggarwal, Curcumin: Getting back to the roots. *Ann NY Acad Sci* **1056**, 206–217 (2005).
7. A. L. Cheng, C. H. Hsu, J. K. Lin, M. M. Hsu, Y. F. Ho, T. S. Shen, J. Y. Ko, J. T. Lin, B. R. Lin, W. Ming-Shiang, H. S. Yu, S. H. Jee, G. S. Chen, T. M. Chen, C. A. Chen, M. K. Lai, Y. S. Pu, M. H. Pan, Y. J. Wang, C. C. Tsai, and C. Y. Hsieh, Phase I clinical trials of curcumin, a chemopreventive agent, in patients with high-risk or pre-malignant lesions. *Anticancer Res* **21**(4B), 2895–2900 (2001).
8. B. B. Aggarwal, A. Kumar, and A. C. Bharti, Anticancer potential of curcumin: Preclinical and clinical studies. *Anticancer Res* **23**(1A), 363–398 (2003).
9. A. Duviox, R. Blasius, S. Delhalle, M. Schnekenburger, F. Morceau, E. Henry, M. Dicato, and M. Diederich, Chemopreventive and therapeutic effects of curcumin, *Cancer Lett* **223**(2), 181–190 (2005).
10. R. A. Sharma, A. J. Gescher, and W. P. Steward, Curcumin: The story so far. *Eur J Cancer* **41**(13), 1955–1968 (2005).

11. A. T. Dinkova-Kostova, C. Abeygunawardana, and P. Talalay, Chemoprotective properties of phenylpropenoids, bis(benzylidene)cycloalkanones, and related Michael reaction acceptors: Correlation of potencies as phase 2 enzyme inducers and radical scavengers. *J Med Chem* **41**(26), 5287–5296 (1998).
12. B. Mutus, J. D. Wagner, C. J. Talpas, J. R. Dimmock, O. A. Phillips, and R. S. Reid, 1-p-chlorophenyl-4,4-dimethyl-5-ethylamino-1-penten-3-one hydrobromide, a sulfhydryl-specific compound which reacts irreversibly with protein thiols but reversibly with smaller molecular weight thiols. *Anal Biochem* **177**(2), 237–243 (1989).
13. S. Mathews and M. N. A. Rao, Interaction of curcumin with glutathione. *Int J Pharm* **76**(3), 257–259 (1991).
14. S. Awasthi, U. Pandya, S. S. Singhal, J. T. Lin, V. Thiviyanathan, W. E. Seifert, Y. C. Awasthi, and G. A. S. Ansara, Curcumin-glutathione interactions and the role of human glutathione S-transferase P1-1. *Chemico-Biol Interact* **128**(1), 19–38 (2000).
15. H. M. Wortelboer, M. Usta, A. E. Van der Velde, M. G. Boersma, B. Spenkelink, J. J. Van Zanden, J. Jelmer, I. M. C. M. Rietjens, P. J. Van Bladeren, and N. H. Cnubben, Interplay between MRP inhibition and metabolism of MRP inhibitors: The case of curcumin. *Chem Res Toxicol* **16**(12), 1642–1651 (2003).
16. Y. J. Wang, M. H. Pan, A. L. Cheng, L. I. Lin, Y. S. Ho, C. Y. Hsieh, and J. K. Lin, Stability of curcumin in buffer solutions and characterization of its degradation products. *J Pharm Biomed Anal* **15**(12), 1867–1876 (1997).
17. M. J. Ansari, S. Ahmad, K. Kohli, J. Ali, and R. K. Khar, Stability-indicating HPTLC determination of curcumin in bulk drug and pharmaceutical formulations. *J Pharm Biomed Anal* **9**(1–2), 132–138 (2005).
18. J. R. Dimmock, P. Kumar, A. J. Nazarali, N. L. Motaganahalli, T. P. Kowalchuk, M. A. Beazely, J. W. Quail, E. O. Oloo, T. M. Allen, J. Szydlowski, E. De Clerq, and J. Balzarini, Cytotoxic 2,6-bis(arylidene)cyclohexanones and related compounds. *Eur J Med Chem* **35**(11), 967–977 (2000).
19. J. R. Dimmock, M. P. Padmanilayam, G. A. Zello, K. H. Nienaber, T. M. Allen, C. L. Santos, E. De Clerq, J. Balzarini, E. K. Manavathu, and J. P. Stables, Cytotoxic analogues of 2,6-bis(arylidene)cyclohexanones. *Eur J Med Chem* **38**(2), 169–177 (2003).
20. J. R. Dimmock, M. P. Padmanilayam, R. N. Puthucode, A. J. Nazarali, N. L. Motaganahalli, G. A. Zello, J. W. Quail, E. O. Oloo, H.-B. Kraatz, J. S. Prisciak, T. M. Allen, C. L. Santos, J. Balzarini, E. De Clerq, and E. K. Manavathu, A conformational and structure-activity relationship study of cytotoxic 3,5-bis(arylidene)-4-piperidones and related N-acryloyl analogues. *J Med Chem* **44**(4), 586–593 (2001).
21. J. R. Dimmock, A. Jha, G. A. Zello, J. W. Quail, E. O. Oloo, K. H. Nienaber, E. S. Kowalczyk, T. M. Allen, C. L. Santos, E. De Clerq, J. Balzerini, E. K. Manavathu, and J. P. Stables, Cytotoxic N-[4-(3-aryl-3-oxo-1-propenyl)phenylcarbonyl]-3,5-bis(phenylmethylene)-4-piperidones and related compounds. *Eur J Med Chem* **37**(12), 961–972 (2002).
22. H. I. El-Subbagh, S. M. Abu-Zaid, M. A. Mahran, F. A. Badria, and A. M. Al-Obaid, Synthesis and biological evaluation of certain α, β-unsaturated ketones and their corresponding fused pyridines as antiviral and cytotoxic agents. *J Med Chem* **43**(14), 2915–2921 (2000).
23. J. R. Dimmock, U. Das, H. I. Gul, M. Kawase, H. Sakagami, Z. Baráth, I. Ocsovsky, and J. Molnár, 3-Arylidene-1(4-nitrophenylmethylene)-3,4-dihydro-1*H*-naphthalen-2-ones and related compounds displaying selective toxicity and reversal of multidrug resistance in neoplastic cells. *Bioorg Med Chem Lett* **15**, 1633–1636 (2005).

24. N. M. Pandya, N. S. Dhalla, and D. D. Santani, Angiogenesis: A new target for future therapy. *Vasc Pharmacol* **44**(5), 265–274 (2006).
25. J. L. Arbiser, N. Klauber, R. Rohan, R. Van Leeuwen, M. Huang, C. Fisher, E. Flynn, and H. R. Byers, Curcumin is an in vivo inhibitor of angiogenesis. *Mol Med* **4**(6), 376–383 (1998).
26. T. P. Robinson, T. Ehlers, R. B. Hubbard, X. Bai, J. L. Arbiser, D. J. Goldsmith, and J. P. Bowen, Design, synthesis, and biological evaluation of angiogenesis inhibitors: Aromatic enone and dienone analogues of curcumin. *Bioorg Med Chem Lett* **13**(1), 115–117 (2003).
27. J. R. Dimmock, N. M. Kandepu, M. Hetherington, J. W. Quail, U. Pugazhenthi, A. M. Sudom, M. Chamankhah, P. Rose, E. Pass, T. M. Allen, S. Halleran, J. Szydlowski, B. Mutus, M. Tannous, E. K. Manavathu, T. G. Meyers, De Clerq, E., and J. Balzarini, Cytotoxic activities of Mannich bases of chalcones and related compounds. *J Med Chem* **41**(7), 1014–1026 (1998).
28. Y. Satomi, Inhibitory effects of 3′-methyl-3-hydroxychalcone on proliferation of human malignant tumor cells and on skin carcinogenesis. *Int J Cancer* **55**(3), 506–514 (1993).
29. L. W. Wattenberg, J. B. Coccia, and A. R. Galbraith, Inhibition of carcinogen-induced pulmonary and mammary carcinogenesis by chalcone administered subsequent to carcinogen exposure. *Cancer Lett* **83**(1–2), 165–169 (1994).
30. M. L. Edwards, D. M. Stemerick, and P. S. Sunkara, Chalcones: A new class of antimitotic agents. *J Med Chem* **33**(7), 1948–1954 (1990).
31. Y. Xia, Z. Yang, P. Xia, K. F. Bastow, Y. Nakanishi, and K. Lee, Antitumor agents. Part 202: Novel 2′-aminochalcones: Design, synthesis, and biological evaluation. *Bioorg Med Chem Lett* **10**(8), 699–701 (2000).
32. T. P. Robinson, R. B. Hubbard, T. J. Ehlers, J. L. Arbiser, D. J. Goldsmith, and J. P. Bowen, Synthesis and biological evaluation of aromatic enones related to curcumin. *Bioorg Med Chem* **13**(12), 4007–4013 (2005).
33. C. M. Ahn, W. Shin, H. B. Woo, S. Lee, and H. Lee, Synthesis of symmetrical bis-alkynyl or alkyl pyridine and thiophene derivatives and their antiangiogenic activities. *Bioorg Med Chem Lett* **14**(15), 3893–3896 (2004).
34. J. S. Shim, J. H. Kim, H. Y. Cho, Y. N. Yum, S. H. Kim, H. Park, B. S. Shim, S. H. Choi, and H. J. Kwon, Irreversible inhibition of CD13/aminopeptidase N by the antiangiogenic agent curcumin. *Chem Biol* **10**(8), 695–704 (2003).
35. E. Hahm, Y. S. Gho, S. Park, C. Park, K. Kim, and C. Yang, Synthetic curcumin analogs inhibit activator protein-1 transcription and tumor-induced angiogenesis. *Biochem Biophys Res Commun* **321**(2), 337–344 (2004).
36. K. Singletary and C. MacDonald, Inhibition of benzo[a]pyrene- and 1,6-dinitropyrene-DNA adduct formation in human mammary epithelial cells by dibenzoylmethane and sulforaphane. *Cancer Lett* **155**(1), 47–54 (2000).
37. H. Ohtsu, Z. Xiao, J. Ishida, M. Nagai, H. Wang, H. Itokawa, C. Su, C. Shih, T. Chiang, E. Chang, Y. Lee, M. Tsai, C. Chang, and K. Lee, Antitumor Agents 217. Curcumin analogues as novel androgen receptor antagonists with potential as anti-prostate cancer agents. *J Med Chem* **45**(23), 5037–5042 (2002).
38. L. Lin, Q. Shi, A. K. Nyarko, K. F. Bastow, C.-C. Wu, C. Y. Su, C. C. Shih, and K. H. Lee, Antitumor Agents 250. Design and synthesis of new curcumin analogues as potential anti-prostate cancer agents. *J Med Chem* **49**(13), 3963–3972 (2006).
39. *Journal of Clinical Oncology*, 2006 ASCO Annual Meeting Proceedings (Post-Meeting Edition). Vol 24, No 18S (June 20 Supplement), 2006: 14151, American Society of Clinical Oncology; http://meeting.jco.org/cgi/content/abstract/24/18_suppl/14151;

cf. *Houston Chronicle*, July 11, 2005, In Cancer fight, a spice brings hope to the table. Available from http://pancreaticalliance.org/news/july2005.html

40. A. C. Bharti, S. Shishodia, J. M. Reuben, D. Weber, R. Alexanian, S. Raj-Vadhan ETALN. Donato, and B. B. Aggarwal, Nulear factor-kappaB and STAT3 are constitutively active in CD138+ cells derived from multiple myeloma patients, and suppression of these transcription factors lead to apoptosis. *Blood* **103**(8), 3175–3184 (2004).
41. S. Singh and B. B. Aggarwal, Activation of transcription factor NF-κB is suppressed by curcumin (diferuloylmethane). *J Biol Chem* **270**(42), 24,995–25,000 (1995).
42. S. M. Plummer, K. A. Holloway, A. Karen, M. M. Manson, R. J. L. Munks, A. Kaptein, S. Farrow, and L. Howells, Inhibition of cyclo-oxygenase 2 expression in colon cells by the chemopreventive agent curcumin involves inhibition of NF-κB activation via the NIK/IKK signaling complex. *Oncogene* **18**(44), 6013–6020 (1999).
43. C. E. Eberhart and R. N. Dubois, Eicosanoids and the gastrointestinal tract. *Gastroenterology* **109**(1), 285–301 (1995).
44. C. Thiemermann, The spice of life: curcumin reduces the mortality associated with experimental sepsis. *Crit Care Med* **34**, 2009–2011 (2006).
45. A. M. Siddiqui, X. Cui, R. Wu, W. Dong, M. Zhou, M. Hu, H. H. Simms, and P. Wang, The anti-inflammatory effect of curcumin in an experimental model of sepsis is mediated by up-regulation of peroxisome proliferator-activated receptor-gamma. *Crit Care Med* **34**, 1874–1882 (2006).
46. M. L. P. S. van Iersel, J. H. T. M. Ploemen, I. Struik, C. van Amersfoort, A. E. Keyzer, J. G. Schefferlie, and P. J. Van Bladeren, Inhibition of glutathione S-transferase activity in human melanoma cells by α,β-unsaturated carbonyl derivatives. Effects of acrolein, cinnamaldehyde, citral, crotonaldehyde, curcumin, ethacrynic acid, and trans-2-hexenal. *Chem-Biol Interact* **102**(2), 117–132 (1996).
47. A. T. Dinkova-Kostova and P. Talalay, Relation of structure of curcumin analogs to their potencies as inducers of Phase 2 detoxification enzymes. *Carcinogenesis* **20**(5), 911–914 (1999).
48. A. T. Dinkova-Kostova, M. A. Massiah, R. E. Bozak, R. J. Hicks, and P. Talalay, Potency of Michael reaction acceptors as inducers of enzymes that protect against carcinogenesis depends on their reactivity with sulfhydryl groups. *Proc Natl Acad Sci USA* **98**(6), 3404–3409 (2001).
49. W. M. Weber, L. A. Hunsaker, S. F. Abcouwer, L. M. Deck, J. Vander, and L. David, Anti-oxidant activities of curcumin and related enones. *Bioorg Med Chem* **13**(11), 3811–3820 (2005).
50. K. M. Youssef and M. A. El-Sherbeny, Synthesis and antitumor activity of some curcumin analogs. *Arch Pharmazie (Weinheim, Germany)* **338**(4), 181–189 (2005).
51. K. M. Youssef, A. M. Ezzo, M. I. El-Sayed, A. A. Hazzaa, A. H. El-Medany, and M. Arafa, Curcumin analogs as anticancer agents: 1) preclinical safety evaluation in mice and rats. 2) Chemopreventive effects in DMH-Induced colon cancer in albino rats model, submitted.
52. B. M. Markaverich, T. H. Schauweker, R. R. Gregory, M. Varma, F. S. Kittrell, D. Medina, and R. S. Rajender, Nuclear type II sites and malignant cell proliferation: Inhibition by 2,6-bisbenzylidenecyclohexanones. *Cancer Res* **52**(9), 2482–2488 (1992).
53. B. K. Adams, E. M. Ferstl, M. C. Davis, M. Herold, S. Kurtkaya, R. F. Camalier, M. G. Hollingshead, G. Kaur, E. A. Sausville, F. R. Rickles, J. P. Snyder, D. C. Liotta, and M. Shoji, Synthesis and biological evaluation of novel curcumin analogs as anti-cancer and anti-angiogenesis agents. *Bioorg Med Chem* **12**(14), 3871–3883 (2004).

54. R. J. Anto, J. George, K. V. Dinesh Babu, K. N. Rajasekharan, and R. Kuttan, Antimutagenic and anticarcinogenic activity of natural and synthetic curcuminoids. *Mutat Res* **370**(2), 127–131 (1996).
55. S. M. McElvain and R. E. McMahon, Piperidine derivatives. XXI. 4-Piperidone, 4-piperidinol, and certain of their derivatives. *J Am Chem Soc* **71**, 901–906 (1949).
56. B. K. Adams, J. Cai, J. Armstrong, M. Harold, Y. J. Lu, A. Sun, J. P. Snyder, D. C. Liotta, D. P. Jones, and M. Shoji, EF24, a novel synthetic curcumin analog, induces apoptosis in cancer cells via a redox-dependent mechanism. *Anti-cancer Drugs* **16**(3), 263–275 (2005).
57. C. Syung-ai, A. L. Kumari, and A. Khar, Effect of curcumin on normal and tumor cells: Role of glutathione and bcl-2. *Mol Cancer Ther* **3**, 1101–1108 (2004).
58. A. Laurent, C. Nicco, C. Chéreau, C. Goulvestre, J. Alexandre, A. Alves, E. Lévy, F. Goldwasser, Y. Panis, O. Soubrane, B. Weill, and F. Batteux, Controlling tumor growth by modulating endogenous production of reactive oxygen species. *Cancer Res* **65**, 948–956 (2005).
59. R. M. Kluck, E. Bossy-Wetzel, D. R. Green, and D. D. Newmeyer, The release of cytochrome c from mitochondria: a primary site for Bcl-2 regulation of apoptosis. *Science* **275**(5303), 1132–1136 (1997).
60. T. Kuwana and D. D. Newmeyer, Bcl-2-family proteins and the role of mitochondria in apoptosis. *Curr Opin Cell Biol* **15**(16), 691–699 (2003).
61. H. Zou, W. J. Henzel, X. Liu, A. Lutschg, and X. Wang, Apaf-1, a human protein homologous to *C. elegans* CED-4, participates in cytochrome c-dependent activation of caspase-3. *Cell* **90**(3), 405–413 (1997).
62. N. A. Thornberry and Y. Lazebnik, Caspases: Enemies within, *Science* **281**(5381), 1312–1316 (1998).
63. P. Li, D. Nijhawan, I. Budihardjo, S. M. Srinivasula, M. Ahmad, E. S. Alnemri, and X. Wang, Cytochrome c and dATP-dependent formation of Apaf-1/caspase-9 complex initiates an apoptotic protease cascade. *Cell* **91**(4), 479–489 (1997).
64. M. H. Pan, W. L. Chang, S. Y. Lin-Shiau, C. T. Ho, and J. K. Lin, Induction of apoptosis by garcinol and curcumin through cytochrome c release and activation of caspase in human leukemia HL-60 cells. *J Agric Food Chem* **49**(3), 1464–1474 (2001).
65. D. Morin, S. Barthelemy, R. Zini, S. Labidalle, and J. Tillement, Curcumin induces the mitochondrial permeability transition pore mediated by membrane protein thiol oxidation. *FEBS Lett* **495**(1–2), 131–136 (2001).
66. R. J. Anto, A. Mukhopadhyay, K. Denning, and B. B. Aggarwal, Curcumin (diferuloylmethane) induces apoptosis through activation of caspase-8, BID cleavage and cytochrome c release: Its suppression by ectopic expression of Bcl-2 and Bcl-xl. *Carcinogensis* **23**(1), 143–150 (2002).
67. L. Ghibelli, S. Coppola, G. Rotilio, E. Lafavia, V. Maresca, and M. R. Ciriolo, Non-oxidative loss of glutathione in apoptosis via GSH extrusion. *Biochem Biophys Res Commun* **216**(1), 462–469 (1995).
68. S. Tan, Y. Sagara, Y. Liu, P. Maher, and D. Schubert, The regulation of reactive oxygen species production during programmed cell death. *J Cell Biol* **141**(16), 1423–1432 (1998).
69. H. Fu, S. Thomas, D. C. Liotta, and J. P. Snyder, in preparation.
70. A. Brown, H. Shim, and J. P. Snyder, in preparation.
71. A. Sun, S. Mao, Y. Lu, M. Shojii, D. C. Liotta, and J. P. Snyder, in preparation.

72. T. Ouchi, E. Yamabe, K. Hara, M. Hirai, and Y. Ohya, Design of attachment type of drug delivery systems by complex formation of avidin with biotinyl drug model and biotinyl saccharide, *J Cont Release* **94**, 281–291 (2004).
73. G. Schoellmann and E. Shaw, Direct evidence for the presence of histidine in the active center of chymotrypsin. *Biochemistry* **2**, 252–255 (1963).
74. N. S. Callander, N. Varki, and L. V. Rao, Immunohistochemical identification of tissue factor in solid tumors. *Cancer* **70**(5), 1194–1201 (1992).
75. J. Contrino, G. Hair, D. L. Kreutzer, and F. R. Rickles, *In situ* detection of tissue factor in vascular endothelial cells: Correlation with the malignant phenotype of human breast disease. *Nature Med* **2**(2), 209–215 (1996).
76. C. B. Hansen, C. Pyke, L. C. Petersen, and L. V. M. Rao, Tissue factor-mediated endocytosis, recycling, and degradation of factor VIIa by a clathrin-independent mechanism not requiring the cytoplasmic domain of tissue factor. *Blood* **97**(6), 1712–1720 (2001).
77. A. Sun, M. Shoji, Y. J. Lu, D. C. Liotta, and J. P. Snyder, Synthesis of EF24-tripeptide chloromethylketone: A novel curcumin-related anticancer drug delivery system. *J Med Chem* **49**(11), 3153–3158 (2006).
78. M. Shoji, A. Sun, W. Kisiel, Yang J. Lu, H. Shim, B. E. McCarey, C. Nichols, E. T. Parker, J. Pohl, A. R. Alizadeh, C. Mosley, D. C. Liotta, and J. P. Snyder, submitted.
79. C. Kettner and E. Shaw, Synthesis of peptides of arginine chloromethyl ketone. Selective inactivation of human plasma kallikrein. *Biochemistry* **17**(1), 4778–4784 (1978).
80. S. A. Buhrow, J. M. Reid, L. Jia, M. Shojii, J. P. Snyder, D. C. Liotta, and M. M. Ames. AACR abstract (2005); part of an NCI subcontract to the Mayo Clinic under the Rapid Access to NCI Discovery (RAND) program sponsored by the NCI.

ANTIOXIDANT AND ANTI-INFLAMMATORY PROPERTIES OF CURCUMIN

Venugopal P. Menon and Adluri Ram Sudheer

Abstract: Curcumin, a yellow pigment from *Curcuma longa*, is a major component of turmeric and is commonly used as a spice and food-coloring agent. It is also used as a cosmetic and in some medical preparations. The desirable preventive or putative therapeutic properties of curcumin have also been considered to be associated with its antioxidant and anti-inflammatory properties. Because free-radical-mediated peroxidation of membrane lipids and oxidative damage of DNA and proteins are believed to be associated with a variety of chronic pathological complications such as cancer, atherosclerosis, and neurodegenerative diseases, curcumin is thought to play a vital role against these pathological conditions. The anti-inflammatory effect of curcumin is most likely mediated through its ability to inhibit cyclooxygenase-2 (COX-2), lipoxygenase (LOX), and inducible nitric oxide synthase (iNOS). COX-2, LOX, and iNOS are important enzymes that mediate inflammatory processes. Improper upregulation of COX-2 and/or iNOS has been associated with the pathophysiology of certain types of human cancer as well as inflammatory disorders. Because inflammation is closely linked to tumor promotion, curcumin with its potent anti-inflammatory property is anticipated to exert chemopreventive effects on carcinogenesis. Hence, the past few decades have witnessed intense research devoted to the antioxidant and anti-inflammatory properties of curcumin. In this review, we describe both antioxidant and anti-inflammatory properties of curcumin, the mode of action of curcumin, and its therapeutic usage against different pathological conditions.

1. CURCUMIN: THE SPICE OF LIFE–UNLOCKING THE SECRETS OF CURCUMIN

–*"Imagine if the key to disease prevention was as close as your kitchen shelf. It's not the product of someone's imagination, but the product of years of medical research. Scientists are beginning to take notice of a well-known spice as a potent new preventive therapy against disease, especially cancer—John C. Martin,* LE Magazine, *September 2001"*

More than one billion people consume curcumin regularly in their diets. Curcumin has long been used in Eastern medicine and is gaining attention in Western medicine, not only as a nonsteroidal anti-inflammatory drug (NSAID) but also for its chemopreventive properties. Essentially, curcumin is believed to possess generalized protective properties.

Curcumin, a yellow pigment from *Curcuma longa*, is a major component of turmeric and is commonly used as a spice and food-coloring material. It exhibits anti-inflammatory,[1] antitumor, and antioxidant[2] properties. Curcumin is a low-molecular-weight polyphenol, first chemically characterized in 1910, with the molecular formula of $C_{21}H_{20}O_6$. It is generally regarded as the most active constituent of and comprises 2–8% of most turmeric preparations. It has long been used as the yellow spice in Indian food and as a naturally occurring medicine for the treatment of inflammatory diseases.[3]

The desirable preventive or putative therapeutic properties of curcumin have also been considered to be associated with its antioxidant property.[4] Because free radical-mediated peroxidation of membrane lipids and oxidative damage of DNA and proteins are believed to be associated with a variety of chronic pathological complications such as cancer, atherosclerosis, neurodegenerative diseases, and aging,[5] curcumin is thought to play a vital role against oxidative-stress-mediated pathological conditions. Hence, the past few decades have witnessed intense research devoted to the antioxidant activity of curcumin. Before pointing out the potential antioxidant property of curcumin, it is worthwhile to outline the role of free radicals and antioxidants in health and disease.

2. ROLE OF FREE RADICALS IN HEALTH AND DISEASE

It has been nearly 50 years since Denham Harman[6] suggested that free radicals produced during aerobic respiration cause cumulative oxygen damage, resulting in aging and death.

Oxygen is an essential molecule for all aerobic forms; however, oxygen plays univalent roles. Although oxygen is indispensable for all cells for chemical energy production (ATP), it is also often transformed into highly reactive forms: reactive oxygen species (ROS), which are often very toxic to the cells.[7,8] Approximately 2% of the oxygen reduced by the mitochondria then forms superoxide ($O_2^{-\cdot}$) or the dismutation product H_2O_2. Superoxide and peroxide reacts with metal ions (Heiber-Weiss and Fenton's reactions) to promote additional radical generation, particularly with the generation of hydroxyl radicals. The hydroxyl radical reacts with all components of the cell, including lipid membrane, DNA and proteins.[9]

Nitric oxide (NO) has an unpaired electron and is therefore a free-radical species. It is a short-lived, lipophilic molecule generated from L-arginine by NO synthase (NOS). NO is involved physiologically in vasorelaxation, neurotransmission, inhibition of platelet aggregation, immune defense, and intracellular signaling. However, NO reacts with $O_2^{-\cdot}$ to form peroxynitrite ($ONOO^-$), which is a powerful oxidant. NO bioactivity is related to the production of many reactive intermediates, but many of these reactive nitrogen species (RNS) are capable of damaging DNA or hindering DNA repair.[10] It is now beyond doubt that oxidants are generated *in vivo* and can cause significant damage to cells.

When an imbalance occurs between oxidants and defense systems, in favor of oxidants, oxidative stress occurs. This oxidative stress in cells results in severe metabolic dysfunctions, including loss of cell integrity, enzyme function, genomic stability, and so forth, which ultimately lead to pathogenesis of many human diseases (e.g., inflammation, ischemia, atherosclerosis, arthritis, cancer, Parkinson's disease, Alzheimer's disease, and so forth).

3. ANTIOXIDANTS: WHY ARE THEY NEEDED?

To deal with the threat of oxidant-induced damage, biological antioxidants were evolved. Cells are equipped with an impressive repertoire of antioxidant enzymes, as well as small antioxidant molecules, the later being mostly ingested from fruits and vegetables. The antioxidant defenses include the following:

1. Superoxide dismutase (SOD), which hastens the dismutation of $O_2^{\cdot-}$ to H_2O_2, catalase, and glutathione peroxidase (GPx), which converts H_2O_2 to water
2. Hydrophilic radical scavengers such as ascorbate, urate, and glutathione (GSH)
3. Lipophilic radical scavengers such as tocopherols, flavonoids, carotenoids, and ubiquinol
4. Enzymes involved in the reduction of oxidized form of small-molecule antioxidants (GSH reductase and ascorbate reductase) or responsible for the maintenance of protein thiols (thioredoxin reductase)

Antioxidant systems are complex and act in concert to decrease ROS load. It also helps to divert ROS to other reaction pathways that form less reactive products, to selectively inactivate (in redox terms) transition metal ions, and, when all of these fails, to provide sacrificial molecules that act as a replicable or recyclable "buffer" to absorb oxidative hits and excess energy.[11]

Human antioxidant defenses are effective, but they are not infallible and oxidative damage to key biological sites occurs, accumulates with age, and contributes to senescence and age-related disease. This means that oxidative stress, which is an oxidant:antioxidant imbalance in favor of oxidation, remains a real and constant threat.

The evolution of our endogenous antioxidant system has not progressed beyond the breakeven point of cost-effectiveness. This has led to the attention of dietary antioxidants. Wood and Brooks[12] suggested that "we are what we ate," and although many geographical and environmental factors undoubtedly determined the evolutionary development of our species, our future health as individuals might well depend on what we eat today.

4. CURCUMIN: ANTIOXIDANT MECHANISM

The antioxidant mechanisms of curcumin have recently been the focus of interest of free-radical chemists and biologists.

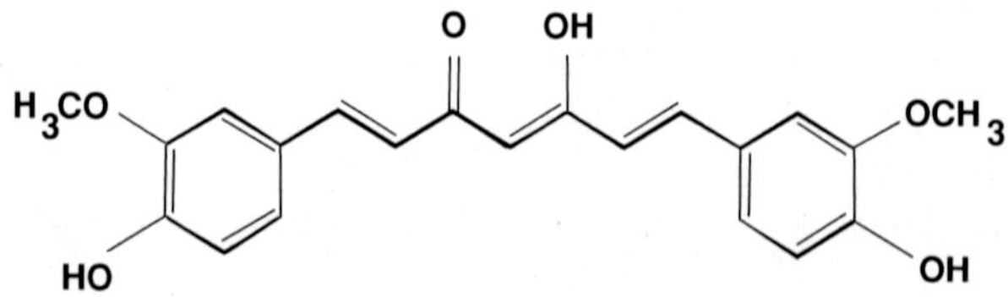

Figure 1. Curcumin.

Curcumin is known to protect biomembranes against peroxidative damage. Peroxidation of lipids is known to be a free-radical-mediated chain reaction, leading to the damage of the cell membranes, and the inhibition of peroxidation by curcumin is mainly attributed to the scavenging of the reactive free radicals involved in the peroxidation. Most of the antioxidants have either a phenolic functional group or a β-diketone group. Curcumin is an unique antioxidant, which contains a variety of functional groups, including the B-diketo group, carbon–carbon double bonds, and phenyl rings containing varying amounts of hydroxyl and methoxy substituents (Figure 1).[13]

The central argument is whether the phenolic or the central methylenic hydrogen in the heptadienone moiety is responsible for its antioxidant activity. Jovanovic and collaborators[14] concluded that curcumin is a superb H-atom donor by donating the H-atom from the central methylenic group rather than from the phenolic group in acidic and neutral aqueous and acetonitrile solutions. On the other hand, Barclay et al.[15] proposed that curcumin is a classical phenolic chain-breaking antioxidant, donating H-atoms from the phenolic group. Priyadarsini et al.[16] have also claimed that the phenolic group is essential for the free-radical-scavenging activity and that the presence of the methoxy group further increased the activity.

Theoretical calculations by the density functional theory (DFT) demonstrated that the enol form of curcumin is significantly more stable than the diketo form and that the bond dissociation enthalpy (BDE) of the phenolic O:H bond is significantly lower than the BDE of the central O:H bond, suggesting that the hydrogen atom abstraction takes place in the phenolic group.[13,16,17] It was also pointed out that the relative contribution of the phenolic group and the central methylenic group on the antioxidant activity depends on the activity of attacking radical and the reaction medium.[13,18] Litwinienko and Ingold[19] recently compared the rate constants of the reaction of 1,1-Diphenyl-2-picrylhydrazyl (DPPH) radical with curcumin in ionizing solvents and nonionizing solvents and resolved the curcumin antioxidant controversy by the mechanism of sequential proton loss electron transfer (SPLET); that is, in solvents that support ionization, curcumin reacts with electrophilic radicals initially at ionized keto-enol moiety and the resulting neutral radicals lose a phenolic photon, thus yielding the same phenoxyl radical, as would have been formed by H-atom transfer (HAT) from the phenolic hydroxyl group of the curcumin anion to the radicals. However, in nonionizing solvent, the SPLET mechanism cannot occur and the reactions involve only HAT from a phenolic hydroxyl group of the neutral curcumin to the radical.[19]

Another important mechanism is that curcumin can degrade into trans-6-(4'-hydroxy-3'-methoxyphenyl)-2,4-dioxo-5-hexanal, ferulic acid, ferruloylmethane, and vanillin at basic pH within 30 min.[3] Among these, ferrulic acid and vannilin are well-established antioxidants.

5. ANTI-INFLAMMATORY PROPERTY OF CURCUMIN

Curcumin was found to have a miraculous power in anti-inflammatory response. The natural anti-inflammatory activity of curcumin is on a par with steroidal drugs and nonsteroidal drugs as indomethacin and phenylbutazone, which have dangerous side effects. Its anti-inflammatory property appears to be mediated through the inhibition of induction of COX-2, LOX, iNOS and production of cytokines such as interferon-γ and tumor necrosis factor, and activation of transcription factors like NF-κB, and AP-1.

5.1. Effect of Curcumin on Cyclooxygenases and Lipoxygenases

The anti-inflammatory properties of curcumin have been attributed, at least in part, to suppression of prostaglandins (PGs) synthesis.[20] The involvement of PGs and other eicosanoids in the development of human cancer has been known for over two decades. Importantly, an increase in PG synthesis might influence tumor growth in human beings and experimental animals, and numerous studies have illustrated the effect of PG synthesis on carcinogen metabolism, tumor cell proliferation, and metastatic potential.[21] Cyclooxygenase (COX) is a key enzyme responsible for the conversion of arachidonic acid to PGs. It consists of two different isoforms, designated COX-1 and COX-2. COX-1 is a constitutive isoform present in most tissues and is generally regarded as a "housekeeping" enzyme[22] and its inhibition results in serious effects such as peptic ulceration or impairment of renal blood flow. In contrast, COX-2 is constitutively expressed only in brain and spinal cord tissue and it can also be induced in a wide variety of normal tissues by the hormones of ovulation and pregnancy, cytokines, growth factors, oncogenes, and tumor promoters.[23] COX-2 overexpression has been implicated in the carcinogenesis of tumors of the colon, rectum, breast, head and neck, lung, pancreas, stomach, and prostate.[24] There is growing evidence that inhibitors of COX-2 activity are useful for treating inflammation and preventing or treating cancer.[25] Therefore, agents that interfere with the signaling mechanisms governing the transcription of COX-2 should also inhibit inflammation and tumorigenesis. Further investigations suggest that arachidonic acid (AA) metabolites derived from lipoxygenase (LOX) pathways play an important role in growth-related signal transduction, implying that intervention through these pathways should be useful for arresting cancer progression.

An expanding body of evidence suggests that curcumin inhibits the expression of COX-2. Kawamori et al.[26] have demonstrated that dietary curcumin significantly inhibits phospholipase A2 in colonic mucosa and tumors, leading to the release of arachidonic acid from phospholipids, alters COX and LOX activities, and modifies

PGE_2 levels. Unlike selective COX-2 inhibitors, which inhibit the catalytic activity of the COX enzyme, curcumin decreases COX-2 expression at the transcriptional level.[27] Several lines of evidence also indicate that the mechanism of action of curcumin is not limited to PG inhibition. They have also observed that dietary curcumin inhibits LOX activity and the production of the LOX metabolites in the colonic mucosa and in tumors. LOX metabolites have also been shown to promote tumor cell adhesion, stimulate the spreading of tumor cells, and augment metastatic potential.[28]

In one study, Zhang and colleagues[27] from the Cornell University campus in New York City, exposed gastrointestinal cells to two known tumor promoters: bile acids (BA) and phorbol esters (PMA). The team found COX-2 to be induced in several of the cell lines, accompanied by a 10-fold increase in the synthesis of inflammatory-causing PGE_2. However, dose-dependent treatment of the cells with curcumin suppressed both BA- and PMA-mediated induction of COX-2 protein, genetic COX-2 expression (as measured by mRNA), and the synthesis of PGE_2. Most impressive, however, was the discovery that curcumin directly inhibited the enzymatic activity of COX-2.

An additional study presented at the 1999 American Association for Cancer Research (AACR) conference also examined the pain-relieving properties of curcumin. Researchers at the University of California, San Diego and the Veterans Administration Medical Center, San Diego[29] investigated whether curcumin could suppress COX-2 expression in human colon cancer cells. After exposing such cells to curcumin, the researchers found the compound not only inhibited cell growth but also reduced the expression of COX-2 mRNA in a time- and dose-dependent manner.

Therefore, curcumin would appear to be a safe, natural COX-2 inhibitor in humans, given its safety profiles and demonstrated anti-inflammatory activity.

5.2. Effect of Curcumin on Inducible Nitric Oxide Synthase

Another enzyme that plays a pivotal role in mediating inflammation is inducible nitric oxide synthase (iNOS). iNOS catalyzes the oxidative deamination of L-arginine to produce NO, a potent pro-inflammatory mediator. NO has multifaceted roles in mutagenesis and carcinogenesis.[30] In addition to acting as an initiator of carcinogenesis, NO is involved in the promotional stage of tumorigenesis or neoplastic transformation. NO is also known to affect tumor progression by regulating the angiogenesis, possibly by stimulating the production of vascular endothelial growth factor (VEGF).[31] NO reacts rapidly with superoxide anion to produce the extremely powerful oxidant peroxynitrite ($ONOO^-$).[32] Peroxynitrite can cause various DNA modifications and other types of cellular injury, contributing to genotoxicity or initiation of multistage carcinogenesis. One of the secondary processes that might follow DNA damage by NO or peroxynitrite includes activation of the tumor suppressor gene *p53* or the DNA repair enzyme poly(ADP-ribose)polymerase (PARP).[33] It is known that activation of *p53* or PARP is often associated with apoptotic cell death. There are numerous reports on the NO- or peroxynitrite-induced apoptosis.[34] Increased expression of iNOS and/or its catalytic activity has been observed in several human tumor tissues and also in chemically induced animal tumors as well as in inflammatory disorders.[35–37] Human breast tumor biopsies in higher grades exhibited elevated levels of iNOS, compared with those in lower-grade ones.[38] Thus, aberrant or excessive expression of iNOS, as in the case of COX-2, is considered to be implicated in the pathogenesis of cancer, and compounds that can selectively inhibit abnormal expression of iNOS can act as a potential candidate for chemoprevention. iNOS has been shown to be involved in the regulation of COX-2 and, hence, the subsequent production of pro-inflammatory PGs.[39]

In addition to COX-2, iNOS also appears to be a target for the anti-inflammatory effect of curcumin. Curcumin is reported to inhibit the NO production and expression of iNOS protein and mRNA in RAW 264.7 cells stimulated with lipopolysaccharides (LPSs) or interferon-γ.[40] Downregulation of the iNOS gene by curcumin in RAW 264.7 cells might be attributed to suppression of c-Jun/AP-1 activation, because it is a known fact that the consensus AP-1-binding sequence is present in the promoter region of the iNOS gene and it can be attenuated by curcumin.[40] Chan et al.[41] have reported that curcumin can inhibit the iNOS gene expression in isolated BALB/c mouse peritoneal macrophages and in the livers of LPS-injected mice.

5.3. Effect of Curcumin on Nuclear Factor-κB

One of the most ubiquitous eukaryotic transcription factors that regulate expression of genes involved in controlling cellular proliferation/growth, inflammatory responses, cell adhesion, and so forth is nuclear factor-κB (NF-κB).[42,43] The functionally active NF-κB exists mainly as a heterodimer consisting of subunits of the Rel family (e.g., Rel A or p65, p50, p52, c-Rel, v-Rel, and Rel B), which is normally sequestered in an inactive cytoplasmic complex by binding to an inhibitory protein, IκB. Exposure of cells to such external stimuli as mitogens, inflammatory cytokines, ultraviolet radiation, ionizing radiation, viral proteins, bacterial LPSs, and ROS causes rapid phosphorylation of IκB with subsequent degradation by proteosomes. Dissociation of IκB from NF-κB allows the activated free dimer to translocate to the nucleus, where it induces, through binding the cis-acting κB element, the transcription of a large variety of target genes that normally encode cytokines, cell adhesion molecules, growth factors, and so forth. The transcriptional activity of NF-κB is regulated via an elaborate series of intracellular signal transduction events in response to external stimuli.[30] In addition to its central roles in mediating inflammation, NF-κB is important in control via cell proliferation, oncogenesis, and cell transformation.[44] Thus, aberrant constitutive activation of NF-κB/Rel has been observed in an array of human and experimentally induced tumors and transformed cells in culture. There is increasing evidence that constitutive activation of this transcription factor is associated with the proliferation and survival of certain tumor cells and causes resistance to apoptosis.[45]

The data from experimental studies have demonstrated that curcumin inhibits the activation of NF-κB in different cancer cell lines.[46] It has been found that oxidative stress activates NF-κB DNA-binding activity. Because curcumin has been known as an antioxidant, its inhibitory effects on oxidative stress might be mediated through the suppression of NF-κB DNA-binding activity. It has been reported that curcumin inhibited IKB kinase (IKK), suppressed both constitutive and inducible NF-κB activation and potentiated tumor necrosis factor (TNF)-induced apoptosis. Curcumin treatment reduced the amount of phosphorylated IKK, which ultimately prevents the translocation of NF-κB to the nucleus. Curcumin also showed strong antioxidant and anticancer properties through regulating the expression of genes that require the activation of activator protein (AP1) and NF-κB.[47]

5.4. Effect of Curcumin on Tumor Necrosis Factor

Tumor necrosis factor (TNF) has been shown to mediate tumor initiation, promotion, and metastasis.[48] Moore et al.[49] have reported that TNF-deficient mice have been shown to be resistant to skin carcinogenesis. The induction of pro-inflammatory genes by TNF has been linked to most diseases. The pro-inflammatory effects of TNF are primarily due to its ability to activate NF-κB. Almost all cell types, when exposed to TNF, activate NF-κB, leading to the expression of inflammatory genes. These include COX-2, LOX-2, cell adhesion molecules, inflammatory cytokines, chemokines, and iNOS. TNF has been found to be a growth factor for most tumor cells.[50]

Because of the critical role of TNF in mediating tumorigenesis, agents that can suppress TNF activity have the potential for therapy of TNF-linked diseases. Curcumin was found to have a spellbound effect in the suppression of TNF production.[51] The constitutive activation of NF-κB in mantle cell lymphoma (MCL) cells is due to autocrine expression of TNF. TNF mRNA is constitutively expressed in the MCL cell lines; curcumin inhibits the expression of both TNF mRNA and TNF protein in MCL cell lines.[51] Suppression of TNF by curcumin led to inhibition of NF-κB and cell proliferation, as was the case when TNF secretion was neutralized using anti-TNF antibody.[51]

Thus, curcumin exerts a protective role against inflammatory diseases through the suppression of COX, LOX, iNOS, NF-κB, TNF and some other inflammatory mediators.

6. THERAPEUTIC USE OF CURCUMIN

6.1. Cancer

Curcumin has proved its credentials as a wonderful chemopreventive agent against a variety of cancers. Oxidative stress induced by ROS has been linked to tumor promotion in mouse skin and other tissues.[52–54] Hydrogen peroxide promotes the transformation of chemically initiated mouse epidermal cells.[52] When different types of tumor promoter were applied topically to mouse skin, there was a distinct increase in the production of hydrogen peroxide in the epidermis, which correlated with their promoting potential.[55] Superoxide anion radicals were also formed in keratinocytes stimulated with the tumor promoter such as 12-*O*-tetradecanoylphorbol-13-acetate (TPA).[56] Further support for the association between pro-oxidant status and tumor promotion derives from the observation that many structurally unrelated antioxidants and radical scavengers exert antipromotional activity.[52,53] Exogenous superoxide dismutase (SOD) inhibited the promotion by TPA of JB6 mouse epidermal cell transformation, providing additional evidence for a critical role of ROS in tumor promotion.[57] Similarly, the lipophilic biomimetic SOD cupric 3,5-diidopropylsalicylic acid inhibited tumor promotion in mouse epidermis.[58] Ornithine decarboxylase (ODC), a rate-limiting enzyme in polyamine biosynthesis, has been utilized as a biochemical marker for tumor promotion. ODC induction was found to be suppressed by SOD and catalase in murine mammary tumor cells and by butylated hydroxytoluene in mouse epidermal cells, suggesting the intermediacy of Reactive oxygen intermediates (ROIs) in the tumor promotion.[52]

Persistent, local inflammation has been considered to contribute to multistage carcinogenesis. ROS produced during the inflammatory tissue damage can trigger a series of reactions responsible for malignant transformation, particularly in the stage of promotion.[52]. There is accumulating evidence that ROS influence the intracellular signaling cascades mediating cell proliferation. Activity or expression of several protein kinases have been shown to be regulated by the pro-oxidant state of cells. ROS are typical by-products of eicosanoid metabolism.[59] ROS are released from the cells of the inflammatory skin infiltrate. A correlation exists between the ability of a compound to induce PG release *in vitro* and its ability to promote papilloma formation in mouse skin *in vivo*. Verma et al.[60] suggests that PGs play a crucial role in the induction of ODC activity and mouse skin tumor promotion by TPA. Both ROI generation by inflammatory cells and skin tumor promotion are efficiently inhibited by protease inhibitors, indicating an interrelationship between the two processes.[54]

Chemopreventive properties of curcumin have been extensively investigated and well documented.[61,62] One of the most plausible mechanisms underlying the chemopreventive effects of curcumin involves suppression of tumor promotion. Thus, topical application of curcumin strongly inhibited TPA-induced inflammation, hyperplasia, proliferation, ODC activity, ODC mRNA expression, generation of ROIs, oxidized DNA base modification, and papilloma formation in mouse skin.[62–63] Curcumin inhibited COX-2 and LOX activities in TPA-treated mouse epidermis.[20] Treatment of several human gastrointestinal cell lines with curcumin suppressed expression of COX-2 protein and mRNA, PGE_2 production, and AP-1 DNA binding induced by TPA or chenodeoxycholate.[27] Suppression of TPA-induced activation of c-Jun/AP-1 in cultured NIH3T3 cells has been proposed to be responsible for the antitumor-promoting activity that curcumin retains.[64]

In conclusion, the anticancer potential of curcumin stems from its ability to suppress proliferation of a wide variety of tumor cells, downregulate transcription factors NF-κB, AP-1, and Egr-1, downregulate the expression of COX-2, LOX, iNOS, MMP-9, uPA, TNF, chemokines, cell surface adhesion molecules, and cyclin D1, downregulate growth factor receptors (such as EGFR and HER2), and inhibit the activity of c-Jun N-terminal kinase, protein tyrosine kinases, and protein serine/threonine kinases. It also influences the free-radical production during the activation of carcinogen and helps in detoxification of carcinogens.

6.2. Atherosclerosis

The most common form of heart disease is atherosclerosis. Atherosclerosis involves the deposition of fatty substances, cholesterol, complex carbohydrates and fibrin (a clotting material in the blood), and so forth in the inner lining of the major artery. The deposition that results (referred to as plaque) can partially or totally block the flow of blood through the artery. This can lead to the formation of a blood clot (thrombus) on the surface of the plaque. If either of these events occurs and blocks the coronary artery, a heart attack might result. Some controllable mechanisms that are involved in the development of atherosclerosis are oxidation of low-density lipoprotein-cholesterol (LDL-C), abnormal platelet aggregation, and inflammation.[65,66] Curcumin has gained importance because of its antioxidant and antiplatelet aggregating qualities. Curcumin's ability to control platelet aggregation appears directly to be related to thromboxane inhibition (a promoter of aggregation) and an increase in prostacyclin activity, an inhibitor of aggregation.[67,68] Curcumin, being a powerful antioxidant, quenches free radicals, thereby decreasing the cellular damage. It also helps in reducing blood lipid levels, particularly cholesterol. Rats fed 0.1% curcumin, along with a cholesterol diet, had about one-half of the blood cholesterol as rats fed equal amounts of cholesterol but without curcumin.[66]

Curcumin, with its potent anti-inflammatory activity, reduces the inflammation, promotes fibrinolysis, and inhibits the leukotriene formation,[69,70] which are important steps in the prevention of atherosclerosis.

6.3. Aging

A number of theories have been proposed to explain the nature of aging and one such is the free-radical theory.[71] According to the free-radical theory of aging, these very reactive species, produced continuously during normal metabolism, eventually accumulates, damaging DNA and other macromolecules. This is due to progressive defects in the defense systems against reactions that generate free radicals. The result is the appearance of degenerative lesions and cellular death. Then the organism ages and, finally dies. Different observations support the hypothesis of the major role of free radicals in aging. The potential life span of a species is determined genetically. However, it can be altered by environmental conditions such as diet. Harman[72,73] has demonstrated, for example, that life expectation of many species is increased by 20% by adding antioxidants to the diet. It has also been shown that animals, which live longer or with longer life spans, present higher SOD levels.[74] Curcumin might have been delaying the aging of human beings with its effective antioxidant property, as it is known that billions of people consume curcumin daily in their diets.

6.4. Neurodegenerative disease

Oxidative stress has been implicated in mechanisms leading to neuronal cell injury in various pathological states of the brain. Alzheimer's disease (AD) is a progressive disorder with cognitive and memory decline, speech loss, personality changes, and synapse loss.[75] Many approaches have been undertaken to understand AD, but the heterogeneity of the etiologic factors makes it difficult to define the clinically most important factor determining the onset and progression of the disease. However, increasing evidence indicates that factors such as oxidative stress and disturbed protein metabolism and their interaction in a vicious cycle are central to AD pathogenesis.

Increasing interest has been focused on identifying dietary compounds that can inhibit, retard, or reverse the multistage pathophysiological events underlying AD pathology. AD, in fact, involves a chronic inflammatory response associated with both brain injury and β-amyloid associated pathology. All of the above evidence suggests that stimulation of various repair pathways by mild stress has significant effects on delaying the onset of various age-associated alterations in cells, tissues, and organisms.

Curcumin has emerged as a strong inducer of the heat shock response. In light of this finding, curcumin supplementation has recently been considered an alternative, nutritional approach to reduce oxidative damage and amyloid pathology associated with AD.

The potential neuroprotective action of curcumin was discovered during a screening of its potential to protect against the adverse effects from high doses of alcohol, which revealed positive results. Since then, studies have indicated potential benefits for AD and Parkinson's disease, based on laboratory animal models; clinical work is now beginning. In addition, studies in animal models of AD indicate a direct effect of curcumin in decreasing the amyloid pathology of AD. As the widespread use of curcumin as a food additive and relatively small short-term studies in humans suggest safety, curcumin is a promising agent in the treatment and/or prevention of AD. Just as in AD, inflammation and oxidative damage play a strong role in the neurodegenerative process of Hodgkin's disease (HD): Oxidative damage helps to degrade nerve cells in the basal ganglia and cerebral cortex and chronic inflammation in the brains of people with HD is believed to play a significant role in the progression of the disease. As shown previously, curcumin was able to reduce inflammation and oxidative damage in mouse models of AD.[76]

6.5. Liver Fibrosis

Liver fibrosis and cirrhosis result from the majority of chronic liver insults and represent a common and difficult clinical challenge of worldwide importance. At present, the only curative treatment for end-stage cirrhosis is transplantation.

Hence, there is a considerable imperative to develop antifibrotic strategies that are applicable to liver fibrosis. Development of liver fibrosis entails major alterations in both the quantity and quality of the extracellular matrix (ECM).[77,78] The remodeling of ECM both in normal and pathological conditions is controlled by a group of enzymes called matrix metalloproteinases (MMPs).[79]

Curcumin has been extensively investigated for its hepatoprotective potential. Normally, oxidative stress and inflammation play an important role in the development of alcohol and heated polyunsaturated fatty acids (PUFAs) and CCL_4-induced liver fibrosis.[80] Curcumin, due to its effective antioxidant and anti-inflammatory properties, inhibits liver fibrosis.[81] Curcumin is also reported to influence the hepatic expression patterns of MMPs.[82] Thus, curcumin emerged as an potent antifibrotic compound.

6.6. Diabetes

Environmental factors play an important role in the etiology and management of diabetes, and antioxidants in food and medicinal plants are potential modulators of diabetes onset, progression, and complications. Among the naturally occurring compounds, curcumin has received the most attention as an antidiabetic compound.

Oxidative stress as a consequence of hyperglycemia and changes in energy metabolism and inflammatory mediators play an important role in the pathophysiology of diabetes in association with depleted cellular antioxidant defense systems and enhanced production of ROS.[82] Oxidative stress associated with hyperglycemia impairs cellular function and alters vascular and neural function. High glucose concentrations promote free-radical production via the following three biochemical pathways: advanced glycation end products (AGEs),[83] protein kinase C activation,[84] and aldose reductase pathway.[85] Another important factor that increases ROS is TNF[86]; it forms a possible connection between obesity and diabetes[87] and has been linked to insulin resistance[88] and diabetic complications.

Curcumin generally improves oxidative status, protects and enhances endogenous defenses, and directly mediates various mechanics of pathology. Many studies have shown that curcumin prevents AGE-induced complications in rats.[89] Curcumin might act by sparing or enhancing the function of the endogenous antioxidants.[90] Antioxidant activities of curcumin might occur synergistically with glucose-lowering activity.[90] The antidiabetic action of curcumin seems to be mediated through (1) stimulation of the pancreas to produce and secrete insulin, (2) interference with dietary glucose absorption, (3) insulin-sparing action of the constituent bioactive compounds, and (4) antioxidant and anti-inflammataory properties of curcumin.

6.7. AIDS

More recently, curcumin has been shown to inhibit HIV replication,[91] and currently it is in clinical trials. Mazumder et al.[92] have shown that curcumin inhibits p24 antigen production and Tat-mediated transcription. They also shown that curcumin inhibits purified HIV-l integrase, suggesting that the anti-HIV activity of curcumin could be due to several mechanisms. Energy minimization studies suggest that the anti-integrase activity of curcumin could be due to an intramolecular stacking of two phenyl rings that brings the hydroxyl groups into close proximity (antioxidant property). The present data suggest that HIV-l integrase inhibition might contribute to the antiviral activity of curcumin. These observations suggest new strategies for antiviral drug development that could be based on curcumin as a lead compound for the development of inhibitors of HIV-l integrase.[92]

6.8. Autoimmune Disease

The immune system has evolved to discriminate self from nonself antigens, thereby protecting the host from microbial pathogens and malignant tumors.[93] Nevertheless, a breakdown in this fundamental immunoregulatory process often results in the development of chronic infectious diseases, malignant tumors, and organ-specific autoimmune diseases. Recent studies have used nutraceuticals (human diets of plant origin, containing many hundreds of biologically active compounds called nutraceuticals) to successfully target organ-specific autoimmune diseases.[94] The pathogenesis of autoimmune diseases involves complex immune mechanisms in which the pro-inflammatory cytokines contribute to manifestation of the diseases.

The anti-inflammatory action of curcumin is associated with its ability to inhibit reactive-oxygen-generating enzymes such as COX, LOX, xanthine dehydrogenase, and iNOS.[95] The involvement of interleukin (IL)-12 in the pathogenesis of rheumatoid arthritis and myocarditis has been well established, and inhibition of IL-12 has decreased the clinical symptoms of these autoimmune diseases.[96] In view of the central role played by IL-12 in Th1 differentiation, the IL-12/Th1 axis has become a novel molecular target in the treatment of Th1 cell-mediated autoimmune diseases.

Many studies suggest that nutraceuticals ameliorate autoimmune diseases by blocking Th1 cell-mediated inflammation and/or by promoting the progenitor cell growth and differentiation in the repair process.

6.9. Psoriasis

Curcumin, by its immune-modulating, anti-inflammatory, and antioxidant properties, shows a favorable effect on a mouse model of psoriasis.[97] Heng et al.[98] showed that topical treatment with curcumin results in resolution of the psoriatic activity as assessed by clinical, histological, and immunological criteria in patients. According to them, this antipsoriatic effect is linked to a curcumin-caused modulation of phosphorylase-kinase (Phk) activity that integrates multiple calcium/calmodulin-dependent signaling pathways. These pathways are coupled to glycogenolysis and ATP-dependent phosphorylation, thus ensuring the required energy supply for cell proliferation and migration.

The effectiveness of curcumin is shown by the fact that the raised Phk activity found in untreated psoriasis, that was decreased by the vitamin D_3 analogue (and indirect inhibitor of Phk) calcipotriol, was even more decreased to near-normal values by the curcumin treatment.

Previous reports suggest additional mechanisms for the antipsoriatic effects of the antioxidants from *C. longa*.[99,100] Thus, *in vitro* exposure of human keratinocytes of the HaCaT line to a hydroalcoholic extract of the rhizome of this plant was as effective for inhibition of the ultraviolet-induced secretion of IL-6 and IL-8 as medium supplementation with betamethasone-17-valerate. This observed downregulation of pro-inflammatory cytokines supports the view that the curcumin might exert a favorable effect on psoriasis-linked inflammation.

Curcumin, by its effective antioxidant and anti-inflammatory properties, was evaluated as a remarkable antipsoriatic compound.

6.10. Drug/Environmental Carcinogen-Induced Toxicity

The two most commonly abused substances in the general population are alcohol and nicotine.

Alcohol is considered to be an important recreational beverage, which is non-toxic at lower concentrations, but it is toxic at higher concentrations. Alcohol-related problems are one of the world's major public health concerns. Alcoholism is seen in all races, ethnic groups, and socioeconomic strata. Oxidative stress plays an important role in the development of alcohol-induced tissue injury.[101] Antioxidant compounds are important in treating this condition. Curcumin received much attention as a hepatoprotective agent. It is well established that curcumin protects experimental rats from chronic alcohol toxicity by its effective antioxidant property.[102]

Nicotine, a major toxic component of cigarette smoke, plays an important role in the development of cardiovascular disease and lung cancer in smokers.[103] Nicotine damages the system through the production of free radicals and inflammatory mediators.[104] Curcumin was found to offer significant protection to the experimental rats from nicotine-induced toxicity.[105]

Benzo(*a*)pyrene [B(*a*)P] is a classical environmental carcinogen present in cigarette smoke and generally regarded to be involved in the development of lung cancer. Curcumin is reported to inhibit the B(*a*)P-induced DNA mutations and thus protects the host organism from the development of lung cancer.[106]

7. CONCLUSION

Curcumin, with its impressive antioxidant and anti-inflammatory properties, was found to be a genuine natural product in treating a wide array of diseases. Its antioxidant property is believed to be due to the presence of different functional groups, including methoxy, phenoxy, and carbon–carbon double bonds in its structure. Its remarkable anti-inflammatory property kept it in the limelight over the decades in treating inflammatory-mediated diseases including cancer, atherosclerosis, diabetes, rheumatoid arthritis, and so forth. Its anti-inflammatory property appears to be mediated through the inhibition of induction of COX-2, LOX, and iNOS and the production of cytokines such as interferon-γ, tumor necrosis factor, and many other transcription factors such as NF-κB.

REFERENCES

1. A. Mukhopadhyay, N. Basu, N. Ghatak, and P. K. Gujral, Antiinflammatory and irritant activities of curcumin analogues in rats. *Agents Actions* **12**, 508–515 (1982).
2. A. C. Reddy and B. R. Lokesh, Effect of dietary turmeric (*Curcuma longa*) on iron induced lipid peroxidation in the rat liver. *Food Chem Toxicol* **32**, 279–283 (1994).
3. R. A. Sharma, A. J. Gescher, and W. P. Steward, Curcumin: The story so far. *Eur J Cancer* **41**, 1955–1968 (2005).
4. C. Kalpana, K. N. Rajasekharan, and V. P. Menon, Modulatory effects of curcumin and curcumin analog on circulatory lipid profiles during nicotine-induced toxicity in Wistar rats. *J Med Food* **8**(2), 246–250 (2005).
5. A. R. Sudheer, C. Kapana, M. Srinivasan, and V P. Menon, Ferulic acid modulates altered lipid profiles and prooxidant/antioxidant status in circulation during nicotine-induced toxicity: A dose-dependent study. *Toxicol Mech Methods* **15**, 1–7 (2005).
6. D. Harman, Aging: A theory based on free radical and radiation chemistry. *J Gerentol* **2**, 298–300 (1956).
7. A. P. Arrigo, Gene expression and thiol redox state. *Free Radical Biol Med* **27**, 936–944 (1983).
8. J. J. Hadad, Antioxidant and prooxidant mechanisms in the regulation of redox (Y)—sensitive factors. *Cell Signal* **14**, 879–897 (1989).
9. B. Halliwell and J. M. C. Gutteridge, *Free Radicals in Biology and Medicine*. Oxford University Press, Oxford, 1989, p. 63
10. J. J. Poderoso, M. C. Carreras, C. Lisdero, N. Riobo, F. Schopfer, and A. Boveris, Nitric oxide inhibits electron transfer and increases superoxide radical production in rat heart mitochondria and submitochondrial particles. *Arch Biochem Biophys* **328**, 85–92 (1996).
11. B. Halliwell and J. M. C. Gutteridge, Antioxidant defenses. In: *Free Radicals in Biology and Medicine*, 3rd ed. Oxford University Press, Oxford, 1999, pp. 105–245.
12. B. Wood and A. Brooks, Human evolution: We are what we ate. *Nature* **400**, 219–220 (1990).
13. J. S. Wright, Predicting the antioxidant activity of curcumin and curcuminoids. *J Mol Struct (Theochem)* **591**, 207–217 (2002).
14. S. V. Jovanovic, S. Steenken, C. W. Boone, and M. G. Simic, H-atom transfer is a preferred antioxidant mechanism of curcumin. *J Am Chem Soc* **121**, 9677–9681 (1999).
15. L. R. C. Barclay, M. R. Vinqvist, K. Mukai, H. Goto, Y. Hashimoto, A. Tokuanga, and H. Uno, The antioxidant mechanism of curcumin: Classical methods are needed to determine antioxidant mechanism and activity. *Org Lett* **2**, 2841–2843 (2000).
16. K. I. Priyadarsini, D. K. Maity, G. H. Naik, M. S. Kumar, M. K. Unnikrishnan, J. G. Satav, and H. Mohan, Role of phenolic O:H and methylene hydrogen on the free

radical reaction and antioxidant activity of curcumin. *Free Radical. Biol Med* **35**, 475–484 (2003).

17. Y. M. Sun, H. Y. Zhang, D. Z. Chen, and C. B. Liu, Theoretical elucidation on the antioxidant mechanism of curcumin: A DFT study. *Org Lett* **4**, 2909–2911 (2002).
18. L. Shen, H. Y. Zhang, and H. F. Ji, Successful application of TD-DFT in transient absorption spectra assignment. *Org Lett* **7**, 243–246 (2005).
19. G. Litwinienko and K. U. Ingold, Abnormal solvent effect on hydrogen atom abstraction: Resolution of the curcumin antioxidant controversy. The role of sequential proton loss electron transfer. *J Org Chem* **64**, 5888–5896 (2004).
20. M. T. Huang, T. Lysz, T. Ferraro, T. F. Abidi, J. D. Laskin, and A. H. Conney, Inhibitory effects of curcumin on in vitro lipoxygenase and cyclooxygenase activities in mouse epidermis. *Cancer Res* **51**, 813–819 (1991).
21. L. Qiao, V. Kozoni, G. J. Tsioulias, M. I. Koutsos, R. Hanif, and S. J. Shiff, Selected eicosaniods increase the proliferation rate of human colon carcinoma cell lines and mouse colonocytes in vivo. *Biochim Biophys Acta* **1258**, 215–223 (1995).
22. C. D. Funk, L. B. Funk, M. E. Kennedy, A. S. Pong, and G. A. FitzGerald, Human platelet/erythroleukemia cell prostaglandin G/H synthase:cDNA cloning, expression and gene chromosomal assignment. *FASEB J* **5**, 2304–2312 (1991).
23. K. Subbaramaiah, N. Telang, J. T. Ramonetti, R. Araki, B. DeVito, B. B. Weksler, and A. J. Dannenberg, Transcription of cyclooxygenase-2 is enhanced in transformed mammary epithelial cells *Cancer Res* **56**, 4424–4429 (1996).
24. H. Y. Fang, T. S. Lin, J. P. Lin, Y. C. Wu, K. C. Chow, and L. S. Wang, Cyclooxygenase-2 in human non-small cell lung cancer. *Eur J Surg Oncol* **29**, 171–177 (2003).
25. J. L. Masferrer, B. S. Zweifel, P. T. Manning, S. D. Hauser, K. M. Leahy, W. G. Smith, P. C. Isakson, and K. Seibert, Selective inhibition of cyclooxygenase-2 in vivo is antiinflammatory and nonulcerogenic. *Proc Natl Acad Sci USA* **91**, 3228–3232 (1994).
26. T. Kawamori, C. V. Rao, K. Seibert, and B. S. Reddy, Chemopreventive activity of celecoxib, a specific cyclooxygenase-2 inhibitor, against colon carcinogenesis. *Cancer Res* **58**, 409–412 (1998).
27. F. Zhang, N. K. Altorki, J. R. Mestre, K. Subbaramaiah, and A. J. Dannenberg, Curcumin inhibits cyclooxygenase-2 transcription in bile acid- and phorbol ester-treated human gastrointestinal epithelial cells. *Carcinogenesis* **20**, 445–451 (1999).
28. J. Hong, M. Bose, J. Ju, J. H. Ryu, X. Chen, and S. Sang, Modulation of arachidonic acid metabolism by curcumin and related beta-diketone derivatives: Effects on cytosolic phospholipase A(2), cyclooxygenases and 5-lipoxygenase. *Carcinogenesis* **25**(9), 1671–1679 (2004).
29. A. Goel, C. R. Boland, and D. P. Chauhan, Specific inhibition of cyclooxygenase-2 (COX-2) expression by dietary curcumin in HT-29 human colon cancer cells. *Cancer Lett* **172**,111–118 (2001)
30. Y. J. Surh, K. S. Chun, H. H. Cha, S. S. Han, Y. S. Keum, K. K. Park, and S. S. Lee, Molecular mechanisms underlying chemopreventive activities of anti-inflammatory phytochemicals: Down-regulation of COX-2 and iNOS through suppression of NF-kB activation. *Mutat Res* **480**, 243–268 (2001).
31. K. Chin, Y. Kurashima, T. Ogura, H. Tajiri, S. Yoshida, and Esumi, Induction of vascular endothelial growth factor by nitric oxide in human glioblastoma and hepatocellular carcinoma cells. *Oncogene* **15**, 437–442 (1997).
32. N. V. Blough and O. C. Zafiriou, Reaction of superoxide with nitric oxide to form peroxynitrite in alkaline aqueous solution. *Inorg Chem* **24**(4), 3502–3505 (1995).

33. C. Szabo and H. Ohshima, DNA damage induced by peroxynitrite: subsequent biological effects. *Nitric Oxide* **1**, 373–385 (1997).
34. Messmer U. K., M. Ankarcrona, P. Nicotera, and B. Brune, p53 expression in nitric oxide-induced apoptosis. *FEBS Lett* **355**, 23–26 (1994).
35. S. R. Goldstein, G. Y. Yang, X. Chen, S. K. Curtis, and C. S. Yang, Studies of iron deposits, inducible nitric oxide synthase and nitrotyrosine in a rat model for esophageal adenocarcinoma. *Carcinogenesis* **19**, 1445–1449 (1998).
36. L. L. Thomsen, D. W. Miles, L. Happerfield, L. G. Bobrow, R. G. Knowles, and S. Moncada, Nitric oxide synthase activity in human breast cancer. *Br J Cancer* **72**, 41–44 (1995).
37. M. Takahashi, K. Fukuda, T. Ohata, and K. Wakabayashi, Increased expression of inducible and endothelial nitric oxide synthases in rat colon tumors induced by azoxymethane. *Cancer Res* **57**, 1233–1237 (1997).
38. D. C. Jenkins, I. G. Charles, L. L. Thomsen, D. W. Moss, L. S. Holmes, S. A. Baylis, P. Rhodes, K. Westmore, P. C. Emson, and S. Moncada, Roles of nitric oxide in tumor growth. *Proc Natl Acad Sci USA* **92**, 4392–4396 (1995).
39. L. M. Landino, B. C. Crews, M. D. Timmons, J. D. Morrow, and L. J. Marnett, Peroxynitrite, the coupling product of nitric oxide and superoxide, activates prostaglandin biosynthesis. *Proc Natl Acad Sci USA* **93**, 15,069–15,074 (1996).
40. I. Brouet and H. Ohshima, Curcumin, an anti-tumor promoter and anti-inflammatory agent, inhibits induction of nitric oxide synthase in activated macrophages. *Biochem Biophys Res Commun* **206**, 533–540 (1995).
41. M. M. Chan, H. I. Huang, M. R. Fenton, and D. Fong, In vivo inhibition of nitric oxide synthase gene expression by curcumin, a cancer preventive natural product with anti-inflammatory properties. *Biochem Pharmacol* **55**, 1955–1962 (1998).
42. F. Chen, V. Castranova, X. Shi, and L. M. Demers, New insights into the role of nuclear factor-kB, a ubiquitous transcription factor in the initiation of diseases. *Clin Chem* **45**, 7–17 (1999).
43. D. Gius, A. Botero, S. Shah, and H. A. Curry, Oxidation/reduction status in the regulation of transcription factors NF-kB and AP-1. *Toxicol Lett* **106**, 93–106 (1999).
44. I. Luque and C. Gelinas, Rel/NF-kB and IkB factors in oncogenesis. *Semin Cancer Biol* **8**, 103–111 (1997).
45. H. Jo, R. Zhang, T. A. McKinsey, J. Shao, R. D. Beauchap, D. W. Ballard, and P. Liang, NF-kB is required for H-ras oncogene-induced abnormal cell proliferation and tumorigenesis. *Oncogene* **19**, 841–849 (2000).
46. S. Singh and B. B. Aggarwal, Activation of transcription factor NF-kB is suppressed by curcumin (diferuloylmethane). *J Biol Chem* **270**, 24,995–25,000 (1995).
47. F. H. Sarkar and Y. Li, Cell signaling pathways altered by natural chemopreventive agents. *Mutat Res* **555**, 53–64 (2004).
48. B. B. Aggarwal, Signalling pathways of the TNF superfamily: A double-edged sword. *Nat Rev Immunol* **3**(9),745–756 (2003).
49. R. J. Moore, D. M. Owens, G. Stamp, C. Arnott, F. Burke, N. East, H. Holdsworth, L. Turner, B. Rollins, M. Pasparakis, G. Kollias, and F. Balkwill, Mice deficient in tumor necrosis factor-alpha are resistant to skin carcinogenesis. *Nature Med* **5**(7), 828–831 (1999).
50. B. J. Sugarman, B. B. Aggarwal, P. E. Hass, I. S. Figari, M .A. Palladino, Jr., and H. M. Shepard, Recombinant human tumor necrosis factor-alpha: Effects on proliferation of normal and transformed cells in vitro. *Science* **230**(4728), 943–945 (1985).

51. S. Shishodia, H. M. Amin, R. Lai, and B. B. Aggarwal, Curcumin (diferuloylmethane) inhibits constitutive NF-kappaB activation, induces G1/S arrest, suppresses proliferation, and induces apoptosis in mantle cell lymphoma. *Biochem Pharmacol* **70**(5), 700–713 (2005).
52. P. A. Cerutti, Prooxidant states and tumor promotion. *Science* **227**, 375–381 (1985).
53. W. J. Kozumbo, M. A. Trush, and T. W. Kensler, Are free radicals invlolved in tumor promotion? *Chem–Biol Interact* **54**, 199–207 (1985).
54. W. Troll and R. Wiesner, The role of oxygen free radicals as a possible mechanism of tumor promotion. *Annu Rev Pharmacol Toxicol* **25**, 509–528 (1985).
55. E. M. Perchellet and J. P. Perchellet, Characterization of the hydroperoxide response observed in mouse skin treated with tumor promoters in vivo. *Cancer Res* **49**, 6193–6201 (1989).
56. J. J. Pence and J. J. Reiners, Murine epidermal xanthine oxidase activity: Correlation with degree of hyperplasia induced by tumor promoters. *Cancer Res* **47**, 6388–6392 (1987).
57. L. Srinivas, T. Gindhart, and N. H. Colburn, Tumor-promoter resistant cells lack trisialoganglioside response. *Proc Natl Acad Sci USA* **79**, 4988–4991 (1982).
58. T. W. Kensler, D. M. Bush, and W. J. Kozumbo, Inhibition of tumor promotion by a biomimetic superoxide dismutase. *Science* **221**, 75–77 (1983).
59. L. J. Marnett, Peroxy free radicals: potential mediators of tumor initiation and promotion. *Carcinogenesis* **8**, 1365–1373 (1987).
60. A. K. Verma, C. L. Ashendel, and R K. Boutwell, Inhibition by prostaglandin synthesis inhibitors of the induction of epidermal ornithine decarboxylase activity, the accumulation of prostaglandins, and tumor promotion caused by 12-O-tetradecanoylphorbol-13-acetate. *Cancer Res* **40**, 308–315 (1980).
61. M. Nagabhushan and S. V. Bhide, Curcumin as an inhibitors of cancer. *J Am Coll Nutr* **11**, 192–198 (1992).
62. Y. P. Lu, R. L. Chang, M. T. Huang, and A. H. Conney, Inhibitory effects of topical application of low doses of curcumin on 12-O-tetradecanoylphorbol-13-acetate-induced increase in ornithine decarboxylase mRNA in mouse epidermis. *Carcinogenesis* **14**, 293–297 (1993).
62a. Y. J. Surh, K. K. Park, K. S. Chun, J. M. Lee, E. Lee, and S. S. Lee, Anti-tumor promoting activities of selected pungent substances present in ginger. *J Environ Toxicol Pathol Oncol* **18**, 131–139 (1999).
63. M. T. Huang, W. Ma, P. Yen, J. G. Xie, J. Han, K. Frenkel, D. Grunberger, and A. H. Conney, Inhibitory effects of topical application of low doses of curcumin on 12-O-tetra-decanoylphorbol-13-acetate-induced tumor promotion and oxidized DNA bases in mouse epidermis. *Carcinogenesis* **18**, 83–88 (1997).
63a. Y. Nakamura, Y. Ohto, A. Murakami, T. Osawa, and H. Ohigashi, Inhibitory effects of curcumin and tetrahydrocurcuminoids on the tumor promoter-induced reactive oxygen species generation in leukocytes in vitro and in vivo. *Jpn J Cancer Res* **89**, 361–370 (1998).
64. T. S. Huang, S. C. Lee, and J. K. Lin, Suppression of c-Jun/AP-1 activation by an inhibitor of tumor promotion in mouse fibroblast cells. *Proc Natl Acad. Sci USA* **88**, 5292–5296 (1991).
65. S. Kilaru, S. G. Frangos, A. H. Chen, D. Gottler, A. Dhadwal, O. Araim, and B. E. Sumpio, Nicotine: A review of its role in atherosclerosis. *J Am Coll Surg* **193**, 538–546 (2001).

66. R. Olszanecki, J. Jawien, M. Gajda, L. Mateuszuk, A. Gebska, M. Korabiowska, S. Chlopicki, and R. Korbut, Effect of curcumin on atherosclerosis in apoE/LDLR-double knockout mice. *J Physiol Pharmacol* **56**(4), 627–635 (2005).
67. R. Srivastava, M. Dikshit, R. C. Srimal, and B. N. Dhawan, Anti-thrombotic effect of curcumin. *Thromb Res* **40**(3), :413–417 (1985).
68. N. Toda, T. Okamura, I. Shimizu, and Y. Tatsuno, Postmortem functional changes in coronary and cerebral arteries from humans and monkeys. *Cardiovasc Res* **11**, 707–713 (1985).
69. R. B. Arora, S. K. Gupta, R. C. Sharma, and H. H. Siddiqui, Isolation and characterization of a sodium retaining substance from pig heart muscle and its role in myocardial infarction. *Indian J Med Res* **59**(3), 483-493 (1971).
70. S. Chandra, S. K. Mukherjee, and N. Sethi, Effect of argemone oil feeding on blood biochemistry and tissue changes in albino rats. *Indian J Med Sci* **26**(5), 308–312 (1972).
71. R L. Saul, P. Gee, and B. N. Ames, Free radicals, DNA damage and aging. In: *Modern Biological Theories of Aging*, H. R. Warner (ed). Raven Press, New York, 1987.
72. D. Harman, Free radical theory of aging: Effect of free radical inhibitors on the mortality rate of male LAP mice, *J Gerontol* **23**, 476–482 (1968).
73. D. Harman, The aging process. *Proc Natl Acad Sci USA* **78**,124–128 (1981).
74. J. M. Talmashoff, T. Ouo, and R. G. Cutter, Superoxide dismutase: Correlation with life-span and specific metabolic rate in primate species. *Proc Natl Acad Sci USA* **77**, 2777–2781 (1980).
75. J. M. Ringman, S. A. Frautschy, G. M. Cole, D. L. Masterman, and J. L. Cummings, A potential role of the curry spice curcumin in Alzheimer's disease. *Curr Alzheimer Res* **22**(2), 131–136 (2005).
76. M. Grundman, M. Grundman, and P. Delaney, Antioxidant strategies for Alzheimer's disease. *Proc Nutr Soc.* **61**(2), 191–202 (2002).
77. R. C. Benyon and M. J. Arthur, Mechanisms of hepatic fibrosis. *J. Pediatr Gastroenterol Nutr* **27**, 75–85 (1998).
78. A. D. Burt, Cellular and molecular aspects of hepatic fibrosis. *J Pathol* **170,** 105–114 (1993).
79. H. Nagase and F. Woessner, Jr., Matrix metallo proteinases. *J Biol Chem* **274**, 21,491–21, 494 (1999).
80. M. J. Arthur, Fibrogenesis II. Metalloproteinases and their inhibitors in liver fibrosis. *Am J Physiol Gastrointest Liver Physiol* **279**(2), G245–G249 (2000).
81. G. Akila, V. Rajakrishnan, P. Viswanathan, K. N. Rajashekaran, and V. P. Menon, Effects of curcumin on lipid profile and lipid peroxidation status in experimental hepatic fibrosis. *Hepatol Res* **11**,147–157 (1998).
82. R. Rukkumani, K. Aruna, P. S. Varma, and V. P. Menon, Curcumin influences hepatic expression patterns of matrix metalloproteinases in liver toxicity. *Italian J Biochem* **53**, 61–66 (2004).
82a. B. Tesfamariam, Free radicals in diabetic endothelial cell dysfunction. *Free Radical Biol Med* **16**, 383–391 (1994).
83. A. Bierhaus, M. A. Hofmann, R. Ziegler, and P. Nawroth, AGE and other interaction with AGE-receptors in vascular disease and diabetes-I. The AGE concept. *Cardiovasc Res* **37**, 586–590 (1998).
84. G. L. King, H. Ishii, and D. Koya, Diabetic vascular dysfunction: A model of excessive activation of protein kinase C. *Kidney Int* **60**, S77–S85 (1997).

85. J. R. Williamson, K. Chang, M. Frangos, K. S. Hasan, Y. Ido, T. Kawamura, J. R. Nyengaard, M. Van Den Enden, C. Kilo, and R. G. Tilton, Hyperglycemic pseudohypoxia and diabetic complications. *Diabetes* **42**, 801–813 (1993).
86. B. Goossens, J. Grooten, K. De Vos, and W. Fiers, Direct evidence for tumor necrosis factor-induced mitochondrial reactive oxygen intermediates and their involvement in cytotoxicity. *Proc Natl Acad Sci USA* **92**(18), 8115–8119 (1995).
87. G. Boden, Role of fatty acids in the pathogenesis of insulin resistance and NIDDM. *Diabetes* **46**(1), 3–10 (1997).
88. C. R. Kahn, Causes of insulin resistance. *Nature* **373**(6513),384–385 (1995).
89. G. B. Sijitlal, P. Chitra, and G. Chandrakasan, Effect of curcumin on the advanced glycation and cross-linking of collagen in diabetic rats. *Biochem Pharmacol* **56**(12), 1067–1074 (1998).
90. T. Mahesh, M. Sri Balasubashini, and V. P. Menon, Photo irradiated curcumin supplementation in streptozotocin-induced diabetic rats: Effect on lipid peroxidation. *Therapie* **59**(6), 639–644 (2004).
91. C. J. Li, L. J. Zhang, B. J. Dezube, C. S. Crumpacker, and A. B. Pardee, Three inhibitors of type 1 human immunodeficiency virus long terminal repeat-directed gene expression and virus replication. *Proc Natl Acad Sci USA* **90**, 1839–1842 (1993).
92. A. Mazumder, K. Raghavan, J. Weinstein, K. W. Kohn, and Y. Pommier, Inhibition of human immunodeficiency virus type-l integrase by curcumin. *Biochem Pharmacol.* **49**(8),1165–1170 (1995).
93. C. A. Janeway, The immune system evolved to discriminate infectious nonself from noninfectious self. *Immunol Today* **13**, 11 (1992).
94. C. Natarajan and J. J. Bright, Curcumin inhibits experimental allergicencephalomyelitis by blocking IL-12 signaling through JAK-STAT pathway in T-cells and differentiation of neural antigen specific ThI cells. *J Immunol* **169**, 6506 (2002).
95. H. C. Huang, T. R. Jan, and S. F. Yeh, Inhibitory effect of curcumin, an anti-inflammatory agent, on vascular smooth muscle cell proliferation. *Eur J Pharmacol* **221**, 381–385 (1992).
96. M. Kitagawa, H. Mitsui, H. Nakamura, S. Yoshino, S. Miyakawa, N. Ochiai, M. Onobori, H. Suzuki, and T. Sumida, Differential regulation of rheumatoid synovial cell interleukin-12 production by tumor necrosis factor alpha and CD 40 signals. *Arthritis Rheum* **42**, 1917–1926 (1999).
97. B. Bosman, Testing of lipoxygenase inhibitors, cycloxigenase inhibitors, drugs with immunomodulating properties and some reference antipsoriatic drugs in the modified mouse tail test, an animal model of psoriasis. *Skin Pharmacol* **7**, 324–334 (1994).
98. M. C. Heng, M. K. Song, J. Harker, and M. K. Heng, Drug-induced suppression of phosphorylase kinase activity correlates with resolution of psoriasis as assessed by clinical, histological and immunohistochemical parameters. *Br J Dermatol* **143**, 937–949 (2000).
99. A. Bernd, C. Theilig, S. Kippenberger, A. Ramirez-Bosca, M. Podda, J. Diaz, J. Miquel, and R. Kaufmann, An extract of *Curcuma longa* exerts anti-oxidative, anti-inflammatory and antiproliferative effects on human keratinocytes in vitro. *J Invest Dermatol* **109**, 460–464 (1997).
100. A. Bernd, C. Theilig, S. Kippenberger, A. Ramirez-Bosca, J. Diaz, J. Miquel, and R. Kaufmann, Effect of *Curcuma longa* extract on the expression of proinflammatory cytokines. *Skin Pharmacol App. Skin Physiol* **13**, 226–234 (2000).
101. I. Kurose, H. Higuchi, S. Kato, S. Miura, N. Watanabe, Y. Kamegaya, K. Tomita, M. Takaishi, Y. Haire, M. Fukuda, K. Mizukami, and H. Ishii, Oxidative stress on

mitochondria and cell membrane of cultured rat hepatocytes and perfused liver exposed to ethanol. *Gastroenterology* **112**, 1331–1343 (1997).
102. V. Rajakrishnan, P. Vishwanathan, K. N. Rajasekharan, G. Gunashekaran, and V. P. Menon, Role of curcumin in alcoholic hepatotoxicity. *Med Sci Res* **26**, 715–719 (1998).
103. A. Helen, K. Krishnakumar, P. L. Vijayammal, and K. T. Augusti, Antioxidant effect of onion oil (*Allium Cepa* linn.) on the damages induced by nicotine in rats as compared to alpha-tocopherol. *Toxicol Lett* **116**, 61–68 (2000).
104. D. Yildiz, N. Ercal, and D. W. Armstrong, Nicotine enantiomers and oxidative stress. *Toxicology* **130**, 155–165 (1998).
105. C. Kalpana and V. P. Menon, Modulatory effects of curcumin on lipid peroxidation and antioxidant status during nicotine-induced toxicity. *Pol J Pharmacol* **56**, 581–586 (2004).
106. K. Polasa, A. N. Naidu, I. Ravindranath, and K. Krishnaswamy, Inhibition of B(a)P induced strand breaks in presence of curcumin *Mutat. Res* **557**, 203–213 (2004).

MODULATION OF TRANSCRIPTION FACTORS BY CURCUMIN

Shishir Shishodia, Tulika Singh, and Madan M. Chaturvedi

Abstract: Curcumin is the active ingredient of turmeric that has been consumed as a dietary spice for ages. Turmeric is widely used in traditional Indian medicine to cure biliary disorders, anorexia, cough, diabetic wounds, hepatic disorders, rheumatism, and sinusitis. Extensive investigation over the last five decades has indicated that curcumin reduces blood cholesterol, prevents low-density lipoprotein oxidation, inhibits platelet aggregation, suppresses thrombosis and myocardial infarction, suppresses symptoms associated with type II diabetes, rheumatoid arthritis, multiple sclerosis, and Alzheimer's disease, inhibits HIV replication, enhances wound healing, protects from liver injury, increases bile secretion, protects from cataract formation, and protects from pulmonary toxicity and fibrosis. Evidence indicates that the divergent effects of curcumin are dependent on its pleiotropic molecular effects. These include the regulation of signal transduction pathways and direct modulation of several enzymatic activities. Most of these signaling cascades lead to the activation of transcription factors. Curcumin has been found to modulate the activity of several key transcription factors and, in turn, the cellular expression profiles. Curcumin has been shown to elicit vital cellular responses such as cell cycle arrest, apoptosis, and differentiation by activating a cascade of molecular events. In this chapter, we briefly review the effects of curcumin on transcription factors NF-κB, AP-1, Egr-1, STATs, PPAR-γ, β-catenin, nrf2, EpRE, p53, CBP, and androgen receptor (AR) and AR-related cofactors giving major emphasis to the molecular mechanisms of its action.

1. INTRODUCTION

Transcription factors are proteins that bind DNA at a specific promoter or enhancer region or site and act like genetic switches to regulate the expression of various genes (Table 1). Transcription factors can be selectively activated or deactivated by other proteins, often as the final step in signal transduction. Eukaryotic transcription factors are nearly always positive regulators of transcription, although they occasionally act as negative regulators. Hundreds of transcription factors with functionally different domains essential for DNA binding and activation have been identified and characterized in several organisms. Some of the transcription factors are important targets for therapeutic intervention in several types of disease (see Refs. 1 and 2 and references therein). For instance, the pro-inflammatory transcription

Table 1. Transcription factors and their regulated genes.

TRANSCRIPTION FACTOR	GENES	REFS.
NF-κB	*COX-2, MMP-9, cyclinD1, IL-6, IL-8, VEGF, uPA, ICAM-1, VCAM, Bcl$_2$, BclX$_L$, survivin, cMyc, TRAF-1, TRAF-2, Bfl-1/A1, FLIP, xIAP, cIAP, HO-1* and others	1
AP-1	*DNMT1, COX-2, MMP-9, HB-EGF, GM-CSF, cyclin D1, WAF-1, p53, ARF, FasL, Fas, Bim, Bcl$_2$, BclX$_L$, VEGF, uPA, uPAR, proliferin, cathepsin, HO-1, IGF-1, hTERT*, collagenase, osteocalcin, oxidative response genes, oxidoreductases, xenobiotic response genes	28, 29
STAT-3	*Cylcin D1, COX-2, MMP-2, MMP-10, MMP-1 p21^{WAF1}, VEGF, BclX$_L$, Mcl-1, c-Myc, survivin, HSP27, adrenomedullin, c-Fos, MEK5*	30–32
β−catenin	*CyclinD1, cMyc, LIF, MMP, connexin 43, PPAR-δ, ITF-2, MDR1, gastrin*	33–35
EpRE	Quinine reductase, genes encoding phase II drug-metabolizing enzymes.	36, 37
Egr-1	*NF-κB, cyclin D1, EGFR, IGF-II, PDGF-A, TGFβ1, p57, KIP2, CBP, TRAIL*, NGFI-A-binding proteins 1 and 2, tyrosine hydroxylase, thrombospondin 1, neuroendocrine-associated genes	38–43
P53	*p21, bax , TFIIB, Apaf-1, SAK* (Snk/Plk-akin kinase)	44, 45
Nrf2	HO-1, γGCS	46
AR	*PSA, IGF-1, Pem, IL-6, Myc, IL-2, IL-4, IL-10, MAPK8, SOCS1, CREB1*, prostatein C3, sex-limited protein (*slp*), spermine-binding protein, prostate-binding protein C2A, cystatin-related protein2, calreticulin, and probasin.	47, 48
PPAR-γ	Adipocyte-specificaP2 gene, *resistin, endothelin-1*	49–51

Abbreviations: ARF, ADP-ribosylation factor; CBP, CREB-binding protein; CREB, cyclic AMP response element-binding protein; COX, cyclooxygenase; DNMT1, DNA methyltransferase-1; EGF, epidermal growth factor; EGFR, epidermal growth factor receptor; FLIP, FLICE-inhibitory protein; GCS, glutamyl cysteine synthetase; GM-CSF, granulocyte macrophage colony-stimulating factor; HO-1, haemoxygenase-1; HSP, heat shock protein; hTERT, human telomerase reverse transcriptase; IAP, inhibitor of apoptosis; ICAM, intercellular adhesion molecule; IGF, insulin-like growth factor; IL, interleukin; MMP, matrix metalloproteinase; ITF-2, immunoglobulin transcription factor-2; LIF, leukemia inhibitory factor; MAPK, mitogen-activated protein kinase; MDR, multidrug resistance; NGF, nerve growth factor; PDGF, platelet-derived growth factor; PPAR, peroxisome proliferator-activated receptor; PSA, prostate-specific antigen; TRAF, tumor necrosis factor receptor-associated factor; TGF, transforming growth factor; TRAIL, tumor necrosis factor-related apoptosis-inducing ligand; SAK, Snk/Plk-akin kinase; SOCS, suppressors of cytokine signaling; TF, transcription factor; uPA, urokinase plasminogen activator; uPAR, urokinase plasminogen activator receptor; VCAM, vascular cell adhesion molecule; VEGF, vascular endothelial growth factor.

factor, nuclear factor (NF)-κB controls the expression of genes that affect important cellular processes, such as cell transformation, proliferation, cell survival, invasion, metastasis, adhesion, angiogenesis, and apoptosis.[3,4] Activator protein (AP)-1 activation is linked to growth regulation, cell transformation, inflammation, and innate immune response. AP-1 has been implicated in regulation of genes involved in apoptosis and proliferation. Signal transducer and activator of transcription (STAT) are dual-function proteins whose constitutive activation by specific classes of oncoproteins with protein tyrosine kinase (PTK) activity contributes to oncogenesis by eliciting permanent alterations in cells' genetic programs. Some of the other transcription factors involved in disease are early growth response-1 (Egr-1), peroxisome proliferator-activated receptor-γ (PPAR-γ), electrophile-response element (EpRE), β-catenin, NF-E2-related factor 2 (Nrf2), and androgen receptor (AR).

The anti-inflammatory and antioxidant compound curcumin targets transcription factors to prevent pathways that are harmful to cells' normal growth. Curcumin has been shown to modulate the signaling pathways responsible for regulating the activation of NF-κB, AP-1, STAT3, Egr-1, PPAR-γ, EpRE, β-catenin, Nrf-2, and AR. In this chapter, we review the effects of curcumin on transcription factors, giving emphasis to the molecular mechanisms of curcumin's activity.

2. NUCLEAR FACTOR-KAPPA B

NF-κB is a transcription factor that is activated in response to variety of stimuli, including cytokines, mitogens, carcinogens, chemotherapeutic agents endotoxin, physical and chemical stresses, radiation, hypoxia, and other inflammatory stimuli.[3,5,6] NF-κB activation has been shown to regulate the expression of over 200 genes that are involved in cellular transformation, proliferation, antiapoptosis, angiogenesis, invasion, and metastasis. Although IκBα kinase is the major kinase, over 60 different protein kinases have been linked to the activation of NF-κB by different agents.

NF-κB is present in all cells in a resting state in the cytoplasm; only when activated and translocated to the nucleus is a sequence of events generated. Under resting conditions, NF-κB consists of a heterotrimer of p50, p65, and IκBα in the cytoplasm. The phosphorylation, ubiquitination, and degradation of IκBα leads to the release of the p50–p65 heterodimer, which then translocates to the nucleus and binds its specific response elements on the DNA of a given gene.

Constitutively active NF-κB is frequently encountered in a wide variety of tumors and in tumor tissues derived from patients[7–9] (e.g., multiple myeloma,[10] acute myelogenous leukemia,[11] T-cell leukemia,[12] acute myeloid leukemia,[13] prostate,[14] and breast[15]). More importantly, suppression of NF-κB in these tumor samples has been shown to inhibit proliferation, and induce cell cycle arrest and apoptosis, indicating the crucial role of NF-κB in the proliferation and survival of cells.[16] The development of a drug that can specifically suppress NF-κB activation requires a full understanding of the mechanism by which NF-κB is activated in response to various stimuli.

Our laboratory has shown that various tumor promoters, including phorbol ester, tumor necrosis factor (TNF), and hydrogen peroxide activate NF-κB and that curcumin downregulates the activation of NF-κB[17] (Figure 1). Curcumin

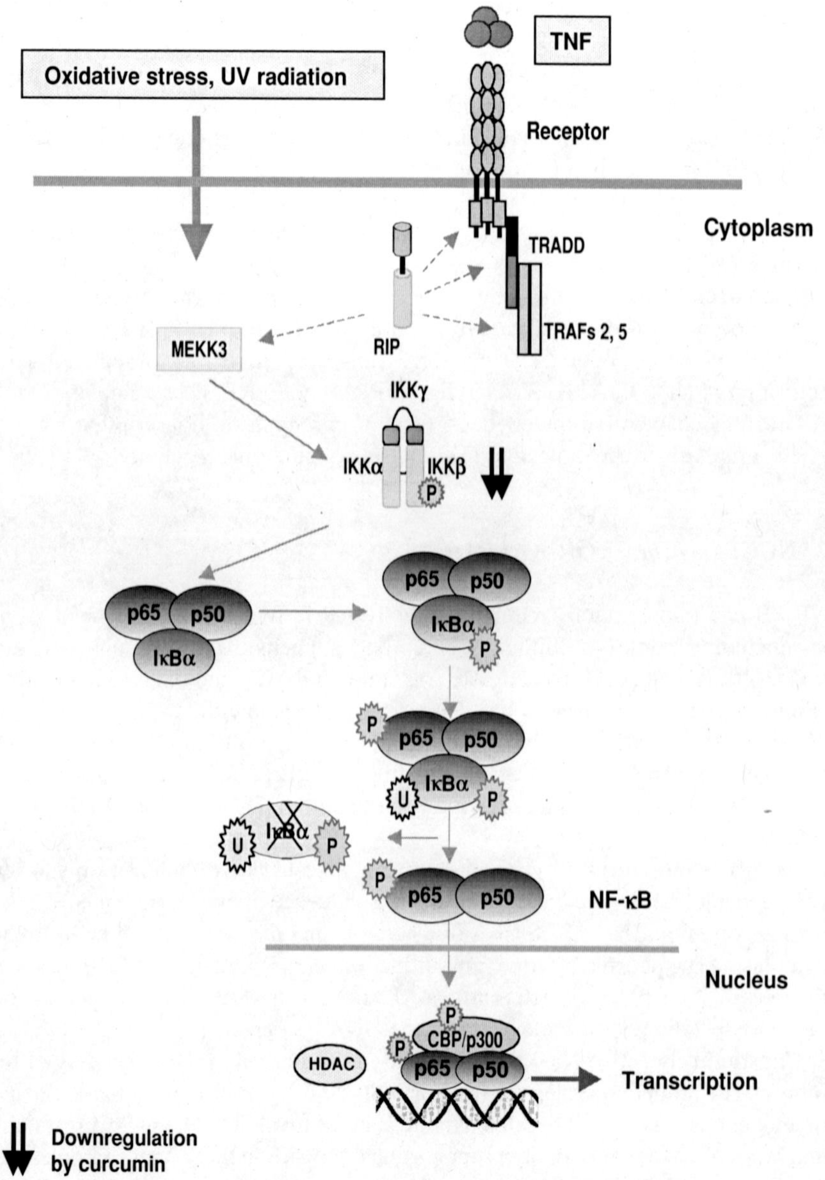

Figure 1. Modulation of NF-κB signaling pathway by curcumin. *Abbreviations*: TRADD, TNF receptor-associated death domain; MEKK, MAPK/ERK kinase kinase; TRAF, TNF receptor-associated factor; RIP, receptor interacting protein; IKK, IκB kinase; IκB, inhibitory κB; CBP, CREB-binding protein; HDAC, histone deacetylase; P, phosphorylation; U, ubiquitination.

downregulates the expression of cell proliferation and antiapoptotic and metastatic gene products through suppression of IκBα kinase and Akt activation.[18] Subsequently, others showed that curcumin-induced downregulation of NF-κB is mediated through suppression of IκBα kinase (IKK) activation.[19,20] IκBα mutations lead to molecular defects in apoptotic signaling in classic Hodgkin's lymphoma, multiple myeloma, and activated B-cell-like diffuse large B-cell lymphoma, which can be circumvented by targeting NF-κB through inhibition of the IκB kinase complex. Thus, targeting IκB kinases might represent an attractive therapeutic approach against these malignancies regardless of the mutational status of IκBα.[21] Recently, we have shown that curcumin downregulated cigarette-smoke-induced NF-κB activation through inhibition of IκBα kinase in human lung epithelial cells.[22] Curcumin suppresses the constitutively active NF-κB activation in mantle cell lymphoma through the suppression of IKK.[23] This led to the downregulation of cyclin D1, cyclooxygenase-2 (COX2) and matrix metalloproteinases (MMP)-9 by curcumin.

Philip and Kundu have recently reported that curcumin downregulates osteopontin (OPN) induced NF-κB-mediated promatrix metalloproteinase-2 activation through IκBα/IKK signaling.[24] Zheng et al. demonstrated that curcumin arrested cell growth at the G_2/M phase and induced apoptosis in human melanoma cells by inhibiting NF-κB activation and, thus, depletion of endogenous nitric oxide.[25] Kim et al. recently reported that curcumin inhibited LPS-induced mitogen-activated protein kinase (MAPK) activation and the translocation of NF-κB p65 in dendritic cells.[26] We have found that curcumin suppresses the paclitaxel-induced NF-κB pathway in breast cancer cells and inhibits lung metastasis of human breast cancer in nude mice.[27]

Downregulation of Notch signaling by curcumin might be a novel strategy for the treatment of patients with pancreatic cancer.[16] The Notch-1 signaling pathway is associated mechanistically with NF-κB activity during curcumin-induced cell growth inhibition and apoptosis of pancreatic cells. A recent report, however, suggested that the curcumin-induced apoptosis is mediated through the impairment of the ubiquitin proteasome system (UPS). Curcumin disrupts the UPS function by directly inhibiting the enzyme activity of the proteasome's 20S core catalytic component. The direct inhibition of proteasome activity causes an increase in the half-life of IκBα that ultimately leads to the downregulation of NF-κB activation. The curcumin-induced proteasomal malfunction might be linked with both antiproliferative and anti-inflammatory activities.[52]

NF-κB has been shown to upregulate the pro-inflammatory genes during cardiopulmonary bypass and cardiac global ischemia and reperfusion. Suppression of NF-κB activation by curcumin led to the downregulation of cardiac pro-inflammatory genes and inhibition of MMPs during cardiopulmonary bypass, thereby lessening the severity of cardiac mechanical dysfunction after global cardiac ischemia/reperfusion injury.[53] The pathogenic bacterium *Neisseria gonorrhoeae* (Ngo) is responsible for an array of diseases ranging from urethritis to disseminated gonococcal infections. Early events in the establishment of infection involved Ngo-induced induction of the pro-inflammatory cytokines TNF, IL-6, and IL-8 through the induction of NF-κB. Curcumin inhibited the NF-κB-mediated cytokine release and the innate immune response. In addition to the inhibition of Ngo-induced signaling, curcumin treatment of cells completely abolished the adherence of bacteria to cells in late infection, underlining the high potential of curcumin as an antimicrobial compound without cytotoxic side effects.[54]

NF-κB is a versatile transcription factor that plays an important role in diverse phenomena, including inflammation, cell proliferation, apoptosis, and oncogenesis. Thus, the capability of curcumin to interfere with the activation of NF-κB is of particular interest to biologists and clinicians. However, the large number of target genes makes it difficult to propose a unique cellular response to NF-κB inhibition.

3. ACTIVATOR PROTEIN 1

Activator protein-1 is another transcription factor that has been closely linked with the proliferation and transformation of tumor cells.[55] The activation of AP-1 is frequently associated with the activation of NF-κB. The activation of AP-1 requires the phosphorylation of c-jun through activation of stress-activated kinase c-Jun N-terminal kinase (JNK).[56] The activation of JNK leads to cellular transformation.[57] Curcumin has been shown to inhibit the activation of AP-1 induced by tumor promoters[58] and JNK activation induced by carcinogens[59] (Figure 2). Bierhaus et al. demonstrated that curcumin caused the inhibition of AP-1 due to its direct interaction with the AP-1 DNA-binding motif.[60] Dickinson et al. have demonstrated that the beneficial effects elicited by curcumin appear to be due to changes in the pool of transcription factors that compose EpRE and AP-1 complexes, affecting gene expression of glutamate–cysteine ligase and other phase II enzymes.[61]

Curcumin suppressed the constitutive AP-1 DNA-binding and transcriptional activity in the HTLV-1-infected T-cell line. Curcumin inhibited the growth of HTLV-1-infected T-cell lines by inducing cell cycle arrest followed by apoptosis. The suppression of the constitutively active AP-1 by curcumin might partly be due to the suppression of JunD expression by curcumin.[62] Infection with high-risk human papillomaviruses (HPVs) leads to the development of cervical carcinoma, predominantly through the action of viral oncoproteins E6 and E7. Curcumin suppressed the expression of viral oncogenes E6 and E7, as well as downregulated the binding of AP-1, an indispensable component for efficient epithelial tissue-specific gene expression of HPV.[63] Hydrogen peroxide stimulates the proliferation and migration of human prostate cancer cells through the activation of AP-1 and upregulation of the heparin affin regulatory peptide (HARP) gene. Curcumin abrogated both hydrogen peroxide-induced HARP expression and LNCaP cell proliferation and migration.[64] Prusty and Das recently reported that curcumin downregulated AP-1-binding activity in tumorigenic HeLa cells.[65] Squires et al. have demonstrated that curcumin suppresses the proliferation of tumor cells through the inhibition of Akt/PKB activation.[66]

Interleukin (IL)-18-induced vascular endothelial growth factor (VEGF) plays an important role in angiogenesis in rheumatoid synoviocytes. The increased level of

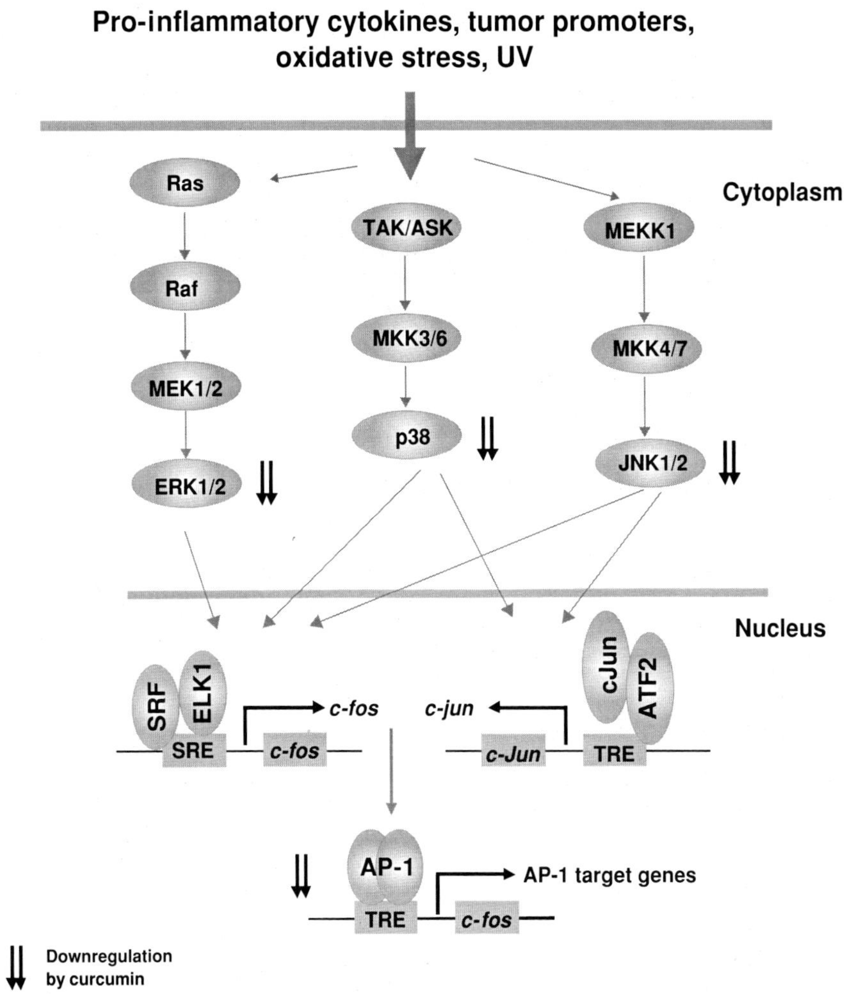

Figure 2. Modulation of AP-1 signaling pathway by curcumin. *Abbreviations*: MEK, MAPK/ERK kinase; MKK, MEK kinase; ERK, extracellular signal-regulated kinase; JNK, cJun N-terminal kinase; TAK/ASK, transforming growth factor-β-activated/apoptosis signal-regulated kinase; SRE, serum response element; SRF, serum response factor; TRE, transforming growth factor-β-response element.

VEGF in fibroblast-like synoviocytes isolated from the patients with rheumatoid arthritis was associated with increased AP-1-binding activity to the VEGF promoter site. Curcumin dose-dependently abrogated the effect of IL-18 on VEGF production. These findings suggest that the downregulation of IL-18 activity or AP-1 signal pathway by curcumin can be a potential therapeutic target for rheumatoid arthritis.[67]

4. EARLY GROWTH RESPONSE-1

The transcription factor Egr-1 is a member of the family of immediate early-response genes and regulates a number of patho-physiologically relevant genes in vasculature that are involved in growth, differentiation, immune response, wound healing, and blood clotting. Pendurthi et al. investigated the effect of curcumin on Egr-1 expression in endothelial cells and fibroblasts.[68] They showed that pretreatment of endothelial cells and fibroblasts with curcumin suppressed TPA and serum-induced Egr-1 binding to the consensus Egr-1-binding site and also to the Egr-1-binding site present in the promoter of the tissue factor gene. Similarly, curcumin inhibits human colon cancer cell growth by suppressing gene expression of EGFR by reducing the trans-activation activity of Egr-1.[38] Curcumin also inhibited TPA-induced *de novo* synthesis of Egr-1 protein in endothelial cells. Suppression of Egr-1 protein expression in curcumin-treated cells stemmed from the suppression of Egr-1 mRNA. Curcumin inhibited serum and TPA-induced expression of tissue factor and urokinase-type plasminogen activator receptor mRNA in fibroblasts. These results showed that curcumin suppresses the induction of Egr-1 and thereby modulates the expression of Egr-1-regulated genes in endothelial cells and fibroblasts. The downregulation of tissue factor by curcumin has also been demonstrated in bovine aortic endothelial cells.[60]

5. SIGNAL TRANSDUCER AND ACTIVATOR OF TRANSCRIPTION

The STAT proteins are signaling molecules with dual functions that were discovered during studies on interferon (IFN)-γ-dependent gene expression.[69] Seven mammalian STAT family members have been molecularly cloned and share common structural elements.[70] They can be activated by phosphorylation through janus kinase (JAK) or cytokine receptors, G-protein-coupled receptors, or growth factor receptors (such as EGFR); by platelet-derived growth factor receptors that have intrinsic tyrosine kinase activity; or by intracellular nonreceptor tyrosine kinase recruitment[31,71] (see Figure 3). Of the seven STAT proteins identified so far, constitutive activation of STAT3 and STAT5 have been implicated in multiple myeloma, lymphomas, leukemias, and several solid tumors, making these proteins logical targets for cancer therapy. These STAT proteins contribute to cell survival and growth by preventing apoptosis through increased expression of antiapoptotic proteins, such as bcl-2 and bcl-X_L. Recently, STAT 3 was shown to be a direct activator of the VEGF gene, which is responsible for increased angiogenesis. Elevated STAT3 activity has been detected in head and neck squamous cell carcinoma,[72] leukemias,[73] lymphomas,[74] and multiple myeloma.[75]

Bharti et al. demonstrated that curcumin inhibited IL-6-induced STAT3 phosphorylation and consequent STAT3 nuclear translocation.[75] Curcumin had no effect on STAT5 phosphorylation but inhibited the interferon α-induced STAT1 phosphorylation. The constitutive phosphorylation of STAT3 found in certain multiple

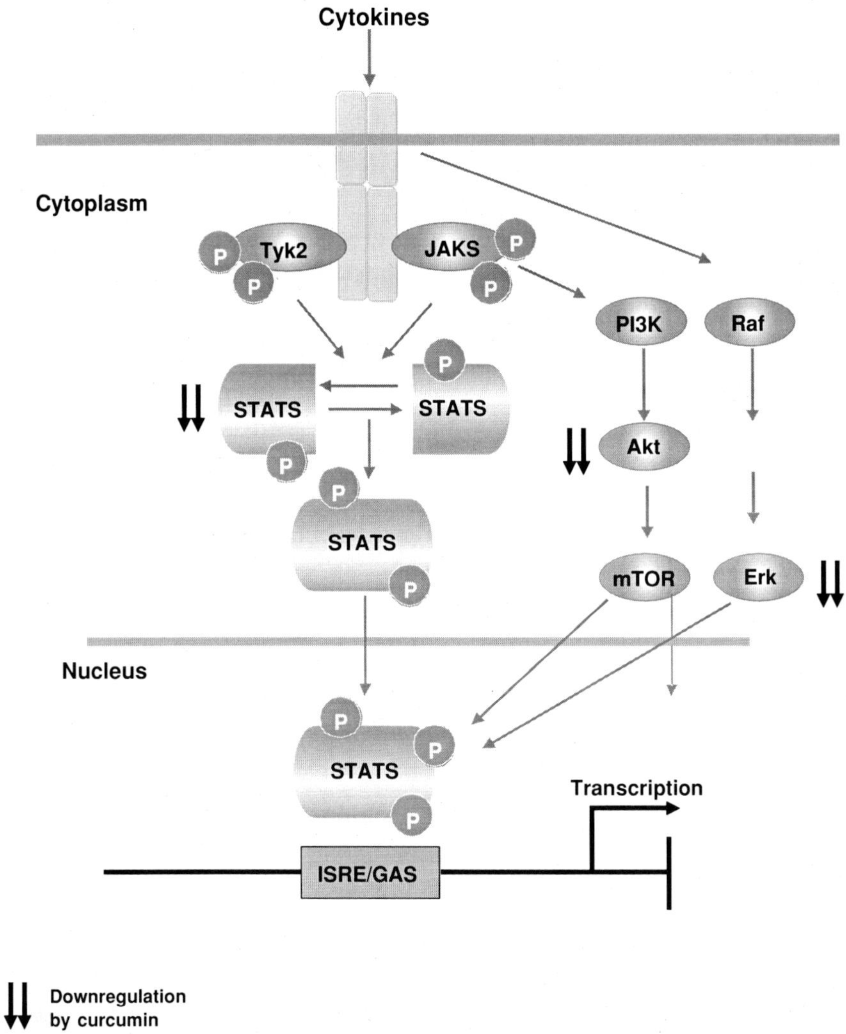

Figure 3. Modulation of STAT-3 signaling pathway by curcumin. *Abbreviations*: PI3K, phosphatidylinositol-3-kinase; ERK, extracellular signal-regulated kinase; ISRE, IFN-stimulated responsive element; GAS, γIFN activation site; mTOR, mammalian target of rapamycin.

myeloma multiple myeloma cells was also abrogated by treatment with curcumin. Curcumin-induced inhibition of STAT3 phosphorylation was reversible. Compared with AG490, a well-characterized JAK2 inhibitor, curcumin was a more rapid (30 min vs. 8 h) and more potent (10 μM vs. 100 μM) inhibitor of STAT3 phosphorylation. In addition, dexamethasone-resistant multiple myeloma cells were found to be sensitive to curcumin. Overall, these results demonstrated that curcumin was a potent inhibitor of STAT3 phosphorylation and this plays a role in curcumin's suppression of the proliferation of multiple myeloma.

Kim et al. investigated the inhibitory action of curcumin on JAK-STAT signaling in the brain. Curcumin markedly inhibited the phosphorylation of STAT1 and STAT3 as well as JAK1 and JAK2 in rat primary microglia activated with gangliosides, lipopolysaccharide, or IFN-γ.[76] Li et al. showed that curcumin suppressed oncostatin-M-stimulated STAT1 phosphorylation, DNA-binding activity of STAT1, and c-Jun N-terminal kinase activation without affecting JAK1, JAK2, JAK3, ERK1/2, and p38 phosphorylation.[77] Curcumin also inhibited OSM-induced MMP-1, MMP-3, MMP-13, and TIMP-3 gene expression.

Natarajan et al. showed that treatment of activated T-cells with curcumin inhibited IL-12-induced tyrosine phosphorylation of JAK 2, tyrosine kinase 2, STAT3, and STAT4 transcription factors.[78] The inhibition of the JAK–STAT pathway by curcumin resulted in a decrease in IL-12-induced T-cell proliferation and Th1 differentiation. STAT5 signaling pathway might be involved in the proliferation of primary CML cells. Curcumin has been shown to inhibit the cellular proliferation and the expression of STAT5 mRNA and to downregulate the activation of STAT5 in primary CML cells[79] and K562 leukemia cells.[80]

6. PEROXISOME PROLIFERATOR-ACTIVATED RECEPTOR-GAMMA

Peroxisome proliferator-activated receptor -γ (PPAR-γ) is a member of the nuclear hormone receptor gene family that is activated by fatty acids and plays a role in insulin sensitivity and adipogenesis. Activation of PPAR-γ inhibits the proliferation of nonadipocytes. The level of PPAR-γ is dramatically diminished along with activation of hepatic stellate cells (HSCs). Xu et al. demonstrated that curcumin dramatically induced the gene expression of PPAR-γ and activated PPAR-γ in activated HSC.[81] Blocking its trans-activating activity by a PPAR-γ antagonist markedly decreased the effects of curcumin on inhibition of cell proliferation. Chen and Xu recently reported that curcumin activation of PPAR-γ inhibited Moser cell (human colon cancer cell line) growth and mediated the suppression of the gene expression of cyclin D1 and EGFR.[82] These results provided a novel insight into the roles and mechanisms of curcumin in inhibition of colon cancer cell growth and potential therapeutic strategies for the treatment of colon cancer.

Hepatic fibrogenesis occurs as a wound-healing process after many forms of chronic liver injury. Hepatic fibrosis ultimately leads to cirrhosis if not treated effectively. During liver fibrogenesis, quiescent HSCs become active, a transformation that is associated with enhanced cell proliferation and overproduction of the extracellular matrix (ECM). Inhibition of cell proliferation and induction of apoptosis are potential strategies to block the activation of HSCs for the prevention and treatment of liver fibrosis. Zheng et al. reported that curcumin stimulated PPAR-γ activity in activated hepatic stellate cells *in vitro*, which was required for curcumin to reduce cell proliferation, induce apoptosis and suppress

ECM gene expression.[83] Activation of PPAR-γ by curcumin also resulted in the interruption of transforming growth factor (TGF)-β signaling by suppressing gene expression of TGF-β receptors, leading to the inhibition of connective tissue growth factor (CTGF) gene expression that is responsible for overproducing the ECM. Inhibition of αI(I)-collagen gene expression by curcumin in activated HSCs results from suppression of CTGF gene expression through increasing cellular GSH contents and interruption of TGF-β signaling. These results provide novel insights into the mechanisms underlying the inhibition of HSC activation by curcumin.[83]

The protective effect of curcumin makes it a strong candidate as a novel therapy for sepsis. The beneficial effect of curcumin appears to be mediated by upregulation of PPAR-γ. Studies with rats indicate that intravenous administration of curcumin before the onset of sepsis attenuated tissue injury, reduced mortality, and decreased the expression of TNF in experimentally induced sepsis. Similar results were also found when curcumin was administered after the onset of sepsis. Moreover, the downregulated PPAR-γ in the liver at 20 h after cecal ligation and puncture (CLP) was significantly improved by curcumin treatment.[84]

7. ELECTROPHILE-RESPONSE ELEMENT

Transcription has been shown in several systems to be mediated through binding of transcription factor complexes to TRE and electrophile-response elements (EpRE). Curcumin exposure has been shown to increase the enzymes responsible for glutathione synthesis, particularly glutamate-cysteine ligase (GCL), and metabolism as well as glutathione content, suggesting the eliciting of an adaptive response to stress. Studies have shown that curcumin caused an increase in binding of proteins to DNA sequences for both cis elements, but, more importantly, it altered the composition and nuclear content of proteins in these complexes. Curcumin exposure increased JunD and c-Jun content in AP-1 complexes and increased JunD while decreasing MafG/MafK in EpRE complexes. Thus, the beneficial effects elicited by curcumin appear to be due to changes in the pool of transcription factors that compose EpRE and AP-1 complexes, affecting gene expression of GCL and other phase II enzymes.[61]

8. TUMOR SUPPRESSOR GENE *P*53

p53 is a tumor suppressor and transcription factor. It is a critical regulator in many cellular processes, including cell signal transduction, cellular response to DNA damage, genomic stability, cell cycle control, and apoptosis. The protein activates the transcription of downstream genes such as $p21^{WAF1}$ and *Bax* to induce the apoptotic process, inhibiting the growth of cells with damaged DNA or cancer cells.[85,86] Mutant *p*53 loses its ability to bind DNA effectively, and, as a consequence, the p21 protein is not made available to regulate cell division.

Thus, cells divide uncontrollably and form tumors. Subjects with only one functional copy of the *p53* gene are predisposed to cancer and usually develop several independent tumors in a variety of tissues in early adulthood. Beyond its effects in these early cancer syndromes, *p53* mutants are found in most tumor types, where they contribute to the complex network of molecular events leading to tumor formation. Curcumin has been shown to be a potent inhibitor of *p53*.[87]

In B-cells, curcumin downregulates the expression of the tumor suppressor gene *p53* as well as the survival genes *egr-1*, *c-myc*, and *bcl-X_L*.[87] Curcumin is known to induce apoptosis in eight melanoma cell lines, four with wild type and four with mutant *p53* without inducing additional expression of *p53*[88]; however, in human breast cancer cells, curcumin induces apoptosis through*p53*-dependent Bax induction.[89,90] Curcumin also inhibits cell cycle progression of immortalized human umbilical vein endothelial cells by upregulating the cyclin-dependent kinase inhibitors, $p21^{WAF1/CIP1}$, $p27^{KIP1}$, and p53.[91] In neuroblastoma, curcumin upregulates p53 expression and induce nuclear translocation of p53, followed by induction of $p21^{WAF-1/CIP-1}$ and Bax expression.[92]

9. BETA-CATENIN

β-Catenin is a central component of the cadherin cell adhesion complex and plays an essential role in the Wingless/Wnt signaling pathway. Signaling through the Wnt/β-catenin pathway is dependent on the levels of β-catenin in the cell. In the absence of a Wnt signal, a multiprotein destruction complex, which includes the adenomatous polyposis coli (APC) protein and a member of the Axin family, facilitates the phosphorylation of β-catenin by glycogen synthase kinase-3 (GSK3). Phosphorylated β-catenin is bound by a component of an E3 ubiquitin ligase and is ubiquitinated and degraded by the proteasomes. Unphosphorlyated β-catenin escapes degradation, accumulates in the cell, and enters the nucleus. In the nucleus, β-catenin interacts with members of the TCF/LEF family of transcription factors to stimulate expression of target genes.

Curcumin treatment impairs both Wnt signaling and cell–cell adhesion pathways, resulting in the cell cycle arrest at G_2/M phase and induction of apoptosis in HCT-116 colon cancer cells.[93] Curcumin induced the activation of caspase-3 that, in turn, mediated the cleavage of β-catenin, decreased the transactivation of β-catenin/Tcf-Lef, decreased promoter DNA-binding activity of the β-catenin/Tcf-Lef complex, and decreased the levels of c-Myc protein. Mahmoud et al., while investigating the efficacy of curcumin for the prevention of tumors in C57BL/6J-Min/+ (Min/+) mice, found that curcumin decreased the expression of the oncoprotein β-catenin in the enterocytes of the Min/+ mouse, which led to its antitumor effect. These animals bear a germ-line mutation in the *Apc* gene and spontaneously develop numerous intestinal adenomas by 15 weeks of age. Curcumin decreased tumor formation in Min/+ mice by over 60%. Tumor prevention by curcumin was associated with increased enterocyte apoptosis.[94]

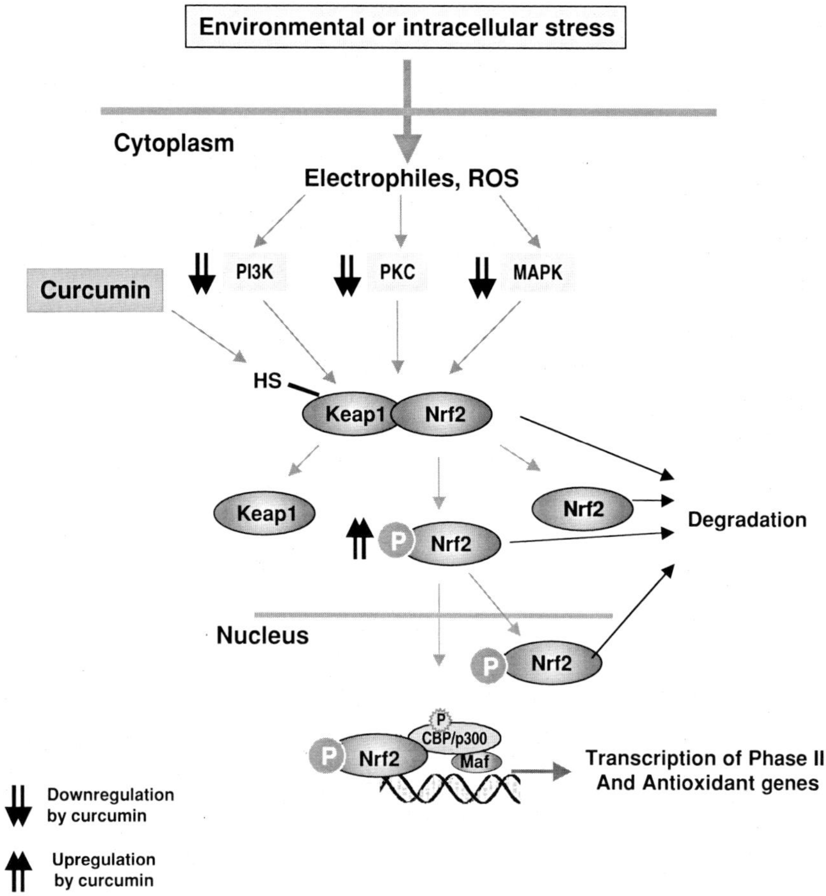

Figure 4. Modulation of Nrf2-mediated gene transcription by curcumin. *Abbreviations*: PI3K, phosphatidylinositol-3-kinase; PKC, protein kinase C; MAPK, mitogen-activated protein kinase; Keap, Kelch-like ECH-associated protein 1; CBP, CREB-binding protein.

10. NF-E2-RELATED FACTOR 2

The transcription factor NF-E2-related factor 2 (Nrf2) normally exists in an inactive state as a result of binding to a cytoskeleton-associated protein Keap1. It can be activated by redox-dependent stimuli. Alteration of the Nrf2–Keap1 interaction enables Nrf2 to translocate to the nucleus, bind to the antioxidant-responsive element (ARE), and initiate the transcription of genes coding for detoxifying enzymes and cytoprotective proteins (Figure 4). The Nrf2/ARE signaling pathway plays a key role in activating cellular antioxidants, including heme oxygenase-1 (HO-1), NADPH quinone oxidoreductase-1 (NQO1), and GSH. This response is also triggered by a class of electrophilic compounds, including curcumin.

Curcumin stimulates the expression of Nrf2 in a concentration- and time-dependent manner in renal epithelial cells. This effect is associated with a significant increase in HO-1 protein expression and hemoxygenase activity. Curcumin stimulates *ho-1* gene activity by promoting inactivation of the Nrf2–Keap1 complex, leading to increased Nrf2 binding to the resident *ho-1* AREs.[95]

Rushworth et al. have shown that curcumin activates ARE-mediated gene expression in human monocytes via PKC delta, upstream of p38 and Nrf2.[96] Gastrointestinal glutathione peroxidase (GI-GPx) has been suggested to act as barrier against hydrogen peroxide absorption and has also been implicated in the control of inflammation and malignant growth. Curcumin has been found to exert its anti-inflammatory and anticarcinogenic effects also by the upregulation of the selenoprotein GI-GPx by activating the Nrf2/Keap1 system.[97]

11. ANDROGEN RECEPTORS AND AR-RELATED COFACTORS

Nakamura et al.[98] have evaluated the effects of curcumin in cell growth, activation of signal transduction, and transforming activities of both androgen-dependent and -independent cell lines. The prostate cancer cell lines LNCaP and PC-3 were treated with curcumin, and its effects on signal transduction and expression of androgen receptor and AR-related cofactors were analyzed. Their results showed that curcumin downregulates transactivation and expression of AR, AP-1, NF-κB, and CREB. It also inhibited the transforming activities of both cell lines as evidenced by reduced colony-forming ability in soft agar. These studies suggest that curcumin has a potential therapeutic effect on prostate cancer cells through downregulation of AR and AR-related cofactors, AP-1, NF-κB, and CBP.[98]

12. cAMP RESPONSE ELEMENT-BINDING PROTEIN

p300/CBP, along with histone acetyltransferases (HATs), have been implicated in cancer cell growth and survival. Acetylation by HAT of specific lysine residues on the N-terminal tail of core histones results in uncoiling of the DNA and increased accessibility to transcription factor binding. In contrast, histone deacetylation by histone deacetylase (HDAC) represses gene transcription by promoting DNA winding, thereby limiting access to transcription factors. CBP and HATs represent novel, therapeutically relevant molecular targets for drug development.

Curcumin is a selective HAT inhibitor.[99] The α and β unsaturated carbonyl groups in the curcumin side chain function as Michael reaction sites and the Michael reaction acceptor functionality of curcumin is required for its HAT-inhibitory activity. In cells, curcumin promotes proteasome-dependent degradation of p300 and the closely related CBP protein. In addition to inducing p300 degradation, curcumin inhibited the acetyltransferase activity of purified p300. Radio-labeled curcumin formed a covalent association with p300; however, tetrahydrocurcumin displayed no p300 inhibitory activity, consistent with a

Michael reaction-dependent mechanism. Curcumin was able to effectively block histone hyperacetylation in both PC3-M prostate cancer cells and peripheral blood lymphocytes induced by the histone deacetylase inhibitor MS-275.

Balasubramanyam et al. found that curcumin is a specific inhibitor of the CBP HAT activity but not of p300/CBP-associated factor, *in vitro* and *in vivo*.[100] Furthermore, curcumin could also inhibit the p300-mediated acetylation of p53 *in vivo*. It specifically represses the p300/CBP HAT activity-dependent transcriptional activation. Curcumin could also inhibit the acetylation of HIV-Tat protein *in vitro* by p300 as well as proliferation of the virus in SupT1 cells. Thus, nontoxic curcumin, which targets p300/CBP, might serve as a lead compound in combinatorial HIV therapeutics. These data thus suggest curcumin as a novel compound for development of possibly therapeutic, p300/CBP-specific HAT inhibitors.

13. CONCLUSION

The evidence presented above clearly demonstrate that curcumin strongly affects the activity of a number of important transcription factors, including NF-κB, AP-1, p53, Egr-1, STAT-3, AR, and AR- related cofactors. Some of these transcription factors are either constitutively expressed or overexpressed in cells leading to diseased conditions. The versatile chemical structure of curcumin enables the molecule to interact with a large number of molecules inside of the cell, leading to a variety of biological effects, like modulation of cell cycle, suppression of growth, induction of differentiation, upregulation of proapoptotic factors, and inhibition of reactive oxygen species production. These activities of curcumin might, in part, explain the molecular basis of the wide and complex effects of this phytochemical; however, further investigations are required to establish its efficacy in humans.

REFERENCES

1. B. B. Aggarwal and S. Shishodia, Molecular targets of dietary agents for prevention and therapy of cancer. *Biochem Pharmacol* **71**, 1397–1421 (2006).
2. S. Shishodia, G. Sethi, and B. B. Aggarwal, Curcumin: getting back to the roots. *Ann NY Acad Sci* **1056**, 206–217 (2005).
3. B. B. Aggarwal, Nuclear factor-kappaB: the enemy within. *Cancer Cell* **6**, 203–208 (2004).
4. S. Shishodia and B. B. Aggarwal, Nuclear factor-kappaB: A friend or a foe in cancer? *Biochem Pharmacol* **68**, 1071–1080 (2004).
5. A. S. Baldwin, Jr., Series introduction: the transcription factor NF-kappaB and human disease. *J Clin Invest* **107**, 3–6 (2001).
6. H. L. Pahl, Activators and target genes of Rel/NF-kappaB transcription factors. *Oncogene* **18**, 6853–6866 (1999).
7. D. K. Giri and B. B. Aggarwal, Constitutive activation of NF-kappaB causes resistance to apoptosis in human cutaneous T cell lymphoma HuT-78 cells. Autocrine role of

tumor necrosis factor and reactive oxygen intermediates. *J Biol Chem* **273**, 14,008–14,014 (1998).
8. H. Lee, M. Wu, F. A. La Rosa, M. P. Duyao, A J. Buckler, and G. E. Sonenshein, Role of the Rel-family of transcription factors in the regulation of c-myc gene transcription and apoptosis of WEHI 231 murine B-cells. *Curr Top Microbiol Immunol* **194**, 247–255 (1995).
9. C. Y. Wang, M. W. Mayo, and A. S.Baldwin, Jr. TNF- and cancer therapy-induced apoptosis: Potentiation by inhibition of NF-kappaB. *Science* **274**, 784–787 (1996).
10. A. C. Bharti, N. Donato, S. Singh, and B. B. Aggarwal, Curcumin (diferuloylmethane) down-regulates the constitutive activation of nuclear factor-kappa B and IkappaBalpha kinase in human multiple myeloma cells, leading to suppression of proliferation and induction of apoptosis. *Blood* **101**, 1053–1062 (2003).
11. R. J. Anto, Mukhopadhyay, K. Denning, and B. B. Aggarwal, Curcumin (diferuloylmethane) induces apoptosis through activation of caspase-8, BID cleavage and cytochrome c release: Its suppression by ectopic expression of Bcl-2 and Bcl-xl. *Carcinogenesis* **23**, 143–150 (2002).
12. N. Mori, M. Fujii, S. Ikeda, Y. Yamada, M. Tomonaga, D. W. Ballard, and N. Yamamoto, Constitutive activation of NF-kappaB in primary adult T-cell leukemia cells. *Blood* **93**, 2360–2368 (1999).
13. C. E. Bueso-Ramos, F. C. Rocha, S. Shishodia, L. J. Medeiros, H. M. Kantarjian, S. Vadhan-Raj, Z. Estrov, T. L. Smith, M. H. Nguyen, and B. B. Aggarwal, Expression of constitutively active nuclear factor-kappa B RelA transcription factor in blasts of acute myeloid leukemia. *Hum Pathol* **35**, 246–253 (2004).
14. T. Dorai, Y. C. Cao, B. Dorai, R. Buttyan, and A. E. Katz, Therapeutic potential of curcumin in human prostate cancer. III. Curcumin inhibits proliferation, induces apoptosis, and inhibits angiogenesis of LNCaP prostate cancer cells in vivo. *Prostate* **47**, 293–303 (2001).
15. H. Nakshatri, P. Bhat-Nakshatri, D. A. Martin, DR. J. Goulet, Jr., and G. W. Sledge, Jr., Constitutive activation of NF-kappaB during progression of breast cancer to hormone-independent growth. *Mol Cell Biol* **17**, 3629–3639 (1997).
16. Z. Wang, Y. Zhang, S. Banerjee, Y. Li, and F. H. Sarkar, Notch-1 down-regulation by curcumin is associated with the inhibition of cell growth and the induction of apoptosis in pancreatic cancer cells. *Cancer* **106**, 2503–2513 (2006).
17. S. Singh and B. B. Aggarwal, Activation of transcription factor NF-kappa B is suppressed by curcumin (diferuloylmethane). *J Biol Chem* **270**, 24,995–25,000 (1995).
18. S. Aggarwal, H. Ichikawa, Y. Takada, S. K. Sandur, S. Shishodia, and B. B. Aggarwal, Curcumin (diferuloylmethane) down-regulates expression of cell proliferation and antiapoptotic and metastatic gene products through suppression of IkappaBalpha kinase and Akt activation. *Mol Pharmacol* **69**, 195–206 (2006).
19. C. Jobin, C. A. Bradham, M. P. Russo, B. Juma, A. S. Narula, D. A. Brenner, and R. B. Sartor, Curcumin blocks cytokine-mediated NF-kappa B activation and proinflammatory gene expression by inhibiting inhibitory factor I-kappa B kinase activity. *J Immunol* **163**, 3474–3483 (1999).
20. S. M. Plummer, K. A. Holloway, M. M. Manson, R. J. Munks, A. Kaptein, S. Farrow, and L. Howells, Inhibition of cyclo-oxygenase 2 expression in colon cells by the chemopreventive agent curcumin involves inhibition of NF-kappaB activation via the NIK/IKK signalling complex. *Oncogene* **18**, 6013–6020 (1999).
21. R. K. Thomas, M. L. Sos, T. Zander, O. Mani, A. Popov, D. Berenbrinker, S. Smola-Hess, J. L. Schultze, and J. Wolf, Inhibition of nuclear translocation of nuclear

factor-kappaB despite lack of functional IkappaBalpha protein overcomes multiple defects in apoptosis signaling in human B-cell malignancies. *Clin Cancer Res* **11**, 8186–8194 (2005).

22. S. Shishodia, P. Potdar, C. G. Gairola, and B. B. Aggarwal, Curcumin (diferuloylmethane) down-regulates cigarette smoke-induced NF-kappaB activation through inhibition of IkappaBalpha kinase in human lung epithelial cells: Correlation with suppression of COX-2, MMP-9 and cyclin D1. *Carcinogenesis* **24**, 1269–1279 (2003).
23. S. Shishodia, H. M. Amin, R. Lai, and B. B.Aggarwal, Curcumin (diferuloylmethane) inhibits constitutive NF-kappaB activation, induces G1/S arrest, suppresses proliferation, and induces apoptosis in mantle cell lymphoma. *Biochem Pharmacol* **70**, 700–713 (2005).
24. S. Philip and G. C. Kundu, Osteopontin induces nuclear factor kappa B-mediated promatrix metalloproteinase-2 activation through I kappa B alpha /IKK signaling pathways, and curcumin (diferulolylmethane) down-regulates these pathways. *J Biol Chem* **278**, 14,487–14,497 (2003).
25. M. Zheng, S. Ekmekcioglu, E. T. Walch, C. H. Tang, and E. R. Grimm, Inhibition of nuclear factor-kappaB and nitric oxide by curcumin induces G2/M cell cycle arrest and apoptosis in human melanoma cells. *Melanoma Res* **14**, 165–171 (2004).
26. K. H. Kim, H. Y. Park, J. H. Nam, J. E. Park, J. Y Kim, M. I. Park, K. O. Chung, K. Y. Park, and J. Y. Koo, The inhibitory effect of curcumin on the growth of human colon cancer cells (HT-29, WiDr) in vitro. *Korean J Gastroenterol* **45**, 277–284 (2005).
27. B. B. Aggarwal, S. Shishodia, Y. Takada, S. Banerjee, R. A. Newman, C. E. Bueso-Ramos, and J. E. Price, Curcumin suppresses the paclitaxel-induced nuclear factor-kappaB pathway in breast cancer cells and inhibits lung metastasis of human breast cancer in nude mice. *Clin Cancer Res* **11,** 7490–7498 (2005).
28. R. Eferl and E. F. Wagner, AP-1: A double-edged sword in tumorigenesis. *Nat Rev Cancer* **3**, 859–868 (2003).
29. M. Takakura, S. Kyo, M. Inoue, W. E. Wright, and J. W. Shay, Function of AP-1 in transcription of the telomerase reverse transcriptase gene (TERT) in human and mouse cells. *Mol Cell Biol* **25**, 8037–8043 (2005).
30. T. Bowman, M. A. Broome, D. Sinibaldi, W. Wharton, W. J. Pledger, J. M. Sedivy, R. Irby, T. Yeatman, S. A. Courtneidge, and R. Jove, Stat3-mediated Myc expression is required for Src transformation and PDGF-induced mitogenesis. *Proc Natl Acad Sci USA* **98**, 7319–7324 (2001).
31. J. F. Bromberg, C. M. Horvath, D. Besser, W.W. Lathem, and J. E. Darnell, Jr., Stat3 activation is required for cellular transformation by v-src. *Mol Cell Biol* **18**, 2553–2558 (1998).
32. F. C. Hsieh, G. Cheng, and J. Lin, Evaluation of potential Stat3-regulated genes in human breast cancer. *Biochem Biophys Res Commun* **335**, 292–299 (2005).
33. J. Coyle-Rink, L. Del Valle, T. Sweet, K. Khalili, and S. Amini, Developmental expression of Wnt signaling factors in mouse brain. *Cancer Biol Ther* **1**, 640–645 (2002).
34. Y. Kim, R. C. Sills, and C. D. Houle, Overview of the molecular biology of hepatocellular neoplasms and hepatoblastomas of the mouse liver. *Toxicol Pathol* **33**, 175–180 (2005).
35. Y. Zhai, R. Wu, D. R. Schwartz, D. Darrah, H. Reed, F. T. Kolligs, M. T. Nieman, E. R. Fearon, and K. K. Cho, Role of beta-catenin/T-cell factor-regulated genes in ovarian endometrioid adenocarcinomas. *Am J Pathol* **160**, 1229–1238 (2002).

36. M. M. Montano, A. K. Jaiswal, and B. S. Katzenellenbogen, Transcriptional regulation of the human quinone reductase gene by antiestrogen-liganded estrogen receptor-alpha and estrogen receptor-beta. *J Biol Chem* **273**, 25,443–25,449 (1998).
37. M. Zhu and W. E. Fahl, Functional characterization of transcription regulators that interact with the electrophile response element. *Biochem Biophys Res Commun* **289**, 212–219 (2001).
38. A. Chen, J. Xu, and A. C. Johnson, Curcumin inhibits human colon cancer cell growth by suppressing gene expression of epidermal growth factor receptor through reducing the activity of the transcription factor Egr-1. *Oncogene* **25**, 278–287 (2006).
39. M. Fu, X. Zhu, J. Zhang, J. Liang, Y. Lin, L. Zhao, M. U. Ehrengruber, and Y. E. Chen, Egr-1 target genes in human endothelial cells identified by microarray analysis. *Gene* **315**, 33–41 (2003).
40. Y. Moon, F. G. Bottone, Jr., M. F. McEntee, and T. E. Eling, Suppression of tumor cell invasion by cyclooxygenase inhibitors is mediated by thrombospondin-1 via the early growth response gene Egr-1. *Mol Cancer Ther* **4**, 1551–1558 (2005).
41. L. Stefano, J. Al Sarraj, O. G. Rossler, C. Vinson, and G. Thiel, Up-regulation of tyrosine hydroxylase gene transcription by tetradecanoylphorbol acetate is mediated by the transcription factors Ets-like protein-1 (Elk-1) and Egr-1. *J Neurochem* **97**, 92–104 (2006).
42. D. Xiao, D. Chinnappan, R. Pestell, C. Albanese, and H. C. Weber, Bombesin regulates cyclin D1 expression through the early growth response protein Egr-1 in prostate cancer cells. *Cancer Res* **65**, 9934–9942 (2005).
43. J. Svaren, T. Ehrig, S. A. Abdulkadir, M. U. Ehrengruber, M. A. Watson, and J. Milbrandt, EGR1 target genes in prostate carcinoma cells identified by microarray analysis. *J Biol Chem* **275**, 38,524–38,531 (2000).
44. J. Li, M. Tan, L. Li, D. Pamarthy, T. S. Lawrence, and Y. Sun, SAK, a new polo-like kinase, is transcriptionally repressed by p53 and induces apoptosis upon RNAi silencing. *Neoplasia* **7**, 312–323 (2005).
45. Y. Sun, p53 and its downstream proteins as molecular targets of cancer. *Mol Carcinog* **45**, 409–415 (2006).
46. A. C. Wild, H. R. Moinova, and R. T. Mulcahy, Regulation of gamma-glutamylcysteine synthetase subunit gene expression by the transcription factor Nrf2. *J Biol Chem* **274**, 33,627–33,636 (1999).
47. J. Kim and G. A. Coetzee, Prostate specific antigen gene regulation by androgen receptor. *J Cell Biochem* **93**, 233–241 (2004).
48. P. Petrusz, D. A. Jeyaraj, and G. Grossman, Microarray analysis of androgen-regulated gene expression in testis: the use of the androgen-binding protein (ABP)-transgenic mouse as a model. *Reprod Biol Endocrinol* **3**, 70 (2005).
49. I. Bogacka, H. Xie, G. A. Bray, and S. R. Smith, The effect of pioglitazone on peroxisome proliferator-activated receptor-gamma target genes related to lipid storage in vivo. *Diabetes Care* **27**, 1660–1667 (2004).
50. P. Delerive, F. Martin-Nizard, G. Chinetti, F. Trottein, J. C. Fruchart, J. Najib, P. Duriez, and B. Staels, Peroxisome proliferator-activated receptor activators inhibit thrombin-induced endothelin-1 production in human vascular endothelial cells by inhibiting the activator protein-1 signaling pathway. *Circ Res* **85**, 394–402 (1999).
51. S. Hummasti and P. Tontonoz, The peroxisome proliferator-activated receptor N-terminal domain controls isotype-selective gene expression and adipogenesis. *Mol Endocrinol* **20**, 1261–1275 (2006).

52. P. Dikshit, A. Goswami, A. Mishra, M. Chatterjee, and N. R. Jana, Curcumin induces stress response, neurite outgrowth and prevent NF-kappaB activation by inhibiting the proteasome function. *Neurotox Res* **9**, 29–37 (2006).
53. C. H. Yeh, Y. M. Lin, Y. C. Wu, and P. J. Lin, Inhibition of NF-kappa B activation can attenuate ischemia/reperfusion-induced contractility impairment via decreasing cardiomyocytic proinflammatory gene up-regulation and matrix metalloproteinase expression. *J Cardiovasc Pharmacol* **45**, 301–309 (2005).
54. S. Wessler, P. Muenzner, T. F. Meyer, and M. Naumann, The anti-inflammatory compound curcumin inhibits Neisseria gonorrhoeae-induced NF-kappaB signaling, release of pro-inflammatory cytokines/chemokines and attenuates adhesion in late infection. *Biol Chem* **386**, 481–490 (2005).
55. M. Karin, Z. Liu, and E. Zandi, AP-1 function and regulation. *Curr Opin Cell Biol* **9**, 240–246 (1997).
56. Y. Xia, C. Makris, B. Su, E. Li, J. Yang, G. R. Nemerow, and M. Karin, MEK kinase 1 is critically required for c-Jun N-terminal kinase activation by proinflammatory stimuli and growth factor-induced cell migration. *Proc Natl Acad Sci USA* **97**, 5243–5248 (2000).
57. C. Huang, J. Li, W. Y. Ma, and Z. Dong, JNK activation is required for JB6 cell transformation induced by tumor necrosis factor-alpha but not by 12-O-tetradecanoylphorbol-13-acetate. *J Biol Chem* **274**, 29,672–29,676 (1999).
58. M. T. Huang, T. Lysz, T. Ferraro, T. F. Abidi, J. D. Laskin, and A. H. Conney, Inhibitory effects of curcumin on in vitro lipoxygenase and cyclooxygenase activities in mouse epidermis. *Cancer Res* **51**, 813–819 (1991).
59. Y. R. Chen and T. H. Tan, Inhibition of the c-Jun N-terminal kinase (JNK) signaling pathway by curcumin. *Oncogene* **17**, 173–178 (1998).
60. A. Bierhaus, Y. Zhang, P. Quehenberger, T. Luther, M. Haase, M. Muller, N. Mackman, R. Ziegler, and P. P. Nawroth, The dietary pigment curcumin reduces endothelial tissue factor gene expression by inhibiting binding of AP-1 to the DNA and activation of NF-kappa B. *Thromb Haemost* **77**, 772–782 (1997).
61. D. A. Dickinson, K. E. Iles, H. Zhang, V. Blank, and H. J. Forman, Curcumin alters EpRE and AP-1 binding complexes and elevates glutamate-cysteine ligase gene expression. *FASEB J* **17**, 473–475 (2003).
62. M. Tomita, H. Kawakami, J. N. Uchihara, T. Okudaira, M. Masuda, N. Takasu, T. Matsuda, T. Ohta, Y. Tanaka, and N. Mori, Curcumin suppresses constitutive activation of AP-1 by downregulation of JunD protein in HTLV-1-infected T-cell lines. *Leuk Res* **30**, 313–321 (2006).
63. C. S. Divya and M. R. Pillai, Antitumor action of curcumin in human papillomavirus associated cells involves downregulation of viral oncogenes, prevention of NFkB and AP-1 translocation, and modulation of apoptosis. *Mol Carcinog* **45**, 320–332 (2006).
64. C. Polytarchou, M. Hatziapostolou, and E. Papadimitriou, Hydrogen peroxide stimulates proliferation and migration of human prostate cancer cells through activation of activator protein-1 and up-regulation of the heparin affin regulatory peptide gene. *J Biol Chem* **280**, 40,428–40,435 (2005).
65. B. K. Prusty and B. C. Das, Constitutive activation of transcription factor AP-1 in cervical cancer and suppression of human papillomavirus (HPV) transcription and AP-1 activity in HeLa cells by curcumin. *Int J Cancer* **113**, 951–960 (2005).
66. M. S. Squires, E. A. Hudson, L. Howells, S; Sale, C. E. Houghton, J. L. Jones, L. H. Fox, M. Dickens, S. A. Prigent, and M. M. Manson, Relevance of mitogen activated protein kinase (MAPK) and phosphotidylinositol-3-kinase/protein kinase B

(PI3K/PKB) pathways to induction of apoptosis by curcumin in breast cells. *Biochem Pharmacol* **65**, 361–376 (2003).

67. M. L. Cho, Y. O. Jung, Y. M. Moon, S. Y. Min, C. H. Yoon, S. H. Lee, S. H. Park, C. S. Cho, D. M. Jue, and H. Y. Kim, Interleukin-18 induces the production of vascular endothelial growth factor (VEGF) in rheumatoid arthritis synovial fibroblasts via AP-1-dependent pathways. *Immunol Lett* **103**, 159–166 (2006).
68. U. R. Pendurthi and L. V. Rao, Suppression of transcription factor Egr-1 by curcumin. *Thromb Res* **97**, 179–189 (2000).
69. J. E. Darnell, Jr., I. M. Kerr, and G. R. Stark, Jak-STAT pathways and transcriptional activation in response to IFNs and other extracellular signaling proteins. *Science* **264**, 1415–1421 (1994).
70. J. E. Darnell, Jr., STATs and gene regulation. *Science* **277**, 1630–1635 (1997).
71. C. L. Yu, D. J. Meyer, G. S. Campbell, A. C. Larner, C. Carter-Su, J. Schwartz, and R. Jove, Enhanced DNA-binding activity of a Stat3-related protein in cells transformed by the Src oncoprotein. *Science* **269**, 81–83 (1995).
72. J. R. Grandis, S. D. Drenning, A. Chakraborty, M. Y. Zhou, Q. Zeng, A.S. Pitt, and D. J. Tweardy, Requirement of Stat3 but not Stat1 activation for epidermal growth factor receptor- mediated cell growth In vitro. *J Clin Invest* **102**, 1385–1392 (1998).
73. N. Carlesso, D. A. Frank, and J. D. Griffin, Tyrosyl phosphorylation and DNA binding activity of signal transducers and activators of transcription (STAT) proteins in hematopoietic cell lines transformed by Bcr/Abl. *J Exp Med* **183**, 811–820 (1996).
74. R. M. Weber-Nordt, C. Egen, J. Wehinger, W. Ludwig, V. Gouilleux-Gruart, R. Mertelsmann, and J. Finke, Constitutive activation of STAT proteins in primary lymphoid and myeloid leukemia cells and in Epstein-Barr virus (EBV)-related lymphoma cell lines. *Blood* **88**, 809–816 (1996).
75. A. C. Bharti, N. Donato, and B. B. Aggarwal, Curcumin (diferuloylmethane) inhibits constitutive and IL-6-inducible STAT3 phosphorylation in human multiple myeloma cells. *J Immunol* **171**, 3863–3871 (2003).
76. H. Y. Kim, E. J. Park, E. H. Joe, and I. Jou, Curcumin suppresses Janus kinase-STAT inflammatory signaling through activation of Src homology 2 domain-containing tyrosine phosphatase 2 in brain microglia. *J Immunol* **171**, 6072–6079 (2003).
77. W. Q. Li, F. Dehnade, and M. Zafarullah, Oncostatin M-induced matrix metalloproteinase and tissue inhibitor of metalloproteinase-3 genes expression in chondrocytes requires Janus kinase/STAT signaling pathway. *J Immunol* **166**, 3491–3498 (2001).
78. C. Natarajan and J. J. Bright, Curcumin inhibits experimental allergic encephalomyelitis by blocking IL-12 signaling through Janus kinase-STAT pathway in T lymphocytes. *J Immunol* **168**, 6506–6513 (2002).
79. W. H. Chen, Y. Chen, G. H. Cui, J. X. Gu, D. Hu, W. K. Chen, and X. G. Li, Effect of curcumin on STAT5 signaling pathway in primary CML cells. *Zhongguo Shi Yan Xue Ye Xue Za Zhi* **12**, 572–576 (2004).
80. W. H. Chen, Y. Chen, J. X. Gu, and J. He, Effect of curcumin on STAT5 signaling molecule in K562 cells. *Zhonghua Xue Ye Xue Za Zhi* **25**, 151–153 (2004).
81. J. Xu, Y. Fu, and A. Chen, Activation of peroxisome proliferator-activated receptor-gamma contributes to the inhibitory effects of curcumin on rat hepatic stellate cell growth. *Am J Physiol Gastrointest Liver Physiol* **285**, G20–G30 (2003).
82. A. Chen and J. Xu, Activation of PPAR{gamma} by curcumin inhibits Moser cell growth and mediates suppression of gene expression of cyclin D1 and EGFR. *Am J Physiol Gastrointest Liver Physiol* **288**, G447–G456 (2005).

83. S. Zheng and A. Chen, Curcumin suppresses the expression of extracellular matrix genes in activated hepatic stellate cells by inhibiting gene expression of connective tissue growth factor. *Am J Physiol Gastrointest Liver Physiol* **290**, G883–G893 (2006).
84. A. M. Siddiqui, X. Cui, R. Wu, W. Dong, M. Zhou, M. Hu, H. H. Simms and P. Wang, The anti-inflammatory effect of curcumin in an experimental model of sepsis is mediated by up-regulation of peroxisome proliferator-activated receptor-gamma. *Crit Care Med* **34**, 1874–1882 (2006).
85. W. S. el-Deiry, T. Tokino, V. E. Velculescu, D. B. Levy, R. Parsons, J. M. Trent, D. Lin, W. E. Mercer, K. W. Kinzler, and B. Vogelstein, WAF1, a potential mediator of p53 tumor suppression. *Cell* **75**, 817–825 (1993).
86. B. Vogelstein and K. W.Kinzler, p53 function and dysfunction. *Cell* **70**, 523–526 (1992).
87. S. S. Han, S. T. Chung, D. A. Robertson, D. Ranjan, and S. Bondada, Curcumin causes the growth arrest and apoptosis of B cell lymphoma by downregulation of egr-1, c-myc, bcl-XL, NF-kappa B, and p53. *Clin Immunol* **93**, 152–161 (1999).
88. J. A. Bush, K. J. Cheung, Jr., and G Li, Curcumin induces apoptosis in human melanoma cells through a Fas receptor/caspase-8 pathway independent of p53. *Exp Cell Res* **271**, 305–314 (2001).
89. T. Choudhuri, S. Pal, M. L. Agwarwal, T. Das, and G. Sa, Curcumin induces apoptosis in human breast cancer cells through p53-dependent Bax induction. *FEBS Lett* **512**, 334–340 (2002).
90. T. Choudhuri, S. Pal, T. Das, and G. Sa, Curcumin selectively induces apoptosis in deregulated cyclin D1-expressed cells at G2 phase of cell cycle in a p53-dependent manner. *J Biol Chem* **280**, 20,059–20,068 (2005).
91. M. J. Park, E. H. Kim, I. C. Park, H. C. Lee, S. H. Woo, J. Y. Lee, Y. J. Hong, C. H. Rhee, S. H. Choi, B. S. Shim, et al., Curcumin inhibits cell cycle progression of immortalized human umbilical vein endothelial (ECV304) cells by up-regulating cyclin-dependent kinase inhibitor, p21WAF1/CIP1, p27KIP1 and p53. *Int J Oncol* **21**, 379–383 (2002).
92. A. Liontas and H. Yeger, Curcumin and resveratrol induce apoptosis and nuclear translocation and activation of p53 in human neuroblastoma. *Anticancer Res* **24**, 987–998 (2004).
93. A. S. Jaiswal, B. P. Marlow, N. Gupta, and S. Narayan, Beta-catenin-mediated transactivation and cell-cell adhesion pathways are important in curcumin (diferuylmethane)-induced growth arrest and apoptosis in colon cancer cells. *Oncogene* **21**, 8414–8427 (2002).
94. N. N. Mahmoud, A. M. Carothers, D. Grunberger, R. T. Bilinski, M. R. Churchill, C. Martucci, H. L. Newmark, and M. M. Bertagnolli, Plant phenolics decrease intestinal tumors in an animal model of familial adenomatous polyposis. *Carcinogenesis* **21**, 921–927 (2000).
95. E. Balogun, M. Hoque, P. Gong, E. Killeen, C. J. Green, R. Foresti, J. Alam, and R. Motterlini, Curcumin activates the haem oxygenase-1 gene via regulation of Nrf2 and the antioxidant-responsive element. *Biochem J* **371**, 887–895 (2003).
96. S. A. Rushworth, R. M. Ogborne, C. A. Charalambos, and M. A. O'Connell, Role of protein kinase C delta in curcumin-induced antioxidant response element-mediated gene expression in human monocytes. *Biochem Biophys Res Commun* **341**, 1007–1016 (2006).
97. A. Banning, S. Deubel, D. Kluth, Z. Zhou, and R. Brigelius-Flohe, The GI-GPx gene is a target for Nrf2. *Mol Cell Biol* **25**, 4914–4923 (2005).

98. K. Nakamura, Y. Yasunaga, T. Segawa, D. Ko, J. W. Moul, S. Srivastava, and J. S. Rhim, Curcumin down-regulates AR gene expression and activation in prostate cancer cell lines. *Int J Oncol* **21**, 825–830 (2002).
99. M. G. Marcu, Y. J. Jung, S. Lee, E. J. Chung, M. J.Lee, J. Trepel, and L. Neckers, Curcumin is an inhibitor of p300 histone acetylatransferase. *Med Chem* **2**, 169–174 (2006).
100. K. Balasubramanyam, R. A. Varier, M. Altaf, V. Swaminathan, N. B. Siddappa, U. Ranga, and T. K. Kundu, Curcumin, a novel p300/CREB-binding protein-specific inhibitor of acetyltransferase, represses the acetylation of histone/nonhistone proteins and histone acetyltransferase-dependent chromatin transcription. *J Biol Chem* **279**, 51,163–51,171 (2004).

CANCER CHEMOPREVENTIVE EFFECTS OF CURCUMIN

Young-Joon Surh and Kyung-Soo Chun

Abstract: Chemoprevention, which is referred to as the use of nontoxic natural or synthetic chemicals to intervene in multistage carcinogenesis, has emerged as a promising and pragmatic medical approach to reduce the risk of cancer. Numerous components of edible plants, collectively termed "phytochemicals" have been reported to possess substantial chemopreventive properties. Curcumin, a yellow coloring ingredient derived from *Curcuma longa* L. (Zingiberaceae), is one of the most extensively investigated and well-defined chemopreventive phytochemicals. Curcumin has been shown to protect against skin, oral, intestinal, and colon carcinogenesis and also to suppress angiogenesis and metastasis in a variety animal tumor models. It also inhibits the proliferation of cancer cells by arresting them in the various phases of the cell cycle and by inducing apoptosis. Moreover, curcumin has a capability to inhibit carcinogen bioactivation via suppression of specific cytochrome P450 isozymes, as well as to induce the activity or expression of phase II carcinogen detoxifying enzymes. Well-designed intervention studies are necessary to assess the chemopreventive efficacy of curcumin in normal individuals as well as high-risk groups. Sufficient data from pharmacodynamic as well as mechanistic studies are necessary to advocate clinical evaluation of curcumin for its chemopreventive potential.

1. INTRODUCTION

Over the past few decades, there has been tremendous progress in our understanding of the molecular biology of cancer. Nonetheless, we have not conquered this dread disease yet. Like the majority of other human disorders, cancer is basically preventable. One of the most promising approaches to reduce the risk of cancer is chemoprevention.[1,2] According to both clinical observations and experimental models, cancer develops in a stepwise fashion starting with a single oncogenic mutation (either of a proto-oncogene or a tumor suppressor gene, or both) in a single cell. Chemoprevention is the attempt to use natural and synthetic compounds to intervene in the early precancerous stages of carcinogenesis, before malignancy manifests. Recently, there have been considerable efforts to search for naturally occurring substances for the intervention of carcinogenesis. Many components derived from dietary or medicinal plants have been found to possess substantial chemopreventive properties.[3] One good example is curcumin, which can act in all stages of multistep carcinogenesis. Chemopreventive properties of curcumin have been extensively investigated and well documented.[4–7]

2. CHEMOPREVENTIVE EFFECTS OF CURCUMIN

2.1. Animal Studies

Animal models are an important component of chemopreventive research. They provide a means of identifying effective compounds, of carrying out fundamental investigations into their mechanisms of action, of determining how they can be used optimally, of evaluating safety and toxicity, and, ultimately, of providing an information base for developing intervention trials in humans.[8,9]

The anticarcinogenic effects of curcumin and its underlying mechanisms have been investigated in several animal tumor systems, including skin, colon, lung, duodenal, stomach, esophageal, and oral carcinogenesis (*vide infra*).

2.1.1. Chemopreventive Effects on Skin Carcinogenesis

The mouse skin carcinogenesis model is an excellent animal model in studying molecular alterations associated with the multistep process of malignant transformation.[10] Topical application of curcumin (1–10 μmol) together with 5 nmol of the prototype tumor promoter 12-*O*-tetradecanoylphorbol-13-acetate (TPA), twice weekly for 20 weeks to female CD-1 mice previously initiated with 7,12-dimethylbenz[*a*]anthracene (DMBA), strongly inhibited papilloma formation.[6,11] In an additional study, topical application of relatively low doses of curcumin (20 or 100 nmol) markedly abrogated TPA-induced tumor promotion.[12] Topical application of commercial-grade curcumin (consisting of approximately 77% curcumin, 17% demethoxycurcumin, and 3% bis-demethoxycurcumin), pure curcumin, or demethoxycurcumin had almost equipotent inhibitory effects on TPA-induced tumor promotion in DMBA-initiated mouse skin carcinogenesis.[13] In other studies, dietary administration of 2% turmeric also significantly inhibited DMBA plus TPA-induced skin tumor formation in female Swiss mice.[14] In a benzo[*a*]pyrene (B[a]P)-initiated and TPA-promoted two-stage skin tumorigenesis model, 3 or 10 μmol curcumin reduced the number of tumors per mouse by 58% or 62%, respectively. The percentage of tumor-bearing mice was decreased by 18–25%.

2.1.2. Chemopreventive Effects on Hepatocarcinogenesis

N-Nitrosodiethylamine (DENA) is a powerful hepatocarcinogen in experimental animals.[15] Phenobarbital acts as a promoter of hepatocarcinogenesis when administered subsequent to an initiating agent like DENA. Curcumin (100 mg/kg/day) attenuated the DENA-initiated and phenobarbital-promoted formation of hepatic hyperplastic nodules, body weight loss, and hypoproteinemia in Wistar rats.[16] The chemopreventive effect of curcumin was also demonstrated in a murine hepatocarcinogenesis model. Five-week-old C3H/HeN mice were injected i.p. with *N*-diethylnitrosamine. A group of mice were fed a 0.2% curcumin-containing diet, starting 4 days before *N*-diethylnitrosamine DEN injection and until termination of the experiment. At the age of 42 weeks, the curcumin group had an 81% reduction in the multiplicity and a 62% reduction in the incidence of hepatocarcinoma.[17] Busquets et al. evaluated the chemopreventive potential of curcumin by treating rats that had been inoculated with the Yoshida AH-130 ascites hepatoma, a fast-growing tumor that causes the death of the animal in about 10 days after inoculation.[18] The results showed that curcumin treatment, significantly decreased tumor growth by 31%, and the tumor volume was not significantly affected.

2.1.3. Chemopreventive Effects on Colorectal Carcinogenesis

The Min/+ mouse has an autosomal dominant heterozygous nonsense mutation of the tumor suppressor *Apc* gene,[19] which is homologous to human germ-line and somatic *APC* mutations. The C57Bl/6J Min/+ inbred mouse model is particularly advantageous for investigating chemopreventive agents, particularly those target early-stage lesions because scores of adenomas grow to a detectable size within a few months on a defined genetic background.[20] Because Min/+ mice develop adenomas as a result of inactivation of the same tumor suppressor gene known to underlie the pathogenesis of most colon cancers in humans, use of this model seems likely to be appropriate to the design of human chemoprevention trials.[21]

Min/+ mice received curcumin mixed with their diet commencing 1 week post-weaning. Animals were killed at the 18th week of age, and the tumor multiplicity as well as the size was inspected. Dietary curcumin (0.2% and 0.5%) significantly reduced the intestinal tumor burden.[22] In another study, dietary administration of 0.1% curcumin for 10 weeks reduced the frequency of intestinal tumors in Min/+ mice by 65%.[23] The azoxymethane (AOM)-induced colorectal tumor model is also widely employed in studying the chemopreventive effects of dietary agents. AOM-induced tumors share many histopathologic characteristics with human tumors.[21] Subcutaneous injections of AOM (10 mg/kg body weight) once a week for 6 consecutive weeks) produced hyperplasia and the focal areas of dysplasia in the female CF-1 mouse colon. Administration of 2% curcumin in the diet inhibited AOM-induced formation of dysplasia.[24] Commercial-grade dietary curcumin also inhibited AOM-induced mouse colon carcinogenesis.[25] In an AOM-induced colon cancer model using male Fischer 344 rats, 2000 ppm of curcumin in the diet significantly lowered the incidence of colon adenocarcinomas and the multiplicity of both invasive and noninvasive adenocarcinomas. Furthermore, dietary curcumin significantly reduced the colon tumor volume by 57% in Fischer F344 rats.[26] Similarly, curcumin (8 and 16 g/kg) in the diet elicited pronounced inhibition of both incidence and multiplicity of adenomas in the same animal model.[27]

2.1.4. Chemopreventive Effects on Oral Carcinogenesis

Male F344 rats fed the diet containing curcumin at a dose of 0.5 g/kg diet (500 ppm) during the initiation and post-initiation stages exhibited 91% reduction in the frequency of 4-nitroquinoline-1-oxide-induced tongue carcinoma by 91%.[28] The incidence of oral preneoplasia was also decreased by curcumin administration. Likewise, dietary curcumin significantly inhibited *N*-nitrosomethylbenzylamine-induced esophageal carcinogenesis in rats when given during the postinitiation phase as well as the initiation phase.[29] In Syrian golden hamsters, curcumin treatment protected against DMBA-induced or methyl-(acetoxymethyl)-nitrosamine-induced oral mucosal tumorigenesis.[30,31]

2.1.5. Chemopreventive Effects on Stomach Carcinogenesis

Ikezaki et al. investigated the modifying effects of curcumin on glandular stomach carcinogenesis in male Wistar rats treated with *N*-methyl-*N′*-nitro-*N*-nitrosoguanisine and sodium chloride.[32] The total incidence of combined atypical hyperplasias and adenocarcinomas produced in the glandular stomach was 10% lower in the 0.05% curcumin-fed group than that observed in the basal diet-fed group. However, the mean number of atypical hyperplasias or adenocarcinomas in the glandular stomach of the curcumin-treated rats was decreased by 45%. Turmeric (2% or 5%) in the diet significantly inhibited the formation of B[a]P-induced forestomach tumors in female Swiss mice, and this response was dose- and time-dependent.[30] Commercial-grade dietary curcumin inhibited B[a]P-induced forestomach carcinogenesis and *N*-ethyl-*N′*-nitro-*N*-nitrosoguanidine-induced duodenal carcinogenesis.[25]

2.1.6. Chemopreventive Effects on Other Malignancies

Intraperitoneal administration of curcumin at the dose of 100 mg/kg or 200 mg/kg significantly decreased the number of palpable mammary tumors and suppressed the production of mammary adenocarcinomas in Sprague–Dawley rats.[33] Similarly, the formation of mammary DMBA–DNA adducts was inhibited by curcumin treatment. In contrast, tumor development for female rats fed diets containing 1.0% curcumin for 14 days prior to dosing with DMBA was not different from that of controls. These data suggest that curcumin, when given by systemic administration, can act as an effective chemopreventative agent toward DMBA-induced rat mammary tumorigenesis.[34] The potential chemopreventive activity of curcumin on experimentally induced pulmonary tumorigenesis was evaluated.[35] Treatment with dietary curcumin (2000 ppm) starting from 1 week after treatment with a mixture of B[a]P and the tobacco-specific nitrosamine 4-(methyl-nitrosamino)-1-(3-pyridyl)-1-butanone (NNK) until termination of the experiment had no effect on lung tumor multiplicity. Long–Evans Cinnamon (LEC) rats, an inbred mutant strain that accumulates copper due to an aberrant copper-transporting ATPase gene, develop acute hepatitis, chronic liver injury, and liver tumors as a result of copper-induced oxidative stress, lipid peroxidation, and DNA damage. Frank et al. investigated the modulating role of curcumin in liver and kidney carcinogenesis in LEC rats.[36] In untreated LEC rats, the frequencies of acute liver failure, hepatomas, and kidney tumors were 32%, 100%, and 10%, respectively, which were not altered by curcumin treatment. However, benign and malignant tumors at other sites that included leukemia, bladder carcinoma, prostate carcinoma, and phaeochromocytoma were only present in the untreated control group but not in the curcumin treatment group.

Six-week-old female SENCAR mice were administered 1 mg of DMBA by gavage once a week for 5 weeks. At the 20th week following the first dose of DMBA, 68% of mice developed mammary tumors and 45% had lymphomas/leukemias. Feeding the 2% curcumin diet had little or no effect on the incidence of mammary tumors, but the incidence of lymphomas/leukemias was reduced by 53%.[11]

2.1.7. Chemopreventive Effects on Angiogenesis and Metastasis

Angiogenesis is now regarded as a critical event to the transition of premalignant lesions in a hyperproliferative state to the malignant phenotype, thus facilitating tumor growth and metastasis.[37] The intensity of angiogenesis, as assessed by counting microvessels in malignant tissues, acts as a prognostic factor for many solid tumors such as breast, prostate, and ovarian cancers and very early cancers of the endometrium and lung.[38]

The antiangiogenic effect of curcumin on tumor neogenesis was investigated by carefully assessing the neocapillary formation induced by hepatocellular carcinoma cells (HepG2) in mice. In the curcumin (3000 mg/kg)-treated group, the increase of tumor neocapillary density was attenuated significantly.[39] Ohashi et al. showed that daily oral administration of curcumin suppressed intrahepatic metastasis of hepatocellular carcinoma CBO140C12 cells, and the growth of implanted tumors was not affected.[40] Hong et al. attempted to characterize the antimetastatic effects of a curcumin-supplemented diet using a prostate xenograft model. All of the mice developed tumors on their foot pads, following the implantation of the prostate carcinoma DU-145 cells.[41] Curcumin markedly reduced the tumor volume in the tumor-bearing site. The metastatic nodules were significantly fewer in the curcumin-treated group than those in the untreated group. Dorai et al. injected subcutaneously into nude mice the androgen-dependent LNCaP prostate cancer cells mixed with matrigel.[42] A significant decrease in the microvessel density as measured by CD31 antigen staining was seen in the 2% curcumin-fed group after the 6th week. Aggarwal and his colleagues investigated the ability of curcumin to modulate human breast cancer metastasis to the lung in a nude mouse xenograft model.[43] They found that curcumin plus paclitaxel (Taxol) suppressed the incidence of breast cancer metastasis in the lung. Treatment with 10 mg/kg paclitaxel alone modestly reduced the incidence of metastasis. However, both curcumin alone and curcumin plus paclitaxel significantly reduced the incidence and the number of metastasized nodules in the lung. Although there was no direct inhibitory effect of curcumin on pulmonary carcinogenesis, curcumin was tested for the inhibition of lung metastasis induced by B16F10 melanoma cells in mice.[44] Oral administration of curcumin at 200 nmol/kg body weight was found to inhibit the pulmonary metastasis as assessed by the reduction in the number of lung tumor nodules (80%). Consequent to the inhibition of the formation of lung tumor nodules, the life span of animals treated with curcumin was substantially increased. The formation of metastases from liver and kidney tumors into the lung and peritoneum in LEC rats was totally prevented by curcumin in spite of the lack of direct chemopreventive activities of curcumin on tumorigenesis toward these organs.[36]

2.2. Epidemiologic and Clinical Studies

An anecdote introducing the potential application of curcumin as a topical treatment for oral cancers and leukoplakia has been reported.[45] According to this report, there was a reduction in the size of the lesions observed in 10% of the 62 patients treated, but there was no control group included.

In a pilot study conducted in Leicester, UK, standardized curcumin extracts containing 36–180 mg curcumin were administered daily for up to 4 months to patients with progressive advanced colorectal cancer refractory to standard chemotherapies.[46] Oral curcuma extract containing up to 180 mg curcumin was well tolerated, and dose-limiting toxicity was not observed. Neither curcumin nor its metabolite was detected in blood or urine, but curcumin was recovered from feces, which reflected the low systemic bioavailability as well as poor gastrointestinal absorption of the compound after oral dosing. Five of 15 patients treated in this study experienced radiologically stable disease staus for 3 months. The venous level of a tumor marker, carcino-embryonic antigen was decreased significantly in one patient.[46] In a subsequent phase 1 clinical trial, patients with progressive advanced colorectal cancer were given tablets containing various doses of curcumin, ranging from 0.45 g to 3.6 g, administered daily for up to 4 months. Again, there was no discernible toxicity observed except mild diarrhea.[47] Curcumin and its glucuronide and sulfate metabolites were detected in plasma and urine. Consumption of a 3.6-g curcumin tablet inhibited the production of prostaglandin E_2 (PGE_2) in blood leukocytes as measured *ex vivo*. Two of 15 patients experienced radiologically stable conditions for 4 months after the curcumin treatment. Based on these observations, a daily oral dose of 3.6 g curcumin has been advocated for evaluation in the prevention or treatment of malignancies at sites other than the gastrointestinal tract.[47] Cheng and colleagues completed a phase I clinical trial of curcumin (1–8 g per day) in patients with pre-invasive malignant or high-risk premalignant conditions of the bladder, skin, cervix, stomach, or oral mucosa in Taiwan.[48] This study demonstrated that curcumin is not toxic to humans up to 8 g per day when taken orally for 3 months, although the quality of life was not recorded. Garcea and colleagues reported that ingestion of 3.6 g of curcumin daily for a week decreased the levels of the oxidative DNA adduct 3-(2-deoxy-β-di-erythro-pentafuranosyl) pyr[1,2-α]-purin-10(3*H*) one in patients with colorectal cancer.[49] The same adduct levels were 2.5-fold higher in malignant tissues compared with normal ones.

Currently, the National Cancer Institute is investigating nearly 40 agents in phase I to III human clinical trials. Curcumin is one of those prospective agents in phase I clinical trials pending its evaluation for the prevention of colon, breast, lung, and prostate cancer. Enthusiasm for curcumin in cancer prevention stems from its performance in pre-clinical settings (http://www.cancer.gov/clinicaltrials/view_clini-caltrials.aspx?version = patient&cdrid = 67916&clickitem = ClinicalTrials SearchResult, NCI).

3. MECHANISMS UNDERLYING CHEMOPREVENTIVE ACTIVITIES OF CURCUMIN

The aforementioned chemopreventive effects of curcumin are likely to be the sum of several distinct mechanisms (Figure 1). The plausible mechanisms by which curcumin exerts its inhibitory effects on multistage carcinogenesis are described below and summarized in Table 1.

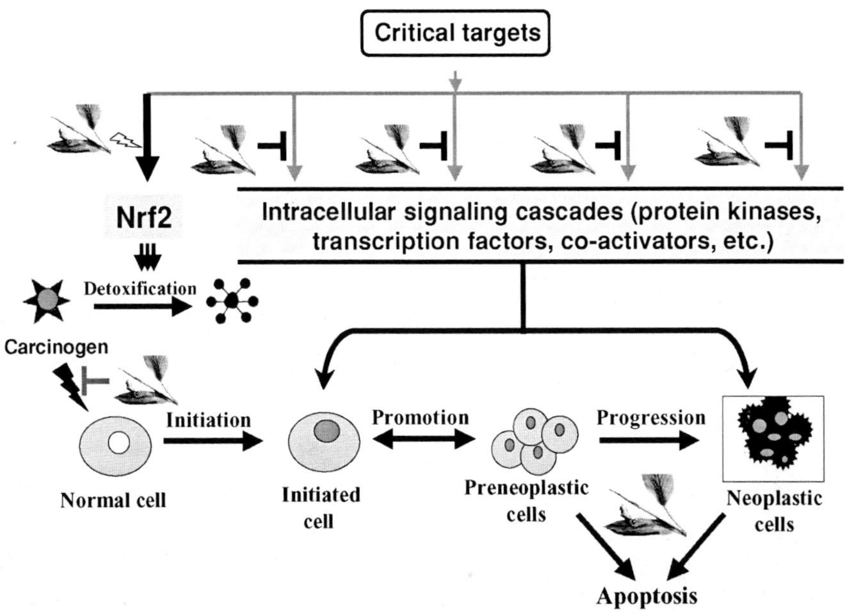

Figure 1. Effects of curcumin in turmeric on multistage carcinogenesis. Curcumin inhibits tumor initiation by blocking the metabolic activation of carcinogens or by stimulating their detoxification. It also exerts antitumor-promoting effects by suppressing inflammatory signaling mainly mediated by COX-2 and iNOS that are under the control of NF-κB and other transcription factors. Curcumin also acts in the progression stage of carcinogenesis by inhibiting metastasis and angiogenesis, which are crucial for the survival and spread of tumor cells. Furthermore, curcumin has antiproliferative effects that are attributed to its capability to induce apoptosis of precancerous and malignant cells or inhibit the cell cycle progression. The mechanistic basis for the aforementioned chemopreventive effects of curcumin is summarized in Table 1 and described more in detail in the Section 3 of this chapter and also in other chapters.

Table 1. Summary of chemopreventive effects of curcumin and underlying mechanisms.

CHEMOPREVENTIVE EFFECTS	MECHANISMS	CELL LINES/ANIMAL MODELS EMPLOYED
Inhibition of carcinogen activation	Inhibition of cytochrome P450 activity/ expression	MCF-7 cells,[50] rat liver microsome,[51] Sprague–Dawley rats,[51] male Swiss mice,[52] female Swiss Webster mice[53]
Stimulation of carcinogen detoxification	Induction of GST activity	Female A/J mice,[54] male ddY mice,[55] male F344 rats,[56] female Swiss Webster mice[53]
	Induction of UGT activity	Male Wistar rats[57]
	Upregulation of HO-1 expression/activity	Human hepatocytes,[58] porcine renal epithelial cells,[59] vascular endothelial cells,[60] human renal proximal tubule cells[61]
Anti-inflammation	Suppression of COX-2 expression or activity	Mouse epidermis,[62] human gastrointestinal cells,[63] human colonic epithelial cells,[64] mouse skin,[65] BV2 microglial cells[66]
	Suppression of iNOS expression or activity	RAW 264.7 cells,[67] BALB/c mouse peritoneal macrophages[68]
Apoptosis induction	Stimulation of caspase activity	K562 cells,[69] human melanom a cells,[70] neuro 2a cells[71]
	Suppression of cell survival signaling	U937 monocytic lymphoma cells,[72,73] human kidney carcinoma cells,[74] DU145 cells[75]
Inhibition of cell cycle progression	G2/M arrest	HL-60 cells,[76] MCF-7 human breast tumor cell line,[77] human colon cancer cells,[78–80] gastric KATO-III cells[80]
	G0/G1 arrest	Human umbilical vein endothelial cells,[81] human kidney carcinoma cells[74]
Inhibition of angiogenesis	Inhibition of VEGF expression	Ehrlich ascites tumor (EAT) cells, NIH 3T3 cells, and HUVEC cells,[82] U937 and Raji cells,[83] MDA-MB-231 cells[84]
	Inhibition of CD13/ aminopeptidase	APN-(+) versus APN-(−) tumor cells[85]
Inhibition of metastasis and invasion	Inhibition of MMP expression or activity	Murine melanoma cells B16F10,[86] human hepatocellular carcinoma SK-Hep-1 cells,[87] hepatocellular carcinoma CBO140C12 cells,[40] MCF10A human breast epithelial cells,[88] human astroglioma cells[89]
	Inhibition of CD13/ aminopeptidase	APN-(+) versus APN-(−) tumor cells[85]
Inhibition of oncogene expression/activation	Decreased c-*ras*, c-*jun*, c-*fos* and c- *myc* expression	CD-1 mouse skin,[90] mouse tumorous skin,[91] SKH-1 mouse skin,[92] mouse epidermal JB6 cells,[92] B lymphoma cells,[93] HCT-116 cells[94]
Potentiation of tumor suppressor function	Increased p53 accumulation or phosphorylation	MCF-7,[95] human breast cancer cell lines TR9–7,[96] human neuroblastoma cell lines,[97] human ovarian cancer cells,[98] colon adenocarcinoma cells HT-29[99]

3.1. Inhibition of Carcinogen Activation

Modulation of enzymes involved in metabolic activation and detoxification/excretion of carcinogens is one of the representative mechanisms of chemopreventive agents.[100] Phase I enzymes, including those that belong to the cytochromes P450 (CYP450) superfamily, are intended, in general, to facilitate the elimination of toxic xenobiotics by the addition of functional groups, rendering these compounds more water soluble. Although phase I functionalization reaction might be necessary for efficient detoxification, induction of phase I xenobiotic metabolizing enzymes might also be detrimental, as it can produce ultimate electrophilic species capable of reacting with DNA, thereby initiating carcinogenesis.[101] The effects of turmeric and curcuminoids on the formation of B[a]P-derived DNA adducts were studied *in vitro* by employing the mouse liver postmitochondrial S9 fraction.[102] A dose-dependent decrease in binding of B[a]P metabolites to calf thymus DNA was observed in the presence of turmeric and curcumin. Demethoxycurcumin and bis-demethoxycurcumin also inhibited the B[a]P–DNA adduct formation dose-dependently.

When MCF-7 human mammary epithelial carcinoma cells were treated with 1 μM of DMBA for 24 h, there was an increase in CYP1A1 enzyme activity. Curcumin competitively inhibited the DMBA-induced CYP1A1 activity in these cells. The CYP1A1 activity of microsomes isolated from DMBA-treated cells was inhibited by 50% with 1 μM curcumin treatment.[50] Curcumin also blocked the metabolic activation of DMBA, as measured by the formation of DMBA–DNA adducts and decreased DMBA-induced cytotoxicity.[50] Thapliyal and Maru investigated the effects of curcumin, demethoxycurcumin, and bis-demethoxycurcumin on the activities of isozymes of CYP1A1, CYP1A2, and CYP2B1, which are predominantly involved in the metabolism of B[a]P.[51] *In vitro* incubation of rat liver microsomes with each of the compounds showed a dose-dependent decrease in carbon monoxide binding to microsomes and also inhibited CYP1A1, CYP1A2, and CYP2B1 activity, which were induced by B[a]P and NNK.[51] Pretreatment of Sprague–Dawley rats with 1% turmeric through the diet resulted in a significant decrease in B[a]P-induced CYP1A1 and CYP1A2 and phenobarbitone (PB)-induced CYP2B1 in the liver, lung, and stomach, although the extent of the inhibition was different.[54] A similar inhibition was also reported in female A/J mice. Thus, the administration of 2% curcumin in the diet to female A/J mice for 2 weeks produced a 25% decrease in the CYP1A catalytic activity but not protein levels.[54] Female Swiss Webster mice given curcumin (200 mg/kg or 400 mg/kg, p.o.) for 2 weeks displayed a 25% decrease in the hepatic CYP1A catalytic activity, whereas the activities of ovarian aromatase, hepatic catechol-*O*-methyltransferase and hepatic UDP-glucuronosyltransferase (UGT) were not altered.[53] Additionally, there was a 20% decrease in the catalytic activity and a 28% decrease in polypeptide levels of CYP3A following 2 weeks of curcumin treatment at 400 mg/kg. In another study, 1% turmeric administered through the diet to male Swiss mice significantly inhibited the activities of CYP1A1 and CYP1A2 in the forestomach, liver, and lung.[52]

3.2. Stimulation of Carcinogen Detoxification

Phase II detoxifying enzymes conjugate the activated compounds to endogenous ligands, such as glutathione (GSH), glucuronic acid, acetic acid, or sulfuric acid, thereby facilitating their excretion. Many of the chemopreventive agents have been shown to increase the levels of a wide variety of phase II drug metabolizing enzymes, such as glutathione-*S*-transferase (GST), NAD(P)H:quinone oxidoreductase 1 (NQO1), UDP-glucuronosyltransferase, aldehyde dehydrogenase, aldo-keto-reductase, microsomal epoxide hydrolase, glutamate cysteine ligase (GCL), glutathione synthetase, γ-glutamyl transpeptidase, heme oxygenase-1 (HO-1), and leukotriene B4 12-hydroxydehydrogenase.[103] The 5′-flanking regions of these genes contain a common cis element, commonly known as the antioxidant-responsive element (ARE) or electrophile-responsove element.[3] The transcription of ARE-driven genes is regulated, at least in part, by nuclear transcription factor erythroid 2p45 (NF-E2)-related factor 2 (Nrf2), which is sequestered in cytoplasm by Kelch-like ECH-associated protein 1 (Keap1). Exposure of cells to ARE inducers results in the dissociation of Nrf2 from Keap1, which allows Nrf2 to translocate into the nucleus.[104]

It is well known that although CYP1A1 is important in the conversion of B[a]P to an ultimate electrophilic and carcinogenic form, (+)-anti-7,8-dihydroxy-9,10-epoxy-7,8,9,10-tetrahydrobenzo[*a*]pyrene, detoxification of this reactive intermediate is accomplished by GST. Feeding of 2% curcumin in the diet to female A/J mice for 14 days caused a 2.3-fold increase in hepatic GST activities.[54] Two percent curcumin in the diet fed to male ddY mice for 30 days enhanced the activities of GST and quinone reductase 1.7 and 1.8 times in the liver and 1.1 and 1.3 times in the kidney, respectively, as compared to the levels in control animals.[55] Male F344 rats fed curcumin by gavage over 5 days exhibited an increased total GST levels and GST-μ enzymatic activities in the prostate.[56] Whereas curcumin treatment for 2 weeks resulted in a 20% increase in GST activity, there was no parallel increase in hepatic stores of GSH.[53] Dietary supplementation of 1% curcumin for 2 weeks significantly enhanced UGT activities in both the liver and intestine.[57] Curcumin disrupts the Nrf2–Keap1 complex, leading to nuclear translocation and increased ARE binding of Nrf2, which was associated with a significant increase in the expression of HO-1[59] and GCL.[105] Likewise, curcumin induced the expression and activity of HO-1 in human hepatocytes,[58] vascular endothelial cells,[60] and human renal proximal tubule cells.[61]

3.3. Suppression of Pro-inflammatory Signaling

Cyclooxygenase-2 (COX-2) and inducible nitric oxide synthase (iNOS) are important enzymes that mediate inflammatory processes. Improper upregulation of COX-2 and iNOS has been associated with the pathophysiology of certain types of human cancer as well as inflammatory disorders. Because chronic inflammation is closely linked to tumor promotion, substances with potent anti-inflammatory activities are anticipated to exert chemopreventive effects on carcinogenesis, particularly in the promotion stage.[106]

The ability of curcumin to inhibit both activity and induced expression of COX-2 has been demonstrated in various cell lines and animal models.[107] Topical application of curcumin strongly inhibited the activities of COX-2 and lipoxygenase, two key enzymes involved in arachidonic acid cascades and in mouse epidermal microsomes and cytosol, respectively[62] Treatment of several human gastrointestinal cell lines with curcumin suppressed the expression of COX-2 protein and mRNA as well as PGE_2 production induced by TPA or chenodeoxycholate.[63] Curcumin inhibited nitric oxide (NO) production and the expression of iNOS protein and mRNA in RAW 264.7 cells stimulated with bacterial lipopolysaccharide (LPS) or interferon (IFN)-γ.[67]

The inhibitory effects of curcumin on pro-inflammatory gene expression are related to its inactivation of activator protein-1 (AP-1) and nuclear factor-κB (NF-κB).[108] Curcumin pretreatment caused marked inhibition of COX-2 mRNA and its protein expression as well as NF-κB DNA-binding activity in tumor necrosis factor (TNF)-α-treated human colonic epithelial cells in culture[64] and mouse skin *in vivo*.[65] Curcumin significantly inhibited LPS-mediated induction of COX-2 expression at both mRNA and protein levels in BV2 microglial cells by blocking the activation of NF-κB and AP-1.[66] Chan et al. have reported the inhibition by curcumin of iNOS gene expression in isolated BALB/c mouse peritoneal macrophages and in the livers of LPS-injected mice.[68] Curcumin was found to inhibit IL-1β-mediated expression of pro-inflammatory genes, such as ICAM-1 and IL-8, in rat intestinal or human colonic epithelial cell lines via the blockade of NF-κB activation.[109]

3.4. Induction of Cancer Cell Apoptosis

The ability of curcumin to induce apoptosis selectively in cancerous and transformed cells contributes to its anticancer potential. Curcumin has been reported to efficiently induce apoptosis in various cell lines, including HL-60, K562, MCF-7, and HeLa.[110] It also induces apoptosis in scleroderma lung fibroblasts without affecting normal lung fibroblasts.[111] Jiang and colleagues reported that curcumin induced cell shrinkage, chromatin condensation, and DNA fragmentation, characteristics of apoptosis, in immortalized mouse embryo fibroblasts (NIH 3T3), erbB2-transformed NIH 3T3 cells, mouse sarcoma S180, human colon cancer HT-29 cells, human kidney cancer cells, and human hepatocellular carcinoma Hep G2 cells.[112]

One of the major signaling pathways involved in apoptotic cell death includes the intracellular caspases, a family of structurally related cysteine proteases.[113] Caspase activity is responsible, either directly or indirectly, for the proteolytic cleavage of certain cellular proteins, which is characteristic of apoptotic cell death. For example, caspases 2, 3, 6, 7, and 9 can cleave poly(ADP ribose)polymerase.[114] Bcl-2 family proteins are one of the well-defined regulators of apoptosis. The ratio of antiapoptotic (e.g., Bcl-2) and proapoptotic (e.g., Bax) proteins determines, in part, how cells respond to apoptotic or survival signals.[115]

The mechanisms responsible for apoptosis induction by curcumin include the release of cytochrome-*c* and modulation of cell survival and death signaling pathways involving Akt, NF-κB, AP-1, or JNK and downregulation of the expression of survival genes and inhibitor of apoptosis (IAP).[116] Curcumin induced apoptosis through Akt dephosphorylation, inhibition of Bcl-2, Bcl-xl, and IAP, cytochrome-*c* release, and caspase-3 activation in human kidney carcinoma cells[74] and U937 monocytic lymphoma cells. Duvoix and colleagues demonstrated that curcumin efficiently induced proteolytic cleavage of pro-caspases 8 and 9 and PARP, leading to apoptosis in K562 cells.[69] Curcumin induced apoptosis through the activation of caspases 8 and 9 in human melanom cells.[70] Jana et al. demonstrated that curcumin inhibited proteasome activity in mouse neuro 2a cells, potentially leading to the induction of apoptosis through caspase-9 activation.[71] The compound has been shown to inhibit the activation of transcription factors NF-κB and AP-1, which regulate the genes responsible for proliferation and antiapoptosis. Curcumin treatment suppressed the constitutive activation of NF-κB and AP-1 in DU145 cells, which, in turn, downregulated endogenous Bcl-2.[75] Recently, Hussain et al. demonstrated that curcumin suppressed proliferation of several T-cell lines via dephosphorylation (inactivation) of constitutively active Akt and GSK3.[117] Aggarwal et al. have also reported that curcumin abrogates TNF–induced Akt activation in U937 cells.[72]

3.5. Cell Cycle Arrest

Cell cycle regulatory proteins and checkpoints are downstream elements of signaling cascades crucial for cell proliferation. Studies in a variety of cell lines have demonstrated that curcumin exerts antiproliferative effects by inducing not only apoptosis but also cell cycle arrest.[116]

In human promyelocytic leukemia (HL-60) cells treated with 25 μM curcumin for 48 h, about 60% cells were initially arrested in the G2/M phase of cell cycle and then in the G0/G1 phase, and as a result, DNA synthesis was halted and apoptosis was induced[76] Curcumin induced cell cycle arrest in the G2/M phase in human colon cancer HT-29 cells,[78] in another colon carcinoma cell line (Lovo),[79] and in the MCF-7 human breast tumor cell line.[77] Curcumin suppressed cell proliferation and induced G2/M arrest in gastric KATO-III and colon HCT-116 cancer cells.[80] The levels of cyclin D and cyclin E declined with curcumin treatment in both cell lines. Curcumin was also found to induce G0/G1 and/or G2/M phase cell cycle arrest, upregulate CDKIs, p21WAF1/CIP1, p27KIP1, and p53, and slightly downregulate cyclin B1 and cdc2 in human umbilical vein endothelial cells (HUVEC).[81] In human colon cancer-derived Moser cells, the curcumin-induced cell cycle arrest has been accompanied by suppression of gene expression of cyclin D1 and epidermal growth factor receptor, which was mediated via stimulation of the trans-activating activity of peroxisome proliferator-activated receptor γ.[118] (A novel curcumin derivative, (4-[3,5-bis-[2-(4-hydroxy-3-methoxy-phenyl)-ethyl]-4,5-dihydro-pyrazol-1-yl]-benzoic acid), exhibits potent inhibitory activities against the proliferation of HCT-15 human colon cancer cells by antagonizing the Ca^{2+}/calmodulin function.[119]

3.6. Inhibition of Angiogenesis and Metastasis

Angiogenesis is now regarded as critical to the transition of premalignant lesions in a hyperproliferative state to the malignant phenotype, thus facilitating tumor growth and metastasis.[37] There is increasing evidence that vascular endothelial growth factor (VEGF) and angiopoietins are the prime regulators of normal and pathological angiogenesis. Curcumin was previously described as a good antiangiogenesis agent.[120] Curcumin has inhibitory effects against VEGF and angiopoietins 1 and 2 in Ehrlich ascites tumor cells, VEGF and angiopoietin 1 in NIH3T3 cells, and tyrosine kinase Flk-1/KDR (VEGF receptor-2) in HUVEC cells.[82] In U937 and Raji cells, expression of VEGF mRNA induced by TNF-α was suppressed by curcumin.[83] In estrogen-negative MDA-MB-231 human breast cancer cells, curcumin inhibited the transcript levels of VEGF and another angiogenesis factor, basic fibroblast growth factor.[84]

Matrix metalloproteinases (MMPs) are a family of zinc-dependent endopeptidases that can degrade components of the extracellular matrix,[121] allowing tumor cells to migrate to their secondary sites of growth via blood and lymphatic vessels. Among the MMPs, a great concern has focused on MMP-2 and MMP-9. These two MMPs have been demonstrated to degrade type IV collagen and are involved in the invasion, angiogenesis, and metastasis of tumor cells[122] Increased expression and activity of MMP-2 and MMP-9 have been demonstrated in many human tumors.[123]

Curcumin treatment inhibited significantly MMP-2 activity and expression of membrane type-1 matrix metalloproteinase (MT1-MMP) and focal adhesion kinase, an important component of the intracellular signaling pathway, in highly metastatic murine melanoma cells (B16F10). This led to a pronounced reduction in cell adhesion to extracellular matrix (ECM) ligands fibronectin and vitronectin[86] and enhancement in the expression of antimetastatic proteins, tissue inhibitor metalloproteinase (TIMP)-2, nonmetastatic gene 23 (Nm23), and E-cadherin.[124] Curcumin, at 10 μM, inhibited cellular migration and invasion via suppression of MMP-9 secretion in the human hepatocellular carcinoma (SK-Hep-1) cell line.[87] It also inhibited the production of MMP-9 and adhesion and migration to fibronectin and laminin in hepatocellular carcinoma CBO140C12 cells.[40] Furthermore, curcumin inhibited phorbol ester-induced upregulation of MMP-9 by blocking ERK1/2 phosphorylation and NF-κB transcriptional activity in MCF10A human mammary epithelial cells[88] and by inhibiting the protein kinase C signaling pathways in human astroglioma cells.[89] CD13/aminopeptidase N (APN) is a membrane-bound, zinc-dependent metalloproteinase that plays a key role in tumor invasion and angiogenesis. Curcumin, by directly binding to APN, irreversibly inhibits its activity.[89] Interestingly, curcumin strongly inhibited APN-positive tumor cell invasion and basic fibroblast growth factor-induced angiogenesis, but it failed to exert such anti-invasive effects in APN-negative tumor cells. In another study, the antiangiogenic activity of demethoxycurcumin, a naturally occurring structural analog of curcumin, was explored by utilizing cDNA microarray-based gene expression analysis in cultured HUVECs using cDNA microarray analysis. Of 1024 human cancer-focused genes arrayed, 9 angiogenesis-related genes were down-regulated over fivefold in demethoxycurcumin-treated cells.[125] In support of the results obtained from cDNA microarray analysis, MMP-9 expression was down-regulated over fivefold by demethoxycurcumin treatment.

3.7. Modulation of Oncogenes and Tumor Suppressor Genes

Oncogenes and tumor suppressor genes control cell cycle and apoptosis. Events leading to carcinogenesis involve mutations in oncogenes, resulting in a dominant gain of function, or mutations in tumor suppressor genes with a resultant loss of their inhibitory action. The Ras signaling cascade constitutes a major pathway that has been demonstrated to be misregulated in about 30% of all human tumors.[126] In fact, *ras* genes are the most frequently mutated oncogenes that can be detected in human tumors.

The Jun family of transcription factors includes c-Jun, JunB, and JunD. Together with Fos family members (FosB, Fra-1, and Fra-2), they form the group of AP-1 proteins that, after dimerization, bind to so-called TPA-responsive elements (TRE) in the promoter and enhancer regions of target genes. AP-1-regulated genes include important regulators of invasion, angiogenesis, metastasis, proliferation, differentiation, and survival. The transcription factor c-Jun cooperates with oncogenic alleles of *ras* in malignant transformation. Member of the *myc* gene family are involved in the regulation of growth and development of normal and cancer cells. In particular, c-*myc*, the cellular homologue of the avian myelocitic leukemia virus, is implicated in a large number of human solid tumors, leukemias, and lymphomas as well as in a variety of animal neoplasias.[127] Deregulated expression of Myc can drive cell proliferation and vasculogenesis, inhibit cell differentiation, and promote metastasis and genomic instability.[128]

Kakar and Roy found that topical application of curcumin (10 μmol) on the dorsal side of the CD-1 mouse skin 30 min before TPA treatment inhibited TPA-induced expression of c-*fos* and c-*jun* by 90% and that of c-*myc* by 60%.[90] Orally administered 1% curcumin significantly decreased protein expression of *ras* and *fos* proto-oncogenes in the tumorous skin.[91] Topical application of 10 μmol curcumin together with 5 nmol TPA once a day for 5 days strongly inhibited epidermal hyperplasia and expression of c-Jun and c-Fos.[92] Moreover, application of 10 μmol curcumin to SKH-1 mouse skin twice a day for 5 days immediately after 180 mJ/cm^2 ultraviolet B (UVB) exposure had a variable inhibitory effect on the UVB-induced expression of c-Fos and c-Jun and on epidermal hyperplasia.[92] These data strongly suggest that curcumin might inhibit tumor promoter or UVB-induced skin cancer through modulation of expression of these oncogenes. Curcumin at 10 nM concentration inhibited the expression of c-Jun induced by TPA in mouse epidermal JB6 cells[92] and downregulated the expression of c-*myc* in B lymphoma cells[93] and HCT-116 cells.[94]

p53, which is the most commonly inactivated tumor suppressor and mutated in about 60% of human cancers, functions as a transcription factor regulating genes involved in cell cycle arrest, apoptosis, and DNA repair.[129] Curcumin induced accumulation of WT p53 and apoptosis in the human breast cancer cell lines MCF-7 and TR9–7,[95,130] human neuroblastoma cell lines,[97] and human ovarian cancer cells.[98] Choudhuri et al. reported that curcumin induced apoptosis in a p53-dependent manner in human mammary epithelial carcinoma cells.[130] In colon adenocarcinoma HT-29 cells, treatment with 50 μM curcumin caused no changes in total p53 expression[99] However, a notable change was observed in the serine phosphorylation level of p53.

In the etoposide-treated human RKO colorectal cancer cell line, curcumin inhibited accumulation of phosphorylated wild-type p53 and induction of G_1 growth arrest, which appears to be attributable to its α,β-unsaturated electrophilic moiety that might disrupt p53 conformation required for its serine phosphorylation.[131] Tsvetkov et al. showed that curcumin inhibited NQO1 activity both *in vitro* and *in vivo* and disrupted the binding of NQO1 to functionally active wild-type p53, induced ubiquitin-independent degradation of p53, and inhibited p53-mediated apoptosis in normal thymocytes and myeloid leukemic cells.[132] The authors suggested that curcumin can promote p53 degradation by an ubiquitin-independent mechanism and thereby protects cells against p53-induced apoptosis.

4. CONCLUDING REMARKS

Chemoprevention by edible phytochemicals is now considered to be an inexpensive, readily applicable, acceptable, and accessible approach to cancer control and management. The optimization of intervention trials of diet-derived putative chemopreventive agents is currently under development in normal populations as well as high-risk groups as critical molecular targets of chemoprevention are being unraveled. Curcumin exerts potential chemopreventive activities in several animal tumor models via multiple underlying molecular mechanisms targeting all stages of multistep carcinogenesis (Figure 1). Curcumin has been considered to be evaluated in intervention trials for its potential use as a cancer chemopreventive agent. Apart from the assessment of efficacy as well as elucidation of underlying molecular mechanisms, measurement of potential biomarkers of pharmacodynamic effects, including absorption and systemic bioavailability, are also important in launching the large-scale intervention trials for evaluating the chemopreventive potential of dietary phytochemicals, including curcumin. Curcumin is poorly absorbed from the intestine, and its systemic bioavailability after oral feeding is relatively low. Nonetheless, curcumin as an ingredient of turmeric has exhibited chemopreventive and other health beneficial effects. It will be important to determine whether curcumin will be more effective in humans as an individual agent or as part of the foodstuffs from which it is derived. Derivatization of curcumin as a lead compound to elevate the bioavailability as well as chemopreventive and therapeutic efficacy might also be considered for future human clinical trials.

REFERENCES

1. P. Greenwald, From carcinogenesis to clinical interventions for cancer prevention. *Toxicology* **166**, 37–45 (2001).
2. G. J. Kelloff, J. A. Crowell, V. E. Steele, R. A. Lubet, W. A. Malone, C. W. Boone, L. Kopelovich, E. T. Hawk, R. Lieberman, J. A. Lawrence, et al., Progress in cancer chemoprevention: development of diet-derived chemopreventive agents. *J Nutr* **130**, 467S–471S (2000).
3. Y.-J. Surh, Cancer chemoprevention with dietary phytochemicals. *Nature Rev Cancer* **3**, 768–780 (2003).
4. A. H. Conney, T. Lysz, T. Ferraro, T. F. Abidi, P. S. Manchand, J. D. Laskin, and M. T. Huang, Inhibitory effect of curcumin and some related dietary compounds on tumor promotion and arachidonic acid metabolism in mouse skin. *Adv Enzyme Regul* **31**, 385–396 (1991).
5. A. Duvoix, R. Blasius, S. Delhalles, M. Schnekenburger, F. Morceau, E. Henry, M. Dicato, M. and M. Diederich, Chemopreventive and therapeutic effects of curcumin. *Cancer Lett* **223**, 181–190 (2005).
6. M. T. Huang, Z. Y. Wang, C. A. Georgiadis, J. D. Laskin, and A. H. Conney, Inhibitory effects of curcumin on tumor initiation by benzo[a]pyrene and 7,12-dimethylbenz[a]anthracene. *Carcinogenesis* **13**, 2183–2186 (1992).
7. M. Nagabhushan and S. V. Bhide, Curcumin as an inhibitor of cancer. *J Am Coll Nutr* **11**, 192–198.
8. L. A. Cohen, A review of animal model studies of tomato carotenoids, lycopene, and cancer chemoprevention. *Exp Biol Med (Maywood)* **227**, 864–868 (2002).
9. S. Das, S. Banerjee, and P. Saha, The models for assessment of chemopreventive agents: single organ models. *Asian Pac J Cancer Prev* **5**, 15–23 (2004).
10. J. DiGiovanni, Multistage carcinogenesis in mouse skin. *Pharmacol Ther* **54**, 63–128 (1992).
11. M. T. Huang, Y. R. Lou, J. G. Xie, W. Ma, Y. P. Lu, P. Yen, B. T. Zhu, H. Newmark, and C. T. Ho, Effect of dietary curcumin and dibenzoylmethane on formation of 7,12-dimethylbenz[a]anthracene-induced mammary tumors and lymphomas/leukemias in Sencar mice. *Carcinogenesis* **19**, 1697–1700 (1998).
12. M. T. Huang, W. Ma, P. Yen, J. G. Xie, J. Han, K. Frenkel, D. Grunberger, and A. H. Conney, Inhibitory effects of topical application of low doses of curcumin on 12-*O*-tetradecanoylphorbol-13-acetate-induced tumor promotion and oxidized DNA bases in mouse epidermis. *Carcinogenesis* **18**, 83–88 (1997).
13. M. T. Huang, W. Ma, Y. P. Lu, R. L. Chang, C. Fisher, P. S. Manchand, H. L. Newmark, and A. H. Conney, Effects of curcumin, demethoxycurcumin, bisdemethoxycurcumin and tetrahydrocurcumin on 12-O-tetradecanoylphorbol-13-acetate-induced tumor promotion. *Carcinogenesis* **16**, 2493–2497 (1995).
14. M. A. Azuine and S. V. Bhide, Chemopreventive effect of turmeric against stomach and skin tumors induced by chemical carcinogens in Swiss mice. *Nutr Cancer* **17**, 77–83 (1992).

15. M. A. Pereira, S. L. Herren-Freund, A. L. Britt, and M. M. Khoury, Effect of coadministration of phenobarbital sodium on N-nitrosodiethylamine-induced gamma-glutamyltransferase-positive foci and hepatocellular carcinoma in rats. *J Natl Cancer Inst* **72**, 741–744 (1984).
16. M. Sreepriya and G. Bali, Chemopreventive effects of embelin and curcumin against N-nitrosodiethylamine/phenobarbital-induced hepatocarcinogenesis in Wistar rats. *Fitoterapia* **76**, 549–555 (2005).
17. S. E. Chuang, M. L. Kuo, C. H. Hsu, C. R. Chen, J. K. Lin, G. M. Lai, C. Y. Hsieh, and A. L. Cheng, Curcumin-containing diet inhibits diethylnitrosamine-induced murine hepatocarcinogenesis. *Carcinogenesis* **21**, 331–335 (2000).
18. S. Busquets, N. Carbo, V. Almendro, M. T. Quiles, F. J. Lopez-Soriano, and J. M. Argiles, Curcumin, a natural product present in turmeric, decreases tumor growth but does not behave as an anticachectic compound in a rat model. *Cancer Lett* **167**, 33–38 (2001).
19. M. B. Thompson, The Min mouse: A genetic model for intestinal carcinogenesis. *Toxicol Pathol* **25**, 329–332 (1997).
20. A. R. Moser, C. Luongo, K. A. Gould, M. K. McNeley, A. R. Shoemaker, and W. F. Dove, ApcMin:A mouse model for intestinal and mammary tumorigenesis. *Eur J Cancer* **31A**, 1061–1064 (1995).
21. D. E. Corpet and F. Pierre, Point: From animal models to prevention of colon cancer. Systematic review of chemoprevention in min mice and choice of the model system. *Cancer Epidemiol Biomarkers Prev* **12**, 391–400 (2003).
22. S. Perkins, R. D. Verschoyle, K. Hill, I. Parveen, M. D. Threadgill, R. A. Sharma, M. L. Williams, W. P. Steward, and A. J. Gescher, Chemopreventive efficacy and pharmacokinetics of curcumin in the min/+ mouse, a model of familial adenomatous polyposis. *Cancer Epidemiol Biomarkers Prev* **11**, 535–540 (2002).
23. N. N. Mahmoud, A. M. Carothers, D. Grunberger, R. T. Bilinski, M. R. Churchill, C. Martucci, H. L. Newmark, and M. M. Bertagnolli, Plant phenolics decrease intestinal tumors in an animal model of familial adenomatous polyposis. *Carcinogenesis* **21**, 921–927 (2000).
24. M. T. Huang, E. E. Deschner, H. L. Newmark, Z. Y. Wang, T. A. Ferraro, and A. H. Conney, Effect of dietary curcumin and ascorbyl palmitate on azoxymethanol-induced colonic epithelial cell proliferation and focal areas of dysplasia. *Cancer Lett* **64**, 117–121 (1992).
25. M. T. Huang, H. L. Newmark, and K. Frenkel, Inhibitory effects of curcumin on tumorigenesis in mice. *J Cell Biochem* **27(Suppl)**, 26–34 (1997).
26. C. V. Rao, A. Rivenson, and B. S. Reddy, Chemoprevention of colon carcinogenesis by dietary curcumin, a naturally occurring plant phenolic compound. *Cancer Res* **55**, 259–266 (1995).
27. M. A. Pereira, C. J. Grubbs, L. H. Barnes, H. Li, G. R. Olson, I Eto, M. Juliana, L. M. Whitaker, G. J. Kelloff, V. E. Steele, and R. A. Lubet, Effects of the phytochemicals, curcumin and quercetin, upon azoxymethane-induced colon cancer and 7,12-dimethylbenz[a]anthracene-induced mammary cancer in rats. *Carcinogenesis* **17**, 1305–1311 (1996).
28. T. Tanaka, H. Makita, M. Ohnishi, Y. Hirose, A. Wang, H. Mori, K. Satoh, A. Hara, and H. Ogawa, Chemoprevention of 4-nitroquinoline 1-oxide-induced oral carcinogenesis by dietary curcumin and hesperidin: Comparison with the protective effect of beta-carotene. *Cancer Res* **54**, 4653–4659 (1994).

29. J. Ushida, S. Sugie, K. Kawabata, Q. V. Pham, T. Tanaka, K. Fujii, H. Takeuchi, Y. Ito, and H. Mori, Chemopreventive effect of curcumin on N-nitrosomethylbenzylamine-induced esophageal carcinogenesis in rats. *Jpn J Cancer Res* **91**, 893–898 (2000).
30. M. A. Azuine and S. V. Bhide, Adjuvant chemoprevention of experimental cancer: Catechin and dietary turmeric in forestomach and oral cancer models. *J Ethnopharmacol* **44**, 211–217 (1994).
31. N Li, X. Chen, J. Liao, G. Yang, S. Wang, Y. Josephson, C. Han, J. Chen, M. T. Huang, and C. S. Yang, Inhibition of 7,12-dimethylbenz[a]anthracene (DMBA)-induced oral carcinogenesis in hamsters by tea and curcumin. *Carcinogenesis* **23**, 1307–1313 (2002).
32. S. Ikezaki, A. Nishikawa, F. Furukawa, K. Kudo, H. Nakamura, K. Tamura, and H. Mori, Chemopreventive effects of curcumin on glandular stomach carcinogenesis induced by N-methyl-N′-nitro-N-nitrosoguanidine and sodium chloride in rats. *Anticancer Res* **21**, 3407–3411 (2001).
33. K. Singletary, C. MacDonald, M. Iovinelli, C. Fisher, and M. Wallig, Effect of the beta-diketones diferuloylmethane (curcumin) and dibenzoylmethane on rat mammary DNA adducts and tumors induced by 7,12-dimethylbenz[a]anthracene. *Carcinogenesis* **19**, 1039–1043 (1998).
34. K. Singletary, C. MacDonald, M. Wallig, and C. Fisher, Inhibition of 7,12-dimethylbenz[a]anthracene (DMBA)-induced mammary tumorigenesis and DMBA-DNA adduct formation by curcumin. *Cancer Lett* **103**, 137–141 (1996).
35. S. S. Hecht, P. M. Kenney, M. Wang, N Trushin, S. Agarwal, A. V. Rao, and P. Upadhyaya, Evaluation of butylated hydroxyanisole, myo-inositol, curcumin, esculetin, resveratrol and lycopene as inhibitors of benzo[a]pyrene plus 4-(methylnitrosamino)-1-(3-pyridyl)-1-butanone-induced lung tumorigenesis in A/J mice. *Cancer Lett* **137**, 123–130 (1999).
36. N. Frank, J. Knauft, F. Amelung, J. Nair, H. Wesch, and H. Bartsch, No prevention of liver and kidney tumors in Long-Evans Cinnamon rats by dietary curcumin, but inhibition at other sites and of metastases. *Mutat Res* **523–524**, 127–135 (2003).
37. J. Folkman, Angiogenesis in cancer, vascular, rheumatoid and other disease. *Nature Med* **1**, 27–31 (1995).
38. C. Chen, S. Parangi, M. J. Tolentino, and J. Folkman, A strategy to discover circulating angiogenesis inhibitors generated by human tumors. *Cancer Res* 55, 4230–4233 (1995).
39. P. Yoysungnoen, P. Wirachwong, P. Bhattarakosol, H. Niimi, and S. Patumraj, Antiangiogenic activity of curcumin in hepatocellular carcinoma cells implanted nude mice. *Clin Hemorheol Microcirc* **33**, 127–135 (2005).
40. Y. Ohashi, Y. Tsuchiya, K. Koizumi, H. Sakurai, and I. Saiki, Prevention of intrahepatic metastasis by curcumin in an orthotopic implantation model. *Oncology* **65**, 250–258 (2003).
41. J. H. Hong, K. S. Ahn, E. Bae, S. S. Jeon, and H. Y. Choi, The effects of curcumin on the invasiveness of prostate cancer in vitro and in vivo. *Prostate Cancer Prostatic Dis* **9**, 147–152 (2006)
42. T. Dorai, Y. C. Cao, B. Dorai, R. Buttyan, and A. E. Katz, Therapeutic potential of curcumin in human prostate cancer. III. Curcumin inhibits proliferation, induces apoptosis, and inhibits angiogenesis of LNCaP prostate cancer cells in vivo. *Prostate* **47**, 293–303 (2001).

43. B. B. Aggarwal, S. Shishodia, Y. Takada, S. Banerjee, R. A. Newman, C. E. Bueso-Ramos, and J. E. Price, Curcumin suppresses the paclitaxel-induced nuclear factor-kappaB pathway in breast cancer cells and inhibits lung metastasis of human breast cancer in nude mice. *Clin Cancer Res* **11**, 7490–7498 (2005).
44. L. G. Menon, R. Kuttan, and G. Kuttan, Inhibition of lung metastasis in mice induced by B16F10 melanoma cells by polyphenolic compounds. *Cancer Lett* **95**, 221–225 (1995).
45. R. Kuttan, P. C. Sudheeran, and C. D. Josph, Turmeric and curcumin as topical agents in cancer therapy. *Tumori* **73**, 29–31 (1987).
46. R. A. Sharma, H. R. McLelland, K. A. Hill, C. R. Ireson, S. A. Euden, M. M. Manson, M. Pirmohamed, L. J. Marnett, A. J. Gescher, and W. P. Steward, Pharmacodynamic and pharmacokinetic study of oral Curcuma extract in patients with colorectal cancer. *Clin Cancer Res* **7**, 1894–1900 (2001).
47. R. A. Sharma, S. A. Euden, S. L. Platton, D. N. Cooke, A. Shafayat, H. R. Hewitt, T. H. Marczylo, B. Morgan, D. Hemingway, S. M. Plummer, M. Pirmohamed, A. J. Gescher, and W. P. Steward, Phase I clinical trial of oral curcumin: Biomarkers of systemic activity and compliance. *Clin Cancer Res* **10**, 6847–6854 (2004).
48. A. L. Cheng, C. H. Hsu, J. K. Lin, M. M. Hsu, Y. F. Ho, T. S. Shen, J. Y. Ko, J. T. Lin, B. R. Lin, W. Ming-Shiang, et al., Phase I clinical trial of curcumin, a chemopreventive agent, in patients with high-risk or pre-malignant lesions. *Anticancer Res* **21**, 2895–2900 (2001).
49. G. Garcea, D. P. Berry, D. J. Jones, R. Singh, A. R. Dennison, P. B. Farmer, R. A. Sharma, W. P. Steward, and A. J. Gescher, Consumption of the putative chemopreventive agent curcumin by cancer patients: assessment of curcumin levels in the colorectum and their pharmacodynamic consequences. *Cancer Epidemiol Biomarkers Prev* **14**, 120–125 (2005).
50. H. P. Ciolino, P. J. Daschner, T. T. Wang, and G. C. Yeh, Effect of curcumin on the aryl hydrocarbon receptor and cytochrome P450 1A1 in MCF-7 human breast carcinoma cells. *Biochem Pharmacol* **56**, 197–206 (1998).
51. R. Thapliyal and G. B. Maru, Inhibition of cytochrome P450 isozymes by curcumins in vitro and in vivo. *Food Chem Toxicol* **39**, 541–547 (2001).
52. R. Thapliyal, S. S. Deshpande, and G. B. Maru, Effects of turmeric on the activities of benzo(a)pyrene-induced cytochrome P-450 isozymes. *J Environ Pathol Toxicol Oncol* **20**, 59–63 (2001).
53. S. P. Valentine, M. J. Le Nedelec, A. R. Menzies, M. J. Scandlyn, M. G. Goodin, and R. J. Rosengren, Curcumin modulates drug metabolizing enzymes in the female Swiss Webster mouse. *Life Sci* **78**, 2391–2398 (2005).
54. S. V. Singh, X. Hu, S. K. Srivastava, M. Singh, H. Xia, J. L. Orchard, and H. A. Zaren, Mechanism of inhibition of benzo[a]pyrene-induced forestomach cancer in mice by dietary curcumin. *Carcinogenesis* **19**, 1357–1360 (1998).
55. M. Iqbal, S. D. Sharma, Y. Okazaki, M. Fujisawa, and S. Okada, Dietary supplementation of curcumin enhances antioxidant and phase II metabolizing enzymes in ddY male mice: possible role in protection against chemical carcinogenesis and toxicity. *Pharmacol Toxicol* **92**, 33–38 (2003).
56. S. B. Jones and J. D. Brooks, Modest induction of phase 2 enzyme activity in the F-344 rat prostate. *BMC Cancer* **6**, 62 (2006).
57. E. M. van der Logt, H. M. Roelofs, F. M. Nagengast, and W. H. Peters, Induction of rat hepatic and intestinal UDP-glucuronosyltransferases by naturally occurring dietary anticarcinogens. *Carcinogenesis* **24**, 1651–1656 (2003).

58. S. J. McNally, E. M. Harrison, J. A. Ross, O. J. Garden, and S. J. Wigmore, Curcumin induces heme oxygenase-1 in hepatocytes and is protective in simulated cold preservation and warm reperfusion injury. *Transplantation* **81**, 623–626 (2006).
59. E. Balogun, M. Hoque, P. Gong, E. Killeen, C. J. Green, R. Foresti, J. Alam, and R. Motterlini, Curcumin activates the haem oxygenase-1 gene via regulation of Nrf2 and the antioxidant-responsive element. *Biochem J* **371**, 887–895 (2003).
60. R. Motterlini, R. Foresti, R. Bassi, and C. J. Green, Curcumin, an antioxidant and anti-inflammatory agent, induces heme oxygenase-1 and protects endothelial cells against oxidative stress. *Free Radical Biol Med* **28**, 1303–1312 (2000).
61. N. Hill-Kapturczak, V. Thamilselvan, F. Liu, H. S. Nick, and A. Agarwal, Mechanism of heme oxygenase-1 gene induction by curcumin in human renal proximal tubule cells. *Am J Physiol Renal Physiol* **281**, F851–F859 (2001).
62. M. T. Huang, T. Lysz, T. Ferraro, T. F. Abidi, J. D. Laskin, and A. H. Conney, Inhibitory effects of curcumin on in vitro lipoxygenase and cyclooxygenase activities in mouse epidermis. *Cancer Res* **51**, 813–819 (1991).
63. F. Zhang, N. K. Altorki, J. R. Mestre, K. Subbaramaiah, and J. A. Dannenberg, Curcumin inhibits cyclooxygenase-2 transcription in bile acid- and phorbol ester-treated human gastrointestinal epithelial cells. *Carcinogenesis* **20**, 445–451 (1999).
64. S. M. Plummer, K. A. Holloway, M. M. Manson, R. J. Munks, A. Kaptein, S. Farrow, and L. Howells, Inhibition of cyclo-oxygenase 2 expression in colon cells by the chemopreventive agent curcumin involves inhibition of NF-κB activation via the NIK/IKK signalling complex. *Oncogene* **18**, 6013–6020 (1999).
65. K. S. Chun, Y. S. Keum, S. S. Han, Y. S. Song, S. H. Kim, and Y.-J. Surh, Curcumin inhibits phorbol ester-induced expression of cyclooxygenase-2 in mouse skin through suppression of extracellular signal-regulated kinase activity and NF-κB activation. *Carcinogenesis* **24**, 1515–1524 (2003).
66. G. Kang, P. J. Kong, Y. J. Yuh, S. Y. Lim, S. V. Yim, W. Chun, and S. S. Kim, Curcumin suppresses lipopolysaccharide-induced cyclooxygenase-2 expression by inhibiting activator protein 1 and nuclear factor κB bindings in BV2 microglial cells. *J Pharmacol Sci* **94**, 325–328 (2004).
67. I. Brouet and H. Ohshima, Curcumin, an anti-tumour promoter and anti-inflammatory agent, inhibits induction of nitric oxide synthase in activated macrophages. *Biochem Biophys Res Commun* **206**, 533–540 (1995).
68. M. M. Chan, H. I. Huang, M. R. Fenton, and D. Fong, In vivo inhibition of nitric oxide synthase gene expression by curcumin, a cancer preventive natural product with anti-inflammatory properties. *Biochem Pharmacol* **55**, 1955–1962 (1998).
69. A. Duvoix, F. Morceau, M. Schnekenburger, S. Delhalle, M. M. Galteau, M. Dicato, and M. Diederich, Curcumin-induced cell death in two leukemia cell lines: K562 and Jurkat. *Ann NY Acad Sci* **1010**, 389–392 (2003).
70. J. A. Bush, K. J. Cheung, Jr., and G. Li, Curcumin induces apoptosis in human melanoma cells through a Fas receptor/caspase-8 pathway independent of p53. *Exp Cell Res* **271**, 305–314 (2001).
71. N. R. Jana, P. Dikshit, A. Goswami, and N. Nukina, Inhibition of proteasomal function by curcumin induces apoptosis through mitochondrial pathway. *J Biol Chem* **279**, 11,680–11,685 (2004).

72. S. Aggarwal, H. Ichikawa, Y. Takada, S. K. Sandur, S. Shishodia, and B. B. Aggarwal, Curcumin (diferuloylmethane) down–regulates expression of cell proliferation and antiapoptotic and metastatic gene products through suppression of IκBα kinase and Akt activation. *Mol Pharmacol* **69**, 195–206 (2006).
73. J. H. Bae, J. W. Park, and T. K. Kwon, Ruthenium red, inhibitor of mitochondrial Ca^{2+} uniporter, inhibits curcumin-induced apoptosis via the prevention of intracellular Ca2+ depletion and cytochrome c release. *Biochem Biophys Res Commun* **303**, 1073–1079 (2003).
74. J. H. Woo, Y. H. Kim, Y. J. Choi, D. G. Kim, K. S. Lee, J. H. Bae, S. Min do, J. S. Chang, Y. J. Jeong, Y. H. Lee, et al., Molecular mechanisms of curcumin-induced cytotoxicity: induction of apoptosis through generation of reactive oxygen species, down-regulation of Bcl-XL and IAP, the release of cytochrome c and inhibition of Akt. *Carcinogenesis* **24**, 1199–1208 (2003).
75. A. Mukhopadhyay, C. Bueso-Ramos, D. Chatterjee, P. Pantazis, and B. B. Aggarwal, Curcumin downregulates cell survival mechanisms in human prostate cancer cell lines. *Oncogene* **20**, 7597–7609 (2001).
76. Y. Wu, Y. Chen, and M. He, The influence of curcumin on the cell cycle of HL-60 cells and contrast study. *J Tongji Med Univ* **20**, 123–125 (2000).
77. A. Simon, D. P. Allais, J. L. Duroux, J. P. Basly, S. Durand-Fontanier, and C. Delage, Inhibitory effect of curcuminoids on MCF-7 cell proliferation and structure–activity relationships. *Cancer Lett* **129**, 111–116 (1998).
78. R. Hanif, L. Qiao, S. J. Shiff, and B. Rigas, Curcumin, a natural plant phenolic food additive, inhibits cell proliferation and induces cell cycle changes in colon adenocarcinoma cell lines by a prostaglandin-independent pathway. *J Lab Clin Med* **130**, 576–584 (1997).
79. H. Chen, Z. S. Zhang, Y. L. Zhang, and D. Y. Zhou, Curcumin inhibits cell proliferation by interfering with the cell cycle and inducing apoptosis in colon carcinoma cells. *Anticancer Res* **19**, 3675–3680 (1999).
80. L. Moragoda, .R. Jaszewski, and A. P. Majumdar, Curcumin induced modulation of cell cycle and apoptosis in gastric and colon cancer cells. *Anticancer Res* **21**, 873–878 (2001).
81. M. J. Park, E. H. Kim, I. C. Park, H. C. Lee, S. H. Woo, J. Y. Lee, Y. J. Hong, C. H. Rhee, S. H. Choi, B. S. Shim, et al., Curcumin inhibits cell cycle progression of immortalized human umbilical vein endothelial (ECV304) cells by up-regulating cyclin-dependent kinase inhibitor, p21WAF1/CIP1, p27KIP1 and p53. *Int J Oncol* **21**, 379–383 (2002).
82. A. E. Gururaj, M. Belakavadi, D. A Venkatesh, D. Marme, and B. P. Salimath, Molecular mechanisms of anti-angiogenic effect of curcumin. *Biochem Biophys Res Commun* **297**, 934–942 (2002).
83. W. H. Chen, Y. Chen, and G. H. Cui, Effects of TNF-alpha and curcumin on the expression of VEGF in Raji and U937 cells and on angiogenesis in ECV304 cells. *Chin Med J (Engl)* **118**, 2052–2057 (2005).
84. Z. M. Shao, Z. Z. Shen, C. H. Liu, M. R. Sartippour, V. L. Go, D. Heber, and M. Nguyen, Curcumin exerts multiple suppressive effects on human breast carcinoma cells. *Int J Cancer* **98**, 234–240 (2002).
85. J. S. Shim, J. H. Kim, H. Y. Cho, Y. N. Yum, S. H. Kim, H. J. Park, B. S. Shim, S. H. Choi, and H. J. Kwon, Irreversible inhibition of CD13/aminopeptidase N by the antiangiogenic agent curcumin. *Chem Biol* **10**, 695–704 (2003).

86. A. Banerji, J. Chakrabarti, A. Mitra, and A. Chatterjee, Effect of curcumin on gelatinase A (MMP-2) activity in B16F10 melanoma cells. *Cancer Lett* **211**, 235–242 (2004).
87. L. I. Lin, Y. F. Ke, Y. C. Ko, and J. K. Lin, Curcumin inhibits SK-Hep-1 hepatocellular carcinoma cell invasion in vitro and suppresses matrix metalloproteinase-9 secretion. *Oncology* **55**, 349–353 (1998).
88. K. W. Lee, J. H. Kim, H. J. Lee, and Y. J. Surh, Curcumin inhibits phorbol ester-induced up-regulation of cyclooxygenase-2 and matrix metalloproteinase-9 by blocking ERK1/2 phosphorylation and NF-kappaB transcriptional activity in MCF10A human breast epithelial cells. *Antioxid Redox Signal* **7**, 1612–1620 (2005).
89. M. S.Woo, S. H. Jung, S. Y. Kim, J. W. Hyun, K. H. Ko, W. K. Kim, and H. S. Kim, Curcumin suppresses phorbol ester-induced matrix metalloproteinase-9 expression by inhibiting the PKC to MAPK signaling pathways in human astroglioma cells. *Biochem Biophys Res Commun* **335**, 1017–1025 (2005).
90. S. S. Kakar and D. Roy, Curcumin inhibits TPA induced expression of c-*fos*, c-*jun* and c-*myc* proto-oncogenes messenger RNAs in mouse skin. *Cancer Lett* **87**, 85–89 (1994).
91. P. Limtrakul, S. Anuchapreeda, S. Lipigorngoson, and F. W. Dunn, Inhibition of carcinogen induced c-Ha-ras and c-fos proto-oncogenes expression by dietary curcumin. *BMC Cancer* **1**, 1 (2001).
92. Y. P. Lu, R. L. Chang, Y. R. Lou, M. T. Huang, H. L. Newmark, K. R. Reuhl, and A. H. Conney, Effect of curcumin on 12-*O*-tetradecanoylphorbol-13-acetate- and ultraviolet B light-induced expression of c-Jun and c-Fos in JB6 cells and in mouse epidermis. *Carcinogenesis* **15**, 2363–2370 (1994).
93. S. S. Han, S. T. Chung, D. A. Robertson, D. Ranjan, and S. Bondada, Curcumin causes the growth arrest and apoptosis of B cell lymphoma by downregulation of egr-1, c-myc, bcl-XL, NF-kB, and p53. *Clin Immunol* **93**, 152–161 (1999).
94. A. S. Jaiswal, B. P. Marlow, N. Gupta, and S. Narayan, Beta-catenin-mediated transactivation and cell–cell adhesion pathways are important in curcumin (diferuylmethane)-induced growth arrest and apoptosis in colon cancer cells. *Oncogene* **21**, 8414–8427 (2002).
95. D. Bech-Otschir, R. Kraft, X. Huang, P. Henklein, B. Kapelari, C. Pollmann, and W. Dubiel, COP9 signalosome-specific phosphorylation targets p53 to degradation by the ubiquitin system. *EMBO J* **20**, 1630–1639 (2001).
96. T. Choudhuri, S. Pal, M. L. Agwarwal, T. Das, and G. Sa, Curcumin induces apoptosis in human breast cancer cells through p53-dependent Bax induction. *FEBS Lett* **512**, 334–340 (2002).
97. A. Liontas and H. Yeger, Curcumin and resveratrol induce apoptosis and nuclear translocation and activation of p53 in human neuroblastoma. *Anticancer Res* **24**, 987–998 (2004).
98. M. Shi, Q. Cai, L. Yao, Y. Mao, Y. Ming, and G. Ouyang, Antiproliferation and apoptosis induced by curcumin in human ovarian cancer cells. *Cell Biol Int* **30**, 221–226 (2006).
99. G. Song, Y. B. Mao, Q. F. Cai, L. M. Yao, G. L. Ouyang, and S. D. Bao, Curcumin induces human HT-29 colon adenocarcinoma cell apoptosis by activating p53 and regulating apoptosis-related protein expression. *Braz J Med Biol Res* **38**, 1791–1798 (2005).
100. T. Prestera and P. Talalay, Electrophile and antioxidant regulation of enzymes that detoxify carcinogens. *Proc Natl Acad Sci USA* **92**, 8965–8969 (1995).

101. E. A. Chiocca and D. J. Waxman, Cytochrome P450-based gene therapies for cancer. *Methods Mol Med* **90**, 203–222 (2004).
102. S. S. Deshpande and G. B. Maru, Effects of curcumin on the formation of benzo[a]pyrene derived DNA adducts in vitro. *Cancer Lett* **96**, 71–80 (1995).
103. J. S. Lee and Y. J. Surh, Nrf2 as a novel molecular target for chemoprevention. *Cancer Lett* **224**, 171–184 (2005).
104. S. Numazawa and T. Yoshida, Nrf2-dependent gene expressions: A molecular toxicological aspect. *J Toxicol Sci* **29**, 81–89 (2004).
105. D. A. Dickinson, K. E. Iles, H. Zhang, V. Blank, and H. J. Forman, Curcumin alters EpRE and AP-1 binding complexes and elevates glutamate-cysteine ligase gene expression. *FASEB J* **17**, 473–475 (2003).
106. Y.-J. Surh, K. S. Chun, H. H. Cha, S. S. Han, Y. S. Keum, K. K. Park, and S. S. Lee, Molecular mechanisms underlying chemopreventive activities of anti-inflammatory phytochemicals: down-regulation of COX-2 and iNOS through suppression of NF-κB activation. *Mutat Res* **480–481**, 243–268 (2001).
107. Y.-J. Surh, Molecular mechanisms of chemopreventive effects of selected dietary and medicinal phenolic substances. *Mutat Res* **428**, 305–327 (1999).
108. C. S. Divya and M. R. Pillai , Antitumor action of curcumin in human papillomavirus associated cells involves downregulation of viral oncogenes, prevention of NFκB and AP-1 translocation, and modulation of apoptosis. *Mol Carcinog* **45**, 320–332 (2006).
109. C. Jobin, C. A. Bradham, M. P. Russo, B. Juma, A. S. Narula, D. A. Brenner, and R. B. Sartor, Curcumin blocks cytokine-mediated NF-κB activation and proinflammatory gene expression by inhibiting inhibitory factor I-κB kinase activity. *J Immunol* **163**, 3474–3483 (1999).
110. M. Roy, S. Chakraborty, M. Siddiqi, and R. K. Bhattacharya, Induction of apoptosis in tumor cells by natural phenolic compounds. *Asian Pacific J Cancer Prev* **3**, 61–67 (2002).
111. E. Tourkina, P. Gooz, J. C. Oates, A. Ludwicka-Bradley, R. M. Silver, and S. Hoffman, Curcumin-induced apoptosis in scleroderma lung fibroblasts: role of protein kinase cepsilon. *Am J Respir Cell Mol Biol* **31**, 28–35 (2004).
112. M. C. Jiang, H. F. Yang-Yen, J. J. Yen, and J. K. Lin, Curcumin induces apoptosis in immortalized NIH 3T3 and malignant cancer cell lines. *Nutr Cancer* **26**, 111–120 (1996).
113. H. R. Stennicke and G. S. Salvesen, Biochemical characteristics of caspases-3, -6, -7, and -8. *J Biol Chem* **272**, 25719–25723 (1997).
114. H. Sakahira, M. Enari, and S. Nagata, Cleavage of CAD inhibitor in CAD activation and DNA degradation during apoptosis. *Nature* **391**, 96–99 (1998).
115. S. N. Farrowand R. Brown, New members of the Bcl-2 family and their protein partners. *Curr Opin Genet Dev* **6**, 45–49 (1996).
116. R. A. Sharma, A. J. Gescher, and W. P. Steward, Curcumin: the story so far. *Eur J Cancer* **41**, 1955–1968 (2005).
117. A. R. Hussain, M. Al-Rasheed, P. S. Manogaran, K. A. Al-Hussein, L. C. Platanias, K. A. Kuraya, and S. Uddin, Curcumin induces apoptosis via inhibition of PI3′-kinase/AKT pathway in acute T cell leukemias. *Apoptosis* **11**, 245–254 (2006).
118. A. Chen and J. Xu, Activation of PPARγ by curcumin inhibits Moser cell growth and mediates suppression of gene expression of cyclin D1 and EGFR. *Am J Physiol Gastrointest Liver Physiol* **288**, G447–G456 (2005).

119. J. S. Shim, J. Lee, H. H. Park, S. J. Park, and H. J. Kwon, A new curcumin derivative, HBC, interferes with the cell cycle progression of colon cancer cells via antagonization of the Ca^{2+}/calmodulin function. *Chem Biol* **11**, 1455–1463 (2004).
120. J. L. Arbiser, N. Klauber, R. Rohan, R. van Leeuwen, M. T. Huang, C. Fisher, E. Flynn, and H. R. Byers, Curcumin is an in vivo inhibitor of angiogenesis. *Mol Med* **4**, 376–383 (1998).
121. L. A. Liotta and W. G. Stetler-Stevenson, Tumor invasion and metastasis: An imbalance of positive and negative regulation. *Cancer Res* **51**, 5054s–5059s (1991).
122. A. John and G. Tuszynski, The role of matrix metalloproteinases in tumor angiogenesis and tumor metastasis. *Pathol Oncol Res* **7**, 14–23 (2001).
123. M. Egeblad and Z. Werb, New functions for the matrix metalloproteinases in cancer progression. *Nature Rev Cancer* **2**, 161–174 (2002).
124. S. Ray, N. Chattopadhyay, A. Mitra, M. Siddiqi, and A. Chatterjee, Curcumin exhibits antimetastatic properties by modulating integrin receptors, collagenase activity, and expression of Nm23 and E-cadherin. *J Environ Pathol Toxicol Oncol* **22**, 49–58 (2003).
125. J. H. Kim, J. S. Shim, S. K. Lee, K. W. Kim, S. Y. Rha, H. C. Chung, and H J. Kwon, Microarray-based analysis of anti-angiogenic activity of demethoxycurcumin on human umbilical vein endothelial cells: crucial involvement of the down-regulation of matrix metalloproteinase. Jpn J *Cancer Res* **93**, 1378–1385 (2002).
126. J. L. Bos, Ras oncogenes in human cancer: A review. *Cancer Res* **49**, 4682–4689 (1989).
127. G. F. Claassen and S. R. Hann, Myc-mediated transformation: the repression connection. *Oncogene* **18**, 2925–2933 (1999).
128. S. Adhikary and M. Eilers, Transcriptional regulation and transformation by Myc proteins. *Nature Rev Mol Cell Biol* **6**, 635–645 (2005).
129. T. Tokino and Y. Nakamura, The role of p53-target genes in human cancer. *Crit Rev Oncol Hematol* **33**, 1–6 (2000).
130. T. Choudhuri, S. Pal, T. Das, and G. Sa, Curcumin selectively induces apoptosis in deregulated cyclin D1-expressed cells at G2 phase of cell cycle in a p53-dependent manner. *J Biol Chem* **280**, 20059–20068 (2005).
131. P. J. Moos, K. Edes, J. E. Mullally, and F. A. Fitzpatrick, Curcumin impairs tumor suppressor p53 function in colon cancer cells. *Carcinogenesis* **25**, 1611–1617 (2004).
132. P. Tsvetkov, G. Asher, V. Reiss, Y. Shaul, L Sachs, and J. Lotem, Inhibition of NAD(P)H:quinone oxidoreductase 1 activity and induction of p53 degradation by the natural phenolic compound curcumin. *Proc Natl Acad Sci USA* **102**, 5535–5540 (2005).

ANTITUMOR, ANTI-INVASION, AND ANTIMETASTATIC EFFECTS OF CURCUMIN

Girija Kuttan, Kuzhuvelil B. Hari Kumar, Chandrasekharan Guruvayoorappan, and Ramadasan Kuttan

Abstract: Curcumin was found to be cytotoxic in nature to a wide variety of tumor cell lines of different tissue origin. The action of curcumin is dependent on with the cell type, the concentration of curcumin (IC_{50}: 2–40 μg/mL), and the time of the treatment. The major mechanism by which curcumin induces cytotoxicity is the induction of apoptosis. Curcumin decreased the expression of antiapoptotic members of the *Bcl-2* family and elevated the expression of *p53*, *Bax*, *procaspases 3*, *8*, and *9*. Curcumin prevents the entry of nuclear factor κB (NF-κB) into the nucleus there by decreasing the expression of cell cycle regulatory proteins and survival factors such as Bcl-2 and survivin. Curcumin arrested the cell cycle by preventing the expression of *cyclin D1*, *cdk-1* and *cdc-25*. Curcumin inhibited the growth of transplantable tumors in different animal models and increased the life span of tumor-harboring animals. Curcumin inhibits metastasis of tumor cells as shown in *in vitro* as well as *in vivo* models, and the possible mechanism is the inhibition of matrix metalloproteases. Curcumin was found to suppress the expression of *cyclooxygenase-2*, *vascular endothelial growth factor*, and *intercellular adhesion molecule-1* and elevated the expression of antimetastatic proteins, the *tissue inhibitor of metalloproteases-2*, *nonmetastatic gene 23*, and *E-cadherin*. These results indicate that curcumin acts at various stages of tumor cell progression.

1. INTRODUCTION

Curcumin (1, 7-bis(4-hydroxy-3-methoxypentyl)-1, 6-hepadiene-3, 5-dione), also known as diferuloylmethane, belongs to the class of polyphenols present in the rhizomes of turmeric (*Curcuma longa* L), and content of curcumin in the dried rhizomes is around 5–10%. Turmeric is extensively used in the Indian subcontinent in culinary art. Turmeric is also used as an indigenous medicine for the treatment of various ailments for many centuries. Most of the biological activities of turmeric are due to curcumin. Curcumin is demonstrated to have a wide spectrum of pharmacological properties. There are several reviews that describe the different pharmacological properties of curcumin.[1–4] In the present chapter, we describe the antiproliferative and antimetastatic activity of curcumin and its implication in cancer prevention and treatment.

2. CYTOTOXIC AND ANTIPROLIFERATIVE ACTIVITY OF CURCUMIN

Curcumin is found to have antiproliferative and differentiation-inducing properties in different types of cell line *in vitro*. The mode of action of curcumin varies considerably with the cell type, the concentration of curcumin, and the time of the treatment. One of the initial reports describing the cytotoxicity of curcumin was against Dalton's lymphoma ascites cells (DLA) *in vitro*, in which curcumin (4 μg/mL) produced 50% cytotoxicity. Curcumin also inhibited the growth of Chinese hamster ovary cells and human leukemic lymphocytes in culture.[5,6]

Curcumin was found to be highly antiproliferative in nature to a variety of human leukemic cell lines. The antiproliferative action of curcumin is best characterized in a human chronic myelogenic leukemia cell line, namely K-562. At a concentration of 20 μg/mL, curcumin produced 50% growth arrest as determined by 3-(4, 5-Dimethyl-2-thiazolyl)-2, 5-diphenyl-2 H tetrazolium bromide (MTT) assay.[7] One of the main mechanisms through which curcumin produces cell arrest is by inducing apoptosis. Curcumin liberates cytochrome-*c* from mitochondria, activates caspases and Bax, and downregulates the Bcl-2 levels. Curcumin also downregulates the expression of Wilms' tumor-1 (*WT-1*) gene, which is highly overexpressed in leukemic blast cells of myeloid and lymphoid origin and serves as a marker for leukemic detection.[8] The downregulation of *WT-1* by curcumin is dependent on concentration. Moreover the expression of *MEK-1*, *c-jun*, and *P210 bcr/abl* were decreased by curcumin, which ultimately retard the ras-mediated signal transduction cascade, thus affecting the process of cell proliferation.[9] There are reports describing the inhibitory effect of curcumin on *STATs* and telomerase.[10] Curcumin also decreases the expression of glutathione-*S*-transferase P1-1 (*GSTP1-1*) in K-562 cells, as seen by reporter gene assay. GSTP1-1 is implicated in the carcinogenesis and the resistance of cancer cells against conventional chemotherapeutic agents. Curcumin also elevates the levels some of the components of the apoptotic pathway such as of *pro-caspases 8* and *9* as well as *poly ADP ribose polymerase*.[11] All of these events contribute to the antiproliferative action of curcumin in *K-562* cells.

Curcumin also suppressed the growth of several T-cell leukemia cell lines in a dose-dependent manner. Curcumin was found to induce apoptosis in these cell lines.[12] Apart from this, curcumin inhibited the action of inhibitory apoptotic proteins (*IAPs*) and downregulated the JAK–STAT (Signal tranducers and activators of transcription) pathway.[13] Curcumin suppressed the binding between AP-1 transcription factor and DNA.[14] Curcumin was shown to induce cell cycle arrest by reducing the expression of *cyclin D1, cdk1, cdc-25*, and *survivin*, thus providing a way for the apoptotic machinery to act.[15] The survival pathway mediated by the *Akt*-PI3K cascade was also inhibited by curcumin.[12]

Curcumin was found to be highly cytotoxic toward several malignant cell lines of colon origin. In the HCT 116 cell line, curcumin prevented the entry of nuclear factor κB (NF-κB) into the nucleus.[16]. NF-κB is a sequence-specific transcription factor and plays a central role in the regulation many types of immune, inflammatory, and carcinogenic response. NF-κB activation is required for normal cell growth; abnormal activation is seen several types of pathological process.[17] The cytotoxic potential of curcumin in the HCT 116 line is a concentration- and time-dependent event, and curcumin either activates or deactivates several pathways that are necessary for the normal growth of the cell. Curcumin activates Janus kinase (*JNK*) and mitogen-activated protein kinase (*MAPK*).[16] Curcumin blocked the entry of the cell cycle from G2 to M by inhibiting the expression of *cdc2/cyclin B*.[18] The proapoptotic members of the Bcl-2 family, such as *Bax*, were activated and antiapototic genes like $Bcl\text{-}_{XL}$ were inhibited by curcumin.[19] It also triggers caspase-3-mediated cell death. Curcumin was found to activate *GADD153*, which, in turn, acts as an activator of apoptosis.[20] Curcumin mediated the degradation of β-catenin, thus affecting the wnt signaling pathway. The cell–cell adhesion pathway mediated by E-cadherin was also blocked by curcumin.[21] In conclusion, curcumin acts as a potent growth suppressive and cytotoxic agent to the HCT116 cell line and the action is mediated by (1) the induction of caspases-mediated apoptosis, (2) the impairment of wnt signaling events, (3) the inhibition of cell–cell adhesion, and (4) the blocking of the transition of the cell cycle from G2 to M.

In the human hepatoma G2 cell line (HepG2), the cytotoxic action of curcumin is mainly by inducing DNA damage of both nuclear and mitochondrial genomes. The extensive mitochondrial damage might be the initial triggering signal for the induction of apoptosis.[22]. Curcumin also inhibited the hepatocyte growth factor (*HGF*) and its receptor c-met (*HGFR*).[23]

Curcumin in a dose- and time-dependent and fashion induces p53-mediated apoptotic death in basal cell carcinoma lines, as evident from electrophoretic mobility shift assay (EMSA) and antisense studies.[24]

Human head and neck squamous cell carcinoma cell lines (HNSCC) is a type of epithelial cancer with a very low survival rate. In the culture, curcumin inhibited the growth of these cells by triggering apoptosis. Curcumin decreased the elevation of *Bcl-2*, matrix metalloproteinase-9 (*MMP-9*), *cyclooxygenase-2* (COX-2), *interleukin* (*IL*)*-6*, and *cyclin D1* and arrested the cell cycle in the G1/S phase. HNSCC cells showed constitutive overexpression of NF-κB and treatment with curcumin inhibited the NF-κB activation, which is the other mechanism through which curcumin acts as a growth-inhibitory substance to HNSCC cells.[25]

Microarray analysis of human breast cancer carcinoma cell line (MCF-7) treated with curcumin revealed that curcumin upregulated 22 genes and downregulated 17 genes, which are components of multiple signaling pathways that control various biological processes. The major upregulated genes were *caspase-3*, *caspase-4*, *JNK*, *proliferating cell nuclear antigen* (*PCNA*), *death-associated protein-6*, *growth arrest and DNA damage inducible protein*, and downregulated genes such as *tumor necrosis factor* (*TNF*)-β, *TNF receptor*, *protein kinase B*, and *caspase-9* precursor.[26]

In several types human melanoma cell, curcumin produces cytotoxicity. In the A375 cell line, curcumin produces cell growth arrest in a time- and concentration-dependent manner and the major mechanism was found to be apoptosis.[27] Curcumin induces apoptosis through the Fas receptor/caspase-8 pathway independent of p53 and suppresses the antiapoptotic gene *XIAP*.[28] Curcumin prevents the translocation of NF-κB to the nucleus; downregulates the expression of *c-myc* and inducible nitric oxide synthase (*iNOS*), and induces G2/M arrest. Curcumin also inhibits the activity of an antiapoptotic gene *XIAP* and elevates the levels of *p53*, *p21*, *p27* (*KIP1*), and *check point kinase 2*.[29]

In some other cell lines, curcumin mediates its cytotoxic action via generating reactive oxygen species (ROS). Although curcumin is a potent scavenger of free radicals, there are reports describing the free-radical-generation potential of curcumin.[30,31] In the human submandibular gland carcinoma (HSG) cell line, at a very low concentration ($>15\ \mu M$), curcumin generates ROS that cause damage to mitochondria, as evident from the decrease in the mitochondrial membrane potential and externalization of phosphatydyl serine, and the whole process eventually ends up in the initiation of apoptosis.[32] In human gingival fibroblast also, treatment with curcumin produces a dose-dependent generation of ROS, and this phenomenon was attributed to the cytotoxic activity.[32] The growth-suppressive effect of curcumin on follicular lymphoma cells is also mediated by the generation of ROS. Flow cytometry and Western blot analysis revealed that curcumin shifted the equilibrium of Bcl-2 family members in favor of apoptosis and initiated a caspases-mediated cell death in these cell lines.[33]

Curcumin act as a cytotoxic agent *in vivo* also. Curcumin inhibited a DLA-induced solid tumor, as seen from the reduction in tumor volume and also inhibited ascites tumor in mice. The administration of curcumin enhanced the percentage of survival of animals bearing ascites.[5,6] When the three naturally occurring curcuminoids (curcumin I, II, and III) were compared for their cytotoxic, antitumor, and inhibition of tumor promotion activities, curcumin III was found to be more active than curcumin I and II.[34,35] Curcumin was found to decrease the Ehrlich's ascites carcinoma (EAC) cell number by the induction of apoptosis in the EAC cells, as seen from flow-cytometric analysis of cell cycle phase distribution of nuclear DNA and oligonucleosomal fragmentation. Moreover, curcumin upregulates the proapoptotic gene *Bax*, induces the release of cytochrome-*c* from the mitochondria, and, finally, induces the activation of caspase-3.[36]

Curcumin inhibited spontaneous tumors developed in C57BL/6J−Min/+ (Min/+) mice. These animals bear a germ-line mutation in the Apc gene and spontaneously develop numerous intestinal adenomas by 15 weeks of age. Curcumin also decreased the expression of β-catenin. Immunohistochemical analysis of the tissues revealed that treatment with curcumin increased the mucosal CD4+ Tcells and Bcells, suggesting that curcumin modulates lymphocyte-mediated immune functions.[37,38] Even though curcumin has been shown to inhibit transplantable and spontaneous tumors, the major action of curcumin is in the inhibition of carcinogenesis. Carcinogenesis is a complex process and involves the following steps; initiation, progression, and promotion. Even though the mechanism of action of curcumin during carcinogenesis is mainly ascribed to the inhibition of the carcinogen metabolism and its increased excretion, it might be possible that curcumin also acts as an inhibitor of cell proliferation during carcinogenesis. Curcumin had an inhibitory effect on both the initiation and promotion stages of carcinogenesis, as seen from the 7, 12-dimethylbenz [*a*]anthracene (DMBA)- and croton oil or 12-*O*-tetradecanoylphorbol-13-acetate (TPA)-induced two-stage skin carcinogenesis.[39–42] The TPA-induced epidermal lipoxygenase and cyclooxygenase activity was decreased by curcumin, thereby inhibiting the arachidonic acid metabolism. Because the critical roles of prostaglandins, lipoxygenase, and COX in the development of cancer are well established now; the inhibition of these enzymes by curcumin might be one of the mechanisms behind its anticarcinogenic potential.[43,44] Curcumin significantly delayed the onset of sarcoma produced by 20-methylcholanthrene in mice.[39] Curcumin has been shown to inhibit B[a]P-induced forestomach papilloma in mice.[41] Curcumin decreased the elevated levels of phase I enzymes, which are involved in the activation of B[a]P in the liver and increased the activities of phase II enzymes, which are involved in the detoxification of metabolites derived from B[a]P.[45] Topical administration of curcumin decreased the formation of B[a] P–DNA adducts on the mouse epidermis.[40] Curcumin was found to inhibit *N*-ethyl-*N′*-nitro-*N*-nitrosoguanidine-induced duodenal tumorigenesis in C57BL/6 mice and azoxymethane (AOM)-induced colon tumorigenesis in CF-1 mice. Histopathological analysis revealed that curcumin inhibited the number of adenomas and adenocarcinomas of the duodenum and colon.[40] The AOM-induced colon carcinogenesis in F344 rats was significantly inhibited by curcumin. Curcumin suppressed AOM-induced prostaglandin (PG) and thromboxane (Tx) formation in the liver, ornithine decarboxylase and tyrosine protein kinase activity in the liver and colon, and the formation of aberrant crypt foci in the colon.[46] Curcumin also inhibited the *N*-nitrosomethylbenzylamine (NMBA)-induced esophageal carcinogenesis in male F344 rats and the administration of curcumin significantly decreased the expression of cell proliferation biomarkers.[47] Curcumin inhibited the hepatocarcinogenesis (HCC) in C3H/HeN mice induced by *N*-nitrosodiethylamine (NDEA), as evident from reduction in multiplicity and incidence of the development of HCC. Curcumin downregulated the levels of *p21(ras), PCNA*, and *CDC2* proteins that were elevated after NDEA administration.[48] The administration of curcumin has been shown to reverse the hepatic damage and hepatocarcinogenesis caused by aflatoxins, a group of potent mycotoxins, in ducklings, as seen by the assay of marker enzymes and histopathological evaluation.[49]

Curcumin also acts as an immunomodulator. Immunomodulators are agents that modify host responses to antigens, with resultant therapeutic effects. Curcumin enhanced the natural killer cell activity. Curcumin has been shown to inhibit mitogen-stimulated cell proliferation. The phagocytic activities of macrophages were increased by curcumin and treatment with curcumin was found to increase the relative organ weights of lymphoid organs such as the spleen and thymus. The administration of curcumin also elevated the total white blood cell count, bone marrow cellularity, and levels of α-esterase-positive (a marker of maturing monocytes) cells. Immunomodulaory action can be another mechanism behind the cytotoxic action of curcumin.[50,51]

Cytokines play an important role in immunomodulation, and curcumin is known to alter the cytokine profiles. Curcumin also inhibited the production of pro-inflammatory cytokines. In the human pancreatic carcinoma cell line SUIT-2, curcumin inhibited the constitutive production of IL-8 in a concentration-dependent manner and reduced the NF-κB activity significantly.[52] In the keratinocyte cell line NCTC 2544, ultraviolet B (UVB) treatment was found to significantly enhance the expression of *TNF-α*, *IL-6*, and *IL-8*, both at the mRNA and protein levels. Curcumin decreased the UVB-mediated overexpression of all of the three pro-inflammatory cytokines in a dose-dependent manner. The UVB-induced cytokine overexpression was accompanied by the activation of NF-κB and AP-1 transcription factors, as assessed by electrophoretic mobility shift assays. The decrease in the expression of cytokines by curcumin is probably mediated by the inhibition of the activation of NF-κB and AP-1.[53] Curcumin potently inhibited the production of IL-12 in a dose-dependent manner from mouse macrophages stimulated with lipopolysaccharide (LPS). The inhibition of IL-12 production might be a key therapeutic strategy for modulating immunological diseases dominated by type 1 cytokine responses.[54] Curcumin, at a dose of 5 μM, inhibited LPS-induced production of TNF-α and IL-1 by a human monocytic macrophage cell line, Mono Mac 6, and reduced the LPS-induced activation of NF-κB and the biological activity of TNF-α in a L929 fibroblast lytic assay (bioassay)[55] Curcumin also inhibited the delayed type of hypersensitivity reaction (type IV) mediated by T_H1 subtypes. T_H1 cells also produce pro-inflammatory cytokines. Inhibition of type IV hypersensitivity reactions by curcumin further substantiates its effect on the inhibition of pro-inflammatory cytokines.[50]

3. ANTIMETASTATIC ACTIVITY OF CURCUMIN

Metastasis is the process by which cancer cells migrate from the tissue of origin to other distant sites through blood flow to form new malignant lesions in other organs. A successful formation of metastatic foci consists of complex and there are several interconnected steps involved. During metastasis, the invasive tumor cells might have to penetrate a number of cellular barriers, which include the basement membrane.[56,57] A group of proteolytic enzymes, namely matrix metalloproteinases (MMPs), are involved in these processes. MMPs are a family of over 28 enzymes and are Zn^{2+}-dependent. MMPs play a key role not only in normal processes of extracellular matrix degradation but also in pathological processes such as tissue remodeling, cancer invasion, and metastasis. Expression of MMPs is generally upregulated in a wide range of malignant tumors.[57,58] The major component of the basement membrane is type IV collagen, and of the several classes of MMPs, two of them (MMP-2 and MMP-9) are the key players in the degradation of type IV collagen. Tissue inhibitors of matrix metalloproteinases (TIMPs) are a class of endogenously present inhibitors of MMPs.

Curcumin was found to be highly antimetastatic in nature. An early report from our laboratory described the antimetastatic activity of curcumin against B16F-10 melanoma cells-induced pulmonary metastasis in C57BL/6 mice. Curcumin inhibited the formation of lung nodules induced by B16F-10 melanoma cells by 89.3% and increased the life span of animals by 143.9%. The invasive property of B16F-10 melanoma cells across the collagen matrix was inhibited by curcumin, as seen from the Boyden chamber assay. Zymographic analysis showed that curcumin inhibited the activities of MMP-2 and MMP-9.[59,60] Curcumin also downregulated the activities of membrane type 1 MMP (*MT1-MMP*) and focal adhesion kinase (*FAK*), which plays a role in the integrin-mediated signal transduction cascade in B16F-10 melanoma cells.[61] Curcumin-treated B16F-10 cells showed a marked reduction in the expression of alpha 5 beta1 and alpha 5 beta3 integrin receptors. Curcumin also enhanced the expression of antimetastatic proteins, *TIMP-2*, non-metastatic gene 23 (*Nm23*), and *E-cadherin*, which reduced the metastatic tendency of the melanoma cells.[62]

Curcumin was also found to be highly antimetastatic against prostrate DU145 cells both *in vitro* as well as *in vivo*. It reduced the metastatic ability of DU145 in a xenograft tumor model. The administration of curcumin produced a marked decrease in the tumor volume and the levels of MMP-2 and MMP-9.[63] In a human breast cancer xenograft model, the administration of curcumin markedly decreased the metastasis to the lung and suppressed the expression of *NF-κB*, *MMP-9*, *COX-2*, vascular endothelial growth factor (*VEGF*), and *intercellular adhesion molecule-1*.[64] 12-*O*-Tetradecanoylphorbol 13-acetate (TPA) is a well-known tumor promoter and in human breast epithelial cells (MCF10A). TPA induces profound expression of COX-2 and MMP-9 and thereby elevated the levels of PGs and the invasive and metastatic tendency of the cells. The increase in the expression is mediated by the upregulation of the signal transducing events mediated by extracellular signal-regulated protein kinase (*ERK1/2*) and *NF-κB*. The treatment of the cells with curcumin inhibits the above cascade, thereby decreasing the expression of *COX-2* and *MMP-9*, which, in turn, alters the invasive and metastatic properties of the cells.[65] Osteopontin (OPN), a type extracellular matrix protein, has been shown to be overexpressed in various types of cancer. OPN plays an important role in the ability of tumor cells to survive and metastasize to other distant organs. OPN stimulated the expression of *proMMP-2* and *MT1-MMP* through a NF-κB-mediated pathway in murine B16F-10 melanoma cells. The OPN-mediated expression of NF-κB, proMMP-2, and MT1-MMP was suppressed by curcumin and it inhibited the OPN-mediated tumor growth in a nude mice model.[66]

Curcumin reduced the metastasis of tumors in Long Evans Cinnamon (LEC) rats, which develop tumors in the kidney and the liver due to an aberrant copper-transporting *ATPase* gene. LEC rats accumulate copper in their body. Treatment with curcumin, although failing to prevent the induction of primary tumors in the kidney and the liver; reduced the metastasis of tumors to other distant sites.[67]

4. CONCLUSION

In conclusion, curcumin shows profound cytotoxic potential against tumor cells both *in vitro* and *in vivo*. The major mechanism of cytotoxicity is the induction of apoptosis. Curcumin either activates or represses several signaling events that are required for the normal cell function. Curcumin modulates these signaling pathways in such a way that the final events will be the execution of death in these cells. In order to execute apoptosis, curcumin follows several routes such as inhibiting antiapoptotic proteins, inhibiting survival pathways, arresting the cell cycle before the mitotic phase, inhibiting the translocation of NF-κB to the nucleus, triggering the generation of ROS, induceing cytochrome-*c* release from mitochondria, and activating the apoptotic machinery. Curcumin inhibits the metastasis of tumor cells both *in vitro* and *in vivo* and the possible mechanism is the inhibition of MMPs. Curcumin also acts as a good immunomodulator. It augments the natural killer cell activity and inhibits the production of cytokines. These biological activities warrant further studies of curcumin in the treatment and prevention of human neoplasm.

REFERENCES

1. Srimal, R. C. and Dhawan B. N. (1973) Pharmacology of diferuloylmethane (curcumin), a non-steroidal anti-inflammatory agent. *J. Pharm Pharmacol.* 25,447–452.
2. Y. J. Surh, Cancer chemoprevention with dietary phytochemicals. *Nat Rev Cancer* **3**, 768–780 (2003).
3. B. B. Aggarwal, A. Kumar, and A. C. Bharti, Anticancer potential of curcumin: Preclinical and clinical studies. *Anticancer Res* **23**, 363–398 (2003).
4. R. K. Maheshwari, A. K. Singh, J. Gaddipati, and R. C. Srimal, Multiple biological activities of curcumin: A short review. *Life Sci* **78**, 2081–2087 (2006).
5. R. Kuttan, P. Bhanumathy, K. Nirmala, and M. C. George, Potential anticancer activity of turmeric (*Curcuma longa*).*Cancer Lett* **29**, 197–202 (1985).
6. K. K. Soudamini. and R. Kuttan, Cytotoxic and tumor reducing properties of curcumin. *Indian J Pharmacol* **20**, 95–101 (1998).
7. S. Anuchapreeda, P. Thanarattanakorn, S. Sittipreechacharn, P. Chanarat, and P. Limtrakul, Curcumin inhibits WT1 gene expression in human leukemic K562 cells. *Acta Pharmacol Sin* **27**, 360–366 (2006).
8. S. Anuchapreeda, P. Limtrakul, P. Thanarattanakorn, S. Sittipreechacharn, and P. Chanarat, Inhibitory effect of curcumin on WT1 gene expression in patient leukemic cells. *Arch Pharm Res* **29**, 80–87 (2006).
9. L. X. Wu, J. H. Xu, X. W. Huang, K. Z. Zhang, C. X. We, and Y. Z. Chen, Downregulation of p210(bcr/abl) by curcumin involves disrupting molecular chaperone functions of Hsp90. *Acta Pharmacol Sin* **27**, 694–699 (2006).
10. S. Chakraborty, U. Ghosh, N. P. Bhattacharyya, R. K. Bhattacharya, and M. Roy, Inhibition of telomerase activity and induction of apoptosis by curcumin in K-562 cells.*Mutat Res* **596**, 81–90 (2006).
11. A. Duvoix, F. Morceau, S. Delhalle, M. Schmitz, M., Schnekenburger, M. M. Galteau, M. Dicato, and M. Diederich, Induction of apoptosis by curcumin: mediation by glutathione S-transferase P1-1 inhibition. *Biochem Pharmacol* **66**, 1475–1483 (2003).
12. A. R. Hussain, M. Al-Rasheed, P. S. Manogaran, K. A. Al-Hussein, L. C. Platanias, K. Al Kuraya, and S. Uddin, Curcumin induces apoptosis via inhibition of PI3′-kinase/AKT pathway in acute T cell leukemias. *Apoptosis* **11**, 245–254 (2006).

13. J. Rajasingh, H. P. Raikwar, G. Muthian, C. Johnson, and J. J. Bright, Curcumin induces growth-arrest and apoptosis in association with the inhibition of constitutively active JAK-STAT pathway in T cell leukemia. *Biochem Biophys Res Commun* **340**, 359–368 (2006).
14. M. Tomita, H. Kawakami, J. N. Uchihara, T. Okudaira, M. Masuda, N. Takasu, T. Matsuda, T. Ohta, Y. Tanaka, and N. Mori, Curcumin suppresses constitutive activation of AP-1 by downregulation of JunD protein in HTLV-1-infected T-cell lines. *Leuk Res* **30**, 313–321 (2006).
15. M. Tomita, H. Kawakami, J. N. Uchihara, T. Okudaira, M. Masuda, N. Takasu, T. Matsuda, T. Ohta, Y. Tanaka, K. Ohshiro,. and N. Mori, Curcumin (diferuloylmethane) inhibits constitutive active NF-kappaB, leading to suppression of cell growth of human T-cell leukemia virus type I-infected T-cell lines and primary adult T-cell leukemia cells. *Int J Cancer* **118**, 765–772 (2006).
16. G. P. Collett and F C.Campbell, Overexpression of p65/RelA potentiates curcumin-induced apoptosis in HCT116 human colon cancer cells. *Carcinogenesis* **27**, 1285–1291 (2006).
17. A. Kumar and B. B. Aggarwal, Nuclear factor-kB: Its role in health and disease. *J Mol Med* **82**, 434–438 (2004).
18. A. S. Jaiswal, B. P. Marlow, N. Gupta, and S. Narayan, Beta-catenin-mediated transactivation and cell-cell adhesion pathways are important in curcumin (diferuylmethane)-induced growth arrest and apoptosis in colon cancer cells. *Oncogene* **21**, 8414–8427 (2002).
19. R. Rashmi, S. Kumar, and D. Karunagaran, Human colon cancer cells lacking Bax resist curcumin-induced apoptosis and Bax requirement is dispensable with ectopic expression of Smac or downregulation of Bcl-XL. *Carcinogenesis* **26**, 713–723 (2005).
20. D. W. Scott and G. Loo, Curcumin-induced GADD153 gene up-regulation in human colon cancer cells. *Carcinogenesis* **25**, 2155–2164 (2004).
21. S. Narayan, Curcumin, a multi-functional chemopreventive agent, blocks growth of colon cancer cells by targeting beta-catenin-mediated transactivation and cell-cell adhesion pathways. *J Mol Histol* **35**, 301–307 (2004).
22. J. Cao, L Jia, H. M. Zhou, Y. Liu, and L. F. Zhong, Mitochondrial and nuclear DNA damage induced by curcumin in human hepatoma G2 *Cells Toxicol Sci* **91**, 476–483 (2006).
23. D. W. Seol, Q. Chen, and R. Zarnegar, Transcriptional activation of the hepatocyte growth factor receptor (c-met) gene by its ligand (hepatocyte growth factor) is mediated through AP-1. *Oncogene* **19**, 1132–1137 (2000).
24. S. H. Jee, S. C. Shen, C. R. Tseng, H. C. Chiu, and M. L. Kuo, Curcumin induces a p53-dependent apoptosis in human basal cell carcinoma cells. *J Invest Dermatol* **111**, 656–661 (1998).
25. S. Aggarwal, Y. Takada, S. Singh, J. N. Myers, and B. B. Aggarwal, Inhibition of growth and survival of human head and neck squamous cell carcinoma cells by curcumin via modulation of nuclear factor-kappaB signaling. *Int J Cancer* **111**, 679–692 (2004).
26. C. Ramachandran, S. Rodriguez, R. Ramachandran, P. K. Raveendran Nair, H. Fonseca, Z. Khatib, E. Escalon, and S. J. Melnick, Expression profiles of apoptotic genes induced by curcumin in human breast cancer and mammary epithelial cell lines. *Anticancer Res* **25**, 3293–3302 (2005).
27. S. Qiu, S. S. Tan, J. A. Zhang, A. Liu, J. Y. Yuan, G. Z. Rao, and W. Y. Wang, Apoptosis induced by curcumin and its effect on c-myc and caspase-3 expressions in human melanoma A375 cell line. *Di Yi Jun Yi Da Xue Xue Bao* **25**, 1517–1521 (2005).

28. J. A. Bush, K. J. Cheung, Jr., and G. Li, Curcumin induces apoptosis in human melanoma cells through a Fas receptor/caspase-8 pathway independent of p53. *Exp Cell Res* **271**, 305–314 (2001).
29. M. Zheng, S. Ekmekcioglu, E. T. Walch, C. H. Tang, and E. A. Grimm, Inhibition of nuclear factor-kappaB and nitric oxide by curcumin induces G2/M cell cycle arrest and apoptosis in human melanoma cells. *Melanoma Res* **14**, 165–171 (2004).
30. K. Elizabeth and M. N. A. Rao, Effect of curcumin on hydroxyl radical generation through Fenton reaction. *Int J Pharm* **57**, 173–176 (1989).
31. K. Elizabeth and M. N. A. Rao, Oxygen radical scavenging activity of curcumin. *Int J Pharm* **58**, 237–240 (1990).
32. T. Atsumi, K. Tonosaki, and S. Fujisawa, Induction of early apoptosis and ROS-generation activity in human gingival fibroblasts (HGF) and human submandibular gland carcinoma (HSG) cells treated with curcumin. *Arch Oral Biol* **51**, 913–921 (2006).
33. J. Skommer, D. Wlodkowic, and J. Pelkonen, Cellular foundation of curcumin-induced apoptosis in follicular lymphoma cell lines. *Exp Hematol* **34**, 463–474 (2006).
34. R. J. Anto, G. Kuttan, and R. Kuttan, Anti-tumour and antioxidant activity of natural curcuminoids. *Cancer Lett* **94**, 79–83 (1995).
35. R. J. Anto, J. George, K. D. Babu, K. N. Rajasekharan, and R Kuttan, Antimutagenic and anticarcinogenic activity of natural and synthetic curcuminoids. *Mutat Res* **370**, 127–131 (1996).
36. S. Pal, T. Choudhuri, S. Chattopadhyay, A. Bhattacharya, G. K. Datta, T. Das, and G. Sa, Mechanisms of curcumin-induced apoptosis of Ehrlich's ascites carcinoma cells. *Biochem Biophys Res Commun* **288**, 658–665 (2001).
37. N. N. Mahmoud, A. M. Carothers, D. Grunberger, R. T. Bilinski, M. R. Churchill, C. Martucci, H. L. Newmark, and M. M. Bertagnolli, Plant phenolics decrease intestinal tumors in an animal model of familial adenomatous polyposis. *Carcinogenesis* **21**, 921–927 (2000).
38. M. Churchill, A. Chadburn, R. T. Bilinski, and M. M. Bertagnolli, Inhibition of intestinal tumors by curcumin is associated with changes in the intestinal immune cell profile. *J Surg Res* **89**, 169–175 (2000).
39. K. K. Soudamini and R. Kuttan, Inhibition of chemical carcinogenesis by curcumin. *J Ethnopharmacol* **27**, 227–233 (1989).
40. M. T. Huang, Y. R. Lou, W. Ma, H. L. Newmark, R. K. Reuhl, and A. H. Conney, Inhibitory effects of dietary curcumin on forestomach, duodenal, and colon carcinogenesis in mice. *Cancer Res* **54**, 5841–5847 (1994).
41. M. Nagabhushan and S.V. Bhide, Curcumin as an inhibitor of cancer. *J Am Coll Nutr* **11**,192–198 (1992).
42. M. T. Huang, R. C. Smart, C. Q. Wong, and A. H. Conney, Inhibitory effect of curcumin, chlorogenic acid, caffeic acid, and ferulic acid on tumor promotion in mouse skin by 12-O-tetradecanoylphorbol-13-acetate. *Cancer Res* **48**, 5941–5946 (1988).
43. M. T. Huang, T. Lysz, T. Ferraro, T. F. Abidi, J. D. Laskin, and A. H. Conney, Inhibitory effects of curcumin on in vitro lipoxygenase and cyclooxygenase activities in mouse epidermis. *Cancer Res* **51**, 813–819 (1991).
44. A. H. Conney, T. Lysz, T. Ferraro, T. F. Abidi, P. S. Manchand, J. D. Laskin, and M. T. Huang, Inhibitory effect of curcumin and some related dietary compounds on tumor promotion and arachidonic acid metabolism in mouse skin. *Adv Enzyme Regul* **31**, 385–396 (1991).

45. S. V. Singh, X. Hu, S. K. Srivastava, M. Singh, H. Xia, J. L. Orchard, and H. A. Zaren, Mechanism of inhibition of benzo[a]pyrene-induced forestomach cancer in mice by dietary curcumin. *Carcinogenesis* **19**, 1357–1360 (1998).
46. C. V. Rao, B. Simi, and B. S. Reddy, Inhibition by dietary curcumin of azoxymethane-induced ornithine decarboxylase, tyrosine protein kinase, arachidonic acid metabolism and aberrant crypt foci formation in the rat colon. *Carcinogenesis* **14**, 2219–2225 (1993).
47. J. Ushida, S. Sugie, K. Kawabata, Q. V. Pham, T. Tanaka, K. Fujii, H. Takeuchi, Y. Ito, and H. Mori, Chemopreventive effect of curcumin on N-nitroso methyl benzylamine-induced esophageal carcinogenesis in rats. *Jpn J Cancer Res* **91**, 893–898 (2000).
48. S. E. Chuang, M. L. Kuo, C. H. Hsu, C. R. Chen, J. K. Lin, G. M. Lai, C. Y. Hsieh, and A. L. Cheng, Curcumin-containing diet inhibits diethylnitrosamine-induced murine hepatocarcinogenesis. *Carcinogenesis* **21**, 331–335 (2000).
49. K. B. Soni, A. Rajan, and R. Kuttan, Reversal of aflatoxin induced liver damage by turmeric and curcumin. *Cancer Lett* **66**, 115–121 (1992).
50. A. Sini, R. Kuttan,. and G. Kuttan, Immunomodulatory activity of curcumin. *Immunol Invest* **28**, 291–303 (1999).
51. V. S. Yadav, K. P. Mishra, D. P. Singh, S. Mehrotra, and V. K. Singh, Immunomodulatory effects of curcumin. *Immunopharmacol Immunotoxicol* **27**, 485–497 (2005).
52. H. Hidaka, T. Ishiko, T. Furuhashi, H. Kamohara, S. Suzuki, M. Miyazaki, O. Ikeda, S. Mita, T. Setoguchi, and M. Ogawa, Curcumin inhibits interleukin 8 production and enhances interleukin 8 receptor expression on the cell surface: impact on human pancreatic carcinoma cell growth by autocrine regulation. *Cancer* **95**, 1206–1214 (2002).
53. G. A. Laquerriere, S. C. Gangloff, R. Le Naour, C. Trentesaux, W. Hornebeck, and M. Guenounou, Relative contribution of NF-kappaB and AP-1 in the modulation by curcumin and pyrrolidine dithiocarbamate of the UVB-induced cytokine expression by keratinocytes.*Cytokine* **18**, 168–177 (2002).
54. B. Y. Kang, S. W. Chung, W. Chung, S. Im, S. Y. Hwang, and T. S. Kim, Inhibition of interleukin-12 production in lipopolysaccharide-activated macrophages by curcumin.*Eur J Pharmaco*l **384**, 191–195 (1999).
55. M. M. Chan, Inhibition of tumor necrosis factor by curcumin, a phytochemical. *Biochem Pharmacol* **49**, 1551–1556n (1995).
56. L. A. Liotta, Tumor invasion and metastasis; role of the basement membrane. *Am.J.Pathol* **117**, 339–348 (1984).
57. L. A. Liotta, Tumor invasion and metastasis; role of extracellular matrix. *Cancer Res* **46**, 1–7 (1986).
58. B. E. Bachmeier, C. M. Iancu, M. Jochum, and A. G. Nerlich, Matrix metalloproteinases in cancer: comparison of known and novel aspects of their inhibition as a therapeutic approach. *Expert Rev Anticancer Ther* **5**, 149–163 (2005).
59. L. G. Menon, R. Kuttan, and G. Kuttan, Inhibition of lung metastasis in mice induced by B16F10 melanoma cells by polyphenolic compounds.*Cancer Lett* **95**, 221–225 (1995).
60. L. G. Menon, R. Kuttan, and G. Kuttan, Anti-metastatic activity of curcumin and catechin. *Cancer Lett* 141,159–165 (1999).
61. A. Banerji, J. Chakrabarti, A. Mitra, and A. Chatterjee, Effect of curcumin on gelatinase A (MMP-2) activity in B16F10 melanoma cells. *Cancer Lett* **211**, 235–242 (2004).
62. S. Ray, N. Chattopadhyay, A. Mitra, M. Siddiqi, and A. Chatterjee, Curcumin exhibits antimetastatic properties by modulating integrin receptors, collagenase activity, and expression of Nm23 and E-cadherin. *J Environ Pathol Toxicol Oncol* **22**, 49–58 (2003).

63. J. H. Hong, K. S. Ahn, E. Bae, S. S. Jeon, and H. Y. Choi, The effects of curcumin on the invasiveness of prostate cancer in vitro and in vivo. *Prostate Cancer Prostatic Dis* **9**, 147–152 (2006).
64. B. B. Aggarwal, S. Shishodia, Y. Takada, S. Banerjee, R. A. Newman, C. E. Bueso-Ramos, and J. E. Price, Curcumin suppresses the paclitaxel-induced nuclear factor-kappaB pathway in breast cancer cells and inhibits lung metastasis of human breast cancer in nude mice. *Clin Cancer Res* **11**, 7490–7498 (2005).
65. K. W. Lee, J. H. Kim, H. J. Lee, and Y. J. Surh, Curcumin inhibits phorbol ester-induced up-regulation of cyclooxygenase-2 and matrix metalloproteinase-9 by blocking ERK1/2 phosphorylation and NF-kappaB transcriptional activity in MCF10A human breast epithelial cells. *Antioxid Redox Signal* **7**, 1612–1620 (2005).
66. S. Philip, A. Bulbule, and G. C. Kundu, Matrix metalloproteinase-2: Mechanism and regulation of NF-kappaB-mediated activation and its role in cell motility and ECM-invasion. *Glycoconj J* **21**, 429–441 (2004).
67. N. Frank, J. Knauft, F. Amelung, J. Nair, H. Wesch, and H. Bartsch, No prevention of liver and kidney tumors in Long-Evans Cinnamon rats by dietary curcumin, but inhibition at other sites and of metastases. *Mutat Res* **523–524**, 127–135 (2003).

CURCUMIN AS AN INHIBITOR OF ANGIOGENESIS

Sulochana S. Bhandarkar and Jack L. Arbiser

Abstract: Angiogenesis, the formation of new blood vessels from host vasculature, is critical for tumor growth and metastases. Curcumin, a novel small-molecular-weight compound, has been shown to inhibit carcinogenesis in different organs and the common link between these actions is its antiangiogenic effect. Curcumin is a direct inhibitor of angiogenesis and also downregulates various proangiogenic proteins like vascular endothelial growth factor and basic fibroblast growth factor. Curcumin's antiangiogenic effect is also in part due to its inhibitory effect on signal transduction pathways, including those involving protein kinase C and the transcription factors NF-κB and AP-1. Curcumin has an inhibitory effect on two groups of proteinases involved in angiogenesis that are the members of the matrix metalloproteinase family and the urokinase plasminogen activator family. Cell adhesion molecules are upregulated in active angiogenesis and curcumin can block this effect, adding further dimensions to curcumin's antiangiogenic effect. Curcumin shows a dose-dependent inhibition on tumor necrosis factor, a versatile cytokine, which has its effect on angiogenesis through the signal transduction pathways, expression of proangiogenic factors, and cell adhesion molecules. Curcumin's effect on the overall process of angiogenesis compounds its enormous potential as an antiangiogenic drug.

1. INTRODUCTION

Angiogenesis, the growth of new capillary blood vessels, is crucial for tumor growth and expansion.[1,2] Tumors require a constant supply of oxygen and nutrients, and diffusion from nearby capillaries can supply adequate nutrition for tumors less than 2 mm^2, but for continued growth, tumors must develop their own blood supply.[3] Tumor masses acquire the ability to produce proangiogenic factors that stimulate the growth of host blood vessels.[4] The acquisition of the proangiogenic factors is mediated by a switch to an angiogenic phenotype that induces angiogenesis and allows rapid expansion of tumor growth.[5,6] Angiogenic tumors also produce positive regulators of angiogenesis and mobilize angiogenic promoters from the extracellular matrix. Antiangiogenic therapy can therefore interfere with any or all of these mechanisms and prevent tumor cells from developing a viable blood supply.[3]

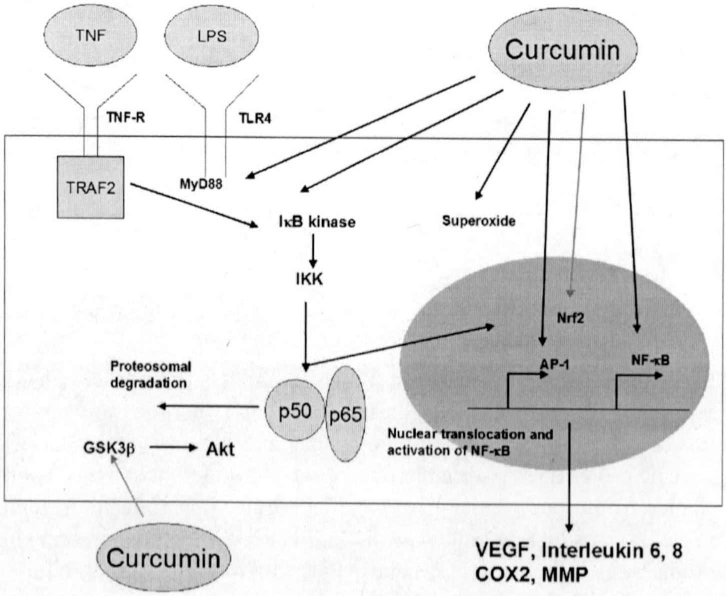

Figure 1. Suppression of angiogenesis pathway by curcumin.

Angiogenesis inhibitors can be divided into two classes. The first class, or direct angiogenesis inhibitors, are relatively specific for endothelial cells and have little effect on tumor cells. Indirect inhibitors might not have direct effects on endothelial cells, but they downregulate the production of angiogenesis stimulators. Curcumin has been shown by Arbiser et al.[7] to be a direct inhibitor of angiogenesis and also plays an important role in the downregulation of proangiogenic proteins (Figure 1). Curcumin's antiangiogenic effect is also in part due to its inhibitory effect on the production of cytokines relevant to tumor growth such as tumor necrosis factor and its antiapoptotic effect: its inhibitory effect on endothelial cell attachment, motility and proliferation.

2. CURCUMIN: INHIBITIOR OF PROANGIOGENIC PROTEINS

The angiogenic switch, which is essential for angiogenesis, is mediated by angiogenic oncogenes,[8] which upregulate the expression of proangiogenic proteins such as VEGF (vascular endothelial growth factor) and bFGF (basic fibroblast growth factor) and reduce the expression of angiogenesis inhibitors.[2,9] Among the proangiogenic proteins, VEGF and bFGF are crucial factors in pathological angiogenesis.[7,10] VEGF-A is considered the most important of the five isoforms. VEGF mediates angiogenic signals through its VEGFR-1, -2, and -3 receptors, which initiate signaling events that begins with dimerization and trans-autophosphorylation of TK residues in the receptors, which, in turn, activate phospholipase C-γ, phospho-inositide 3 (PI3) kinase (PI3-K), GTPase activating protein (GAP), mitogen-activated protein kinase (MAPK), and others.[10] Cui et al. showed that VEGF secretion by U937 and Raji cells is increased by tumor necrosis factor-α (TNF-α) treatment and suppressed by curcumin treatment.[11] TNF-α augments the expression of VEGF165 and VEGF121 mRNA and curcumin reduces the expression. Angiogenesis was tested by network formation of endothelial cells on Matrigel and no networks or cords formed in control and curcumin groups and there was tube formation on matrigel in the supernatants of the Raji culture group and the supernatants groups treated by the VEGF group and TNF-α in Raji cells. The study concluded that expressions of VEGF mRNA in U937 and Raji cells were increased by TNF-α and suppressed by curcumin. VEGF and TNF-α can induce angiogenesis, and curcumin can inhibit angiogenesis in ECV304 cells and can, therefore, inhibit potential mechanisms controlling tumor neovascularization.

Overexpression of VEGF and cyclooxygenase-2 (COX-2) has been demonstrated in the HepG2 cell line (hepatocellular carcinoma cell line).[12,13] COX-2 supports tumor angiogenesis both directly and indirectly, as shown by Millianta et al.[14] and directly stimulates the production of angiogenic factors from tumor cells. Antiangiogenic activity of curcumin is further compounded by its ability to reduce the tumor-induced overexpression of VEGF and COX-2 as, shown by Yoysungnoen et al.[15] HepG2 cells were inoculated onto the upper layer of the skin-fold chamber and curcumin solutions were orally fed to the HepG2 cell-implanted nude mice. The tumor neocapillary density (NCD) was evaluated using a digital image analysis and demonstrated the NCD of HepG2-groups were significantly increased on day 7 and 14, compared to the aged-matched controls and this was attenuated by daily treatment of curcumin solution (3000 mg/kg BW). Curcumin treatment thereby inhibits tumor angiogenesis by reduction of angiogenic biomarkers such as VEGF and COX-2 and this inhibition also occurs with liposomal curcumin as shown by Li et al.[16]through attenuation of the NF-κB mechanism.

Basic fibroblast growth factor (bFGF) is another potent angiogenic factor and stimulates both endothelial proliferation and migration. The activity of bFGF on endothelial cells might be in part through stimulation of protein kinase C.[7,17] Curcumin and its analogues results in potent inhibition of bFGF-induced corneal neovascularization assessed by measuring vessel length and density in the normally avascular cornea. Intraperitoneal administration of curcumin at doses up to 300 mg/kg BW did not inhibit corneal neovascularization, which might be due to the well-known rapid metabolism of curcumin. Curcumin inhibits the proliferation of primary endothelial cells in the presence and absence of bFGF and also inhibits proliferation of an immortalized endothelial cell line, as shown by Arbiser et al.[7]

3. EFFECT ON ENDOTHELIAL CELL MIGRATION AND INVASION

The role of endothelial cell migration is a crucial step in angiogenesis. The effect of curcumin on endothelial cell migration, attachment, and tube formation was studied on Matrigel by Aggarwal and Natarajan.[18] Curcumin had no effect on endothelial cell migration or attachment to either plastic of Matrigel, but caused a dose-dependent inhibition of tube formation when the cells were treated before plating or at the time of plating on Matrigel. Curcumin treatment inhibited angiogenesis in a subcutaneous Matrigel plug model in mice and caused the preformed tubes to break down.

During angiogenesis, extracellular proteolysis has been implicated in different steps such as provisional matrix remodeling, basement membrane degradation, cell migration, and invasion.[19–21] The group of proteinases involved in extracellular matrix (ECM) remodeling comprises four different families based on the nature of the chemical group responsible for catalytic activity: the serine, cysteine, aspartic, and metalloproteinases.[22] Curcumin's antiangiogenic property is due in part by its inhibitory action on metalloproteinases and the serine proteinase family, the urokinase plasminogen activator system (uPA). uPA interacts with a specific receptor (uPAR) via the epidermal growth factor (EGF)-like domain in the urokinase amino-terminal fragment (ATF).[23] Its angiogenic effect is due to its effect on the migration of endothelial cells and through the activation and/or release of several angiogenic factors such bFGF, transforming growth factor (TGF), TNF, hepatocyte growth factor (HGF), and VEGF.

In mouse keratinocytes, uPA expression/secretion is increased by TGF-β1. Curcumin decreases the uPA levels induced by TGF-β1 in transformed keratinocytes; inhibits the TGF-β-induced synthesis of fibronectin, an early response gene to the growth factor; and reduces TGF-β-stimulated cell migration and invasiveness.[24,25] Curcumin also inhibits EGF-stimulated urokinase production, although not statistically significant. Curcumin modulates the EGF-stimulated uPA production, which involves the activation of the extracellular signal-regulated kinases 1/2 and JNK signaling pathways and also inhibits the phosphorylation of the EGF receptor.[26] In another study by Parra et al.,[27] uPA induced by N-methyl-N-nitro-N-nitrosoguanidine (MNNG) was inhibited by curcumin. Curcumin acted at the (1) AP-1 binding to the uPA enhancer element, (2) uPA transcriptional activity, and (3) uPA mRNA expression to abrogate the uPA secretion. The multifunctional properties render the uPA–uPAR system an attractive target for curcumin as antiangiogenic therapy.

The other family of proteinases involved in ECM remodeling is the metalloproteinases. Endothelial cell attachment to the extracellular matrix, detachment, and migration/invasion are functions of matrix metalloproteinases (MMPs), which have been clearly implicated in angiogenesis by Collins et al.[28] and Stetler-Stevenson et al.[29]Gelatinase A (MMP-2) and gelatinase B (MMP-9) are metalloproteinases that cause the formation of new capillaries by activating growth factors and it was shown that curcumin inhibits the gelatinolytic activities of secreted 53- and 72-kDa MMP and suppresses the expression and transcription of the 72-kDa MMP, indicating its inhibitory effect at both the transcriptional and posttranscriptional level. Gelatinase-B expression is induced by the transcription factor AP-1, which, in turn, is regulated by FGF-2 and this expression is inhibited by curcuminoids. Using corneal implantation pellets, it has been shown that the FGF-2 pellet was inhibited by coimplantation of curcuminoid pellet and this correlates with the inhibition of endogenous gelatinase-B expression. These results provide evidence that curcuminoids inhibit expression of gelatinase-B and target the FGF-2 angiogenic signaling pathway and that curcumin acts as an angiogenesis inhibitor by modulating MMPs.[24]

4. EFFECT ON ADHESION MOLECULES

Cell adhesion molecules play a determining role in tumor metastasis and curcumin can downregulate their expression. Adhesion molecules such as intracellular adhesion molecule-1 (ICAM-1, also called CD54), vascular cell adhesion molecule-1 (VCAM-1), and endothelial leukocyte adhesion molecule-1 (ELAM-1, also called E-selectin) are induced by TNF.[30–32] ICAM-1, VCAM-1, and ELAM-1 are 95-kDa, 110-kDa, and 115-kDa proteins, respectively, and are expressed in endothelial cells. Treatment of endothelial cells with TNF augmented the adhesion of monocytes to endothelial cells and this adhesion was due to increased expression of ICAM-1, VCAM-1, and ELAM-1 in the endothelial cell. Curcumin completely blocked the adhesion of monocytes[32] with endothelial cells as well as the cell surface expression of ICAM-1, VCAM-1, and ELAM-1. TNF-induced expression of adhesion molecules on human umbilical vein endothelial cells has been reported by others.[31] Although curcumin inhibited adhesion even when administered 1 h after TNF treatment, maximum inhibition occurred when added either 1 h before or at the same time as TNF. The downregulation of adhesion molecules by curcumin might contribute to its anticancer properties.

These properties of curcumin were further studied by Aggarwal et al.[33] and showed that curcumin also acts through the TNF-induced NF-κB-dependent pathway. VEGF, MMP-9, and ICAM-1 are regulated by NF-κB, and Western blot analysis revealed that curcumin blocked TNF-induced VEGF, ICAM-1, and MMP-9 protein expression in a time-dependent manner. These results suggest that curcumin plays a role in suppressing angiogenesis and metastasis through various different pathways.

5. EFFECT ON APOPTOSIS OF MELANOMA CELLS

The NF-κB transcription factor plays a central role in the pathogenesis of melanoma.[34] NF-κB activity is inhibited in part under conditions in which melanoma cells undergo apoptosis and previous reports have described NF-κB inhibition with exposure to higher concentrations of curcumin (60 μM) for shorter periods (6 h) in melanoma cell lines.[35] Therefore, the NF-κB machinery is suppressed both by short exposures to high concentrations of curcumin and by longer exposures to lower concentrations of curcumin. Liposomal curcumin was also shown by Li and Kurzrock[16] to have an inhibitory effect on NF-κB. Under apoptosis-inducing conditions, IKK, the upstream regulator of NF-κB, is inhibited strongly by curcumin. Partial inhibition of NF-κB but strong inhibition of IKK by curcumin suggests that, in melanoma cells, signaling molecules other than IKK can regulate NF-κB activity. The ERK1/2-mediated pathway and an Akt-mediated pathway contribute to NF-κB activation independent of IKK[36,37] and play a role in melanoma cell proliferation or survival,[38–40] but under apoptosis-inducing conditions, neither the B-Raf/ERK pathway nor the Akt pathway was inhibited by curcumin. This shows that curcumin's proapoptotic acivity is associated with inhibition of the IKK/NF-κB transcriptional machinery, but not the B-Raf/ERK or Akt pathways, which implies that suppression of the viability of melanoma cells can occur despite the continued activation of the B-Raf/MEK/ERK and Akt pathways.[34] Interleukin (IL)-8, a pleiotropic chemokine that previously has been shown by Hoffmann et al.[41] to play a role in the promotion of malignant cell proliferation as well as in angiogenesis, is regulated in part by NF-κB, and although curcumin can cause NF-κB inhibition, it has surprisingly increased IL-8 levels in the high-secreting IL-8 melanoma cell lines. Therefore, it appears that IL-8 secretion is independent of curcumin-induced NF-κB inhibition and this might be due to IL-8 expression is regulated transcriptionally by AP-1 and C/EBP as well as NF-κB.[42] Curcumin also represses TNF-induced NF-κB-dependent antiapoptotic gene products. NF-κB regulates the expression of antiapoptotic proteins IAP1/2, XIAP, Bcl-2, Bcl-xL, Bfl-1/A1, and FLIP induced by TNF. Also, curcumin was shown to block the expression of these TNF-induced antiapoptotic proteins as well.[33]

6. EFFECT ON CELL MOTILITY AND PROLIFERATION

Cell motility is essential for a wide range of cellular activities, including angiogenesis.[24] In the highly invasive SK-Hep-1 cell line of human hepatocellular carcinoma (HCC), an *in vitro* assay, without or with the Matrigel matrix, was used to quantitate cellular migration and invasion. Curcumin, at 10 μM, inhibited cellular migration and invasion of SK-Hep-1. This cell line also showed a higher secretion of MMP-9, which was inhibited by curcumin in a dose-dependent fashion. Curcumin, therefore, has a significant anti-invasion activity in SK-Hep-1 cells, and this effect is associated with its inhibitory action on MMP-9 secretion.[43]

Ras has been implicated as a direct regulator of endothelial cell differentiation and microinjection of oncogenic H-Ras proteins into endothelial cells stimulates random motility.[44,45] Ras functions upstream of MAPK families, which include extracellular-signal regulated kinase (ERK) and c-Jun N-terminal kinase (JNK). JNK's role in endothelial cell motility has been proven by using a stable transfectant (DAR-ECV) of ECV304 endothelial cells expressing previously established oncogenic H-Ras. DAR-ECV cells showed a twofold increase in angiogenic potential compared to ECV-304 cells. Pretreatment with curcumin decreased the basal motility of DAR-ECV cells in a dose-dependent manner and suppressed the motility stimulated by known JNK agonists such as TNF-α and anisomycin. These results suggest that curcumin has an inhibitory effect on the

Ras-SEK-1-JNK pathway, which regulates the motility of endothelial cells during angiogenesis.[45]

7. EFFECT ON CYCLIN D1

Cyclin D1 is a proto-oncogene that is overexpressed as a result of the amplification or translocation in many cancers, including the breast, esophagus, lung, liver, head and neck, colon, and prostate.[24] Curcumin treatment of prostate cancer, breast cancer, and multiple myeloma cell lines correlates with the downregulation of cyclin D1 protein.[46,47] The suppression of cyclin D1 by curcumin led to inhibition of CDK-4-mediated phosphorylation of retinoblastoma protein. Curcumin-induced downregulation of cyclin D1 was inhibited by lactacystin, an inhibitor of 26S proteosome, suggesting that curcumin represses cyclin D1 expression by promoting proteolysis. Curcumin also downregulated mRNA expression and inhibited the activity of cyclin D1 promoter-dependent reporter gene expression. Thus, curcumin downregulates cylclin D1 expression through the activation of both transcriptional and posttranscriptional mechanisms, and this might contribute to the antiproliferative effects of curcumin.[24]

The precise molecular target of curcumin remains unknown. Curcumin and curcumin derivatives are an attractive pharmacophore because curcumin affects many targets, making resistance to curcumin less likely. A novel target of curcumin is the cop9 signalosome, which is a multiunit protein (at least eight units to date) that is involved in the proteolytic degradation of p53. Mdm2, the natural antagonist of p53, and an ubiquitin ligase, targets p53 for cop9/26S proteasomal-mediated degradation of p53. Proteasome inhibition has become an attractive strategy for tumor therapy, with velcade, and an inhibitor of the 26S proteasome, becoming first-line therapy for multiple myeloma. Curcumin has been found to inhibit the function of the cop9 signalosome. Inhibition of the cop9 signalosome has far-reaching effects on the angiogenic switch. p53 levels are upregulated, and p53 has been shown to be a negative regulator of VEGF. Inhibition of the cop9 signalosome also leads to downregulation of the angiogenic transcription factors Id1 and Id3. Mice deficient in Id1 and Id3 have a decreased ability to accept tumor xenografts because these mice have a decreased number of bone-marrow-derived endothelial stem cells.

The cop9 signalosome is regulated by several proteins, including casein kinase 2 and inositol 1,3,4-trisphosphate 5/6-kinase (5/6-kinase). Inositol 1,3,4-trisphosphate 5/6-kinase (5/6-kinase) phosphorylates many of the same substrates, such as Ikb, as does cop9, and it is possible that the 5/6 kinase provides some of the specificity of the cop9 signalosome and, like the cop9 signalosome, is inhibited by curcumin. Intriguingly, inositol 1,3,4-trisphosphate 5/6-kinase prevents apoptosis due to TNF-α, which might thus allow survival and hypertrophy of tissue under inflammatory conditions, a phenomenon that can be readily appreciated in inflammatory conditions such as psoriasis, rheumatoid arthritis, and inflammatory bowel disease.[48] Topical curcumin has demonstrated efficacy in psoriasis, a

TNF-mediated inflammatory disorder, as well as cutaneous metastases of internal tumors.

8. SUMMARY

Curcumin affects the overall process of angiogenesis by its downregulation of transcription factors such as NF-κB, proangiogenic factors such as VEGF, bFGF, and COX-2; inhibition of cell motility, cellular adhesion molecules, endothelial cell migration, invasion, and extracellular proteolysis. It also has antiproliferative and proapoptotic effects on tumor cells. All of these studies and the lack of toxicity of curcumin point toward curcumin's enormous potential as an antiangiogenic drug.

REFERENCES

1. J. Folkman, Tumor angiogenesis: Therapeutic implications. *New Engl J Med* **285**, 1182–1186 (1971).
2. J. Folkman, Angiogenesis and apoptosis. *Semin Cancer Biol* **13**(2), 159–167 (2003).
3. T. P. Robinson, T. Ehlers, R. B. Hubbard IV, X. Bai, J. L. Arbiser, D. J. Goldsmith, and J. P. Bowen, Design, synthesis, and biological evaluation of angiogenesis inhibitors: Aromatic enone and dienone analogues of curcumin. *Bioorg Med Chem Lett* **13**(1), 115–117 (2003).
4. M. S. Furness, T. P. Robinson, T. Ehlers, R. B. Hubbard 4th, J. L. Arbiser, D. J. Goldsmith, and J. P. Bowen, Antiangiogenic agents: Studies on fumagillin and curcumin analogs. *Curr Pharm Des* **11**, 357–373 (2005).
5. J. Folkman and D. Hanahan, Patterns and emerging mechanisms of the angiogenic switch during tumorigenesis. *Cell* **86**, 353–364 (1996).
6. J. L. Arbiser, M. A. Moses, C. A. Fernandez, N. Ghiso, Y. Cao, N. Klauber, D. Frank, M. Brownlee, E. Flynn, S. Parangi, H. R. Byers, and J. Folkman, Oncogenic H-ras stimulates tumor angiogenesis by two distinct pathways. *Proc Natl Acad Sci USA* **94**(3), 861–866 (1997).
7. J. L. Arbiser, N. Klauber, R. Rohan, R. van Leeuwen, M. T. Huang, C. Fisher E. Flynn, and H. R. Byers, Curcumin is an in vivo inhibitor of angiogenesis. *Mol Med* **4**, 376–383 (1998).
8. R. Klafer and J. L. Arbiser, Regulation of angiogenesis and tumorigenesis by signal transduction cascades: lessons from benign and malignant endothelial tumors. *J Investig Dermatol Symp Proc* **5**(1), 79–82 (2000).
9. G. N. Naumov, E. Bender, D. Zurakowski, S. Y. Kang, D. Sampson, E. Flynn, R. S. Watnick, O. Straume, L. A. Akslen, J. Folkman, and N. Almog, A model of human tumor dormancy: An angiogenic switch from the nonangiogenic phenotype. *J Natl Cancer Inst* **98**(5), 316–325 (2006).
10. R. Bianco, D. Melisi, F. Ciardiello, G. Tortora, Key cancer cell signal transduction pathways as therapeutic targerts, *Eur J Cancer* **42**, 290–294 (2006).
11. W. H. Chen, Y. Chen, and G. H. Cui, Effects of TNF-alpha and curcumin on the expression of VEGF in Raji and U937 cells on angiogenesis in ECV304 cells. *Chin Med J (Engl)* **118**(24), 2052–2057 (2005).

12. S. Sawant, S. Aparicio, A. R. Tink, N. Lara, C. J. Barnstable, and J. Tombran-Tink, Regulation of factors controlling angiogenesis in liver development: A role for PEDF in the formation and maintenance of normal vasculature. *Biochem Biophys Res Commun.* **325**(2), 408–413 (2004).
13. K. Q. Hu, C. H. Yu, Y. Mineyama, J. D. McCracken, D. J. Hillebrand, and M. Hasan, Inhibited proliferation of cyclooxygenase-2 expressing human hepatoma cells by NS-398, a selective COX-2 inhibitor, *Int J Oncol* **22**(4), 757–763 (2003).
14. F. Millanta, S. Citi, D. Della Santa, M. Porciani, and A. Poli, COX-2 expression in canine and feline invasive mammary carcinomas: Correlation with clinicopathological features and prognostic molecular markers. *Breast Cancer Res Treat.* **98**(1), 115–120 (2006) Mar 15; [Epub ahead of print]
15. P. Yoysungnoen, P Wirachwong, P Bhattarakosol, H. Niimi, and S. Patumraj, Effects of curcumin on tumor angiogenesis and biomarkers, COX-2 and VEGF, in hepatocellular carcinoma cell-implanted nude mice. *Clin Hemorheol Microcirc* **34**(1–2), 109–115 (2006).
16. L. Li, F. S. Braiteh, and R. Kurzrock, Liposome-encapsulated curcumin: In vitro and in vivo effects on proliferation, apoptosis, signaling, and angiogenesis. *Cancer* **104**(6), 1322–1331 (2005).
17. K. C. Kent, S. Mii, E. O. Harrington, J. D. Chang, S. Mallette, and J. A. Ware, Requirement for protein kinase C activation in basic fibroblast growth factor-induced human endothelial cell proliferation. *Cir Res* **77**, 231–238 (1995).
18. B. B. Aggarwal and K. Natarajan, Tumor necrosis factor: Developments during the last decade. *Eur Cytokine Netw* **7**, 93–124 (1996).
19. A. Noel, C. Gilles, K. Bajou, L. Devy, F. Kebers, J. M. Lewalle, et al., Emerging roles for proteinases in cancer. *Invasion Metastasis* **17**, 221–239 (1997).
20. D. B. Rifkin, R. Mazzieri, J. S. Munger, I. Noguera, and J. Sung, Proteolytic control of growth factor availability. *APMIS* **107**, 80–85 (1999).
21. M. Egeblad and Z. Werb, New functions for the matrix metalloproteinases in cancer progression, *Nature Rev* **2**, 161–174 (2002).
22. S. S. Twining, Regulation of proteolytic activity in tissues. *Crit Rev Biochem Mol Biol* **29**, 315–383 (1994).
23. H. Li, C. Soria, F. Griscelli, P. Opolon, J. Soria, P. Yen, C. Legrand, J. P. Vannier, D. Belin, M. Perricaudet, and H. Lu, Amino-terminal fragment of urokinase inhibits tumor cell invasion in vitro and in vivo: Respective contribution of the urokinase plasminogen activator receptor-dependent or -independent pathway. *Hum Gene Ther* **16**(10), 1157–1167 (2005).
24. B. B. Aggarwal, A. Kumar, and A. C. Bharti, Anticancer potential of curcumin: Preclinical and clinical studies. *Anticancer Res* **23**(1A), 363–398 (2003).
25. J. F. Santibanez, M. Quintanilla, and J. Martinez, Genistein and curcumin block TGF-beta 1-induced u-PA expression and migratory and invasive phenotype in mouse epidermal keratinocytes. *Nutr Cancer* **37**(1),49–54 (2000).
26. P. C. Smith, J. F. Santibanez, J. P. Morales, and J. Martinez, Epidermal growth factor stimulates urokinase-type plasminogen activator expression in human gingival fibroblasts. Possible modulation by genistein and curcumin. *J Periodontal Res* **39**(6), 380–387 (2004).
27. M. Parra, F. Lluis, F. Miralles, C. Caelles, and P. Munox-Canoves, The cJun N-terminal kinase (JNK) signaling pathway mediates induction of urokinase-type plasminogen activator (uPA) by the alkylating agent MNNG. *Blood* **96**(4), 1415–1424 (2000).

28. T. Collins, M. A. Read, A. S. Neish, M. Z. Whitley, D. Thanos, and T. Maniatis, Transcriptional regulation of endothelial cell adhesion molecules: NF-kappa B and cytokine-inducible enhancers. *FASEB J* **9**(10), 899–909 (1995).
29. W. G. Stetler-Stevenson, Matrix metalloproteinases in angiogenesis: A moving target for therapeutic intervention. *J Clin Invest* **103**(9),1237–1241 (1999).
30. A. Kumar, S. Dhawan, N. J. Hardegen, and B. B. Aggarwal, Curcumin inhibition of tumor necrosis factor (TNF)-mediated adhesion of monocytes to endothelial cells by suppression of cell surface expression of adhesion molecules and of nuclear factor-kappa B activation. *Biochem Pharamcol* **55**, 775–783 (1998).
31. B. Gupta and B. Ghosh, *Curcuma longa* inhibits TNF-alpha induced expression of adhesion molecules on human umbilical vein endothelial cells. *Int J Immunopharmacol.* **21**, 745–757 (1999).
32. Aggarwal, D. Thaloor, A. K. Singh, G. S. Sidhu, P. V. Prasad, H. K. Kleinman, R. K. Maheshwari, Inhibition of angiogenic differentiation of human umbilical vein endothelial cells by curcumin. *Cell growth Differ* **9**(4), 305–312 (1998).
33. S. Aggarwal, H. Ichikawa, Y. Takada, S. K. Sandur, S. Shishodia, and B. B. Aggarwal, Curcumin (diferuloylmethane) down-regulates expression of cell proliferation and antiapoptotic and metastatic gene products through suppression of IκBα Kinase and Akt activation. *Mol Pharmacol* **69**(1), 195–206 (2006).
34. D. R. Siwak, S. Shishodia, B. B. Aggarwal, and R. Kurzrock, Curcumin-induced antiproliferative and proapoptotic effects in melanoma cells are associated with suppression of IkappaB kinase and nuclear factor kappaB activity and are independent of the B-Raf/mitogen-activated/extracellular signal-regulated protein kinase pathway and the Akt pathway. *Cancer* **104**(4), 879–890 (2005).
35. J. A. Bush, K.-J. Cheung, Jr., and G. Li, Curcumin induces apoptosis in human melanoma cells through a Fas receptor/easpase 8 pathway independent of p53. *Exp Cell Res* **271**, 305–314 (2001).
36. P. Dhawan and A. Richmond, A novel NF-κB-inducing kinase-MAPK kinase signaling pathway up-regulates NF-κB activity in melanoma cells. *J Biol Chem* **277**(10), 7920–7928 (2002).
37. L. V. Madrid, C. Y. Wang, D. C. Guttridge, A. J. Schottelius, A. S. Baldwin, Jr., and M. W. Mayo, Akt suppresses apoptosis by stimulating the transactivation potential of the RelA/p65 subunit of NF–κB. *Mol Cell Biol* **20**, 1626–1638 (2000).
38. M. Fujita, D. A. Norris, H. Yagi, et al., Overexpression of mutant *ras* in human melanoma increases invasiveness, proliferation, and anchorage-independent growth in vitro and induces tumour formation and cachexia in vivo. *Melanoma Res* **9**, 279 –291 (1999).
39. J. M. Stahl, M. Cheung, A. Sharma, N. R. Trivedi, S. Shanmugam, and G. P. Robertson, Loss of PTEN promotes tumor development in malignant melanoma. *Cancer Res* **63**, 2881–2890 (2003).
40. A. Jetzt, J. A. Howe, M. T. Horn, E. Maxwell, Z. Yin, D. Johnson, C. C. Kumar, Adeno et al., Adenoviral-mediated expression of a kinase-dead mutant of Akt induces apoptosis selectively in tumor cells and suppresses tumor growth in mice. *Cancer Res* **63**, 6697–6706 (2003).
41. E. Hoffmann, O. Dittrich-Breiholtz, O. Holtmann, and M. Kracht, Multiple control of interleukin-8 gene expression. *J Leuk Biol* **72**, 847–855 (2002).
42. X. Le, Q. Shi, B. Wang, et al., Molecular regulation of constitutive expression of interleukin-8 in human pancreatic adenocarcinoma. *J Interferon Cytokine Res* **20**, 935–946 (2000).

43. L. I. Lin, Y. F. Ke, Y. C. Ko, and J. K. Lin, Curcumin inhibits SK-Hep-1 hepatocellular carcinoma cell Invasion in vitro and suppresses matrix metalloproteinase-9 secretion *Oncology*. **55**, 349–353 (1998).
44. P. L. Fox, PG. Sa, S. F. Dobrowolski, and D. W. Stacey, The regulation of endothelial cell motility by p21 ras. *Oncogene* **9**(12), 3519–3526 (1994).
45. E. Y. Shin, S. Y. Kim, E. G. Kim, C-Jun N-terminal kinase is involved in motility of endothelial cell. *Exp Mol Med* **33**(4), 276–83 (2001).
46. A. C. Bharti, N. Donato, S. Singh, and B. B. Aggarwal, Curcumin (diferuloylmethane) down-regulates the constitutive activation of nuclear factor-kappa B and IkappaBalpha kinase in human multiple myeloma cells, leading to suppression of proliferation and induction of apoptosis. *Blood* **101**(3), 1053–1062 (2002).
47. A. Mukhopadhyay, S. Banerjee, L. J. Stafford, C. Xia, M. Liu, B. B. Aggarwal. Curcumin-induced suppression of cell proliferation correlates with down-regulation of cyclic D_1 expression and CDK4-medicated retinoblastoma protein phosphorylation. *Oncogene* **21**(57), 8852–8861 (2002.)
48. Y. Sun, M. P. Wilson, and P. W. Majerus, Inositol 1,3,4-trisphosphate 5/6-kinase associates with the COP9 signalosome by binding to CSN1. *J Biol Chem* **277**(48), 45759–4564 (2002).

NEUROPROTECTIVE EFFECTS OF CURCUMIN

Greg M. Cole, Bruce Teter, and Sally A. Frautschy

Abstract: Neurodegenerative diseases result in the loss of functional neurons and synapses. Although future stem cell therapies offer some hope, current treatments for most of these diseases are less than adequate and our best hope is to prevent these devastating diseases. Neuroprotective approaches work best prior to the initiation of damage, suggesting that some safe and effective prophylaxis would be highly desirable. Curcumin has an outstanding safety profile and a number of pleiotropic actions with potential for neuroprotective efficacy, including anti-inflammatory, antioxidant, and anti-protein-aggregate activities. These can be achieved at submicromolar levels. Curcumin's dose–response curves are strongly dose dependent and often biphasic so that *in vitro* data need to be cautiously interpreted; many effects might not be achievable in target tissues *in vivo* with oral dosing. However, despite concerns about poor oral bioavailability, curcumin has at least 10 known neuroprotective actions and many of these might be realized *in vivo*. Indeed, accumulating cell culture and animal model data show that dietary curcumin is a strong candidate for use in the prevention or treatment of major disabling age-related neurodegenerative diseases like Alzheimer's, Parkinson's, and stroke. Promising results have already led to ongoing pilot clinical trials.

1. INTRODUCTION

Many neurodegenerative diseases of aging involve the accumulation of protein aggregates, oxidative damage, and inflammation. Curcumin has multiple desirable characteristics for a neuroprotective drug, including anti-inflammatory, antioxidant, and anti-protein-aggregate activities that we have previously reviewed.[1,2] Because of its pluripotency, oral safety, long history of use, and inexpensive cost, curcumin has great potential for the prevention of multiple neurological conditions for which current therapeutics are less than optimal. Examples reviewed include Alzheimer's, Parkinson's, Huntingtin's, head trauma, aging, and stroke. Despite the widely held belief that curcumin's poor systemic bioavailability precludes therapeutic utility outside of the colon,[3,4] there is ample animal model evidence for very effective neuroprotection in a variety of disease models. Conversely, many of curcumin's reported toxic effects are achieved only at doses that will not be reached in systemic tissues with oral dosing. One of the key obstacles with curcumin, as with other compounds lacking adequate patent protection, is that there has been no push for development from the private sector. What is needed is preclinical and clinical development support from either government or philanthropic support. One example of this might be support from the US NIH Aging and Complementary and Alternative Medicine Institutes. In the following pages, we review some of the literature supporting the neuroprotective utility for curcumin, beginning with Alzheimer's disease.

2. BIPHASIC OR DOSE-DEPENDENT RESPONSES TO CURCUMIN

As reviewed elsewhere in this volume, curcumin is both a potent antioxidant and anti-inflammatory agent and has a long history of use, both as a food preservative and in traditional Indian and Asian medicine, often as an oral or topical extract for conditions where Western medicine might employ a nonsteroidal anti-inflammatory drug and/or vitamin E. Curcumin's activity and structure–function relation as a radical scavenger, metal chelator, and antioxidant has received considerable attention . It clearly reduces mRNA production for pro-inflammatory mediators, including cytokines, inducible nitric oxide synthase (iNOS), and cyclooxygenase (COX)-2.[5,6] Apparently, this is due to limiting activator protein (AP)-1 and nuclear factor (NF)-κB-mediated gene transcription[7,8]; however, the direct molecular targets at low doses are not entirely clear. Curcumin inhibition of AP-1 and NF-κB-mediated transcription occurs at relatively low (<100 nM) doses and might be due to inhibition of histone acetylase (HAT) or activation of histone deacetylase (HDAC) activity.[9] At high doses (>3 μM) that are relevant to colon cancer but unlikely achievable with oral delivery in plasma and tissues outside of the gut, curcumin can act as an alkylating agent,[10] a phase II enzyme inducer,[11] and stimulate antioxidant response element-mediated protective gene expression.[12] Some of the effects of curcumin at high *in vitro* doses are clearly toxic and undesirable beyond its use in cancer therapy. For example, inhibition of proteasomal function and potentiation of huntingtin toxicity can be achieved with dosing >3 μM *in vitro*.[13] Proteasomal inhibition would clearly be undesirable in neurodegenerative disorders, which often have protein aggregate accumulation, whereas proteasome stimulation would be protective. However, the dose dependence of curcumin's effects on the proteasome is biphasic with doses up to 1 μM (e.g., achievable *in vivo*) causing 46% increased proteasomal activity and higher doses leading to proteasome inhibition.[14] Proteasome activation would presumably be a useful response in neurodegenerative diseases with accumulating aggregates.

As discussed below, many protective effects, including anti-amyloid, antioxidant, and anti-inflammatory activities, can be obtained with doses at or below 1 μM. For example, low-nanomolar doses can inhibit histone acetyltransferase[15] and JNK-stimulated AP-1 activity, suggesting that these functions are likely central to many *in vivo* effects,[16,17] including central nervous system (CNS) neuroprotective activity.[18] Also as reviewed below, low-dose curcumin can limit the aggregation of multiple forms of amyloid-forming peptides that lead to intraneuronal or extracellular aggregates in a variety of neurodegenerative diseases.

3. ALZHEIMER'S DISEASE

Alzheimer's (AD) is the most prevalent form of age-related dementia, with AD risk doubling every 5 years after age 65. Thus, AD risk for persons living into their eighties rises to 20–40% depending on the population. There are millions of AD patients in the United States today and this number is expected to double and double again with the demographic shift toward a more aged population, leading to over 10 million expected cases, unless preventive measures can be achieved.[19] The classical pathology of AD involves neurodegeneration and the accumulation of protein aggregates to form two major lesions: neurofibrillary tangles (NFTs) and senile plaques. The senile plaques consist of abnormal neuronal proceses ("dystrophic neurites") and activated glial cells surrounding and penetrating a more central proteinaceous deposit of amyloid fibrils made up of β-amyloid (Aβ) peptide. The Aβ peptide is typically 40–42 amino acids in length and is derived from a larger single membrane spanning "amyloid precursor protein" (APP) by endoproteolytic cleavage. The N-terminus is exoplasmic and cut by a rate-limiting β-secretase enzyme (BACE 1). The final secreted amyloid peptide product is amphipathic with the 12–14-amino-acid C-terminal hydrophobic amino acid tail cut from within the membrane by a "γ-secretase" enzyme complex. Aβ peptide is, thus, normally rapidly produced and equally rapidly degraded. However, at elevated concentrations, it has a strong tendency to self-aggregate to form poorly degradable, β-pleated sheet-rich oligomers, protofilaments, and, finally, filaments that have the histochemical staining properties of amyloid. These Aβ filaments deposited in plaques can be visualized with the amyloid dyes thioflavin S and Congo red. The 2-amino-acid longer Aβ1–42, typically a minor species, forms aggregates more than a thousand times faster than Aβ1–40. A large number of different autosomal-dominant AD mutations have been found in APP and the "presenilin" component of the γ-secretase complex and all of these cause more Aβ1–42 to be made, resulting in early-onset AD. Thus, the genetics of AD clearly implicate an etiopathogenic role for increased Aβ1–42. Further, because mutations in Aβ itself can also increase the aggregation rate and cause AD, most researchers are convinced that Aβ aggregates initiate pathogenesis.[20,21] Transgenic mouse models that overexpress human mutant APP develop neuritic amyloid plaques that closely resemble the senile plaques in AD patients,[22,23] but although they show hyperphosphorylated tau, they do not develop neurofibrillary tangles. More recently, tangle pathology has been achieved by expressing high levels of mutant human tau or wild-type human tau on a mouse tau knockout background, but curcumin effects have not been reported on in these models.

3.1. Amyloid Reduction

We initially tested curcumin in a mutant APP transgenic plaque-forming animal model and found that it not only reduced indices of oxidative damage and inflammation, but it also reduced amyloid plaques and accumulated Aβ.[24] We also found that curcmin reduced oxidative damage, inflammation, and cognitive deficits in rats receiving CNS infusions of toxic Aβ.[25] Tests on cultured HEK or 293 cells transfected with human APP and producing measurable Aβ failed to show any evidence of secretase inhibition and reduced Aβ production. However, because curcumin structurally resembles the amyloid-binding dye Congo red, we tested the ability of curcumin to bind amyloid and inhibit Aβ aggregation and found that it dose-dependently blocked Aβ aggregation at submicromolar concentrations.[1] A more extensive report on these observations showed that curcumin not only stained plaques and inhibited Aβ aggregation and fibril formation *in vitro*; but curcumin also inhibited the formation of Aβ oligomers and their toxicity and readily entered the brain to label plaques *in vivo*.[26] More significantly, we found that curcumin appeared to reduce preformed amyloid *in vitro* and to markedly suppress Aβ accumulation and plaques *in vivo* even when the drug treatment was begun when the mice were old enough to already have well-established amyloid burdens at levels similar to AD patients. This efficacy in late stages of amyloid deposition is in marked contrast to other antioxidants and other treatments that fail to reduce amyloid in the same Tg2576 model mice when treatments are begun late.[27,28] Curcumin's *in vivo* capacity to reduce β-amyloid accumulation might derive from multiple activities beyond this first mechanism: (1) direct binding inhibition of Aβ aggregate formation. Amyloid formation has been shown to be limited by five additional mechanisms: (2) metal chelation,[29] (3) the antioxidant vitamin E,[28] (4) lowering cholesterol[30,31] and reducing expression of the β-secretase enzyme BACE1 by reducing its induction by both (5) pro-inflammatory cytokines [interleukin (IL)-1β and tumor necrosis factor (TNF)-α)[32] and (6) the lipid peroxidation product 4 hydroxynonenal acting on JNK-mediated transcription.[33] Curcumin might work to limit amyloid production by direct inhibition of aggregates and control of all five of these pathways, including chelating metals,[34] limiting oxidative damage better than vitamin E,[35,36] lowering cholesterol,[37,38] reducing pro-inflammatory cytokines,[24,38,39] lipid peroxidation,[25]and protein oxidation[24] and JNK-mediated transcription[39] to control BACE1 expression. For example, treatment with curcumin reduced BACE1 mRNA in cultured primary rat neurons and in aging Tg2576 (Morihara, Ma, and Cole, unpublished data). Iron chelation is another activity that also has *in vivo* support.[40] Further, although it is not clear whether it can do so *in vivo* because the dosing seems to require $>3\ \mu M$,[11] curcumin can induce phase II enzymes in astrocytes and heme oxygenase-1 in neurons *in vitro*.[41] Two additional mechanisms might contribute to amyloid reduction: (7) Amyloid aggregates can be cleared via phagocytosis by brain macrophages, curcumin at dosing as low as the 100–500-nM range can stimulate microglial phagocytosis, and clearance of amyloid *in vitro* and curcumin appears to promote phagocytosis *in vivo* (Yang, et al., unpublished data). (8) Finally, one of the major defenses against intraneuronal protein aggregate formation is the induction of heat shock proteins (HSPs) that function as molecular chaperones to block protein aggregate formation.[42] Increased HSP expression from transgenes clearly protects from neurotoxicity arising from intraneuronal protein aggregates.[43] Like several other nonsteroidal anti-inflammatory drugs (NSAIDs), curcumin can potentiate the production of HSPs in response to cellular stress *in vitro* and *in vivo*[44]. Curcumin potentiates the *in vitro* and *in vivo* HSP response to infused (*in vivo)* or applied soluble Aβ aggregates on neurons in culture (Frautschy et al., unpublished results). Thus, there are eight known ways for curcumin to limit β-amyloid accumulation and protect against amyloid peptide-mediated toxicity.

A very recent report using direct *in vivo* multiphoton microscopy to repeatedly observe the same amyloid plaques in AD model mice showed the ability of curcumin to enter the brain, bind plaques, and reduce amyloid plaque size by 30%, and to significantly reduce soluble Aβ *in vivo*.[45] These data encourage the continued development of curcumin as an anti-amyloid agent and efforts to understand its mechanisms of action.

3.2. Inhibition of Amyloid Toxicity

The mechanisms by which β-amyloid peptide aggregates act to cause AD remain unclear, but they appear to include induction of oxidative damage[46,47] as well as inflammation[4,49] and neurotoxicity, the latter mediated through JNK activation.[50,51] Thus, curcumin might act not only by limiting amyloid aggregates but also by suppressing their pro-oxidant, pro-inflammatory, and JNK-mediated toxic amyloid aggregate effects. Further, high doses of curcumin can also inhibit amyloid toxicity *in vitro* and neurotoxic p75 neurotrophin receptor signaling.[52] AD pathogenesis also involves the accumulation of other protein aggregates, including intraneuronal tau amyloid as NFTs and α-synuclein aggregates (discussed below), which curcumin could potentially suppress. Tau dimerization is initiated by oxidative damage[53] and at least some tau kinases, notably mitogen-activated protein kinase (MAPK), are activated by oxidative damage.[54] Further, tau pathology appears to induce oxidative stress and mitochondrial dysfunction, suggesting antioxidants might protect.[55] Finally, like all amyloids, tau aggregates contain a core β-sheet domain that plays a central role in aggregation and might be blocked by natural and synthetic amyloid-binding dyes, potentially including curcumin.

In summary, curcumin's known activities target at least eight anti-amyloid mechanisms relevant to AD pathogenesis, suggesting that it might be useful in preventing or treating AD. Although there is no epidemiology isolating curcumin intake as a variable, age-adjusted AD prevalence and incidence in an area with high curcumin intake (rural India) was surprisingly low compared to the United States and other Western countries.[56] Collectively, available evidence warrants the exploration of curcumin in clinical trials for AD treatment and prevention; a pilot trial in early AD evaluating dosing and efficacy with clinical end points and biomarkers is currently underway at UCLA's Alzheimer's Disease Center.[2]

4. PARKINSON'S DISEASE

Another prevalent, age-related neurodegenerative condition, the movement disorder Parkinson's disease (PD), involves relatively selective vulnerability to the neuromelanin-bearing dopaminergic neurons of the pars compacta region of the substantia nigra and their terminals in the striatum. In Western populations, significant age-related loss of pigmented neuromelanin-bearing neurons commonly occurs in the this region, but symptoms of PD do not manifest until 60–80% neuron loss.

4.1. Oxidative Damage and Inflammation

Of the age-related neurodegenerative conditions, PD has long had the strongest associations with elevated oxidative damage, including that associated with auto-oxidative dopamine breakdown and related semiquinone metabolism to superoxide, as well as monoamine oxidase production of hydrogen peroxide.[57] Low doses of curcumin can inhibit dopamine toxicity *in vivo*.[18] More recently, mitochondrial electron transport defects at complex I and increased free-radical production have been identified in PD brain and peripheral sites, whereas oxidative damage to vulnerable dopaminergic neurons and a PD syndrome can be produced in human and animal models by the MPTP toxin (reviewed in Ref. 58). MPTP toxicity is mediated by MPP+, and curcumin can directly inhibit MPP+ toxicity to the PC12 neuronal cell line.[16]

Further, support for a free-radical role in PD comes from evidence that selective neuron loss, aggregation of α-synuclein, and clinical symptoms resembling PD can be produced by the pesticide toxin rotenone that targets mitochondrial electron transport and causes increased free-radical production.[59] Although not as closely associated with inflammation as AD, recent studies have shown chronic microglial activation in PD and that a single pro-inflammatory stimulus results in sustained microglial activation around dopaminergic neurons that can contribute to their loss in animal models.[60] These data provide some rationale for protection from PD with the polyphenolic antioxidant/ NSAID curcumin.

4.2. Synuclein Aggregation

Although rare, some genetic cases of PD are linked to mutations in a synaptic protein called α-synuclein that was originally identified from smaller peptides isolated in amyloid-containing fractions of AD brains.[61] The α-synuclein protein is another aggregating, fibril-forming protein that is a major component of the Lewy body lesions characteristic of PD as well as certain cases of AD and several other neurodegenerative conditions. Synuclein aggregates show evidence of nitration-based oxidative damage[62] that might play a critical role in aggregate formation.[63] Recent studies have shown that curcumin can reduce the aggregation of α-synuclein,[64] and administration to cultured cells with α-synuclein aggregate formation results in fewer aggregates.[65]

5. OTHER NEURODEGENERATIVE DISEASES WITH PROTEIN MISFOLDING

5.1. Mad Cow Disease

"Mad cow" involves the aggregation of infectious prion proteins that form protease-resistant toxic species with a β-sheet core. Low doses ($IC_{50} \sim 10$ nM) of curcumin effectively inhibited protease-resistant prion protein aggregation and accumulation in neuroblastoma cells *in vitro*, but an initial trial to delay scrapie pathogenesis *in vivo* was unsuccessful.[66] The reasons for the failure in the animal model remain unclear and should be further explored, but one likely explanation would be the failure to obtain adequate curcumin blood levels with oral administration.

5.2. Huntington's Disease and Other CAG Repeat Diseases

These diseases have extended C-terminal CAG repeats coding for polyglutamine, which causes protein aggregates to form at a rate determined by the repeat length. Because curcumin resembles Congo red and its chrysamine G homologue Congo red, its anti-amyloid-binding protein properties are generic and should extend to other protein-misfolding diseases with a β-pleated sheet, including the polyglutamine dieases like Huntington's disease (HD).[67] Evidence for a protective effect in an HD transgenic model has been recently obtained by a UCLA investigator,[68] leading to a pilot curcumin clinical trial with HD patients at UCLA. Marie-Charcot Tooth disorder is another example of a similar protein-misfolding neuropathy and curcumin protects against this disorder *in vitro*[69] (and *in vivo* in a transgenic model (Lupski, personal communication).

5.3. Tauopathies

Aggregates of the microtubule-associated protein tau are present in neurofibrillary tangles in AD and tau mutations have been genetically linked to neurodegeneration in some forms of frontotemporal dementia (FTD), which can be modeled in FTD mutant tau transgenic mice.[70] There is currently intense interest in the neurotoxicity of soluble tau aggregates because of a recent report using a doxycycline-regulated tau transgenic that showed that turning off tau transgene expression in older tangle-bearing mice fails to reduce tangles, but markedly protects against neurodegeneration.[71] Curcumin might protect against the formation of these soluble tau aggregates because the initial tau dimerization step can be driven by oxidative damage, notably lipid peroxidation[53] or redox-regulated disulfides.[72] Further, as discussed earlier, the induction of HSPs should also protect against aggregates. Although an abstract report claimed that curcumin can reduce tau pathology in one of the tau transgenic models, as far as we are aware there are no peer-reviewed publications on this topic. Nevertheless, in a model of CNS Aβ infusion into genomic wild-type tau transgenic mice, dietary curcumin appeared to limit Aβ infusion and tau transgene-related cognitive deficits (Frautschy et al., unpublished data). Based on this suggestive data, ongoing studies are further examining the impact of curcumin on tau pathology.

6. CEREBROVASCULAR DISEASE AND STROKE

Cerebrovascular and cardiovascular disease risk factors overlap AD risk factors and many dementia cases are mixed. Therefore, if confirmed in larger trials, curcumin's reported ability to lower total cholesterol and raise high-density lipoprotein (HDL) cholesterol in humans should be relevant to dementia prevention.[73] To provide another example, homocysteinuria appears to be an important risk factor for both AD and cardiovascular disease.[74] Curcumin effectively protects against homocysteine-induced endothelial damage.[75] Free-radical damage and inflammation contribute to ischemic damage after a stroke. Prior and even delayed curcumin treatment reduces this damage. For example, curcumin injections i.p. reduced damage to vulnerable hippocampal CA1 and preserved antioxidant enzymes and glutathione, even when initiated 3 and 24 h after ischemia[76] Further, curcumin has been shown to protect in a standard middle cerebral artery occlusion rat model for stroke.[77].

7. HEAD TRAUMA

Head trauma is a stringent test of neuroprotective activity and a validated environmental risk factor for AD.[78,79] ;epeated head trauma is also the cause of boxer's dementia (dementia pugilistica), which involves both tangles[80] and Aβ42 deposition.[81] ApoE4, the major genetic risk factor for AD and brain trauma, synergistically increases the risk of AD and Aβ deposition.[82] Further, in an APP transgenic animal model for AD, brain trauma and the APP transgene act synergistically to increase both cognitive deficits and neurodegeneration.[83] Thus, protection against head trauma by curcumin, as shown in an animal model,[84] is another mechanism for potential AD prevention by curcumin.

8. ALCOHOL-INDUCED NEUROTOXICITY

Ethanol-induced toxicity involves lipid peroxidation, inflammation, and other well-established curcumin targets. Not surprisingly, curcumin can effectively protect against ethanol-induced oxidative damage, inflammation, and resulting liver damage[6] and ethanol-induced CNS neurodegeneration *in vivo*.[85] These reports show that despite claims of poor bioavailability, properly delivered , curcumin or its metabolites are effective in protecting tissues from oxidative damage outside of the gastrointestinal tract.

9. THE AGING BRAIN

Curcumin is one of the few drugs likely to slow aging rates, as evidenced by the ability of its major metabolite, tetrahydrocurcumin, to increase the life span in middle-aged mice.[86] Evidence for an impact on aging brain has been recently produced in aging rats, where chronic curcumin treatment was shown to result in reduced lipid peroxidation and accumulation of the age-pigment lipofuscin and to increase the antioxidant defense enzymes glutathione peroxidase and superoxide dismutase as well as sodium potassium ATPase, which normally declines.[87] Curcumin resembles another biphenolic antioxidant, resveratrol, that is believed to have antiaging activity via induction of sirtuins and HDAC activtion, so curcumin's ability to limit HAT and promote neurogenesis[15] might also impact longevity, promoting a sirtuin-like effect on HAT-regulated transcription. These results are intriguing, consistent with other measures of normal brain aging, including protection against CNS oxidative damage, and support the hypothesis that curcumin might slow normal aging of the brain and presumably other tissues in which age-related oxidative damage is an issue.

10. STEM CELL NEURODIFFERENTIATION AND ADULT NEUROGENESIS

Although still controversial, adult neurogenesis appears to be both modulatable and therapeutically significant.[88] It would be of obvious utility to functionally replace lost neurons in neurodegenerative diseases. Curcumin has been reported to stimulate neuronal differentiation of stem cells *in vitro* and adult neurogenesis *in vivo*, notably in the striatum.[15] Although this is a single report that needs confirmation and extension, it shows additional potential for curcumin in conditions with CNS injury and neurodegeneration.

11. OBSTACLES FACING THE CLINICAL DEVELOPMENT OF CURCUMIN

As summarized in Table 1, curcumin has at least ten neuroprotective effects and it can apparently act at nanomolar or even picomolar doses. For example, curcumin's ki for amyloid binding is 200 picomolar.[89]

Curcumin is neuroprotective in multiple animal models and has great potential for the prevention or treatment of age-related dementia arising from AD or cardiovascular disease, Parkinson's disease, other diseases of aging, and aspects of aging itself. Like any drug, it needs preclinical development to establish dosing, formulation, pharmacokinetics, therapeutic windows, and potential toxicity. Normally, these issues are the concern of drug companies, but in the absence of patent protection, this is unlikely to occur. Further, large clinical trials will be required to establish efficacy for any of curcumin's many disease indications. Primary

Table 1. Ten neuroprotective effects of curcumin.

LIMITS	MECHANISM
Pro-inflammatory cytokine induction	Inhibition of AP-1, NF-κB, HAT, HDAC stimulation?
Reactive oxygen species (ROS)	Scavenger, metal/Fe chelator, induces AO defense enzymes
Aβ production	Suppresses cholesterol, BACE1 induction
Amyloid aggregates	Congo red mimetic/aggregate inhibitor
Misfolded protein accumulation	Potentiates HSPS
Neurotoxicity	JNK pathway
Excitotoxicity	COX-2 induction via AP-1, NF-κB
Toxicity	Phase II inducer, HO-1
Particulate toxins	Increases phagocyte clearance
Neuron loss	Stimulates neurogenesis

Note: References are reviewed in the text.

prevention of age-related neurodegeneration would be the eventual goal, but this is even less likely to ever be tested in clinical trials. Government or philanthropic support will likely be required to realize curcumin's potential for ameliorating age-related neurodegeneration and other debilitating conditions with enormous personal and economic costs.

12. CONCLUSION: RATIONALE FOR MULTITARGET APPROACH TO AGE-RELATED DISEASE

Most chronic age-related conditions are not caused by foreign pathogens, but the failure to repair or resist chronic age-related lesions arising from naturally occurring damage or imbalances. They involve prolonged multistep cascades that induce slow degeneration that would best be dealt with by long-term prevention with very inexpensive and safe interventions rather than new drugs with unknown or unacceptable costs and side-effect profiles. This is a huge issue because in the absence of a foreign pathogen, most of the targets will involve essential physiological pathways, where major inhibition will predictably lead to sideeffects. With modest efficacy from multiple beneficial activities, a pleiotropic drug like curcumin can be efficacious without side effects. Further, the original cause might be superseded by subsequent steps in the cascade and no single pathway might be responsible for ongoing degeneration. For example, most of these diseases involve inflammation and oxidative damage, which are known curcumin targets. Atherosclerosis and stroke, colon cancer, and Alzheimer's are prime examples. Furthermore, Alzheimer's and other neurodegenerative diseases of aging typically involve amyloidogenic protein misfolding and aggregation that can be directly combated by curcumin's anti-amyloid activity and possibly by potentiating HSP synthesis. Curcumin's favorable effects on cholesterol metabolism are also likely to reduce vascular disease and mixed dementia that cause dementia and frequently overlap AD. There are other likely beneficial effects. Finally, stimulation of neurogenesis might facilitate functional replacement of lost neurons, and curcumin has been reported to stimulate adult neurogenesis. With so much potential, the argument for curcumin's further development for neurodegenerative and other diseases of aging is compelling.

REFERENCES

1. G. M. Cole, F. Yang, G. P. Lim, J. L. Cummings, D. L. Masterman, and S. A. Frautschy, A rationale for curcuminoids for the prevention or treatment of Alzheimer's disease. *Curr Med Chem-Immun, Endoc, & Metab Agents* **3**, 15–25 (2003).
2. J. M. Ringman, S. A. Frautschy, G. M. Cole, D L. Masterman, and J. L. Cummings, A potential role of the curry spice curcumin in Alzheimer's disease. *Curr Alzheimer Res* **2**, 131–136 (2005).
3. G. Garcea, D. J. Jones, R. Singh, A. R. Dennison, p. B. Farmer, R. A. Sharma, W. P. Steward, A. J. Gescher, and D. P. Berry, Detection of curcumin and its metabolites in hepatic tissue and portal blood of patients following oral administration. *Br J Cancer* **90**, 1011–1015 (2001).
4. C. D. Lao, M. T. T. Ruffin, D. Normolle, D. D. Heath, S. I. Murray, J. M. Bailey, M. E. Boggs, J. Crowell, C. L. Rock, and D. E. Brenner, Dose escalation of a curcuminoid formulation. *BMC Complement Altern Med* **6**, 10 (2006).
5. H. Y. Hsu and M. H. Wen, Lipopolysaccharide-mediated reactive oxygen species and signal transduction in the regulation of interleukin-1 gene expression. *J Biol Chem* **277**, 22,131–22,139 (2002).
6. A. A. Nanji, K. Jokelainen, G. L. Tipoe, A. Rahemtulla, P. Thomas, and A. J. Dannenberg, Curcumin prevents alcohol-induced liver disease in rats by inhibiting the expression of NF-kappa B-dependent genes. *Am J Physiol Gastrointest Liver Physiol* **284**, G321–327 (2003).
7. F. A. Al-Omar, M. N. Nagi, M. M. Abdulgadir, K. S. Al Joni, and A. A. Al-Majed, Immediate and delayed treatments with curcumin prevents forebrain ischemia-induced neuronal damage and oxidative insult in the rat hippocampus. *Neurochem Res* **31**, 611–618 (2006).

7a. S. Shishodia, P. Potdar, C. G. Gairola, and B. B. Aggarwal, Curcumin (diferuloylmethane) down-regulates cigarette smoke-induced NF-kappaB activation through inhibition of IkappaBalpha kinase in human lung epithelial cells: Correlation with suppression of COX-2, MMP-9 and cyclin D1. *Carcinogenesis* **24**, 1269–1279 (2005).

8. G. Kang, P. J. Kong, Y. J. Yuh, S. Y. Lim, S. V. Yim, W. Chun, and S. S. Kim, Curcumin suppresses lipopolysaccharide-induced cyclooxygenase-2 expression by inhibiting activator protein 1 and nuclear factor kappab bindings in BV2 microglial cells. *J Pharmacol Sci* **94**, 325–328 (2004).
9. I. Rahman, J. Marwick, and P. Kirkham, Redox modulation of chromatin remodeling: Impact on histone acetylation and deacetylation, NF-kappaB and pro-inflammatory gene expression. *Biochem Pharmacol* **68**, 1255–1267 (2004).
10. J. Fang, J. Lu, an d A. Holmgren, Thioredoxin reductase is irreversibly modified by curcumin: A novel molecular mechanism for its anticancer activity. *J Biol Chem* **280**, 25,284–25,290 (2005).

11. A. T. Dinkova-Kostova and P. Talalay, Relation of structure of curcumin analogs to their potencies as inducers of Phase 2 detoxification enzymes. *Carcinogenesis* **20**, 911–914 (1999).
12. E. Balogun, M. Hoque, P. Gong, E. Killeen, C. J. Green, R. Foresti, J. Alam, and R. Motterlini, Curcumin activates the haem oxygenase-1 gene via regulation of Nrf2 and the antioxidant-responsive element. *Biochem J* **371**, 887–895 (2003).
13. P. Dikshit, A. Goswami, A. Mishra, N. Nukina, and N. R. Jana, Curcumin enhances the polyglutamine–expanded truncated N-terminal huntingtin-induced cell death by promoting proteasomal malfunction. *Biochem Biophys Res Commun* **342**, 1323–1328 (2006).
14. R. E. Ali and S. I. Rattan, Curcumin's biphasic hormetic response on proteasome activity and heat-shock protein synthesis in human keratinocytes. *Ann NY Acad Sci* **1067**, 394–399 (2006).
15. S. K. Kang, S. H. Cha, and H. G. Jeon, Curcumin-induced histone hypoacetylation enhances caspase-3-dependent glioma cell death and neurogenesis of neural progenitor cells. *Stem Cells Dev* **15**, 165–174 (2006).
16. M. M. Chan, H. I. Huang, M. R. Fenton, and D. Fong, In vivo inhibition of nitric oxide synthase gene expression by curcumin, a cancer preventive natural product with anti-inflammatory properties. *Biochem Pharmacol* **55**, 1955–1962 (1998).
17. F. E. Parodi, D. Mao, T. L. Ennis, M. B. Pagano, and R. W. Thompson, Oral administration of diferuloylmethane (curcumin) suppresses proinflammatory cytokines and destructive connective tissue remodeling in experimental abdominal aortic aneurysms, *Ann Vasc Surg* **20**, 360–368 (2006).
18. Y. Luo, A. Hattori, J. Munoz, Z. Qin, and G. Roth, Intrastriatal dopamine injection induces apoptosis through oxidation-involved activation of transcription factors ap-1 and nf-kappab in rats. *Mol Pharmacol* **56**, 254–264 (1999).
19. R. Brookmeyer, S. Gray, and C. Kawas, Projections of Alzheimer's disease in the United States and the public health impact of delaying disease onset. *Am J Public Health* **88**, 1337–1342 (1998).
20. J. Hardy, Amyloid, the presenilins and Alzheimer's disease . *Trends Neurosci* **20**, 154–159 (1997).
21. D. J. Selkoe, Alzheimer's disease: Genotypes, phenotypes, and treatments. *Science* **275**, 630–631 (1997).
22. D. Games, D. Adams, R. Alessandrini, R. Barbour, P. Berthelette, C. Blackwell, T. Carr, J. Clemens, T. Donaldson, F. Gillespie, T. Guido, S. Hagoplan, K. Johnson-Wood, K. Khan, M. Lee, P. Leibowitz, I. Lieberburg, S. Little, E. Masliah, L. McConlogue, M. Montoya-Zavala, L. Mucke, L. Paganini, E. Penniman, M. Power, D. Schenk, P. Seubert, B. Snyder, F. Soriano, H. Tan, J. Vitale, S. Wadsworth, B. Wolozin, and J. Zhao, Alzheimer-type neuropathology in transgenic mice overexpressing V717F b-amyloid precursor protein. *Nature* **373**, 523–527 (1995).
23. K. Hsiao, P. Chapman, S. Nilsen, C. Eckman, Y. Harigaya, S. Younkin, F. Yang, and G. Cole, G., Correlative memory deficits, Ab elevation and amyloid plaques in transgenic mice. *Science* **274**, 99–102 (1996).
24. G. P. Lim, T. Chu, F. Yang, W. Beech, S. A. Frautschy, and G. M. Cole, The curry spice curcumin reduces oxidative damage and amyloid pathology in an Alzheimer transgenic mouse. *J Neurosci* **21**, 8370–8377 (2001).
25. S. A. Frautschy, W. Hu, S. A. Miller, P. Kim, M. E. Harris-White, and G. M. Cole, Phenolic anti-inflammatory antioxidant reversal of Aβ-induced cognitive deficits and neuropathology. *Neurobiol Aging* **22**, 991–1003 (2001).

26. F. Yang, G. P. Lim, A. N. Begum, O. J. Ubeda, M. R. Simmons, S. S. Ambegaokar, P. P. Chen, R. Kayed, C. G. Glabe, S. A. Frautschy, and G. M. Cole, Curcumin inhibits formation of amyloid beta oligomers and fibrils, binds plaques, and reduces amyloid in vivo. *J Biol Chem* **280**, 5892–5901 (2005).
27. P. Das, M. P. Murphy, L. H. Younkin, S. G. Younkin, and T. E. Golde, Reduced effectiveness of Abeta1-42 immunization in APP transgenic mice with significant amyloid deposition. *Neurobiol Aging* **22**, 721–727 (2001).
28. S. Sung, Y. Yao, K. Uryu, H. Yang, V. M. Lee, J. Q. Trojanowski, and D. Pratico,. Early vitamin E supplementation in young but not aged mice reduces Abeta levels and amyloid deposition in a transgenic model of Alzheimer's disease. *FASEB J* **18**, 323–325 (2004).
29. X. Huang, C. S. Atwood, R. D. Moir, M. A. Hartshorn, R. E. Tanzi, and A. I. Bush, Trace metal contamination initiates the apparent auto-aggregation, amyloidosis, and oligomerization of Alzheimer's Abeta peptides. *J Biol Inorg Chem* **9**, 954–960 (2004).
30. K. Fassbender, M. Simons, C. Bergmann, M. Stroick, D. Lutjohann, P. Keller, H. Runz, S. Kuhl, T. Bertsch, K. von Bergmann, M. Hennerici, K. Beyreuther, and T. Hartmann, Simvastatin strongly reduces levels of Alzheimer's disease beta -amyloid peptides Abeta 42 and Abeta 40 in vitro and in vivo. *Proc Natl Acad Sci USA* **98**, 5856–5861 (2001).
31. L. M. Refolo, M. A. Pappola, J. LaFrancois, B. Malester, S. D. Schmidt, T. Thomas-Bryant, G. S. Tint, R. Wang, M. Mercken, S. S. Petanceska, and K. E. Duff, A cholesterol-lowering drug reduces β-amyloid pathology in a transgenic mouse model of Alzheimer's disease. *Neurobiol Dis* **8**, 890–899 (2001).
32. M. Sastre, I. Dewachter, G. E. Landreth, T. M. Willson, T. Klockgether, F. van Leuven, and M. T. Heneka, Nonsteroidal anti-inflammatory drugs and peroxisome proliferator-activated receptor-gamma agonists modulate immunostimulated processing of amyloid precursor protein through regulation of beta-secretase. *J Neurosci* **23**, 9796–9804 (2003).
33. M. Tabaton, Oxidative stress and beta-APP proteolytic processing. *Neurobiol Aging* **25 (S2)**, S69 (S64-02-03) (2004).
34. E. Tamagno, M. Parola, P. Bardini, A. Piccini, R. Borhi, M. Gugliemotto, G. Santoro, A. Davit, O. Danni, M. A. Smith, G. Perry, M. Tabaton, Beta-site APP cleaving enzyme up-regulation induced by 4-hydroxynonenal is mediated by stress-activated protein kinasses pathways. *J Neurochem* **92**, 628–636 (2005).
35. L. Baum and A. Ng, Curcumin interaction with copper and iron suggests one possible mechanism of action in Alzheimer"s disease animal models. *J Alzheimers Dis* **6**, 367–377; discussion 443–469 (2004).
36. K. B.Suni and R. Kuttan, Effect of oral curcumin administration on serum peroxides and cholesterol levels in human volunteers. *Indian J Physiol Pharmacol* **36**, 273–275 (1992).
37. M. N. Sreejayan Rao, Curcuminoids as potent inhibitors of lipid peroxidation. *J Pharm Pharmacol* **46**(12), 1013–1016 (1994).
38. P. Venkatesanand and M. N. Structure-activity relationships for the inhibition of lipid peroxidation and the scavenging of free radicals by synthetic symmetrical curcumin analogues. *J Pharm Pharmacol* **52**, 1123–1128 (2000).
39. D. Peschel, R. Koerting, and N. Nass, Curcumin induces changes in cxpression of genes involved in cholesterol homeostasis. *J Nutr Biochem* (2006).
40. Y. Abe, S. Hashimoto, and T. Horie, Curcumin inhibition of inflammatory cytokine production by human peripheral blood monocytes and alveolar macrophages. *Pharmacol Res* **39**, 41–47 (1999).

41. Y. Jiao, J. T. Wilkinson, E. Christine Pietsch, J. L. Buss, W. Wang, R. Planalp, F. M. Torti, and S. V. Torti, Iron chelation in the biological activity of curcumin. *Free Radical Biol Med* **40**, 1152–1160 (2006).
42. G. Scapagnini, C. Colombrita, M. Amadio, V. D'Agata, E. Arcelli, M. Sapienza, A. Quattrone, and V. Calabrese, Curcumin activates defensive genes and protects neurons against oxidative stress. *Antioxid Redox Signal* **8**, 395–403 (2006).
43. K. Ohtsukaand T. Suzuki, Roles of molecular chaperones in the nervous system. *Brain Res Bull* **53**, 141–146 (2000).
44. C. J. Cummings, Y. Sun, P. Opal, B. Antalffy, R. Mestril, H. T. Orr, W. H. Dillmann, and Y. Zoghbi, Over-expression of inducible HSP70 chaperone suppresses neuropathology and improves motor function in SCA1 mice. *Hum Mol Genet* **10**, 1511–1518 (2001).
45. K. Kato, H. Ito, K. Kamei, and I. Iwamoto, Stimulation of the stress-induced expression of stress proteins by curcumin in cultured cells and in rat tissues in vivo. *Cell Stress Chaperones* **3**, 152–160 (1998).
46. M. S. Garcia-Alloz, L. Dodwell, A. Borelli, S. Raju, and B J. Backskai, In vivo reduction of plaque size in APPswe/PS1D9 mice treated with curcumin (P4-342). *Alzheimer's and Dementia* **2(Suppl)**, S617 (2006).
47. C. Behl, J. Davis, G. M. Cole, and D. Schubert, Vitamin E protects nerve cells from amyloid β-protein toxicity. *Biochem Biophys Res Commun* **186**, 944–950 (1992).
48. C. Behl, J. B. Davis, R. Lesley, and D. Schubert, Hydrogen peroxide mediates amyloid β-protein toxicity. *Cell* **77**, 817–827 (1994).
49. R. E. Mrakand W. S. Griffin, Interleukin-1, neuroinflammation, and Alzheimer's disease. *Neurobiol Aging* **22**, 903–908 (2001).
50. Z. Xie, M. Wei, T. E. Morgan, P. Fabrizio, D. Han, C. E. Finch, and V. D. Longo, Peroxynitrite mediates neurotoxicity of amyloid beta-peptide1-42- and lipopolysaccharide-activated microglia. *J Neurosci* **22**, 3484–3492 (2002).
51. W. Wei, X. Wang, and J. W. Kusiak, Signaling events in amyloid beta-peptide-induced neuronal death and insulin-like growth factor I protection. *J Biol Chem* **277**, 17,649–17,656 (2002).
52. A. M. Minogue, A. W. Schmid, M. P. Fogarty, A. C. Moore, V. A. Campbell, C. E. Herron, and M. A. Lynch, Activation of the c-Jun N-terminal kinase signaling cascade mediates the effect of amyloid-beta on long term potentiation and cell death in hippocampus: A role for interleukin-1beta? *J Biol Chem* **278**, 27,971–27,980 (2003).
53. P. Kuner, R. Schubenel, and C. Hertel, Beta-amyloid binds to p57NTR and activates NFkappaB in human neuroblastoma cells. *J Neurosci Res* **54**, 798–804 (1998).
54. T. C. Gamblin, M. E. King, J. Kuret, R. W. Berry, and L. I. Binder, Oxidative regulation of fatty acid-induced tau polymerization. *Biochemistry* **39**, 14,203–14,210 (2000).
55. X. Zhu, H. G. Lee, A. K. Raina, G. Perry, and M. A. Smith, The role of mitogen-activated protein kinase pathways in Alzheimer's disease, *Neurosignals* **11**, 270–281 (2002).
56. D. C. David, S. Hauptmann, I. Scherping, K. Schuessel, U. Keil, P. Rizzu, R. Ravid, S. Drose, U. Brandt, W. E. Muller, A. Eckert, and J. Gotz, Proteomic and functional analyses reveal a mitochondrial dysfunction in P301L tau transgenic mice. *J Biol Chem* **280**, 23,802–23,814 (2005).
57. V. Chandra, R. Pandav, H. H. Dodge, J. M. Johnston, S. H. Belle, S. T. DeKosky, and M. Ganguli, Incidence of Alzheimer's disease in a rural community in India, the Indo-US study, *Neurology* **57**, 985–989 (2001).

58. R. J. Mehlhornand G. M. Cole, The free radical theory of aging: A critical review. *Adv Free Radical Biol Med* **1**, 165–223 (1985).
59. W. Duan and M P. Mattson, Dietary restriction and 2-deoxyglucose administration improve behavioral outcome and reduce degeneration of dopaminergic neurons in models of Parkinson's disease. *J Neurosci Res* **57**, 195–206 (1999).
60. Y. Zhou, G. Gu, D. R. Goodlett, T. Zhang, C. Pan, T. J. Montine, K. S. Montine, R. H. Aebersold, and J. Zhang,. Analysis of alpha -synuclein-associated proteins by quantitative proteomics, *J Biol Chem* **279**, 39155–39164 (2004).
61. J. S. Hong, Role of inflammation in the pathogenesis of Parkinson's disease: Models, mechanisms, and therapeutic interventions. *Ann NY Acad Sci* **1053**, 151–152 (2005).
62. K. Uéda, H. Fukushima, E. Masliah, Y. Xia, A. Iwai, D. Otero, J. Kondo, Y. Ihara, and T. Saitoh, Molecular cloning of a novel amyloid component in Alzheimer's disease. *Proc Natl Acad Sci USA* **90**, 11,282–11,286 (1993).
63. B. I. Giasson, J. E. Duda, I. V. MurrayQ. Chen, J. M. Souza, H. I. Hurtig, H. Ischiropoulos, J. Q. Trojanowski, and V. M. Lee, Oxidative damage linked to neurodegeneration by selective alpha-synuclein nitration in synucleinopathy lesions. *Science* **290**, 985–989 (2000).
64. T. Takahashi, H. Yamashita, T. Nakamura, Y. Nagano, and S. Nakamura, Tyrosine 125 of alpha-synuclein plays a critical role for dimerization following nitrative stress. *Brain Res* **938**, 73–80 (2002).
65. K. Onoand M. Yamada, Antioxidant compounds have potent anti-fibrillogenic and fibril-destabilizing effects for alpha-synuclein fibrils in vitro. *J Neurochem* **97**, 105–115 (2006).
66. N. Pandey and J. E. Galvin, Curcumin prevents aggregation of alpha-synuclein. *Soc Neurosci* **31**, abs 1007.9 (2005).
67. B. Caughey, L. D. Raymond, G. J. Raymond, L. Maxson, J. Silveira, and G. S. Baron, Inhibition of protease-resistant prion protein accumulation in vitro by curcumin. *J Virol* **77**, 5499–5502 (2003).
68. N. F. Bence, R. R. Sampat, and R. R. Kopito, Impairment of the ubiquitin-proteasome system by protein aggregation. *Science* **292**, 1552–1555 (2001).
69. C. Zhu, M. A. Hickey, K. Gallant, M. S. Levine, M. F. Chesselet. Differential effects of curcumin and coenzyme Q10 treatment on huntingtin aggregate in CAG 140 knock-in mouse model of Huntington's disease Soc. *Neurosci* **32**, Abs 472.8 (2006).
70. M. Khajavi, K. Inoue, W. Wiszniewski, T. Ohyama, G. J. Snipes, and J. R. Lupski, Curcumin treatment abrogates endoplasmic reticulum retention and aggregation-induced apoptosis associated with neuropathy-causing myelin protein zero-truncating mutants. *Am J Hum Genet* **77**, 841–850 (2005).
71. M Hutton, J. Lewis, D. Dickson, S. H. Yen, and E. McGowan, Analysis of tauopathies with transgenic mice. *Trends Mol Med* **7**, 467–470 (2001).
72. K. Santacruz, J. Lewis, T. Spires, J. Paulson, L. Kotilinek, M. Ingelsson, A. Guimaraes, M. DeTure, M. Ramsden, E. McGowan, C. Forster, M. Yue, J. Orne, C. Janus, A. Mariash, M. Kuskowski, B. Hyman, M/Hutton, and K. H. Ashe, Tau suppression in a neurodegenerative mouse model improves memory function. *Science* **309**, 476–481 (2005).
73. K. Bhattacharya, K. B. Rank, B. D. Evans, and S. K. Sharma, S.K., Role of cysteine-291 and cysteine-322 in the polymerization of human tau into Alzheimer-like filaments. *Biochem Biophys Res Commun* **285**, 20–26 (2001).
74. B. E. Dwyer, A. K. Raina, G. Perry, and M. A. Smith, Homocysteine and Alzheimer's disease: A modifiable risk? *Free Radical Biol Med* **36**, 1471–1475 (2004).

75. G. Ramaswami, H. Chai, Q. Yao, P. H. Lin, A. B. Lumsden, and C. Chen, Curcumin blocks homocysteine-induced endothelial dysfunction in porcine coronary arteries. *J Vasc Surg* **40**, 1216–1222 (2004).
76. J. Chen, X. O. Tang, J. L. Zhi, Y. Cui, H. M. Yu, E. H. Tang, S. N. Sun, J. Q. Feng, nd P. X. Chen, Curcumin protects PC12 cells against 1-methyl-4-phenylpyridinium ion-induced apoptosis by bcl-2-mitochondria-ROS-iNOS pathway. *Apoptosis* **11**, 943–953 (2006).
77. Q. Wang, A. Y. Sun, A. Simonyi, M. D. JensenP.B. Shelat, G. E., Rottinghaus, R. S. MacDonald, D. K. Miller, D. E., Lubahn, G. A. Weisman, and G. Y. Sun, Neuroprotective mechanisms of curcumin against cerebral ischemia-induced neuronal apoptosis and behavioral deficits. *J Neurosci Res* **82**, 138–148 (2005).
78. J. A. Mortimer, C. M. Duijn, V. Chandra, L. Fratiglioni, A. B. Graves, A. Heyman, A. F. Jorm, E. Kokmen, K. Kondo, W. A. Rocca, S. L. Shalat, H. Soininen, and A. Hofman, Head trauma as a risk factor for Alzheimer's disease, A collaborative re-analysis of case-control studies. *Int J Epidemiol* **20**, S28–S35 (1991).
79. J. L. Cummings, JH. V. Vinters, G. M. Cole, and Z. S. Khachaturian, Alzheimer's disease: Etiologies, pathophysiology, cognitive reserve, and treatment opportunities. *Neurology* **51**, S2–17; discussion S65–S17 (1998).
80. H. Wisniewski, H. K. Narang, J. Corsellis, and R. D. Terry, Ultrastructural studies of the neuropil and neurofibrillary tangles in Alzheimer's disease and post-traumatic dementia. *J Neuropathol Exp Neurol* **35**, 367 (1976).
81. S. M. Gentleman, B. D. Greenberg, M. J. Savage, M. Noori, S. J. Newman, G. W. Roberts, S. T. Griffin, and D. Graham, Ab42 is the prominant form of amyloid b-protein in the brains of short-term surviviors of head injury. *Neuroreport* **8**, 1519–1522.
82. J. A. Nicoll, G. W. Roberts, and D. I. Graham, Amyloid beta-protein, APOE genotype and head injury. *Ann NY Acad Sci* **777**, 271–275 (1996).
83. D. H. Smith, M. Nakamura, T. K. McIntosh, J. Wang, A. Rodriguez, X. H. Chen, R. Raghupathi, K. E. Saatman, J. Clemens, M. L. Schmidt, V. M. Lee, and J. Q. Trojanowski, Brain trauma induces massive hippocampal neuron death linked to a surge in beta-amyloid levels in mice overexpressing mutant amyloid precursor protein. *Am J Pathol* **153**, 1005–1010 (1998).
84. A. Wu, Z. Ying, and F. Gomez-Pinilla, Dietary curcumin counteracts the outcome of traumatic brain injury on oxidative stress, synaptic plasticity, and cognition. *Exp Neurol* **197**, 309–317 (2006).
85. V. Rajakrishnan, P. Viswanathan, K. Rajasekharan, and V. Menon, Neuroprotective role of curcumin from *Curcuma longa* on ethanol-induced brain damage. *Phytother Res* **13**, 571–574 (1999).
86. K. Kitani, T. Yokozawa, and T. Osawa, Interventions in aging and age-associated pathologies by means of nutritional approaches. *Ann NY Acad Sci* **1019**, 424–426 (2004).
87. K. Bala, B. C. Tripathy, and D. Sharma, Neuroprotective and anti-ageing effects of curcumin in aged rat brain regions. *Biogerontology* **7**, 81–89 (2006).
88. G. Kempermann, H. G. Kuhn, and F. H. Gage, More hippocampal neurons in adult mice living in an enriched environment. *Nature* **386**, 493–495 (1997).
89. E. K. Ryu, Y. S. Choe, K.-H. Lee, Y. Choi, and B.-T. Kim. Curcumin and dehydrozingerone derivatives: synthesis, radiolabeling, and evaluation for β-amyloid plaque imaging. *J Med Chem* **49**, 6111–6119 (2006).

REGULATION OF COX AND LOX BY CURCUMIN

Chinthalapally V. Rao

Abstract: Turmeric (*Curcuma longa*) is extensively used as a household remedy for various diseases. For the last few decades, work has been done to establish the biological activities and pharmacological actions of curcumin, the principle constituent of turmeric. Curcumin has proven to be beneficial in the prevention and treatment of a number of inflammatory diseases due to its anti-inflammatory activity. Arachidonic acid-derived lipid mediators that are intimately involved in inflammation are biosynthesized by pathways dependent on cyclooxygenase (COX) and lipoxygenase (LOX) enzymes. The role of LOX and COX isoforms, particularly COX-2, in the inflammation has been well established. At cellular and molecular levels, curcumin has been shown to regulate a number of signaling pathways, including the eicosanoid pathway involving COX and LOX. A number of studies have been conducted that support curcumin-mediated regulation of COX and LOX pathways, which is an important mechanism by which curcumin prevents a number of disease processes, including the cancer. The specific regulation of 5-LOX and COX-2 by curcumin is not fully established; however, existing evidence indicates that curcumin regulates LOX and COX-2 predominately at the transcriptional level and, to a certain extent, the posttranslational level. Thus, the curcumin-selective transcriptional regulatory action of COX-2, and dual COX/LOX inhibitory potential of this naturally occurring agent provides distinctive advantages over synthetic COX/LOX inhibitors, such as nonsteroidal anti-inflammatory drugs. In this review, we discuss evidence that supports the regulation of COX and LOX enzymes by curcumin as the key mechanism for its beneficial effects in preventing various inflammatory diseases.

1. INTRODUCTION

Currently, chronic diseases are, by far, the leading cause of death in the world, and their impact is steadily growing. Approximately 17 million people die prematurely each year as a result of the global epidemic of chronic disease.[1,2] Elimination of risk factors as an approach for the prevention of any given chronic disease has so far had limited success in reducing burden of many chronic diseases such as cardiovascular, cancer, and diabetes. For example, it has been known for decades that tobacco smoking, excessive alcohol drinking, and unhealthy dietary habits and lifestyles are major risk factors for many of these chronic diseases,[3,4] but, to date, not much progress has been made in eliminating these high-risk factors. In the prevention of chronic diseases, one key new approach appears to be modulation of the inflammatory cascade, as research is expanding that links many chronic diseases to inflammatory events. Synthetic nonsteridoal anti-inflammatory drugs (NSAIDs) are widely used to treat many chronic inflammatory diseases; these agents primarily act on the suppression of the formation of pro-inflammatory eicosanoids. Evidence is presented that suggests that the inhibition of cyclooxygenase (COX) and lipoxygenase (LOX) are equally important in the inhibition of chronic inflammation and cancer[5–8] currently available NSAIDs are predominately COX inhibitors. Based on the hypothesis that natural agents are capable of modulating both LOX and COX activities will provide several advantages over synthetic drugs in the prevention and treatment of chronic diseases, including cancer.[9–11] In this review, we discuss evidence that supports the regulation of COX and LOX enzymes by curcumin as a key mechanism for its beneficial effects in the prevention of inflammatory diseases. Also, we discuss advantages of the use of curcumin in the regulation of COX and LOX pathways when compared to synthetic inhibitors of COX or LOX in cancer prevention and treatment.

2. TURMERIC: AN ANTI-INFLAMMATORY AGENT

The use of medicinal plants or their crude extracts in the prevention and/or treatment of several chronic diseases has been traditionally practiced in various different ethnic societies worldwide. In South and Southeast Asia, including India, turmeric, the powdered rhizome of *Curcuma longa* L. has been used extensively in foods for both its flavor and color. Turmeric has a long tradition of use in the Ayurvedic and Chinese systems of medicine, particularly as an anti-inflammatory agent and for the treatment of flatulence, jaundice, menstrual difficulties, hematuria, hemorrhage, and colic. Turmeric can also be applied topically in poultices to relieve pain and inflammation.[12–14] Current researchas focused on turmeric's principle component—curcumin—and its potential beneficial effects in various disease preventions and treatments. Curcumin (Figure 1), which has been identified as the major pigment in turmeric, possesses both anti-inflammatory and antioxidant properties. Curcumin's antioxidant, and anti-inflammatory activities

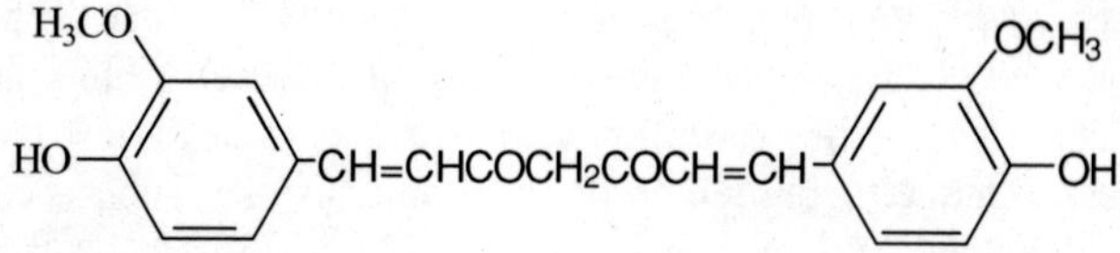

Figure 1. Chemical structure of curcumin [diferuloylmethane; 1,7-bis-(4-hydroxy-3-methoxyphenyl)-1,6-heptadine-3,5-dione].

help in the prevention and treatment of diseases associated with oxidative stress and inflammation.

It is noteworthy that turmeric/curcumin has been known for its broad spectrum of health beneficial effects over thousands of years, but the biological action of this agent has been understood only in recent times. The anti-inflammatory effects of turmeric were supported by many laboratory investigations. Arora et al. reported anti-inflammatory activity in different fractions of the petroleum ether extract of *C. longa*.[15] The total extracts of the rhizome of turmeric were evaluated for their anti-inflammatory activity in albino rats (180–200 g) and compared with that of hydrocortisone acetate and phenylbutazone. It was found that the anti-inflammatory activity of turmeric extracts was almost as active as hydrocortisone acetate in the inflammation induced by the cotton pellet method. Curcumin isolated from the alcoholic extract of turmeric has been shown to be a potent anti-inflammatory agent. Recently, the anti-inflammatory activity of curcumin has been demonstrated in acute and chronic models of inflammation in rats and mice.[11,16] In rats with Freud׳s adjuvant-induced arthritis, administration of curcumin significantly reduced the inflammatory swelling compared to the control[16] In instances of acute inflammation, oral administration of curcumin was found to be as effective as cortisone or phenylbutazone, and even in chronic inflammation, it was shown to be effective[17] Curcumin might also be applied topically to animal skin to counteract inflammation and irritation associated with inflammatory skin conditions and allergies.[17] Further, clinical studies also confirmed the anti-inflammatory property of curcumin in patients with postoperative inflammation.[18] Further, oral administration of curcumin to rats, at a dose of 3 mg/kg body weight, and sodium curcumin, at a dose of 0.1 mg/kg, inhibited formalin-induced arthritis in the animals. In fact, curcumin, once again, was comparatively as effective as phenylbutazone in this application. In a double-blind trial of 49 patients diagnosed with rheumatoid arthritis, when curcumin was given at a dose of 1200 mg/day for 5–6 weeks, there was an overall improvement in morning stiffness and physical endurance; this yielded comparable effects to those obtained with phenylbutazone.[12] These studies suggest that curcumin has been shown to regulate signaling pathways associated with inflammation.

3. INHIBITION OF EICOSANOID PATHWAY

Similar to nonsteroidal anti-inflammatory agents, curcumin might act via a single mechanism or in a combination of any of the mechanisms involving inhibition of arachidonic acid metabolism, inhibition of COX, inhibition of the prostaglandin (PG) synthesis, inhibition of LOX, inhibition of cytokines [interleukin (IL), tumor necrosis factor (TNF), etc.], release of steroidal hormones from the adrenals, stabilization of lysosomal membrane, uncoupling of oxidative phosphorylation, and so forth.[9,19–22] Accumulating evidence suggests that the inhibition of eicosanoid pathways (Figure 2) as a predominant mechanism for most anti-inflammatory actions

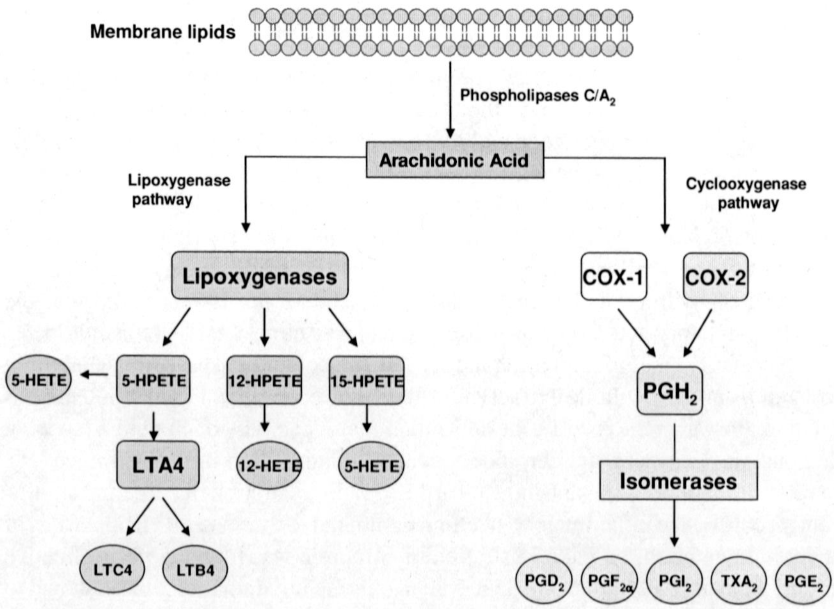

Figure 2. Products and enzymes of arachidonic acid metabolism involved in the inflammatory process. PG, prostaglandin; LT, leukotrienes, HETE, hydroxy eicosatetraenoic acid.

is induced by curcumin.[19–21] It is noteworthy that, so far, more than 100 research articles were published implicating the inhibitory effects of COX and LOX pathways by curcumin.

4. INHIBITION OF COX PATHWAY

Cyclooxygenase catalyzes the transformation of arachidonic acid into prostaglandin $H_2(PGH_2)$ as a first step; in turn, it is transformed into a series of final active products in different cell types and with different expressions. Two COX isoforms, COX-1 and COX-2, have been identified. They are encoded by two different genes,[23] and it has been postulated that whereas COX-1 is expressed in mammalian cells, particularly in endothelium, platelets, and the kidneys in physiological conditions, COX-2 is inducible in pathological conditions by inflammatory stimulation.[24,25] it has, therefore, been suggested that constitutive COX-1 is involved in homoeostatic processes, whereas COX-2 is the isoform that plays a major part in the inflammatory process and the pain associated with it. Nevertheless, there is accumulating evidence that COX-1 and COX-2 have overlapping actions. The inhibitory effect of curcumin on COX has been shown by a number of laboratories utilizing *in vitro* and *in vivo* model systems.

Srivastava et al. demonstrated that curcumin inhibited the incorporation of [^{14}C]-arachidonic acid (AA) into platelet phospholipids and inhibited the deacylation of

AA-labeled phospholipids (liberation of free AA) on stimulation with the calcium ionophore.[26,27] Rat peritoneal macrophages preincubated with 10 μM curcumin for 1 h inhibited the incorporation of AA into membrane lipids by 82% and that of PGE_2 by 45%.[28] Curcumin appears to block the synthesis of certain PGs through inhibition of the COX enzyme.[29,30] Preclinical experiments conducted in laboratories also showed that dietary curcumin (0.2%) suppressed chemically induced colon cancers and this inhibition was associated with decreased activities of colonic mucosal and tumor phospholipase A_2 (50%) and phospholipase Cgγ1 (40%) and the formation of PGs such as PGE_2, $PGF_2\alpha$, PGD_2, 6-keto $PGF_1\alpha$, and thromboxane B_2.[31,32] similarly, chemopreventive properties of curcumin against skin carcinogenesis could be related to the potent inhibitory effect of this agent on arachidonic acid-induced inflammation and on arachidonic acid metabolism through the COX pathway in the mouse epidermis.[33,34] in an *in vivo* study, the PGE_2 content in the inflammatory exudates of control rats with inflammation was 7.29 μg/mL. Treatment of the animals with curcumin (200 mg/kg) for 4 days reduced the PGE_2content of the exudates by 45%.[33] *In vitro* studies revealed that curcumin decreased phorbol ester-induced PGE_2 production down to almost preinduction level.[35] In a confirmatory Western analysis using a COX-2 monoclonal antibody, curcumin was shown to reduce phorbol ester-induced COX-2 protein expression consistently by 60–70%. In contrast, curcumin metabolites tetrahydrocurcumin, hexahydrocurcumin, and curcumin sulfate interfered with COX-2 protein inhibition only weakly.[35] Similarly, Ramsewak et al. demonstrated that curcumin was active against the COX-2 enzyme compared to the COX-1 enzyme.[36] Thus, curcumin was found to be effective in inhibiting PG synthesis in inflammatory exudates as well as in the *in vitro* and *in vivo* tumors by modulating the COX-pathway.

5. MOLECULAR MECHANISM OF COX-2 REGULATION

The exact molecular mechanism by which curcumin produces the suppression of COX isoforms is not yet fully understood, but available information leans more toward pleiotropic mechanisms. Zhang et al. investigated whether curcumin inhibited chenodeoxycholate (CD)- or phorbol ester (phorbol 12-myristate 13-acetate, PMA)-mediated induction of COX-2 in several gastrointestinal cell lines (SK-GT-4, SCC450, IEC-18, and HCA-7).[37] Treatment with curcumin suppressed CD- and PMA-mediated induction of COX-2 protein and synthesis of PGE_2. In the same study, curcumin also suppressed the induction of COX-2 mRNA by CD and PMA.[37] To investigate the effect of curcumin on COX-2 expression, HT-29 human colon cancer cells were treated with various concentrations of curcumin. Curcumin inhibited the cell growth of HT-29 cells in a concentration- and time-dependent manner. There was a marked inhibition of mRNA and protein expression of COX-2, but not COX-1.[38] Kim et al. demonstrated that the inhibitory action of curcumin on Janus kinase (JAK)–STAT (signal transducer and activator of transcription) signaling could contribute to its anti-inflammatory activity in the brain.[39] In both rat primary microglial and murine BV2 microglial cells, curcumin effectively suppressed the ganglioside, lipopolysaccharide (LPS)- or interferon (IFN-γ)-stimulated induction of COX-2.

Recent studies have demonstrated that eukaryotic transcription factor nuclear factor-κB (NF-κB) was involved in the regulation of COX-2. Surh et al. studied the molecular mechanism underlying the anti-inflammatory activity of curcumin.[40] They suggested the downregulation of COX-2 through suppression of NF-κB. Repression of the degradation of the inhibitory unit IκBα, which hampers subsequent nuclear translocation of the functionally active subunit of NF-κB, might be responsible for the inhibition of NF-κB by curcumin.[40,41] Han et al. demonstrated that curcumin inhibited the 12-*O*-tetradecanoylphorbol-13-acetate (TPA)-induced NF-κB activation by preventing the degradation of the inhibitory protein IκBα and the subsequent translocation of the p65 subunit in cultured human promyelocytic leukemia (HL-60) cells.[42] Alternatively, curcumin repressed the TPA-induced activation of NF-κB through direct interruption of the binding of NF-κB to its consensus DNA sequences. It is important to note that studies carried out by Shishodia et al. showed that curcumin downregulates cigarette-smoke-induced NF-κB activation through the inhibition of IκBκ and it correlated with the suppression of COX-2.[43]. Chun et al. demonstrated the effect of curcumin on TPA-induced expression of COX-2 in a female mouse.[44] Immunohistochemical analysis of TPA-treated mouse skin revealed an enhanced expression of COX-2 localized primarily in the epidermal layer, which was markedly suppressed by a curcumin pretreatment. Curcumin treatment attenuated TPA-stimulated NF-κB activation in mouse skin, which was associated with its blockade of the degradation of the inhibitory protein IκBα and of the subsequent translocation of the p65 subunit to the nucleus.[45]

Another important molecular mechanism by which curcumin regulates the COX-2 is through the modulating of inducible nitric oxide synthase (iNOS) expression. Several *in vitro* and *in vivo* studies have provided evidence for interplay (cross-talk) between COX-2 and iNOS expression and activities.[46–52] Salvemini et al. showed that NO enhanced the activity of COX enzymes through mechanisms independent of cGMP in the RAW 264.7 macrophage cell line.[50] Tetsuka et al. found that COX-2 activation by NO increased the production of PGE_2in rat mesangial cells.[47] Similarly, Mei et al. observed that NO induced COX-2 expression in conditionally immortalized murine colonic epithelial cells.[48] *In vivo* studies using iNOS-deficient mice revealed that a significant reduction of urinary and peritoneal macrophage PGE_2 levels correlated with nitrite and nitrate levels.[49] In animal models of colon cancer, we have shown that the administration of the colon-specific carcinogen azoxymethane (AOM) increased both iNOS and COX-2 activities in colonic mucosa of F334 rats.[53–55] Importantly, AOM-induced iNOS and COX-2 activities have been significantly suppressed in animals fed a diet of curcumin and iNOS-selective inhibitors.[53,54] The mechanism(s) by which NO/iNOS regulates COX-2 activity is not fully understood. Some studies suggested that COX-mediated arachidonic acid oxygenation (by free-radical mechanisms) requires NO/peroxynitrite for the catalysis.[46,49,51] It is possible that curcumin might directly act on the arachidonic acid oxygenation or snaring reactive nitrogen species due to its antioxidant property. However, recent investigations

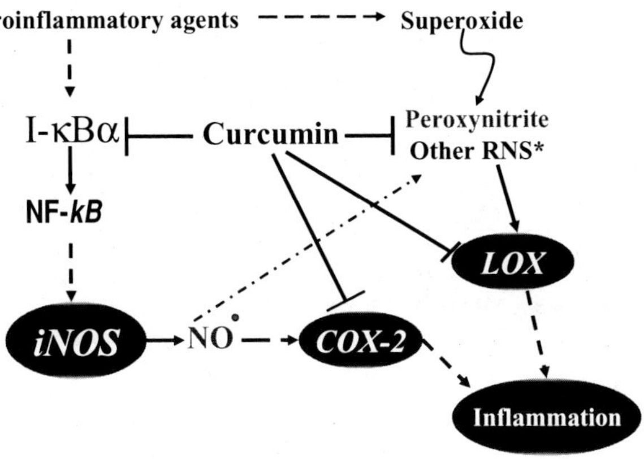

Figure 3. Mechanism of modulation of COX-2 and LOX by curcumin. *RNS, reactive nitrogen species.

indicated that the regulation of COX-2 by iNOS/NO is more complex and also involves a β-catenin/TCF-LEF-mediated transcriptional pathway.[56] Surely, iNOS seems to play a critical role in modulating COX-2 activities that, in turn, influence inflammatory processes, as depicted in Figure 3.

6. REGULATION OF LOX PATHWAY

Lipoxygenases, like COX, are enzymes implicated in the arachidonic acid metabolism catalyzing the insertion of oxygen into various positions in arachidonic acid. Among various LOX enzymes, 5-LOX-mediated arachidonic acid metabolites play an important role in the inflammation.[57,58] The final and biologically active metabolites of the 5-LOX cascade are hydroxyeicosatetraenoic acids (HETEs), leukotrienes (LTs), and lipoxins. Leukotrienes are potent mediators of inflammation; these include LTB_4 and the so-called cysteinyl LTs (LTC_4, LTD_4, and LTE_4).[59] The target of the biological effects of LTB_4 has been found to be primarily inflammatory cells: $LTB_4$is a potent stimulator of leukocyte activation, and adhesion of these cells to vascular endothelium elicits chemokinetic and chemotactic responses.[60] Furthermore, LTB_4 has been shown to be involved in the pathogenesis of a variety of inflammatory diseases.[61] It has been observed that this LT stimulates the production and release of pro-inflammatory cytokines from macrophages and lymphocytes and, recently, from synovial membrane. Inhibitory effects of curcumin on LOX have been shown by number of laboratories utilizing *in vitro* and *in vivo* model systems. Srivastava et al. demonstrated that curcumin inhibited the formation of LTs in rat peritoneal macrophages induced by the calcium ionophore.[26,27] Rat peritoneal macrophages preincubated with 10 μM curcumin for 1 h inhibited the formation of LTB_4 by 61% and LTC_4 by 34%.

Flynn et al. studied the inhibitory activities of curcuminoids on the 5-HETE. Various diaryl-heptonoids, including curcumin, were found to be potent inhibitors of 5-HETE productions by intact human neutrophils with IC_{50} values ranging from 4 to 8 μM.[62] Huang et al. showed that topical application of curcumin markedly inhibited TPA- and arachidonic acid-induced epidermal inflammation (ear edema) in mice, but chlorogenic acid, caffeic acid, and ferulic acid were only weakly active or inactive.[34] The *in vitro* addition of 3, 10, 30, or 100 μM curcumin to cytosol from homogenates of mouse epidermis inhibited the metabolism of arachidonic acid to 5-HETE by 40%, 60%, 66%, or 83%, respectively, and the metabolism of arachidonic acid to 8-HETE was inhibited by 40%, 51%, 77%, or 85%, respectively, whereas chlorogenic acid, caffeic acid, or ferulic acid (100 μM) inhibited the metabolism of arachidonic acid to 5-HETE by 36%, 10%, or 16%, respectively.[34] Similarly, our laboratory experiments carried out in rats showed that the administration of 0.2% curcumin suppressed AOM-induced colonic mucosal and tumor 5(S)-, 8(S)-, 12(S)- and 15(S)-HETEs formation up to 68%[32] Also, Ammon et al. showed the inhibitory effect of curcumin on the 5-LOX in rat peritoneal neutrophils and 12-LOX activities in human platelets.[29] Recently, Hong et al. showed that curcumin and related β-diketone derivatives released arachidonic acid and its metabolites in the murine macrophage RAW264.7 cells and in HT-29 human colon cancer cells.[63] They examined their effects on the catalytic activities and protein levels of 5-LOX. At 10 μM, dibenzoylmethane, trimethoxydibenzoylmethane, tetrahydrocurcumin, and curcumin effectively inhibited the release of arachidonic acid and its metabolites in LPS-stimulated RAW cells and A23187-stimulated HT-29 cells. Curcumin significantly inhibited the activity of human recombinant 5-LOX, showing estimated IC_{50} values of 0.7 μM.[63] The results suggest that curcumin inhibits 5-LOX activities in both *in vitro* and *in vivo* models.

The exact mechanism by which curcumin regulates 5-LOX is not yet explored and thus far very few studies are reported.[58,64–66] One of these studies suggested that LOX catalyzed the oxygenation of curcumin and that curcumin can act as a lipoxygenase substrate. By the use of X-ray diffraction and mass spectrometry, they found an unoccupied electron mass that appeared to be an unusual degradation product of curcumin (4-hydroxyperoxy-2-methoxyphenol) located near the soybean LOX catalytic site.[64] In another study, West et al. showed that curcumin might regulate 5-LOX through the inhibition of transforming growth factor (TGF)-α signaling.[65] Understanding how curcumin inhibits LOX might help in the development of novel drugs used for treatment where LOXs are involved.

7. ADVANTAGES OF DUAL COX/LOX INHIBITION BY CURCUMIN

There are a number of advantages of curcumin application in the prevention and treatment of various inflammatory diseases and cancer when compared to the conventional NSAIDs and the selective COX-2 inhibitors because of its dual

COX/LOX inhibitory property. COX-2 inhibitors primarily exert their activity by reducing the production of PGs induced in the inflammatory process. In recent years, it has been clarified that PG synthesis is only one part of the arachidonic acid pathway, this precursor being a substrate that gives rise to many other lipid mediators, such as the LTs and the lipoxins (LXs). Leukotrienes themselves have a major role in the development and persistence of the inflammatory process, and it is now clear that PGs and LTs have complementary effects, whereas the production of LXs can counteract the inflammatory actions of LTs. Also, it is important to note that the effect of COX-2 inhibitors on the incidence of cardiovascular diseases raised several concerns[66]; particularly, COX-2-specific inhibitors could potentially increase the thrombotic risk, because they block the production of PGI_2, a potent antiaggregating agent,[67] but curcumin does not have such side effects. Shah et al. studied the mechanism of platelet aggregation by curcumin.[69] They showed that curcumin inhibited platelet aggregation mediated by the platelet agonists' epinephrine (200 μM), platelet activating factor (PAF, 800 nM), collagen (20 μg/mL), and AA (0.75 mM). Curcumin preferentially inhibited PAF and AA-induced aggregation at low concentrations (IC_{50}; 20–25 μM), and at higher concentrations, it inhibited aggregation induced by other platelet agonists. In view of these concepts, it has been suggested that blocking both LOX and COX pathways by curcumin might have synergistic effects and achieve optimal anti-inflammatory activity. In addition, taking into account the roles of LTB_4 and cysteinyl LTs (against which neither selective nor nonselective NSAIDs are effective) in the inflammatory process, dual inhibition of the COX and 5-LOX pathways by curcumin could produce a wider spectrum of anti-inflammatory effects.

8. SUMMARY

Curcumin is a major pigment in turmeric, a powdered rhizome of *Curcumin longa* Linn, which has been extensively used as a coloring and flavoring agent in foods and for the treatment of a variety of inflammatory conditions and other chronic diseases. There is convincing evidence that curcumin inhibits AA-mediated COX and LOX pathways. A number of studies demonstrated that curcumin suppresses the COX-2 expression and activities more preferentially when compared to COX-1. Similarly, many studies also suggest that curcumin inhibits the 5-LOX and other LOXs. The exact molecular mechanism by which curcumin suppresses COX and LOX isoforms is not yet fully understood, but available information leans more toward pleiotropic mechanisms involving transcriptional as well as posttranslational modifications. Based on dual COX-2 and LOX inhibitory activities, curcumin has several advantages over the other synthetic agents that inhibit either COX or LOX. Furthermore, the lack of toxicity and side effects as well as its availability in large quantities as a natural product that has been used in population groups for several centuries are additional advantages in applying this agent for the prevention and treatment of inflammatory diseases.

9. ACKNOWLEDGMENTS

We thank Ms. Julie DeVore and Dr. Suresh Guruswamy for helping in the preparation of this manuscript. This work was in part supported by NCI grants R01CA-80003 and R01-CA-94962.

REFERENCES

1. *World Health Organization, Health Report, 2006*. Lyon: WHO Publications Press, 2006.
2. Chronic Disease Prevention. Center for Disease Control, Annual Report (2006).
3. W.C. Willett, M. J. Stampfer, B. A. Colditz, G. A. Rosner, and F. E. Speizer, Relation of meat, fat and fiber intake to the risk of colon cancer in a prospective study among women. *N Engl J Med* **323**, 1664–1672 (1990).
4. J. D. Potter and K. Steinmatz, Vegetables, fruits and phytoestrogens as preventive agents. *IARC Sci Publ* **139**, 61–90 (1995).
5. C. V. Rao and B. S. Reddy, NSAIDs and chemoprevention. *Curr Cancer Drug Targets* **4**, 29–44 (2004).
6. J. P. Collet, C. Sharpe, E. Belzile, J. F. Boivin, J. Hanley, and L. Abenhaim, Colorectal cancer prevention by non-steroidal anti-inflammatory drugs: Effects of dosage and timing. *Br J Cancer* **81**, 62–68 (1999).
7. S. R. Maxwell, R. A. Payne, G. D. Murray, and D. J. Webb,. Selectivity of NSAIDs for COX-2 and cardiovascular outcome. *Br J Clin Pharmacol* **62**(2), 243–245 (2006).
8. D. M. Schreinemachers and R. B. Everson, Aspirin use and lung, colon, and breast cancer incidence in a prospective study.*Epidemiology* **5**, 138–146 (1994).
9. K. Kohli, J. Ali, M. J. Ansari, and Z. Raheman, Curcumin: A natural anti-inflammatory agent. *Indian J Pharmacol* **37**, 141–147 (2005).
10. H. H. Tonnesen, Chemistry of curcumin and curcuminoids. In: C.-T. Ho, C. Y. Lee, and M-T. Haung, eds. *Phenolic Compounds in Food and their Effect of Health. Vol. 1: Analysis, Occurrence and Chemistry*. ACS Symposium Series No. 506, pp. 143–153, Washington, DC: American Chemical Society, 1992. pp. 143–153.
11. R. C. Srimal and B. N. Dhawan, Pharmacology of diferuloylmethane (curcumin), a non-steroidal anti-inflammatory agent. *J Pharm Pharmacol* **25**, 447–452 (1973).
12. H. P. T. Ammon and M. A. Wahl, Pharmacology of *Curcuma longa*. *Planta Med* **57**, 1–7 (1991).
13. B. B. Aggarwal, A. Kumar, and A. C. Bharti, Anticancer potential of curcumin, preclinical and clinical studies. *Anticancer Res*, **23**, 363–398 (2003).
14. I. Chattopadhyay, K. Biswas, U. Bandyopadhyay, and R. K. Banerjee, Turmeric and curcumin: Biological actions and medicinal applications. *Cur Sci* **87**, 44–53 (2004).
15. R. Arora, N. Basu, and V. Kapoor, Anti-inflammatory studies on *Curcuma longa* (turmeric). *Indian J Med Res* **59**, 1289–1295 (1971).
16. R. C. Srimal, N. M. Khanna, and B. N. Dhawan, A preliminary report on anti inflammatory activity of curcumin. *Int J Pharm* **3**, 10–13 (1971).
17. A. Mukhopadhyay, N. Basu, and N. Ghatak, Anti-inflammatory and irritant activities of curcumin analogues in rats. *Agents Actions* **12**, 508–515 (1982).
18. R. R. Satoskar, S. J. Shah, and S. G. Shenoy, Evaluation of anti-inflammatory property of curcumin in patients with postoperative inflammation. *Int J Clin Pharmacol Ther Toxicol* **24**, 651–654 (1986).

19. R. Maheshwari, A. K. Singh, J. Gaddopati, and R. C. Srimal, Multiple biological activities of curcumin: A short review. *Life Sci* **78,** 2081–2087 (2006).
20. T. H. Leu and M. C. Maa, The molecular mechanisms for the antitumorigenic effect of curcumin. *CurrMed Chem Anticancer Agents* **2**, 357–370 (2002).
21. S. Shishodia, H. M. Amin, R. Lai, and B. B. Aggarwal, Curcumin (diferuloylmethane) inhibits constitutive NF-κB activation, induces G1/S arrest, suppresses proliferation, and induces apoptosis in mantle cell lymphoma. *Biochem Pharmacol* **70**(5), 700–713 (2005).
22. R. L. Thangapazham, S. Sharma, and R. Maheshwari, Multiple molecular targets in cancer chemoprevention by curcumin. *AAPS J* **8**, 443–449 (2006).
23. W. L. Smith, R. M. Garavito, and D. L. DeWitt, Prostaglandin endoperoxide H synthases (cyclooxygenases)-1 and -2. *J Biol Chem* **271**, 33,157–33,160 (1996).
24. J. Y. Jouzeau, B. Terlain, A. Abid, E. Nedelec, and P. Netter, Cyclo-oxygenase isoenzymes. How recent findings affect thinking about nonsteroidal anti-inflammatory drugs. *Drugs* **53**, 563–582 (1997).
25. J. Y. Fu, . J. L. Masferrer, K. Seibert, A. Raz, and P. Needleman, The induction and suppression of prostaglandin H2 synthase (cyclooxygenase) in human monocytes. *J Biol Chem* **265**, 16,737–16,740 (1990).
26. K. C. Srivastava, A. Bordia, and S. K. Verma, Curcumin, a major component of food spice turmeric (*Curcuma longa*) inhibits aggregation and alters eicosanoid metabolism in human blood platelets. *Prostaglandins Leukot Essent Fatty Acids* **52**, 223–227 (1995).
27. A. H. Conney, T. Lysz, T Ferraro, T. F. Abidi, P. S. Manchand, J. D. Laskin, and M. T. Huang, Inhibitory effect of curcumin and some related dietary compounds on tumor promotion and arachidonic acid metabolism in mouse skin. *Adv Enzyme Regul* **31**, 385–396 (1991).
28. B. Joe and B. R. Lokesh, Effect of curcumin and capsaicin on arachidonic acid metabolism and lysosomal enzyme secretion by rat peritoneal macrophages. *Lipids* **32**, 1173–1180 (1997).
29. H. P. Ammon, H. Safayhi, T. Mack, and J. Sabieraj, Mechanism of anti-inflammatory actions of curcumin and bowsellic acids. *J Ethnopharmacol* **38**, 113–119 (1993).
30. R. Srivastava, Inhibition of neutrophil response by curcumin. *Agents Actions* **28**, 298–303 (1989).
31. C. V. Rao, B. Simi, and B. S. Reddy, Inhibition by dietary curcumin of azoxymethane-induced ornithine decarboxylase, tyrosine protein kinase, arachidonic acid metabolism and aberrant crypt foci formation in the rat colon. *Carcinogenesis* **14**, 2219–2225 (1993).
32. C. V. Rao, A. Rivenson, B. Simi, and B. S. Reddy, Chemoprevention of colon carcinogenesis by dietary curcumin, a naturally occurring plant phenolic compound. *Cancer Res* **55**, 259–266 (1995).
33. M.-T. Huang, T. Lysz, T. Ferraro, T. F. Abidi, J. D. Laskin, and A. H. Conney, Inhibitory effects of curcumin on in vitro lipoxygenase and cyclooxygenase activities in mouse epidermis. *Cancer Res* **51**, 813–819 (1991).
34. M.-T. Huang, R. C. Smart, C.-Q. Wong, and A. H. Cooney, Inhibitory effect of curcumin, chlorogenic acid, caffeic acid, and ferulic acid on tumor promotion in mouse skin by 12-O-tetradecanoylphorbol-13-acetate. *Cancer Res* **48**, 5941–5946 (1998).
35. C. Ireson, S. Orr, D. J. Jones, R. Verschoyle, C. K. Lim, J. L. Luo, et al., Characterization of metabolites of the chemopreventive agent curcumin in human and rat hepatocytes and in the rat in vivo, and evaluation of their ability to inhibit phorbol ester-induced prostaglandin E2 production. *Cancer Res* **61**, 1058–1064 (2001).

36. R. S. Ramsewak, D. L. DeWitt, and M. G. Nair, Cytotoxicity, antioxidant and anti-inflammatory activities of curcumins I-III from *Curcuma longa*. *Phytomedicine* **7**, 303–308 (2000).
37. F. Zhang, N. K. Altorki, J. R. Mestre, K. Subbaramaiah, and A. J. Dannenberg, Inhibition of cyclo-oxygenase 2 expression in colon cells by the chemopreventive agent curcumin involves inhibition of NF-kappaB activation via the NIK/IKK signaling complex. *Carcinogenesis* **20**, 445–451 (1999).
38. A. Goel, C. R. Boland, and D. P. Chauhan, Specific inhibition of cyclooxygenase-2 (COX-2) expression by dietary curcumin in HT-29 human colon cancer cells. *Cancer Lett* **172**, 111–118 (2001).
39. H. Y. Kim, E. J. Park, E. H. Joe, and I. Jou, Curcumin suppresses Janus kinase–STAT inflammatory signaling through activation of Src homology 2 domain-containing tyrosine phosphatase 2 in brain microglia. *J Immunol* **171**, 6072–6079 (2003).
40. Y. J. Surh, K. S. Chun, H. H. Cha, S. S. Han, Y. S. Keum, K. K. Park KK, et al., Molecular mechanism underlying chemopreventive activities of anti-inflammatory phytochemicals: down regulation of COX-2 and iNOS through suppression of NF-kappa B activation. *Mutation Res* **480**, 243–268 (2001).
41. A. A. Nanji, K. Jokelainen, G. L. Tipoe, A. Rahemtulla, P. Thomas, and A. J. Dannenberg, Curcumin prevents alcohol-induced liver disease in rats by inhibiting the expression of NF-kappa B-dependent genes. *Am J Physiol Gastrointest Liver Physiol* **284**, 321–327 (2003).
42. S. S. Han, Y. S. Keum, H. I. Seo, and Y. L. Surh, Curcumin suppresses activation of NF-κB and AP-1 induced by phorbol ester in cultured human promyelocytic leukemia cells. *J Biochem Mol Biol* **35**, 337–342 (2002).
43. S. Shishodia, H. M. P. Potdar, C. G. Gairola, and B. B. Aggarwal, Curcumin (diferuloylmethane) down-regulates cigarette smoke-induced NF-κB activation through inhibition of IkappaBalpha kinase in human lung epithelial cells: Correlation with suppression of COX-2, MMP->9 and cyclin D1. *Carcinogenesis* **24**(7), 1269–1279 (2003).
44. K. S. Chun, Y. S. Keum, S. S. Han, Y. S. Song, S. H. Kim, and Y. J. Surh, Curcumin inhibits phorbol ester-induced expression of cyclooxygenase-2 in mouse skin through suppression of extracellular signal-regulated kinase activity and NF-kappaB activation. *Carcinogenesis* **24**, 1515–1524 (2003).
45. A. C. Bharti, N. Donato, S. Singh, and B. B. Aggarwal, Curcumin (diferuloylmethane) down-regulates the constitutive activation of nuclear factor-kappa B and IkappaBalpha kinase in human multiple myeloma cells, leading to suppression of proliferation and induction of apoptosis. *Blood* **10**, 1053–1062 (2003).
46. D. Salvemini, S. L. Settle, J. L. Masferrer, K. Seibert, M. G. Currie, and P. Needleman, Regulation of prostaglandin production by nitric oxide: An in vivo analysis. *Br J Pharmacol* **114**, 1171–1178 (1995).
47. T. Tetsuka, D. D. Iken, B. W. Miler, Z. Guan, L. D. Baier, and A. R. Morrison, Nitric oxide amplifies interleukin 1-induced cyclooxygenase-2 expression in rat mesangial cells. *J Clin Invest* **97**, 2051–2055 (1996).
48. J. M. Mei, N. G. Hord, D. F. Winterstein, S. P. Donald, and J. M. Phang, Expression of prostaglandin endoperoxide H synthase-2 induced by nitric oxide in conditionally immortalized murine colonic epithelial cells. *FASEB J* **14**, 1188–1192 (2000).
49. L. J. Marnett, T. L. Wright, B. C. Crews, S. R. Tannenbaum, and J. D. Morrow, Regulation of prostaglandin biosynthesis by nitric oxide is revealed by targeted deletion of inducible nitric oxide synthase. *J Biol Chem* **275**, 13,427–13,421 (2000).

50. D. Salvemini, T. P. Misko, J. L. Masferrer, K. Seibert, M. G. Currie, and P. Needleman, Nitric oxide activates cyclooxygenase enzymes. *Proc Natl Acad Sci USA* **90**, 7240–7245 (1993).
51. L. J. Marnett, S. W. Rowlinson, D. C. Goodwin, A. S. Kalgutkar, and C. A. Lanzo, Arachidonic acid oxygenation by COX-1 and COX-2. *J Biol Chem* **274**, 22,903–22,906 (1999).
52. V. B. O'Donell, B. Coles, M. J. Lewis, B. C. Crews, L. J. Marnett, and B. A. Freeman, Catalytic consumption of nitric oxide by prostaglandin H synthase regulates platelet function. *J Biol Chem* **275**, 38,239–38,243 (2000).
53. C. V. Rao, C. Indranie, B. Simi, P. T. Manning, J. R. Connor, and B. S. Reddy, Chemopreventive properties of a selective inducible nitric oxide synthase inhibitor in colon carcinogenesis, administered alone or in combination with celecoxib, a selective cyclooxygenase inhibitor. *Cancer Res* **62**, 165–170 (2002).
54. C. V. Rao, T. Kawamori, R. Hamid, and B. S. Reddy, Chemoprevention of colonic aberrant crypt foci by an inducible nitric oxide synthase-selective inhibitor. *Carcinogenesis* **20**, 641–644 (1999).
55. C. V. Rao, I. Cooma, M. V. Swamy, B. Simi, and B. S. Reddy, Modulation of inducible nitric oxide synthase and cyclooxygenase activities by curcumin during different stages of experimental colon carcinogenesis. *Proc Am Assoc Cancer Res* **39**, 3084 (2001).
56. Y. Liu, G. L. Borchert, and J. M. Phang, PEA3, an Ets transcription factor, mediates the induction of cyclooxygenase-2 by nitric oxide in colorectal cancer cells. *J Biol Chem* **279**, 18,694–18,700 (2004).
57. A. Sala, S. Zarini, and M. Bolla, Leukotrienes, lipid bioeffectors of inflammatory reactions. *Biochemistry (Mosc)* **63**, 84–92 (1998).
58. O. P. Radmark, The molecular biology and regulation of 5-lipoxygenase. *Am J Respir Crit Care Med* **161**, S11–S25 (2000).
59. J. F. Penrose, K. F. Austen, and B. K. Lam, Leukotrienes: Biosynthetic pathways, release and receptor-mediated actions with relevance to disease states. In: J. L.Gallin and R. Snyderman, eds. *Inflammation Basic Principles And Clinical Correlates*. Philadelphia: Lippincott Williams & Wilkins, 1999, pp. 361–372.
60. M. A. Bray, A. W. Ford-Hutchinson, and M. J. Smith, Leukotriene B4: An inflammatory mediator in vivo. *Prostaglandins* **22**, 213–222 (1981).
61. R. A. Lewis, K. F. Ansten, and R. J. Soberman, Leukotrienes and other products of the 5-lipoxigenase pathway. Biochemistry and relation to pathobiology in human diseases. *N Engl J Med* **192**, 439–446 (2000).
62. D. L. FlynnM. F. Rafferty, and A. M. Boctor, Inhibition of 5-hydroxyeicosatetraenoic acid (5-HETE) formation in intact human neutrophils by naturally occurring diarylheptanoids: Inhibitory activities of curcuminoids and yakuchinones. *Leukotrienes Med* **22**, 357–360 (1986).
63. J. Hong, M. Bose, J. Ju, J.H. Ryu, X. Chen, S. Sang, M. J. Lee, and C. S. Yang, Modulation of arachidonic acid metabolism by curcumin and related beta-diketone derivatives; effects of cytosolic phospholipase A2, cyclooxygenases and 5-lipoxygenase. *Carcinogenesis* **25**, 1671–1679 (2004).
64. J. E. Skrzypczak, N. P. McCabe, S. H. Selman, and J. Jankun, Curcumin inhibits lipoxygenase by binding to its central cavity: Theoretical and X-ray evidence. *Int J Mol Med* **6**, 521–526 (2000).
65. M. West, M. Mhatre, A. Ceballos, R. A. Floyd, P. Grammas, S. P. Gabbita, L. Hamdheydari, T. Mai, Z. Zemlan, and K. Hensley, The arachidonic acid 5-lipoxygenase inhibitor

nordihydroguaiaretic acid inhibits tumor necrosis factor alpha activation of microglia and extends survival of G93A-SOD1 transgenic mice. *J Neurochem* **91**(1), 133–143 (2004).

66. N. S. Prasad, R. Raghavendra, B. R. Lokesh, and K. A. Naidu, Spice phenolic inhibits human PMNL 5-lipoxygenase. *Prostaglandins Leukot Essent Fatty Acids* **70**, 521–528 (2004).
67. B. F. McAdam, F. Catella-Lawson, I. A. Mardini, S. Kapoor, J. A. Lawson, and G. A. FitzGerald, Systemic biosynthesis of prostacyclin by cyclooxygenase (COX)-2: The human pharmacology of a selective inhibitor of COX-2. *Proc Natl Acad Sci USA* **96**, 272–277 (1999).
68. B. H. Shah, Z. Nawaz, S. A. Pertani, A. Roomi, H. Mahmood, S. A. Saeed, et al., Inhibitory effect of curcumin, a food spice from turmeric, on platelet-activating factor- and arachidonic acid-mediated platelet aggregation through inhibition of thromboxane formation and Ca^{2+} signaling. *Biochem Pharmacol* **58**, 1167–1172 (1999).

MOLECULAR TARGETS OF CURCUMIN

Jen-Kun Lin

Abstract: Curcumin possesses anti-inflammatory activity and is a potent inhibitor of reactive-oxygen-generating enzymes such as lipoxygenase/cyclooxygenase, xanthine dehydrogenase/oxidase, and inducible nitric oxide synthase (iNOS); it is an effective inducer of heme oxygenase-1. Curcumin is also a potent inhibitor of protein kinase C (PKC), EGF-receptor tyrosine kinase, and IκB kinase. Subsequently, curcumin inhibits the activation of NF-κB and the expressions of oncogenes including c-jun, c-fos ,c-myc, NIK, MAPKs, ERK, ELK, PI3K, Akt, CDKs, and iNOS. It is considered that PKC, mTOR, and EGFR tyrosine kinase are the major upstream molecular targest for curcumin intervention, whereas the nuclear oncogenes such as c-jun, c-fos, c-myc, CDKs, FAS, and iNOS might act as downstream molecular targets for curcumin actions. It is proposed that curcumin might suppress tumor promotion through blocking signal transduction pathways in the target cells. The oxidant tumor promoter TPA activates PKC by reacting with zinc thiolates present within the regulatory domain, whereas the oxidized form of cancer chemopreventive agent such as curcumin can inactivate PKC by oxidizing the vicinal thiols present within the catalytic domain. Recent studies indicated that proteasome-mediated degradation of cell proteins play a pivotal role in the regulation of several basic cellular processes, including differentiation, proliferation, cell cycling, and apoptosis. It has been demonstrated that curcumin-induced apoptosis is mediated through the impairment of the ubiquitin–proteasome pathway.

1. INTRODUCTION

Chemoprevention is the attempt to use dietary factors, synthetic pharmacological agents, and changes in lifestyle to intervene in the precancerous stages of carcinogenesis before the invasive disease begins.[1] It has been suggested that diet has an impact on cancer incidence and that daily consumption of vegetables and fruits decreases the risk for human cancer.[2,3] Recently, efforts have been focused on identifying dietary phytochemicals, which have the ability to inhibit the processes of carcinogenesis. Among these phytochemicals, curcumin has been demonstrated to be a promising cancer chemopreventive agent in animal systems.[4,5] Curcumin has been listed as the third generation of cancer chemopreventive agents by the Institute of Cancer Chemoprevention, NCI, NIH of the United States. A study of the clinical application of curcumin as a chemopreventive agent was intensively carried out at the NCI.[6] Our recent study on phase I clinical trial of curcumin in patients with high-risk or premalignant lesions has demonstrated that curcumin is not toxic to humans up to 8000 mg/day when taken for 3 months and has a promising biologic effect in the chemoprevention of several types of cancer.[7] Turmeric is widely used as a spice and coloring agent in several foods such as curry, mustard, bean cake, cassava paste, and potato chips as well as cosmetics and drugs. Another species, namely *C. wenyujin*, has been used for centuries in Chinese traditional medicine for the treatment of a variety of inflammatory conditions such as hepatitis and bile duct disorders.[8] Curcumin has been demonstrated to have potent antioxidant[9–11] and anti-inflammatory activities,[4,5,12,13] and it inhibits the carcinogen–DNA adduct.[14] and tumorigenesis in several animal models.[15–18]

2. CANCER CHEMOPREVENTION BY CURCUMIN

Several studies have demonstrated that curcumin inhibited chemical carcinogenesis in different tissue sites in several experimental animal models. Curcumin inhibited the tumor initiation by benzo[*a*]pyrene (BaP) and 7,12-dimethylbenz[*a*]anthracene (DMBA) in mouse epidermis.[14] The topical application of curcumin strongly inhibited tumor promotion in the skin of DMBA-initiated mice.[4,15,17] Feeding 0.5–2.0% curcumin in the diet decreased BaP-induced forestomach tumors per mouse by 51–53% when administered during the initiation period and 47–67% when administered during the postinitiation period.[16] Further studies indicated that curcumin might inhibit BaP-induced forestomach cancer in mice by affecting both activation as well as inactivation pathways of BaP metabolism in the liver.[19] Feeding curcumin in the diet decreased the number of *N*-ethyl-*N*′-nitro-*N*-nitrosoguanidine (ENNG)-induced duodenal tumors per mouse.[16] Administration of curcumin in the diet decreased the number of azoxymethane (AOM)-induced colon tumors in mice[16] and rats.[18]

Curcumin is an effective agent for chemoprevention action at the radiation-induced initiation stage of mammary carcinogenesis.[20] A recent study in our laboratory also indicated that curcumin effectively inhibits diethylnitrosamine-induced hepatocarcinogenesis in mice.[21] It is suggested that the feasibility of using curcumin in the chemoprevention of human hepatocellular carcinoma should be further explored.[7] Recent studies have indicated the combined inhibitory effects of curcumin and phenethylisothiocyanate on the growth of PC-3 prostate xenografts in immunodeficient mice.[22]

3. METABOLISM OF CURCUMIN

The pharmacokinetic properties of curcumin have been investigated in mice.[23] After intraperitoneal adminstration of curcumin (0.1 g/kg) to mice, about 2.25 μg/mL of curcumin appeared in the plasma in the first 15 min. One hour after administration, the levels of curcumin in the intestine, spleen, liver, and kidneys were 177, 26, 27, and 7.5 μg/g, respectively. Only traces (0.41 μg/g) were observed in the brain at 1 h. To clarify the nature of the metabolites of curcumin, the plasma was analyzed by reversed-phase high-performane liquid chromatography (HPLC), and two putative conjugates were observed. Further treatment of the plasma with β-glucuronidase resulted in a decrease in the levels of these two putative conjugates and the concomitant appearance of the tetrahydrocurcumin and curcumin, respectively. To investigate the nature of these glucuronide conjugates *in vivo*, the plasma was analyzed by electrospray. The chemical structures of these metabolites were determined by mass spectrometry–mass spectrometry (MS/MS) analysis.[23] The experimental results suggested that curcumin was first biotransformed to dihydrocurcumin and tetrahydrocurcumin and these compounds subsequently were converted to monoglucuronide conjugates. These results suggest that curcumin–glucuronide, dihydrocurcumin–glucuronide, tetrahydrocurcumin–glucuronide, and tetrahydrocurcumin are major metabolites of curcumin in mice.

The bioavailability of parent curcumin is low,[7] so its pharmacological activity can be mediated, in part, by curcumin metabolites. The major products of curcumin biotransformation by hepatocytes occur only at low abundance in rat plasma after curcumin administration and metabolism of curcumin by reduction or conjugation generate species with reduced ability to inhibit cyclooxygenase (COX)-2 expression.[24] Because the gastrointestinal tract seems to be exposed more prominently to unmetabolized curcumin than any other tissue, the results support the clinical evaluation of curcumin as a colorectal cancer chemopreventive agent.

Curcumin glucuronide was identified in intestinal and hepatic microsomes, and curcumin sulfate, tetrahydrocurcumin, and hexahydrocurcumin were found as curcumin metabolites in intestinal and hepatic cytosol from humans and rats .The extent of curcumin conjugation was much greater in intestinal fractions from humans than in those from rats, whereas curcumin conjugation was less extensive in hepatic fractions from humans than in those from rats. The curcumin-reducing ability of cytosol from human intestinal and liver tissue exceeded that observed with the corresponding rat tissue by factors of 18 and 5, respectively.[25] Curcumin sulfate was identified in the incubation of curcumin with intact rat gut sacs. Curcumin was sulfated by human phenol sulfotransferase isoenzymes SULT1A1 and SULT1A3.

4. MAJOR TARGETS FOR THE BIOLOGICAL ACTIONS OF CURCUMIN

4.1. Antioxidative Effects Through Modulating Related Enzyme Systems

Curcumin possesses anti-inflammatory activity and is a potent inhibitor of reactive-oxygen-generating enzymes, such as lipoxygenase (LOX)/cyclooxygenase (COX), xanthine dehydrogenase/oxidase, and inducible nitrogen oxide synthase (iNOS).[26] Simultaneous administration of 2 and 10 μM curcumin with 100 ng/mL trifluoroacetic acid (TPA) inhibits TPA-induced increases in xanthine oxidase activity measured 30 min later by 22.7% and 36.5%, respectively.[27] Based on these findings, induction of xanthine oxidase activity is deemed to be one of the major causative elements in TPA-mediated tumor promotion, and the major inhibitiory mechanism of curcumin on TPA-induced increases in xanthine dehydrogenase/oxidase enzyme activities is through direct inactivation in the protein level.[27]

It is interesting to note that curcumin induces heme oxygenase-1 (HO-1) and protects endothelial cells against oxidative stress.[28] Exposure of bovine aortic endothelial cells to curcumin (5–15 μM) resulted in both a concentration- and time-dependent increase in HO-1 mRNA, protein expression, and heme oxygenase activity. Interestingly, prolonged incubation (18 h) with curcumin in normoxic or hypoxic conditions resulted in enhanced cellular resistance to oxidative damage. This cytoprotective effect was considerably attenuated by tin protoporphyrin IX, an inhibitor of heme oxygenase activity. Regulation of HO-1 expression by curcumin and other polyphenols is evoked by a distinctive mechanism that is not necessary linked to changes in glutathione but might depend on redox signals sustained by specific and targeted sulfhydryl groups.[29]

Curcumin is a potent scavenger of a variety of reactive oxygen species (ROS), including superoxide anion,[9] hydroxyl radical, singlet oxygen,[10] nitric oxide, and peroxynitrite. Curcumin has the ability to protect lipids, hemoglobin, and DNA against oxidative degradation. Pure curcumin has more potent superoxide anion scavenging activity than demethoxycurcumin or bisdemethoxycurcumin.[9] Curcumin is a potent inhibitor of ROS-generating enzymes cyclooxygenase and lipoxygenase in mouse epidermis.[5]

Supplementation with *Curcuma longa* extract reduces oxidative stress and attenuates the development of fatty streaks in male New Zealand white rabbits fed a high-cholesterol diet (1.3%).[30] Many studies have shown the capacity of curcumin to prevent lipid peroxidation, a key process in the onset and progression of many diseases, including atherosclerosis. It has been observed that curcumin reduces plasma lipid peroxides, reduces the susceptibility of LDL to oxidation, inhibits the proliferation of vascular smooth muscle cells, has an antithrombotic effect, and inhibits platelet aggregation *in vivo* and *ex vivo*.

Curcumin prevents colon cancer in rodent models. It inhibits lipid peroxidation and cyclooxygenase-2 (COX-2) expression and induces glutathione-*S*-transferase (GST). The total GST activity and adducts of malondialdehyde with DNA in colon mucosa, liver, and blood leukocytes were significantly inhibited by curcumin.[31]

4.2. Metabolic Enzyme Induction

Metabolic studies showed that curcumin significantly inhibited CYP1A1-mediated benzo(a)pyrene diol bioactivation in both oral squamous cell carcinoma cells and intact oral mucosa.[32] Because CYP1A1 is one of the primary carcinogen-activating enzymes in oral mucosa, the use of curcumin as an oral cavity chemopreventive agent have significant clinical impact via its ability to inhibit carcinogen bioactivation. Curcumin exhibits anticancer activity in rodents and in humans. Its efficacy appears to be related to induction of GST enzymes, inhibition of prostaglandin E_2 (PGE_2) production, or suppression of oxidative DNA adduct formation.[33] Curcumin and a number of naturally occurring and synthetic analogues are phase II enzyme inducers, as demonstrated by their ability to elevate the enzyme activity of quinone reductase in murine hepatoma cells. It is reasonable to assume that phase II enzyme induction plays a significant role in the chemopreventive and antioxidant activities of these curcuminoids.[34]

It has been demonstrated that coordinate induction of phase II proteins and elevation of glutathione protect cells against the toxic and carcinogenic effects of electrophiles and oxidants. All inducers react covalently with thiols at rate that are closely related to their potencies. Inducers disrupt the cytoplasmic complex between the actin-bound protein Keap1 and the transcription factor Nrf2, thereby releasing Nrf2 to migrate to the nucleus where it activates the antioxidant response element (ARE) of phase II genes and accelerate their transcription.[35] This finding suggests that reaction of cysteine thiols is followed by rapid formation of protein disulfide linkages. The most reactive residues of Keap1 (C^{257}, C^{273}, C^{288}, and C^{297}) were identified by mapping the hexamethasone-modified cysteines by mass spectrometry of tryptic peptides. The residues are located in the intervening region between BTB and Kelch repeat domains of Keap1 and probably are the direct sensors of inducers of the phase II enzyme system.[35]

4.3. Induction of Apoptosis

We have demonstrated that curcumin (30 μM) induces apoptosis in several tumor cell lines.[36] The curcumin-induced apoptosis is highly dependent on the origin and malignancy of cell lines. It appears that the typical apoptosis can only be induced in immortalized mouse embryo fibroblast NIH 3T3, erbB2 oncogene-transformed NIH 3T3, mouse Sarcoma 180, human colon cancer cell HT29, human kidney cancer cell 293, and human hepatocellular carcinoma HepG2 cells; but not in primary cultures of mouse embryonic fibroblast C3H 10T1/2, rat embryonic fibroblast, and human foreskin fibroblast cells.[36] Treatment of NIH 3T3 cells with the protein kinase C (PKC) inhibitor staurosporine, the tyrosine kinase inhibitor herbimycin A, or arachidonic acid metabolism inhibitor quinacrine induces typical apoptosis. These results suggest that blocking the cellular signal transduction in immortalized or transformed cells might trigger the induction of apoptosis.

We have also demonstrated that curcumin (3.5 μg/mL) induces human promyelocytic HL-60 cells. The apoptosis-inducing activity of curcumin appeared in a dose- and time-dependent manner.[37] Flow-cytometric analysis showed that the hypodiploid DNA peak of propidium iodide-stained nuclei appeared at 4 h after 7-μg/mL curcumin treatment. The action mechanism has been demonstrated to be through cytochrome-*c* release and activation of caspases.[38] The antioxidants

N-acetyl-L-cysteine (NAC), L-ascorbic acid, α-tocopherol, catalase, and superoxide dismutase effectively prevented curcumin-induced apoptosis.

The combined treatment of LNCaP prostate cancer cells with curcumin (10 μM) and tumor necrosis factor- related apoptosis-inducing ligand (TRAIL, 20 ng/mL) remarkably induced apoptosis through inducing cleavage of procaspase-3, procaspase-8, and procaspase-9, truncation of Bid and release of cytochrome-*c* from the mitochondria, indicating that both the extrinsic (receptor-mediated) and intrinsic (chemical-induced) pathways of apoptosis are triggered in prostate cancer cells treated with a combination of curcumin and TRAIL. These findings define a potential use of curcumin to sensitize prostate cancer cells for TRAIL-mediated immunotherapy.[39]

4.4. Inhibition of Mammalian Target of Rapamycin

Curcumin inhibits proliferation/growth, induces apoptosis, and suppresses motility of cells by inhibition of mammalian targets of rapamycin (mTOR)-mediated signaling pathways in rhabdomyosarcoma cells. Recently, two mTOR complexes (m-TORmLST8-raptor and mTOR-mLST8-rictor) have been identified. The former is rapamycin-sentsitive, whereas the latter is rapamycin-insensitive. Treatment with curcumin promoted dissociation of raptor from mTOR in a concentration-dependent manner, but did not affect association of mLST8 with mTOR. Furthermore, unlike rapamycin, curcumin also induced the dissociation of rictor from mTOR. Therefore, curcumin might represent a novel class of mTOR inhibitor.[40] It is found that curcumin induced sustained hyperphosphorylation of c-jun and activation of JNK, which might be responsible for curcumin-induced apoptosis of the rhabdomyosarcoma cells.

The conserved checkpoint protein kinase mTOR is a key regulator of cell growth and proliferation and increasing evidence supports that the mTOR pathway plays a central role in the genesis of cancer. Curcumin has been shown to inhibit carcinogenesis and tumor cell proliferation. Curcumin inhibits PKB/Akt, one of the upstream regulators of mTOR; furthermore, curcumin also activates AMPK1, which senses cellular ATP levels and inhibits mTOR indirectly. Inhibition of mTOR signaling by curcumin is mediated by the inhibition of Akt and activation of AMPK, and the inhibition of mTOR signaling is important for the inhibition of PC-3 cell proliferation by curcumin.[41]

Human tumors that overexpressed ErbB2, which have been previously shown to have higher VEGF expression, showed significantly higher p70S6K phosphorylation as well. Increased vascular epithelial growth factor (VEGF) expression also significantly correlated with higher levels of Akt and mTOR phosphorylation. Additionally, patients with tumors having increased p70S6K phosphorylation showed a trend for worse disease-free survival and increased metastasis. These findings show that ErbB2 increases VEGF protein production by activating p70S6K in cell lines, xenografts, and human cancers and suggest that these signaling molecules might serve as targets for antiangiogenic and antimetastatic therapies.[42]

4.5. Suppression of Akt Signaling

Althugh curcumin has several different molecular targets within the MAPK and PI3K/PKB signaling pathways that could contribute to inhibition of proliferation and induction of apoptosis, inhibition of basal activity of Akt/PKB, but not ERK, might facilitate apoptosis in the tumor cell line.[43] Recent studies have demonstrated that curcumin downregulates nuclear factor (NF-κB) through inhibiting IκBα, Bcl-2, Bcl-xL, cyclin D1, and interleukin (IL)-6 in human multiple myeloma cells leading to the suppression of proliferation and induction of apoptosis, thus providing the molecular basis for the treatment of multiple myeloma patients with this pharmacologically safe agent.[44] Curcumin causes dose-dependent apoptosis and DNA fragmentation of Caki cells, which is preceded by the sequential dephosphorylation of Akt, downregulation of the antiapoptotic Bcl-2, Bcl-xL, and IAP proteins, release of cytochrome-*c*, and activation of caspase-3, cyclosporin A, as well as caspase inhibitor, specifically inhibit curcumin-induced apoptosis in Caki cells. Pretreatment with *N*-acetylcysteine markedly prevented dephosphorylation of Akt and cytochrome-*c* release, and cell death, suggesting a role for ROS in this process.[45]

4.6. Inhibition of MDM2 Oncogene Action

The anti-cancer effect of curcumin might be through direct inhibition of the MDM2 oncogene action. It is well established that the MDM2 oncogene plays a major role in human cancer development and progression. Its tumorigenic properties are associated with both p53-dependent and p53-independent pathways. It has been demonstrated that in a dose-dependent manner, curcumin inhibited DMD2 expression in various human cancer cell lines with different p53 backgrounds. Curcumin inhibited MDM2 expression at the transcription level, affecting DMD2 promoter activity. The levels of p21 and Bax were increased and E2F1 and Bcl2 decreased. Curcumin induced apoptosis and inhibited proliferation in PC3 cells. The chemosensitization and radiosensitization effects of curcumin *in vitro* and *in vivo* are tightly associated with its MDM2 inhibitory effects.[46]

4.7. Suppression of c-jun and c-fos Expression

In 1991, we have made an interesting finding that the phorbol ester TPA-induced transcriptional factor c-jun/AP-1 in mouse fibroblast cells is suppressed by curcumin.[47] Elevated expression of gene transcriptionally induced by TPA is among the events required for tumor promotion. Functional activation of transcriptional factor c-Jun/AP-1 is believed to play an important role in signal transduction of TPA-induced tumor promotion. Suppression of the c-jun/AP-1 activation by curcumin (10 μM) is observed in mouse fibroblast cells. These findings show for the first time that the effect of curcumin on TPA-induced inflammation/tumor promotion could be studied at the molecular level.

Curcumin also inhibits the TPA- and ultraviolet B (UVB) light-induced expression of c-jun and c-fos in JB6 cells and in mouse epidermis.[48] Recent studies indicated that curcumin treatment attenuated TPA-stimulated NF-κB activation in mouse skin, which was associated with its blockade of degradation of the inhibitory protein IκBα and also of subsequent translocation of the p65 subunit to nucleus.[49] TPA treatment resulted in rapid activation via phosphorylation of ERK1/2 and p38 MAP kinases, which are upstream of NF-κB. The MEK1/2 inhibitor UO126 strongly inhibited NF-κB activation, whereas the p38 inhibitor SB203580 failed to block TPA-induced NF-κB activation in mouse skin. It is suggested that curcumin inhibits the catalytic activity of ERK1/2 in mouse skin and its suppression of COX-2 expression by inhibiting ERK activity and NF-κB activation might provide the molecular basis for the antitumor-promoting effects of curcumin in mouse skin carcinogenesis.[49,50] Curcumin has potent antineoplastic activity in several tumor types and is thought to exert anti-inflammatory effects in part through inhibition of NF-κB, which has been linked to inhibition of the cytokine IL-8, production of PGE_2, and VEGF. IL-8, PGE_2, and VEGF are expressed in ovarian cancers and are associated with increased angiogenesis and poor prognosis. Curcumin is cytotoxic in ovarian cancer cells and exerts its effects by inhibiting NF-κB activation and decreasing levels of IL-8 and VEGF. The inhibition of these angiogenic factors along with the induction of apoptosis might have broad clinical benefits to ovarian cancer patients.[51]

4.8. Inhibition of Protein Kinase C

Treatment with 15 or 20 μM curcumin for 15 min inhibited TPA-induced PKC activity in particulate fractions by 26% or 60% and did not affect the level of PKC protein. However, the inhibitory effect of curcumin was reduced after preincubation with the thiol compounds.[52]

Dietary antioxidants are important in cancer prevention. The conventional view held for a long time is that antioxidants act by scavenging free radicals. Although these actions of antioxidants are certainly important in preventing promutagenic DNA damage caused by oxidants, other actions of antioxidants, particularly those influencing cell signaling mechanisms, have also recently come to light. Antioxidants are believed to induce their own effects on cell signaling, such the PKC pathway in the precancer cells to decrease tumor promotion, a critical stage in carcinogenesis.[53] By having different oxidation susceptible regions, PKC can respond to both oxidant tumor promoters and cancer preventive antioxidants to elicit apposite cellular responses. The oxidant tumor promoter (such as TPA) activates PKC by reacting with zinc thiolates present within the regulatory domain. In contrast, the oxidized forms of some cancer preventive agents, such as polyphenolics (curcumin, ellagic acid, and 4-hydroxytamoxifen) and seleno compounds, can inactivate PKC by oxidizing the vicinal thiols present within the catalytic domain. This brings an efficient counteractive mechanism to block the signal transduction induced by the tumor promoter at the first step itself.[54]

4.9. Suppression of EGF Receptor Tyrosine Kinase Activity

Curcumin (10 μM) inhibits EGF receptor kinase activity up to 90% in a dose- and time-dependent manner and also inhibits EGF-induced tyrosine phosphorylation of EGF-receptors in A431 cells.[55] Treatment of NIH 3T3 cells with a saturating concentration of EGF for 5–15 min induced increased EGF-R tyrosine phosphorylation by 4– to 11-fold and this was inhibited by curcumin, which also inhibited the growth of EGF-stimulated cells.[56] Curcumin has been shown to suppress the expression of iNOS *in vivo*.[57] The EGF is a well-known mitogen, but it paradoxically induces apoptosis in cells that overexpress its receptor. It has been demonstrated that the EGF-induced apoptosis is accelerated if NF-κB is inactivated by curcumin and sodium salicylate.[58] Under the NF-κB inactivated condition, A431 cells were more sensitive to EGF with decreased cell viability and increased externalization of phosphatidylserine on the cell surface, DNA fragmentation, and activation of caspases (3 and 8 but not 9), typical features of apoptosis. These results were further supported by the potentiation of the growth inhibitory effects of EGF by chemical inhibitors of NF-κB (such as curcumin and sodium salicylate) and the protective role of Rel A evidenced by the resistance of A431-Rel A cells (stably transfected with Rel A) to EGF-induced apoptosis.[58]

4.10. Proteasome System in Cell Proliferation and Apoptosis

It has been proposed that curcumin mediates growth arrest in breast cancer cells by targeting the proteasome[59] Evidence points to curcumin as a potent antioxidant and anti-inflammatory agent, both desirable properties that have been reported to play key roles in inhibiting carcinogenesis. Using a panel of three breast cancer cell culture models (MDA-MB-231, MDA-MB-436, and Hs578T), it has been shown that curcumin mediates its cell cycle inhibitory activities by blocking the chymotrypsin-like activity of the proteasome *in vitro*. It is suggested that curcumin might mediate G1 arrest and possible cytostasis and apoptosis by blocking the proteasome activity and upregulating the p21 protein in breast cancer cells.[59] In addition to the mechanisms by which the growth factors exhibit both stimulatory and inhibitory activity in a single cell depending on the context of the other signal molecules present, the final outcome is presumably influenced by a host of regulatory molecules other than the growth factors and their receptors.[60] It is thus clearly important to recognize that a potent mitogen like EGF also sends out apoptotic signals and identifies conditions in which these signals are regulated. NF-κB inhibition makes A431 cells more susceptible to EGF-induced apoptosis, whereas Rel A protects them against it. EGF stimulation in A431 cells enhances the degradation of IκBα, but not IκBβ, and proteasome inhibitors such as ALLN or MG132 block EGF-mediated NF-κB activation, indicating that EGF-induced NF-κB activation requires proteasome-dependent IκB degradation. Furthermore, the EGF-induced DNA-binding complex of NF-κB in A431 was found to be composed of p50/Rel A heterodimers, but not c-Rel.[61]

It has been demonstrated that curcumin-induced apoptosis is mediated through the impairment of the ubiquitin–proteasome system. Exposure of curcumin to the mouse neuro2a cells causes a dose-dependent decrease in proteasome activity and an increase in ubiquitinated proteins. Curcumin exposure also decreases the turnover of the destabilized enhanced green fluorescence protein, a model substrate for proteasome and cellular p53 protein.[62] In our laboratory, a similar effect was observed in another polyphenolic: pentagalloylglucose (5GG). It is interesting to note that 5GG induces G1 arrest and apoptosis in human Jurkat T-cells through inhibiting proteasome activity and elevating $p27^{kip1}$, $P21^{cip1/WAF1}$, and Bax proteins.[63]

Proteasome-mediated degradation of cell proteins plays a pivotal role in the regulation of several basic cellular processes, including differentiation, proliferation, cell cycling, apoptosis, gene expression, and signal transduction. Imbalances in proteasome-mediated protein degradation contribute to various human diseases such as cancer and neurodegenerative and myodegenerative disases, suggesting that the proteasome might be a novel target for anticancer therapy.[64]

4.11. Modulation of Ca^{2+} and Cellular p53 Protein

When COLO205 colorectal carcinoma cells were treated with curcumin (60 μM), the appearance of apoptotic DNA ladders was delayed about 5 h and G1 arrest was detected.[65] The reduction of p53 gene expression was accompanied by the induction of HSP70 gene expression in the curcumin-treated cells. These findings suggest that curcumin might induce the expression of the HSP70 gene through the initial depletion of intracellular Ca^{2+} followed by the suppression of p53 gene function in the target cells.[65]

4.12. Suppression of Hepatocellualr Carcinoma Invasion by Inhibiting MMP-9

An *in vitro* assay, without or with the Matrigel matrix, was used to quantitate cellular migration and invasion. Gelatin-based zymography was adapted to assay the secretion of matrix metalloproteinase-9 (MMP-9). We found that curcumin at 10 μM inhibited 17.4% and 70.6% of cellular migration and invasion of SK-Hep-1 cells, respectively. Compared with a less invasive human hepatocellular carcinoma cell line Huh 7, SK-Hep-1 showed a much higher MMP-9 secretion. Furthermore, parallel with its anti-invasion activity, curcumin inhibited MMP-9 secretion in SK-Hep-1 in a dose-dependent fashion. We conclude that curcumin has a significant anti-invasion activity in SK-Hep-1 cells and that this effect is associated with its inhibitory action on MMP-9 secretion.[66]

Osteopontin (OPN) is a member of the extracellular matrix protein; it is a non-collagenous, sialic acid-rich, and glycosylated phosphoprotein. OPN stimulates tumor growth and activation of promatrix metalloproteinase-2 (ProMMP-2) through NF-κB-mediated induction of membrane type-1 matrix metalloproteinase (MT1-MMP) in murine melanoma cells.[67] Recently, it has been shown that curcumin inhibited the OPN-induced IκBα phosphorylation and degradation by inhibiting the IKK activity, Moreover, curcumin inhibited the OPN-induced translocation of p65, NF-κB DNA-binding, and NF-κB transcriptional activity. Curcumin also inhibited OPN-induced cell proliferation, cell migration, extracellular matrix invasion, and synergistically induced apoptotic morphology with OPN in these cells.[68]

5. MECHANISM OF ACTION OF CURCUMIN IN CHEMOPREVENTION

Multiple evidences have been indicated that many dietary constituents are chemopreventive in animal models, and experiments with cultured cells are revealing various potential action mechanisms. Several compounds classified as blocking agents can prevent, or greatly reduce, initiation of carcinogenesis, wereas suppressing agents affect later stages of the promoting process by reducing cell proliferation. Many naturally occurring compounds such as curcumin, catechins, theaflavins, and others have both types of activity. These compounds exhibit their blocking mechanisms through alteration of phase II drug-metabolizing activities and scavenging of ROS in the target tissue. Meanwhile, these compound might act as suppressing agents to suppress carcinogenesis involving modulation of signal transduction that leads to altered gene expression, cell cycle arrest, or apoptosis.[69] Although curcumin alone had little or no effect on cellular differentiation, when it was combined with all-trans retinoic acid or 1α-25-dihydroxyvitamin D_3, a synergistic effect was observed. It is possible that many dietary chemicals in fruits, vegetables, and other edible plants can prevent cancer by synergizing with endogenously produced stimulators of differentiation such as all-trans retinoic acid, 1α-25-dihydroxyvitamin D_3, and butyrate.[70]

Recent intensive studies on the action mechanisms of curcumin in various biological systems have indicated that this compound has engaged in multiple antitumor-promoting pathways.[48,71] It has been demonstrated that the TPA-induced tumor promotion is significantly inhibited by curcumin.[4,12,18] Angiogenesis is a key component of cancer metastasis. ErbB2- overexpressing breast tumors tend to be more angiogenic than other breast tumors. One of the most potent inducers of angiogenesis is VEGF, which induces endothelial cell proliferation and migration. In human breast tumors, overexpression of ErbB2 is correlated with increased VEGF expression and the transcription factor hypoxia-inducible factor-1α has been postulated to mediate the ErbB2 upregulation of VEGF.[22] It is conceivable that the molecular mechanism of action of curcumin is quite complicated and dispersed. The primary target of curcumin could be on the plasma membrane where the activity of PKC is first inhibited.[52] In addition, the activity of EGF receptor tyrosine kinase is also inhibited.[55] Some PKC-mediated nuclear protein factors, such as IκB kinase and NF-κB, are then inhibited through various signal transduction pathways. The TRE-binding activity of c-Jun/AP-1 is then repressed[47] and, finally, the transcription of genes essential for cell proliferation are suppressed, as indicated by the inhibition of related enzymes such as ornithine decarboxylase, PKC, COX and LOX. It appears that activation of calcium-dependent protein kinases (such as PKC) or inhibition of protein phosphatases results in tumor promotion.[72] In the case of tumor promoters, it appears that a common final effect is to increase the phosphorylation of the protein substrate on serine or threonine residues. It appears that when any essential component of a signal transduction pathway is rendered hyperactive or autonomous, it might acquire the ability to drive the cell into unchecked proliferation and lead to tumor promotion. Curcumin might attenuate or suppress the hyperactivity of these components of signal transduction and maintain the normal cell function.[73]

ACKNOWLEDGMENTS

This study was supported by the National Science Council (NSC-94-2300-B-002-118 and NSC94-2320-B-002-019). This author would like to express his sincere appreciation to his associates, Professor Lin-Shiau Shoei-Yn, Professor M. H. Pan, Professor C. Y. Liu, and Professor T. S. Huang, and others for their continuous and important contribution to this work.

REFERENCES

1. G. J. Kelloff, C. W. Boone, V. E. Steele, et al., Progress in cancer chemoprevention: Prospective on agent selection and short term clinical intervention trials.*Cancer Res* **54(Suppl)**, 2015s–2024s (1994).
2. B. Armstrongand R. Doll, Environmental factors and cancer incidence and mortality in different countries, with special reference to dietary practices. *Int J Cancer*,**15**, 617–631 (1975).
3. R. L. Phillips, Role of life-style and dietary habits in risk of cancer among Seventh-Day Adventists. *Cancer Res* **35**, 3513–3522 (1975).
4. M. T. Huang, R. C. Smart, C. Q. Wong, and A. H. Conney, Inhibitory effect of curcumin, chlorogenic acid, caffeic acid and ferulic acid on tumor promotion in mouse skin by 12-O-tetradecanoylphorbol-13-acetate. *Cancer Res* **48**, 5941–5946 (1988).
5. M. T. Huang, T. Lysz, T. Ferraro, T. F. Abidi, J. D. Laskin, and A. Conney,. Inhibitory effects of curcumin in vivo lipoxygenase and cyclooxygenase activities in mouse epidermis. *Cancer Res* **51**, 813–819 (1991).
6. G. J. Kelloff, J. A. Crowell, E. T. Hawk, et al., Stragegy and planning for chemopreventive drug–development: Clinical development. *J Cell Biochem.* **26(Suppl),** 54–71 (1996).
7. A. L. Cheng, C. H. Hsu, J. K. Lin, M. M. Hsu, Y. F. Ho, T. S. Shen, TJ. Y. Ko, J. T. Lin, B. R. Lin, W. Ming-Shiang, H. S. Yu, S. H. Jee, G. S. Chen, T. M. Chen. C. A. Chen, M. K. Lai, . Y. S. Pu, M. H. Pan, Y. J. Wang, C. C. Tsai, and C. Y. Hsieh, Phase I clinical trial of curcumin, a chemopreventive agent, in patients with high risk or pre-malignant lesion. *Anti-cancer Res* **21**(4B), 2895–2900 (2001). .
8. C. M. Chen and H. C. Fang, chemical analysis of the active principles of curcuma species In: C. Y. Sung, ed. *Modern Treatise on Chinese Herbal Medicines*. Beijing:

The Institute of Pharmaceutical Sciences, Medical Academia, 1997,Vol. III, pp 95–105.

9. E. Kunchandyand M. N. A. Rao, Oxygen scavenging activity of curcumin. *Int J Pharm* **38**, 239–240 (1990) .
10. M. Subramanian, M. N. A. Sreejayan Rao, T. P. A. Devasagyam, and B. B. Singh, Diminution of singlet oxygen induced DNA-damage by curcumin and related antioxidants. *Mutat Res* **311**, 249–255 (1994).
11. M. N. A. Sreejayan Rao, Curcuminoids as potent inhibitors of lipid peroxidation. *J Pharm Pharmacol* **46**, 1013–1016 (1994).
12. M. T. Huang, W. Ma, P. Yen, J. G. Xie, J. Han, K. D. Fenkel, K. D. Grunberger, and Conney, Inhibitory effects of topical application of low doses of curcumin on TPA-induced tumor promotion and oxidized DNA bases in mouse epidermis. *Carcinogenesis* **18**, 83–88 (1997).
13. C. A. Shih and J. K.Lin, Inhibition of 8-hydroxydeoxyguanosine formation by curcumin in mouse fibroblast cells. *Carcinogenesis* **14**, 709–712 (1994).
14. A. H. Conney, T. Lysz, T. Ferraro, T. F. Abidi, P. S. Manchand, J. D. Laskin, and M. T. Huang, Inhibitory effect of curcumin and some related dietary compounds on tumor promotion and arachidonic acid metabolism in mouse skin. *Adv Enzyme Regul* **31**, 385–389 (1991).
15. M. T. Huang, Z. Y. Wang, C. A. Georgiadis, J. D. Laskin, and A. H. Conney, Inhibitory effect of curcumin on tumor initiation by benzo[a]pyrene and 7,12- dimethylbenz[a]anthracene. *Carcinogenesis* **13**, 947–954 (1992).
16. M. T. Huang, Y. R. Lou, YW. Ma, H. L. Newmark, K. R. Reuhl, and A, H. Conney, Inhibitory effect of dietary curcumin on forestomach, duodenal and colon carcinogenesis in mice. *Cancer Res* **54**, 5841–5847 (1994).
17. M. T. Huang, W. Ma, Y. P. Lu, YR. L. Chang, C. Fischer, P. S. Manchand, H. L. Newmark, and H. H. Conney, Effects of curcumin, demethoxy- curcumin,bisdemethoxycurcumin and tetrahydrocurcumin on TPA-induced tumor promotion. *Carcinogenesis* **16**, 2493–2497 (1995).
18. C. V. Rao, A. B. Riven, A. B.Simi, and B. S. Reddy, Chemoprevention of colon carcinogenesis by dietary curcumin, a naturally occurring plant phenolic compound.*Cancer Res* **55**, 259–266 (1995).
19. S. V. Singh, X. Hu, S. K. Srivastava, M. Singh, H. Xia, J. L. Orchard, and H. A. Zaren, Mechanism of inhibition of benzo[a]pyrene-induced forestomach cancer in mice by dietary curcumin. *Carcinogenesis* **19**(8), 1357–1360 (1998).
20. H. Inano, M. Onoda, N. Inafuku, et al., Potent protective action of curcumin on radiation-induced initiation of mammary tumorigenesis in rats. *Carcinogenesis* **21**(10), 1835–1841 (2000).
21. S. E. Chuang, M. L. Kuo, C. H. Hsu, C. R. Chen, J. K. Lin, G. M. Lai, C. Y. Hsieh, and A. L. Cheng, Curcumin-containing diet inhibits diethylnitrosamine- induced murine hepatocarcinogenesis.*Carcinogenesis* **21**(2), 331–335 (2000).
22. K. S. Klos, S. L. Wyszomierski, M. Sun, et al., ErbB2 increases vascular endothelial growth factor protein synthesis via activation of mammalian target of rapamycin/p70S6K leading to increased angiogenesis andspontaneous metastasis of human breast cancer cells. *Cancer Res* **66**, 2028–2037 (2006).
23. M. H. Pan, T. M. Huang, Tand J. K. Lin, Biotransformation of curcumin through reduction and glucuronizationm in mice. *Drug Metab Dispos* **27**, 486–494 (1999).
24. C. Ireson, S. Orr, D. J. Jones, R. Verschoyle, C. K. Lim, J. L. Luo, L. Howells, S. Plummer, R. Jukes, M. Williams, W. P. Steward, and A.Gescher, Characterization of

metabolites of the chemopreventive agent curcumin in human and rat hepatocytes and in the rat in vivo, and evaluation of their ability to inhibit phorbol ester-induced prostaglandin E2 production. *Cancer Res* **61**(3), 1058–1064 (2001).

25. C. Ireson, D. J. Jones, S. Orr, M. W. Coughtrie, D. J. Hoocock, M. L. Williams, P. B. Farmer, W. P. Steward, and A. Gescher, Metabolism of the cancer chemopreventive agent curcumin in human and rat intestine. *Cancer Epidermiol Biomarkers Prev* **11**(1), 105–111 (2002).
26. J. K.Lin and S. Y. Lin-Shiau, Cancer chemoprevention by curcumin. *Proc Natl Sci Counc Repub China B* **25**(2), 59–66 (2001) .
27. J. K. Lin, T. S. HuangC. A. Shih, and J. L. Liu, Molecular mechanism of action of curcumin. In: C. T. Ho, T. Osawa, M. T. Huang, and R. T. Rosen, eds. *Food Phyto-Chemicals for Cancer Prevention II.* ACS Symposium Series 547. Washington, DC: Amercian Chemical Society, Washington, 1994**,** pp.196–203.
28. R. Motterlin, R. Foresti, R.,\ Bassi, and C. J. Green, Curcumin, an antioxidant and anti-inflammatory agent, induces heme oxygenase-1 and protects endothelial cells against oxidative stress. *Free Radical Biol Med* **28**(8), 1303–1312 (2000).
29. G. Scapagnini, R Foresti, V. Calabrese, A. M. Giuffrida Stella, C. J. Green, and R. Motterlin, Caffeic acid phenethyl ester and curcumin: A novel class of heme oxygenase-1 inducers. *Mol Pharmacol* **3**, 554–561 (2002).
30. J. L.Quiles, M. Dolores Mesa, C. L. Ramirez-Tortosa, C. M. Anguilera, M. Battino, A. Gil, and M. Carmen Ramirez-Tortosa, Curcuma longa extract supplementation induces oxidative stress and attenuates aortic fatty streak development in rabbits. *Arterioscler Thromb Vasc Biol* **22**, 1225–1231 (2002).
31. R. A. Sharma, C. R. Ireson, R. D. Verschoyle, K. A. Hill, M. L. Williams, C. Leuratti, M. M. Manson, L. J. Marett, W. P. Steward, and A. Gescher, Effect of dietary curcumin on glutathione S-transferase and malondialdehyde-DNA adducts in rat liver and colon mucosa: Relationship with drug levels. *Clin Cancer Res* **7**, 1452–1458 (2001).
32. A. L. Rinaldi, M. A. Monge, H. W. Fields, et al., Curcumin activates the arylhydrocarbon receptor yet significantly inhibits (-)-benzo(a)pyrene-7R-trans-7,8-dihydrodiol bioactivation in oral squamous cell carcinoma cells and oral mucosa. *Cancer Res* **62**, 5451–5456 (2002).
33. R. A. Sharma, S. A. Euden, S. L. Platton, et al., Phase I clinical trial of oral curcumin: Biomarkers of systematic activity and compliance. *Clin Cancer Res* **10**, 6847–6854 (2004).
34. A. Dinkova-Kostova and P. Talalay, Relation of structure of curcumin alalogs to their potencies as inducers of phase 2 detoxification enzymes. *Carcinogenesis* **20**(5), 911–914 (1999).
35. A. Dinkova-Kostova, W. D. Holtzclaw, R. N. Cole, K. Itoh, N. Wakabayashi, Y. Katoh, M. Yamamoto, and P. Talalay, Direct evidence that sulfhydryl groups of Keap 1 are the sensors regulating induction of phase 2 enzymes that protect against carcinogens and antioxidants. *Proc Natl Acad Sci USA* **99**(18), 11,908–11,913 (2002).
36. M. C. Jiang, H. F. Yang-Yen, J. J. Yen, and J. K. Lin, Curcumin induces apoptosis in immortalized NIH 3T3 and malignant cancer cell lines. *Nutr Cancer* **26**, 111–120 (1996).
37. M. L. Kuo, T. S. Huang, and J. K. Lin, Curcumin, an antioxidant and anti-tumor promoter, induced apoptosis in human leukemia cells.*Biochim Biophys Acta* **1317**, 95–100 (1996).

38. M. H. PanW. L. Chang, S. Y. Lin-Shiau, C. T. Ho, and J. K. Lin, Induction of apoptosis by garcinol and curcumin through cytochrome c release and activation of caspases in human leukemia HL-60 cells. *J Agric Food Chem* **49**(3), 1464–1474 (2001).
39. D. Deeb, Y. X. Xu, H. Jiang, X. Gao, N. Janakiraman, R. A. Chapman, and S. C. Gautam, Curcumin enhances tumor necrosis factor-related apoptosis-inducing-ligand-induced apoptosis in LNCaP prostate cancer cells. *Mol Cancer Ther* **2**(1), 95–103 (2003).
40. C. S. Beevers and S. Huang, Curcumin disrupts the complexes of mammalian target of rapamycin. *Proc Am Assoc Cancer Res* **7**, 356 (2006)
41. S. Yu, G. Shen, and T. A. Kong, Curcumin inhibits mTOR signaling by inhibiting protein kinase B/Akt and activating AMP-activated protein kinase (AMPK) in prostate cancer cell line PC-3.*Proc Am Assoc Cancer Res* **47**, 538 (2006).
42. T. O. Khor, Y. S. Keum, W. S. Lin, et al., Combined inhibitory effects of curcumin and phenethylisothiocyanate on the growth of human PC-3 prostate xenografts in immunodeficient mice. *Cancer Res* **66**(2), 613–621 (2006).
43. M. S. Squires, E. A. Hidson, L. Howells, S. Sale, C. E. Houghton, J. L. Jones, L. H. Fox, M. Dickens, S. A. Prigent, and M. M. Manson, Relevance of mitogen activated protein kinase (MAPK) and phosphatidylinositol 3-kinase/protein kinase B (PI3K/PKB) pathways to induction of apoptosis by curcumin in breast cells. *Biochem Pharmacol* **65**(3), 361–376 (2003).
44. A. C. Bharti, N. Donato, S. Singh, and B. B.Aggarwal, Curcumin down-regulates the constitutive activation of nuclear factor κB and IκBα kinase in human multiple myeloma cells leading to suppression of proliferation and induction of apoptosis. *Blood* **101**, 1053–1062 (2003).
45. J. H. Woo, Y. H. Kim, Y. J. Choi, D. G. Kim, K. S. Lee, J. H. Hae, D. S. Min, J. S. Chang, Y. J. Jeong, Y. S. Lee, J. W. Park, and J. K. Kwon, Molecular mechanisms of curcumin-induced cytotoxicity: Induction of apoptosis through generation of reactive oxygen species, down-regulation of Bcl-XL and IAP , the release of cytochrome c and inhibition of Akt. *Carcinogenesis* **24**(7), 1199–1208 (2003).
46. M. Li, H. Wang, Z. Zhang, et al., Curcumin, a multifunctional chemopreventive agent, inhibits MDM2 oncogene which is associated with its anti-cancer, chemosensitization and radiosensitization effects. *Proc Am Assoc Cancer Res* **47**, 538 (2006).
47. T. S. Huang, S. C. Lee, and J. K. Lin, Suppression of c-Jun/AP-1 activation by an inhibitor of tumor promotion *Proc Natl Acad Sci USA* **88**, 5292–5296 (1991).
48. Y. P. Lu, R. L. Cahng, Y. R. Lou, M. T. Huang, H. L. Newmark, K. R. Reuhl, and A. H. Conney, Effect of curcumin on TPA- and ultraviolet B light induced expression of c-jun and c-fos in JB6 cells and in mouse epidermis. *Carcinogenesis* **15**, 2363–2370 (1994).
49. K. S. Chun, Y. S. Keum, S. S. Han, Y. S. Song, S. H. Kim, and Y. J. Surh, Curcumin inhibits phorbal ester-induced expression of cyclooxygenase-2 in mouse skin through expression of extracellular signal-regulated kinase activity and NFκB activation. *Carcinogenesis* **24**(9), 1515–1524 (2003) .
50. S. Singhand B. B. Aggarwal, Activation of transcription factor NFκB is suppressing by curcumin.*J Biol Chem* **270**(42), 24,995–25,000 (1995).
51. S. K. Fogoros, M. Choi, and J. R. Liu, Curcumin mediates angiogenic factors through inhibition of NFκκn ovarian cancer cells. *Proc Am Assoc Cancer Res* **47**, 312 (2006).
52. J. Y. Liu, S. J. Lin, and J. K. Lin, Inhibitory effects of curcumin on protein kinase C activity induced by TPA in NIH 3T3 cells. *Carcinogenesis* **14**, 857–861 (1993).

53. R. Gopalakrishna and S. Jaken, Protein kinase C signaling and oxidative stress. *Free Radical Biol Med* **28**, 1349–1361 (2000).
54. R. Gopalakrishna and U. Gundimeda, Antioxidant regulating protein kinase C in cancer prevention.*J Nutr* **132**, 3819s–3823s (2002).
55. L. Korutla and R. Kumar, Inhibitory effects of curcumin on epidermal growth factor receptor kinase activity in A431 cells.*Biochim Biophys Acta* **1224**, 597–600 (1994) .
56. L. Korutla, J. Y. Cheung, J. Mendelsohn, and R. Kumar, Inhibition of ligand–induced activation of epidermal growth factor receptor tyrosine phosphorylation by curcumin.*Carcinogenesis* **16**, 1741–1745 (1995).
57. M. M. Y. .Chan, H. I. Huang, M. R. Fenton, and D. Fong, In vivo inhibition of nitric oxide synthase gene expression by curcumin, a cancer preventive natural product with anti-inflammatory properties.*Biochem Pharmacol* **55**, 1955–1962 (1998).
58. R. J. Anto, M. Venkatraman, and D. Karunagaran, Inhibition of NFκB sensitizes A431 cells to epidermal growth factor-induced apoptosis, whereas its activation by ectopic expression of Rel A confers resistance. *J Biol Chem* **278**(28), 25,490–25,498 (2003).
59. E. T. Efuet and K. Keyomarsi, Curcumin and simvastatin mediate growth arrest in breast cancer cells by targeting proteasome. *Proc Am Assoc Cancer Res* **47**, 1092 (2006).
60. M. B. Sporn and A. B. Roberts, Peptide growth factors are multifunctional. *Nature* **332**, 217–219 (1998).
61. L. Sun and G. Carpenter, Epidermal growth factor activation of NF-κB is mediated through IκBα degradation and intracellular free calcium. *Oncogene* **16**, 2095–2102 (1998).
62. N. R. Jana, P. Dikshit, A. Goswami, and W. Nukina, Inhibition of proteasomal function by curcumin induces apoptosis through mitochondrial pathway. *J Biol Chem* 279(12), 11,680–11,685 (2004) .
63. W. J. Chen and J. K. Lin, Induction of G1 arrest and apoptosis in human Jurkat T cells by pentagalloylglucose through inhibiting proteasome activity and elevating p27, p21 and bax protein. *J Biol Chem* **279**(14), 13,496–13,525 (2004).
64. C. Naujokat and S. Hoffmann, Role and function of 26S proteasome in proliferation and apoptosis. *Lab Invest* **82**, 965–980 (2002).
65. Y. C. Chen, T. C. Kuo, S. Y. Lin-Shiau, and J. K. Lin, Induction of HSP70 gene expression by modulation of calcium ion and cellular p53 protein by curcumin in colorectal carcinoma cells.*Mol Carcinog* **17**, 224–234 (1996).
66. L. I. Lin, Y. F. Ke, Y. C. Ko, and J. K. LinCurcumin inhibits SK-Hep-1 hepatocellular carcinoma cell invasion in vitro and suppresses matrix metalloproteinase 9 secretion. *Oncology* **55**, 349–353 (1998).
67. S. Phillip, A. Bulbule, and G. C. Kundu, Osteopontin stimulates tumor growth and activation of promatrix metalloproteinase-2 through nuclear factor-κB mediated induction of membrane type 1 matrix metalloproteinase in murine melanoma cells. *J Biol Chem* **276**, 44,926–44,935 (2001).
68. S. Phillip and G. C. Kandu, Osteopontin induces nuclear factor κB-mediated promatrix metalloproteinase-2 activation through IκBα/IKK signaling pathways and curcumin down regulate these pathways. *J Biol Chem* **278**(16), 14,487–14,497 (2003).
69. M. M. Manson, A. Gescher, E. A. Hudson, S. M. Plummer, M. S. Squires, and S. A. Prigent, Blocking and suppressing mechanisms of chemoprevention by dietary constituents. *Toxicol Lett* **112–113**, 499–505 (2000) .

70. A. H. Conney, Y. R. Lou, J. G. Xie, T. Osawa, H. L. Newmark, Y. Liu, R. L. Chang, and M. Huang, Some perspectives on dietary inhibition of carcinogenesis: Studies with curcumin and tea. *Proc Soc Exp Biol Med* **216**(2), 234–245 (1997) .
71. J. K. Lin and C.A. Shih, Inhibitory effect of curcumin on xanthine dehydrogenase/oxidase induced by TPA in NIH 3T3 cells. *Carcinogenesis* **15**, 1717–1721 (1994).
72. R. A. Haystead, A. T. Sim, and D. Carling, Effects of the tumor promoter okadaic acid on intracellular protein phosphorylation and metabolism. *Nature* **337**, 78–81 (1989).
73. A. H. Conney, Enzyme induction and dietary chemicals as approaches to cancer chemoprevention: The Seventh Dewitt Goodman Lecture. *Cancer Res* **63**, 7005–7031 (2003).

CELL GROWTH REGULATION

Devarajan Karunagaran, Jeena Joseph and
Thankayyan R. Santhosh Kumar

Abstract: Curcumin, the active ingredient of turmeric (*Curcuma longa*) used in culinary and medical practices in Asia, has immense potential for being used in cancer chemotherapy because of its control over the cell growth regulatory mechanisms. The present chapter throws light on the role of curcumin in modulating the various phases of the cell cycle and its apoptosis-inducing effects. This is followed by a discussion on the implications of these effects of curcumin for its use as a chemotherapeutic agent in cancer. Curcumin affects various cell cycle proteins and checkpoints involving downregulation of some of the cyclins and cyclin-dependent kinases, upregulation of cdk inhibitors, and inhibition of DNA synthesis. In addition, curcumin also exerts indirect control over cell division such as inhibition of telomerase activity. Remarkably, some studies point toward a selective growth-inhibitory effect of curcumin on transformed cell lines compared to non-transformed cell lines. Curcumin has also been demonstrated to have proapoptotic effects in several *in vitro* studies, mostly through the mitochondria-mediated pathway of apoptosis. Curcumin-mediated regulation of apoptosis involves caspases, Bcl2 family members, inhibitors of apoptosis proteins, and heat shock proteins. The accumulating data on the *in vitro* and *in vivo* actions of curcumin together with the ongoing human clinical trials will provide a better understanding of curcumin-mediated cell growth regulation, ultimately catering to the needs of human welfare.

1. INTRODUCTION

In multicellular organisms, mutations in somatic cells affecting genes that control cell proliferation and apoptosis often lead to the development of tumors. Even though tumors are diverse and heterogeneous, they have commonly lost or modified the mechanisms that normally regulate cell proliferation and death. Thus, a logical approach to ensure selectivity of antitumor agents is to target critical factors that regulate cell growth, proliferation, and apoptosis. Increasing evidence suggests that a variety of plant-derived natural compounds can target the basic machinery of DNA synthesis and cell division responsible for deregulated proliferation and apoptosis in cancer cells. Curcumin is the yellow pigment and active ingredient of turmeric (*Curcuma longa* rhizomes) that is commonly used as a spice in Asia. Turmeric is well known for its medicinal properties in Indian and

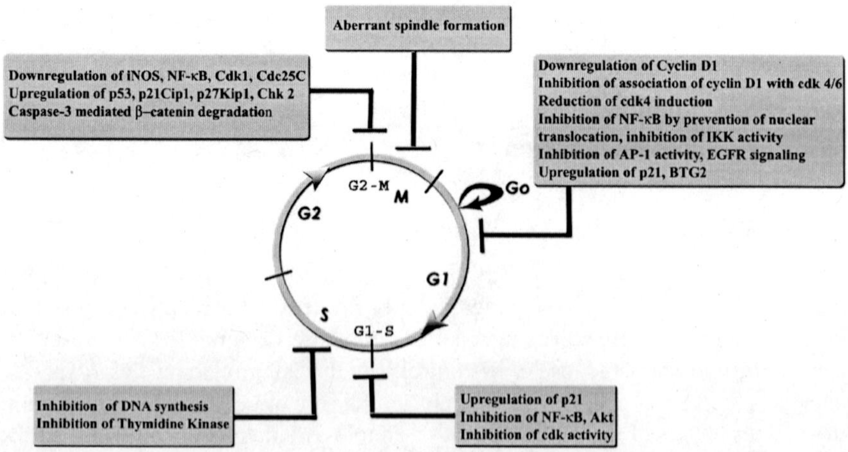

Figure 1. Molecular mechanisms involved in the regulation of the cell cycle by curcumin.

Chinese systems of medicine and has been widely used for the treatment of several diseases.[1] Epidemiological observations, although inconclusive, suggest that turmeric consumption might reduce the risk of some forms of cancer and might render other protective biological effects in humans. These biological effects of turmeric have been attributed to its constituent curcumin that has been widely studied for its anti-inflammatory, antiangiogenic, antioxidant, wound-healing, and anticancer effects.[2] The present chapter throws light on the role of curcumin in modulating the various phases of the cell cycle and its apoptosis-inducing effects. This is followed by a discussion on the implications of these effects of curcumin for its use as a chemotherapeutic agent in cancer. Curcumin-mediated effects on the cell cycle are outlined in Figure 1.

2. CURCUMIN AND THE CELL CYCLE

Curcumin affects many cell cycle proteins and checkpoints, which are crucial in regulating cell proliferation. Curcumin is said to influence cell cycle by the downregulation of cyclin D1, inhibition of DNA synthesis, upregulation of p53 and Kip family cdk inhibitors like $p21^{WAF1/CIP1}$ and $p27^{KIP1}$ and inhibition of NF-κB.

2.1. Direct Targets of Curcumin in the Cell Cycle

2.1.1. G0/G1 Phase

The proliferation of rat calvarial osteoblastic cells was markedly inhibited by curcumin at 5–10 μM and it did not induce apoptosis but arrested cells at the

G1 phase of the cell cycle. In addition, curcumin stimulated the expression of mRNA for p21$^{WAF1/CIP1}$, which inhibits the activity of cyclin-dependent kinases, and inhibited the phosphorylation of histone H1.[3] HBC (4-[3,5-bis-[2-(4-hydroxy-3-methoxy-phenyl)-ethyl]-4,5-dihydro-pyrazol-1-y l]-benzoic acid) is a recently developed curcumin derivative that exhibits potent inhibitory activities against the proliferation of several tumor cell lines. HBC induced prolonged phosphorylation of ERK1/2 and activated p21$^{WAF1/CIP1}$ expression, resulting in the induction of G0/G1 cell cycle arrest in HCT15 colon cancer cells.[4] Curcumin reversibly inhibits normal mammary epithelial cell cycle progression by downregulating cyclin D1 expression and blocking its association with Cdk4/Cdk6 as well as by inhibiting phosphorylation and inactivation of retinoblastoma protein. In addition, curcumin significantly upregulates p21$^{WAF1/CIP1}$ in normal cells and arrests them in the G0 phase of the cell cycle that facilitates these cells to escape from curcumin-induced apoptosis at the G2 phase. Interestingly, these processes remain unaffected by curcumin in carcinoma cells that express high levels of cyclin D1. Similarly, in an ectopically overexpressed system, curcumin cannot downregulate cyclin D1 and thus block cell cycle progression. Hence, these cells progress into the G2 phase and undergo apoptosis.[5] In U937 cells, the levels of cyclin D1 mRNA and protein were decreased upon treatment with curcumin, by preventing nuclear translocation of nuclear factor (NF)-κB and by the upregulation of BTG2, a negative transcriptional regulator of cyclin D1.[6] Activation of PPAR-γ (peroxisome proliferator-activated receptor-γ) by curcumin inhibits cell growth in a human colon cancer-derived cell line (Moser cells) by suppressing cyclin D1 and EGFR (epidermal growth factor receptor) expression.[7] Curcumin induces G1 phase arrest in embryonal carcinoma (PCC4) cells, followed by their differentiation.[8] In silica-induced injury on human embryonic lung fibroblasts, inhibition of activator protein (AP)-1 activity by cucumin resulted in a reduced induction of cyclin D1 and CDK4.[9] Curcumin prevents cell growth of HTLV-I-infected T-cell lines and primary ATL cells but not of normal peripheral blood mononuclear cells by inhibiting phosphorylation of IκB-α kinase (IKK), thus suppressing the action of NF-κB. Cell cycle arrest, with the reduction in the expression of cyclin D1, Cdk1, and Cdc25C was observed.[10] Other observations on curcumin-induced G1 phase arrest and/or cyclin D1 downregulation through inhibition of NF-κB have been reported in bladder cancer cells,[11] human lung epithelial cells,[12] and human multiple myeloma cells.[13]

2.1.2. G1/S Phase

In human umbilical vein endothelial cells (HUVEC), curcumin effectively blocked the cell cycle progression during the S phase by inhibiting the activity of thymidine kinase. A [^{3}H]-thymidine incorporation assay revealed a curcumin-mediated inhibition of DNA synthesis and accumulation of >46% cells in the early S phase. Pulse-labeling studies with [3H]-thymidine demonstrated that curcumin affected cells that were actively undergoing DNA synthesis.[14] Similarly in rat aortic smooth muscle cells (A7r5), curcumin treatment (1–100 μM) significantly decreased the uptake of [^{3}H]-thymidine in a concentration-dependent manner.[15] Loss of thymidine kinase activity might be one of the possible mechanism(s) for the inhibition of DNA synthesis activity by curcumin. In human Burkitt's lymphoma (CA46) cells, curcumin inhibits proliferation, associated with a decrease in the cell population in the S phase.[16] In head and neck squamous cells (HNCCs), curcumin arrested the cell cycle in the G1/S phase by the inhibition of IKK and NF-κB that resulted in suppression in the expression of various cell survival and cell proliferative genes, including Bcl-2, cyclin D1, interleukin (IL)-6, cyclooxygenase (COX)-2, and matrix metalloprotenase (MMP)-9.[17] Curcumin-induced arrest of mantle cell lymphoma cells in the G1/S phase by inhibiting constitutive NF-κB activation has also been reported.[18] Human colon cancer cells (Lovo), upon curcumin treatment, exhibited cell cycle arrest largely in the S and G2/M phases, preventing the entry into the next phase of the cell cycle.[19]

2.1.3. G2/M Phase

There are several reports of curcumin-induced G2/M phase arrest, the exact mechanism of which is, however, not well understood. Curcumin induces G2/M phase arrest in colon cancer (HCT-116) cells with a decline in the levels of cyclin D and E, whereas the level of cyclin B remains unaffected.[20] Studies by Grimm and colleagues show that curcumin induces melanoma cell apoptosis and cell cycle arrest at the G2/M phase, which is associated with the downregulation of inducible nitric oxide synthase (iNOS), NF-κB, and DNA-dependent protein kinase catalytic subunit expression and upregulation of p53, p21$^{WAF1/CIP1}$, p27^{KIP1}, and checkpoint kinase 2.[21] In a gene expression profiling study on two human colon cancer cell lines (HT29 and Caco-2), short-term exposure (3–6 h) to curcumin induced expression changes of many genes, including those that control cell cycle progression. Especially in HT29 cells, several genes regulating the transition through the G2/M phases were modulated. FACS analysis showing cell cycle arrest at the G2/M transition phase arrest augments these data.[22] Curcumin treatment causes p53- and p21$^{WAF1/CIP1}$-independent G2/M phase arrest and apoptosis in HCT 116 cells. Further investigation revealed that curcumin induced caspase-3-mediated degradation of β-catenin, thereby decreasing promoter DNA-binding activity of the β-catenin/Tcf-Lef complex, thereby decreasing Cdc2/cyclin B1 kinase activity, which explains the G2/M phase arrest.[23] In a recent study on squamous cell carcinoma, curcumin (3.5 μM) has been reported to induce S/G2M phase cell cycle arrest, the stages at which chemotherapy is more effective, thus sensitizing these cells to the ionizing effects of radiation.[24] In breast cancer (MCF-7) cells, curcumin (10–20 μM, 24–48 h) treatment arrested cell growth at the G2/M phase. Although this concentration is not sufficient to induce apoptosis, it is highly effective in inhibiting cell proliferation for over 6 days. The arrested mitotic cells exhibit monopolar spindles, and chromosomes do not undergo normal anaphase movements. After 48 h, most cells eventually leave the M phase, and many form multiple micronuclei instead of individual daughter nuclei.[25] These observations indicate that curcumin induces the assembly of aberrant, monopolar spindles that are impaired in their ability to segregate chromosomes. The production of cells with extensive micronucleation after curcumin treatment suggests that at least some of the cytostatic effects of this phytochemical are due to its ability to disrupt normal mitosis. This raises the possibility that curcumin might promote genetic instability under some circumstances. This shows that curcumin induces a mitotic catastrophelike death in this cell line. This observation can have potential implications in improvising the efficacy of existing anticancer therapy. This is because, in contrast to treatment-induced physiological antiproliferative responses such as apoptosis and senescence, mitotic catastrophe is potentiated by cellular changes that occur in neoplastic transformation. Therefore, it is indeed a promising area for further studies towards achieving tumor selectivity in anticancer therapy.

2.2. Indirect Targets of Curcumin in Cell Cycle

Apart from controlling the molecules directly involved in the cell cycle, curcumin affects several signaling pathways and enzymes that indirectly have a role in regulating the cell cycle. A gene expression profiling study, aimed at identifying novel targets of curcumin among the 12,625 probes analyzed, using MDA 1986 cell line made the following observations: Among other genes affected, two negative regulators of the cell cycle—*mad* (antagonizes *myc* transcriptional activity) and $p27^{Kip1}$ (a cdk inhibitor)—were induced 68- and 3-fold, respectively. Two additional genes involved in growth control—*K-sam* and *HER3*—and that of *E2F-5* (a transcription factor that regulates genes controlling cell proliferation) were also upregulated.[26] It was reported that curcumin inhibits telomerase activity in human breast cancer cells (MCF-7) in a dose-dependent manner, with 93.4% inhibition of activity at 100 μM curcumin. Such a growth regulatory effect of curcumin has been found to be exerted by downregulating the expression of human telomerase reverse transcriptase (hTERT).[27] It is not clear how exactly curcumin targets telomerase activity and whether telomerase inhibition is a cause or consequence of apoptotic cell death. As constitutive telomerase activity is a characteristic feature of tumor cells, further studies on this unique activity of curcumin might unravel a selective antitumor potential of this phytochemical. Such compounds are expected to show specificity toward cancer cells, highlighting their superiority in anticancer therapy compared to their counterparts. Curcumin also inhibits the activity of enzymes like ornithine decarboxylase,[28,29] protein kinase C (PKC),[15,30] COX and lipoxygenase (LOX),[31] and oxidative enzymes such as β-carotene-15,15′-dioxygenase.[32] Curcumin inhibited ErbB-2 autophosphorylation and transphosphorylation *in vitro* in a concentration-dependent manner and depleted the $p185^{neu}$ protein *in vivo* in human breast cancer cells.[33] In one study, it was reported that curcumin inhibited cellular growth of both transformed and nontransformed cell lines in a nonselective manner and inhibition of cell growth was not always associated with apoptosis.[34] However, other studies on nontransformed cell lines support the finding that curcumin has no effect on their growth.[35] Similarly, curcumin did not inhibit growth in primary cultures of mouse embryonic fibroblasts and human foreskin fibroblasts when it did so in tumor cell lines.[36] Curcumin inhibits proliferation and induces apoptosis in human B-cell non-Hodgkin's lymphoma (Raji cells) while showing no apparent inhibition of proliferation and apoptosis induction in normal peripheral blood mononuclear cells.[37] Improved understanding of this differential activity in transformed and nontransformed cells is very essential in the light of the clinical application of curcumin in anticancer therapy. Further studies are needed to draw a general conclusion about its selectivity to cancer cells.

3. CURCUMIN-INDUCED APOPTOSIS

Apoptosis is a highly regulated cell death process, which efficiently eliminates unwanted cells in multicellular organisms in a programmed manner. The relative expression levels and/or activities of the proapoptotic (Bax, Bid, Bak, p53, active caspases) and antiapoptotic (Bcl-2, IAPs, heat shock proteins) molecules in the intracellular milieu determine the survival status of a cell. Curcumin is a potent inducer of apoptosis and induces typical features of apoptosis, such as phosphatidyl serine exposure, cell shrinkage, chromatin condensation, and DNA fragmentation in several cell lines. Curcumin-mediated cell death involves the downregulation of cell survival genes such as *NF-κB, egr-1, c-myc*, and *Bcl-XL* or upregulation of cell-death-promoting genes such as *p53* and *Bax*. There are reports of apoptosis induction by curcumin through both the extrinsic and intrinsic pathways.[38]

3.1. Effects on the Extrinsic Pathway

In human melanoma cells, curcumin activated caspase-3 and caspase-8, but not caspase-9, and it induced Fas receptor aggregation in a FasL-independent manner. Several conditions, including the presence of caspase-8 inhibitors, low-temperature incubation (known to prevent receptor aggregation), and expression of dominant negative Fas associated protein with death domain (FADD), prevented curcumin-mediated cell death induction, strongly supporting the involvement of extrinsic pathway in curcumin-mediated apoptosis in human melanoma cells.[39] In gastric (KATO-III) and colon cancer (HCT-116) cells, curcumin stimulates the activity of caspase-8 and the cleavage of caspase-3 and poly (ADP) ribose polymerase (PARP) (a caspase-3 substrate).[40] Curcumin was reported to inhibit cell proliferation and induce apoptosis in human Burkitt's lymphoma cells (CA46) by upregulating Fas receptor expression, among its other effects.[16] However, evidence of curcumin-mediated induction of apoptosis by the death receptor pathway are limited. Also, to show conclusively that curcumin acts through Fas signaling, synthesis of FasL should be evident before the cells become apoptotic and cytotoxicity should be inhibited by manipulations that block FasL–Fas and FADD–procaspase-8 interactions and cells that lack Fas should fail to respond to the drug. Unfortunately, such rigorous evidence of the involvement of Fas signaling in response to curcumin is not available.

3.2. Effects on the Intrinsic Pathway

Mitochondria play a pivotal role in the induction of apoptosis through the intrinsic pathway. DNA damage, ischemia, and oxidative stress are examples of apoptotic signals that lead to cell death through the mitochondria. The mitochondrial pathway of apoptosis involves the permeabilization of the mitochondrial outer membrane, opening of the permeability transition pore, and release of apoptotic factors like cytochrome-*c*, Smac, and apoptosis including factor (AIF) from the mitochondrial intermembrane space. This ultimately results in the activation of caspases, the executioners of cell death, and cleavage of their myriad substrate proteins like PARP and lamins. Curcumin induces the loss of $\Delta\psi_m$, changes in mitochondrial permeability, and release of apoptogenic factors from the mitochondrial intermembrane space in many cell lines, thus proving to be a potent proapoptotic signal. There have also been several reports of curcumin-mediated regulation of proapoptotic and antiapoptotic molecules, shifting the balance toward cell death. Our laboratory reported that metastatic colon cancer cells were more sensitive to curcumin-induced apoptosis than their counterparts from the primary sites of tumor of the same patient.[41] In primary effusion lymphoma (PEL) cells, curcumin induces a loss of $\Delta\psi_m$, subsequent cytochrome-*c* release, caspase-3 activation, and PARP cleavage, suggesting the involvement of the intrinsic pathway of apoptosis.[42] In rat liver mitochondria, curcumin induced an increase in membrane permeability, resulting in swelling, $\Delta\psi_m$, and inhibition of ATP synthesis, and these effects of curcumin were found to be due to the opening of the permeability transition pore.[20] Curcumin induces apoptosis through the mitochondrial pathway involving caspase-8 (blocked by dominant negative FADD-like-IL-1beta-converting enzyme (FLICE), but not by dominant negative FADD), Bcl-2 interacting domain (BID) cleavage, cytochrome-*c* release, and caspase-3 activation in HL-60 cells.[43] In HL-60 cells, the Bcl-2 protein level decreased to 30% after 6 h of curcumin treatment and a further 6 h of treatment subsequently reduced it to 20%.[44] Overexpression of Bcl-2 or Bcl-XL in HL-60 cells results in either a blockage or a delay in curcumin-induced cytochrome-*c* release, activation of caspase-8 and caspase-3, and cleavage of BID and PARP.[43] Studies from our laboratory have shown that curcumin induces activation of caspase-3 and capase-9 and AIF release in human colon cancer cells (SW480) and these effects were reversed by the overexpression of Hsp70 and restored by the introduction of antisense Hsp.[41,45] Further, our studies revealed that downregulation of Bcl-XL and Ku70 (known to suppress the apoptotic translocation of Bax) increased their sensitivity toward curcumin-induced nuclear condensation and cell death.[46] Also, overexpression of these proteins resulted in protection by inhibition of release of cytochrome-*c*, Smac (second mitochondria-derived activator of caspases), and AIF, as well as the activation of caspase-9, caspase-8, and caspase-3.[46] In a human ovarian cancer cell line (Ho8910), curcumin treatment (40 μM) suppresses cell growth and induces apoptosis, accompanied by a decrease in the expression of Bcl-2, Bcl-XL and pro-caspase-3 and an increase in the levels of p53 and Bax.[47] In human lung cancer cell lines (A549 and H1299), a decrease in the expression of Bcl-2 and Bcl-XL protein occurs with a 12-h exposure of

40 μM curcumin.[48] In smooth muscle cells (A7r5), a significant reduction in Bcl-2 mRNA was detected with curcumin.[15] In contrast, in AK-5 cells[49] and Ehrlich's ascites carcinoma cells,[50] curcumin treatment does not seem to affect the levels of Bcl-2. However, in the latter case, in spite of Bcl-2 levels remaining unaffected, the intracellular protein level of Bax is found to be upregulated upon the introduction of curcumin, suggesting that curcumin-mediated cytochrome-*c* release and apoptosis in Ehrlich's ascites carcinoma cells bypass the Bcl-2 checkpoint overriding its protective effects. Our laboratory reported an important role for Bax in the curcumin-induced mitochondrial apoptosis pathway. In HCT116 cells lacking Bax (Bax$^{-/-}$), curcumin-induced activation of caspase-9 and caspase-3 was blocked, whereas that of caspase-8 remained unaltered. Further, Bax$^{-/-}$ cells resisted curcumin-induced release of cytochrome-*c*, AIF, and Smac. Reintroduction of Bax, downregulation of Bcl-XL, and overexpression of Smac highly sensitized the Bax$^{-/-}$ cells to curcumin-induced apoptosis.[51] In human colon adenocarcinoma (HT-29), incubation of cells with 50 μM curcumin (0–24 h) resulted in a time-dependent downregulation of Bcl-2 as well as upregulation of Bax, pro-caspase-9 and pro-caspase-3 gene expression, in a p53-dependent manner, suggesting the involvement of the intrinsic pathway.[52] Inhibitors of apoptosis proteins (IAPs) bind and inhibit caspases, and studies on human melanoma cells reported suppression of XIAP upon curcumin treatment.[39] A study on HTLV-1-infected T-cell lines reported that curcumin induced cell cycle arrest, followed by apoptosis by reducing the expression of XIAP and survivin.[10] Binding of Smac (released from mitochondria) to IAPs can block their inhibition over caspases. Zheng et al reported that transfection of extrinsic Smac gene sensitized gastric cancer cells (MKN-45) to curcumin-induced apoptosis.[53] Major mechanisms by which curcumin is known to induce extrinsic and intrinsic pathways of apoptosis are listed in Table 1.

3.3. Protective Effects of Curcumin on Apoptosis

In contrast to the proapoptotic effects of curcumin, some reports suggest that it inhibits apoptosis induced by other agents. Curcumin (30 mg/kg body weight i.p. injections or 2 g/kg of diet for 2 months) exhibited neuroprotective effects against cerebral ischemia-induced neuronal apoptosis and behavioral deficits in experiments on Mongolian gerbils. Curcumin administration decreases lipid peroxidation, mitochondrial dysfunction, and apoptotic indices such as a decrease in $\Delta\psi_m$, Δcytochrome-*c* release, and caspase-3 activation.[54] In human proximal tubule epithelial cells, curcumin exhibited a cytoprotective effect against shiga toxin-induced injury, which was proposed by the authors to be probably due to increased expression of Hsp70.[55] An antiapoptotic chondroprotective role of curcumin was observed in articular chondrocytes by reversing the effects of IL-1β. IL-1 β is a proinflammatory cytokine that mediates cartilage degradation in osteoarticular disorders like osteoarthritis and rheumatoid arthritis, thereby activating matrix-degrading enzymes, leading to chondrocyte apoptosis. Curcumin treatment (50 μM) for 30 min relieved IL-1β-induced matrix degradation and caspase-3 activation.[56]

Table 1. Apoptotic pathways and mechanisms induced by curcumin.

PATHWAYS OF APOPTOSIS	CELLS/TISSUES	MECHANISMS	REFS.
Extrinsic pathway	Human melanoma	Caspase-3 and caspase-8 activation Fas receptor aggregation	39
	Gastric (KATO-III) Colon (HCT-116)	Caspase-3 and caspase8 activation PARP cleavage	40
	Human Burkitt's lymphoma (CA46)	Fas receptor expression	16
Intrinsic pathway	Rat liver mitochondria	Mitochondrial swelling, increase in permeability, loss of $\Delta\psi_m$	20
	Human acute myelogenous leukemia (HL-60)	Caspase-8 activaion (not blocked by FADD), caspase-3 activation, cytochrome c release, BID and PARP cleavage	43
		Decrease in Bcl-2 protein	44
	Human colon cancer (SW480)	Caspase-3 and caspase-9 activation, AIF release	41
	Human ovarian cancer (Ho8910)	Decrease in Bcl-2, Bcl-XL expression, Increase in p53 and Bax expression	47
	Human lung cancer (A549 and H1299)	Decrease in Bcl-2, Bcl-XL expression (p53 independent)	48
	Primary effusion lymphoma (PEL)	Loss of $\Delta\psi_{m,}$, Cyt c release, Caspase-3 activation, PARP cleavage	42
	Vascular smooth muscle (A7r5) cells	Decrease in Bcl-2 mRNA	15
	Ehrlich's ascites carcinoma	Bax protein upregualtion	50
	Human colon cancer (HCT116)	Caspase-3 and caspase-9 activation, caspase-8 unaffected (blocked in Bax knockouts)	51
	Human colon adenocarcinoma (HT-29)	Downregulation of Bcl-2, upregulation of Bax, caspase-3 and caspase-9 gene expression (p53 dependent)	52
	Human melanoma	Suppression of XIAP	39
	HTLV-1 infected T-cell leukemia	Suppression of XIAP and survivin expression	10
	Rat histiocytoma (AK-5)	ROS generation, sytochrome-*c* release	64

3.4. Influence of Oxidation Status of the Cells on the Actions of Curcumin

One of the well-known biological activities of curcumin is its antioxidant activity, by virtue of which it protects the cells from oxidative damage induced by reactive oxygen intermediates. The antioxidant effects are due to its ability to be a potent scavenger of a variety of reactive oxygen species (ROS), including the superoxide anion, the hydroxyl radical, the singlet oxygen, nitric oxide, and peroxynitrite by forming stable curcumin dimers upon radical binding[57,58] and inhibiting ROS-induced DNA damage.[59,60] Turmeric lowers lipid peroxidation by enhancing the activities of antioxidant enzymes, superoxide dismutase, catalase, and glutathione peroxidase.[61] The ability of curcumin to enhance intracellular glutathione levels is also attributed to its antioxidant property.[62] Interestingly, a pro-oxidant property has also been attributed to curcumin, which explains its proapoptotic activity in many cases. The pro-oxidant DNA cleaving activity is reported to be due to binding of Cu(II) at three proposed sites in the curcumin molecule.[63] In human leukemia cells (HL 60), curcumin-induced apoptosis has been reported to be effectively abrogated by the antioxidants *N*-acetyl-L-cysteine (NAC), L-ascorbic acid, α-tocopherol, catalase, and superoxide dismutase.[44] In rat histiocytoma (AK-5), there is evidence of curcumin-mediated apoptosis via generation of ROS with subsequent release of cytochrome-*c*.[64] Moreover, curcumin being lipophilic, it can collapse the $\Delta\psi_m$ and increase the permeability of the mitochondrial membrane for protons, thus interfering with the energy-coupling system in the mitochondria. This leads to the generation of abnormal levels of superoxide radicals, which has been reported to correlate with the sensitivity towards curcumin-induced apoptosis.[65,66] In fact, dietary curcumin can cause oxidative stress in Asian patients with acute vitilago.[67] Studies on colon cancer cells show that curcumin induces apoptosis via a ROS-associated mechanism that converges on JNK activation and, to a lesser extent, via a parallel ceramide-associated pathway.[68] In a study on the influence of intracellular glutathione levels on curcumin-induced apoptosis, Syng-Ai et al. reported that depletion of glutathione levels (which resulted in the increased generation of ROS) sensitizes tumor cells to curcumin-induced apoptosis.[69] There is a report on the inhibition of drug-induced apoptosis by curcumin through the inhibition of ROS formation and consequent blockade of JNK function in human breast cancer cells.[70] Similarly, another antioxidant, resveratrol, also inhibits anticancer drug-induced caspase activation, DNA fragmentation, and translocation of cytochrome-*c* in human leukemia cells.[71] In addition, curcumin (50 μM) inhibits apoptosis in dexamethasone-treated rat thymocytes accompanied by partial suppression of AP-1 activity.[72] An obvious question that follows this observation is whether a reverse approach of using pro-oxidants in combination with anticancer drug therapy can enhance the sensitivity of these anticancer drugs. Such studies would reveal valuable insights into developing combination therapies in cancer treatment to combat factors such as multidrug resistance. Presumably, the microenvironment in a cell that alters the oxidation status of a cell at a given time point and conditions determines the pro-oxidant or antioxidant mode of action of curcumin *in vivo*. Thus, it is possible that, the redox action of curcumin in inducing apoptosis is determined by the intracellular oxidation status and manipulating this status could possibly have implications in the sensitization of curcumin-based therapies in tumor cells.

4. INFLUENCE OF CURCUMIN ON SIGNALING PATHWAYS

4.1. NF-κB Signaling

Nuclear factor-κB, an important intracellular target of curcumin, is actually a family of dimeric transcription factors that, upon activation, translocate to the nucleus and controls over 200 genes that regulate the immune system, cell growth, proliferation, angiogenesis, metastasis and inflammation.[73] Curcumin was shown to inhibit the NF-κB activation pathway at a step before IκB-α phosphorylation induced by various stimuli.[74,75] A recent study has shown that curcumin suppresses tumor necrosis factor (TNF)-α-induced activation of NF-κB (by inhibiting the nuclear translocation of the p65 subunit) and Akt (an activator of NF-κB).[76] This is bound to affect the expression of its downstream targets associated with the cell cycle (cyclin D1) and apoptosis (antiapoptotic molecules IAP-1, IAP-2, XIAP, Bcl-2, Bcl-XL, Bfl-1/A1, TRAF1, and FLIP-like inhibitory protein), among others. Studies on melanoma cell lines show that curcumin exhibits antiproliferative and proapoptotic effects by suppressing IκB kinase and NF-κB activity, independent of the B-Raf/MEK/ERK and Akt pathways.[77] Glutathione-*S*-transferase P1-1 (GSTP1-1), a cytosolic enzyme involved in the detoxification of xenobiotics, is implicated in tumor progression and drug resistance. Curcumin induces apoptosis in K562 leukemia cells by downregulating GSTP1-1 mRNA and protein levels by inhibiting TNF-α and phorbol ester-induced AP-1 and NF-κB transcription factor binding to the GSTP1-1 gene promoter.[78] (In a human breast cancer xenograft model, dietary administration of curcumin significantly decreased the incidence of breast cancer metastasis to the lung and suppressed the expression of NF-κB.[79] We stably transfected mouse fibrosarcoma (L-929) cells with *relA* gene (a subunit of NF-κB) and found that the transfected cells were resistant to varying doses of curcumin, whereas the untransfected parent cells were sensitive to curcumin-induced apoptosis in a concentration-dependent manner.[80] Deeb et al. reported that curcumin enhanced the sensitivity of prostate tumor cells to TNF related apoptosis-inducing ligand (TRAIL) by inhibiting NF-κB activation through the blockade of phosphorylation of IκB-α and its degradation.[81,82] We reported that curcumin sensitizes cervical cancer cells (HeLa) to Taxol-induced apoptosis by the down-regulation of NF-κB via inhibition of Akt, a Ser/Thr kinase.[83]

4.2. p53 Signaling

The intracellular activity of p53, often referred to as the "guardian of genome" with separable cell cycle arrest and proapoptotic activities, is regulated by phosphorylation and other posttranslational modifications. Many studies assign a cell-type-dependent role for curcumin in stabilizing and upregulating intracellular levels of this tumor suppressor protein. In colon adenocarcinoma (HT29) and human ovarian carcinoma (Ho-8910) cells, curcumin upregulates the level of phospho-p53 (Ser 15) in a time- and concentration-dependent manner, associated with the regulation of apoptosis-related proteins Bax and Bcl-2.[47,52,84] CK2 and PKD are protein kinases, which are associated with a complex called CSN (COP9 signalosome). CSN is involved in the phosphorylation of p53 at Thr-155, directing it to 26S proteasome, resulting in the ubiquitin-mediated degradation of p53. It is reported that curcumin inhibits CK2 and PKD activities, thus stabilizing p53 levels inside the cell.[85] A recent study on mammary epithelial cells that express deregulated levels of cyclin D1 reveals the involvement of curcumin in selectively inducing apoptosis of the cancerous cells at the G2 phase in a p53-dependent manner.[5] However, p53-independent induction of apoptosis by curcumin in human lung cancer cell lines (A549 and H1299) has also been reported recently.[27] Similarly, curcumin treatment causes p53- and p21-independent apoptosis in HCT-116 cells.[23]

4.3. JAK–STAT Signaling

In PEL cells that exhibit a constitutively active JAK–STAT pathway, curcumin induces growth suppression and apoptosis by inhibiting the JAK–STAT pathway.[42] Curcumin-mediated apoptosis occurs through the suppression of the JAK–STAT pathway in a dose-dependent manner in HTLV-1 transformed T-cell leukemia (MT-2, HuT-102, SLB-1).[86] Curcumin inhibits IL-6-mediated as well as constitutive phosphorylation of STAT3 and its nuclear translocation in multiple myeloma cells.[87]

4.4. Other Signaling Intermediates

Other intracellular factors that influence the course of curcumin-induced apoptosis are transcription factors like AP-1, several adhesion molecules, kinases such as receptor tyrosine kinases, mitogen-activated protein kinasea (MAPKs), and c-jun N-terminal kinases (JNKs). In a recent study on human papilloma virus (HPV) infection-associated cervical cancer cells, it has been reported that curcumin plays a role in downregulating the viral oncogenes and preventing NF-κB and AP-1 translocation, followed by induction of apoptosis.[88] In HTLV-1-transformed T-cell leukemia, curcumin suppressed the constitutive activation of AP-1 (a transcription factor playing a critical role in the oncogenesis of this cancer) by downregulation of JunD protein.[89] In contrast, another study reported AP-1 as an inducer of bile acid-induced liver cell apoptosis and curcumin is shown to inhibit AP-1 transcriptional activity.[90] Chen et al. reported that curcumin inhibits the MEKK1–JNK signaling pathway, thus revealing a possible mechanism of its inhibition of AP-1 and NF-κB signaling by curcumin [91] Recently, studies on acute T-cell leukemia revealed that curcumin inhibits proliferation and induces apoptosis in these cells by constitutively suppressing the activated targets of the PI3K/AKT pathway.[92] One of the mechanisms of curcumin-mediated inhibition of prostate cancer cells might be via inhibition of Akt.[93] In a cDNA array-based differential gene expression to identify novel genes regulated by curcumin upon treatment of MDA 1986 cells (squamous cell carcinoma of the oral cavity cell line), Yan et al. reported that of the

12,695 probes analyzed, close to 700 mRNAs showed a response to curcumin.[26] Of this, a proapoptotic activating transcription factor 3 (ATF3) was induced greater than fourfold. Two negative regulators of growth control—Mad (antagonizer of myc transcriptional activity) and $p27^{KIP1}$ (a CDK inhibitor)—were induced 68-fold and 3-fold, respectively. K-sam (keratinocyte growth factor receptor), HER3 (an EGFR family receptor), and E2F-5 transcription factor (which regulates cell proliferation) were downregulated. Moreover, Frizzled-1 (Wnt receptor) was most strongly repressed, suggesting a role for curcumin in the regulation of the prosurvival Wnt signaling pathway. In fact, an earlier study reported caspase-3-mediated cleavage and impairment of β-catenin (a downstream element of Wnt signaling) upon curcumin treatment, suggesting its action upon the Wnt signaling pathway.[23] Korutla and Kumar reported curcumin-induced reduction in the EGF receptor tyrosine kinase activity by 90% in a time- and dose-dependent manner in human epidermoid carcinoma (A431) cells, accompanied by inhibition of EGF-induced phosphorylation of EGF receptors.[94] Studies on human colon cancer cells suggest a role for curcumin in the suppression of transcription of EGFR by reducing the transactivation activity of the transcription factor early growth response-1 (egr-1) and also by reducing the expression of *egr-1*.[95] In HCT116 cells, curcumin treatment leads to transcriptional induction of GADD153 protein, an early-response gene to cellular stress. This is followed by induction of apoptosis that occurred independent of MAPKs.[96] In another study on the same cell line Collett and Campbell reported a MAPK-dependent curcumin-mediated induction of apoptosis.[97] They report that JNK, not p38 MAPK or ERK, is primarily responsible for the induction of apoptosis in HCT116, which is preceded by sustained phosphorylation of c-jun, increased AP-1 transcriptional activity, and inhibition of the transcriptional activity of NF-κB. Curcumin stimulated PPAR-γ activity in activated HSC (hepatic stellate cells) *in vitro*, which was required for curcumin to reduce cell proliferation, induce apoptosis, and suppress ECM (extracellular matrix) gene expression. Further experiments demonstrated that curcumin suppressed the gene expression of TGF-β (transforming growth factor-β) receptors and interrupted the TGF- β signalling pathway in activated HSC, which was mediated by PPAR-γ activation.[98,99] A study on ovarian cancer cells (A2780) reported that the growth regulatory effects of curcumin are probably the result of downregulation of bcl-2 and p53.[100] The proapoptotic activity of curcumin in vascular smooth muscle cells has been proposed to be mediated, at least partly, through inhibition of protein tyrosine kinase activity, PKC activity, c-myc, and bcl-2 mRNA expression.[15] Various signaling intermediates known to be targeted by curcumin are shown in Figure 2.

5. IMPLICATIONS FOR CANCER THERAPY

The main aim of cancer therapy is to kill cancer cells selectively by apoptosis, without affecting the normal cells. Apoptosis is favored over necrosis because it is not associated with inflammation and does not damage the neighboring tissues. It

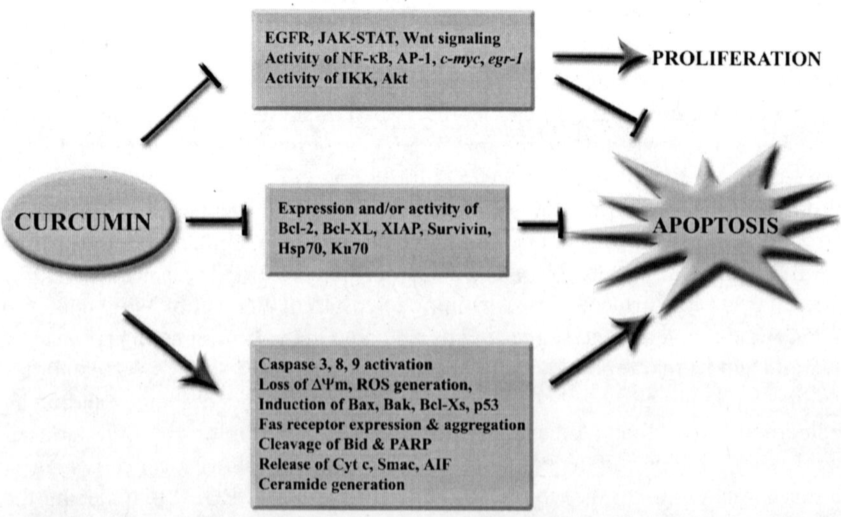

Figure 2. The molecular players involved in the induction of apoptosis by curcumin.

is therefore important to understand the mechanisms responsible for drug-induced apoptosis. Tumor cells evade apoptosis by one of the following mechanisms: upregulation of antiapoptotic factors like bcl-2, bcl-XL, NF-κB, Ku 70, IAPs, and heat shock proteins or downregulation or mutation of proapoptotic factors like Bax and p53. Tumor progression might also be associated with upregulation of cell cycle proteins like cyclins, cdks, or oncogenes like *c*-myc and egr-1. Thus, an effective strategy of cancer chemotherapy is to target these lesions either by upregulation of proapoptotic molecules or targeted downregulation of antiapoptotic proteins. Our studies on human colon cancer (SW480) cells suggests that heat shock protects these cells from curcumin-mediated apoptosis by blocking the release of cytochrome-*c* and Smac and reducing activation of caspases 9 and caspase-3, but not caspase-8.[41] From another study, we also reported that ectopic expression of Hsp70 protects these cells from curcumin-induced death. Antisense DNA to Hsp70 renders them highly sensitive to curcumin, with loss of $\Delta\psi_m$ and a greater release of cytochrome-*c*, AIF, and Smac. Hsp70 partly prevents the release of AIF, but not that of the other proteins.[45] Thus, it is evident that heat shock proteins (Hsps) interfere with the mitochondrial pathway to make apoptosis induced by curcumin less effective. These studies highlight the role of Hsps in conferring chemoresistance in colon cancer (SW480) cells and in using antisense approaches as an effective means to make curcumin-based therapy effective in treating multidrug resistance in cancer. A combination of 5 μM curcumin and 5 nM Taxol induces cytotoxicity and inhibits DNA synthesis more effectively than when Taxol is used alone in human cervical cancer cells. Interestingly, this synergism is not seen in normal cervical cells. Also, factors such as tubulin polymerization and Cdc2 activation induced by Taxol alone remain unaffected.[83]

Recently, Aggarwal has patented his findings showing that a combination of curcumin and paclitaxel (Taxol) can be used to treat and inhibit metastasis of breast tumors (US2004002499). In squamous cell lung carcinoma (H520), curcumin induced sensitization to Vinorelbine (a chemotherapeutic agent)-induced apoptosis, accompanied by upregulation of proapoptotic molecules (Bax and Bcl-Xs) and downregulation of Bcl-2 and Bcl-XL, induction of cytochrome-*c* release, as well as caspase-9 and caspase-3 activation. Curcumin, together with Vinorelbine, resulted in apoptosis of 61.3% cells, whereas alone they caused 23.7% and 38% apoptosis, respectively.[101] In androgen-independent prostate cancer cells (PC-3 and DU 145), curcumin enhances the cytotoxicity of chemotherapeutic agents by the inducton of $p21^{WAF1/CIP1}$ and C/EBP beta expression.[102] Production of IL-6 has been shown to be responsible for poor prognosis and drug resistance in cisplatin-induced apoptosis in certain ovarian cancer cells (CAOV3 and SKOV3). Curcumin has been found to enhance the sensitivity of these cells to cisplatin-induced apoptosis by reducing the autologous production of IL-6, thus proving to be an effective chemotherapeutic in combination with cisplatin.[103] Combining curcumin with 5-flurouracil significantly increased the growth inhibition of AGS human gastric carcinoma cells compared to either curcumin or 5-flurouracil alone, suggesting synergistic actions of the two drugs.[104] Similar synergistic inhibitory effects have also been reported in human colon cancer (HT29) cells.[105] Another study on multiple myeloma cells reported that curcumin-induced downregulation of NF-κB (a factor that has been implicated in chemoresistance) is also associated with conferring chemosensitivity to vincristine and melphalan.[13] Overexpression of a P-glycoprotein (Pgp-170, an MDR-1 gene product), a cell membrane protein, has been implicated in many cases of multidrug resistance (MDR). It is responsible for enhanced drug efflux leading to decreased intracellular drug accumulation. A study on the modulation of MDR-1 gene expression by curcuminoids shows that they decrease MDR-1 gene expression in a dose-dependent manner, with bisdemethoxycurcumin having the maximum effect. Treatment of drug-resistant KB-V1 cells with curcumin increased their sensitivity to vinblastine, which was consistent with a decreased Pgp expression on the cell plasma membrane.[106] Activation of NF-κB upon treatment with anticancer drugs has been found to be a general response in some tumor cell lines. However, treatments of these cells with certain common biologic modulators such as tamoxifen, dexamethasone, and curcumin that regulate NF-κB activation attenuate the doxorubicin (an anticancer drug)-induced NF-κB activation.[107] Very recently, a group has reported the synergistic action of curcumin and celecoxib (a proapoptotic agent) in cell growth inhibition and apoptosis in osteoarthritis synovial arthritis cells. This action has been proposed to be mediated through the involvement of inhibition of COX-2 activity, thus suggesting the potential for developing a combinatorial therapy based on these two drugs to treat rheumatological disorders.[108] PARP cleavage, caspase-3 activation, and apoptotic cell death in prostate cancer (PC-3) cells has been by synergistic inhibitory action of curcumin and β-phenylethylisothyocyanate via inhibition of EGFR signaling.[109,110] Curcumin is an arachidonic acid cascade inhibitor (AACI). Arachidonic acid metabolism intermediates have been reported to influence normal and malignant cell growth. Spingarn et al tested various combinations of AACIs to study their effects on head and neck squamous cell carcinoma.[111] They report that among the many combinations of chemicals analyzed, combination of curcumin with 13-*cis*-retinoic acid has a superior effective synergistic growth inhibition than when they are used alone. Similarly, in HL-60 cells, a combination of curcumin and retinoic acid proved to be a potent inhibitor of cell growth.[112] Chendil et al. studied the effect of a combination of radiation and curcumin (2–4 μM) in PC-3 cells.[113] It was found that such a combination approach results in the downregulation of NF-κB and Bcl-2 protein levels, although the levels of Bax remain unchanged. There was also a significant activation of caspase-9, caspase-3, and cytochrome-*c* release. This resulted in an altered Bcl-2:Bax ratio and enhanced chemosensitizing effect. Thus, compared to irradiation alone, curcumin enhanced the induction of clonogenic inhibition and apoptosis in these cells. In a recent study on squamous cell carcinoma, curcumin (3.5 μM) has been reported to induce S/G2M phase cell cycle arrest, the stages at which chemotherapy is more effective,[24] thus sensitizing these cells to the ionizing effects of radiation.

6. CONCLUSIONS AND FUTURE DIRECTIONS

It is clear that curcumin acts through multiple pathways to bring about its effects on cell growth and apoptotic signaling. Despite our extensive knowledge of the multiple biological effectors/targets of curcumin, more candidates remain to be explored. This is evident from the fact that several genes not known to be affected by curcumin previously are now identified through microarray-based gene expression profiling studies.[26,114] These novel approaches might provide potential new leads to genes and pathways that could play a role in cancer prevention and/or treatment by curcumin. There have been a few recent reports that surveyed apoptotic, antitumor, and antiangiogenic activities of some structural analogues of curcumin.[115–119] Identification of novel curcumin analogues that bypass some of the problems encountered with the use of curcumin, such as poor bioavailability, water solubility, and absorption, is needed. Enhancing bioavailability and ascertaining clinically attainable levels in specific target tissues are attractive strategies to be considered in developing curcumin as an effective anticancer drug. Although conventional chemotherapeutic drugs were not designed to induce apoptosis in cells, the fact that they indirectly do so raises the importance of this form of cell death in cancer chemotherapy. At the same time, there is also a possibility that many of the chemotherapy-induced cell death cannot be classified as apoptosis. The accumulating data on the *in vitro* and *in vivo* actions of curcumin together with the ongoing human clinical trials will provide a better understanding of curcumin-mediated cell growth regulation, ultimately catering to the needs of human welfare.

ACKNOWLEDGMENTS

Work in the authors' laboratory was supported by funding from the Kerala State Council for Science, Technology and the Environment, the Department of Science and Technology (to D. K.), the Life Sciences Research Board, Defence Research Development Organization, Government of India (to D. K. and S. K.), and a Junior Research Fellowship (to J. J.) of the Council of Scientific and Industrial Research, Government of India.

REFERENCES

1. S. Shishodia, G. Sethi, and B. B. Aggarwal, Curcumin: getting back to the roots. *Ann NY Acad Sci* **1056**, 206–217 (2005).
2. R. K. Maheshwari, A. K. Singh, J. Gaddipati, and R. C. Srimal, Multiple biological activities of curcumin: A short review. *Life Sci* **78**, 2081–2087 (2006).
3. M. Notoya, H. Nishimura, J. T. Woo, K. Nagai, Y. Ishihara, and H. Hagiwara, Curcumin inhibits the proliferation and mineralization of cultured osteoblasts. *Eur J Pharmacol* **534**, 55–62 (2006).
4. J. S. Shim, J. Lee, H. J. Park, S. J. Park, and H. J. Kwon, A new curcumin derivative, HBC, interferes with the cell cycle progression of colon cancer cells via antagonization of the Ca2+/calmodulin function. *Chem Biol* **11**, 1455–1463 (2004).
5. T. Choudhuri, S. Pal, T. Das, and G. Sa, Curcumin selectively induces apoptosis in deregulated cyclin D1-expressed cells at G2 phase of cell cycle in a p53-dependent manner. *J Biol Chem* **280**, 20,059–20,068 (2005).
6. Y. K. Kwon, J. M. Jun, S. W. Shin, J. W. Cho, and S. I. Suh, Curcumin decreases cell proliferation rates through BTG2-mediated cyclin D1 down-regulation in U937 cells. *Int J Oncol* **26**, 1597–1603 (2005).
7. A. Chen and J. Xu, Activation of PPAR{gamma} by curcumin inhibits Moser cell growth and mediates suppression of gene expression of cyclin D1 and EGFR. *Am J Physiol Gastrointest Liver Physiol* **288**, G447–G456 (2005).
8. B. K. Batth, R. Tripathi and U. K. Srinivas, Curcumin-induced differentiation of mouse embryonal carcinoma PCC4 cells. *Differentiation* **68**, 133–140 (2001).
9. F. Shen, X. Fan, B. Liu, X. Jia, H. Du, B. You, M. Ye, C. Huang,. and X. Shi, Overexpression of cyclin D1-CDK4 in silica-induced transformed cells is due to activation of ERKs, JNKs/AP-1 pathway. *Toxicol Lett* **160**, 185–195 (2006).
10. M. Tomita, H. Kawakami, J. N. Uchihara, T. Okudaira, M. Masuda, N. Takasu, T. Matsuda, T. Ohta, Y. Tanaka, K. Ohshiro, and N. Mori, Curcumin (diferuloylmethane) inhibits constitutive active NF-kappaB, leading to suppression of cell growth of human T-cell leukemia virus type I-infected T-cell lines and primary adult T-cell leukemia cells. *Int J Cancer* **118**, 765–772 (2006).
11. M. Sun, Y. Yang, H. Li, B. Su, Y. Lu, Q. Wei, and T. Fan, [The effect of curcumin on bladder cancer cell line EJ in vitro]. *Zhong Yao Cai* **27**, 848–850 (2004).
12. S. Shishodia, P. Potdar, C. G. Gairola, and B. B. Aggarwal, Curcumin (diferuloylmethane) down-regulates cigarette smoke-induced NF-kappaB activation through inhibition of IkappaBalpha kinase in human lung epithelial cells: correlation with suppression of COX-2, MMP-9 and cyclin D1. *Carcinogenesis* **24**, 1269–1279 (2003).

13. A. C. Bharti, N. Donato, S. Singh, and B. B. Aggarwal, Curcumin (diferuloylmethane) down-regulates the constitutive activation of nuclear factor-kappa B and IkappaBalpha kinase in human multiple myeloma cells, leading to suppression of proliferation and induction of apoptosis. *Blood* **101**, 1053–1062 (2003).
14. A. K. Singh, G. S. Sidhu, T. Deepa, and R. K. Maheshwari, Curcumin inhibits the proliferation and cell cycle progression of human umbilical vein endothelial cell. *Cancer Lett* **107**, 109–115 (1996).
15. H. W. Chen and H. C. Huang, Effect of curcumin on cell cycle progression and apoptosis in vascular smooth muscle cells. *Br J Pharmacol* **124**, 1029–1040 (1998).
16. Y. Wu, Y. Chen, J. Xu, and L. Lu, Anticancer activities of curcumin on human Burkitt's lymphoma. *Zhonghua Zhong Liu Za Zhi* **24**, 348–352 (2002).
17. S. Aggarwal, Y. Takada, S. Singh, J. N. Myers, and B. B. Aggarwal, Inhibition of growth and survival of human head and neck squamous cell carcinoma cells by curcumin via modulation of nuclear factor-kappaB signaling. *Int J Cancer* **111**, 679–692 (2004).
18. S. Shishodia, H. M. Amin, R. Lai, and B. B. Aggarwal, Curcumin (diferuloylmethane) inhibits constitutive NF-kappaB activation, induces G1/S arrest, suppresses proliferation, and induces apoptosis in mantle cell lymphoma. *Biochem Pharmacol* **70**, 700–713 (2005).
19. H. Chen, Z. S. Zhang, Y. L. Zhang, and D. Y. Zhou, Curcumin inhibits cell proliferation by interfering with the cell cycle and inducing apoptosis in colon carcinoma cells. *Anticancer Res* **19**, 3675–3680 (1999).
20. D. Morin, S. Barthelemy, R. Zini, S. Labidalle, and J. P. Tillement, Curcumin induces the mitochondrial permeability transition pore mediated by membrane protein thiol oxidation. *FEBS Lett* **495**, 131–136 (2001).
21. M. Zheng, S. Ekmekcioglu, E. T. Walch, C. H. Tang, and E. A. Grimm, Inhibition of nuclear factor-kappaB and nitric oxide by curcumin induces G2/M cell cycle arrest and apoptosis in human melanoma cells. *Melanoma Res* **14**, 165–171 (2004).
22. M. J. Van Erk, E. Teuling, Y. C. Staal, S. Huybers, P. J. Van Bladeren, J. M. Aarts, and B. Van Ommen, Time- and dose-dependent effects of curcumin on gene expression in human colon cancer cells. *J Carcinog* **3**, 8 (2004).
23. A. S. Jaiswal, B P. Marlow, N. Gupta, and S. Narayan, Beta-catenin-mediated transactivation and cell-cell adhesion pathways are important in curcumin (diferuylmethane)-induced growth arrest and apoptosis in colon cancer cells. *Oncogene* **21**, 8414–8427 (2002).
24. A. Khafif, R. Hurst, K. Kyker, D. M. Fliss, Z. Gil, and J. E. Medina, Curcumin: A new radio-sensitizer of squamous cell carcinoma cells. *Otolaryngol Head Neck Surg* **132**, 317–321 (2005).
25. J. M. Holy, Curcumin disrupts mitotic spindle structure and induces micronucleation in MCF-7 breast cancer cells. *Mutat Res* **518**, 71–84 (2002).
26. C. Yan, M. S. Jamaluddin, B. Aggarwal, J. Myers, and D. D. Boyd, Gene expression profiling identifies activating transcription factor 3 as a novel contributor to the proapoptotic effect of curcumin. *Mol Cancer Ther* **4**, 233–241 (2005).
27. C. Ramachandran, H. B. Fonseca, P. Jhabvala, E. A. Escalon, and S. J. Melnick, Curcumin inhibits telomerase activity through human telomerase reverse transcritpase in MCF-7 breast cancer cell line. *Cancer Lett* **184**, 1–6 (2002).
28. K. Mehta, P. Pantazis, T. McQueen, and B. B.Aggarwal, Antiproliferative effect of curcumin (diferuloylmethane) against human breast tumor cell lines. *Anticancer Drugs* **8**, 470–481 (1997).

29. E. L. White, L. J. Ross, S. M. Schmid, G. J. Kelloff, V. E. Steele, and D. L. Hill, Screening of potential cancer-preventing chemicals for inhibition of induction of ornithine decarboxylase in epithelial cells from rat trachea. *Oncol Rep* **5**, 717–722 (1998).
30. M. Hasmeda and G. M. Polya, Inhibition of cyclic AMP-dependent protein kinase by curcumin. *Phytochemistry* **42**, 599–605 (1996).
31. M. T. Huang, T. Lysz, T. Ferraro, T. F. Abidi, J. D. Laskin, and A, H, Conney, Inhibitory effects of curcumin on in vitro lipoxygenase and cyclooxygenase activities in mouse epidermis. *Cancer Res* **51**, 813–819 (1991).
32. A. Nagao, M. Maeda, B. P. Lim, H. Kobayashi, and J. Terao, Inhibition of beta-carotene-15,15′-dioxygenase activity by dietary flavonoids. *J Nutr Biochem* **11**, 348–355 (2000).
33. R. L. Hong, W. H. Spohn, and M. C. Hung, Curcumin inhibits tyrosine kinase activity of p185neu and also depletes p185neu. *Clin Cancer Res* **5**, 1884–1891 (1999).
34. S. C. Gautam, Y. X. Xu, K. R. Pindolia, N. Janakiraman, and R. A. Chapman, Non-selective inhibition of proliferation of transformed and nontransformed cells by the anticancer agent curcumin (diferuloylmethane). *Biochem Pharmacol* **55**, 1333–1337 (1998).
35. A. Khar, A. M. Ali, B. V. Pardhasaradhi, C. H. Varalakshmi, R. Anjum, and A. L. Kumari, Induction of stress response renders human tumor cell lines resistant to curcumin-mediated apoptosis: role of reactive oxygen intermediates. *Cell Stress Chaperones* **6**, 368–376 (2001).
36. M. C. Jiang, H. F. Yang-Yen, J. J. Yen, and J. K. Lin, Curcumin induces apoptosis in immortalized NIH 3T3 and malignant cancer cell lines. *Nutr Cancer* **26**, 111–120 (1996).
37. C. Sun, X. Liu, Y. Chen, and F. Liu, Anticancer effect of curcumin on human B cell non-Hodgkin's lymphoma. *J Huazhong Univ Sci Technolog Med Sci* **25**, 404–407 (2005).
38. D. Karunagaran, R. Rashmi, and T. R. Kumar, Induction of apoptosis by curcumin and its implications for cancer therapy. *Curr Cancer Drug Targets* **5**, 117–129 (2005).
39. J. A. Bush, K. J. Cheung, Jr., and G. Li, Curcumin induces apoptosis in human melanoma cells through a Fas receptor/caspase-8 pathway independent of p53. *Exp Cell Res* **271**, 305–314 (2001).
40. L. Moragoda, R. Jaszewski, and A. P. Majumdar, Curcumin induced modulation of cell cycle and apoptosis in gastric and colon cancer cells. *Anticancer Res* **21**, 873–878 (2001).
41. R. Rashmi, T. R. Santhosh Kumar, and D. Karunagaran, Human colon cancer cells differ in their sensitivity to curcumin-induced apoptosis and heat shock protects them by inhibiting the release of apoptosis-inducing factor and caspases. *FEBS Lett* **538**, 19–24 (2003).
42. S. Uddin, A. R. Hussain, P. S. Manogaran, K. Al-Hussein, L. C. Platanias, M. I. Gutierrez, and K. G. Bhatia, Curcumin suppresses growth and induces apoptosis in primary effusion lymphoma. *Oncogene* **24**, 7022–7030 (2005).
43. R. J. Anto, A. Mukhopadhyay, K. Denning, and B. B. Aggarwal, Curcumin (diferuloylmethane) induces apoptosis through activation of caspase-8, BID cleavage and cytochrome c release: its suppression by ectopic expression of Bcl-2 and Bcl-xl. *Carcinogenesis* **23**, 143–150 (2002).
44. M. L. Kuo, T. S. Huang, and J. K. Lin, Curcumin, an antioxidant and anti-tumor promoter, induces apoptosis in human leukemia cells. *Biochim Biophys Acta* **1317**, 95–100 (1996).

45. R. Rashmi, S. Kumar, and D. Karunagaran, Ectopic expression of Hsp70 confers resistance and silencing its expression sensitizes human colon cancer cells to curcumin-induced apoptosis. *Carcinogenesis* **25**, 179–187 (2004).
46. R. Rashmi, S. Kumar, S. and Karunagaran, D. (2004a) Ectopic expression of Bcl-XL or Ku70 protects human colon cancer cells (SW480) against curcumin-induced apoptosis while their down-regulation potentiates it. *Carcinogenesis* **25**, 1867–1877 (2004).
47. M. Shi, Q. Cai, L. Yao, Y. Mao, Y. Ming, and G. Ouyang, Antiproliferation and apoptosis induced by curcumin in human ovarian cancer cells. *Cell Biol Int* (2006).
48. G. Radhakrishna Pillai, A. S. Srivastava, T. I. Hassanein, D. P. Chauhan, and E. Carrier, Induction of apoptosis in human lung cancer cells by curcumin. *Cancer Lett* **208**, 163–170 (2004).
49. R. Anjum and A. Khar, Differential regulation of apoptosis in AK-5 tumor cells by the proto-oncogene Bcl-2: presence of Bcl-2 dependent and independent pathways. *FEBS Lett* **499**, 166–170 (2001).
50. S. Pal, T. Choudhuri, S. Chattopadhyay, A. Bhattacharya, G. K. Datta, T. Das, and G. Sa, Mechanisms of curcumin-induced apoptosis of Ehrlich's ascites carcinoma cells. *Biochem Biophys Res Commun* **288**, 658–665 (2001).
51. R. Rashmi, S. Kumar, and D. Karunagaran, Human colon cancer cells lacking Bax resist curcumin-induced apoptosis and Bax requirement is dispensable with ectopic expression of Smac or downregulation of Bcl-XL. *Carcinogenesis* **26**, 713–723 (2005).
52. G. Song, Y. B. Mao, Q. F. Cai, L. M. Yao, G. L. Ouyang, and S. D. Bao, Curcumin induces human HT-29 colon adenocarcinoma cell apoptosis by activating p53 and regulating apoptosis-related protein expression. *Braz J Med Biol Res* **38**, 1791–1798 (2005).
53. L. D. Zheng, Q. S. Tong, K. X. Tao, L. Wang, and B. Zhang, [Effects of Smac gene overexpression on chemotherapeutic sensitivity of gastric cancer cell line MKN-45]. *Ai Zheng* **23**, 361–366 (2004).
54. Q. Wang, A. Y. Sun, A. Simonyi, M. D. Jensen, P. B. Shelat, G. E. Rottinghaus, R. S. MacDonald, D. K. Miller, D. E. Lubahn, G. A. Weisman, and G. Y. Sun, Neuroprotective mechanisms of curcumin against cerebral ischemia-induced neuronal apoptosis and behavioral deficits. *J Neurosci Res* 82, 138–148 (2005).
55. A. Sood, R. Mathew, and H. Trachtman, Cytoprotective effect of curcumin in human proximal tubule epithelial cells exposed to shiga toxin. *Biochem Biophys Res Commun* **283**, 36–41 (2001).
56. M. Shakibaei, G. Schulze-Tanzil, T. John, and A. Mobasheri, Curcumin protects human chondrocytes from IL-l1beta-induced inhibition of collagen type II and beta1-integrin expression and activation of caspase-3: An immunomorphological study. *Ann Anat* **187**, 487–497 (2005).
57. S. Fujisawa, T. Atsumi, M. Ishihara, and Y. Kadoma, Cytotoxicity, ROS-generation activity and radical-scavenging activity of curcumin and related compounds. *Anticancer Res* **24**, 563–569 (2004).
58. A. C. Reddy and B. R. Lokesh, Studies on the inhibitory effects of curcumin and eugenol on the formation of reactive oxygen species and the oxidation of ferrous iron. *Mol Cell Biochem* **137**, 1–8 (1994).
59. V. K. Shalini and L. Srinivas, Lipid peroxide induced DNA damage: Protection by turmeric (*Curcuma longa*). *Mol Cell Biochem* **77**, 3–10 (1987).
60. M. Subramanian, M. N. Sreejayan, Rao, T. P. Devasagayam, and B. B. Singh, Diminution of singlet oxygen-induced DNA damage by curcumin and related antioxidants. *Mutat Res* **311**, 249–255 (1994).

61. A. C. Reddyand B. R. Lokesh, Effect of dietary turmeric (*Curcuma longa*) on iron-induced lipid peroxidation in the rat liver. *Food Chem Toxicol* **32**, 279–283 (1994).
62. E. L. White, L. J. Ross, S. M. Schmid, G. J. Kelloff, V. E. Steele, and D. L. Hill, Screening of potential cancer preventing chemicals for induction of glutathione in rat liver cells. *Oncol Rep* **5**, 507–512 (1998).
63. H. Ahsan, N. Parveen, N. U. Khan, and S. M. Hadi, Pro-oxidant, anti-oxidant and cleavage activities on DNA of curcumin and its derivatives demethoxycurcumin and bisdemethoxycurcumin. *Chem Biol Interact* **121**, 161–175 (1999).
64. S. Bhaumik, R. Anjum, N. Rangaraj, B. V. Pardhasaradhi, and A. Khar, Curcumin mediated apoptosis in AK-5 tumor cells involves the production of reactive oxygen intermediates. *FEBS Lett* **456**, 311–314 (1999).
65. U. Jacobi, T. Tassopoulos, C. Surber, and J. Lademann, Cutaneous distribution and localization of dyes affected by vehicles all with different lipophilicity. *Arch Dermatol Res* **297**, 303–310 (2006).
66. A. Paradkar, A. A. Ambike, B. K. Jadhav, and K R. Mahadik,) Characterization of curcumin-PVP solid dispersion obtained by spray drying. *Int J Pharm* **271**, 281–286 (2004).
67. K. U. Schallreuter and H. Rokos, Turmeric (curcumin): A widely used curry ingredient, can contribute to oxidative stress in Asian patients with acute vitiligo. *Indian J Dermatol Venereol Leprol* **72**, 57–59 (2006).
68. M. Moussavi, K. Assi, A. Gomez-Munoz, and B. Salh, Curcumin mediates ceramide generation via the de novo pathway in colon cancer cells. *Carcinogenesis* **27**, 1636–1644 (2006).
69. C. Syng-Ai, A. L. Kumari, and A. Khar, Effect of curcumin on normal and tumor cells: role of glutathione and bcl-2. *Mol Cancer Ther* **3**, 1101–1108 (2004).
70. S. Somasundaram, N. A. Edmund, D. T. Moore, G. W. Small, Y. Y. Shi, and R. Z. Orlowski, Dietary curcumin inhibits chemotherapy-induced apoptosis in models of human breast cancer. *Cancer Res* **62**, 3868–3875 (2002).
71. M. S. Ahmad, Sheeba, and M Afzal, Amelioration of genotoxic damage by certain phytoproducts in human lymphocyte cultures. *Chem Biol Interact* **149**, 107–115 (2004).
72. E. Sikora, A. Bielak-Zmijewska, K. Piwocka, J. Skierski, J. and E. Radziszewska, Inhibition of proliferation and apoptosis of human and rat T lymphocytes by curcumin, a curry pigment. *Biochem Pharmacol* **54**, 899–907 (1997).
73. B. B. Aggarwal, Y. Takada, and O. V. Oommen, From chemoprevention to chemotherapy: common targets and common goals. *Expert Opin Invest Drugs* **13**, 1327–1338 (2004).
74. C. Jobin, C. A. Bradham, M. P. Russo, B. Juma, A. S. Narula, D. A. Brenner, and R. B. Sartor, Curcumin blocks cytokine-mediated NF-kappa B activation and proinflammatory gene expression by inhibiting inhibitory factor I-kappa B kinase activity. *J Immunol* **163**, 3474–3483 (1999).
75. S. Singh and B. B. Aggarwal, Activation of transcription factor NF-kappa B is suppressed by curcumin (diferuloylmethane) [corrected]. *J Biol Chem* **270**, 24,995–25,000 (1995).
76. S. Aggarwal, H. Ichikawa, Y. Takada, S. K. Sandur, S. Shishodia, and B. B. Aggarwal, Curcumin (diferuloylmethane) down-regulates expression of cell proliferation and antiapoptotic and metastatic gene products through suppression of IkappaBalpha kinase and Akt activation. *Mol Pharmacol* **69**, 195–206 (2006).

77. D. R. Siwak, S. Shishodia, B. B. Aggarwal, and R. Kurzrock, Curcumin-induced antiproliferative and proapoptotic effects in melanoma cells are associated with suppression of IkappaB kinase and nuclear factor kappaB activity and are independent of the B-Raf/mitogen-activated/extracellular signal-regulated protein kinase pathway and the Akt pathway. *Cancer* **104**, 879–890 (2005).
78. A. Duvoix, F. Morceau, S. Delhalle, M. Schmitz, M. Schnekenburger, M. M. Galteau, M. Dicato, and M. Diederich, Induction of apoptosis by curcumin: mediation by glutathione S-transferase P1-1 inhibition. *Biochem Pharmacol* **66**, 1475–1483 (2003).
79. B. B. Aggarwal, S. Shishodia, Y. Takada, S. Banerjee, R. A. Newman, C. E. Bueso-Ramos, and P. E. Price, Curcumin suppresses the paclitaxel-induced nuclear factor-kappaB pathway in breast cancer cells and inhibits lung metastasis of human breast cancer in nude mice. *Clin Cancer Res* **11**, 7490–7498 (2005).
80. R. J. Anto, T. T. Maliekal, and D. Karunagaran, L-929 cells harboring ectopically expressed RelA resist curcumin-induced apoptosis. *J Biol Chem* **275**, 15,601–15,604 (2000).
81. D. Deeb, H. Jiang, X. Gao, M. S. Hafner, H. Wong, G. Divine, R. A. Chapman, S. A. Dulchavsky, andS. C. Gautam, Curcumin sensitizes prostate cancer cells to tumor necrosis factor-related apoptosis-inducing ligand/Apo2L by inhibiting nuclear factor-kappaB through suppression of IkappaBalpha phosphorylation. *Mol Cancer Ther* **3**, 803–812 (2004).
82. D. Deeb, Y. X. Xu, H. Jiang, X. Gao, N. Janakiraman, R. A. Chapman, and S. C. Gautam, Curcumin (diferuloyl-methane) enhances tumor necrosis factor-related apoptosis-inducing ligand-induced apoptosis in LNCaP prostate cancer cells. *Mol Cancer Ther* **2**, 95–103 (2003).
83. S.V. Bava, V. T. Puliappadamba, A. Deepti, A. Nair, D. Karunagaran, and R. J. Anto, Sensitization of taxol-induced apoptosis by curcumin involves down-regulation of nuclear factor-kappaB and the serine/threonine kinase Akt and is independent of tubulin polymerization. *J Biol Chem* 280, 6301–6308 (2005).
84. M. Shi, Q. Cai, L. Yao, Y. Mao, Y. Ming,.and G. Ouyang, Antiproliferation and apoptosis induced by curcumin in human ovarian cancer cells. *Cell Biol Int* **30**, 221–226 (2006).
85. W. Henke, K. Ferrell, D. Bech-Otschir, M. Seeger, R. Schade, P. Jungblut, M. Naumann, and W. Dubiel,) Comparison of human COP9 signalsome and 26S proteasome lid'. *Mol Biol Rep* **26**, 29–34 (1999).
86. J. Rajasingh, H. P. Raikwar, G. Muthian, C. Johnson, and J. J. Bright, Curcumin induces growth-arrest and apoptosis in association with the inhibition of constitutively active JAK-STAT pathway in T cell leukemia. *Biochem Biophys Res Commun* **340**, 359–368 (2006).
87. A. C. Bharti, N. Donato, and B. B. Aggarwal, Curcumin (diferuloylmethane) inhibits constitutive and IL-6-inducible STAT3 phosphorylation in human multiple myeloma cells. *J Immunol* **171**, 3863–3871 (2003).
88. C. S. Divya and M. R. Pillai, Antitumor action of curcumin in human papillomavirus associated cells involves downregulation of viral oncogenes, prevention of NFkB and AP-1 translocation, and modulation of apoptosis. *Mol Carcinog* **45**, 320–332 (2006).
89. M. Tomita, H. Kawakami, J. N. Uchihara, T. Okudaira, M. Masuda, N. Takasu, T. Matsuda, T. Ohta, Y. Tanaka, and N. Mori, Curcumin suppresses constitutive activation of AP-1 by downregulation of JunD protein in HTLV-1-infected T-cell lines. *Leuk Res* **30**, 313–321 (2006).

90. C. Bernt, T. Vennegeerts, U. Beuers, and C. Rust, The human transcription factor AP-1 is a mediator of bile acid-induced liver cell apoptosis. *Biochem Biophys Res Commun* **340**, 800–806 (2006).
91. Y. R. Chen and T. H. Tan, Inhibition of the c-Jun N-terminal kinase (JNK) signaling pathway by curcumin. *Oncogene* **17**, 173–178 (1998).
92. A. R. Hussain, M. Al-Rasheed, P. S. Manogaran, K. A. Al-Hussein, L. C. Platanias, K. A. Kuraya, and S. Uddin, Curcumin induces apoptosis via inhibition of PI3'-kinase/AKT pathway in Acute T cell Leukemias. *Apoptosis* **11**, 245–254 (2006).
93. L. R. Chaudhary and K. A. Hruska, Inhibition of cell survival signal protein kinase B/Akt by curcumin in human prostate cancer cells. *J Cell Biochem* **89**, 1–5 (2003).
94. L. Korutla and P. Kumar, Inhibitory effect of curcumin on epidermal growth factor receptor kinase activity in A431 cells. *Biochim Biophys Acta* **1224**, 597–600 (1994).
95. A. Chen, J. Xu, and A. C. Johnson, Curcumin inhibits human colon cancer cell growth by suppressing gene expression of epidermal growth factor receptor through reducing the activity of the transcription factor Egr-1. *Oncogene* **25**, 278–287 (2006).
96. D. W. Scott and G. Loo, Curcumin-induced GADD153 gene up-regulation in human colon cancer cells. *Carcinogenesis* **25**, 2155–2164 (2004).
97. G P. Collett and F. C. Campbell, Curcumin induces c-jun N-terminal kinase-dependent apoptosis in HCT116 human colon cancer cells. *Carcinogenesis* **25**, 2183–2189 (2004).
98. S. Zheng and A. Chen, Activation of PPARgamma is required for curcumin to induce apoptosis and to inhibit the expression of extracellular matrix genes in hepatic stellate cells in vitro. *Biochem J* **384**, 149–157 (2004).
99. S. Zheng and A. Chen, Curcumin suppresses the expression of extracellular matrix genes in activated hepatic stellate cells by inhibiting gene expression of connective tissue growth factor. *Am J Physiol Gastrointest Liver Physiol* **290**, G883–893 (2005).
100. L. Zheng, Q. Tong, and C. Wu, Growth-inhibitory effects of curcumin on ovary cancer cells and its mechanisms. *J Huazhong Univ Sci Technolog Med Sci* **24**, 55–58 (2004).
101. S. Sen, H. Sharma, and N. Singh, Curcumin enhances Vinorelbine mediated apoptosis in NSCLC cells by the mitochondrial pathway. *Biochem Biophys Res Commun* **331**, 1245–1252 (2005).
102. T. C. Hour, J. Chen, C. Y. Huang, J. Y. Guan, S. H. Lu, and Y. S. Pu, Curcumin enhances cytotoxicity of chemotherapeutic agents in prostate cancer cells by inducing p21(WAF1/CIP1) and C/EBPbeta expressions and suppressing NF-kappaB activation. *Prostate* **51**, 211–218 (2002).
103. M. M. Chan, D. Fong, K. J. Soprano, W. F. Holmes, and H. Heverling, Inhibition of growth and sensitization to cisplatin-mediated killing of ovarian cancer cells by polyphenolic chemopreventive agents. *J Cell Physiol* **194**, 63–70 (2003).
104. J. Y. Koo, H. J. Kim, K. O. Jung, and K. Y. Park, Curcumin inhibits the growth of AGS human gastric carcinoma cells in vitro and shows synergism with 5-fluorouracil. *J Med Food* **7**, 117–121 (2004).
105. B. Du, L. JiangQ. Xia, and L. Zhong, Synergistic inhibitory effects of curcumin and 5-fluorouracil on the growth of the human colon cancer cell line HT-29. *Chemotherapy* **52**, 23–28 (2006).
106. P. Limtrakul, S. Anuchapreeda, and D. Buddhasukh, Modulation of human multidrug-resistance MDR-1 gene by natural curcuminoids. *BMC Cancer* **4**, 13 (2004).
107. S. E. Chuang, P. Y. Yeh, Y. S. Lu, G. M. Lai, C. M. Liao, M. Gao, and A. L. Cheng, Basal levels and patterns of anticancer drug-induced activation of nuclear factor-kappaB

(NF-kappaB), and its attenuation by tamoxifen, dexamethasone, and curcumin in carcinoma cells. *Biochem Pharmacol* **63,** 1709–1716 (2002).

108. S. Lev-Ari, L. Strier, D. Kazanov, O. Elkayam, D. Lichtenberg, D. Caspi, and N. Arber, Curcumin synergistically potentiates the growth-inhibitory and pro-apoptotic effects of celecoxib in osteoarthritis synovial adherent cells. *Rheumatology (Oxford)* **45**, 171–177 (2006).
109. T. O. Khor, Y. S. Keum, W. Lin, J. H. Kim, R. Hu, G. Shen, C. Xu, A. Gopalakrishnan, B. Reddy, X. Zheng, A. H. Conney, and A. N. Kong, Combined inhibitory effects of curcumin and phenethyl isothiocyanate on the growth of human PC-3 prostate xenografts in immunodeficient mice. *Cancer Res* **66**, 613–621 (2006).
110. J. H. Kim, C. Xu, Y. S. Keum, B. Reddy, A. Conney, and A. N. Kong, Inhibition of EGFR signaling in human prostate cancer PC-3 cells by combination treatment with {beta}-phenylethyl isothiocyanate and curcumin. *Carcinogenesis* **27**, 475–482 (2006).
111. A. Spingarn, P. G. Sacks, D. Kelley, A. J. Dannenberg, and S. P. Schantz,) Synergistic effects of 13-cis retinoic acid and arachidonic acid cascade inhibitors on growth of head and neck squamous cell carcinoma in vitro. *Otolaryngol Head Neck Surg* **118**, 159–164 (1998).
112. Y. Liu, R. L. Chang, X. X. Cui, H. L. Newmark, and A. H. Conney, Synergistic effects of curcumin on all-trans retinoic acid- and 1 alpha,25-dihydroxyvitamin D3-induced differentiation in human promyelocytic leukemia HL-60 cells. *Oncol Res* **9**, 19–29 (1997).
113. D. Chendil, R. S. Ranga, D. Meigooni, S. Sathishkumar, and M. M. Ahmed, Curcumin confers radiosensitizing effect in prostate cancer cell line PC-3. *Oncogene* **23**, 1599–1607 (2004).
114. C. Ramachandran, S. Rodriguez, R. Ramachandran, P. K. Raveendran Nair, H. Fonseca, Z. Khatib, E. Escalon, and S. J. Melnick, Expression profiles of apoptotic genes induced by curcumin in human breast cancer and mammary epithelial cell lines. *Anticancer Res* **25**, 3293–3302 (2005).
115. B. K. .Adams, J. Cai, J. Armstrong, M. Herold, Y. J. Lu, A. Sun, J. P. Snyder, D. C. Liotta, D.P. Jones, and M. Shoji, EF24, a novel synthetic curcumin analog, induces apoptosis in cancer cells via a redox-dependent mechanism. *Anticancer Drugs* **16**, 263–275 (2005).
116. M. S. Furness, T. P. Robinson, T. Ehlers, R. B. T. Hubbard, J. L. Arbiser, D. J. Goldsmith, and J. P.Bowen, Antiangiogenic agents: Studies on fumagillin and curcumin analogs. *Curr Pharm Des* **11**, 357–373 (2005).
117. Q. S. Tong, L. D. Zheng, P. Lu, F. C. Jiang, F. M. Chen, F. Q. Zeng, L. Wang, and J. H. Dong, Apoptosis-inducing effects of curcumin derivatives in human bladder cancer cells. *Anticancer Drugs* **17**, 279–287 (2006).
118. W. M. Weber, L. A. Hunsaker, S. F. Abcouwer, L. M. Deck, and D. L.Vander Jagt, Antioxidant activities of curcumin and related enones. *Bioorg Med Chem* **13**, 3811–3820 (2005).
119. K. M. Youssef and M. A. El-Sherbeny, Synthesis and antitumor activity of some curcumin analogs. *Arch Pharm (Weinheim)* **338**, 181–189 (2005).

CURCUMIN AS CHEMOSENSITIZER

Pornngarm Limtrakul

Abstract: This overview presents curcumin as a significant chemosensitizer in cancer chemotherapy. Although the review focuses on curcumin and its analogues on multidrug resistance (MDR) reversal, the relevance of curcumin as a nuclear factor (NF)-κB blocker and sensitizer of many chemoresistant cancer cell lines to chemotherapeutic agents will also be discussed. One of the major mechanisms of MDR is the enhanced ability of tumor cells to actively efflux drugs, leading to a decrease in cellular drug accumulation below toxic levels. Active drug efflux is mediated by several members of the ATP-binding cassette (ABC) superfamily of membrane transporters, which have now been subdivided into seven families designated A through G. Among these ABC families, the classical MDR is attributed to the elevated expression of ABCB1 (Pgp), ABCC1 (MRP1), and ABCG2 (MXR). The clinical importance of Pgp, MRP1, and MXR for MDR and cancer treatment has led to the investigation of the inhibiting properties of several compounds on these transporters. At present, due in part to the disappointing results associated with the many side effects of synthetic modulators that have been used in clinical trials, current research efforts are directed toward the identification of novel compounds, with attention to dietary natural products. The advantage is that they exhibit little or virtually no side effects and do not further increase the patient's medication burden.

1. INTRODUCTION

This chapter deals with the theoretical background of drug resistance in cancer chemotherapy and clinical significance of the search for chemosensitizers of multidrug resistance (MDR) in cancer. After discussing the basic features of the drug transporter proteins P-glycoprotein (Pgp), multidrug resistance protein-1(MRP1), and mitoxantrone resistance protein (MXR) responsible for this phenomenon, the possible mechanism of action of MDR chemosensitizers is reviewed. This chapter also discusses apoptosis and chemotherapy resistance.

This overview presents curcumin as a significant chemosensitizer in cancer chemotherapy. Although the review focuses on curcumin and its analogues on MDR reversal, the relevance of curcumin as a nuclear factor (NF)-κB blocker and sensitizer of many chemoresistant cancer cell lines to chemotherapeutic agents will also be discussed.

One of the major mechanisms of MDR is the enhanced ability of tumor cells to actively efflux drugs, leading to a decrease in cellular drug accumulation below toxic levels. Active drug efflux is mediated by several members of the ATP-binding cassette (ABC) superfamily of membrane transporters, which have now been subdivided into seven families designated A through G.[1] Among these ABC families, the classical MDR is attributed to the elevated expression of ABCB1 (Pgp), ABCC1 (MRP1), and ABCG2 (MXR).

The clinical importance of Pgp, MRP1, and MXR for MDR and cancer treatment has led to the investigation of the inhibiting properties of several compounds on these transporters. The calcium channel blocking agent verapamil was the first drug described as an inhibitor of the Pgp efflux mechanism.[2] After this discovery, several other compounds have been studied for their inhibitory effects (e.g., valspodar, GF120918, and LY335979).[3,4] Although these agents are effective, one of the major problems with most of them is that the *in vivo* plasma concentrations required to inhibit Pgp are too high and result in severe toxic side effects. At present, due in part to the disappointing results associated with the many side effects of modulators that have been used in clinical trials, current research efforts are directed toward the identification of novel compounds, with attention to dietary natural products. The advantage is that they exhibit little or virtually no side effects and do not further increase the patient's medication burden.

Curcumin has been described as a potent antioxidant and anti-inflammatory agent. The compound has been found to be pharmacologically safe: Human clinical trials indicated no doses-limiting toxicity when administered at doses up to 10 g/day.[5] All of these studies suggest that curcumin has enormous potential in the prevention and therapy of cancer. However, a better understanding of the mechanism would enhance the therapeutic potential of curcumin either alone or in combination with chemotherapy. The study reported by our group demonstrated that curcumin (curcumin I), demethoxycurcumin (curcumin II), and bisdemethoxycurcumin (curcumin III) are potent chemosensitizers of Pgp, MRP1, and MXR, and curcumin I was the most effective form. Tetrahydrocurcumin (THC), a major metabolite of curcumin, is also a good chemosensitizer of Pgp, MRP1, and MXR, and it is able to extend the MDR-reversing properties of curcuminoids *in vivo*.

There is increasing evidence that the inability of the cells to undergo apoptosis might critically contribute to the genesis and progression of cancer and represents an important cause of tumor drug resistance.[6] Tumor cells often evade apoptosis by overexpressing antiapoptotic proteins such as Bcl-2, NF-κB, Akt, and so forth, which give them a survival advantage. Some conventional chemotherapeutic drugs in low concentrations cause upregulation of survival signals, thereby necessitating increments of the effective dose of treatment. Activation of NF-κB has been shown to block apoptosis and promote proliferation; therefore, NF-κB activation induces resistance to chemotherapeutic agents. Thus, agents that induce apoptosis and stimulate NF-κB activity might be effective if given in combination with agents that could inhibit NF-κB. Evolving interest in recent years has focused on phytochemicals augmenting apoptosis as possible candidates for the evaluation of their synergistic efficacy in combination with chemotherapeutic agents. Common biological modulators, including curcumin, have been researched in order to block NF-κB activation as described in this chapter.

2. MECHANISMS OF DRUG RESISTANCE IN CANCER CHEMOTHERAPY

2.1. Drug Resistance in Cancer

The presence or development of resistance to anticancer drugs is the main cause of failure of chemotherapy in the majority of the most common forms of cancer (e.g. lung, colon, breast, and cervix). Resistance to chemotherapeutic drugs has already been present at diagnosis or it can develop after chemotherapy treatment. These two forms of drug resistance are respectively called intrinsic and acquired resistance.[7,8] Intrinsic resistance or *de novo* resistance of cancer cells can be present before chemotherapy, resulting in initial treatment failure such as Hodgkin's disease, testicular cancer, and acute childhood leukemia, but acquired resistance can develop in response to chemotherapeutic intervention, eventually leading to early disease progression despite an initial treatment response (e.g., lymphoma and breast cancer).[9] In both intrinsic and acquired resistance, tumors are often found to be refractory to a variety of drugs with different structures and functions. A similar experimental phenomenon has been termed *multidrug resistance*. MDR can be the result of a variety of mechanisms that are not fully understood.[10] The most important among them are the following: (1) altered membrane transport either by decreased drug uptake or by increased drug efflux[11]; (2) perturbed expression of target enzymes or altered target enzymes[12]; (3) altered drug activation or degradation[13]; (4) enhanced DNA repair[14]; and (5) failure to undergo apoptosis.[15,16] Some of these mechanisms of drug resistance might coexist; however, the most widely implicated mechanism is that concerned with altered membrane transport in tumor cells. This mechanism is often referred to as typical or classical MDR.

2.2. Multidrug Resistance and Drug Transporter Proteins

The human *MDR1* gene product, P-glycoprotein, was the first ATP-dependent system discovered that was implicated in MDR. P-Glycoprotein (also known as Pg-170, Pgp, P-170, or ABCB1) was isolated[17] and proposed to be the transporter protein that pumps out the antitumor agents.[18,19] The overexpression of P-glycoprotein is not the only cause of MDR. Another member of the ATP-binding cassette (ABC) superfamily, which is involved in MDR, is the 190-kDa multidrug resistance-associated protein (MRP1 or ABCC1), encoded by the MRP1 gene. MRP1 is simlar to P-glycoprotein in its capability of decreasing intracellular levels of drugs and is ATP-dependent.[20] The most recently discovered

ABC drug efflux transporter is breast cancer resistance protein (BCRP, MXR, or ABCG2).[21]

Among these ABC families, the classical MDR is attributed to the elevated expression of ABCB1 (Pgp), ABCC1 (MRP1), and ABCG2 (BCRP or MXR).[22,23]

2.2.1. P-Glycoprotein

In various cancer types, such as acute myeloid leukemia, various childhood tumors and locoregionally advanced breast cancer, overexpression of *MDR1*-Pgp has been found to correlate with poor outcome in patients treated with chemotherapy.[24–28] These findings have been interpreted as an indication of Pgp-mediated drug resistance. Various clinical studies have suggested that Pgp-positivity is associated with more aggressive tumor behavior. In colon cancer, Pgp was found to be expressed predominantly in the tumor cells at the invading edge of primary tumors, and Pgp-positivity in primary tumors was associated with a higher incidence of lymph node metastases.[29] In renal cell carcinoma, Pgp-positivity was found significantly more often in invasive than in noninvasive tumors.[30] In primary breast cancer, overexpression of *MDR*/Pgp seems to be more common in advanced locoregional disease than it is in small tumors.[31,32]

Increased expression of Pgp, the product of the human *MDR1* gene, is a well-characterized mechanism used by cancer cells to evade the cytotoxic action of anticancer drugs. Twenty years ago, Juliano and Ling discovered that Pgp was the most ubiquitous marker in MDR cells.[17] P-Glycoprotein (P refers to its proposed role in modulating cellular permeability to drugs) is a high-molecular-weight integral plasma membrane glycoprotein that confers MDR to mammalian cells by acting as an energy-dependent drug efflux pump. P-Glycoproteins are encoded by small gene families, with two members in humans and three in rodents.[33] Despite a high amino acid sequence identity of more than 70% among all Pgps, *MDR* gene products are subdivided into two classes: class I and class II. Overexpression of class I P-glycoprotein causes cancer cells to become resistant to a variety of anticancer drugs (e.g., vinblastine, vincristine, daunorubicin, etoposide, teniposide, and paclitaxel) as well as many other cytotoxic agents, including colchicines, emetine, ethidium bromide, puromycin, and mithramycin.[34] Class II *MDR* gene products are predominantly expressed in the liver bile canaliculi.[35–37]

2.2.1a. Structure of P-Glycoprotein

Mammalian P-glycoproteins are single-chain proteins and consist of approximately 1280 amino acid residues (170 kDa). P-Glycoproteins are composed of 43% sequence homology; between the two halves, there is a hydrophobic, membrane-associated domain (approximately 250 amino acid residues) followed by a hydrophilic nucleotide-binding fold (approximately 300 amino acid residues).[38] These two halves are connected by a linked peptide of approximately 75 amino acids defined as amino acids 633–709 in human Pgp. This peptide conjugated, commonly called the linker region, is highly charged and contains the *in vivo* sites of phosphorylation.

Both the N-terminal membrane-associated domains and the C-terminal membrane-associated domains of human P-glycoprotein harbor six predicted trans-membrane (TM) regions. The N-terminus, the C-terminus as well as the nucleotide-binding folds are located intracellularly. The first extracellular loop is glycosylated. This 12-TM-region model of P-glycoprotein is supported experimentally by cellular epitope localization data obtained from antibodies that specifically recognize the N- or C-terminus of P-glycoprotein, its first and fourth extracellular loop, or the two ATP-binding sites. The two halves of Pgp are essential for activity of the transporter as measured by its ability to confer drug resistance or drug-stimulated ATPase activity. Both transmembrane domains 5, 6 and 11, 12 and the extracellular loops connecting them were determined by photoaffinity labeling, with the Pgp substrate analogues being the major sites of drug interaction. These transmembranes are important determinants in the drug-binding site(s), but they do not offer any insight into whether these sites are autonomous or interdependent.

2.2.1b. Mechanism of Pgp-Mediated Drug Resistance

Discovery of the molecular mechanisms by which Pgp exert its action has been one of the major tasks of research in the field of MDR. Pgp substrates can structurally be very different; however, the physical properties shared by many of them include high hydrophobicity, an amphiphilic nature, and a net positive charge, although neutral compounds, among them hydrophobic peptides, have also been described as substrates of Pgp.[39]

Whereas Pgp fulfills critical functions in transport processes involved in normal physiology, overexpression of this protein in tumor cells results in reduced intracellular accumulation of anticancer agents due to increased drug efflux. Most models of Pgp suggest that it transports drugs across cell membranes in a manner analogous to that defined for active transport proteins. This model predicts that substrates (cytotoxic drugs) bind to specific domains of the protein, which subsequently undergoes an energy-dependent conformational change. This change allows the substrate to be released on the exterior side of the membrane. Complementary models have been proposed, suggesting that (1) Pgp interacts directly with substrates in the plasma membrane (the "hydrophobic vacuum cleaner" model[38] or (2) Pgp might be involved in the transport of drugs from the inner leaflet to outer leaflet of the plasma membrane, from which they diffuse (the flippase model).[40] Identification and characterization of Pgp segments responsible for drug recognition and binding indicate that Pgp interacts directly with drug molecules. Efforts to map the drug-binding domains of Pgp by photoaffinity drug analogues and site-directed mutagenesis indicate that Pgp contains multiple nonoverlapping or partially overlapping drug-binding sites, each having different affinities for different drugs or classes of drugs.[41–43] The two nucleotide-binding domains (NBDs) are a critical feature of Pgp. Reconstitution studies with purified Pgp have shown that transport of hydrophobic substrates against a concentration gradient is coupled to ATP hydrolysis.[44] However, the mechanism by which Pgp couples ATP energy to translocation and efflux of a diverse range of substrates is a largely unresolved debate.[45] Both NBDs can hydrolyze nucleotides, and their ATPase activity, that can be blocked by vanadate, is necessary for drug transport.[46]

Finally, Pgp is phosphorylated by protein kinase C (PKC), and PKC blockers reduce Pgp phosphorylation and increase drug accumulation. These observations suggest that phosphorylation of Pgp stimulates drug transport. However, there is evidence that PKC inhibitors directly interact with Pgp and inhibit drug transport by a mechanism independent of Pgp phosphorylation.[47,48]

2.2.2. Multidrug Resistance Protein-1

The MRP family entered drug resistance in 1992 when Susan Cole and Roger Deeley cloned the multidrug resistance-associated protein gene, now known as MRP1 and was classified to ABCC1.[20] Since then 13 genes for ABCC family have been reported and designated ABCC1 to ABCC13. In 2002, Yabuuchi et al. reported that ABCC13 is predicted to encode a nonfunction protein.[49] As a result, the ABCC family contains only 12 functional proteins. The discovery of the MRP family has considerably broadened the study of MDR in tumor cells and has led to widespread interest in the possible function(s) of the members of this family in normal metabolism.

MRP1 is broadly expressed in the epithelial cells of multiple tissues, including the digestive, urogenital, and respiratory tracts, endocrine glands, and the hematopoietic system.[50] MRP1 expression has been demonstrated in multiple tumor tissues and has been implicated as a component of the MDR phenomenon in leukemia and cancers of the lung, colon, breast, bladder, and prostate.

2.2.2a. Structure of MRP1

The MDR-associated protein (MRP1) is a 190-kDa protein encoded by the *mrp1* gene and is constituted by 1531 amino acids presenting N-linked glycosylation sites.[51] Although the human genome encodes only two Pgps, it contains many genes related to MRP.[52] The protein is predominantly localized to the plasma membrane in drug-resistant cells, with detectable levels present in intracellular membrane compartments of some cell types.[53] Whereas Pgp transports neutral and positively charged molecules in their unmodified form, MRP1 overexpression is associated with an increased ATP-dependent glutathione (GSH) *S*-conjugate transport activity. MRP1 is able to transport a range of substrates as such or conjugated to GSH, glucuronide, and sulfate.[54–56] The predicted topology of MRP1 and several related ABCC proteins differs from that of most eukaryotic ABC transporters, which are composed of two membrane-spanning domains (MSDs), each containing six transmembrane (TM) domains with two NBD sites. MRP1 has an additional NH_2-terminal domain, MSD1, with five TMs and an extracellular NH_2-terminus. Thus, MRP1 is predicted to contain three MSDs with $5 + 6 + 6$ TM helices.[57,58]

2.2.2b. Mechanism of MRP1-Mediated Drug Resistance

It has been proposed that a physiological function of MRP1 is the extrusion of endogenously formed GSH-dependent detoxification products to prevent cellular damage.[55] The generation of *Mrp1 −/−* knockout mice has significantly contributed to the understanding of the physiological role of MRP1. Similar to most ABC transporters, MRP1 requires ATP hydrolysis for its transport; the interaction of ATP with MRP1 was studied by photoaffinity labeling and vanadate-induced trapping experiments using ^{32}P-labeled 8-azido-ATP. The two NBDs show cooperativity in the binding and trapping of the nucleotide.[59,60] Experiments with membrane vesicles from MRP1-overexpressing cells demonstrated that MRP1 is a transporter for the unmodified anticancer drugs vincristine and daunorubicin, but only in the presence of physiological amounts of GSH.[54,61] These results extend the earlier observations that GSH is a critical factor in MRP1-mediated drug resistance. MRP1 transports a wide variety of subtrates that include drugs conjugate with GSH (GS-X pump), glucuronide, and sulfate and some anticancer drugs such as anthracyclines, vinca alkaloids, and epipodophyllotoxin. It has been proposed recently that MRP1 might interact with GSH by at least four different mechanisms.[62] First, GSH might be a direct low-affinity substrate for MRP1 ($K_m \sim 10$ mM). Second, GSH is required for the cotransport of certain MRP1 substrates (e.g., in the case of daunorubicin, vincristine, and aflatoxin) ($K_m \sim 0.1$ mM). Third, GSH stimulates the transport of certain compounds on MRP1, but it is not transported; finally, the transport of GSH is accelerated by certain compounds that are not themselves a substrate for MRP1.

2.2.3. Mitoxantrone Resistance Protein

ABCG2 was first cloned and sequenced from mitoxantrone-resistant S1-M1-80 human colon carcinoma cells and from MCF-7 AdrVp human breast cancer cells selected in doxorubicin (adriamycin).[63,64] This gene is designated ABCG2 by the new nomenclature system but is also referred to as BCRP (breast cancer resistance protein),[64] MXR (mitoxantrone resistance protein),[63] or ABCP (placenta-specific ABC transporter).[65]

2.2.3a. Structure of MXR

The human *ABCG2* gene is located on chromosome 4q22 and encodes a 655-amino-acid polypeptide with a predicted molecular weight of 72 kDa. Therefore, ABCG2 is proposed to be a half-transporter, containing only one set of six TM domains and one NBD site.[65]

Recently, it has been reported that amino acid 482 is an important determinant of substrate recognition by ABCG2.[66] For example, wild-type MXR with an Arg at position 482 does not transport daunorubicin, rhodamine123, and lyso-tracker green; however, these compounds can be transported by mutants with a Thr (T) or Gly (G) at this position.[67] On the other hand, substances such as mitoxantrone, bodipy-prazosin, and Hoechst 33342 are substrates of both wild-type MXR and the two mutants.[67,68] Recently, Miwa et al. generated a large number of mutants in the TM segments and examined the effect of these amino acid substitutions on drug resistance conferred by ABCG2.[69] They found that amino acid substitutions of Glu at position 446, which is predicted to be located within or proximal to the TM2 of ABCG2, resulted in a complete loss of drug resistance to SN-38 and mitoxantrone. Cells transfected with mutant ABCG2 cDNA with substitution of Asn residue at position 557 to Asp (N557D) exhibited comparable resistance to mitoxantrone but significantly reduced resistance to SN-38 relative to wild-type protein. Position 557 is predicted to be located within or proximal to the TM5 segment. These data again provided strong evidence that the drug-binding sites are likely located in the MSD; therefore, amino acids in or proximal to the TM segments are important for substrate recognition by ABCG2. Alternatively, amino acid substitutions in the TM segments might alter the substrate recognition and/or translocation pathway of the protein. Position 557 is a putative N-glycosylation site of ABCG2. Whether glycosylation is important for ABCG2 function is not known at the present time.

2.2.3b. Mechanism of ABCG2-Mediated Drug Resistance

ABCG2 is endogenously expressed at high levels in human placenta and to a lesser extent in the liver, small intestine and colon, ovary, vein and capillary endothelia, kidney, adrenal, and lung, with little to no expression in the brain, heart, stomach, prostate, spleen, and cervix.[23,64,65] Based on its localization, it has been suggested that the physiological roles of ABCG2 might be to protect cells from potentially toxic substances and to prevent absorption of xenobiotics ingested in our diet by actively transporting compounds from cells.

ABCG2 confers resistance to several Pgp substrates such as mitoxantrone, the anthracyclines such as daunorubicin and doxorubicin, the camptothecins, bisantrene, topotecan, rhodamine123, prazosin, and SN-38.[23,63,64,70] In contrast, ABCG2 does not efflux other known Pgp substrates such as taxol, colchicine, verapamil, vinblastine, and calcein-AM, nor the MRP substrates calcein and GSH-conjugated monochlorobimane.[23] Substrates of ABCG2 are reviewed and summarized[71] as follows: I(1). anthracyclines (e.g., daunorubicin, epirubicin, anthracene, mitoxantrone, bisantrane), (2). camptothecin (e.g., SN-38, 9-aminocamtothecin, irrenotecan, diflomotecan, topotecan), (3). nucleoside analogs (AZT, AZT 5′ monophosphate, lamivudine), (4). fluorophores (bodipy-prazosin, Hoechst 33342, rhodamine 123, lyso-tracker green), and (5) polyglutamates (e.g., methotrexate)

2.3. Apoptosis and Chemotherapy Resistance

The determinants of cell survival and cell death are both extrinsic and intrinsic to the cell. All cells are in the default position of being able to undergo apoptosis but are prevented from doing so by extracellular signals within a multicellular organism.[72] These signals arise through cell-to-cell contacts, from the extracellular matrix to which cells are attached[73,74] and from circulating survival factors, such as insulin-like growth factor (IGF)-I and nerve growth factor.[75] Many of these survival components and their downstream effectors, such as BCR-ABL, RAS, and the IGF-1 receptor, are altered in malignancy. In a metastatic tumor cell, survival must be independent of the normal topological context of a tissue. This implies that to become metastatic, intrinsic mechanisms of survival must be initiated to allow survival away from normal controls. The implications of this for cytotoxic drug therapy are that tumor cells might be intrinsically more resistant to undergoing cell death than many normal cell types. Alterations in apoptosis pathways have been shown to be involved in resistance to a variety of cytotoxic agents. Thus, it seems appropriate to refer to apoptosis-related chemotherapy resistance as a type of MDR.

There are external signals that engage apoptosis: Ligation of the APO-1/fas receptor initiates a discrete cell death signaling cascade, presumably by removing the action of internal inhibitors of the default position of cell death.[76] The expression levels of both the death-promoting ligand and its receptor will again determine a hierarchy among different cells as to whether they might readily engage apoptosis. The intrinsic determinants of a survival/death hierarchy are epitomized by members of the BCL-2 family or antiapoptotic genes. It is important for malignant tumors arising from some epithelia that the cell has a relatively high survival potential determined by the expression of antiapototic genes, sufficient for it to survive DNA damage without deletion by apoptosis. These common, high-death-threshold tumors would be resistant to chemotherapy, whereas those rarer tumors arising from hematopoietic cells might be more amenable to the engagement of cell death following cytotoxic therapy. The genes that determine survival and death thresholds might determine intrinsic drug sensitivity and resistance.[77,78]

2.3.1. The BCL-2 Family of Proteins

The BCL-2 gene was identified as a translocated product in follicular lymphoma. Expression of BCL-2 suppressed the apoptosis stimulated by the withdrawal of serum survival factors.[79] A number of gene homologues of BCL-2 have now been discovered that encode both suppressors of apoptosis and accelerators of the process (see the review in Ref. 80). Bcl-2 is the archetypal member of a family of proteins that undergo homodimerizations and heterodimerizations to each other via binding through conserved BH1, BH2, and BH3 domains.[81] The isolation of the Bcl-2 homologue Bax as a protein that immunoprecipitated with Bcl-2 and the finding that its expression accelerated apoptosis suggested a model whereby Bax-Bax homodimers promote apoptosis, whereas the Bcl-2-Bax heterodimer inhibits apoptosis by limiting Bax-Bax homodimerization.[80,81] Knowledge of the family of BCL-2-like genes has been expanding with recent discoveries of sequence-related promoters of apoptosis (bad, bak, bcl-X_s) and inhibitors of apoptosis (bcl-X_L).[82–85] In a variety of cellular backgrounds, BCL-2 and BCL-$X_L$expression has been shown to delay the onset of apoptosis induced by almost all classes of cytotoxic drugs. It could be claimed that the expression of BCL-2 or bcl-$X_L$provides a genuine multidrug or pleotropic resistance, because its inhibition of drug-induced apoptosis crosses the entire spectrum of the pharmacopoeia. Whether ectopic expression of BCL-2 universally provides pleiotropic drug resistance, associated with the long-term survival of cells, is complicated by findings that certain types do not appear to be protected by BCL-2 apoptosis, but are, instead, protected by the expression of its homologue BCL-X_L.[86.87] On the other hand, ectopic expression of the death-promoting Bcl-Xs protein in BCL-2-expressing MCF-7 human breast carcinoma cells sensitized them to the cytotoxicity of both etoposide and paclitaxel[88] - Strategies like this, delivering apoptotic accelerators such as bax and bcl-X_s or inhibitors of bcl-2 or bcl-X_L by expression of mimetics, which prevent proapoptotic homologues from binding to bcl-2 protein, would seem to offer an important route for pleotropic drug resistance.

2.3.2. The Role of TP53 in Determining Drug Sensitivity and Resistance

Many anticancer drugs damage DNA, either directly or indirectly. This damage per se is not lethal but has to be "sensed" by the cell, and, coupled with the execution of apoptosis, this suggests that the failure of sensors could lead to drug resistance. The tumor suppressor TP53 has been suggested to be a direct sensor of DNA damage. Loss of functional p53 might promote pleiotropic drug resistance to DNA-damaging agents. The importance of p53 in promoting DNA damage-induced apoptosis was demonstrated by studies of immature thymocytes *in vitro* or intestinal epithelia *in vivo* from homozygous TP53 null animals.[89–91] Cells that had been γ-irradiated did not undego apoptosis in comparison with those that were homozygously TP53 positive. TP53-null thymocytes also failed to undergo apoptosis after treatment with the topoisomerase II inhibitor etoposide treatment with the non-DNA-damaging corticosteroid dexamethasone, suggesting that the non-DNA-damage-induced pathway was discrete and p53 independent.

3. CHEMOSENSITIZERS FOR CANCER CHEMOTHERAPY

As soon as Pgp and sister proteins were recognized as the main reason of MDR, blocking the efflux of drugs by inhibition of the functions of these transporters has become a realistic way to circumvent MDR.[92] Several chemicals, already known or used as drugs for other purposes, have been tested *in vitro* and *in vivo* on resistant tumor cells. Verapamil, a calcium channel antagonist, was the first compound found active in reversing MDR,[2] and after it, many other compounds have been found effective in the resensitization of resistant malignant cells (see the review in Ref. 93). The compounds are called chemosensitizers, MDR modulators, or MDR-reversing agents.

3.1. Chemosensitizers of Pgp

The process of chemosensitization involves administering a Pgp inhibitor (MDR modulator) and an anticancer drug to cause enhanced intracellular anticancer drug accumulation by impairing the Pgp function. Numerous compounds have been shown to inhibit the drug efflux function of Pgp and, therefore, reverse cellular resistance. In general, they have been classified as MDR modulators belonging to the first, second, or third generation.[94]

The history on the studies of MDR modulators began more than two decades ago with the discovery by Tsuruo and co-workers that the calcium channel blocker verapamil can reverse MDR.[2] Later, it was reported that verapamil inhibits Pgp activity via direct competition with Pgp substrates.[95] Other first-generation MDR modulators include the antimalarial drug quinidine, the calmodulin antagonist trifluoperazine, and the immunosuppressant cyclosporin A.[96] The Cyclosporin A is proved to compete with Pgp substrates for binding to a common drug-binding site of Pgp[96] There were promising results in phase I clinical trials with some of the firstgeneration MDR modulators, but most required high doses,[97] and nonspecific side effects were noted. As a result, their clinical applications in cancer patients have been limited, and this has led to the discovery of so-called second- and third-generation MDR chemosensitizers.

The second-generation MDR modulators include dexverapamil, PSC 833, dexniguldipine, and VX-710. Among these, most of the studies are with PSC833 and VX-710.

PSC833 (valspodar) is an analogue of cyclosporin D, and the results to date suggest that PSC 833 acts as a noncompetitive inhibitor by binding to site(s) other than the substrate-binding site to alter the conformation of Pgp.[98] Numerous studies have been reported for its clinical trials, including phase III clinical trials. Although PSC833 exhibited increased potency, and thus required lower doses to achieve effective *in vivo* plasma concentrations to modulate MDR, it retained some properties that limited its clinical usefulness. VX-710 (biricodar) is an amido-ketopipecolinate derivative that has been shown to block both Pgp and MRP activity.[99] However, similar to PSC833, the use of VX-710 is limited by its unpredictable pharmacokinetic interactions with cytotoxic agents. Most of the third-generation MDR modulators have been developed based on structure–activity relationships and combinatorial chemistry, in the hope of overcoming limitations exhibited by the second-generation molecules.[100] The third-generation MDR modulators, which are currently in clinical development, are LY335979, XR9576, laniquidar (R101933), GF120918, and ONT-093.[100] Both LY335979[101] and XR9576[102] are among the most studied in this group of modulators.

3.2. Chemosensitizers of MRP1

Most MRP1 substrates, as well as inhibitors, are anionic compounds that enter cells poorly, thus making it difficult to obtain a good inhibitor for MRP1 compared with Pgp. A variety of MRP1 inhibitors have been reported.[103] For instance, general inhibitors of organic transport are probenecid, sulfinpyrazone, and indomethacin; inhibitors of MRP-related transporters are the LTD_4 analogue MK571, ONO-1078, glibenclamide,[104] and some GSH conjugates[105]; inhibitors of MRP1 and Pgp are VX-710, agosterol A, PAK-104, verapamil, cyclosporin A, genistein, and quercetin; and GSH-dependent inhibitors of MRP1 are LY 475776 and LY 402913. These compounds are mostly not specific to MRP1 and they need to be used at relatively high concentrations to overcome the MDR mediated by MRP1.[58]

3.3. Chemosensitizers of MXR

A variety of MXR inhibitors have been identified.[71,106] It has been reported recently that GF120918, a third-generation Pgp inhibitor, is also a potent inhibitor of MXR.[107] Various studies showed that GF120918 can be tolerated in humans and animals at concentrations sufficient to inhibit MXR.[106,108] The natural product fumitremorgin C (FTC) secreted from the fungi *Aspergillus fumigatus* was another potent modulator of ABCG2 that was able to completely reverse mitoxantrone resistance and topotecan resistance in ABCG2-overexpressing cells at 1–5-μM concentrations[109] Many studies showed that this compound is highly specific to ABCG2 and did not reverse Pgp- or MRP1-mediated drug resistance. Recently, several FTC analogues such as Ko132 and Ko134 have been developed.[106] These compounds could potentially be further developed as clinically useful ABCG2 inhibitors because they were more potent than FTC; the IC_{50}s are in the range of 85–270 nM. Several of the tyrosine kinase inhibitors (e.g., CI1033) have also been shown to be potent inhibitors of MXR that inhibit the MXR-mediated efflux of topotecan and SN-38 at low micromolar concentrations. Recently, HIV protease inhibitors ritonavir, saquinavir, and nelfinavir also have been found to be effective inhibitors of MXR.[110] Collectively, although a large number of MXR inhibitors has been described, whether any of these compounds are clinically useful in reversing MXR-mediated MDR has yet to be determined.

3.4. Mechanism of Action of MDR Chemosensitizers

Multidrug resistance chemosensitizers might function in two major ways: They can modify either the function or the expression of the proteins involved in MDR.

3.4.1. Modulation of the Function of MDR1/MRP

The idea of finding chemosensitizers that inhibit the function of the drug transporters and thereby reverse MDR has grown in parallel with the biochemical and clinical investigations of the molecular mechanism and regulation of these proteins. The compounds that inhibit MDR might be categorized according to their mode of action on the targeted transporter proteins. The first category involves analogues of the transported (drug) substrates that either competitively or noncompetitively inhibit drug extrusion through MDR1 or MRP. These agents interact with the transporters on their drug-binding sites with significantly higher affinity than any cytotoxic drugs and might be either efficiently transported or not transported by the pumps. In the former case (which is probably true, e.g., for verapamil and for several calmodulin inhibitors), stimulation of the pump turnover might greatly increase ATP consumption in the MDR tumor cells.[111] This might result in an advantageous collateral sensitivity of the tumor cells to the modulating agent. In the latter form (this is probably the case with PSC833), the transporter becomes locked by a substrate analogue that cannot be pumped, thus cannot be cleared from the binding sites.[112]

In the case of MRP, which transports various glutathione *S*-conjugates, agents inducing cellular GSH-depletion might be good candidates for substrate-dependent reversal of drug resistance.[113,114] Moreover, certain prostaglandins (PGA1), tyrosine kinase inhibitors (genistein), and inhibitors of uric acid transport (e.g., benzbromarone) seem to be effective substrate-analogue MRP chemosensitizers.[115,116] The second category of MDR chemosensitizers includes inhibitors of ATP binding, or ATP utilization in the drug pumps. Various non-hydrolyzable or covalently reacting ATP analogues (such as azido-ATP) or compounds like NBD chloride react with crucial lysines at or near the ATP-binding sites.[117,118] Certainly, very little specificity against MDR can be expected from such compounds, as most ATP-binding proteins, including, for example, ion pumps or protein kinases, will be affected as well. The MDR1 protein has two cysteines located in the two highly conserved ATP-binding regions, which can be modified by alkylating agents (e.g., NEM). Alkylation of these cysteine residues irreversibly blocks the function of MDR1, whereas the presence of ATP protects these sites from NEM.[117,118] Recent reports indicate that flavonoids like quercetin might inhibit drug pumps by reacting preferentially with their ATP-binding domains.[119] A dream compound of this kind would be specifically recognized by MDR1 or MRP as a toxic product to be eliminated, and then the compound would irreversibly modify ATP binding or hydrolysis in the same proteins.

The third category of MDR chemosensitizers includes specific antibodies interfering with the function of drug transporters. There are several monoclonal antibodies that react with intracellular functional domains of MDR1 or MRP, but their *in vivo* application is not considered, as they do not enter tumor cells. However, some of the antibodies thatt recognize extracellular epitopes block the conformational changes required for drug transport function and might be good candidates for medical application. The frist such anti-MDR1 monoclonal antibody, MRK16, was developed by Hamada and Tsuruo[14] and shown to inhibit ATP-dependent drug extrusion and to modulate drug resistance.

The last category of mode of action of MDR chemosensitizers would include all other possible drug pump inhibitors that cannot be easily separated by their mode of action. Oligomycin, an effective inhibitor of both MDR1 and MRP, does not seem to be a substrate analogue but might directly block ATP hydrolysis, although its action is certainly not selective.[117] Various detergents seem to inhibit MDR pumps at the site(s) of hydrophobic interactions in or near the membrane lipid bilayer[120,121] with little selectivity.

3.4.2. Modulation of the Expression of MDR1/MRP

Chemosensitizers of the drug transporter transcription might become useful inhibitors, and potential promoter regions of these proteins were identified and characterized in detail.[122,123] Most previous studies on the regulation of *MDR1* gene expression have concentrated on identifying transcription factors involved in the induction of *MDR1* gene promoter activity in drug-resistant cancer cell lines.

The human *MDR1* gene promoter contains a number of regulation sites for SP1, NF-Y, and YB-1 transcription factors.[123–125] These transcription factors have been shown to upregulate *MDR1* gene promoter activity. Recently, it was reported that *MDR1* gene promoter activity might be linked to the cyclic AMP-dependent protein kinase signal pathway, which plays a key role in activating SP1.[126] Activation of Ras and PKC has also been shown to stimulate Jun and Fos families, forming the activator protein-1 (AP-1). AP-1-responsive genes are important in DNA synthesis, DNA repair, and drug detoxification. The promoter/enhancer element of the *MDR1* gene contains the AP-1-binding-site sequence. Because the transcription efficiency of the *MDR1* gene appears to be regulated by AP1,[122] the activation of Fos and Jun might lead to increased expression of the *MDR1* gene. Fos is thought to mediate its effects through transcriptional activation, after it interacts with the Jun protein to form AP-1. Therefore, overexpression of Fos might cause the MDR phenotype by modulation of *MDR1* gene expression.

Overall, it is important to note that the MDR1 promoter is responsive to cellular stress triggered by anticancer drugs, carcinogens, heavy metals, ultraviolet light (UV), heat shock, serum stravation, phosphatase inhibitors, and phorbol esters.[127–129] These regulations occur probably in a species- and cell-specific fashion[130] thus, any effort for their clinical modulation seems to be a long shot. Rather, the expression of the drug pumps MDR1 or MRP might be efficiently modulated by chemically stabilized antisense oligonucleotides[131] or synthetic catalytic RNAs (ribozymes).[132] The most critical issues for their therapeutic use will be increased stability and effective delivery to the target cancer cells.

Inhibitors of protein processing and posttranslational modifications, in principle, might also be used to block the expression of a functional form of MDR1 and MRP drug pumps, as both proteins are posttranslationally modified by N-glycosylation and phosphorylation. However, the inhibitors of their processing are basically unaffected or is not an efficient way to modulate drug transport.[133–135]

4. CHEMOSENSITIZING ACTIVITIES OF CURCUMIN AND ITS ANALOGUES

Curcuminoids are natural phenolic coloring compounds found in the rhizomes of *Curcuma longa* Linn., commonly known as turmeric. The rhizomes contain three major pigments of curcuminoids: curcumin I (diferuloylmethane), curcumin II (demethoxycurcumin), and curcumin III (bisdemethoxycurcumin).[136,137] their chemical structures are illustrated elsewhere.[138,139] All three impart the hallmark yellow pigmentation to the *Curcuma longa* plant and particularly to its rhizome. Ongoing experimental and clinical studies indicate that turmeric and its curcuminoid components exhibit unique antioxidant,[140] anti-inflammatory,[141] and antitumorigenesis properties.[142–144] Their potential use in the prevention of cancer and in the treatment of human immunodeficiency virus (HIV) infection is also a subject of intensive research.[136]

Curcumin has been found to be safe, with no dose-limiting toxicity, when administered at doses up to 10 g/day in humans.[145] However, curcumin undergoes rapid and extensive metabolism in the liver and intestine[146] and demonstrates poor bioavailability, thereby limiting its usefulness as a potent chemopreventive agent. To date, curcumin-glucuronide, dihydrocurcumin-glucuronoside, THC-glucuronoside, and THC (tetrahydrocurcumin) have been demonstrated as the major curcumin metabolites *in vivo*.[144,147,148]

4.1. Effect on Pgp

Due to its wide range of biological and pharmacological effects, lack of toxicity, cyclicity, and lipophilicity, curcumin was examined to determine possible interactions with Pgp expression and function.[149] The commercial grade of curcumin, which contain approximately 77%, 17%, and 3% curcumin I, II, and III, respectively, was used in this study. Curcumin (1–10 μM) downregulated Pgp expression and reduced Pgp-mediated efflux in drug-resistant human cervical carcinoma cells (KB-V1). Curcumin increased rhodamine 123 accumulation in a concentration-dependent manner (1–55 μM) and inhibited the efflux of rhodamine 123 from these cells but had no effect on the wild-type drug-sensitive KB-3-1 cells, which do not express Pgp. Because the time of exposure of cells to curcumin in these experiments was short (1–2 h), it is unlikely that curcumin acted by downregulating MDR1 gene expression, resulting in a reduced level of cellular Pgp. However, the effect of curcumin on the expression of Pgp at the protein (Western blotting) and mRNA [reverse transcription–polymerase chain reaction (RT-PCR)] levels was examined. There was no difference in Pgp expression in KB-V1 cells when treated with curcumin for 1–2 h. Treatment of drug-resistant KB-V1 cells with curcumin increased their sensitivity to vinblastine, but not in wild-type KB-3-1 cells. In addition, curcumin inhibited verapamil-stimulated ATPase activity and the photoaffinity labeling of Pgp with the prazosin analogue iodoarylazidoprazosin in a concentration-dependent manner, indicating direct interaction of curcumin with Pgp and possible binding to the same site as other agents such as prazosin and verapamil. These findings suggest that curcumin might represent a new reversal agent for the chemosensitization of cancer cells.

In another study, curcumin inhibited vinblastine induced Pgp level in a dose- and time-dependent manner in the vinblastine-resistant subline KB-V0.1.[150] Another report from the same group demonstrated that three major curcuminoids modulated Pgp function using human MDR KB-V1 cells and crude membranes of Pgp-overexpressing HighFive insect cells.[138] The IC_{50} of curcumin I, II, and III is not statistically different compared to KB-V1 (expressing high levels of Pgp) and KB-3-1 cells (parental drug sensitive), suggesting that Pgp does not confer resistance to curcumin I, II, or III; in other words, these curcuminoids most likely are not transported by Pgp. Treating the cells with nontoxic doses of curcuminoids increased their sensitivity to vinblastine only in the Pgp-expressing drug-resistant cell line KB-V1, and curcumin I retained the drug in KB-V1 cells more effectively than curcumin II and III. Effects of curcumin I, II, and III on rhodamine 123, calcein AM, and bodipy- FL vinblastine accumulation confirmed these findings. Curcumin I, II and III increased the accumulation of fluorescent substrates in a dose-dependent manner, and at 15 μM, curcumin I was the most effective. These results demonstrated that this effect is not specific to a particular substrate; curcuminoids affected the accumulation of all three substrates in the same manner. The inhibitory effect in a concentration-dependent manner of curcuminoids on verapamil-stimulated ATPase activity and photoaffinity labeling of Pgp with the [^{125}I}-iodoarylazidoprazosin offered additional support that curcumin I was the most potent modulator. Thus, these biochemical results demonstrate that curcuminoids interact directly with Pgp and possibly bind to the same binding sites as other agents such as prazosin, vinblastine, and verapamil. Chemical structure of curcumin I might make it more suitable for binding to the drug-binding site of Pgp than that of curcumin II and III, because curcumin I has a balance of two hydroxyl and methoxyl groups on each side, and the presence of two methoxyl groups in the curcumin I molecule might help its inhibitory activity on the Pgp function.

In another study bisdemethoxycurcumin has been demonstrated to be the most active form of the curcuminoids present in turmeric for modulation of MDR1 gene expression in MDR KB-V1 cells by Western blot and RT-PCR analysis.[151] The nuclear protein was identified by competitive electrophoretic mobility shift assay (EMSA) using unlabeled SP1, AP1, AP2, OCT1, NF-κB, and cAMP-responsive element binding (CREB) oligomers (200 M excess). The result demonstrated that the CREB consensus sequence can compete more completely with the nuclear factor that binds to the labeled probe (MDR1 gene promoter –84 to –65 DNA fragment) than other unlabeled probes[149] This result indicates that CREB is the transcription factor that binds to the MDR1 gene promoter in residues –84 to –65, and this result was confirmed by supershift assay using an anti-CREB antibody. In additional studies, pretreatment of KB-V1 cells with curcuminoids significantly decreased the activity of the MDR1 gene promoter, and bisdemethoxycurcumin produced the maximum inhibitory effect.[149] As tetrahydrocurcumin is the ultimate metabolite of the curcumins *in vivo*, we recently extended our investigation to assess whether THC is able to retain the MDR-reversing activity (manuscript in preparation). Two types of cell line were used for Pgp study: human cervical carcinoma KB-3-1 (wild type) and KB-V-1 and human breast cancer MCF-7 (wild type) and MCF-7 MDR, respectively. The results by flow-cytometry assay indicated that THC is able to inhibit the function of Pgp and thereby significantly increase the accumulation of rhodamine and calcein AM in KB-V-1 cells. The result was confirmed by the effect of THC on [^{3}H]-vinblastine accumulation and efflux in MCF-7 and MCF-7MDR. THC significantly increased the accumulation and inhibited the efflux of [^{3}H]-vinblastine in MCF-7 MDR in a concentration-dependent manner. This effect was not found in the wild type MCF-7 cell line. The interaction of THC with the Pgp molecule was clearly indicated by ATPase assay and photoaffinity labeling of Pgp with the transport substrate. THC stimulated Pgp ATPase activity and inhibited the incorporation of [^{125}I]-iodoarylazidoprazosin (IAAP) into Pgp in a concentration-dependent manner. The MDR-reversing properties of THC on Pgp was determined by 3-(4,5 dimethylthiazol-2-yl)-2,5-diphenyltetrazolium bromide (MTT) assay. THC at 25 μM significantly increased the sensitivity of vinblastine in drug-resistant KB-V-1 cells. This effect was not found in respective drug-sensitive parental cell lines. Taken together, the present study clearly showed that THC inhibits the efflux function of Pgp and it is able to extend the MDR-reversing activity of curcuminoids *in vivo*. Additional *in vivo* studies are required to determine if curcumin has potential as an effective and safe chemosensitizer for treating cancers expressing Pgp.

In another study by Romiti et al., using primary cultures of rat hepatocytes expressing high levels of Pgp after 72 h culture, curcumin (commercial grade) inhibited rhodamine 123 efflux in a dose dependent manner.[152] Western blot analysis indicated that curcumin decreased the protein levels of Pgp in cultures. In photoaffinity labeling studies, curcumin competed with azidopine for binding to Pgp, suggesting a direct interaction with glycoprotein. These results suggest that curcumin is able to modulate *in vitro* both expression and function of hepatic Pgp.

4.2. Effect on MRP-1

The inhibitory effects of a mixture of curcumin I, II, and III on MRP1-mediated transport using isolated membrane vesicles of MRP1-expressing Sf9 cells was recently reported.[153] However, the mechanism of inhibition remains unknown. Moreover, it is unknown whether each curcumin form in the curcumin mixture exhibits the same effect. In another study by Chearwae et al., curcumin mixture and three major curcuminoids purified from turmeric (curcumin I, II, and III) were tested for their ability to modulate the function of MRP1 using HEK293 cells stably transfected with MRP1-pcDNA3.1 and pcDNA3.1 vector alone.[139] The IC_{50} of curcuminoids in these cell lines ranged from 14.5 to 39.3 μM. Results indicated that curcuminoids might not be MRP1 substrate because the IC_{50} values were almost identical in both parental and MRP1-transfected cells. Upon treating the cells with etoposide, in the presence of 10 μM curcuminoids the sensitivity of etoposide was increased several-fold only in MRP1-expressing and not in pcDNA3.1-HEK 293 cells. Western blot analysis showed that the total cellular level of the MRP1 protein level was not affected by treatment with 10 μM curcuminoids for 3 days. The modulatory effect of curcuminoids on MRP1 function was confirmed by the inhibition of efflux of two fluorescent substrates: calcein-AM and fluo4-AM. Although all three curcuminoids increased the accumulation of fluorescent substrates in a concentration-dependent manner, curcumin I was the most effective inhibitor. The potency of curcumin I was comparable to MK-571, which is known to inhibit MRP1-mediated transport with high affinity. In addition, curcuminoids did not affect 8azido[α-^{32}P]ATP binding; however, they did stimulate the basal ATPase activity and inhibited the quercetin-stimulated ATP hydrolysis of MRP1, demonstrating the interaction of curcuminoids most likely at the substrate-binding site(s) on this multidrug transporter. In summary, these results demonstrate that curcuminoids effectively inhibit MRP1-mediated transport, and among curcuminoids, curcumin I, a major constituent of curcumin mixture, is the best modulator.

Recently, other workers have reported the modulation of MRP1 and MRP2 function by curcumin mixture.[153,154] In addition, the curcumin mixture appears to affect the trafficking of ΔF508 mutant of cystic fibrosis transmembrane regulator (CFTR),[155] which also belongs to the ABCC subfamily (ABCC7), similar to MRP1 (ABCC1) and MRP2 (ABCC2). The curcumin mixture has been reported to stimulate the chloride channel activity of wild-type CFTR.[156] Further extensive work of interest is whether curcumin I concentrations achieved *in vivo* are sufficient to inhibit MRP1 function and or expression, and extensive pharmacokinetics studies with curcumin I will be required to know the steady-state levels of phytochemical reached in blood and tissue after its administration at pharmacological doses. However, recent work suggested that the curcumin mixture, and all three pure forms of curcumin (I, II, and III) inhibit the function of MRP1.[139] Curcumin I was the most effective form as an inhibitor of MRP1, similar to previous results with Pgp.[138] These agents thus might have a beneficial effect on cancer chemotherapy with respect to the possibility of long-term use without concerns regarding MRP1 or MDR1 activation.

We recently extended our investigation to assess whether THC, a major metabolite of curcumin, is able to modulate MRP1 function using pcDNA 3.1 and pcDNA3.1-MRP1 transfected HEK293 cells.[157] The efflux of a fluorescent substrate calcein AM was inhibited effectively by THC; thereby, the accumulation of calcein was increased in MRP1-HEK 293 and not its parental pcDNA3.1-HEK 293 cells. The MDR-reversing properties of THC on MRP1 were determined by MTT assay. THC (20–25 μM) significantly increased the sensitivity of etoposide in MRP1-HEK 293 cells. This effect was not found in respective drug-sensitive parental cell lines. A consistent finding was reported in MDCKII cells transfected with MRP1; THC significantly increased 3[H]-EGCG in MDCKII/MRP1-overexpressing cells.[154] Taken together, these studies clearly showed that THC inhibits the efflux function of MRP1 and it is able to extend the MDR-reversing activity of curcuminoids *in vivo*.

4.3. Effect on MXR or ABCG2

As reported earlier, the curcumin mixture and purified curcuminoids (curcumin I, II, and III) could reverse the MDR in cells expressing Pgp and MRP1 by inhibiting the functions mediated by these transporters.[138,139,149] It was also shown that curcumin I, which is a major constituent (70–75%) of a curcumin mixture, was most potent among the purified curcuminoids in inhibiting the activity of both of the transporters. Purified curcuminoids were further evaluated for their modulating effects on the function of either the wild-type 482R or mutant 482T ABCG2 transporter, stably expressed in human embryonic kidney 293 cells and drug-selected MCF7FLV1000 and MCF7AdrVp3000 cells.[158]

It has been reported previously that the amino acid at position 482 has a crucial role in the substrate and inhibitor specificity of ABCG2 and that mutants R482→T/G exhibit altered drug resistance profiles and substrate specificity of MXR.[67,68] Therefore, we decided to investigate the modulating effects of curcuminoids on ABCG2 activity in both wild-type R482 stably expressed in HEK 293 cells, and the mutant 482T overexpressed in MCF7AdrVp3000 cells. The drug-selected MCF7FLV1000 and MCF7AdrVp3000, which overexpressed the wild-type R482 and the mutant 482T ABCG2, respectively, were chosen because the protein was overexpressed in these breast cancer cell lines under its own promoter at higher levels[159] in sufficient quantity for biochemical characterization. Curcumin I, II, III and the curcumin mixture inhibited the efflux of ABCG2 substrates and the presence of nontoxic concentrations of curcuminoids (10 μM) increased (threefold to eightfold) the sensitivity of ABCG2-expressing cells to anticancer drugs, including mitoxantrone, topotecan, SN-38, and doxorubicin. This reversal was not due to reduced expression, becauseABCG2 protein levels were unaltered by treatment with 10 μM of curcuminoids for 3 days. In addition, [^{3}H]-curcumin-I transport assays demonstrate that the curcuminoids are not transported by ABCG2. Curcuminoids stimulated (2.4- to 3.3-fold) ATPase activity of ABCG2 at very low concentrations (7–18 nM) and inhibited both the photolabeling of ABCG2 with two photoaffinity analogues, [^{125}I]-IAAP and [^{3}H]-azidopine, and also the transport of these agents.

Curcuminoids interact at the drug–substrate binding sites on drug transporters with very high affinity and inhibit ABCG2-mediated drug resistance. Taken together, our previous work with Pgp and MRP1 and this study with ABCG2 suggest that curcumin I is a very effective modulator, which should be considered as a potential compound for development of reversal agents designed to overcome MDR mediated by these three major ABC drug transporters.

In another study by our group using MXR-overexpressing MCF7AdrVp3000 or MCF7FL1000 and its parental MCF-7, we assessed whether THC, a major metabolite of curcumin, is able to modulate MXR function.[157] The binding of [^{125}I]-IAAP to MXR was also inhibited by THC, suggesting that THC interacted with the drug-binding site of the transporter. THC dose-dependently inhibited the efflux of mitoxantrone and pheophorbide A from MXR-expressing cells (MCF7AdrVp3000 and MCF7FL1000). THC at 25 μM significantly increased the sensitivity of mitoxantrone in drug-resistance MCF7AdrVp3000 cells. This effect was not found in MCF-7 drug sensitive parental cell lines.

4.4. Effect on NF-κB and Inhibitor Apoptotic Proteins

Nuclear factor-κB has been implicated in both carcinogenesis and the development of drug resistance in cancer cells.[160,161] Most reports suggest that NF-κB mediates survival signals that counteract apoptosis.[156a,162] NF-κB-activated expression of genes that inhibit apoptosis, such as A20, IAPs (inhibitor apoptotic proteins), and TRAFs, is probably involved in the mediation.[163,164] Upon activation, NF-κB dissociates from the inhibitory IκBα and translocates from the cytoplasm to the nucleus, where it binds to the promoter elements and transactivates gene expression.[165] This general activation of NF-κB by anticancer drugs can be attenuated by pretreatment with common biologic modulators. Chuang et al. demonstrated that NF-κB can be activated by all four of the anticancer drugs in three cancer cell lines (liver, uterine cervix, and urinary bladder) examined.[166] Each of the four anticancer drugs used (doxorubicin, 5-FU, cisplatin, and paclitaxel) possesses distinct modes of action that cause different types of damage to cancer cells. However, universal NF-κB activation was observed. These results suggest the existence of a common set of cellular elements that sense the challenge by these drugs as a type of stress and transmits this signal to NF-κB. When cells were pretreated with common biologic modulators such as tamoxifen, dexamethasone, and curcumin, the doxorubicin-induced NF-κB activation was attenuated significantly. This inhibition might play a role in sensitizing cancer cells to chemotherapeutic drugs.

In another study using human hepatic cancer cells and the combination of curcumin with cisplatin or doxorubicin, the levels of NF-κB were lower than those predicted from the effects of the single agents.[6] Except for Bcl-2, the human hepatic cancer cells expressed different other genes, including the IAPs, implicated in cell proliferation and survival. Curcumin determined early changes in cyclooxygenase (COX)-2 and c-myc mRNAs, which were downregulated, and in livin mRNA, which was upregulated. Later it decreased Bcl-X_L mRNA and increased Bcl-X_s and c-IAP-2 mRNAs. Cisplatin and doxorubicin exerted distinct effects on gene expression. The cytotoxic interactions between curcumin and these agents were accompanied by synergistic or additive effects of decrease in the expression of different genes, including c-myc, Bcl-X_L, c-IAP-2, NAIP, and XIAP. The expression of XIAP and other IAPs can be upregulated by NF-κB.[167,168] Thus, the inhibition of NF-κB by curcumin might be of help to antagonize the IAPs as well as other NF-κB target genes (e.g., COX-2, Bcl-X_L and c-myc) involved in the adverse biology of cancer. Singh and Aggarwal showed that curcumin could suppress NF-κB activation induced by TNF, phorbol ester, and H_2O_2 through suppression of IκBα degradation.[169] Recently, Aggarwal et al. demonstrated that curcumin inhibits the tumor necrosis factor (TNF)-induced IκBα kinase complex and Akt activation, which blocks phosphorylation of IκBα and p65, leading to suppression of events required for NF-κB gene expression.[170]

In human cervical carcinoma cells, curcumin sensitizes tumor cells more efficiently to the therapeutic effect of paclitaxel.[171] Paclitaxel is the best anticancer agent that has ever been isolated from plants, but its major disadvantage is its dose-limiting toxicity. Furthermore, tumors tend to acquire resistance to cytotoxic chemotherapeutic agents, including paclitaxel. A combination of 5 nM paclitaxel with 5 μM curcumin augments anticancer effects more efficiently than paclitaxel alone, as evidenced by increased cytotoxicity and reduced DNA synthesis in HeLa cells. This synergistic effect was not observed in normal cervical cells in which paclitaxel downregulates NF-κB. Evaluation of signaling pathways common to paclitaxel and curcumin reveals that this synergism was in part related to downregulation of NF-κB and serine/threonine kinase Akt pathways by curcumin. An electrophoretic mobility shift assay revealed that activation of NF-κB induced by paclitaxel is downregulated by curcumin. Curcumin-downregulated paclitaxel induced phosphorylation of the serine/threonine kinase Akt, a survival signal regulated by NF-κB. Tubulin polymerization and cyclin-dependent kinase

Cdc2 activation induced by paclitaxel was not affected by curcumin. These results lead to the conclusion that the synergistic effect of Taxol and curcumin in inducing apoptosis in cervical cancer cells follows a pathway that is independent of tubulin polymerization and cell cycle arrest, at least at lower concentrations of curcumin.

5. CONCLUSIONS AND FUTURE PERSPECTIVES

Many studies have been performed with the aim of developing effective resistance modulators to overcome the MDR of human cancers. Potent MDR modulators are being investigated in clinical trials. Many current studies are focused on herbal constituents because these have been used for centuries without producing any harmful side effects. Curcuminoids (curcumin, demethoxycurcumin, and bisdemethoxycurcumin) purified from turmeric are able to modulate the efflux function of Pgp, MRP1 and mitoxantrone resistance protein (MXR), and curcumin I, a major constituent of curcumin mixture, was the most effective. Tetrahydrocurcumin, a major metabolite form of curcuminoids in humans, inhibits the efflux function of these three major ABC drug transporters and it is able to extend the MDR-reversing activity of curcuminoids *in vivo*.

Curcuminoids were reported to modulate Pgp, MRP1, and MXR by interacting directly with drug- or substrate-binding site(s). This might involve competitive binding to the substrate-binding site or binding to other drug-binding sites and altering molecular conformation, as indicated by the altered photoaffinity labeling. Curcuminoids did not affect the ATP-binding site; however, they did stimulate the basal ATPase activity and inhibited verapamil-stimulated ATP hydrolysis of Pgp or quercetin hydrolysis of MRP1. Curcuminoids also stimulated ATPase activity of MXR or ABCG2 at very low concentrations (7–18 nM). Curcumin is the most potent inhibitor for all three drug transporters, as the chemical structure of curcumin I might make it more suitable for binding to the drug-binding site of Pgp than that of curcumin II and III, because curcumin I has a balance of two hydroxyl and methoxyl groups on each side, and the presence of two methoxyl groups in the curcumin I molecule might help its inhibitory activity on the Pgp function.

Bisdemethoxycurcumin, or curcumin III, can upregulate MDR1 gene expression. Western blot and RT-PCR analysis indicated that bisdemethoxycurcumin decreased the protein and mRNA levels of Pgp in cultures. The EMSA demonstrated that CREB is the transcription factor that binds to the MDR1 gene promoter in residues –84 to –65. However, curcuminoids do not change protein and RNA levels of MRP1 and MXR drug transporters. The mechanism for herbal modulation of the *MDR1* gene is largely undetermined.

The inhibition of Pgp, MRP1, and MXR by curcumin might provide a novel approach for reversing MDR in tumor cells. Additional *in vivo* studies are required to determine if curcumin has potential as an effective and safe chemosensitizer for treating cancers expressing Pgp. Phase II and III clinical trials of many known MDR modulators might soon yield informative results that should help to decide whether the chemosensitizer works in clinical oncology. In addition, many ABC transporters have not yet been identified and characterized. As more information on these proteins becomes available, we might be able to more effectively design drug combinations that will provide increased selectivity of action at the desired tissue site.

REFERENCES

1. H. Lage, ABC-transporters: Implications on drug resistance from microorganisms to human cancers. *Int J Antimicrob Agents* **22**, 188 (2003).
2. T. Tsuruo, H. Iida, S. Tsukagoshi, and Y. Sakurai, Overcoming of vincristine resistance in P388 leukemia in vivo and in vitro through enhanced cytotoxicity of vincristine and vinblastine by verapamil. *Cancer Res* **41**, 1967 (1981).
3. H. A. Bardelmeijer, J. H. Beijnen, K. R. Brouwer, H. Rosing, W. J. Nooijen, J. H. Schellens, Jand O. van Tellingen, Increased oral bioavailability of paclitaxel by GF120918 in mice through selective modulation of P-glycoprotein. *Clin Cancer Res* **6**, 4416 (2000).
4. L. J. Green, P. Marder, and C. A. Slapak, Modulation by LY335979 of P-glycoprotein function in multidrug-resistant cell lines and human natural killer cells. *Biochem Pharmacol* **61**, 1393 (2001).
5. A. L.Cheng, C. H. Hsu, J. K. Lin, M. M. Hsu, Y. F. Ho, T. S. Shen, J. Y. Ko, J. T. Lin, B. R. Lin, W. Ming-Shiang, H. S. Yu, S. H. Jee, G. S. Chen, T. M. Chen, C. A. Chen, M. K. Lai, Y. S. Pu, M. H. Pan, Y. J. Wang, C. C. Tsai, and C. Y. Hsieh, Phase I clinical trial of curcumin, a chemopreventive agent, in patients with high-risk or pre-malignant lesions. *Anticancer Res* **21**, 2895 (2001).
6. M Notarbartolo, P. Poma, D. Perri, L. Dusonchet, M. Cervello, and N. D'Alessandro, Antitumor effects of curcumin, alone or in combination with cisplatin or doxorubicin, on human hepatic cancer cells. Analysis of their possible relationship to changes in NF-kB activation levels and in IAP gene expression. *Cancer Lett* **224**, 53 (2005).
7. B. Hill, Drug resistance, and overview of the current state of the art. *Int J Oncol* **9**, 197 (1996).
8. P. S. Lacombe, J. A. G. Vicente, J. G. Pages, and P. L. Morselli, Causes and problems of nonresponse or poor response to drugs. *Drugs* **51**, 552 (1996).
9. L. J. Goldstein, MDR1 gene expression in solid tumours. *Eur J Cancer* **32A**(6), 1039 (1996).
10. M. Volmand J. Mattern, Resistance mechanisms and their regulation in lung cancer. *Crit Rev Oncog* **7**, 227 (1996).
11. M. Dietel, What's new in cytostatic drug resistance and pathology. *Pathol Res Pract* **187**, 892 (1991).
12. W. T. Beck, Mechanisms of multidrug resistance in human tumor cells. The roles of P-glycoprotein, DNA topoisomerase II, and other factors. *Cancer Treat Rev* **17**(Suppl A), 11 (1990).
13. C. S. Morrow and K. H. Cowan, Glutathione S-transferases and drug resistance. *Cancer Cells* **2**, 15 (1990).
14. J. R. Hammond, R. M. Johnstone, and P. Gros, Enhanced efflux of [3H]vinblastine from Chinese hamster ovary cells transfected with a full-length complementary DNA clone for the mdr1 gene. *Cancer Res* **49**, 3867 (1989).

15. Y. A. Hannun, Apoptosis and the dilemma of cancer chemotherapy. *Blood* **89**, 1845 (1997).
16. Y. Y. Liu, T. Y. Han, A. E. Giuliano, and M. C. Cabot, Ceramide glycosylation potentiates cellular multidrug resistance. *FASEB J* **15**, 719 (2001).
17. R. L. Juliano and V. Ling, A surface glycoprotein modulating drug permeability in Chinese hamster ovary cell mutants. *Biochim Biophys Acta* **455**, 152 (1976).
18. D. Nielson, C. Maare, and T. Skovsgaard, Influx of daunorubicin in multidrug resistance Erlich ascites tumor cells, correlation to expression of P-glycoprotein and efflux. Influence of verapamil. *Biochem Pharmacol* **50**, 443 (1995).
19. A. R. Safa, Photoaffinity labeling of p-glycoprotein in multidrug resistance cells. *Cancer Invest* **10**, 295 (1992).
20. R. G. Deeley and S. P. Cole, Function, evolution and structure of multidrug resistance protein (MRP). *Semin Cancer Biol* **8**,193 (1997).
21. D. D. Ross, W. Yang, L. V. Abruzzo, W. S. Dalton, E. Schneider, H. Lage, M. Dietel, L. Greenberger, S. P. Cole, and L. A. Doyle, Atypical multidrug resistance: Breast cancer resistance protein messenger RNA expression in mitoxantrone-selected cell lines. *J Natl Cancer Inst* **91**, 429 (1999).
22. Z. E. Sauna, M. M. Smith, M. Muller, K. M. Kerr, and S. V. Ambudkar, The mechanism of action of multidrug-resistance-linked P-glycoprotein. *J Bioenerg Biomembr* **33**, 481 (2001).
23. T. Litman, T. E. Druley, W. D. Stein, and S. E. Bates, From MDR to MXR: New understanding of multidrug resistance systems, their properties and clinical significance. *Cell Mol Life Sci* **58**, 931 (2001).
24. J. C. Leighton, Jr. and L. J. Goldstein, P-Glycoprotein in adultsolid tumors. Expression and prognostic significance. *Hematol Oncol Clin North Am* **9**, 251 (1995).
25. J. P. Marie, P-Glycoprotein in adult hematologic malignancies. *Hematol Oncol Clin North Am* **9**, 239 (1995).
26. R. J. Arceci, Clinical significance of P-glycoprotein in multidrug resistance malignancies. *Blood* **81**, 2215 (1993).
27. R. Pirker, J. Wallner, K. Geissler, W. Linkesch, O. A. Haas, P. Bettelheim, M. Hopfner, R. Scherrer, P. Valent, L. Havelec, et al., MDR1 gene expression and treatment outcome in acute myeloid leukemia. *J Natl Cancer Inst.* **83**, 708 (1991).
28. H. S. Chan, P. S. Thorner, G. Haddad, and V. Ling, Immunohistochemical detection of P-glycoprotein: Prognostic correlation in soft tissue sarcoma of childhood. *J Clin Oncol* **8**, 689 (1990).
29. R. S. Weinstein, S. M. Jakate, J. M. Dominguez, M. D. Lebovitz, G. K. Koukoulis, J. R. Kuszak, L. F. Klusens, T. M. Grogan, T. J. Saclarides, I. B. Roninson, et al., Relationship of the expression of the multidrug resistance gene product (P-glycoprotein) in human colon carcinoma to local tumor aggressiveness and lymph node metastasis. *Cancer Res.* **51**, 2720 (1991).
30. S. W. Tobe, S. E. Noble-Topham, I. L. Andrulis, R. W. Hartwick, K. L. Skorecki, and E. Warner, Expression of the multiple drug resistance gene in human renal cell carcinoma depends on tumor histology, grade, and stage. *Clin Cancer Res* **1**, 1611 (1995).
31. G. Giaccone, S. C. Linn, and H. M. Pinedo, Multidrug resistance in breast cancer, mechanisms, strategies. *Eur J Cancer* **31A**(Suppl 7), S15 (1995).
32. H. M. Pinedo and G. Giaccone, P-Glycoprotein: A marker of cancer-cell behavior. *N Engl J Med* **333**, 1417 (1995).
33. J. L. Biedler, Genetic aspects of multidrug resistance. *Cancer* **70**, 1799 (1992).

34. M. Lehnert, Clinical multidrug resistance in cancer: A multifactorial problem. *Eur J Cancer* **32A**, 912 (1996).
35. E. Buschman, R. J. Arceci, J. M. Croop, M. Che, I. M. Arias, D. E. Housman, and P. Gros, mdr2 encodes P-glycoprotein expressed in the bile canalicular membrane as determined by isoform-specific antibodies. *J Biol Chem* **267**, 18,093 (1992).
36. C. Cordon-Cardo, J. P. O'Brien, J. Boccia, D. Casals, J. R. Bertino, and M. R. Melamed, Expression of the multidrug resistance gene product (P-glycoprotein) in human normal and tumor tissues. *J Histochem Cytochem* **38**, 1277 (1990).
37. J. J. Smit, A. H. Schinkel, C. A. Mol, D. Majoor, W. J. Mooi, A. P. Jongsma, C. R. Lincke, and P. Borst, Tissue distribution of the human MDR3 P-glycoprotein. *Lab Invest* **71**, 638 (1994).
38. M. M. Gottesman and I. Pastan, Biochemistry of multidrug resistance mediated by the multidrug transporter. *Annu Rev Biochem* **62**, 385 (1993).
39. F. Frezard, E. Pereira-Maia, P. Quidu, W. Priebe, and A. Garnier-Suillerot, P-Glycoprotein preferentially effluxes anthracyclines containing free basic versus charged amine. *Eur J Biochem* **268**, 1561 (2001).
40. U. Brinkmann, I. Roots, and M. Eichelbaum, Pharmacogenetics of the human drug-transporter gene MDR1: Impact of polymorphisms on pharmacotherapy. *Drug Discov Today* **6**, 835 (2001).
41. P. Gros and C. Shustik, Multidrug resistance: A novel class of membrane-associated transport proteins is identified. *Cancer Invest* **9**, 563 (1991).
42. M. M. Cornwell, I. Pastan, and M. M.. Gottesman, Certain calcium channel blockers bind specifically to multidrug-resistant human KB carcinoma membrane vesicles and inhibit drug binding to P-glycoprotein. *J Biol Chem* **262**, 2166 (1987).
43. T. W. Loo and D. M. Clarke, Functional consequences of glycine mutations in the predicted cytoplasmic loops of P-glycoprotein. *J Biol Chem* **269**, 7243 (1994).
44. F. J. Sharom, X. Yu, and C. A. Doige, Functional reconstitution of drug transport and ATPase activity in proteoliposomes containing partially purified P-glycoprotein. *J Biol Chem* **268**, 24,197 (1993).
45. P. M. Jones and A. M. George, A new structural model for P-glycoprotein. *J Membr Biol* **166**, 133 (1998).
46. K. Ueda, A. Yoshida, and T. Amachi, Recent progress in P-glycoprotein research. *Anticancer Drug Des* **14**, 115 (1999).
47. A. F. Castro, J. K. Horton, C. G. Vanoye, and G. A. Altenberg, Mechanism of inhibition of P-glycoprotein-mediated drug transport by protein kinase C blockers. *Biochem Pharmacol* **58**, 1723 (1999).
48. G. Conseil, J. M. Perez-Victoria, J. M. Jault, F. Gamarro, A. Goffeau, J. Hofmann, and A. Di Pietro, Protein kinase C effectors bind to multidrug ABC transporters and inhibit their activity. *Biochemistry* **40**, 2564 (2001).
49. H. Yabuuchi, S. Takayanagi, K. Yoshinaga, N. Taniguchi, H. Aburatani and T. Ishikawa, ABCC13, an unusual truncated ABC transporter, is highly expressed in fetal human liver. *Biochem Biophys Res Common* **299**, 410 (2002).
50. A. C. Lockhart, R. G. Tirona, and R. B. Kim, Pharmacogenetics of ATP-binding cassette transporters in cancer and chemotherapy. *Mol Cancer Ther* **2**, 685 (2003).
51. S. P. Cole, G. Bhardwaj, J. H. Gerlach, J. E. Mackie, C. E. Grant, K. C. Almquist, A. J. Stewart, E. U. Kurz, A. M.. Duncan, and R. G. Deeley, Overexpression of a transporter gene in a multidrug-resistant human lung cancer cell line. *Science* **258**, 1650 (1992).

52. P. Borst, R. Evers, M. Kool, and J. Wijnholds, A family of drug transporters: The multidrug resistance-associated proteins. *J Natl Cancer Inst* **92**, 1295 (2000).
53. D. R. Hipfner, R. G. Deeley, and S. P. Cole, Structural, mechanistic and clinical aspects of MRP1. *Biochim Biophys Acta* **1461**, 359 (1999).
54. J. Renes, E. G. de Vries, E. F. Nienhuis, P. L. Jansen, and M. Muller, ATP- and glutathione-dependent transport of chemotherapeutic drugs by the multidrug resistance protein MRP1. *Br J Pharmacol* **126**, 681 (1999).
55. G. Rappa, A. Lorico, R. A. Flavell, and A. C. Sartorelli, Evidence that the multidrug resistance protein (MRP) functions as a co-transporter of glutathione and natural product toxins. *Cancer Res* **57**, 5232 (1997).
56. L. Manciu, X. B. Chang, J. R. Riordan, and J. M. Ruysschaert, Multidrug resistance protein MRP1 reconstituted into lipid vesicles: Secondary structure and nucleotide-induced tertiary structure changes. *Biochemistry* **39**, 13,026 (2000).
57. E. M. Leslie, R. G. Deeley, and S. P. Cole, Multidrug resistance proteins, role of P-glycoprotein, MRP1, MRP2, and BCRP (ABCG2) in tissue defense. *Toxicol Appl Pharmacol* **204**, 216 (2005).
58. A. Haimeur, G. Conseil, R. G. Deeley, and S. P. Cole, The MRP-related and BCRP/ABCG2 multidrug resistance proteins: Biology, substrate specificity and regulation. *Curr Drug Metab* **5**, 21 (2004).
59. M. Gao, H. R. Cui, D. W. Loe, C. E. Grant, K. C. Almquist, S. P. Cole, and R. G. Deeley, Comparison of the functional characteristics of the nucleotide binding domains of multidrug resistance protein 1. *J Biol Chem* **275**, 13,098 (2000).
60. K. Nagata, M. Nishitani, M. Matsuo, N. Kioka, T. Amachi, and K. Ueda, Nonequivalent nucleotide trapping in the two nucleotide binding folds of the human multidrug resistance protein MRP1. *J Biol Chem* **275**, 17,626 (2000).
61. K. Barnouin, I. Leier, G. Jedlitschky, A. Pourtier-Manzanedo, J. Konig, W. D. Lehmann, and D. Keppler, Multidrug resistance protein-mediated transport of chlorambucil and melphalan conjugated to glutathione. *Br J Cancer* **77**, 201 (1998).
62. N. Ballatori, C. L. Hammond, J. B. Cunningham, S. M. Krance, and R. Marchan, Molecular mechanisms of reduced glutathione transport: Role of the MRP/CFTR/ABCC and OATP/SLC21A families of membrane proteins. *Toxicol Appl Pharmacol* **204**, 238 (2005).
63. K. Miyake, L. Mickley, T. Litman, Z. Zhan, R. Robey, B. Cristensen, M. Brangi, L. Greenberger, M. Dean, T. Fojo, and S. E. Bates, Molecular cloning of cDNAs which are highly overexpressed in mitoxantrone-resistant cells: Demonstration of homology to ABC transport genes. *Cancer Res* **59**, 8 (1999).
64. L. A. Doyle and D. D. Ross, Multidrug resistance mediated by the breast cancer resistance protein BCRP (ABCG2). *Oncogene* **22**, 7340 (2003).
65. R. Allikmets, L. M. Schriml, A. Hutchinson, V. Romano-Spica, and M. Dean, A human placenta-specific ATP-binding cassette gene (ABCP) on chromosome 4q22 that is involved in multidrug resistance. *Cancer Res* **58**, 5337 (1998).
66. S. E. Bates, R. Robey, K. Miyake, K. Rao, D. D. Ross, and T. Litman, The role of half-transporters in multidrug resistance. *J Bioenerg Biomembr* **33**, 503 (2001).
67. Y. Honjo, C. A. Hrycyna, Q. W. Yan, W. Medina-Perez, R. W. Robey, A. van de Laar, T. Litman, M. Dean, and S. E. Bates, Acquired mutations in the MXR/BCRP/ABCP gene alter substrate specificity in MXR/BCRP/ABCP-overexpressing cells. *Cancer Res* **61**, 6635 (2001).
68. R. W. Robey, Y. Honjo, K. Morisaki, T. A. Nadjem, S. Runge, M. Risbood, M. S. Poruchynsky, and S. E. Bates, Mutations at amino-acid 482 in the ABCG2 gene affect substrate and antagonist specificity. *Br J Cancer* **89**, 1971 (2003).

69. M. Miwa, S. Tsukahara, E. Ishikawa, S. Asada, Y. Imai, and Y. Sugimoto, Single amino acid substitutions in the transmembrane domains of breast cancer resistance protein (BCRP) alter cross resistance patterns in transfectants. *Int J Cancer* **107**(5), 757 (2003).
70. S. Kawabata, M. Oka, K. Shiozawa, K. Tsukamoto, K. Nakatomi, H. Soda, M. Fukuda, Y. Ikegami, K. Sugahara, Y. Yamada, S. Kamihira, L. A. Doyle, D. D. Ross, and S. Kohno, Breast cancer resistance protein directly confers SN-38 resistance of lung cancer cells. *Biochem Biophys Res Commun* **280**, 1216 (2001).
71. Q. Mao and J. D. Unadkat, Role of the breast cancer resistance protein (ABCG2) in drug transport. *AAPS J* **7**, E118 (2005).
72. M. C. Raff, Social controls on cell survival and cell death. *Nature* **356**, 397 (1993).
73. E. Ruoslahti and J. C. Reed, Anchorage dependence, integrins, and apoptosis. *Cell* **77**, 477 (1994).
74. S. M. Frisch and H. Francis, Disruption of epithelial cell–matrix interactions induces apoptosis. *J Cell Biol* **124**, 619 (1994).
75. E. A. Harrington M. R. Bennett, A. Fanidi, and G. I. Evan, c-Myc-induced apoptosis in fibroblasts is inhibited by specific cytokines. *EMBO J* **13**, 3286 (1994).
76. S. Nagata, Apoptosis regulated by a death factor and its receptor, Fas ligand and Fas. *Phil Trans R Soc Lond B: Biol Sci* **345**, 281 (1994).
77. C. Dive and J. A. Hickman, JDrug-target interactions: Only the first step in the commitment to a programmed cell death? *Br J Cancer* **64**,192 (1991).
78. D. E. Fisher, Apoptosis in cancer therapy: Crossing the threshold. *Cell* **78**, 539–542 (1994).
79. D. Hockenbery, G. Nunez, C. Milliman, R. D. Schreiber, and S. I. Korsmeyer, Bcl-2 is an inner mitochondrial membrane protein that blocks programmed cell death. *Nature* **348**, 334 (1990).
80. J. C. Reed, Bcl-2 and the regulation of programmed cell death. *J Cell Biol* **124**, 1 (1994).
81. X. M. Yin, Z. N. Oltvai, and S. J. Korsmeyer, BH1 and BH2 domains of Bcl-2 are required for inhibition of apoptosis and heterodimerization with Bax. *Nature* **369**, 321 (1994).
82. E. Yang, J. Zha, J. Jockel, L. H. Boise, C. B. Thompson, and S. J. Korsmeyer, Bad, a heterodimeric partner for Bcl-XL and Bcl-2, displaces Bax and promotes cell death. *Cell* **80**, 285 (1995).
83. M. C. Kiefer, M. J. Brauer, V. C. Powers, J. J. Wu, S. R. Umansky, L. D. Tomei, and P. J. Barr, Modulation of apoptosis by the widely distributed Bcl-2 homologue Bak. *Nature* **374**, 736 (1995).
84. S. N. Farrow, J. H. White, I. Martinou, T. Raven, K. T. Pun, C. J. Grinham, J. C. Martinou, and R. Brown, Cloning of a bcl-2 homologue by interaction with adenovirus E1B 19K. *Nature* **374**, 731 (1995).
85. T. Chittenden, E. A. Harrington, R. O'Connor, C. Flemington, R. J. Lutz, G. I. Evan, and B. C. Guild, Induction of apoptosis by the Bcl-2 homologue Bak. *Nature* **374**, 733 (1995).
86. D. T. Chao, G. P. Linette, L. H. Boise, L. S. White, C. B. Thompson, and S. J. Korsmeyer, Bcl-XL and Bcl-2 repress a common pathway of cell death. *J Exp Med* **182**, 821 (1995).
87. A. R. Gottschalk, L. H. Boise, C. B. Thompson, and J. Quintans, Identification of immunosuppressant-induced apoptosis in a murine B-cell line and its prevention by bcl-x but not bcl-2. *Proc Natl Acad Sci USA* **91**, 7350 (1994).

88. V. N. Sumantran, M. W. Ealovega, G. Nunez, M. F. Clarke, and M. S. Wicha, Overexpression of Bcl-XS sensitizes MCF-7 cells to chemotherapy-induced apoptosis. *Cancer Res* **55**, 2507 (1995).
89. A. R. Clarke, C. A. Purdie, D. J. Harrison, R. G. Morris, C. C. Bird, M. L. Hooper, and A. H. Wyllie, Thymocyte apoptosis induced by p53-dependent and independent pathways. *Nature*. **362**, 849 (1993).
90. S. W. Lowe, E. M. Schmitt, S. W. Smith, B. A. Osborne, and T. Jacks, p53 is required for radiation-induced apoptosis in mouse thymocytes. *Nature* **362**, 847 (1993).
91. A. J. Merritt, C. S. Potten, C. J. Kemp, J. A. Hickman, A. Balmain, D. P. Lane, and P. A. Hall, The role of p53 in spontaneous and radiation-induced apoptosis in the gastrointestinal tract of normal and p53-deficient mice. *Cancer Res* **54**, 614 (1994).
92. V. Sandor, T. Fojo, and E. Bates, Future perspectives for the development of P-glycoprotein modulators. *Drug Resist Up-dates*. **1**, 190 (1998).
93. F. Gualtieri, Drugs reverting multidrug resistance (chemosenzitizers). *Chim. Ind* **78**, 1233 (1996).
94. T. J. Lampidis, A. Krishan, L. Planas, and H. Tapiero, Reversal of intrinsic resistance to adriamycin in normal cells by verapamil. *Cancer Drug Deliv* **3**, 251 (1986).
95. E. C. Spoelstra, H. V. Westerhoff, H. M. Pinedo, H. Dekker, and J. Lankelma, The multidrug-resistance-reverser verapamil interferes with cellular P-glycoprotein-mediated pumping of daunorubicin as a non-competing substrate. *Eur J Biochem* **221**, 363 (1994).
96. N. J. Chao, M. Aihara, K. G. Blume, and B. I. Sikic, Modulation of etoposide (VP-16) cytotoxicity by verapamil or cyclosporine in multidrug-resistant human leukemic cell lines and normal bone marrow. *Exp Hematol* **18**, 1193 (1990).
97. C. Avendano and J. C. Menendez, Inhibitors of multidrug resistance to antitumor agents (MDR). *Curr Med Chem* **9**, 159 (2002).
98. P. Atadja, T. Watanabe, H. Xu, and D. Cohen, PSC-833, a frontier in modulation of P-glycoprotein mediated multidrug resistance. *Cancer Metastasis Rev* **17**, 163 (1998).
99. U. A. Germann, D. Shlyakhter, V. S. Mason, R. E. Zelle, J. P. Duffy, V. Galullo, D. M. Armistead, J. O. Saunders, J. Boger, and M. W. Harding, Cellular and biochemical characterization of VX-710 as a chemosensitizer: Reversal of P-glycoprotein-mediated multidrug resistance in vitro. *Anticancer Drugs* **8**(2), 125 (1997).
100. R. Krishna and L. D. Mayer, Multidrug resistance (MDR) in cancer. Mechanisms, reversal using modulators of MDR and the role of MDR modulators in influencing the pharmacokinetics of anticancer drugs. *Eur J Pharm Sci* **11**, 265 (2000).
101. A. H. Dantzig, K. L. Law, J. Cao, and J. J. Starling, Reversal of multidrug resistance by the P-glycoprotein modulator, LY335979, from the bench to the clinic. *Curr Med Chem* **8**, 39 (2001).
102. A. Stewart, J. Steiner, G. Mellows, B. Laguda, D. Norris, and P. Bevan, Phase I trial of XR9576 in healthy volunteers demonstrates modulation of P-glycoprotein in CD56+ lymphocytes after oral and intravenous administration. *Clin Cancer Res* **6**, 4186 (2000).
103. B. Tan, D. Piwnica-Worms, and L. Ratner, Multidrug resistance transporters and modulation. *Curr Opin Oncol* **12**, 450 (2000).
104. L. Payen, L. Delugin, A. Courtois, Y. Trinquart, A. Guillouzo, and O. Fardel, The sulphonylurea glibenclamide inhibits multidrug resistance protein (MRP1) activity in human lung cancer cells. *Br J Pharmacol* **132**, 778 (2001).

105. D. Burg, P. Wielinga, N. Zelcer, T. Saeki, G. J. Mulder, and P. Borst, Inhibition of the multidrug resistance protein 1 (MRP1) by peptidomimetic glutathione-conjugate analogs. *Mol Pharmacol* **62**, 1160 (2002).
106. J. D. Allen, S. C. Van Dort, M. Buitelaar, O. van Tellingen, and A. H. Schinkel, Mouse breast cancer resistance protein (Bcrp1/Abcg2) mediates etoposide resistance and transport, but etoposide oral availability is limited primarily by P-glycoprotein. *Cancer Res* **63**, 1339 (2003).
107. M. de Bruin, K. Miyake, T. Litman, R. Robey, and S. E. Bates, Reversal of resistance by GF120918 in cell lines expressing the ABC half-transporter, MXR. *Cancer Lett* **146**, 117 (1999).
108. C. M. Kruijtzer, J. H. Beijnen, H. Rosing, W. W. ten Bokkel Huinink, M. Schot, R. C. Jewell, E. M. Paul, and J. H. Schellens, Increased oral bioavailability of topotecan in combination with the breast cancer resistance protein and P-glycoprotein inhibitor GF120918. *J Clin Oncol* **20**, 2943 (2002).
109. S. K. Rabindran, D. D. Ross, L. A. Doyle, W. Yang, and L. M. Greenberger, Fumitremorgin C reverses multidrug resistance in cells transfected with the breast cancer resistance protein. *Cancer Res* **60**, 47 (2000).
110. A. Gupta, Y. Zhang, J. D. Unadkat, and Q. Mao, HIV protease inhibitors are inhibitors but not substrates of the human breast cancer resistance protein (BCRP/ABCG2). *J Pharmacol Exp Ther* **310**, 334 (2004).
111. H. J. Broxterman, H. M. Pinedo, G. J. Schuurhuis, and J. Lankelma, Cyclosporin A and verapamil have different effects on energy metabolism in multidrug-resistant tumour cells. *Br J Cancer* **62**, 85 (1990).
112. D. Cohen, Modulation of resistance and P-glycoprotein function in tumor cells. 'ATP Binding cassette (ABC) Transporter: From multidrug resistance to genetic disease. *Abstract Book*, 1997, p. 46.
113. S. P. Cole, K. E. Sparks, K. Fraser, D. W. Loe, C. E. Grant, G. M. Wilson, and R. G. Deeley, Pharmacological characterization of multidrug resistant MRP-transfected human tumor cells. *Cancer Res* **54**, 5902 (1994).
114. G. Jedlitschky, I. Leier, U. Buchholz, M. Center, and D. Keppler, ATP-dependent transport of glutathione S-conjugates by the multidrug resistance-associated protein. *Cancer Res* **54**, 4833 (1994).
115. H. J. Broxterman, G. Giaccone, and J. Lankelma, Multidrug resistance proteins and other drug transport-related resistance to natural product agents. *Curr Opin Oncol*.**7**, 532 (1995).
116. P. R. Twentyman and C. H. Versantvoort, Experimental modulation of MRP (multidrug resistance-associated protein)-mediated resistance. *Eur J Cancer* **32A**, 1002 (1996).
117. M. K. al-Shawi and A. E. Senior, Characterization of the adenosine triphosphatase activity of Chinese hamster P-glycoprotein. *J Biol Chem* **268**, 4197 (1993).
118. M. K. al-Shawi, I. L. Urbatsch, and A. E. Senior, Covalent inhibitors of P-glycoprotein ATPase activity. *J Biol Chem* **269**, 8986 (1994).
119. A. B Shapiro and V. Ling, Effect of quercetin on Hoechst 33342 transport by purified and reconstituted P-glycoprotein. *Biochem Pharmacol* **53**, 587 (1997).
120. S. V. Ambudkar, Purification and reconstitution of functional human P-glycoprotein. *J Bioenerg Biomembr* **27**, 23 (1995).
121. D. M. Woodcock, S. Jefferson, M. E. Linsenmeyer, P. J. Crowther, G. M. Chojnowski, B. Williams, and I. Bertoncello, Reversal of the multidrug resistance phenotype with

cremophor EL, a common vehicle for water-insoluble vitamins and drugs. *Cancer Res* **50**, 4199 (1990).

122. K. Ueda, I. Pastan, and M. M. Gottesman, Isolation and sequence of the promoter region of the human multidrug-resistance (P-glycoprotein) gene. *J Biol Chem* **262**, 17,432 (1987).
123. Q. Zhu and M. S. Center, Cloning and sequence analysis of the promoter region of the MRP gene of HL60 cells isolated for resistance to adriamycin. *Cancer Res* **54**, 4488 (1994).
124. R. Sundseth, G. MacDonald, J. Ting, and A. C. King, DNA elements recognizing NF-Y and Sp1 regulate the human multidrug-resistance gene promoter. *Mol Pharmacol* **51**, 963 (1997).
125. T. Ohga, T. Uchiumi, Y. Makino, K. Koike, M. Wada, M. Kuwano, and K. Kohno, Direct involvement of the Y-box binding protein YB-1 in genotoxic stress-induced activation of the human multidrug resistance 1 gene. *J Biol Chem* **273**, 5997 (1998).
126. C. Rohlff and R. I. Glazer, Regulation of the MDR1 promoter by cyclic AMP-dependent protein kinase and transcription factor Sp1. *Int J Oncol* **12**, 383 (1998).
127. K. Kohno, S. Sato, H. Takano, K. Matsuo, and M. Kuwano, The direct activation of human multidrug resistance gene (MDR1) by anticancer agents. *Biochem Biophys Res Commun* **165**, 1415 (1989).
128. N. Kioka, Y. Yamano, T. Komano, and K. Ueda, Heat-shock responsive elements in the induction of the multidrug resistance gene (MDR1). *FEBS Lett* **301**, 37 (1992).
129. K. V. Chin, S. Tanaka, G. Darlington, I. Pastan, and M. M. Gottesman, Heat shock and arsenite increase expression of the multidrug resistance (MDR1) gene in human renal carcinoma cells. *J Biol Chem* **265**, 221 (1990).
130. K. V. Chin, K. Ueda, I. Pastan, and M. M. Gottesman, Modulation of activity of the promoter of the human MDR1 gene by Ras and p53. *Science* **255**, 459 (1992).
131. C. Cucco and B. Calabretta, In vitro and in vivo reversal of multidrug resistance in a human leukemia-resistant cell line by mdr1 antisense oligodeoxynucleotides. *Cancer Res* **56**, 4332 (1996).
132. J. Bertram, K. Palfner, M Killian, W. Brysch, K. H. Schlingensiepen, W. Hiddemann, and M. Kneba, Reversal of multiple drug resistance in vitro by phosphorothioate oligonucleotides and ribozymes. *Anticancer Drugs* **6**, 124 (1995).
133. A. H. Schinkel, S. Kemp, M. Dolle, G. Rudenko, and E. Wagenaar, *N*-Glycosylation and deletion mutants of the human MDR1 P-glycoprotein. *J Biol Chem* **268**, 7474 (1993).
134. H. Lis and N. Sharon, Protein glycosylation. Structural and functional aspects. *Eur J Biochem* **218**, 1 (1993).
135. H. R. Goodfellow, A. Sardini, S. Ruetz, R. Callaghan, P. Gros, P. A. McNaughton, and C. F. Higgins, Protein kinase C-mediated phosphorylation does not regulate drug transport by the human multidrug resistance P-glycoprotein. *J Biol Chem* **271**, 13,668 (1996).
136. H. P. Ammon and M. A. Wahl, Pharmacology of *Curcuma longa*. *Planta Med* **57**, 1 (1991).
137. R. S. Ramsewak, D. L. DeWitt, and M. G. Nair, Cytotoxicity, antioxidant and anti-inflammatory activities of curcumins I–III from *Curcuma longa*. *Phytomedicine* **7**, 303 (2000).
138. W. Chearwae, S. Anuchapreeda, K. Nandigama, S. V. Ambudkar, and P. Limtrakul, Biochemical mechanism of modulation of human P-glycoprotein (ABCB1) by cur-

cumin I, II, and III purified from Turmeric powder. *Biochem Pharmacol* **68**, 2043 (2004).

139. W. Chearwae, C. P. Wu, H. Y. Chu, T. R. Lee, S. V. Ambudkar, and P. Limtrakul, Curcuminoids purified from turmeric powder modulate the function of human multidrug resistance protein 1 (ABCC1). *Cancer Chemother Pharmacol* **57**, 376 (2006).
140. L. M. Antunes, M. C. Araujo, J. D. Darin, and M. L. Bianchi, Effects of the antioxidants curcumin and vitamin C on cisplatin-induced clastogenesis in Wistar rat bone marrow cells. *Mutat Res* **465**, 131 (2000).
141. T. Kawamori, R. Lubet, V. E. Steele, G. J. Kelloff, R. B. Kaskey, C. V. Rao, and B. S. Reddy, Chemopreventive effect of curcumin, a naturally occurring anti-inflammatory agent, during the promotion/progression stages of colon cancer. *Cancer Res* **59**, 597 (1999).
142. M. L. Kuo, T. S. Huang, and J. K. Lin, Curcumin, an antioxidant and anti-tumor promoter, induces apoptosis in human leukemia cells. *Biochim Biophys Acta* **1317**, 95 (1996).
143. P. Limtrakul, S. Lipigorngoson, O. Namwong, A. Apisariyakul, and F. W. Dunn, Inhibitory effect of dietary curcumin on skin carcinogenesis in mice. *Cancer Lett* **116**, 197 (1997).
144. B. B. Aggarwal, A. Kumar, and A. C. Bharti, Anticancer potential of curcumin: Preclinical and clinical studies. *Anticancer Res* **23**, 363 (2003).
145. N. Chainani-Wu, Safety and anti-inflammatory activity of curcumin: A component of tumeric (*Curcuma longa*). *J Altern Complement Med.* **9**, 161 (2003).
146. C. R. Ireson, D. J. Jones, S. Orr, M. W. Coughtrie, D. J. Boocock, M. L. Williams, P. B. Farmer, W. P. Steward, and A. J. Gescher, Metabolism of the cancer chemopreventive agent curcumin in human and rat intestine. *Cancer Epidemiol Biomarkers Prev* **11**, 105 (2002).
147. M. H. Pan, T. M. Huang, and J. K. Lin, Biotransformation of curcumin through reduction and glucuronidation in mice. *Drug Metab Dispos* **27**, 486 (1999).
148. J. K. Lin, M. H. Pan, and S. Y. Lin-Shiau, Recent studies on the biofunctions and biotransformations of curcumin. *Biofactors* **13**, 153 (2000).
149. S. Anuchapreeda, P. Leechanachai, M. M. Smith, S. V. Ambudkar, and P. N. Limtrakul, Modulation of P-glycoprotein expression and function by curcumin in multidrug-resistant human KB cells. *Biochem Pharmacol* **64**,573 (2002).
150. P. Waiwut, S. Anuchapreeda, and P. Limtrakul, Curcumin inhibits the P-glycoprotein level in carcinoma of cervix cells (KB-carcinoma cell lines) induced by vinblastine, *Chiang Mai Med Bull* **41**, 135 (2002).
151. P. Limtrakul, S. Anuchapreeda, and D. Buddhasukh, Modulation of human multidrug-resistance MDR-1 gene by natural curcuminoids. *BMC Cancer* **4**, 13 (2004).
152. N. Romiti, R. Tongiani, F. Cervelli, and E. Chieli Effects of curcumin on P-glycoprotein in primary cultures of rat hepatocytes. *Life Sci* **62**(25), 2349 (1998).
153. H. M. Wortelboer, M. Usta, A. E. van der Velde, M. G. Boersma, B. Spenkelink, J. J. van Zanden, I. M. Rietjens, P. J. van Bladeren, and N. H. Cnubben, Interplay between MRP inhibition and metabolism of MRP inhibitors, the case of curcumin. *Chem Res Toxicol* **16**, 1642 (2003).
154. J. Hong, J. D. Lambert, S. H. Lee, P. J. Sinko, and C. S. Yang, Involvement of multidrug resistance-associated proteins in regulating cellular levels of (-)-epigallocatechin-3-gallate and its methyl metabolites. *Biochem Biophys Res Commun* **310**, 222 (2003).

155. M. E. Egan, M. Pearson, S. A. Weiner, V. Rajendran, D. Rubin, J. Glockner-Pagel, S. Canny, K. Du, G. L. Lukacs, and M. J. Caplan, Curcumin, a major constituent of turmeric, corrects cystic fibrosis defects. *Science* **304**(5670)**,** 600 (2004).
156. A. L. Berger, C. O. Randak, L. S. Ostedgaard, P. H. Karp, D. W. Vermeer, and M. J. Welsh, Curcumin stimulates cystic fibrosis transmembrane conductance regulator Cl^- channel activity. *J Biol Chem* **280**, 5221 (2005).
156a. R. C. Bargou, F. Emmerich, D. Krappmann, K. Bommert, M. Y. Mapara, W. Arnold, H. D. Royer, E. Grinstein, A. Greiner, C. Scheidereit, and B. Dorken, Constitutive nuclear factor-kappaB-RelA activation is required for proliferation and survival of Hodgkin's disease tumor cells. *J Clin Invest* **100**, 2961 (1997).
157. P. Limtrakul, W. Chearwae, S. Shukla, C. Phisalphong and S. Ambudkar S, Modulation of function of three ABC drug transporters, P-glycoprotein (ABCB1), mitoxantrone resistance protein (ABCG2)) and multidrug resistance protein 1 (ABCC1) by tetrahydrocurcumin, a major metabolite of curcumin. *Molecular and Cellular Biochemistry* (Sep 8), 2006; Epub ahead of print).
158. C. Chearwae, S. Shukla, P. Limtrakul, S. Ambudkar, Modulation of the function of the multidrug resistance linked ATP-binding cassette transporter ABCG2 by cancer chemopreventive agent curcumin. *Molecular Cancer Therapeutic* **5**(8), 1995–2006 (2006).
159. R. W. Robey, K. Steadman, O. Polgar, K. Morisaki, M. Blayney, P. Mistry, and S. E. Bates, Pheophorbide a is a specific probe for ABCG2 function and inhibition. *Cancer Res* **64**, 1242 (2004).
160. D. Hanahan and R. A. Weinberg, The hallmarks of cancer. *Cell* **100**, 57 (2000).
161. N. D. Perkins, The Rel/NF-kappa B family: Friend and foe. *Trends Biochem Sci* **25**,434 (2000).
162. C. Y. Wang, J. C. Cusack, Jr., R. Liu, and A. S. Baldwin, Jr., Control of inducible chemoresistance: Enhanced anti-tumor therapy through increased apoptosis by inhibition of NF-kappaB. *Nat Med* **5**, 412 (1999).
163. M. Barkett and T. D. Gilmore, Control of apoptosis by Rel/NF-kappaB transcription factors. *Oncogene* **18**, 6910 (1999).
164. S. Y. Foo and G. P. Nolan, NF-kappaB to the rescue, RELs, apoptosis and cellular transformation. *Trends Genet* **15**, 229 (1999).
165. P. A. Baeuerle and D. Baltimore, I kappa B, a specific inhibitor of the NF-kappa B transcription factor. *Science* **242**, 540 (1988).
166. S. E. Chuang, P. Y. Yeh, Y. S. Lu, G. M. Lai, C. M.. Liao, M. Gao, and A. L. Cheng, Basal levels and patterns of anticancer drug-induced activation of nuclear factor-κB (NF-κB), and its attenuation by tamoxifen, dexamethasone, and curcumin in carcinoma cells. *Biochem Pharmacol.* **63**, 1709 (2002).
167. C. Y. Wang, M. W. Mayo, R. G. Korneluk, D. V. Goeddel, and A. S. Baldwin, Jr., NF-kappaB antiapoptosis: Induction of TRAF1 and TRAF2 and c-IAP1 and c-IAP2 to suppress caspase-8 activation. *Science* **281**, 1680 (1998).
168. N. Mitsiades, C. S. Mitsiades, V. Poulaki, D. Chauhan, P. G. Richardson, T. Hideshima, N. Munshi, S. P. Treon, and K. C. Anderson, Biologic sequelae of nuclear factor-kappaB blockade in multiple myeloma: Therapeutic applications. *Blood* **99**, 4079 (2002).
169. S. Singh and B. B. Aggarwal, Activation of transcription factor NF-κB is suppressed by curcumin (diferuloylmethane) [corrected]. *J Biol Chem* 270, 24,995 (1995).

170. S. Aggarwal, H. Ichikawa, Y. Takada, S. K. Sandur, S. Shishodia, and B. B. Aggarwal, Curcumin (diferuloylmethane) down-regulates expression of cell proliferation and antiapoptotic and metastatic gene products through suppression of IκBα kinase and Akt activation. *Mol Pharmacol* **69**, 195 (2006).
171. S. V. Bava, V. T. Puliappadamba, A. Deepti, A. Nair, D. Karunagaran, and R. J.Anto, Sensitization of taxol-induced apoptosis by curcumin involves down-regulation of nuclear factor-kappaB and the serine/threonine kinase Akt and is independent of tubulin polymerization. *J Biol Chem* **280**, 6301 (2005).

RADIOPROTECTION AND RADIOSENSITIZATION BY CURCUMIN

Ganesh C. Jagetia

Abstract: This chapter gives an overview of the radioprotective and radiosensitizing effect of curcumin. Ionizing radiations interact with biological molecules inducing radiolytic products like e_{aq}, $^{\bullet}OH$, $^{\bullet}H$, ^{-}OH, ^{+}H, O_2, and peroxides. These free radicals damage important biomolecules and subsequently inflict deleterious effects in the organism. Whole-body exposure to ionizing radiations results in central nervous system, gastrointestinal tract, and bone marrow syndromes, whereas chronic irradiation causes cancer, birth anomalies, erythema, and dysfunctions to almost all organ of the body depending on the total dose and site of irradiation. Curcumin (diferuloyl methane), a yellow pigment present in the rhizomes of turmeric, has been used in Southeast Asia to give yellow color and flavor to curries. Turmeric has been used to treat various ailments in the Ayurvedic system of medicine in India. Recently, it has been evaluated for its radioprotective and radiosensitizing activities. Curcumin has been found to exert a dual mode of action after irradiation depending on its dose. It has been reported to protect various study systems against the deleterious effects induced by ionizing radiation and to enhance the effect of radiation. Therefore, curcumin can be very useful during radiotherapy of cancer. Administration of curcumin in patients will be able to kill the tumor cells effectively by enhancing the effect of radiation and, at the same time, protect normal cells against the harmful effects of radiation. The available information on curcumin suggests that the radioprotective effect might be mainly due to its ability to reduce oxidative stress and inhibit transcription of genes related to oxidative stress and inflammatory responses, whereas the radiosensitive activity might be due the upregulation of genes responsible for cell death.

1. INTRODUCTION

1.1. Interaction of Ionizing Radiations with Biological Systems

Ionizing radiations like X-rays and γ-rays, β particles, α particles, and neutrons have sufficient energy to knock out an electron and ionize atoms of the medium. The ability to ionize radiations to produce ionization is responsible for biological damage. Ionizing radiations interact with the biological molecules in two ways: the direct effect and the indirect effect. High-LET (linear energy transfer) radiations interact with biological systems by directly producing free radicals in the

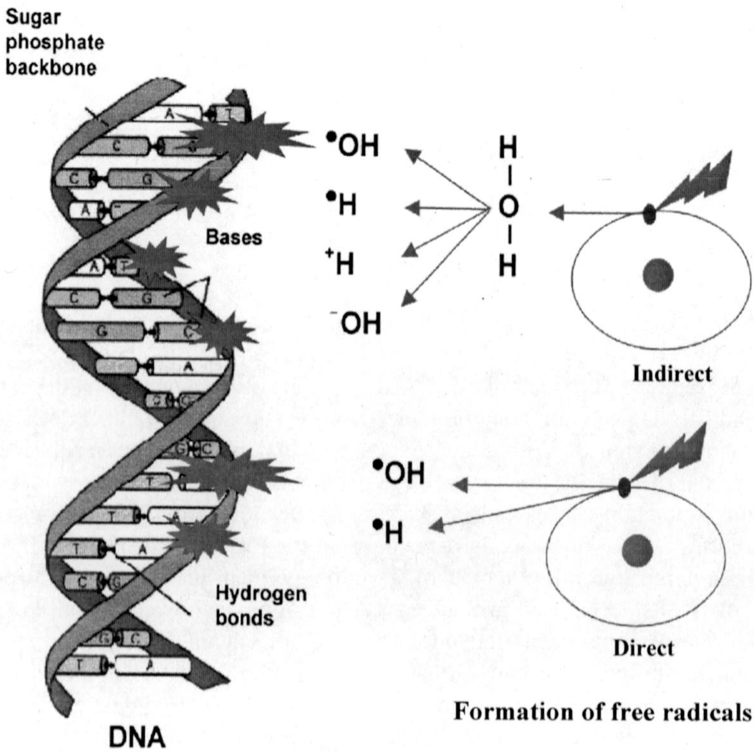

Figure 1. Mechanism of radiation-induced damage.

biomolecules and causing cellular damage (Figure 1). Because biological systems contain 75–90% water, the predominant effect by which ionizing radiations cause damage to the important biomolecules is through radiolysis of water (the indirect effect; Figure 1), where ionizing radiation interacts with water molecule(s) to produce a wide range of reactive oxygen species (ROS) such as e_{aq}, •OH, •H, $^{-}$OH, $^{+}$H, O_2, and peroxides.[1] Among these hydroxyl (•OH) and hydrogen (•H) radicals are free radicals (containing an unpaired electron in the outermost orbit and are highly reactive). The •OH radical is a highly reactive and oxidizing species that can react virtually with all cell constituents at diffusion-controlled rates.[2,3] Among the entire cell constituents, DNA, lipids, and proteins are the principal targets for the attack of •OH radicals.[4] As a result of the interaction of •OH radicals with the cellular genome, a cascade of events is initiated leading to various pathophysiological disorders and eventually to cell death.[5–7]

1.2. Harmful Effects of Ionizing Radiation

The whole-body exposure to ionizing radiations induces radiation sickness and syndromes depending on the exposure dose. If the dose of radiation is very high (i.e., ≥50 Gy), it causes death within minutes and up to 48 h and is called the central nervous system syndrome. It is characterized by ataxia, uncoordinated aberrant movements and tremors, response to stimuli with convulsions, vomiting, repeated defecation, watery diarrhea, nystagmus, meningismus, tremors, and intermittent seizures, which might be found singly or in combination. Whole-body irradiation to doses between 5 and 10 Gy cause the gastrointestinal syndrome, and death of the individual occurs between 3 and 10 days due to failure of the gastrointestinal tract. It is characterized by fever, vomiting, anorexia, diarrhea, infection, dehydration, weight loss, diminishing food and water intake, and decrease in gastric retention and intestinal absorption. If the exposure dose is still lower (i.e., 2–10 Gy), the death of exposed individual takes a longer time (i.e., within 30 days) and is due to the failure of bone marrow, and the characteristic features include onset of chills, fatigue, petechial hemorrhages in the skin and ulceration of mouth and pharynx, and epilation. Cutaneous injury from thermal or radiation burns is characterized by the loss of epidermis and, at times, dermis. Injuries to the skin might cover small areas but extend deeply into the soft tissue, even reaching underlying muscle and bone.[8,9] They might be accompanied by profound local edema and place the patient at risk for a compartment syndrome. Patients presenting with burns immediately after exposure have thermal rather than radiation burns. Significant injuries to the integument decrease the $LD_{50/60}$ and amplify the risk for death at any radiation exposure dose.

Apart from acute doses, exposure to low and chronic doses of radiation has been reported to cause cancer, birth anomalies, erythema, and dysfunctions to almost all organ of the body depending on the total dose. Local body irradiation at low as well as high doses of radiation cause pathological changes in almost all organs and these changes are radiation dose dependent. Some of the organs like the brain; bone, muscles, thyroid, pituitary, adrenals, and liver are radioresistant, whereas others like the lymphoid organs, reproductive organs, bone marrow, and intestinal crypts are radiosensitive. Some of the effects require a definite dose to express the changes (e.g., erythema) and are known as nonstochastic or deterministic effects, whereas for others like the genetic effect and cancer, no threshold dose exists and this is known as the stochastic effect. In humans, therapy-related adverse effects are erythema, xerostomia, mucositis, loss of taste, infection of wounds, and fistulas.

1.3. Therapeutic Uses of Ionizing Radiation

The therapeutic uses of radiation date back to 1896 when the dermatologist Freund used it to treat a hairy nevus. In the early days of radiation, it was primarily used for dermatological treatment. Emil Grubbe, a Chicago electrician and metallurgist, first treated the recurrent breast cancer of a 55-year-old woman in the last days of January 1896—only weeks after the announcement of Roentgen's discovery. In 1901, Frands Williams published work on the X-ray cure of a cancer of the lower lip.[10,11] Radiation therapy is useful in cases where surgical removal of the cancer is not possible or when surgery might debilitate the patient (e.g., when tumors are located close to the spinal cord). Together with image-guided treatment planning, radiation therapy is a powerful tool in the treatment of cancer, particularly when the cancer is detected at an early stage. Radiation therapy is used in the treatment of as many as 50% of all cancer patients. Cancer patients receive radiation therapy each year, either alone or in conjunction with surgery, chemotherapy, or other forms of cancer therapy.

Radiation therapy can be used following surgery to destroy any cancer cells that were not removed by surgery, or prior to surgery to "shrink" a previously inoperable tumor to a manageable size to enable surgical excision. Radiation can also be used to destroy any remaining cancer cells after surgery. Chemotherapy and radiation therapy can also be used together to effectively treat the cancer. Radiation can also be used to help relieve symptoms of advanced cancer (such as bleeding or pain), even if a cure is not possible.

2. RADIATION PROTECTION BY CURCUMIN

The discovery of X-rays by Roentgen in 1895 and radioactivity by Bacquerel in 1896 can be considered as the turning point in human health care, as the X-ray allowed one to see inside the human body. Although harmful effects of ionizing radiation was reported within months of the discovery of X-rays, its real magnitude was not known. The study of occupational workers like physicians and scientists handling radioactivity gave a clear picture of the harmful effects of ionizing radiations, which was further strengthened after the study of Japanese atomic bomb survivors of 1945. The use of chemicals to protect against the harmful effects of radiation was attempted after World War II with the realization of the need to safeguard humans against the military use of atomic weapons. Patt and his co-workers were the first to investigate the effect of the amino acid cysteine in rats exposed to lethal doses of X-rays.[12] They found that pretreatment of rats protected them against radiation-induced lethality. With the recognition that normal tissue protection during radiotherapy is as important as the destruction of the cancer cells, the focus of protection research became more therapy oriented. Thereafter, several chemical compounds and their analogues have been screened for their radioprotective ability; however, their high toxicity at optimum protective doses precluded their clinical use.[13,14] The other major drawback of these compounds was that they were unable to provide postirradiation protection. A good chemical protector should be able to protect against the deleterious effect of radiation during therapeutic procedures as well as during nuclear accidents, space flight, and so forth. An ideal radioprotector should be inexpensive, have no toxic implications, and can be orally administered, with rapid absorption and a reasonably good dose reduction factor.

Dietary agents that are already consumed by humans have not received the attention they deserve for their potential radioprotective effect. It is likely that if such agents are radioprotective, they might be more easily and safely used in patients undergoing radiotherapy than other radioprotective chemicals. Patients might tolerate dietary agents better than other drugs because humans consume many of these dietary ingredients daily.[15] Turmeric, *Curcuma longa* (family: Zingiberaceae), is an ancient spice, a native of Southeast Asia, used from antiquity as a dye and a condiment. It is still used in rituals in the Hindu religion and as a dye for holy robes, being natural, unsynthesized, and inexpensive. Its use dates back nearly 4000 years, to the Vedic culture in India, where it was used as a culinary spice and had some religious significance. Turmeric, as an additive, improves the palatability, aesthetic appeal, and shelf life of perishable food items. The use of turmeric became more popular when it was found to act as a therapeutic agent for various illnesses. In the Ayurvedic system of medicine, turmeric is used as a tonic and as a blood purifier. Its role in the treatment of skin diseases and its ability to soften rough skin resulted in the prolific use of turmeric in topical creams and bath soaps in India. Turmeric is also used in home remedies in the treatment of cuts, wounds, bruises, and sprains. Its use as an anti-inflammatory and antimicrobial agent has been recognized for more than a century. The importance of turmeric in medicine took a new twist when it was discovered that the dried rhizome of *Curcuma longa* is very rich in phenolics, whose structures have been identified as curcuminoids, which are chemically related to its principal ingredient, curcumin. Curcumin (diferuloyl methane), the natural yellow pigment in turmeric, is isolated from the rhizomes of the plant *Curcuma longa*. It constitutes about 3–4% of the composition of turmeric. In the south and southeast tropical Asian countries, turmeric has been used for centuries as a spice to give the specific flavor and yellow color to curry.[16,17]

Oral administration of 5, 10, and 20 mg/kg body weight (b. wt.) curcumin to mice significantly reduced the frequencies of micronucleated polychromatic erythrocytes (MPCEs) exposed to whole-body exposure to 1.15 Gy or 0.05 Gy/s γ-radiation at 24, 30, and 48 h postirradiation. This effect was observed after a single administration either 2 h before or immediately after irradiation.[18] In another study, oral administration of 400 μmol to mice reduced the radiation-induced chromosomal aberrations and micronuclei in polychromatic erythrocytes of mice exposed to 1.5–3 Gy γ-radiation.[19] Pregnant rats exposed to 1.5 or 2.6 Gy γ-radiation on day 20 of gestation developed mammary tumors, whereas feeding of 1% curcumin diet to rats between 11 and 23 days of gestation consistently reduced the radiation-induced tumors, indicating that curcumin treatment protected rats against radiation-induced mammary tumors.[20,21] Irradiation of rats to 13.5, 15.5, 16, or 18 Gy γ-irradiation resulted in a dose-dependent increase in the radiation-induced mucositis in rats, whereas curcumin treatment at a dose of 200 mg/kg daily until the end of the experiment protected rats against the radiation-induced mucositis.[22]

Curcumin was evaluated for its efficacy in radioprotection using *Saccharomyces cerevisiae* cells. Curcumin (1, 10, or 100 mM) treatment protected normal yeast cells from damage induced by γ-radiation. Using rad52 mutants, which lack a recombinational DNA repair pathway, it was found that protection was solely brought about by reducing DNA damage rather than by interfering with DNA repair. These results were further confirmed with DNA repair polymerase studies.[23] Irradiation was found to enhance inducible nitric oxide synthase (iNOS) and nitric oxide (NO) production in mouse macrophages, whereas treatment of macrophages with 10 μM reduced the NO production and nitration of proteins.[24] Protein kinase C (PKC) activity has been reported to be elevated in mice liver cytosol after 5 Gy irradiation, whereas treatment of liver cytosol with curcumin reduced the radiation-induced PKC activity.[25] The ability of curcumin and its metabolite tetrahydrocurcumin (THC) was found to inhibit radiation-induced lipid peroxidation in rat liver microsomes. Irradiation was found to increase lipid peroxidation in a dose-dependent manner and curcumin or THC treatment reduced the radiation-induced lipid peroxidation in liver microsomes.[26]

The glyoxalase enzyme system is present during embryogenesis and tissue maturation and continues to persist until cell death.[27] It regulates cell division and differentiation.[28–30] The glyoxalase system is reported to be radiosensitive,[31] particularly glyoxalase I, which could be used as a biochemical indicator for radiomodification studies.[32,33] Irradiation of mice liver to 0–6 Gy have been reported to increase the activity of glyoxalase I in a dose-dependent manner up to 4 Gy, whereas glyoxalase II activity was inhibited after irradiation. Treatment of mice with 5, 25, or 50 mg/kg b.wt. curcumin before exposure to different doses of γ-radiation resulted in the restoration of glyoxalase I activity to the control level at all curcumin doses. Curcumin pretreatment also helped to reduce the radiation-induced decline in glyoxalase II activity.[34]

The author and his group have studied the radioprotective effect of curcumin by using various study systems *in vivo* and *in vitro*. The use of the excision wound model has consistently proved the radioprotective activity of curcumin. A full-thickness wound was created on the dorsum of the irradiated mice and the progression of wound contraction was monitored by capturing video images of the wound at various postirradiation days. Whole-body irradiation of mice to 6 Gy of γ-radiation caused a significant delay in the wound contraction and wound-healing time. Oral administration of mice with 25, 50, 100, 150, or 200 mg/kg b.wt. curcumin before irradiation resulted in a curcumin dose-dependent elevation in the wound contraction and the highest contraction was observed at 100 mg/kg. The wound contraction was significantly greater at 3 ($p < 0.009$), 6 ($p < 0.05$), and 9 ($p < 0.05$) days postirradiation with 100 mg/kg curcumin. The complete healing of the wound was achieved by day 22 postirradiation in the curcumin-treated irradiation group.

Extended studies have shown that treatment of mice with 100 mg/kg curcumin before whole-body exposure to 2, 4, 6, or 8 Gy protected mice against the radiation-induced delay in the repair and regeneration of the wound.[35,36]

Hemi-body exposure of mice to 2, 4, 6, or 8 Gy resulted in a delayed wound-healing, whereas oral administration of 100 mg/kg curcumin protected mice against the radiation-induced delay in the wound repair and reduced the mean wound-healing time when compared to irradiated wounds.[37] The effect of radiotherapy-related exposure was also evaluated where mice were exposed daily to 2 Gy per day for varying times and the assessment of wound-healing was done after exposure to 10, 20, or 40 Gy of hemi-body radiation. This regimen resulted in a radiation dose-dependent inhibition of wound repair and a maximum inhibition of repair was observed in those mice receiving 40 Gy of partial-body γ-radiation. When the animals were given 100 mg curcumin daily before each fraction of radiation, it resulted in the enhancement of repair and faster healing of wounds, as evident by

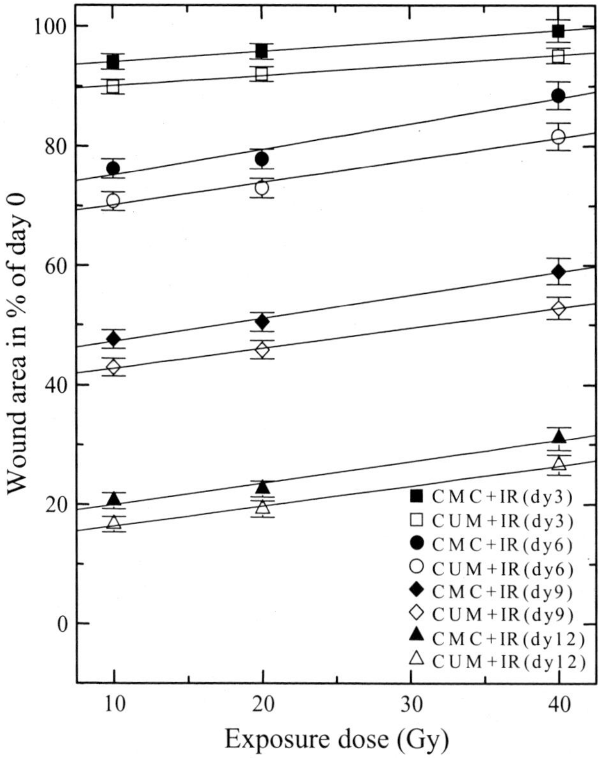

Figure 2. Effect of 100 mg/kg curcumin on the healing of an excision wound exposed to different doses of fractionated γ-radiation.

a greater contraction of the wound (Figure 2) and reduction in the mean wound-healing time (Figure 3). The protection against the radiation–induced changes in wound-healing was due to the higher synthesis of DNA, collagen, hexosamine, nitrite, and nitrate during the healing process in the granulation tissue, which allowed early wound modeling.[36] The histological examination of wound biopsies periodically showed increased fibroblasts, angiogenesis, and collagen deposition.[36]

Lessons from the experience with radioprotectors worldwide are that animal studies with survival as the end point are the most confirmatory, as 30-day survival after lethal whole-body irradiation clearly indicates the capacity of the drug to facilitate recovery, and regeneration. Regeneration of the gastrointestinal epithelium and hemopoietic progenitor cells in the bone marrow, the two most radiosensitive organs, which are essential for sustenance of life.[38] Therefore, survival studies can be considered as the most effective parameter for testing the radioprotective ability of any pharmacological agent after irradiation. Treatment of mice with 100 mg/kg curcumin for 5 consecutive days before exposure to 7, 8, 9, or 10 Gy increased the survival of mice and a dose reduction factor of 1.15 was reported. A similar study was done after infliction of the wound. The combined injuries reduced the $LD_{50/30}$

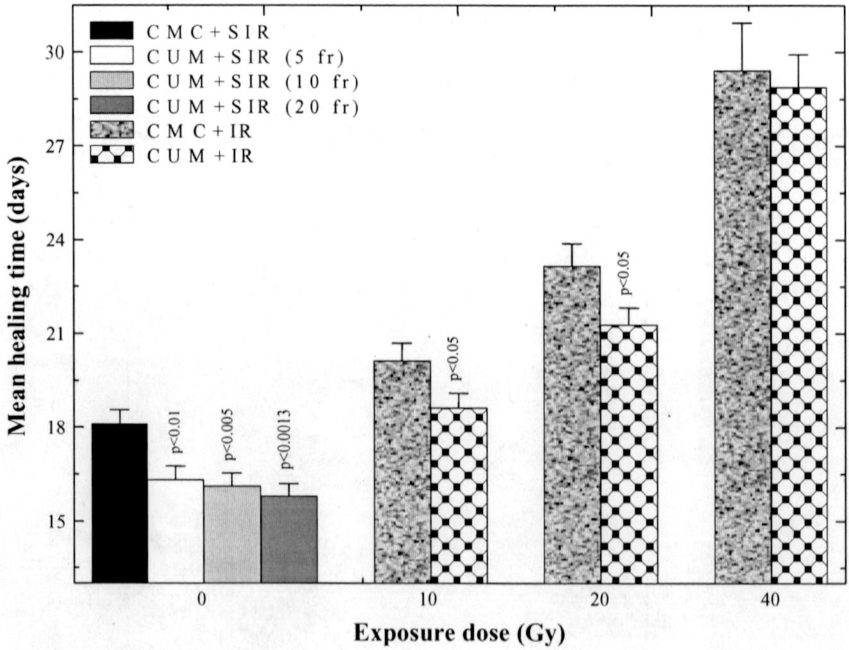

Figure 3. Effect of 100 mg/kg curcumin on the mean wound-healing time in mice exposed to different doses of fractionated γ-radiation.

of mice, whereas curcumin treatment for 5 consecutive days before irradiation and infliction of additional injury in the form of a wound increased the survival of mice (Figure 4), indicating that curcumin has a protective effect even in cases of combined injuries.

The effect of 100 mg/kg curcumin on the alteration in the antioxidant status of mice partially (hemi-body) exposed to 10, 20, or 40 Gy of fractionated (2 Gy/day) γ-radiation was studied. Irradiation of animals resulted in a dose-dependent decline in the activities of superoxide dismutase, glutathione peroxidase, and glutathione concentration at various postirradiation times. Curcumin pretreatment resulted in a significant elevation in the activities of both the enzymes and glutathione in the irradiated mouse skin (Figure 5). Conversely, lipid peroxidation increased in a dose-dependent manner in both of the groups, reaching a peak level by 3 h postirradiation, whereas curcumin pretreatment inhibited the radiation-induced elevation in lipid peroxidation in the skin of mice exposed to different doses of fractionated γ-radiation.

The role of various concentration of curcumin was studied on the radiation-induced genotoxicity in cultured human peripheral blood lymphocytes (HPBLs) exposed to 3 Gy γ-radiation. Treatment of HPBLs to different concentrations of curcumin did not significantly alter the spontaneous frequency of micronucleated binucleate cells (MNBNCs). The irradiation of HPBLs to 3 Gy caused

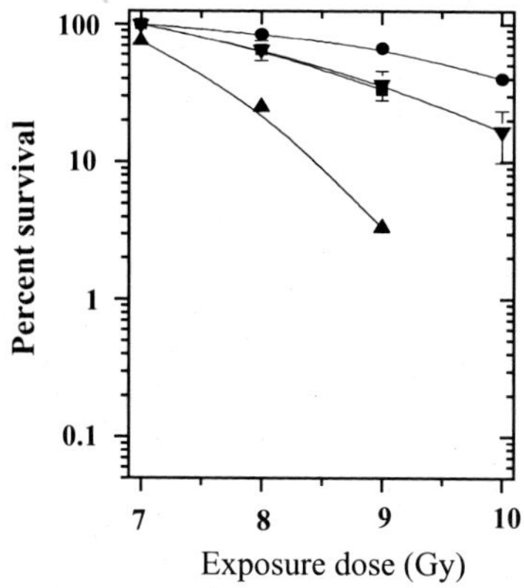

Figure 4. Effect of oral administration of curcumin on the 30-day survival of mice exposed to different doses of whole-body irradiation before infliction of the excision wound. ▲: irradiation + wound; ■: irradiation alone; ▼: curcumin + irradiation + wound; •: curcumin + irradiation

a significant elevation in the frequency of MNBNCs. Treatment of HPBLs with different doses of curcumin before exposure to 3 Gy γ-radiation caused a significant decline in the radiation-induced micronuclei formation up to 2 μg/mL when compared with the non-drug-treated 3-Gy-exposed cultures. A further increase in the curcumin concentration resulted in the elevation in the frequency of MNBNCs, and the highest frequency of MNBNC was observed at 50 μg/mL curcumin when compared with 3 Gy alone. Of all of the concentrations of curcumin screened, the lowest frequency of micronuclei was observed for 0.5 μg/mL (Figure 6).

3. CURCUMIN AND UV RADIATION

Exposure of excessive sunlight is an important etiologic factor in the development of acute inflammation, characterized by erythema, edema, and immunosuppression and is thus linked to the progression of skin cancer.[39,40] Ultraviolet B (UVB) in the 290–320-nm range is a well-known major risk factor for the development of acute inflammation as well as nonmelanoma skin cancer in the epidermis.[41–43] The sequence of carcinogenic events includes UVB-induced p53 mutation, formation and expansion of p53 mutant clones, creation of precancers, and malignant conversion of precancerous lesions to carcinoma *in situ* and SCC (sequamous cell carcinoma).[41,44] Accumulating data indicate that UVB exerts its detrimental effect mainly through the induction of direct DNA damage or the production of reactive oxygen species (ROS).[45–47] Direct DNA damage or ROS often triggers some signaling pathways such as mitogen-activated protein kinases (MAPKs), which are known to be involved in the proliferation and survival of the cells.[48,49] Effect of

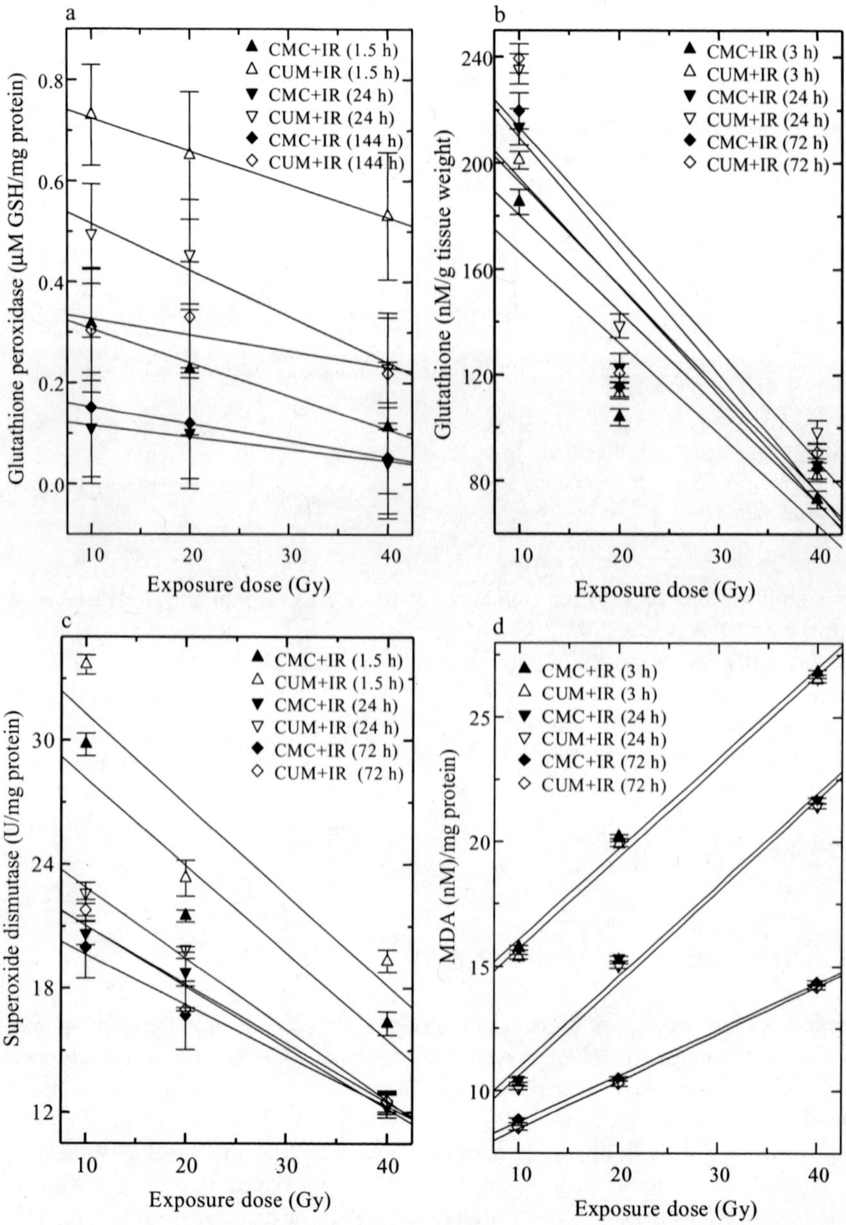

Figure 5. Effect of curcumin on various antioxidants and lipid peroxidation in mouse skin exposed to fractionated doses of γ-radiation: (a) glutathione peroxidase; (b) glutathione; (c) superoxide dismutase, and (d) lipid peroxidation.

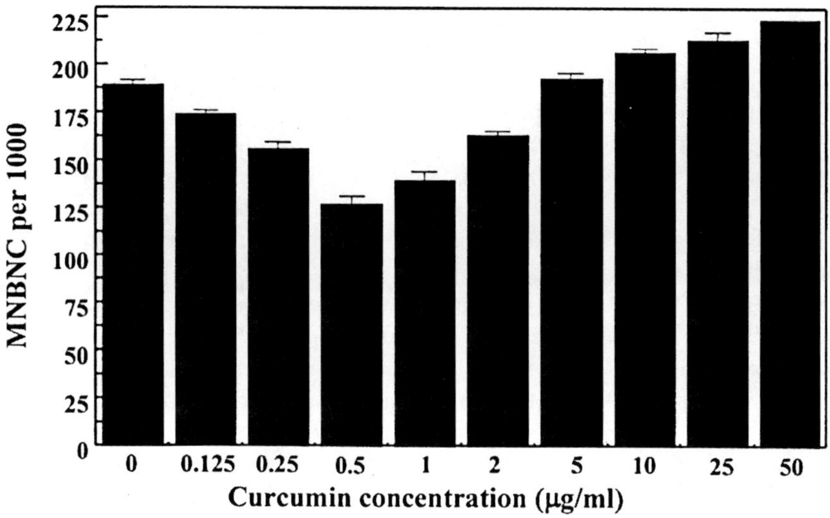

Figure 6. Effect of various concentrations of curcumin on the micronuclei formation in cultured HPBLs exposed to 3 Gy γ-radiation.

UV-induced oxidative stress and apoptosis was studied in A431 cells treated or not with curcumin. UV radiation was found to increase the ROS generation and cause a time-dependent decline in the cell viability, whereas curcumin treatment arrested this trend. Curcumin pretreatment significantly reduced UV-induced ROS, where 50 μM curcumin arrested ROS generation up to 35% of the control. Curcumin treatment was found to reduce matrix mitochondrial membrane potential (MMP) after 1 and 2 h of UV exposure, whereas curcumin pretreatment completely blocked this reduction in MMPs. Curcumin also reduced UV-induced apoptotic changes, cytochrome-*c* release and c-Jun N-terminal kinase (JNK) activation, loss of MMP, mitochondrial release of caspase-3 activation, and cleavage/activation of PAK2 in A431 cells.[50] Curcumin has also been reported to inhibit SOS gene (*umuC*) expression and UV-radiation-induced mutagenesis in *Salmonella typhimurium TA1535/pSKlO02* and *Escherichia coli* K-12 strains in a concentration-dependent manner.[51]

In another study the expression of cycloxygenase (COX)-2 and associated molecular mechanisms were studied in curcumin-treated UVB irradiated HaCaT cells. The expression of COX-2 mRNA and protein were upregulated in UVB-irradiated HaCaT cells in a dose- and time-dependent manner, and treatment of HaCaT cells with curcumin inhibited UVB-irradiation-induced COX-2 mRNA and protein expression in a concentration-dependent manner. Curcumin also inhibited UVB-induced activations of p38 MAPK and JNK in HaCaT cells. The DNA-binding activity of activator protein-1 (AP-1) transcription factor was markedly reduced by curcumin treatment in UVB-irradiated HaCaT cells.[52]

4. RADIOSENSITIZATION BY CURCUMIN

Radiosensitization has been studied extensively with various chemotherapy drugs (cisplatinum, 5-FU, taxols) prior to irradiation or as concomitant treatment for patients with head and neck SCC. The side effects of these drugs, however, prevent their routine use for all irradiated patients; thus, the chemotherapy–radiation protocols are currently used only as a part of clinical trials for organ preservation or for patients with unresectable cancers. Local tumor control, or radiocurability, after curative radiotherapy is contingent on the successful killing of all clonogenic cells capable of tumor regrowth. This is influenced by resistance factors acting at a cellular level, such as intrinsic cellular radiosensitivity, the repair of ionizing radiation-induced DNA damage, and microenvironmental resistance factors at the three-dimensional level of the stromal tissues and tumors. The latter includes tumor hypoxia and its reoxygenation and tumor cell proliferation with the potential for accelerated repopulation of surviving tumor clonogens.[53,54] Specifically, modification of the intrinsic cellular radiosensitivity by resistance factors, as indicated by an increased surviving fraction at 2 Gy within the clonogenic radiation cell survival curve, is an important concept. Small increases in this parameter can lead to large, logarithmic decreases in final cell killing after therapeutic doses of fractionated radiotherapy. The studies on the intrinsic radiosensitivity of some human malignancies based on pretreatment biopsies have been shown to predict for clinical radiocurability and patient outcome.[55,56]

Treatment of the p53 mutant prostate cancer cell line with 2 or 4 μM curcumin before irradiation to 5 Gy resulted in the enhancement of the effect of radiation, as evident by a decline in the cell survival when compared to irradiation alone. Curcumin pretreatment also enhanced the apoptosis at 24 and 48 h after irradiation when compared to radiation alone. Irradiation of PC-3 cells upregulated tumor necrosis factor (TNF)-α protein, leading to an increase in nuclear factor (NF)-κB activity and Bcl-2 protein. Pretreatment of PC-3 cells with curcumin inhibited TNF-α mediated NF-κB activity and expression of Bcl-2 protein without altering the expression of Bax protein. Apart from that, curcumin pretreatment increased the activation of cytochrome-*c* and caspase-9 and caspase-3 enzymes.[57]

Irradiation of squamous cell carcinoma SSC2 resulted in a radiation dose-dependent decline in cell survival as well as cell count. Whereas treatment of SSC2 cells with 3.5 μM curcumin for 48 h enhanced the radiation-induced decline in the cell survival and cell counts. These results indicate that curcumin enhanced the effect of radiation *in vitro*.[58] In another study, the effect of turmeric and curcumin on the frequencies of chromosome aberrations in Chinese hamster ovary (CHO) cells exposed to 2.5 Gy γ-radiation was studied. Treatment of CHO cells with 100, 250, or 500 mg/mL turmeric or 2.5, 5, or 10 μg/mL curcumin, before exposure to 2.5 Gy during different phases of the cell cycle increased the frequencies of chromosome aberrations. Turmeric at 500 μg/mL elevated the frequency of chromosome aberrations during the G_2/S phase, whereas curcumin at 10 μg/mL increased these frequencies during S and G_2/S phases of the cell cycle. These results clearly indicate the exacerbated effect of turmeric and curcumin on radiation-induced clastogenicity, suggesting that these antioxidants are also potentiating agents, depending on the experimental conditions. Turmeric was not clastogenic by itself, whereas curcumin at 10 μg/mL enhanced the chromosomal damage frequency.[59]

5. MECHANISM OF RADIOPROTECTION

It is a well-established fact that radiation induces ROS that follows a cascade of events leading to DNA damage that includes single- or double-strand breaks (dsb), base damage, and DNA–DNA or DNA–protein cross-links, and these lesions cluster as complex, local, multiply-damaged sites. The DNA dsbs are considered the most lethal events after ionizing radiation and have been found to be the main target of cell killing by radiation. The putative mechanisms of radioprotection by curcumin are shown in Figure 7. The radioprotective activity of curcumin might not be due to a single mechanism but to several mechanisms. The scavenging of radiation-induced free radicals and the elevation of cellular antioxidants by curcumin in irradiated systems could be one of the main leading mechanisms. Upregulation of enzymes like catalase, glutathione transferase (GST), glutathione peroxidase (GSHpx), superoxide dismutase (SOD), and their mRNAs might be another mechanism of radioprotection by curcumin. Reduction in lipid peroxidation and elevation in glutathione (GSH) and increase in sulphydryl groups might also contribute to some extent for its radioprotective activity. Inhibition of

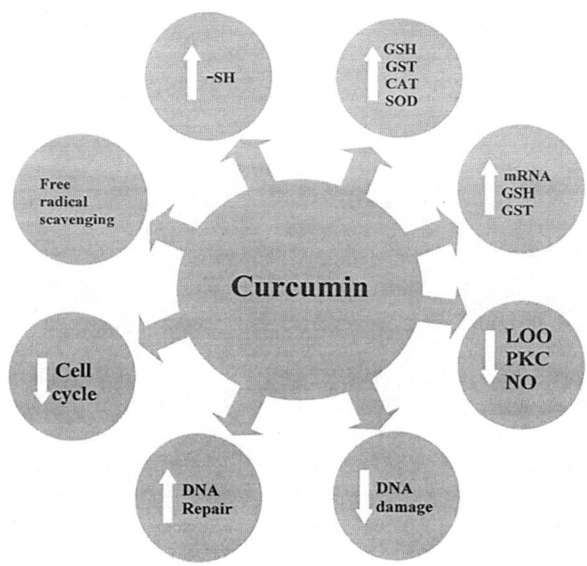

Figure 7. Mechanism of radioprotection by curcumin. ↑ indicates increase or upregulation and ↓ indicates decrease or inhibition.

activation of PKC, MAPK, and NO by curcumin might also provide protection against radiation-induced damage. Curcumin has been reported to scavenge free radicals, increase antioxidant status, inhibit lipid peroxidation, and elevate GST, GSHpx, SOD, GSH, and sulphydryl groups.[25,26,60–65] Curcumin has been reported to enhance GSH content and acid-soluble sulphydryl groups.[66]

6. MECHANISM OF RADIOSENSITIZATION

Ionizing radiation can initiate cellular damage through the cell plasma membrane by sphingomyelin–ceramide or Fas-mediated pathways or as clustered damage within the nuclear DNA. Induction of transcription factors including c-MYC, p53, c-FOS/c-JUN, and NF-κB proteins follows thereafter with binding to specific DNA sequences responsible for the transcription of cytokine-, growth factor-, and cell cycle-related genes. Ionizing radiation can also modify intracellular signaling through modification of the activity of tyrosine kinases, mitogen-activated protein kinases (MAPK), stress-activated protein kinases (SAPK), and RAS associated proteins. EGFR (*EGR-1*) and other genes,[67] such as cell cycle-related proteins like *GADD45*, *p21CIP1/WAF*, *cyclin B*, *p*53, growth factors and their receptors, or cytokines such as platelet-derived growth factor, transforming growth factor-β, epidermal growth factor, TNF, and interleukins leading to radioresistance during radiotherapy of tumor.[68,69] Nuclear damage created by ionizing radiation causes a number of DNA lesions, including single- or double-strand breaks, base damage, and DNA–DNA or DNA–protein cross-links, and these lesions cluster as complex, local, multiply-damaged sites. Persistence of these DNA lesions can be curative in cancer patients, if repair is optimized in normal versus tumor tissues, thus improving the therapeutic ratio.

Curcumin can radiosensitize cells by more than one defined pathway (Figure 8). It can inhibit/downregulate the TNF-α and NF-κB activation pathways.[70–72] Curcumin inhibits COX, lipoxygenases, prostaglandin E_2 and leukotrienes B4 and C4.[73–75] It can also downregulate Bcl-2, Bcl-X_L TNF-α, PKC, cell cycle cytokines, cytochrome P450, and poly (ADP-ribose) polymerase (PARP). Curcumin upregulates Bcl-xs and Bax and activates caspase-9 and caspase-3.[57,76,77] The caspase-9 activation is effected by cytochrome-*c* release from mitochondria.[76] Further, curcumin has been reported to arrest cell proliferation in the G_2-G_0-S-phase.[78–80] The inhibition of cell cycle by curcumin is mediated by the activation of PPAR-γ suppression of the cyclin D1 a critical protein in the cell cycle. The curcumin also blocks epithelial growth factor (EGF) signaling by inhibition of EGF receptor (EGFR) tyrosine phosphorylation and suppressing the gene expression of EGFR.[81,82] It has also been reported to inhibit p38, MAPK, and JNK.[52]

7. CONLCLUSIONS

Ionizing radiation inflicts damage in biological systems by inducing free radicals and ROS that lead to a cascade of molecular events like elevation in c-MYC, p53,

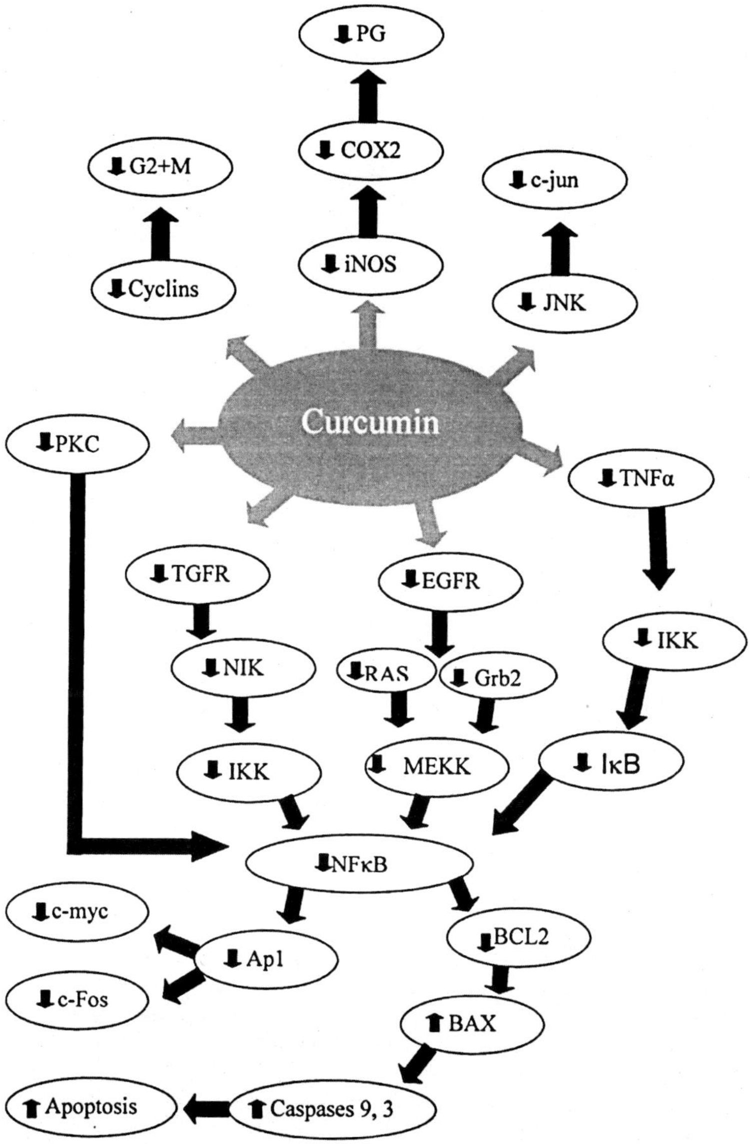

Figure 8. Mechanism of radiosensitization by curcumin. ↑ indicates increase or upregulation and ↓ indicates down regulation or inhibition.

c-FOS/c-JUN, NF-κB cytokine, growth factors, MAPK, SAPK, RAS-associated proteins, EGFR, and cyclins and decline in antioxidants resulting in DNA damage. Whole-body irradiation cause central nervous system, gastrointestinal, bone marrow, and cutaneous syndromes, whereas, chronic irradiation leads to pathogenesis of various organs, mutagenesis, and carcinogenesis. Curcumin, a yellow pigment present in turmeric, is a versatile molecule, which can protect against radiation-induced damage at low doses, whereas it might increase the effect of radiation at higher doses. It is nontoxic up to 12 g/day in humans. The mechanism of radioprotection and radiosensitization are several. The radioprotection of curcumin is due to scavenging of free radicals, elevation in antioxidants, upregulation of mRNAs for GSH, GST, SOD, and catalase enzymes. It also inhibits NO production. The increased radiosensitivity could be due to inhibition of radiation-induced elevation of growth factors, cytokines, cyclins, NF-κB, PKC, TNF-α and inhibition of cell cycle at the $G_2 + M$ phase, increased apoptosis, and some other unknown mechanisms.

REFERENCES

1. C. Von Sonntag, *The Chemical Basis of Radiation Biology*. London: Taylor and Francis, 1987.
2. G. V. Buxton, C. L. Greenstock, W. P. Helman, and A. B. Ross, Critical review of rate constants for reactions of hydrated electrons, hydrogen atoms and hydroxyl radical in aqueous solution. *J Phys Chem Ref Data* **17**, 513–886 (1988).
3. M. C. R. Symons and Gutteridge, *Free Radicals and Ion, Chemistry Biology and Medicine.* Oxford: Oxford University Press, 1998, pp. 40–60.
4. W. A. Pryor, Cancer and free radicals. *Basic Life Sci* **39**, 45–59 (1986).
5. S. S. Wallace, S. S., 1988, Detection and repair of DNA base damages produced by ionizing radiation. *Environ Mol Mutagen* **12,** 431–477 (1988).
6. H. Esterbauer, Estimation of peroxidative damage, a critical review. *Pathol Biol Paris* **44**, 25–28 (1996).
7. B. Halliwell and J. M. C. Gutteridge, *Free Radicals in Biology and Medicine*. Oxford: Oxford University Press, 1997.
8. R. I. Walker and R. J. Cerveny, (eds.), *Medical Consequences of Nuclear Warfare*. Falls Church, VA: Office of the Surgeon General, 1989.
9. A. V. Barabanova, Acute radiation syndrome with cutaneous syndrome. In: R. C. Ricks, M. E. Berger, and F. M. O'Hara, eds. *The Medical Basis for Radiation Accident Preparedness, The Clinical Care of Victims*. New York: Parthenon, 2002, pp. 217–224.
10. H. D. Kogelnik, [100 years radiotherapy. On the birth of a new specialty]. *Wien Klin Wochenschr.* **110(9)**, 313–320 (1998).
11. H. D. Kogelnik, The history and evolution of radiotherapy and radiation oncology in Austria. *Int J Radiat Oncol Biol Phys* **35(2)**, 219–226 (1996).
12. H. M. Patt, E. B. Tyree, R. L. Straube, and D. E. Smith, D. E., 1949, Cysteine protection against X-irradiation, *Science* **110**, 213–214 (1949).
13. T. R. Sweeney, *Survey of Compounds from the Antiradiation Drug Development Program of the U.S. Army Medical Research and Development Command.* Washington, DC: U.S.Government Printing Office, 1979. pp. 308–318.
14. J. R. Maisin, J. R., 1998, Bacq and Alexander Award lecture: Chemical radioprotection, past, present, and future prospects. *Int J Radiat Biol* **73**, 443–450 (1998).
15. G. C. Jagetia and V. A.Venkatesha, Effect of mangiferin on radiation-induced micronucleus formation in cultured human peripheral blood lymphocytes. *Environ Mol Mutagen* **46**, 12–21 (2005).

16. H. P. Ammon and M. A. Wahl, Pharmacology of *Curcuma longa*, *Planta Med* **57(1)**, 1–7 (1991).
17. D. Eigner and D. Scholz, Ferula asa-foetida and *Curcuma longa* in traditional medical treatment and diet in Nepal *J. Ethnopharmacol* **67**, 1–6 (1999).
18. S. K. Abraham, L. Sarma, and P. C. Kesavan, Protective effects of chlorogenic acid, curcumin and beta-carotene against gamma-radiation-induced in vivo chromosomal damage. *Mutat Res* **303(3)**, 109–112 (1993).
19. K. C. Thresiamma, J. George, and R. Kuttan, Protective effect of curcumin, ellagic acid and bixin on radiation induced genotoxicity, *J Exp Clin Cancer Res* **17**, 431–434 (1998).
20. H. Inano, M. Onoda, N. Inafuku, M. Kubota, Y. Kamada, T. Osawa, H. Kobayashi, and K. Wakabayashi, Chemoprevention by Curcumin during the promotion stage of tumorigenesis of mammary gland in rats irradiated with X-rays. *Carcinogenesis* **20(6),** 1011–1018 (1999).
21. H. Inano, M. Onoda, K. Suzuki, H. Kobayashi, and K.Wakabayashi, Radiation-induced mammary tumors in virgin and parous rats administered contraceptive steroids, 17 α-ethibnylestradiol and norethisterone. *Carcinogenesis* **21(5)**, 1043–1050 (2000).
22. M. Rezvani and G. A.Ross, Modification of radiation-induced acute oral mucositis in the rat, *Int J Radiat Biol* **80(2)**, 177–182 (2004).
23. P. Nemavarkar, B. K. Chourasia, and K. Pasupathy, Evaluation of radioprotective action of compounds using Saccharomyces cerevisiae. *J Environ Pathol Toxicol Oncol* **23(2)**, 145–151 (2004).
24. H. Narang and M. Krishna, Inhibition of radiation induced nitration by Curcumin and nicotinamide in mouse macrophages. *Mol Cell Biochem* **276**, 7–13 (2005).
25. P. Varadkar, P. Dubey, M. Krishna, and N. C. Verma, Modulation of radiation-induced protein kinase C activity by phenolics, *J Radiol Prot* **21** , 361–370 (2001).
26. S. M. Khopde, K. I. Priyadarsini, S. N. Guha, J. G. Satav, P. Venkatesan, and M. N. Rao, Inhibition of radiation-induced lipid peroxidation by Tetrahydrocurcumin. Possible mechanisms by pulse radiolysis. *Biosci Biotechnol Biochem* **64(3)**, 503–509 (2000).
27. A. C. McLellan and P. J. Thornalley, Glyoxalase activity in human red blood cells fractionated by age. *Mech Ageing Dev* **48**, 63–71 (1989).
28. A. Szent-Györgyi, Bioelectronics of cancer. *Bioenergetics* **4,** 533–562 (1973).
29. A. Szent-Györgyi, Protein radicals, regulation and cancer. *Int J Quantum Chem* **QBS4**, 179–184 (1997).
30. N. I. Hooper, M. J. Tisdale, and P. J. Thornalley, Glyoxalase activity during differentiation of human leukaemic cells in vitro. *Leuk Res* **11**, 1141–1148 (1987).
31. R. Sharma and R. K. Kale, Effect of radiation on glyoxalase I and glyoxalase II activities in spleen and liver of mice. *Int J Radiat Biol* **63(2),** 233–238 (1993).
32. R. Sharma-Luthra and R. K. Kale, Inhibition of radiationinduced changes of glyoxalase I activity in mouse spleen and liver by phenothiazines. *Int J Radiat Biol* **67(4),** 403–410 (1995).
33. R. K. Kale, Exploitation of hypoxia for radiation therapy: A lesson from phenothiazines. *Med Hypothes* **47**, 107–110 (1996).
34. D. Choudhary, D. Chandra, and R. K. Kale, Modulation of radioresponse of glyoxalase system by curcumin. *J Ethnopharmacol* **64**, 1–7 (1999).
35. G. C. Jagetia and G. K. Rajanikant, Effect of various doses of curcumin on the radiation-impaired healing of excision wound in mice: A preliminary study. *J Wound Care* **13(3)**, 107–109 (2004).

36. G. C. Jagetia and G. K. Rajanikant, Role of curcumin, a naturally occurring phenolic compound of turmeric in accelerating the repair of excision wounds in mice whole-body exposed to various doses of γ-radiation. *J Surg Res* **120**, 127–138 (2004).
37. G. C. Jagetia and G. K. Rajanikant, Curcumin treatment enhances the repair and regeneration of wounds in mice hemi-body exposed to γ-radiation, *Plast Reconstr Surg* **115(2)**, 515–528 (2005).
38. G. C. Jagetia, P. Venkatesh, and M. S. Baliga, Fruit extract of *Aegle marmelos* protects mice against radiation-induced lethality. *Integr Cancer Ther* **3(4)**, 323–332 (2004).
39. R. D. Granstein and M. S. Matsui, UV radiation-induced immunosuppression and skin cancer. *Cutis* **74(5)**, 4–9 (2004).
40. Y. Matsumura and H. N. Ananthaswamy, Toxic effects of ultraviolet radiation on the skin. *Toxicol Appl Pharmacol* **195,** 298–308 (2004).
41. D. Grossman and D. J. Leffell, The molecular basis of nonmelanoma skin cancer: A new understanding. *Arch Dermatol* **133**, 1263–1270 (1997).
42. E. C. De Fabo, F. P. Noonan, T. Fears, and G. Merino, Ultraviolet B but not ultraviolet A radiation initiates melanoma. *Cancer Res* **64**, 6372–6376 (2004).
43. J. Ramos, J. Villa, A. Ruiz, R. Armstrong, and J. Matta, UV dose determines key characteristics of nonmelanoma skin cancer. *Cancer Epidemiol Biomarkers Prev* **13**, 2006–2011 (2004).
44. A. Ziegler, A. S. Jonason, D. J. Leffell, J. A. Simon, H. W. Sharma, J. Kimmelman, L. Remington, T. Jacks, and D. E. Brash, Sunburn and p53 in the onset of skin cancer. *Nature* **372,** 773–776 (1994).
45. F. R. de Gruijl, Photocarcinogenesis, UVA *vs.* UVB radiation. *Skin Pharmacol Appl Skin Physiol* **15**, 316–320 (2002).
46. D. Kulms, E. Zeise, B. Poppelmann, and T. Schwarz, DNA damage, death receptor activation and reactive oxygen species contribute to ultraviolet radiation-induced apoptosis in an essential and independent way. *Oncogene* **21**, 5844–5851 (2002).
47. D. E. Heck, A. M. Vetrano, T. M. Mariano, and J. D. Laskin, UVB light stimulates production of reactive oxygen species, unexpected role for catalase. *J Biol Chem* **278**, 22,432–22,436 (2003).
48. S. J. Rhee, Redox signaling, hydrogen peroxide as intracellular messenger. *Exp Mol Med* **31,** 53–59 (1999).
49. M. Torres and H. J. Forman, Redox signaling and the MAP kinase pathways. *Biofactor* **17**, 287–296 (2003).
50. W.-H. Chan, C.-C. Wu, J.-S. and Yu, Curcumin inhibits UV irradiation-induced oxidative stress and apoptotic biochemical changes in human epidermoid carcinoma A431 cells. *J Cell Biochem* **90**, 327–338 (2003).
51. Y. Oda, Inhibitory effect of curcumin on SOS functions induced by UV irradiation. *Mutat Res* **348(2)**, 67–73 (1995).
52. J.-W.Cho, K. Park, G. R. Kweon, B.-C. Jang, W.-K. Baek, M.-H. Suh, C.-W. Kim, K.-S. Lee, and S.-I.Suh, Curcumin inhibits the expression of COX-2 in UVB–irradiated human keratinocytes (HaCaT) by inhibiting activation of AP-1, p38 MAP kinase and JNK as potential upstream targets. *Exp Mol Med* **37(3)**, 186–192 (2005).
53. G. G. Steel and M. J. Peckham, Exploitable mechanism in combined radiotherapy-chemotherapy: The concept of additivity. *Int J Radiat Oncol Biol Phys* **5**, 85–91 (1979).
54. R. K. Schmidt-Ullrich, J. N. Contessa, P. Dent, R. B. Mikkelsen, K. Valerie, D. B. Reardon, G. Bowers, and P. S. Lin, Molecular mechanisms of radiation-induced accelerated repopulation. *Radiat Oncol Invest* **7**, 321–330 (1999).

55. J. Deacon, M. J. Peckham, and G. G. Steel, The radioresponsiveness of human tumours and the initial slope of the cell survival curve. *Radiother Oncol* **2**, 317–323 (1984).
56. C. M. West, S. E. Davidson, S. A. Roberts, and R. D. Hunter, The independence of intrinsic radiosensitivity as a prognostic factor for patient response to radiotherapy of carcinoma of the cervix. *Br J Cancer* **76**, 1184–1190 (1997).
57. D. Chendil, R. S. Ranga, D. Meigooni, S. Sathishkumar, and M. M. Ahmed, Curcumin confers radiosensitizing effect in prostate cancer cell line PC-3 *Oncogene* **23**, 1599–1607 (2004).
58. R. Khafif, K. Hurst, D. M. Kyker, Z. Fliss, Z. Gil, and J. E. Medina, J. E., 2005, Curcumin: A new radiosensitizer of squamous cell carcinoma cells. *Otolaryngol Head Neck Surg* **132**, 317–321 (2005).
59. M. C. P. Araujo, F. L. Dias, and C. S.Takahashi, Potentiation by turmeric and curcumin of γ–radiation-induced chromosome aberrations in Chinese Hamster ovary cells. *Teratogen Carcinogen Mutagen* **19**, 9–18 (1999).
60. M. Subramanian, M. N. A. Sreejayan Rao, T. P. A. Devasagyam, and B. B. Singh, Diminution of singlet oxygen-induced DNA-damage by J.K. Lin and S.Y. Lin-Shiau curcumin and related antioxidants. *Mutat Res* **311**, 249–255 (1994).
61. B. Joe and B. R. Lokesh, Role of capsaicin, curcumin and dietary n-3 fatty acids in lowering the generation of reactive oxygen species in rat peritoneal macrophages. *Biochim Biophys Acta* **1224**, 255–263 (1994).
62. N. Sreejayan and M. N. Rao, Free radical scavenging activity of curcuminoids. *Arzneimittelforschung* **46(2)**, 169–171 (1996).
63. A. C. Reddy and B. R. Lokesh, Effect of curcumin and eugenol on ironinduced hepatic toxicity in rats. *Toxicology* **107**, 39–45 (1996).
64. R. G. Bristow and R. P. Hill, Molecular and cellular basis of radiotherapy. In: H. R. Tia, ed. *The Basic Science of Oncology.* New York: McGraw-Hill, New York, 1998, pp 295–322..
65. S. K. Biswas, D. McClure, L. A. Jimenez, I. L. Megson, and I. Rahman, Curcumin induces glutathione biosynthesis and inhibits NF-kappaB activation and interleukin-8 release in alveolar epithelial cells, mechanism of free radical scavenging activity. *Antioxid Redox Signa.* **7(1–2)**, 32–41 (2005).
66. B. Joe, M. Vijaykumar, and B. R. Lokesh, Biological properties of Curcumin: Cellular and molecular mechanisms of action. *Crit Rev Food Sci Nutr* **44,** 97–111 (2004).
67. F. Bonte, M. S. Noel-Hudson, J. Wepierre, and A. Meybeck, Protective effect of curcuminoids on epidermal skin cells under free oxygen radical stress. *Planta Med* **63**, 265–266 (1997).
68. D. E. Hallahan, D. R. Spriggs, M. A. Beckett, D. W. Kufe, and R. R. Weichselbaum, Increased tumor necrosis factor alpha mRNA after cellular exposure to ionizing radiation. *Proc Natl Acad Sci USA* **86(24)**,10,104–10,107 (1989).
69. C. N. Coleman, Radiation oncology: Linking technology and biology in the treatment of cancer. *Acta Oncol* **41**, 6–13 (2002).
70. S. Singh and B. B. Aggarwal, Activation of transcription factor NF-kappa B is suppressed by Curcumin (diferuloylmethane). *J Biol Chem* **270**, 24,995–25,000 (1995).
71. B. B. Aggarwal, A. Kumar, and A. C. Bharti, Anticancer potential of Curcumin: Preclinical and clinical studies. *Anticancer Res* **23**, 363–398 (2003).
72. A. K. Garg, T. A. Buchholz, and B. B. Aggarwal, Chemosensitization and radiosensitization of tumors by plant polyphenols. *Antioxid Redox Signal* **7(11–12)**, 1630–1647 (2005).

73. C. V. Rao, A. Riven, B. Simi, and B. S. Reddy, Chemoprevention of colon carcinogenesis by dietary Curcumin: A naturally occurring plant phenolic compound. *Cancer Res* **55**, 259–266 (1995).
74. M. T. Huang, H. L. Newmark, and K. Frenkel, Inhibitory effects of Curcumin on tumorigenesis in mice. *J Cell Biochem* **27**, 26–34 (1997).
75. B. Joe and B. R. Lokesh, Effect of Curcumin and capsaicin on arachidonic acid metabolism and lysosomal enzyme secretion by rat peritoneal macrophages. *Lipids* **32,** 1173–1180 (1997).
76. J. C. Reed, JCytochrome c, can't live with it—can't live without it. Cell **91(5)**, 559–562 (1997).
77. S. Sen, H. Sharma, and N. Singh, Curcumin enhances Vinorelbine mediated apoptosis in NSCLC cells by the mitochondrial pathway. *Biochem Biophys Res Commun* **331**, 1245–1252 (2005).
78. R. Hanif, L. Qiao, S. J. Shiff, and B. Rigas, Curcumin, a natural plant phenolic food additive, inhibits cell proliferation, and induces cell cycle changes in colon adenocarcinoma cell lines by a prostaglandin–independent pathway. *J Lab Clin Med* **130**, 576–584 (1997).
79. H. W. Chen and H. C. Huang, Effect of curcumin on cell cycle progression and apoptosis in vascular smooth muscle cells. *Br J Pharmacol* **124**, 1029–1040 (1998).
80. H. Chen, Z. S. Zhang, Y. L. Zhang, and D. Y. Zhou, Curcumin inhibits cell proliferation by interfering with the cell cycle and inducing apoptosis in colon carcinoma cells. *Anticancer Res* **19**, 3675–3680 (1999).
81. A. Chen and J. Xu, J., 2005, Activation of PPARγ by curcumin inhibits Moser cell growth and mediates suppression of gene expression of cyclin D1 and EGFR. *Am J Physiol Gastrointest Liver Physiol* **288,** G447–G456 (2005).
82. L. Korutla, J. Y. Cheung, J. Mendelsohn, and R. Kumar, Inhibition of ligand-induced activation of epidermal growth factor receptor tyrosine phosphorylation by Curcumin. *Carcinogenesis* **16**, 1741–1745 (1995).

IMMUNOMODULATION BY CURCUMIN

Subhash C. Gautam, Xiaohua Gao, and Scott Dulchavsky

Abstract: Turmeric, the bright yellow spice extracted from the tuberous rhizome of the plant *Curcuma longa*, has been used in traditional Indian and Chinese systems of medicine for centuries to treat a variety of ailments, including jaundice and hepatic disorders, rheumatism, anorexia, diabetic wounds, and menstrual difficulties. Most of the medicinal effects of turmeric have been attributed to curcumin, the principal curcumanoid found in turmeric. Recent evidence that curcumin exhibits strong anti-inflammatory and antioxidant activities and modulates the expression of transcription factors, cell cycle proteins, and signal transducing kinases has prompted the mechanism-based studies on the potential of curcumin to primarily prevent and treat cancer and inflammatory diseases. Little work has been done to study the effect of curcumin on the development of immune responses. This review discusses current knowledge on the immunomodulatory effects of curcumin on various facets of the immune response, including its effect on lymphoid cell populations, antigen presentation, humoral and cell-mediated immunity, and cytokine production.

1. INTRODUCTION

Turmeric is a bright yellow spice derived from the rhizomes of *Curcuma longa* Linn, which is widely cultivated in India, China, and Indonesia. *Curcuma longa* is a perennial herb of the Zingerberaceae (ginger) family with pulpy, orange tuberous roots or rhizome and oblong pointed leaves and it bears funnel-shaped yellow flowers. Turmeric is what gives curry powder its unique spicy flavor and color. In addition to its culinary appeal, turmeric powder has long been used for medicinal purposes in traditional Indian (Ayurvedic) and Chinese systems of medicine, particularly as an anti-inflammatory agent. A paste of powdered rhizome and slaked lime or turmeric ground in poultices applied locally is an ancient remedy to relieve pain and inflammation caused by sprain and injury. Turmeric is extensively used to treat indigestion, jaundice, menstrual difficulties, urinary infections, arthritis, and gallstones and as a carminative. In addition to medicinal uses, turmeric is also used in cosmetics and fabric coloring.

The active constituents of turmeric are curcumanoids and volatile oils, including tumerone, atlantone, and zingiberene. The bright yellow color of turmeric is due to polyphenolic pigments, known as curcumanoids. The major curcumanoids present in turmeric are curcumin, demethoxycurcumin, and bisdemethoxycurcumin;

together these curcumanoids comprise 3–6% of turmeric powder. Curcumin is the most researched curcumanoid and it makes up 70–75% of the curcumanoids, demethoxycurcumin about 15–20%, and bisdemethoxycurcumin about 3%. Pure curcumin is an orange-yellow crystalline powder, which is insoluble in water. The molecular formula of curcumin is $C_{21}H_{20}O_6$ and its molecular weight is 368.39 Daltons. Recent investigations have shown curcumin to be a potent antioxidant, anti-inflammatory, and anticarcinogenic agent that has been entered into Phase I clinical trials for chemoprevention of cancers. In experimental studies, curcumin has also shown a wide range of therapeutic effects, including cardioprotective, neuroprotective, hepatoprotective, anti-HIV, and anti-Alzheimer activity. The bioavailability of oral curcumin is low because 40–65% of curcumin passes through the gastrointestinal tract unchanged. Most of the absorbed curcumin is metabolized via glucuronidation to glucuronide and glucuronide/sulfate metabolites in the intestinal mucosa and liver.

The immune system is organized into the incredibly intricate arrangement of central (bone marrow, thymus) and peripheral (lymph nodes, spleen, blood) lymphoid organs and tissues. Various lymphoid cell populations and molecules secreted by them are responsible for protection against infectious agents and cancers as well as destruction of organ transplants and harmful effects of autoimmune diseases. A healthy individual has two levels of defense against foreign agents: innate (natural) immunity and adaptive or acquired immunity. The effector mechanisms of innate immunity are already present before an encounter with microbes and are rapidly activated before the development of adaptive immune responses. The principal effector cells of the innate immunity are phagocytes (neutrophils, monocytes, macrophages) and natural killer cells (NK cells), which are both tissue-borne and wander in the circulation to encounter and destroy microbes that have breached the epithelial barriers as well as self cells that have aged and died or transformed into cancerous cells. The principal mechanism by which neutrophils and other phagocytic cells capture and destroy foreign invaders and effete or transformed cells is through the production of highly toxic free radicals, such as hydrogen peroxide, superoxide anion ($O2^-$). and nitric oxide. Free radicals are generated by lysosomal NADPH oxidases in a process known as respiratory burst.

2. IMMUNOMODULATION BY CURCUMIN

The immunological effects of curcumin have been described in a limited number studies. In this chapter, we present an overview of the current knowledge on the modulatory effects of curcumin on different components of the immune system.

2.1. Anti-inflammatory Effects of Curcumin

Inflammation plays a crucial role in the pathogenesis of a wide array of diseases ranging from the development of cancer to the occurrence of autoimmune disorders and organ damage by infectious agents. Inflammation results from a complex series of actions and reactions triggered by the body's immunological response to tissue damage caused by various disease processes or surgical procedures. Because turmeric has been used in traditional medicine to control inflammatory conditions, several groups of investigators have investigated the anti-inflammatory activity of curcumin or its analogues in acute and chronic models of inflammation in rodents. In rats with Freund's adjuvant-induced arthritis, oral administration of *Curcuma longa* was shown by Srimal and Dhawan to significantly reduce inflammation.[1] These investigators also demonstrated that curcumin was as potent as phenylbutazone in the carrageenan edema test in the rat but was only half as potent in chronic tests. Mukhopadhyay et al. compared the anti-inflammatory activity of curcumin analogues and showed that sodium curcuminate, diacetyl curcumin, triethyl curcumin, and tetrahydrocurcumin were less potent in inhibiting acute and chronic inflammatory responses than curcumin.[2] Marked inhibition of 12-*O*-tetradecanoylphorbol-13-acetate (TPA) and arachidonic acid (AA)-induced epidermal inflammation (ear edema) in mice by curcumin was shown by Huang et al.[3] There are only a few controlled clinical trials in which the efficacy of curcumin in the treatment of inflammatory diseases have been tested. Deodhar and colleagues (1980) compared curcumin with phenylbutazone in 18 rheumatoid arthritis patients; they reported that improvements in morning stiffness, walking time, and joint swelling after 2 weeks of curcumin supplementation (1200 mg/day) were comparable to those by phenylbutazone therapy (300 mg/day).[4] Satoskar and colleagues that 5 days of oral curcumin supplementation (1200 mg/day) was effective in reducing postsurgical edema, tenderness, and pain and was comparable to phenylbutazone therapy.[5] In two uncontrolled studies by Lal et al., oral curcumin (1125 mg/day) for 12 weeks to 22 months was found to improve chronic anterior uveitis and idiopathic inflammatory orbital pseudotumor, inflammatory conditions of the eye.[6,7] Clearly, larger randomized controlled trials are needed to determine whether curcumin supplementation is efficacious in the treatment of inflammatory diseases, such as rheumatoid arthritis, ulcerative colitis, or multiple sclerosis.

Activated monocytes and macrophages release proinflammatory cytokines [tumor necrosis factor (TNF), interleukin (IL)-1, etc.)] to induce inflammation in the tissue. The inflammatory process involves the production of prostaglandins, thromboxanes, and leukotrienes, collectively known as eicosanoids, by macrophages and neutrophils. These mediators of inflammation are synthesized through enzymatic degradation of AA by cylcooxygenase-2 (COX-2) or liopooxygenase (LOX). COX-2 is induced by a variety of inflammatory insults, including an encounter with endotoxins, cytokines, growth factors, tumor promoters, and stress. COX-2 catalyzes the synthesis of series-2 prostaglandins (e.g., PGE_2, $PGF_2 - \alpha$, PGI_2, PGD_2) and thromboxanes (e.g., TXA_2, TXB_2) by mononuclear phagocytes, endothelial cells, polymorphonuclear leukocytes, and platelets. Eicosanoid byproducts of AA are also produced via a LOX group of enzymes, which metabolize AA to produce series-4 leukotriens and various hydroxyeicosatetraenoic acids (e.g., 5-HETE, 12-HETE, and 15-HETE). Together, these AA-derived compounds contribute to pain, inflammation, swelling, vasoconstriction, and thrombosis. Eicosanoids are also elevated in most human cancers and potentially play a significant role in tumor cell proliferation, angiogenesis, spread of cancers, and suppression of immune response.

Hong et al. showed the inhibition of AA generation by curcumin through the prevention of phosphorylation of phospholipase A_2, which catalyzes the hydrolysis of membrane phospholipids into AA.[8] Huang et al. showed the inhibition of the metabolism of AA *in vitro* by curcumin through inhibition of both LOX and COX activities.[3] Zhang et al. demonstrated the inhibition of the chenodeoxycholate- or phorbol-12-myristate-13-acetate (PMA)-induced induction of COX-2 mRNA and protein expression by curcumin in gastrointestinal cell lines.[9] Goel et al. showed that curcumin inhibited the growth of the HT-29 colon cancer cell line through the inhibition of COX-2 mRNA and protein expression without altering the expression of COX-1.[10] In another study by Chun et al., curcumin was shown to suppress PMA-induced expression of COX-2 through the inhibition of extracellular signal-regulated kinase (ERK) activity.[11] Flynn et al. demonstrated the strong inhibition of 5-HETE production by curcumin in human neutrophils.[12] Work by Kim et al. suggests that in the inhibition of the Janus kinase (JAK)–STAT signaling cascade via the effect on the src homology 2 domain-containing protein tyrosine phosphatases (SHP)-2 could contribute to its anti-inflammatory activity.[13] Furthermore, curcumin has been shown to inhibit the production of superoxide and nitric oxide by inflammatory cells, which could also contribute to its anti-inflammatory activity because these free radicals play a significant role in inflammatory processes.[14,15] In addition, as we will discuss later, curcumin has been shown to inhibit several proinflammatory cytokines [e.g., TNF, IL-1, interferon (IFN)] that play an important role in the initiation and progression of inflammatory responses. Thus, there are numerous molecular targets through which curcumin can mediate its anti-inflammatory effects.

2.2. Effect on Lymphoid Cell Populations

Little is known about how curcumin affects central and peripheral lymphoid tissues of the immune system. In a study by Antony et al., treatment of Balb/c mice with curcumin was shown to increase the cellularity and α-esterase-positive cells in the marrow.[16] These investigators also reported significant increase in total white blood cells (WBCs) in the peripheral blood in mice treated with curcumin. Yasni et al. found that in mice fed with *Curcuma xanthorrhiza* for 3–5 weeks, splenic T-lymphocytes were increased throughout the experimental period, but exerted a variable effect on B-lymhocytes and and T-cell subsets.[17] They observed elevation of B-cells at 3 weeks and of T helper cells at 4 weeks without any change in T suppressor cells. The effect of *C. xanthorrhiza* on macrophages in the spleen and peripheral blood was inconsistent. Churchill et al. showed that inhibition of intestinal tumors in C57BL/6J-Min/+ mice by curcumin was associated with changes in the intestinal immune cell profile.[18] Immunohistochemical staining of resident intestinal immune effector cells revealed that although there was no change in total intraepithelial lymphocyte number, treatment with curcumin significantly increased $CD3^+$ T-cell numbers in the small intestinal mucosa. Staining for CD4 and CD8 cells indicated that the increase in $CD3^+$ T-cells was due to changes in the $CD4^+$ subset of T-cells. On the other hand, treatment with curcumin had no effect on the distribution of T cell receptor (TCR) subtypes (i.e., TCRαß or TCRγδ). Curcumin also increased the numbers of B-cells in the small intestinal mucosa. These data suggested that curcumin exerts immunomodulatory effects on the mucosal immune system. Pal et al. (2005) showed amelioration of inhibition of lymphoid cell populations caused by growth of Ehrlich's ascites carcinoma.[19] Growth of these tumors decreased the numbers of bone marrow progenitors as well as thymic and splenic mononuclear cells. Oral administration of curcumin every other day for 21 days increased instead of decreased the cellularity of bone marrow, thymus, and spleen to the normal levels, indicating that curcumin prevents tumor-induced immunosuppression. Restoration of immune cell numbers by curcumin was attributed to the inhibition of tumor-induced apoptosis of lymphoid populations. Because curcumin has been shown to inhibit purified HIV-1 integrase, HIV-1 and HIV-2 proteases, and HIV-1 long terminal repeat directed gene expression of HIV-1 infected cells, it has been evaluated for HIV-infected $CD4^+$ cell replication. In a clinical trial, administration of 2 g/day of curcumin to 18 HIV-infected patients for approximately 29 weeks resulted in a significant increase in the CD4 and CD8 lymphocyte counts compared to the placebo-receiving patients. The CD4 cell count before the treatment ranged from 5 to 615 cells/mL of blood, and after treatment, the range was 283–1467 CD4 cells/mL of blood. The subsequent phase I/II study using doses of 2.7–4.8 g/day of curcumin failed to show any benefit on viral loads or CD4 counts in HIV-positive patients.[20] The negative results of this trial were attributed to poor bioavailability of curcumin. Overall, results of these studies demonstrate that curcumin moderately increases the number of T- and B-cells without altering the numbers of phagocytic macrophages.

2.3. Effect on Antigen-Presenting Cells

The production of an acquired immune response involves the capturing, processing, and presentation of antigens by antigen-presenting cells (APCs), such as macrophages and dendritic cells or B-lymphocytes. Curcumin might modulate immune responses by interfering with antigen presentation by APCs. Curcumin has been studied for its effect on phagocytosis and antigen presentation by APCs in a small number of studies. In a study by Antony et al., treatment with curcumin significantly increased the phagocytic activity of macrophages.[16] Enhancement of the phagocytic activity of peritoneal macrophages by curcumin was also recently reported by Li and Liu.[21] Cole et al. examined the mechanisms by which curcumin and turmeric extract prevents Alzheimer pathogenesis.[22] These investigators demonstrated that in addition to the inhibition of amyloid oligomer and fibril formation and amyloid deposition, curcumin promoted phagocytosis of amyloid by microglial cells *in vitro*. In contrast to these few studies in which curcumin was shown to promote phagocytosis, the opposite effect of curcumin has been shown in others. Lipopolysaccharide (LPS)-elicited hepatic microvascular inflammatory response is characterized by a significant increase in phagocytic activity of cenrilobular Kupffer cells, the number of leucocytes adhering to the sinusoidal wall, and swollen endothelial cells in the periportal and centrilobular regions. Lukita-Atmadja et al. showed that intragastric administration of curcumanoids before intravenous injection of LPS significantly reduced the phagocytic activity of Kupffer cells, the number of adhering leukocytes, and swollen endothelial cells, demonstrating that curcuminoids are effective in inhibiting the hepatic microvascular inflammatory response by suppressing phagocytosis.[23] Dendritic cells (DCs) are the professional APCs that play a key role in the initiation of T-cell responses against infectious microbes and tumors. Until recently, the effect of curcumin on the maturation and functional activity of DCs was unknown. In a study published in the *Journal of Immunology* in 2005, Kim et al. examined the effect of curcumin on surface molecule expression, cytokine production, and signaling pathways in bone marrow-derived murine DCs.[13] Curcumin was found to suppress surface expression of costimulatory molecules CD80 and CD86 and major histocompatibility complex (MHC) II but not MHC class I. Curcumin also impaired the production of IL-12, IL-1β, IL-6, and TNF-α by DCs. Although curcumin-treated DCs were highly efficient in capturing Ag via the mannose receptor-mediated endocytosis, they were very poor stimulators of Th1 and cell-mediated responses. Impairment of some of these DC functions might be related to the inhibition of mitogen-activated protein kinase (MAPK) activation and nuclear translocation of nuclear factor (NF)-κB by curcumin. The inhibition of several of functional attributes of DCs treated with curcumin indicates that curcumin might downregulate the production of T-cell-mediated immune responses by interfering with antigen handling and antigen presentation by DCs.

2.4. Induction of Apoptosis by Curcumin

Apoptosis or programmed cell death plays an essential role in the normal development of multicellular organisms and in maintaining tissue homeostasis.[23a] Aberration of apoptosis has been implicated in tumor development and resistance to cancer therapies.[24] Chemotherapy and radiation therapy destroys tumor cells in part by inducing apoptosis; However, most cancers have high levels of constitutively active NF-κB, which renders them resistant to the induction of apoptosis by anticancer therapies; therefore, the promotion of apoptosis in cancer cells potentially can lead to the regression and improved prognosis of refractory cancers. Curcumin has been extensively studied for the induction or promotion of apoptosis to prevent or inhibit tumor growth and metastasis. *In vitro* treatment with curcumin inhibited cell proliferation or induced apoptosis in leukemia,[25] lymphoma,[26] breast,[27] pancreatic,[28] lung,[29] prostate,[30] melanoma,[31] gastric,[32] colon,[33] and brain[10] tumor cell lines. In most cases, induction of apoptosis by curcumin involved triggering of both the caspase-8 to caspase-3 activation pathway (extrinsic pathway) and mitochondrial pathway (intrinsic pathway) of apoptosis in which cytochrome-*c* released from mitochondria leads to the activation of caspase-9 and caspase-3.[34,35] The molecular mechanisms involved in the induction of apoptosis by curcumin have ranged from a decrease in cellular levels of antiapoptotic Bcl-2, Bcl-XL, and cIAP proteins to an increase in levels of proapoptotic Bax.[36] Rajasingh et al. showed that curcumin induced growth arrest and apoptosis in T-cell leukemia in association with the inhibition of constitutively active JAK–STAT pathways.[37] Activation of MAPK and PI3k/PKB has been shown by Squires et al. to play a role in the inhibition of the proliferation and induction of apoptosis in breast cancer cells by curcumin.[38] Others have suggested the generation of reactive oxygen species (ROS) might be necessary for curcumin's apoptotic effect. A comparison of $p53^{+}$ and $p53^{-}$ human melanoma cell lines by Bush et al. showed that curcumin induced apoptosis in these cell lines through a Fas receptor/caspase-8 pathway independent of p53.[39]

Curcumin has been demonstrated to augment the cytotoxic effects of chemotherapy and radiation therapy.[40,41] There is some evidence that subtoxic concentration of curcumin might also promote apoptosis by death ligands such as tumor necrosis factor-related apoptosis-inducing ligand (TRAIL). Deeb et al. have demonstrated that the combined treatment of prostate cancer cell lines with subtoxic concentrations of curcumin and TRAIL induced apoptosis through the inhibition of NF-κB and the activation of extrinsic and intrinsic pathways of apoptosis.[42–44] The same sort of effect was shown by Gao et al. when human neuroblastoma cell line U87 but not U251 was treated with subtoxic concentration of curcumin and TRAIL together,[45] suggesting that at a nontoxic concentration, curcumin can sensitize prostate and brain tumor cells to TRAIL-induced apoptosis. Jung et al. later showed that curcumin sensitization of renal carcinoma cells to TRAIL-induced apoptosis was mediated through upregulation of death receptor 5 (DR5) by ROS.[46] Taken together, published data demonstrated that curcumin can induce or promote apoptosis by chemotherapeutic drugs and TRAIL through activation of a number of intracellular signal transduction pathways.

3. EFFECT ON SPECIFIC TYPES OF IMMUNE RESPONSES

There are two types of effector mechanism that mediate specific immune responses: (1) those mediated by antibodies produced by B-lymphocytes (humoral immunity) and (2) those mediated by specifically sensitized T-lymphocytes (cell-mediated immunity). Neither of these two forms of immunity has been adequately investigated for modulation by curcumin. A brief overview of the currently available information is summarized below.

3.1. Humoral Immunity

In a study by Kuramoto et al., investigators examined the effect of low and high concentrations of natural food colorings on immunoglobulin production by rat spleen lymphocytes *in vitro*.[47] Both water-soluble and water-insoluble (including curcumin) natural colorings inhibited IgE production at 10 and 20 μM, respectively. Although many of these colorings were found to inhibit the production of IgG and IgM at high concentrations, the water-insoluble colorings actually enhanced IgM production at a low concentration of 1 μM. These investigators concluded that natural colorings possess the capacity to modulate the production of immunoglobulins. South and colleagues examined dietary curcumin for its effect on antibody response in rats *in vivo*.[48] They showed that after 5 weeks of dietary exposure to curcumin at 40 mg/kg, IgG levels were significantly increased. Rats receiving lower dietary concentrations of 1 or 20 mg/kg curcumin showed no change in IgG levels. Effects of curcumin on production of antibodies in response to immunization with antigen were evaluated in two published reports. Antony et al. showed that the administration of curcumin to mice increased the circulating antibody titer after immunization with sheep red blood cells.[16] Curcumin also increased the number of antibody-forming cells against sheep red blood cells (SRBC) in the spleen. In another study, Odot et al. reported a significant decrease in melanoma tumor size and an increase the median survival time of mice immunized with soluble protein extract of mouse B16-R melanoma tumor cells in conjunction with curcumin administration.[49] The tumor inhibitory effect of combination therapy was attributed to the enhancement of antisoluble B16-R protein antibody response by curcumin. Taken together, published data tend to suggest that curcumin augments the production of humoral immune responses.

3.2. Cell-Mediated Immunity

Cell-mediated immunity (CMI) in which antigen-sensitized T-lymphocytes mediate the immune function plays an important role in defense against intracellular pathogens, rejection of organ transplants, antitumor immunity, and development of autoimmune diseases. CMI is evaluated by measuring T-cell proliferation, T-cell-mediated cytotoxicity, and production of cytokines by Tcells. Several recent reports suggested that curcumin is capable of modulating many of these T-cell-mediated immune functions. A brief account of curcumin's effect on each of these T-cell activities is presented below.

3.2.1. T-Cell Proliferation

Antigen presentation to T-lymphocytes by the APCs leads to their activation, proliferation, and clonal expansion. The inhibition or augmentation of the proliferative response of lymphocytes is frequently used to evaluate potential therapeutic agents for their immunomodulatory effects. Curcumin has been investigated for its effect on mitogen or alloantigen-induced proliferation of T-cells *in vitro* and *in vivo*. Yasni et al. demonstrated increased mitogenic responses of spleen cells from rats fed *C. xanthorrhiza* for 3 weeks.[17] However, in many other studies, curcumin was shown to inhibit the proliferation of several mouse and human leukemia cell lines and mitogen- and antigen-induced proliferation of mouse, rat, and human lymphocytes. Gao et al. investigated the effect of curcumin on the proliferative response of murine splenic T-cells stimulated with conconavalin A (Con A), IL-2, or allogeneic cells.[50] They found a significant increase in Con A-induced proliferation of splenic cells at 6.25 μmol/L curcumin, followed by a significant decrease in the proliferation at 12.5 μmol/L curcumin (Figure 1A). The proliferative response

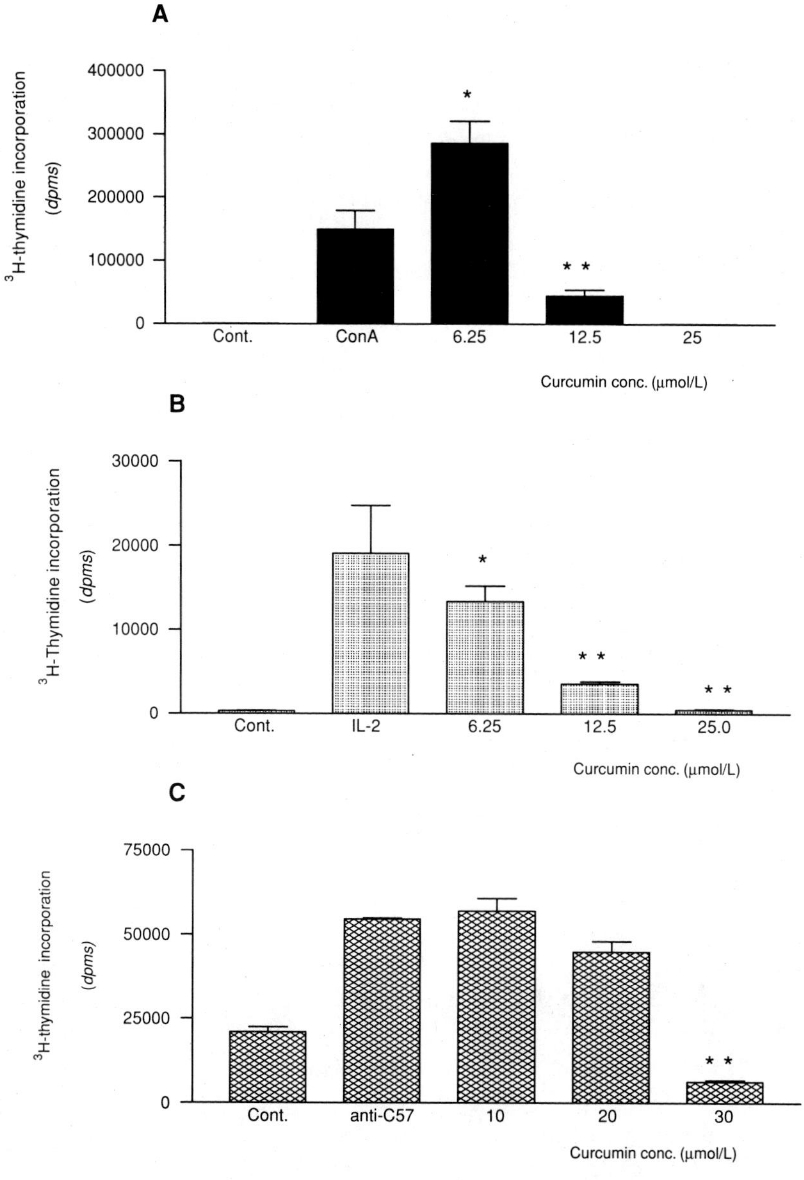

Figure 1. Effect of curcumin on the proliferation of spleen cells. C3H spleen cells (1×10^6 cells/mL) were stimulated with Con A (1 μg/mL) (A) or IL-2 (100 ng/mL) (B), or irradiated C57BL/6 spleen cells (1:1) (C) for 4 days in the absence or the presence of curcumin at concentrations as indicated. 2×10^5 viable cells from each culture were transferred to the wells of a 96-well microtiter tissue culture plate in triplicate. Cultures were pulsed with ^{3}H-thymidine (0.25 μCi/well) for 8 h. ^{3}H-thymidine incorporation was determined by liquid scintillation spectrometry. Data are presented as mean dissociation counts per minute (dpms) ±SEM of three to four experiments.

was completely blocked at 25 μmol/L curcumin. Curcumin also inhibited the IL-2-induced proliferation of splenic cells. The inhibition of IL-2-induced proliferation of spleen cells was dose dependent, because an increasingly suppressive effect was observed at increasing concentrations of curcumin from 6.25 to 25 μmol/L (Figure 1B). The effect of curcumin on the alloantigen-induced proliferation of spleen cells was modest at 20 μmol/L curcumin. However, at 30 μmol/L curcumin, the proliferation of cells was significantly suppressed. These data demonstrate that at 12.5 μmol/L and above, curcumin significantly inhibits the mitogen- and IL-2 induced proliferation of splenic cells, but the alloantigen-induced proliferation of splenic lymphocytes is suppressed only at higher concentration of curcumin (30 μmol/L). These investigators also showed that the inhibitory effect of curcumin on the proliferation of lymphocytes is irreversible, as removal of curcumin from cultures failed to restore responsiveness of treated cells to Con A, IL-2, or allogeneic cells.

The effect of curcumin on human T-cell proliferation has been investigated in two studies. Cipriani et al. investigated the effect of curcumin on the antigen-induced proliferation of γδ T-cells isolated from healthy donors.[51] They demonstrated that curcumin inhibited the isopentenyl pyrophosphate-induced proliferation of these cells in a dose-dependent manner. At 30 μM curcumin, phosphoantigen-induced proliferation of γδ T-cells was completely abolished. Ranjan et al. (2004) examined the effect of curcumin on the mitogen-induced proliferation of lymphocytes isolated from the human spleen.[52] Curcumin at 2.5 μg/mL significantly inhibited the proliferation of these cells by Con A, phytohemagglutinin (PHA) or PMA. Curcumin also inhibited the proliferation of lymphocytes induced by IL-2.

3.2.2. Cell-Mediated Cytotoxicity

Cytotoxic T-lymphocytes (CTLs), a major effector arm of CMI, play a crucial role in destroying virally infected cells, organ transplants, and tumor cells. Curcumin was shown by Shoskes et al. to reduce ischemia–reperfusion injury and rejection of renal transplants synergistically with mycophenolate mofetil in rodent models.[53,54] In a randomized placebo controlled trial by Shoskes et al., bioflavonoid therapy with curcumin and quercetin improved early function in cadaveric renal transplantation, acute graft rejection, and neurotoxicity.[55] Curcumin was also found by Chueh et al. to enhance the immunosuppressive activity of cyclosporin in a heterotopic cardiac transplant model.[56] However, whether modulation of antigen-specific cell-mediated cytotoxity by curcumin played a role in the protection of transplants from rejection was not addressed in these studies. Gao et al. investigated the regulation of development of alloantigen-specific and broadly nonspecific lymphokine-activated killer (LAK) cells by curcumin *in vitro*.[50] At 10 μmol/L curcumin, the generation of C3H/HeN (H-2^k) anti-C57/BL6 (H-2^b) CTLs was only insignificantly reduced (Figure 2A), but was significantly reduced at 20 μmol/L curcumin. The response was, however, completely abrogated at 30 μmol/L curcumin. These investigators further showed that at 10–20 μmol/L curcumin has little effect on the generation of IL-2-induced LAK cells (Figure 2B); however, at 30 μmol/L

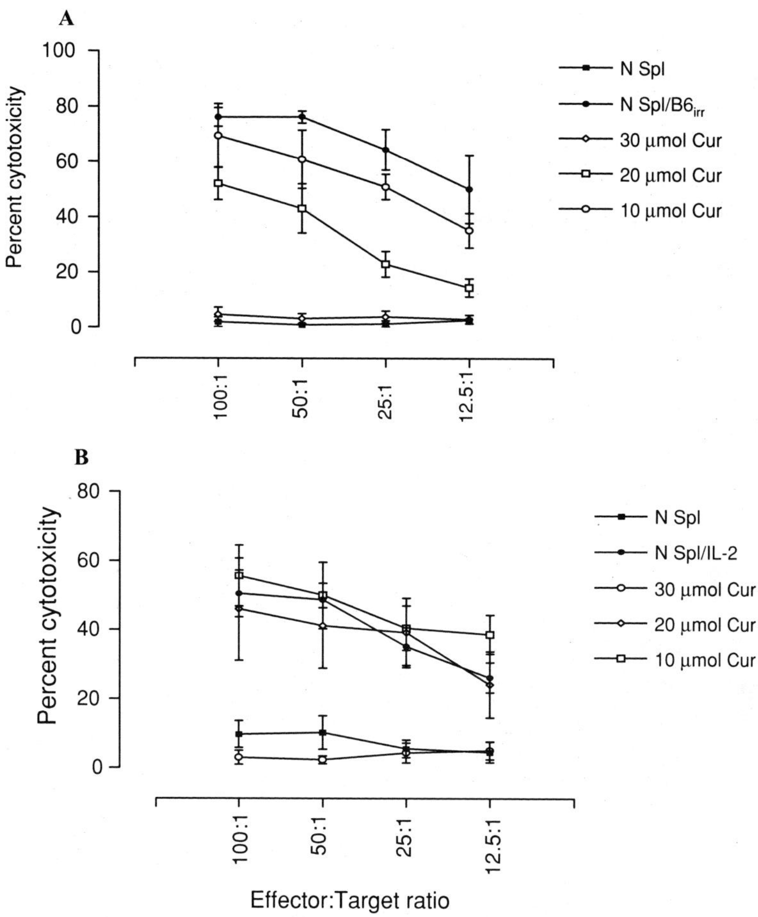

Figure 2. Effect of curcumin on the development of cell-mediated cytotoxicity. For the effect of curcumin on the generation of CTLs, 10^7 C3H spleen cells were cocultured with an equal number of irradiated C57BL/6 spleen cells in 10 mL RPMI-1640 medium for 5 days in the absence or the presence of curcumin (10, 20, and 30 μmol/L). Cytotoxicity of the viable effector cells against EL-4 target cells of C57BL/6 origin was determined in a 4-h ^{51}Cr-release assay (A). The effect of curcumin on LAK cell generation was examined by incubating C3H spleen cells (10^6 cells/mL) with IL-2 (150 ng/mL) for 3 days in the absence or the presence of curcumin (10, 20, and 30 μmol/L). The cytotoxicity of effector cells against YAC-1 target cells was measured in a 4-h ^{51}Cr-release assay (B). In each panel, the results are presented as mean percent cytotoxicity ± SD of three to four experiments.

curcumin, LAK cell production was significantly inhibited. Investigators concluded that although both CTL and LAK cell generation is significantly inhibited at high-dose curcumin (30 μmol/L); only CTL generation is downregulated at lower concentrations of curcumin. Furthermore, effective doses of curcumin were found to irreversibly impair the development of both CTLs and LAKs. Contrary to the downregulation of CTL and LAK cell generation by curcumin, Yadav et al. demonstrated augmentation of NK cell activity by curcumin, indicating an obvious need for more research on curcumin's effects on the development and functionality of these important effector immune cells.[57]

3.2.3. Effect on Cytokine Production

Cytokines are polypeptides produced by lymphocytes, monocytes, and a variety of other cells, including endothelial and epithelial cells. Cytokines are usually produced in response to microbes, noninfectious antigens, or physiological stress. The development of cell-mediated immune responses involves a complex network of cytokine signals generated by APCs and T helper (Th1and Th2) cells. Activated monocytes/macrophages release proinflammatory cytokines (monokines), such as TNF-α, IL-1, and IL-6 that play a prominent role in inflammatory responses. Th1 cells secrete IL-2 and gamma interferon (IFN-γ), which predominantly promote cell-mediated immunity. Th2 cells secrete IL-4, IL-5, IL-6, and transforming growth factor (TGF-β) that upregulate humoral immunity and negatively regulate cell-mediated immunity. The regulation of cytokine and chemokine production by curcumin has been addressed in a number of publications. Chan showed for the first time the inhibition of LPS-induced production of TNF-α by the human monocytic cell line Mono Mac 6 by curcumin at a low concentration of 5 μM.[58] Curcumin also inhibited the biological activity of TNF-α, but at a higher concentration of 50 μM. Kang et al. showed the inhibition of IL-12 production by murine macrophages stimulated with LPS or *Listeria* monocytogenes.[59] Macrophages treated with curcumin reduced the ability of antigen-primed $CD4^+$ T-cells to produce IFN-γ but increased the production of IL-4. Furthermore, macrophages from mice treated with curcumin were also defective for IL-12 induction and their ability to stimulate production of IFN-γ by $CD4^+$ T cells. Gao et al. showed that curcumin irreversibly inhibited the expression of IL-2, IFN-γ by mitogen-stimulated spenic T-cells, and IL-12 production by peritoneal macrophages (Figure 3).[50] In another study by Kim et al., curcumin was shown to inhibit proinflammatory IL-1β, IL-6, and TNF-α by bone marrow-derived DCs stimulated with LPS and demonstrated that suppression of MAPKs and translocation of NF-κB by curcumin was responsible for the impairment of cytokine production.[13]

Curcumin has also been investigated for its effect on cytokine production in disease-specific conditions. Literat et al. examined the effect of curcumin on proinflammatory cytokine expression in lung inflammatory cells from preterm newborns at risk for the development of chronic lung disease (CLD).[60] Significant inhibition of IL-1β and IL-8 but minimal inhibition of TNF-α production by curcumin was observed in preterm lung inflammatory cells stimulated with LPS. The authors of the study concluded that curcumin might be effective as a therapeutic agent in the attenuation of CLD. Kobayashi et al. showed the inhibition of IL-5 and granulocyte-macrophage-colony stimulating factor (GM-CSF) production by curcumin in lymphocytes from bronchial asthmatics.[61] Gaddipati et al. studied the effect of

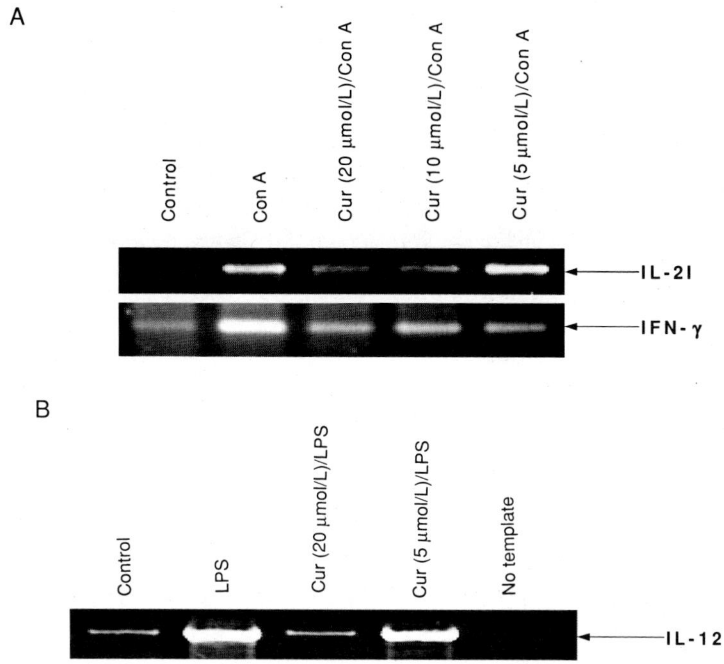

Figure 3. Effect of curcumin on cytokine gene expression. Treatment of spleen cells with curcumin (5,10, or 20 μmol/L) was started 1 h prior to stimulating cells with Con A (1 μg/mL) for 1 h. Similarly, macrophage monolayers were treated with curcumin at 5 or 20 μmol/L starting 1 h prior to stimulation with LPS (500 ng/mL) for 1 h. Total cellular RNA was isolated and reverse transcribed using random primers to generate cDNAs. cDNA (1 μg) was amplified by polymerase chain reaction (PCR) using gene-specific primers (mIL-2, mIFN-γ, or mIL-12, p40). PCR products were separated by 1% agarose gel electophoresis and visualized by ethidium bromide staining. Expected amplified gene products of sizes 294 bp (IL-2), 267 bp (IFN-γ), and 1025 bp (IL-12, p40) were obtained. Similar results were obtained in two separate experiments.

curcumin on the expression of cytokines in the liver of rats after hemorrhage/resuscitation.[62] They showed that pretreatment with curcumin for 7 days inhibited the mRNA transcripts of IL-1α, IL-β, IL-2, IL-6, IL-10, and TNF-α 2 and 24 h after hemorrhage/resuscitation, suggesting that curcumin might be beneficial in protecting against hemorrhagic liver injury. Natarajan and Bright showed that curcumin inhibits experimental allergic encephalomyelitis (EAE) by blocking IL-12 signaling through the JAK–STAT pathway in myelin-specific T-lymphocytes.[63]

Chemotactic cytokines (chemokines) play an important role in the chemotaxis of inflammatory cells to the site of injury and expression of adhesion molecules on endothelial and inflammatory cells. Xu et al. showed downregulation of monocyte chemotactic protein-1 (MCP-1) and interferon-inducible protein-10 (IP-10) by curcumin in bone marrow stromal cells stimulated with IL-1α, IFN-γ, TNF-α, or LPS.[64] The inhibition of chemokines by curcumin was at the level of gene transcription. Abe et al. investigated chemotactic IL-8, monocyte inflammatory protein-1 (MIP-1α), and MCP-1 expression in human peripheral blood monocytes and alveolar macrophages and showed inhibition of these chemokines by curcumin in both cell popupations stimulated with LPS or PMA.[65] Hidaka et al. showed that suppression of IL-8 production by curcumin was associated with the simultaneous increase in the expression of IL-8 receptors, CXCR1 and CXCR2, indicating that curcumin inhibits IL-8 induced internalization of IL-8 receptors.[66] The authors concluded that curcumin not only inhibited IL-8 production but also inhibited signal transduction through IL-8 receptors.

3.3. Effect on NF-κB Activation

Nuclear factor-κB is a member of the Rel family of transcription factors that regulate transcription of numerous genes involved in immune and inflammatory responses, cell proliferation, apoptosis, oncogenesis, and atherosclerosis.[67] Under normal conditions, NF-κB is sequestered in the cytoplasm, as a heterodimer of Rel proteins p50 and p65, by an inhibitory protein IκBα.[68] In response to cytokines, bacterial and viral products, free radicals, ultraviolet (UV) light, and chemotherapeutic agents, IκBα is rapidly phosphorylated and degraded by cytosolic proteosomes. Active NF-κB translocates to the nucleus, where it regulates the expression of target genes containing κB regulatory elements. The anti-inflammatory and anticarcinogenic activity of curcumin might be related to its inhibition of NF-κB, which plays a central role in many inflammatory disease processes and oncogenesis. Indeed, downregulation of constitutive or inducible NF-κB activity by curcumin has been extensively published. Singh and Aggarwal showed that inhibition of TNF-α, PMA, or H_2O_2-induced activation of NF-κB by curcumin in a human myelomonoblastic cell line was associated with the prevention of phosphorylation and degradation of IκBα and translocation of p65 subunit to the nucleus.[69] Xu et al. reported that the inhibition of IL-1α or TNF-α induced MCP-1/JE in bone marrow stromal cells by curcumin-involved suppression of NF-κB and activator protein (AP)-1 by curcumin.[70] Jobin et al. demonstrated that the inhibition of IL-1β-mediated ICAM and IL-8 gene expression involved the suppression of cytokine-induced NF-κB DNA-binding activity, p65 nuclear import, IκBα degradation, and inhibition of IκBα kinase (IKK).[71] Similarly, inhibition of the LPS-induced production of IL-12 by macrophages[72] or IL-12, IL-1β, IL-6, and TNF-α production by DCs involved suppression of LPS-induced nuclear translocation of NF-κB p65.[51]

COX-2 plays a critical role in mediating the inflammatory process and is overexpressed in a number of cancers, including colon, breast, prostate, and lung cancers. Recent studies have shown that COX-2 expression is regulated by NF-κB. Plummer et al. showed that inhibition of COX-2 expression in colon cells by curcumin is through the inhibition of NF-κB activation by blocking the phosphorylation of IκB through the inhibition of IKK.[73] Chun et al. assessed the effect of curcumin

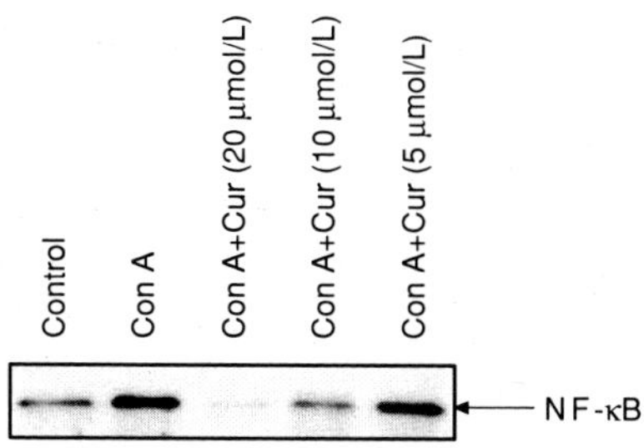

Figure 4. Effect of curcumin on activation of NF-κB. Spleen cells were pretreated with curcumin (5–30 µmol/L) for 1 h before stimulating with Con A (1 µg/mL) for 45 min. Nuclear extracts were prepared from control and treated cells and NF-κB was analyzed analyzed by immunoblotting using anti-NF-κB (p65) antibody.

on TPA-induced expression of COX-2 in mouse skin.[11] Topical application of curcumin prior to the application of TPA inhibited the expression of COX-2 in the epidermal layer. The suppression of COX-2 by curcumin was associated with the inhibition of TPA-stimulated NF-κB activation, IκB degradation, and translocation of p65 subunit to the nucleus.

Gao et al. demonstrated that mitogen induced IL-2 and IFN-γ production by splenic T-cells was associated with the activation of NF-κB.[50] Curcumin inhibited both the production of cytokines and NF-κB activation (Figure 4). In another study by Cipriani et al., curcumin inhibition of antigen-induced chemokines MIP-1α, MIP-1β, and regulated on activation, normal T expressed and secreted (RANTES) produced by γδ T-cells was linked to the suppression of antigen-induced NF-κB and AP-1 activation.[51] Collectively, published data indicate that curcumin modulates NF-κB activation in a cell-specific and stimulus-dependent manner and that the inhibition of NF-κB and NF-κB-dependent gene products constitute part of the molecular basis for the immunomodulatory effects of curcumin.

4. CONCLUSION

The anecdotal and experimental evidence favoring the anti-inflammatory effects of curcumin is compelling, and insufficient clinical evidence exists to warrant further clinical testing and development of this phytochemical as a safe nutraceutical drug for chronic human inflammatory diseases. The evidence favoring curcumin as a potent modulator of humoral or cell-mediated immune responses is weak. Limited experimental data suggest enhancement of the antibody production by curcumin. In contrast, curcumin inhibits the production of cytokines by macrophages and lymphocytes, especially the cytokines with the proinflammatory effects, as well as proliferative and cytotoxic responses of T-lymphocytes *in vitro*. Whether the *in vitro* inhibitory effects of curcumin on T-cell activities would translate into the prevention of transplant rejection or diminish the severity of T-cell-mediated inflammatory diseases *in vivo* remain to be investigated. It appears that the inhibition of transcription factors NF-κB and AP-1 by curcumin plays a central role in inhibition of T-cell proliferation, cytokine production, and inflammation. A better understanding of the mechanism by which curcumin blocks different signaling pathways, including the NF-κB signaling pathway involved in immune and inflammatory responses, would shed light on its mechanism of action and further testing of this novel agent for clinical applications as an anti-inflammatory agent.

ACKNOWLEDGMENT

The authors thank Deborah Dulchavsky for careful editing of this chapter.

REFERENCES

1. R. C. Srimal and B. N. Dhawan, Pharmacology of diferuloyl methane (curcumin), a non-steroidal anti-inflammatory agent. *J Pharm Pharmacol* **25**, 447–452 (1973).
2. A. Mukhopadhyay, N. Basu, N. Ghatak, and P. K. Gujral, Matory and irritant activities of curcumin. *Agents Actions* **12**, 508–515 (1982).
3. M. T Huang, T. Lysz, T. Ferraro, T. F. Abidi, J. D. Laskin, and A. H. Conney, Inhibitory effects of curcumin on in vitro lipoxygenase and cyclooxygenase activities in mouse epidermis. *Cancer Res* **51**, 813–819 (1991).
4. S. D. Deodhar, R. Sethi, and R. C. Srimal, Preliminary study on antirheumatic activity of curcumin (diferuloyl methane). *Indian J Med Res* **71**, 632–634 (1980).
5. R. R. Satoskar, S. J. Shah, and S. G. Shenoy, Evaluation of anti-inflammatory property of curcumin (diferuloyl methane) in patients with postoperative inflammation. *Int J Pharmacol Ther Toxicol* **24**, 651–654 (1986).
6. B. Lal, A. K. Kapoor, O. P. Asthana, P. K. Agrawal, R. Prasad, and P. Kumar, Efficacy of curcumin in the management of chronic anterior uveitis. *Phytother Res* **13**, 318–322 (1999).
7. B. Lal, A. K. Kapoor , O. P. Asthana, and R. C. Srimal, Role of curcumin in idiopathic inflammatory orbital pseudotumours. *Phytother Res* **14**, 443–447 (2000).
8. J. Hong, M. Bose, J. Ju, J. H. Ryu, X. Chen, S. Sang, M. J. Lee, , and C. S. Yang, Modulation of arachidonic acid metabolism by curcumin and related beta-diketone derivatives: Effects on cytosolic phospholipase A(2), cyclooxygenases and 5-lipoxygenase. *Carcinogenesis* **25**, 1671 (2004).
9. F. Zhang, N. K. Altorki, J. R. Mestre, K.Subbaramaiah, and A. J. Dannenberg, Curcumin inhibits cyclooxygenase-2 transcription in bile acid- and phorbol ester-treated human gastrointestinal epithelial cells.*Carcinogenesis* **20**, 445 (1999).
10. A. Goel, C. R. Boland, and D. P. Chauhan, Specific inhibition of cyclooxygenase-2 (COX-2) expression by dietary curcumin in HT-29 human colon cancer cells. *Cancer Lett* **172**, 111–118 (2001).
11. K. S. Chun, Y. S. Keum, S. S. Han, Y. S. Song, S. H. Kim, and Y. J. Surh, Curcumin inhibits phorbol ester-induced expression of cyclooxygenase-2 in mouse skin through suppression of extracellular signal-regulated kinase activity and NF-kappaB activation. *Carcinogenesis* **24**, 1515 (2003).

12. D. L. Flynn, M. F. Rafferty, and A. M. Boctor, Inhibition of human neutrophil 5-lipooxygenase activity by gingerdione, shogaol, capsaicin and related pungent compounds. *Prostaglandins Leukot Med* **24**, 195–198 (1986).
13. G. Y. Kim, K. H. Kim, S. H. Lee, M. S. Yoon, H. J. Lee, D. O. Moon, C. M. Lee, S. C. Ahn, Y. C. Park, and Y. M. Park, Curcumin inhibits immunostimulatory function of dendritic cells: MAPKs and translocation of NF-kappa B as potential targets. *J Immunol* **174**, 8116 (2005).
14. S. Bhaumik, M. D. Jyothi, and A. Khar, Differential modulation of nitric oxide production by curcumin in host macrophages and NK cells. *FEBS Lett* **483**, 78–82 (2000).
15. I. Brouet and H. Ohshima, Curcumin, an anti-tumour promoter and anti-inflammatory agent, inhibits induction of nitric oxide synthase in activated macrophages. *Biochem Biophys Res Commun* **206**, 533–540 (1995).
16. S. Antony, R. Kuttan, and G. Kuttan, Immunomodulatory activity of curcumin. *Immunol Invest* **28**, 291 (1999).
17. S. Yasni, K. Yoshiie, H. Oda, M. Sugano, and K. Imaizumi, Dietary *Curcuma xanthorrhiza* Roxb. increases mitogenic responses of splenic lymphocytes in rats, and alters populations of the lymphocytes in mice. *J Nutr Sci Vitaminol (Tokyo)* **39**, 345 (1993).
18. M. Churchill, A. Chadburn, R. T. Bilinski, and M. M. Bertagnolli, Inhibition of intestinal tumors by curcumin is associated with changes in the intestinal immune cell profile. *J Surg Res* **89**, 169 (2000).
19. S. Pal, S. Bhattacharyya, T. Choudhuri, G. K. Datta, T. Das, and G. Sa, Amelioration of immune cell number depletion and potentiation of depressed detoxification system of tumor-bearing mice by curcumin. *Cancer Detect Prev* **29**, 470 (2005).
20. J. S. James, Curcumin: Clinical trial finds no antiviral effect. *AIDS Treat News* **242**, 1 (1996).
21. X. Li X. Liu, Effect of curcumin on immune function of mice. *J Huazhong Univ Sci Technolog Med Sci* **25**, 137 (2005).
22. G. M. Cole, T. Morihara, G. P. Lim, F. Yang, A. Begum, and S. A. Frautschy, NSAID and antioxidant prevention of Alzheimer's disease: Lessons from in vitro and animal models. *Ann NY Acad Sci* **1035**, 68 (2004).
23. W. Lukita-Atmadja, Y. Ito, G. L. Baker, and R. S. McCuskey, Effect of curcuminoids as anti-inflammatory agents on the hepatic microvascular response to endotoxin. *Shock* **17**, 399 (2002).
23a. H. Steller, Mechanisms and genes of cellular suicide. *Science* **267**, 1445 (1995).
24. M. W. Mayo and A. S. Baldwin, The transcription factor NF-κB: control of oncogenesis and cancer therapy resistance. *Biochim Biophys Acta* **1470**, M55–M62 (2000).
25. A. C. Bharti, N. Donato, S. Singh, and B. B. Aggarwal, Curcumin (diferuloylmethane) down-regulates the constitutive activation of nuclear factor-kB and IkBa kinase in human multiple myeloma cells, leading to suppression of proliferation and induction of apoptosis. *Blood* **101**, 1053 (2003).
26. S. Uddin, A. R. Hussain, P. S. Manogaran, K. Al-Hussein, L. C. Platanias, M. I. Gutierrez, and K. G. Bhatia, Curcumin suppresses growth and induces apoptosis in primary effusion lymphoma. *Oncogene* **24**, 7022 (2005).
27. T. Choudhuri, S. Pal, M. L. Agwarwal, T. Das, and G. Sa, Curcumin induces apoptosis in human breast cancer cells through p53-dependent Bax induction. *FEBS Lett* **512**, 334 (2002).
28. L. Li, B. B. Aggarwal, S. Shishodia, J. Abbruzzese, and R. Kurzrock, Nuclear factor-kappaB and IkappaB kinase are constitutively active in human pancreatic cells, and

their down-regulation by curcumin (diferuloylmethane) is associated with the suppression of proliferation and the induction of apoptosis. *Cancer* **101**, 2351 (2004).

29. P. G. Radhakrishna, A. S.Srivastava, T. I. Hassanein, D. P. Chauhan, and E. Carrier, Induction of apoptosis in human lung cancer cells by curcumin. *Cancer Lett* **208**, 163 (2004).
30. T. Dorai, Y.-C. Cao, B. Dorai, R. Buttyan, and A. E. Katz, Therapeutic potential of curcumin in human prostate cancer. III. Curcumin inhibits proliferation, induces apoptosis, and inhibits angiogenesis of PC3 prostate cancer cells in vivo. *Prostate* **47**, 293 (2001).
31. M. Zheng, S. Ekmekcioglu, E. T. Walch, C. H. Tang, and E. A. Grimm, Inhibition of nuclear factor-kappaB and nitric oxide by curcumin induces G2/M cell cycle arrest and apoptosis in human melanoma cells. *Melanoma Res* **14**, 165 (2004).
32. L. Moragoda, R. Jaszewski, and A. P. Majumdar, Curcumin induced modulation of cell cycle and apoptosis in gastric and colon cancer cell lines. *Oncogene* **20**, 7597 (2001).
33. A. S. Jaiswal, B. P. Marlow, N. Gupta, and S. Narayan, Beta-catenin-mediated transactivation and cell-cell adhesion pathways are important in curcumin (diferuylmethane)-induced growth arrest and apoptosis in colon cancer cells. *Oncogene* **21**, 8414 (2002).
34. A. Mukhopadhyaya, C. Bueso-Ramos, D. Chatterjee, P. Pantazis, and B. Aggarwal, Curcumin downregulates cell survival mechanisms in human prostate cancer cell lines. *Oncogene* **20**, 759 (2001).
35. R. J. Anto, A. Mukhopadadhyay, K. Denning, and B. B. Aggarwal, Curcumin (diferuloylmethane) induces apoptosis through activation of caspase-8, BID cleavage and cytochrome c release: its suppression by ectopic expression of Bcl-2 and Bcl-xL. *Carcinogenesis* **23**, 143 (2002).
36. S. Aggarwal, H. Ichikawa, Y. Takada, S. K. Sandur, S. Shishodia, and B. B. Aggarwal, Curcumin (diferuloylmethane) down-regulates expression of cell proliferation and antiapoptotic and metastatic gene products through suppression of IkappaBalpha kinase and Akt activation. *Mol Pharmacol* **69**, 195 (2006).
37. J. Rajasingh, H. P. Raikwar, G. Muthian, C. Johnson, and J. J. Bright, Curcumin induces growth-arrest and apoptosis in association with the inhibition of constitutively active JAK-STAT pathway in T cell leukemia. *Biochem Biophys Res Commun* **340**, 359 (2006).
38. M. S. Squires, E. A. Hudson, L. Howells, S. Sale, C. E. Houghton, J. L. Jones, L. H. Fox, M. Dickens, S. A. Prigent, and M. M.. Manson, Relevance of mitogen activated protein kinase (MAPK) and phosphotidylinositol-3-kinase/protein kinase B (PI3K/PKB) pathways to induction of apoptosis by curcumin in breast cells. *Biochem Pharmacol* **65**, 361 (2003).
39. J. A. Bush, K. J. Cheung, Jr., and G. Li, Curcumin induces apoptosis in human melanoma cells through a Fas receptor/caspase-8 pathway independent of p53. *Exp Cell Res* **271**, 305 (2001).
40. B. B. Aggarwal, S. Shishodia, Y. Takada, S. Banerjee, R. A. Newman, C. E. Bueso-Ramos, and J. E. Price, Curcumin suppresses the paclitaxel-induced nuclear factor-kappaB pathway in breast cancer cells and inhibits lung metastasis of human breast cancer in nude mice. *Clin Cancer Res* **11**, 7490 (2005).
41. T.-C.Hour, J. Chen, J., Huang, J.-Y. Guan, S.-H. Lu, and Y.-S. Pu, Curcumin enhances cytotoxicity of chemotherapeutic agents in prostate cancer cells by inducing

p21WAFI/CIPI and C/EBPβ expressions and suppressing NF-kB activation. *Prostate* **51**, 211 (2002).

42. D. Deeb, H. Jiang, X. Gao, M. S. Hafner, H. Wong, G. Divine, R. A. Chapman, S. A. Dulchavsky, and S. C. Gautam, Curcumin sensitizes prostate cancer cells to tumor necrosis factor-related apoptosis-inducing ligand/Apo2L by inhibiting nuclear factor-kappaB through suppression of IkappaBalpha phosphorylation. *Mol Cancer Ther* **3**, 803 (2004).
43. D. Deeb, Y. X. Xu, H. Jiang, X. Gao, N. Janakiraman, R. A. Chapman, and S. C. Gautam, Curcumin (diferuloyl-methane) enhances tumor necrosis factor-related apoptosis-inducing ligand-induced apoptosis in LNCaP prostate cancer cells. *Mol Cancer Ther* **2**, 95 (2003).
44. D. D. Deeb, H. Jiang, X. Gao, G. Divine, S. A. Dulchavsky, and S. C. Gautam, Chemosensitization of hormone-refractory prostate cancer cells by curcumin to TRAIL-induced apoptosis. *J Exp Ther Oncol* **5**, 81 (2005).
45. X. Gao, D. Deeb, H. Jiang, Y. B. Liu, S.A. Dulchavsky, and S. C. Gautam, Curcumin differentially sensitizes malignant glioma cells to TRAIL/Apo2L-mediated apoptosis through activation of procaspases and release of cytochrome c from mitochondria. *J Exp Ther Oncol* **5**, 39 (2005).
46. E. M. Jung, J. H. Lim, JT. J. Lee, J. W. Park, K. S. Choi, and T. K. Kwon, Curcumin sensitizes tumor necrosis factor-related apoptosis-inducing ligand (TRAIL)-induced apoptosis through reactive oxygen species-mediated upregulation of death receptor 5 (DR5). *Carcinogenesis* **26**, 1905 (2005).
47. Y. Kuramoto, K. Yamada, O. Tsuruta, and M. Sugano, Effect of natural food colorings on immunoglobulin production in vitro by rat spleen lymphocytes. *Biosci Biotechnol Biochem* **60**, 1712 (1996).
48. E. H. South, J. H. Exon, and K. Hendrix, Dietary curcumin enhances antibody response in rats. *Immunopharmacol Immunotoxicol* **19**, 105 (1997).
49. J. Odot, P. Albert, A. Carlier, M. Tarpin, J. Devy, and C. Madoulet, In vitro and in vivo anti-tumoral effect of curcumin against melanoma cells. *Int J Cancer* **111**, 381 (2004).
50. X. Gao, J. Kuo, H. Jiang, D. Deeb, Y. Liu, G. Divine, R. A. Chapman, S. A. Dulchavsky, and S. C. Gautam, Immunomodulatory activity of curcumin: Suppression of lymphocyte proliferation, development of cell-mediated cytotoxicity, and cytokine production in vitro. *Biochem Pharmacol* **68**, 51 (2004).
51. B. Cipriani, G. Borsellino, H. Knowles, D. Tramonti, F. Cavaliere, G. Bernardi, L. Battistini, andC. F. Brosnan, Curcumin inhibits activation of Vgamma9Vdelta2 T cells by phosphoantigens and induces apoptosis involving apoptosis-inducing factor and large scale DNA fragmentation. *J Immunol* **167**, 3454 (2001).
52. D. Ranjan, C. Chen, T. D. Johnston, H. Jeon, and M. Nagabhushan, Curcumin inhibits mitogen stimulated lymphocyte proliferation, NFkappaB activation, and IL-2 signaling. *J Surg Res* **121**, 171 (2004).
53. D. A. Shoskes, Effect of bioflavonoids quercetin and curcumin on ischemic renal injury: A new class of renoprotective agents. *Transplantation* **66**, 147 (1998).
54. D. A. Shoskes, E. A. Jones, and A. Shahed, Synergy of mycophenolate mofetil and bioflavonoids in prevention of immune and ischemic injury. *Transplant Proc* **32**, 798 (2000).
55. D. Shoskes, C. Lapierre, M. Cruz-Corerra, N. Muruve, R. Rosario, B. Fromkin, M. Braun, and J. Copley, Beneficial effects of bioflavonoids curcumin and quercetin on

early function in cadaveric renal transplantation: A randomized placebo controlled trial. *Transplantation* **80**, 1556 (2005).
56. S. C. Chueh, M. K. Lai, I. S. Liu, F. C. Teng, and J. Chen, Curcumin enhances the immunosuppressive activity of cyclosporine in rat cardiac allografts and in mixed lymphocyte reactions. *Transplant Proc* **35**, 1603 (2003).
57. V. S. Yadav, K. P. Mishra, D. P. Singh, S. Mehrotra, and V. K. Singh, Immunomodulatory effects of curcumin. *Immunopharmacol Immunotoxicol* **27**, 485 (2005).
58. M. M.-Y. Chan, Inhibition of tumor necrosis factor by curcumin, a phytochemical. *Biochem Pharmacol* **49**, 1551 (1995).
59. B. Y. Kang, S. W. Chung, W. Chung, S. Im, S. Y. Hwang, and T. S. Kim, Inhibition of interleukin-12 production in lipopolysaccharide-activated macrophages by curcumin. *Eur J Pharmacol* **384**, 191 (1999).
60. A. Literat, F. Su, M. Norwicki, M. Durand, R. Ramanathan, C. A. Jones, P. Minoo, and K. Y. Kwong, Regulation of pro-inflammatory cytokine expression by curcumin in hyaline membrane disease (HMD). *Life Sci* **70**, 253 (2001).
61. T. Kobayashi, S. Hashomoto, and T. Horie, Curcumin inhibition of Dermatophagoides farinea-induced interleukin-5 (IL-5) and granulocyte macrophage-colony stimulating factor (GM-CSF) production by lymphocytes from bronchial asthmatics. *Biochem Pharmacol* **54**, 819 (1997).
62. J. P. Gaddipati, S. V. Sundar, J. Calemine, P. Seth, G. S. Sidhu, and R. K. Maheshwari, Differential regulation of cytokines and transcription factors in liver by curcumin following hemorrhage/resuscitation. *Shock* **19**, 150 (2003).
63. C. Natarajan and J. J. Bright, Curcumin inhibits experimental allergic encephalomyelitis by blocking IL-12 signaling through Janus kinase-STAT pathway in T lymphocytes. *J Immunol* **168**, 6506 (2002).
64. Y. X. Xu, K. R. Pindolia, N. Janakiraman, C. J. Noth, R. A. Chapman, and S. C. Gautam, Curcumin, a compound with anti-inflammatory and anti-oxidant properties, down-regulates chemokine expression in bone marrow stromal cells. *Exp Hematol* **25**(5), 413 (1997).
65. Y. Abe, S. Hashimoto, and T. Horie, Curcumin inhibition of inflammatory cytokine production by human peripheral blood monocytes and alveolar macrophages. *Pharmacol Res* **39**, 41 (1999).
66. H. Hidaka, T. Ishiko, T. Furuhashi, H. Kamohara, S. Suzuki, M. Miyazaki, O. Ikeda, S. Mita, T. Setoguchi, and M. Ogawa, Curcumin inhibits interleukin 8 production and enhances interleukin 8 receptor expression on the cell surface: Impact on human pancreatic carcinoma cell growth by autocrine regulation. *Cancer* **95**, 1206 (2002).
67. S. Ghosh, Regulation of inducible gene expression by the transcription factor NF-kappaB. *Immunol Res* **19**, 183 (1999).
68. M. Karin, How NF-kB is activated: The role of the IkB kinase (IKK) complex? *Oncogene* **18**, 6867 (1999).
69. S. Singh and B. B. Aggarwal, Activation of transcription factor NF-kappa B is suppressed by curcumin (diferuloylmethane). *J Biol Chem* **270**, 24,995 (1995).
70. Y. X. Xu, K. R. Pindolia, N. Janakiraman, R. A. Chapman, and S. C. Gautam, Curcumin inhibits IL1 alpha and TNF-alpha induction of AP-1 and NF-kB DNA-binding activity in bone marrow stromal cells. *Hematopathol Mol Hematol* **11**, 49 (1997).

71. C. Jobin, C. A. Bradham, M. P. Russo, B. Juma, A. S. Narula, D A. Brenner, and R. B. Sartor, Curcumin blocks cytokine-mediated NF-kappa B activation and proinflammatory gene expression by inhibiting inhibitory factor I-kappa B kinase activity. *J Immunol* **163**, 3474 (1999).
72. B. Y. Kang, Y. J. Song, K. M. Kim, Y. K. Choe, S. Y. Hwang, and T. S. Kim, Curcumin inhibits Th1 cytokine profile in CD4+ T cells by suppressing interleukin-12 production in macrophages. *Br J Pharmacol* **128**, 380 (1999).
73. S. M. Plummer, K. A. Holloway, M. M. Manson, R. J. Munks, A. Kaptein, S. Farrow, and L. Howells, Inhibition of cyclo-oxygenase 2 expression in colon cells by the chemopreventive agent curcumin involves inhibition of NF-kappaB activation via the NIK/IKK signalling complex. *Oncogene* **18**, 6013 (1999).

BENEFICIAL ROLE OF CURCUMIN IN SKIN DISEASES

Rajesh L. Thangapazham, Anuj Sharma and Radha K. Maheshwari

Abstract: In recent years, considerable interest has been focused on curcumin a compound, isolated from turmeric. Curcumin is used as a coloring, flavoring agent and has been traditionally used in medicine and cuisine in India. The varied biological properties of curcumin and lack of toxicity even when administered at higher doses makes it attractive to explore its use in various disorders like tumors of skin, colon, duodenum, pancreas, breast and other skin diseases. This chapter reviews the data on the use of curcumin for the chemoprevention and treatment of various skin diseases like scleroderma, psoriasis and skin cancer. Curcumin protects skin by quenching free radicals and reducing inflammation through nuclear factor-κB inhibition. Curcumin treatment also reduced wound-healing time, improved collagen deposition and increased fibroblast and vascular density in wounds thereby enhancing both normal and impaired wound-healing. Curcumin has also been shown to have beneficial effect as a proangiogenic agent in wound-healing by inducing transforming growth factor-β, which induces both angiogenesis and accumulation of extracellular matrix, which continues through the remodeling phase of wound repair. These studies suggest the beneficial effects of curcumin and the potential of this compound to be developed as a potent nontoxic agent for treating skin diseases.

1. INTRODUCTION

Use of dietary supplements for the treatment of diseases is a part of traditional medicine and is practiced in all parts of the world. Turmeric (*Curcuma longa*), a coloring and flavoring agent, is a dietary additive and has been traditionally used in medicine and cuisine in India. Curcumin (diferuloylmethane), a major active component of turmeric, is a crystalline compound and turmeric has been widely used for centuries in indigenous medicine for the treatment of a variety of inflammatory conditions and other diseases. Preclinical studies have revealed the chemopreventive potential of curcumin in several different animal bioassay models both *in vitro* and *in vivo*, demonstrating the antioxidant, anti-inflammatory, antitumorigenic, and antiangiogenic properties of curcumin. Curcumin is quite safe and the varied biological properties and the lack of toxicity even at higher doses makes it attractive to explore its use in various disorders like tumors of skin, colon, duodenum, pancreas, breast, and other skin diseases.

Skin, considered the largest organ of the body, is made up of layers of tissues that protects underlying muscles and organs. As an interface with the surroundings, it plays the most important role in protecting our body against heat, light, infection, and injury. Hence, skin, the first line of defense, is vulnerable to various adverse conditions like rashes, burns, injuries, infections and disorders like scleroderma, dermatitis, psoriasis and cancer. Over the last decade, there has been increasing interest in turmeric and its medicinal properties in treating various skin conditions. Turmeric is well documented for its medicinal properties in Indian and Chinese systems of medicine. In Ayurvedic medicine, turmeric was prescribed for the treatment of many conditions, including rheumatic pain and various skin disorders. Turmeric and curcumin have traditionally been used for pain and inflammation associated with acne, skin rashes and warts. Skin damage is a serious side effect for patients receiving radiation for treating their skin tumors. In a study with mice exposed to radiation, curcumin-treated ones had fewer blisters or burns when compared to the control mice. Numerous studies reveal the protective effect of curcumin against various chemicals and environmental pollutants, showing the benefits of curcumin against agents that cause skin damage. In this chapter, we review the beneficial roles of curcumin/turmeric in various skin diseases.

2. SCLERODERMA

Derived from the Greek words "sklerosis," meaning hardness and "derma," meaning skin, scleroderma literally means hard skin. Although the main cause is generally unknown, scleroderma is suspected to be induced by a combination of several factors like autoimmunity, environmental exposure, genetics and infections.[1,2] Scleroderma, alternatively known as systemic sclerosis (SSc), is characterized by aberrations of extracellular matrix deposition, severe and often progressive cutaneous and visceral fibrosis, pronounced alterations in the microvasculature and numerous cellular and humoral immune abnormalities.[3] Scleroderma can cause major lung pathology like pulmonary fibrosis, which is a major cause of morbidity and mortality in scleroderma patients. Curcumin is found to be beneficial in scleroderma and other associated organ pathologies. Curcumin exerts the protective effect by mainly modulating the protein kinase C (PKC) pathway. PKC-dependent signal transduction pathways have been found to regulate many intracellular events in fibroblasts involved in the development of fibrosis. Two of the novel PKCs, δ and ε, play important roles in scleroderma. PKC δ is present at elevated levels in scleroderma dermal fibroblasts that might be involved in the dysregulation of collagen gene expression that is characteristic of SSc.[4] Thrombin-induced p21Cip1/WAF1 is involved in both cell survival and DNA synthesis, providing supplementary support for the important roles of thrombin-induced signaling in the emergence, proliferation and persistence of the myofibroblast phenotype critical for the development and progression of pulmonary fibrosis. PKC isoforms α and ε play a major role in the regulation of p21Cip1/WAF.[5] PKC ε is deficient in scleroderma lung fibroblasts (SLF) when compared to normal lung fibroblasts (NLF).[6]

This differential expression makes SLF susceptible to curcumin-induced apoptosis, whereas NLF tolerates the same. Increasing PKC ε expression in SLF provides protection against curcumin and decreasing PKC ε expression or activity in NLF causes the cells to become sensitive to curcumin.[7] This selectivity might be due to the low levels of phase 2 enzymes, which is a consequence of the downregulation of PKC in SLF. This shows that curcumin might be used in the treatment of scleroderma by inducing apoptosis selectively to SLF and not to NLF. Moreover, heme oxygenase 1 (HO-1), a phase 2 enzyme, is being produced at higher level in skin fibroblasts isolated from patients with SSc as a feedback mechanism of hypoxia and free radicals.[8] Curcumin induces the expression of phase 2 detoxification enzymes such as HO-1 and glutathione-*S*-transferase (GST), which are regulated by the antioxidant response element (ARE).[9,10] ARE is regulated and activated by the inactivation of the Nrf2-Keap1 complex.[10] (NF-E2)-related factor 2 (Nrf2) is a transcription factor that regulates the expression of conjugating enzymes like GST and HO-1 via ARE.[11] Nrf2 activity is regulated by its sequestration in the cytoplasm by Kelch-domain-containing protein, Keap1 (*K*elch-like *E*CH-*a*ssociated *p*rotein 1). Keap 1 releases Nrf-2 in the presence of oxidants and chemoprotective agents, thereby leading to the activation of ARE and expression of phase 2 enzymes.[12] Curcumin has been shown to promote the dislodging of Nrf-2 from the Nrf2-Keap1 complex, leading to increased Nrf2 binding to the resident HO-1 AREs, resulting in upregulation of HO-1 expression.[10] Moreover, phosphorylation of Nrf2 by PKC also promotes its dissociation from Keap1 and thereby increases the expression of phase 2 detoxification enzymes.[13] Curcumin induces the Nrf2/ARE pathway, increases HO-1 expression and stimulates Nrf2 binding to the ARE via PKC.[14] These details suggest the important role of PKC in scleroderma, and curcumin, being a PKC modulator as well as antioxidant enzyme inducer, might be of therapeutic value in scleroderma.

3. PSORIASIS

Psoriasis is a noncontagious hyperproliferative skin disease caused by faulty signals in the immune system and is generally considered an autoimmune disease mediated by T-cells.[15] It results in thick, silvery flakes of scale on raised pinkish red skin with well-defined margins. Hence, agents can be screened on the basis of antiproliferative and/or anti-inflammatory properties. Curcumin has the ability to be developed as an antipsoriatic drug because of its ability to curtail keratinocyte proliferation[16] and was found to be effective in the mouse tail animal model of psoriasis.[17] Curcumin decreases the expression of proinflammatory cytokines like interleukin 6 and IL-8 in human keratinocytes.[18] These cytokines are both proinflammatory and are growth factors for keratinocytes. Hence, their inhibition by curcumin might reduce psoriasis-linked inflammation as well as psoriasis-related keratinocyte hyperproliferation. Moreover, HO-1 is being produced at higher levels in the skin of psoriatic patients and might be involved in heme degradation and in the protection of cells from the toxic effects of reactive oxygen species (ROS).[19] Curcumin regulation of HO-1 and counteracting oxidative stress was a major highlight in its protection against scleroderma as well.

Topical therapies have a stronghold in human psoriasis treatment. The most common therapy for psoriasis is topical glucocorticoids due to their anti-inflammatory activity.[20] Even though glucocorticoid treatment is effective, discontinuation is prompted due to its side effect of dermal atrophy. Dermal atrophy is due to the inhibitory effect of glucocorticoids on collagen synthesis and deposition. Therefore, suitable alternatives are pivotal for treating psoriasis. The polyphenolic compound curcumin and related nontoxic antioxidants from the rhizome of the *Curcuma longa* have a favorable effect on psoriasis. Phosphorylase kinase (PK), a calmodulin-containing enzyme, is involved in regulating calcium-dependent phosphorylation events in human epidermis. PK is present in both psoriatic and normal epidermis. However, PK is expressed at significantly higher levels in psoriasis. Higher levels of PK activity, glycogenolysis and phosphorylation are associated with increased psoriatic activity.[21] Curcumin is a selective PK inhibitor and was shown to reduce the PK levels in psoriatic subjects to PK levels as in normal skin.[22] Curcumin was found to be more effective in reducing the PK levels when compared with calcipotriol, a vitamin D_3 analogue in psoriasis.[22] The resolution of psoriasis in the curcumin-treated group was far better than the calcipotriol or untreated group.

Other than topical therapies, phototherapy is also considered effective and safe for treating psoriasis. Curcumin proves phototoxic for *Salmonella typhimurium* and *Escherichia coli* on irradiation with visible light.[23] This potential photosensitizing principle of curcumin makes it applicable in the phototherapy of psoriasis. The combination of phototherapy with curcumin might accelerate the clinical response and might even diminish the exposure load.

4. SKIN CANCER

Skin cancer is a disease in which cancer (malignant) cells are found in the outer layers of skin. Most skin cancers are classified as nonmelanoma, usually occurring in either basal cells or squamous cells, and melanoma, cancer that begins in the melanocytes (cells that produce the skin color or pigment known as melanin).[24] Skin cancer is mainly caused by the interaction of genes and the environment. Exposure to sun is the leading environmental cause of skin cancer. Skin cancer also tends to be genetic (hereditary) and occurs very frequently in certain ethnic groups, especially those with fair complexions. Other possible causes of skin cancer include X-rays,[25] trauma, and exposure to certain chemicals like arsenic.[26]

Murine models are used to study chemical- and radiation-induced skin cancer. One of the reliable models of chemical carcinogensis is the two-stage model in which the tumor is initated by chemicals like 7,12-dimethylbenz[a]anthracene (DMBA), benzo[*a*]pyrene (BAP), and 3-methyl cholanthrene and promoted by treatment with tetradecanoylphorbol-13-acetate (TPA) the following week until papillomas form. Curcumin inhibited and decreased the number of tumors in the

DMBA-induced and TPA-promoted model of chemical carcinogenesis.[27–29] Ornithine decarboxylase (ODC), the rate-limiting enzyme in polyamine biosynthesis, is elevated in mouse epidermis after TPA administration and this induction was inhibited by intraperitoneal injection of curcumin.[30] Topical application of curcumin inhibited TPA-induced epidermal hyperplasia, which might be due to inhibition of c-jun and c-fos expression.[31] Pretreatment with curcumin significantly reduced the ODC levels induced by ultraviolet A (UVA) irradiation, thereby alleviating the TPA-induced dermatitis.[32]

The incidence of skin cancer by radiation is rising and a frequently used preclinical animal model of photocarcinogenesis is the SKH-1 hairless mouse strain.[33] DNA damage and mutation in crucial genes are prominent reasons for the initiation of skin tumor by ultraviolet B (UVB) light. Application of curcumin inhibited the UVB-induced epidermal hyperplasia.[31] Curcumin also controlled the proinflammatory cytokines induced by UVB in keratinocytes by nuclear factor (NF)-κB inhibition.[34] Exposure of the human skin to UVB causes acute inflammation and cyclooxygenase (COX)-2 plays a prominent role in mediating inflammation. Curcumin treatment in UVB-irradiated keratinocytes largely abrogated COX-2 expression by suppressing p38 Mitogen-activated protein kinase (MAPK) and Jun N-terminal kinase (JNK).[35] The anti-inflammatory property of curcumin can be recognized by its ability to decrease TPA-induced COX-2 expression in mouse skin through blocking extracellular signal-regulated kinases (ERK) and NF-κB signaling cascades.[36]

Oxidative stress plays a crucial role in skin carcinogenesis. Activating oxygen can produce compounds called free radicals that cause oxidative stress on cells. Both UV and chemicals have the ability to induce free radicals and such stress could ultimately lead to cancer. Antioxidant plant phenolics like curcumin might be used as an effective chemopreventive agent for mouse skin carcinogenesis by their antioxidant property.[37] All of the experimental evidences provide a molecular mechanism for suppression of tumor promotion as well as for the anti-inflammatory activity exerted by this phytochemical in mouse skin.

5. WOUND-HEALING

Every wound initiates bodily mechanisms that are designed to regenerate the same or almost the same tissue as the original one. Wound-healing proceeds in three interrelated dynamic phases with overlapping time courses irrespective of the wound type and the degree of tissue damage.[38] According to morphological changes in the course of the healing process, three phases are clinically distinguished: (1) the inflammatory or exudative phase, for detachment of the deteriorated tissues and wound cleansing; (2) the proliferative phase, for the development of granulation tissue; and (3) the differentiation phase or a regeneration phase, for maturation, scar formation and epithelialization.

The inflammatory response is characterized by homeostasis and inflammation. Damaged tissue releases adenosine diphosphate (ADP) and a vasoconstrictor, resulting in platelet aggregation and blood vessel constriction, thus sealing the damaged blood vessels and reducing blood flow, respectively.[39] Degranulating platelets release a number of chemokines. Neutrophils, activated endothelial cells, leukocytes, macrophages and epithelial cells, migrate in the wound matrix under chemotactic signals.[38] These cells further secrete chemotactic mediators and enzymes. This helps in removing the wound debris, promoting angiogenesis and stimulatiing fibroblasts and keratinocytes.[40,41] Once the inflammatory cells are active, they become susceptible to transforming growth factor (TGF)-β1-mediated suppression to reverse the inflammatory substance. IL-4 also dampens the inflammatory response and promotes collagen synthesis. This step marks the transition from the inflammatory phase into the next phase of tissue reconstruction—the proliferative phase.

The proliferative phase is marked by reepithelialization, granulation and angiogenesis in the wound tissue.[38] Epithelial cells begin to proliferate, migrate and cover the exposed area to restore the functional integrity of the tissue, serving both the purpose of wound contraction and reformation of the cutaneous barrier.[42] Simultaneous to this process, granulation and angiogenesis start. Granulation tissue was described by Letterer as a temporary primitive tissue until after having fulfilled its function and is subjected to regression and for the most part is gradually converted into the scar tissue.[43] Granulation tissue forms below the epithelium and is composed of inflammatory cells involving leukocytes, histocytes, plasma cells, mast cells and fibroblasts. Nourishment of the nascent tissue is ensured by growing capillaries. Within the granulation tissue, angiogenesis is potentiated by hypoxia, nitric oxide (NO), vascular endothelial growth factor (VEGF), fibroblast growth factor-2 (FGF-2), chemokines and macrophage inflammatory protein (MIP-1α).[44] With the generation of new vasculature, matrix-generating cells move into the granulation tissue. The fibroblast degrades the provisional matrix and synthesizes the new extracellular matrix (ECM). TGF-β stimulates fibroblast to synthesize collagen I, III, and V, proteoglycans, fibronectin and other ECM components. As the repair progresses, the fibroblast displays increased expression levels of adhesion molecules and assumes a myofibroblast phenotype to facilitate wound contraction. In the last phase of wound-healing, which continues for years, fibroblast proliferation ceases and the ECM matures to provide a connective tissue structure that is both strong and flexible and adds to the tensile strength of the wound.

Turmeric has been used since ancient times in India for the treatment of wounds. Topical application of turmeric as a household remedy, for several conditions is described by Nadkarni.[45] In a recent review, Maheshwari et al. described the multiple biological activities of curcumin.[46] Several studies have shown the beneficial effect of curcumin in the enhancement of wound-healing. Curcumin acts at multiple points in wound healing. The anti-inflammatory and ROS scavenger activities of curcumin are regarded as the mechanisms helping in enhancing wound-healing. In the gastric ulcer, curcumin accelerates healing by reepithelialization, preventing glutathione depletion, lipid peroxidation, protein oxidation, attenuating matrix metalloprotein (MMP)-9 and inhibiting MMP-2 activity.[47] When incorporated in a collagen matrix, curcumim resulted in enhanced wound-healing.[48] Wound reduction resulted due to enhanced cell proliferation and efficient free-radical scavenging by curcumin. Curcumin incorporation also resulted in the higher thermostability of collagen. Oxidative damage to keratinocytes, skin cells, and fibroblasts is also reversed by curcumin.[49] At a dose of 10 μg/mL, curcumin protects keratinocytes from the oxidative damage of hydrogen peroxide. Early growth response-1 gene (Egr-1), a transcription factor that regulates genes involved in various pathophysiological processes, including vasculature and wound-healing. Egr-1 is suggested to contribute to the complex series of cellular and thrombogenic events in the development of vascular occlusive lesions.[50] Curcumin suppresses the induction of Egr-1 in endothelial cells and fibroblasts and thus could have a potential therapeutic value, particularly in suppressing thrombogenic events associated with various pathological conditions.[51] The oral and topical administration of curcumin on punch biopsy wounds resulted in faster healing. Curcumin treatment resulted in faster reepithelialization of the epidermis, increased migration of myofibroblasts, fibroblasts, and macrophages in the wound bed, extensive neovascularization and greater collagen deposition.[52] This enhanced wound-healing activity is attributed to increased TGF-β1 expression by curcumin. These studies show that curcumin can be effectively used in therapeutics for enhancing wound-healing.

6. METABOLIC IMPAIRED WOUND-HEALING

The relationship between wound and patient is complex and depends on the overall condition of the afflicted person. It depends on a number of endogenous factors, including age and metabolic condition. In the normal healthy host, wound-healing is usually uncomplicated and proceeds at a rapid rate. In contrast, most healing failures are associated with some form of host impairment, including diabetes, infection, immunosuppression, obesity or malnutrition. Impaired wound-healing is a significant source of morbidity for the surgical patient and might result in complications such as wound dehiscence, anastomotic breakdown and chronic nonhealing ulcers.[53] Multiple factors contribute towards the impairment of wound healing in diabetes. These factors include inadequate blood supply due to venous insufficiency and cuffing of microvessels,[54] decreased proliferative potential of fibroblasts,[55] decreased inflammatory changes, and failure of migrating macrophages to show activation markers.[56] The cost associated with such complications might be extreme due to prolonged hospitalization and increased time away from work. Several studies have shown a beneficial activity of curcumin in diabetic conditions such as blood glucose, cataract, hyperalgesia, and oxidative stress.[57–60] Curcumin has been shown to enhance woundhealing in dexamethasone-induced diabetes in rats.[61] These animals exhibit delayed healing without any treatment. Wounds of animals treated with curcumin showed earlier reepithelialization, improved neovascularization, increased migration of dermal myofibroblasts, fibroblasts, and macrophages into the wound bed, and a higher collagen content. Curcumin treatment resulted in increased expression of TGF-β1 and its receptor TGF-β type I (tIrc) and type II (tIIrc) and induced nitric oxide species (iNOS).[62] Apoptosis was also delayed in diabetic wounds compared to curcumin-treated wounds. These studies show that curcumin has a positive impact on wound repair in diabetic-impaired healing and could be developed as a pharmacological agent for treating diabetic wounds.

7. RADIATION-INDUCED IMPAIRED WOUND-HEALING

Humans are increasingly exposed to radiations, especially in the use of radiation against cancer therapeutic or incidental. Exposure of wounds to radiation results in the impairment of healing and also has malignant potentials.[63] Exposure of the wound tissue to ionizing radiation results in the disruption of the normal healing and thus longer recovery and detrimental effect on the outcome of injury. Wounds, when exposed to radiation, result in severe inhibition of inflammatory response, slow maturation of granulation tissue, reduced collagen and hexosamine synthesis, and delayed reepithelialization.[64] Fibroblasts, the key players in granulation tissue formation and maturation, isolated from the irradiated wound and from the wound after total-body irradiation (TBI) have significantly reduced attachment, adhesion, and colony formation ability and have longer doubling time compared to nonirradiated fibroblasts.[65,66] Systemic irradiation after the creation of wounds results in decreased phagocytic function of wound macrophages. The release of TNF-α and IL-1 from wound macrophages, the number of macrophages in the wound, and the wound breaking strength (WBS) decrease after exposure to irradiation.[67] Oral administration of curcumin before irradiation enhanced the synthesis of collagen, hexosamine, DNA, and nitrate.[68] Curcumin pretreatment results in reduced wound healing time, improved collagen deposition, increase in fibroblast and vascular density. Curcumin pretreatment treatment resulted in a dose dependent increase in the contraction of irradiated wound.[69,70] Therefore, curcumin can be effectively used in the treatment of irradiated wound for improving the delayed healing.

8. CURCUMIN AND ANGIOGENESIS IN SKIN DISEASES

Angiogenesis, the formation of new blood vessels, plays a significant role in physiological processes like growth and development, wound-healing and reproductive functions in adults. Contrary to these beneficial effects, angiogenesis can be detrimental in several pathological conditions, which include malignant diseases like skin cancer as well as nonmalignant conditions like psoriasis, atopic dermatitis and diabetic retinopathy. Astonishingly, VEGF, a proangiogenic factor is overexpressed in the skin of patients with SSc despite insufficient angiogenesis.[71] Angiogenesis appears to be a fundamental inflammatory response early in pathogenesis, and significant abnormalities of vascular morphology and angiogenic growth factors have been described in psoriasis.[72] Proteasome inhibitors have been recently investigated for treating psoriasis and other inflammatory disorders. Compounds like curcumin, which not only inhibit angiogenesis but also inhibit proteasome activity, might provide a suitable alternative to treat psoriasis.[73] Antiangiogenic therapies represent a powerful addition to traditional cancer therapies and other skin diseases. Curcumin is found to hamper tumor angiogenesis[74] and it might be a reason in part for its activity in inhibiting carcinogenesis in skin. Endothelial cell proliferation and differentiation are sequential events involved in capillary formation. The growth of human umbilical vein endothelial cells (HUVECs) stimulated with fibroblast growth factor (FGF) and endothelial cell growth supplement (ECGS) was found to be inhibited by curcumin, wherein it effectively blocked cell cycle progression during the S phase by inhibiting the activity of TK enzyme.[75] Curcumin treatment resulted in inhibiting tube formation as well as reduced the migration of endothelial cells in a Matrigel plug model, thus showing their inhibitory nature on angiogenic differentiation of endothelial cells.[76] All of the cited experimental evidence indicate the antiangiogenesis effect of curcumin and its potential as a valuable therapeutic agent. However, curcumin has shown a beneficial effect as a pro angiogenic agent in wound-healing by inducing TGF-β. During the inflammation that follows injury, TGF-β induces both the angiogenesis and accumulation of the ECM, which continue through the remodeling phase of wound repair. Moreover TGF-β can attract inflammatory and connective tissue cells, which, in turn, control angiogenesis. The preclinical evidence suggests that curcumin might exert a dual effect in angiogenesis and its action mainly depends on the environment in which it is acting.

9. CLINICAL TRIALS WITH CURCUMIN FOR VARIOUS SKIN DISORDERS

Although preclinical data of curcumin in various animal models of skin disease is very encouraging, clinical trials on the pharmacokinetics, pharmacodynamics, and mechanism-based markers of efficacy would help in developing curcumin as a therapeutic agent. Some of the clinical trials on curcumin pertinent to skin diseases are discussed in brief. A phase I study for Bowen's disease (squamous cell carcinoma *in situ*) of the skin was conducted for curcumin and the study reports that there was no treatment-related toxicity up to 8000 mg/day for up to 3 months. Histologic improvement of precancerous lesions was seen in around 33% of patients with Bowen's disease.[77] Another clinical trial in the University of California was carried out for patients with psoriasis. An alcoholic gel preparation containing 1% curcumin was compared with calcipotriol ointment (Dovonexw, 0.005% calcipotriol) in a cohort of 10 patients. Untreated patients with psoriasis were the control group. Curcumin-treated psoriasis patient had much better resolution of the disease within a shorter period of time when compared to the calsipotriol ointment. In a separate experiment, when curcumin gel was compared with the vehicle alone (alcoholic gel), curcumin-treated plagues improved by 25 ± 70%, whereas there was no improvement in 33% of the patient and the condition worsened in 66% of the vehicle-treated plaques.[22] Neem (*Azadirachta indica)* and turmeric were used

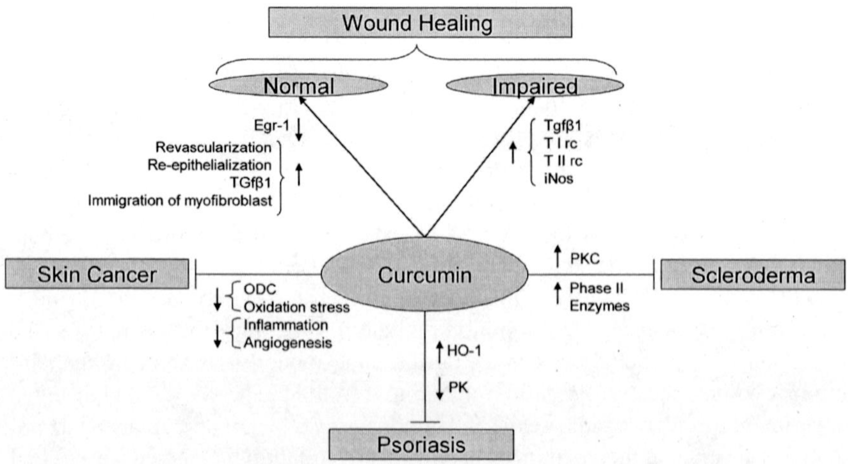

Figure 1. Schematic representation of multiple targets involved in curumin's beneficial effects in various skin diseases and wound-healing.

as a paste for the treatment of chronic ulcers and scabies. Curcumin cured around 97% of the people in a cohort of 814 people suffering from scabies [itchy condition of the skin caused by a tiny mite (*Sarcoptes scabiei*)] within 3–15 days.[78]

10. CONCLUSIONS

There have been a number of encouraging studies on skin benefits of curcumin (Figure 1). Preclinical trials with experimental animal models and few of the human clinical trials conducted so far showed the protection rendered by curcumin from skin diseases like psoriasis. Moreover, it has been shown to be nontoxic in large doses even at 8 g/day. Curcumin appears to protect skin by quenching free radicals and reducing inflammation, and the primary target of curcumin is found to be NF-κB inhibition. These studies suggest the beneficial effect of curcumin and the potential of this compound to be developed as a potent nontoxic agent for treating skin diseases.

ACKNOWLEDGMENTS

Some of these studies were supported by a grant (G174KT) from the NCCAM, National Institute of Health, Bethesda and US-INDIA Foreign Currency Fund from US Department of State to USUHS. The opinions or assertions contained herein are the private views of the authors and should not be construed as official or necessarily reflecting the views of the Uniformed Services University of the Health Sciences or the Department of Defense, USA.

REFERENCES

1. A. Maitre, M. Hours, V. Bonneterre, J. Arnaud, M. T. Arslan, P. Carpentier, A. Bergeret, and R. de Gaudemaris, Systemic sclerosis and occupational risk factors: Role of solvents and cleaning products. *J Rheumatol* **31**, 2395 (2004).
2. M. Bovenzi, F. Barbone, F. E. Pisa, A. Betta, L. Romeo, A. Tonello, D. Biasi, and P. Caramaschi, A case-control study of occupational exposures and systemic sclerosis. *Int Arch Occup Environ Health* **77**, 10 (2004).
3. C. T. Derk and S. A. Jimenez, Systemic sclerosis: Current views of its pathogenesis. *Autoimmun Rev* **2**, 181 (2003).
4. S. A.Jimenez, S. Gaidarova, B. Saitta, N. Sandorfi, D. J. Herrich, J. C. Rosenbloom, U. Kucich, W. R. Abrams, and J Rosenbloom, Role of protein kinase C-delta in the regulation of collagen gene expression in scleroderma fibroblasts. *J Clin Invest* **108**, 1395 (2001).
5. G. S. Bogatkevich, E. Gustilo, J. C. Oates, C. Feghali-Bostwick, R. A. Harley, R. M. Silver, and A. Ludwicka-Bradley, Distinct PKC isoforms mediate cell survival and DNA synthesis in thrombin-induced myofibroblasts. *Am J Physiol Lung Cell Mol Physiol* **288**, L190 (2005).
6. E. Tourkina, P. Gooz, J. Pannu, M. Bonner, D. Scholz, S. Hacker, R. M. Silver, M. Trojanowska, and S. Hoffman, Opposing effects of protein kinase Calpha and protein kinase Cepsilon on collagen expression by human lung fibroblasts are mediated via MEK/ERK and caveolin-1 signaling, *J Biol Chem* **280**, 13,879 (2005).
7. E. Tourkina, P. Gooz, J. C. Oates, A. Ludwicka-Bradley, R. M. Silver, and S. Hoffman, Curcumin-induced apoptosis in scleroderma lung fibroblasts: Role of protein kinase cepsilon. *Am J Respir Cell Mol Biol* **31**, 28 (2004).
8. M. V. Panchenko, H. W. Farber, and J. H. Korn, Induction of heme oxygenase-1 by hypoxia and free radicals in human dermal fibroblasts. *Am J Physiol: Cell Physiol* **278**, C92 (2000).
9. A. T. Dinkova-Kostova and P. Talalay, Relation of structure of curcumin analogs to their potencies as inducers of Phase 2 detoxification enzymes. *Carcinogenesis* **20**, 911 (1999).
10. E. Balogun, M. Hoque, P. Gong, E. Killeen, C. J. Green, R. Foresti, J. Alam, and R. Motterlini, Curcumin activates the haem oxygenase-1 gene via regulation of Nrf2 and the antioxidant-responsive element. *Biochem J* **371**, 887 (2003).
11. K. Itoh, T. Chiba, S. Takahashi, T. Ishii, K. Igarashi, Y. Katoh, T. Oyake, N. Hayashi, K. Satoh, I. Hatayama, M. Yamamoto, and Y. Nabeshima, An Nrf2/small Maf heterodimer mediates the induction of phase II detoxifying enzyme genes through antioxidant response elements. *Biochem Biophys Res Commun* **236**, 313 (1997).
12. B. Pool-Zobel, S. Veeriah, and F. D. Bohmer, Modulation of xenobiotic metabolising enzymes by anticarcinogens-focus on glutathione S-transferases and their role as targets of dietary chemoprevention in colorectal carcinogenesis. *Mutat Res* **591**, 74 (2005).
13. D. A. Bloom and A. K. Jaiswal, Phosphorylation of Nrf2 at Ser40 by protein kinase C in response to antioxidants leads to the release of Nrf2 from INrf2, but is not required for Nrf2 stabilization/accumulation in the nucleus and transcriptional activation of antioxidant response element-mediated NAD(P)H:quinone oxidoreductase-1 gene expression. *J Biol Chem* **278**, 44,675 (2003).
14. S. A. Rushworth, R. M. Ogborne, C. A. Charalambos, and M. A. O'Connell, Role of protein kinase C delta in curcumin-induced antioxidant response element-mediated gene expression in human monocytes. *Biochem Biophys Res Commun* **341**, 1007 (2006).

15. M. Kastelan, L. P. Massari, and I. Brajac, IApoptosis mediated by cytolytic molecules might be responsible for maintenance of psoriatic plaques. *Med Hypotheses* **21**, 21 (2006).
16. A. Pol, M. Bergers, and J. Schalkwijk, Comparison of antiproliferative effects of experimental and established antipsoriatic drugs on human keratinocytes, using a simple 96-well-plate assay. *In Vitro Cell Dev Biol Anim* **39**, 36 (2003).
17. B. Bosman, Testing of lipoxygenase inhibitors, cyclooxygenase inhibitors, drugs with immunomodulating properties and some reference antipsoriatic drugs in the modified mouse tail test, an animal model of psoriasis. *Skin Pharmacol* **7**, 324 (1994).
18. J. Miquel, A. Bernd, J. M. Sempere, J. Diaz-Alperi, and A. Ramirez, The curcuma antioxidants: Pharmacological effects and prospects for future clinical use. A review. *Arch Gerontol Geriatr* **34**, 37 (2002).
19. C. Hanselmann, C. Mauch, and S. Werner, Haem oxygenase-1: A novel player in cutaneous wound repair and psoriasis? *Biochem J* **353**, 459 (2001).
20. M. Lebwohl, Innovations in the treatment of psoriasis. *J Am Acad Dermatol* **51**, S40 (2004).
21. M. C. Heng, M. K. Song, and M. K. Heng, Elevated phosphorylase kinase activity in psoriatic epidermis: Correlation with increased phosphorylation and psoriatic activity. *Br J Dermatol* **130**, 298 (1994).
22. M. C. Heng, M. K. Song, J. Harker, and M. K. Heng, Drug-induced suppression of phosphorylase kinase activity correlates with resolution of psoriasis as assessed by clinical, histological and immunohistochemical parameters. *Br J Dermatol* **143**, 937 (2000).
23. H. H. Tonnesen, H. de Vries, J. Karlsen, and G. Beijersbergen van Henegouwen, Studies on curcumin and curcuminoids. IX: Investigation of the photobiological activity of curcumin using bacterial indicator systems. *J Pharm Sci* 76, 371 (1987).
24. L. Kondapalli, K. Soltani, and M. E. Lacouture, The promise of molecular targeted therapies: Protein kinase inhibitors in the treatment of cutaneous malignancies. *J Am Acad Dermatol* **53**, 291 (2005).
25. C. C. Ramirez, D. G. Federman, and R. S. Kirsner, Skin cancer as an occupational disease: The effect of ultraviolet and other forms of radiation. *Int J Dermatol* **44**, 95 (2005).
26. W. Ding, L. G. Hudson, and K. J. Liu, Inorganic arsenic compounds cause oxidative damage to DNA and protein by inducing ROS and RNS generation in human keratinocytes. *Mol Cell Biochem* **279**, 105 (2005).
27. M. A. Azuine and S. V. Bhide, Chemopreventive effect of turmeric against stomach and skin tumors induced by chemical carcinogens in Swiss mice. *Nutr Cancer* **17**, 77 (1992).
28. M. Nagabhushan and S. V. Bhide, Curcumin as an inhibitor of cancer. *J Am Coll Nutr* **11**, 192 (1992).
29. M. T. Huang, Z. Y. Wang, C. A. Georgiadis, J. D. Laskin, and A. H. Conney, Inhibitory effects of curcumin on tumor initiation by benzo[a]pyrene and 7,12-dimethylbenz[a]anthracene. *Carcinogenesis* **13**, 2183 (1992).
30. Y. P. Lu, R. L. Chang, M. T. Huang, and A. H. Conney, Inhibitory effect of curcumin on 12-O-tetradecanoylphorbol-13-acetate-induced increase in ornithine decarboxylase mRNA in mouse epidermis. *Carcinogenesis* **14**, 293 (1993).
31. Y. P. Lu, R. L. Chang, Y. R. Lou, M. T. Huang, H. L. Newmark, K. R. Reuhl, and A. H. Conney, Effect of curcumin on 12-O-tetradecanoylphorbol-13-acetate- and ultraviolet

B light-induced expression of c-Jun and c-Fos in JB6 cells and in mouse epidermis. *Carcinogenesis* **15**, 2363 (1994).
32. C. Ishizaki, T. Oguro, T. Yoshida, C. Q. Wen, H. Sueki, and M. Iijima, Enhancing effect of ultraviolet A on ornithine decarboxylase induction and dermatitis evoked by 12-O-tetradecanoylphorbol-13-acetate and its inhibition by curcumin in mouse skin. *Dermatology* **193**, 311(1996).
33. G. T. Bowden, Prevention of non-melanoma skin cancer by targeting ultraviolet-B-light signalling. *Nat Rev Cancer* **4**, 23 (2004).
34. A. Grandjean-Laquerriere, S. C. Gangloff, R. Le Naour, C. Trentesaux, W. Hornebeck, and M. Guenounou, Relative contribution of NF-kappaB and AP-1 in the modulation by curcumin and pyrrolidine dithiocarbamate of the UVB-induced cytokine expression by keratinocytes. *Cytokine* **18**, 168 (2002).
35. J. W. Cho, K. Park, G. R. Kweon, B. C. Jang, W. K. Baek, M. H. Suh, C. W. Kim, K. S. Lee, and S. I. Suh, Curcumin inhibits the expression of COX-2 in UVB-irradiated human keratinocytes (HaCaT) by inhibiting activation of AP-1: p38 MAP kinase and JNK as potential upstream targets. *Exp Mol Med* **37**, 186 (2005).
36. K. S. Chun, Y. S. Keum, S. S. Han, Y. S. Song, S. H. Kim, and Y. J. Surh, Curcumin inhibits phorbol ester-induced expression of cyclooxygenase-2 in mouse skin through suppression of extracellular signal-regulated kinase activity and NF-kappaB activation. *Carcinogenesis* **24**, 1515 (2003).
37. N. Ahmad, S. K. Katiyar, and H. Mukhtar, Antioxidants in chemoprevention of skin cancer. *Curr Probl Dermatol* **29**, 128 (2001).
38. I. Aukhil, Biology of wound healing. *Periodontology* **22**, 44 (2000).
39. J. Slavin, The role of cytokines in wound healing. *J Pathol* 178, 5–10 (196).
40. J. Folkman and Y. Shing, Angiogenesis. *J Biol Chem* **267**, 10,931 (1992).
41. J. Folkman and Y. Shing, Control of angiogenesis by heparin and other sulfated polysaccharides. *Adv Exp Med Biol* **313**, 355 (1992).
42. G. S. Ashcroft, T. Greenwell-Wild, M. A. Horan, S. M. Wahl, and M. W. Ferguson, Topical estrogen accelerates cutaneous wound healing in aged humans associated with an altered inflammatory response. *Am J Pathol* **55**, 1137 (1999).
43. E. Letterer, Morphological manifestations of allergic-hyperergic processes during infectious diseases. *Acta Allergol Suppl (Copenh)* **3**, 79 (1953).
44. J. A. Belperio, M. P. Keane, D. A. Arenberg, C. L. Addison, J. E. Ehlert, M. D. Burdick, and R. M. Strieter, CXC chemokines in angiogenesis. *J Leukocyte Biol* **68**, 1 (2000).
45. K. M. Nadkarni, *Curcuma longa*. In: K. M. Nadkarni, ed. *Indian Materia Medica*, Bombay: Popular Prakashan Publishing, 1976.
46. R. K. Maheshwari, A. K. Singh, J. Gaddipati, and R. C. Srimal, Multiple biological activities of curcumin: A short review. *Life Sci* **78**, 2081 (2006).
47. S. Swarnakar, K. Ganguly, P. Kundu, A. Banerjee, P. Maity, and A. V. Sharma, Curcumin regulates expression and activity of matrix metalloproteinases-9 and -2 during prevention and healing of indomethacin-induced gastric ulcer. *J Biol Chem* **280**, 9409 (2005).
48. D. Gopinath, M. R. Ahmed, K. Gomathi, K. Chitra, P. K. Sehgal, and R. Jayakumar, Dermal wound healing processes with curcumin incorporated collagen films. *Biomaterials* **25**, 1911 (2004).
49. T. T. Phan, P. See, S. T. Lee, and S. Y. Chan, Protective effects of curcumin against oxidative damage on skin cells in vitro: Its implication for wound healing. *J Trauma* **51**, 927 (2001).

50. L. M. Khachigian, V. Lindner, A. J. Williams, and T. Collins, Egr-1-induced endothelial gene expression: A common theme in vascular injury. *Science* **271**, 1427 (1997).
51. U. R. Pendurthi and L. V. Rao, Suppression of transcription factor Egr-1 by curcumin. *Thromb Res* **97**, 179 (2000).
52. G. S. Sidhu, A. K. Singh, D. Thaloor, K. K. Banaudha, G. K. Patnaik, R. C. Srimal, and R. K. Maheshwari, Enhancement of wound healing by curcumin in animals. *Wound Repair Regen* **6**, 167 (1998).
53. W. H.Goodson 3rd. and T. Hunt, Wound healing and the diabetic patient. *Surg Gynecol Obstet* **149**, 600 (1997).
54. M. P. Cohen, V. Y. Wu, and J. A. Cohen, Glycated albumin stimulates fibronectin and collagen IV production by glomerular endothelial cells under normoglycemic conditions. *Biochem Biophys Res Commun* **239**, 91 (1997).
55. J. S. Vande Berg, M. C. Robson, and R. J. Mikhail, Extension of the life span of pressure ulcer fibroblasts with recombinant human interleukin-1 beta. *Am J Pathol* **146**, 1273 (1995).
56. K. Moore, F. Ruge, and K. G. Harding, T lymphocytes and the lack of activated macrophages in wound margin biopsies from chronic leg ulcers. *Br J Dermatol* **137**, 188 (1997).
57. L. Pari and P. Murugan, Effect of tetrahydrocurcumin on blood glucose, plasma insulin and hepatic key enzymes in streptozotocin induced diabetic rats. *J Basic Clin Physiol Pharmacol* **16**, 257 (2005).
58. P. Suryanarayana, M. Saraswat, T. Mrudula, T. P. Krishna, K. Krishnaswamy, and G. B. Reddy, Curcumin and turmeric delay streptozotocin-induced diabetic cataract in rats. *Invest Ophthalmol Vis Sci* **46**, 2092 (2005).
59. S. Sharma, S. K. Kulkarni, J. N. Agrewala, and K. Chopra, Curcumin attenuates thermal hyperalgesia in a diabetic mouse model of neuropathic pain. *Eur J Pharmacol* **536**, 256 (2006).
60. T. Osawa and Y. Kato, Protective role of antioxidative food factors in oxidative stress caused by hyperglycemia. *Ann NY Acad Sci* **1043**, 440 (2005).
61. G. S. Sidhu, H. Mani, J. P. Gaddipati, A. K. Singh, P. Seth, K. K. Banaudha, G. K. Patnaik, and R. K. Maheshwari, Curcumin enhances wound healing in streptozotocin induced diabetic rats and genetically diabetic mice. *Wound Repair Regen* **7**, 362 (1999).
62. H. Mani, G. S. Sidhu, R. Kumari, J. P. Gaddipati, P. Seth, and R. K. Maheshwari, Curcumin differentially regulates TGF-beta1, its receptors and nitric oxide synthase during impaired wound healing. *Biofactors* **16**, 29 (2002).
63. R. E. Shore, Overview of radiation-induced skin cancer in humans. *Int J Radiat Biol* **57**, 809 (1990).
64. Q. Gu, D. Wang, C. Cui, Y. Gao, G. Xia, and X. Cui, Effects of radiation on wound healing. *J Environ Pathol Toxicol Oncol* **17**, 117 (1998).
65. R. Rudolph, J. Vande Berge, J. A. Schneider, J. C. Fisher, and W. L. Poolman, Slowed growth of cultured fibroblasts from human radiation wounds. *Plast Reconstr Surg* **18**, 669 (1998).
66. J. F. Qu, T. M. Cheng, L. S. Xu, C. M. Shi, and X. Z. Ran, Effects of total body irradiation injury on the participation of dermal fibroblasts in tissue repair. *Sheng Li Xue Bao* **54**, 395 (2002).
67. S. Song and T. Cheng, The effect of systemic and local irradiation on wound macrophage and repair promoting action of phenytion sodium. *Zhonghma Yi Xue Za Zhi* **77**, 54 (1997).

68. G. C. Jagetia and G. K. Rajanikant, Curcumin treatment enhances the repair and regeneration of wounds in mice exposed to hemibody gamma-irradiation. *Plast Reconstr Surg* **115**, 515 (2005).
69. G. C. Jagetia and G. K. Rajanikant, Effect of curcumin on radiation-impaired healing of excisional wounds in mice. *J Wound Care* 13, 107 (2004).
70. G. C. Jagetia and G. K. Rajanikant, Role of curcumin, a naturally occurring phenolic compound of turmeric in accelerating the repair of excision wound, in mice whole-body exposed to various doses of gamma-radiation. *J Surg Res* **120**, 127 (2004).
71. J. H. Distler, J. R. Kalden, S.Gray, and O. Distler, [Vascular changes in the pathogenesis of systemic sclerosis]. *Z Rheumatol* **63**, 446 (2004).
72. T. T. Leong, U. Fearon, and D. J. Veale, Angiogenesis in psoriasis and psoriatic arthritis: Clues to disease pathogenesis. *Curr Rheumatol Rep* **7**, 325 (2005).
73. J. L. Arbiser, X. C. Li, C. F. Hossain, D. G. Nagle, D. M. Smith, P. Miller, B. Govindarajan, J. DiCarlo, K. R. Landis-Piwowar, and Q. P. Dou, Naturally occurring proteasome inhibitors from mate tea (*Ilex paraguayensis*) serve as models for topical proteasome inhibitors. *J Invest Dermatol* **125**, 207 (2005).
74. M. H. Oak, J. El Bedoui, and V. B. Schini-Kerth, Antiangiogenic properties of natural polyphenols from red wine and green tea. *J Nutr Biochem* **16**, 1 (2005).
75. A. K. Singh, G. S. Sidhu, T. Deepa, and R. K. Maheshwari, Curcumin inhibits the proliferation and cell cycle progression of human umbilical vein endothelial cell. *Cancer Lett* **107**, 109 (1996).
76. D. Thaloor, A. K. Singh, G. S. Sidhu, P. V. Prasad, H. K. Kleinman, and R. K. Maheshwari, Inhibition of angiogenic differentiation of human umbilical vein endothelial cells by curcumin. *Cell Growth Differ* **9**, 305 (1998).
77. A. L. Cheng, C. H. Hsu, J. K. Lin, M. M. Hsu, Y. F. Ho, T. S. Shen, J. Y. Ko, J. T. Lin, B. R. Lin, W. Ming-Shiang, H. S. Yu, S. H. Jee, G. S. Chen, T. M. Chen, C. A. Chen, M. K. Lai, Y. S. Pu, M. H. Pan, Y. J. Wang, C. C. Tsai, and C. Y. Hsieh, Phase I clinical trial of curcumin, a chemopreventive agent, in patients with high-risk or pre-malignant lesions. *Anticancer Res* **21**, 2895 (2001).
78. V. Charles and S. X. Charles, The use and efficacy of *Azadirachta indica* ADR ('Neem') and *Curcuma longa* ('Turmeric') in scabies. A pilot study. *Trop Geogr Med* **44**, 178 (1992).

CARDIOPROTECTIVE EFFECTS OF CURCUMIN

Sumitra Miriyala, Manikandan Panchatcharam, and
Puvanakrishnan Rengarajulu

Abstract: Curcumin, a major active component of turmeric, is extracted from the powdered dry rhizome of *Curcuma longa* Linn (Zingiberaceae) and it has been used for centuries in indigenous medicine. We have shown that curcumin has a protective role against myocardial necrosis in rats. The antioxidant activity of curcumin could be attributed to the phenolic and methoxy groups in conjunction with the 1,3-diketone-conjugated diene system, for scavenging of the oxygen radicals. In addition, curcumin is shown to enhance the activities of detoxifying enzymes such as glutathione-*S*-transferase *in vivo.* We have also shown that oxygen free radicals exacerbate cardiac damage and curcumin induces cardioprotective effect and it also inhibits free-radical generation in myocardial ischemia in rats. This chapter on the cardioprotective effects of curcumin covers the following aspects: (1) the history of curcumin and its discovery as a potent drug with relevance to cardiovascular diseases; (2) mechanistic role of curcumin *in vitro,* emphasizing the antiplatelet and anticoagulant effects; (3) cardiovascular properties of curcumin; (4) application of curcumin in different animal models (viz. myocardial ischemia, myocardial infarction, cardiomyopathy, and arrhythmia *in vitro* and *in vivo*); (5) curcumin free-radical scavenging activity, particularly against $O_2^{\cdot -}$ radical and depletion of the oxidative stress.

1. INTRODUCTION

In recent years, many spices have made a cross-cultural penetration that, in turn, has generated scientific curiosity about their biological effects and modes of action. One such wonder is the "turmeric" spice made from the root of the plant *Curcuma longa* (Zingiberaceae), in the ginger family. In the Ayurvedic system of medicine, turmeric has been prescribed for the treatment of common colds, coughs, jaundice, and upper respiratory disorders.[1] "The next time you feel a cold coming on, try this old Indian remedy: open a capsule or two of curcumin into a small amount of honey. Mix it and eat. You'll feel better almost immediately."

Since ancient times, many properties have been ascribed to extracts of *Curcuma longa.* The plant has been applied for the prevention of skin diseases, hepatic disorders, ulcers and digestive disturbances. It has also been used in the treatment of intestinal parasites and as a remedy for poisoning, snakebites, and various other

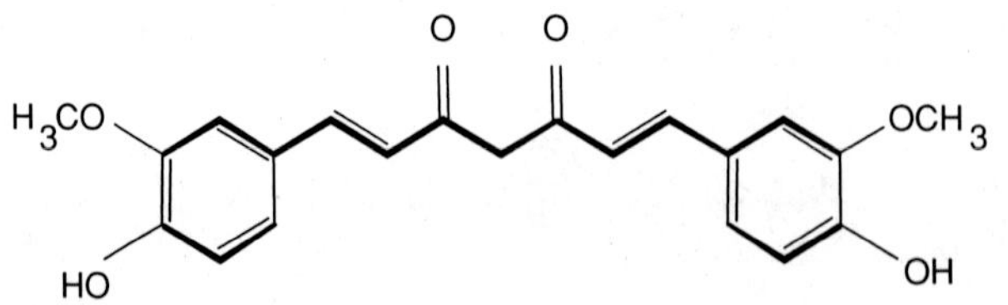

Figure 1. Chemical structure of curcumin (1,7-bis (4-hydroxy-3 methoxyphenyl)-1,6-heptadiene-3,5-dione).

complaints.[2] It is considered to be a magical plant because of its organoleptic properties and undoubted therapeutic and protective effects, especially for the skin and liver.[3] According to a WHO/FAO report (1974), the dietary intake of turmeric in Indian populations ranged between 2 and 2.5 g per person (approximate weight = 60 kg) per day. Keeping in view the amount of curcumin (3–5%) present in dry rhizome, an adult individual might be consuming 60–100 mg of curcumin per day.

Curcumin (Figure 1), a major active component of turmeric, has been used for centuries in indigenous medicine for the treatment of a variety of inflammatory conditions and other diseases.[1] Curcumin contains curcuminoids (including demethoxycurcumin and bisdemethoxycurcumin), which are antioxidants.[4]

Extensive scientific research on curcumin has demonstrated its potent antioxidant properties. This chapter details the cardioprotective effects of curcumin as follows: (1) the history of curcumin and its discovery as a potent drug with relevance to cardiovascular diseases; (2) mechanistic role of curcumin *in vitro,* emphasizing the antiplatelet and anticoagulant effects; (3) cardiovascular properties of curcumin; (5) application of curcumin in different animal models (viz. myocardial ischemia, myocardial infarction, cardiomyopathy, and arrhythmia *in vitro* and *in vivo*); (5) curcumin free-radical scavenging activity, particularly against $O_2^{\cdot -}$ radical and depletion of the oxidative stress.

2. DISCOVERY AS A POTENT DRUG

Curcumin is thought to be the primary pharmacological agent in turmeric. In numerous studies, curcumin's anti-inflammatory effects have been shown to be comparable to the potent drugs hydrocortisone and phenylbutazone as well as over-the-counter anti-inflammatory agents such as Motrin. Unlike the drugs, which are associated with significant toxic effects (ulcer formation, decreased white blood cell count, intestinal bleeding), curcumin produces no toxicity. Curcumin, also known as turmeric root, is gaining attention for its positive impact on a number of diseases, including cholesterol reduction. Scientific evidence has been building since the mid-1980s of curcumin's potential cholesterol-lowering capabilities.

Several studies have reported that curcumin is beneficial in lowering low-density lipoprotein (LDL) and raising high-density lipoprotein (HDL) or good cholesterol while reducing lipid peroxidation. Ten human volunteers were given 500 mg of curcumin for 7 days during a controlled trial at Amala Cancer Research Centre in India. After 7 days, they noted a 29% increase in good cholesterol (HDL) and a reduction of 11.6% in total cholesterol. Lipid peroxidation was also reduced by 33%. Another study published in *Atherosclerosis* in Dececember 1999 by the Faculty of Pharmacy at the University of Granada reported that curcumin was effective in inhibiting LDL oxidation and lowering LDL cholesterol as well as triglycerides. According to the latest research, curcumin reduced cholesterol by interfering with intestinal cholesterol uptake, increasing the conversion of cholesterol into bile acids and increasing the excretion of bile acids.

In conclusion, supplementation with *Curcuma longa* extract reduced oxidative stress and attenuated the development of fatty streaks in rabbits fed a high-cholesterol diet.[5] However, because we have data showing that curcumin acts as a potent drug in cardiovascular disorders, every individual question will be "What is the mechanistic role ?"

3. ANTIPLATELET EFFECT

As pointed out by Pilgeram,[6] "no disease in the history of mankind exacts a greater toll in morbidity than heart or blood vessel disease." However, the physiopathological mechanisms responsible for this disease are not fully understood. Dating back to 1846, Von Rokitansky was the first to propose that the accumulation of a fibrous substance on the endothelial wall of arteries is the fundamental cause of coronary heart disease,[7] and at about the same time, Anitschow showed that diets high in cholesterol and fat resulted in the formation of atheromatous plaques.[8]

Platelets have a key role in atherosclerosis, thrombosis, and acute coronary syndromes. Platelet adhesion, the first step in the process of homeostasis, was triggered by damage to the vessel wall and local exposure of the sub endothelial matrix. Coverage of the exposed site by platelets depended on the recognition of adhesive proteins by specific platelet-membrane glycoproteins, many of which are integrins.[9,10] After the platelet monolayer is formed over the endothelial lesion, specific agonists induce platelet vesicle secretion and aggregation. Platelet aggregation was mediated by the GPIIb/IIIa receptor, a member of the integrin superfamily of membrane-bound adhesion molecules. Integrins are defined as subunit receptors composed of an α subunit (i.e., GPIIb) and a β subunit (i.e., GPIIIa) capable of mediating adhesive interactions between cells or matrix. Although integrins were distributed widely throughout the vasculature, where they were expressed on endothelial cells, smooth muscle cells, and leukocytes, expression of the GPIIb/IIIa integrin is restricted to platelets.[11]

Extracts from several spices possessed antiaggregatory properties and they were shown to alter eicosanoid biosynthesis.[12,13] An ether extract of turmeric inhibited arachidonate-induced platelet aggregation and showed inhibitory effects at several steps of the arachidonic acid (AA) cascade in platelets. We report here the effects of curcumin [1,7-bis (4-hydroxy-3-methoxyphenyl)-l,6-heptadiene-3,5-dione] on platelet aggregation and eicosanoid metabolism.[14]

Curcumin inhibited the incorporation of AA into platelet phospholipids and also inhibited the deacylation of the latter, which would result in reduced amounts of free AA. Thus, it was observed that curcumin exerted its effects at several steps of the AA cascade in platelets (Figure 2). In this respect, curcumin's mechanism of action on the AA cascade in platelets resembled that of garlic and some of its components.[15]

Whether these effects and, more particularly, the inhibition of phospholipids deacylation were specific for platelets or was common to other cell types needs to be determined. If the release of AA was inhibited by curcumin in other cells also (e.g., polymorphonuclear leukocytes), it might explain its (as well as turmeric's) anti-inflammatory activity.

Platelet aggregation occurred when fibrinogen molecules bound to the activated GPIIb/IIIa receptor and connected platelets to one another.[16] Antiplatelet therapy prevented potential thrombolytic-induced platelet aggregation, coronary artery reocclusion, and reinfarction.[11] Curcumin might inhibit the cyclooxygenase (COX) pathway by blocking the GPIIb/IIIa receptor, inhibiting platelet aggregation leading to the formation of blood clots.[17] The IC_{50} shown by the antiplatelet aggregation of curcumin in this study might be of great help to develop potent antiplatelet drugs.[18]

Curcumin, a spice food pigment quite commonly used in natural medicine, exerted its potent antiplatelet activity through inhibition of COX activity and blockade of calcium signaling. The sensitivity of platelet agonists to the inhibitory effect of curcumin varied, platelet-activating factor (PAF) and AA-mediated aggregation being the most sensitive. Both of these agonists were directly or indirectly involved in the stimulation of thromboxane A_2 (TXA_2) production. TXA_2 interacted with its receptors on platelets in an autocrine fashion to activate Gq protein[19] and PAF was known to activate the Gq-phospholipase C signaling pathway.[20,21] Stimulation of Gq protein led to the generation of the second-messenger Inositol phosphate (IP3) and diacylglycerol, which, in turn, induced the release of Ca^{2+} from the dense tubular system and the activation of protein kinase C (PKC), respectively.[22–24] Multiple isoforms of both IP3 receptor and PKC were present in platelets.[25–27]

Defects at any of the signaling cascade, receptor or Gq protein activation, IP3 formation, or Ca^{2+} release could impair the platelet response to agonists. Curcumin inhibited platelet aggregation induced by calcium ionophore A-23187 and other agonists known to increase cytosolic Ca^{2+}. A rise in cytosolic Ca^{2+} levels accompanied platelet activation through the stimulation of enzymes that were not fully functional at the low Ca^{2+} concentration present in the resulting platelets[23] In platelets, multiple signaling mechanisms occurred because of an increase in cytosolic Ca^{2+}. These included the stimulation of phospholipase C, activation of

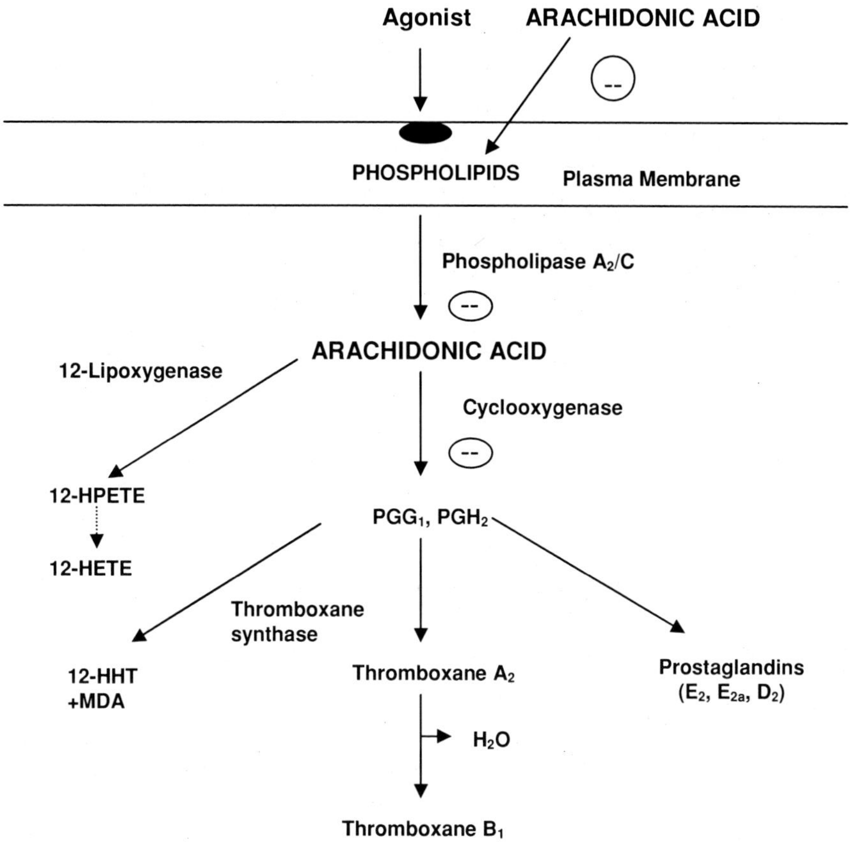

Figure 2. Metabolism of AA in blood platelets and the steps where curcumin shows an inhibitory effect.

inhibitory G-protein (G_i), opening of receptor-operated calcium channels, and mobilization of intracellular calcium.[22,23,28,29]

The common point in the action of the platelet agonists (ADP, epinephrine, collagen, A- 23187, PAF) was the increase in the levels of cytosolic Ca^{2+}, either due to its release from internal stores or through Ca^{2+} influx.[23,30] Because curcumin inhibited the aggregation induced by these agents, it is likely that it could interfere with the Ca^{2+} signaling influx in activated platelets. However, the agonists that were linked with the activation of the COX pathway were more sensitive to the inhibitory effect of curcumin. It seems that curcumin exerted its inhibitory effects at points distal to any specific receptor of G-protein, most likely at COX- and Ca^{2+}-activated signaling cascades. Therefore, curcumin might exhibit preferential inhibitory effects on aggregation mediated by agonists such as (PAF and AA), which activated either one or both of the signaling targets.[22,31,32]

Cells possessed the capabilities of maintaining milieu *inte'rieur through* regulation and desensitization of signaling molecules, which might include receptors, G-proteins, or second messengers.[33] PKC played an important role in signal transduction pathways including platelets, as it directly regulated adenylyl cyclase activity through phosphorylation and inactivation of G_i.[34] PKC acted in synergy with Ca^{2+} mobilization for the activation of platelets.[25] Curcumin (up to 500 mM) did not block the aggregation induced by phorbol myristate acetate (PMA). This excluded the possible role of PKC in curcumin-mediated inhibition of aggregation. These findings were in contrast to previous studies suggesting the inhibitory activity of curcumin on PKC in NIH 3T3 cells,[35,36] but they were in accordance with those that compared to AA and PAF (25–30 mM). Epinephrine, through interaction with α_2-adrenoceptors, activated G_i and inhibited adenylyl cyclase activity, thus causing a decrease in intracellular cAMP levels in platelets.

Agents that decreased cAMP levels stimulated platelet aggregation.[28,29] In fact, the effects of epinephrine in platelets might be mediated through multiple pathways: a decrease in intracellular cAMP levels, stimulation of phospholipase Cb and thus IP3 production by G-protein subunits, an increase in Ca^{2+} influx, and activation of other proteins such as Syk and related adhesion focal tyrosine kinase (RAFTK).[37,38] In addition to epinephrine, other platelet agonists, such as ADP, collagen, and A-23187, which were inhibited by curcumin ($IC_{50} = 650$, 450, and 100 mM, respectively) also induced RAFTK phosphorylation. It might be interesting to examine the effect of curcumin on Syk and RAFTK. The effect of curcumin on ADP- and collagen-mediated aggregation was not well pronounced. ADP caused a transient increase in IP3-mediated Ca^{2+} levels, which might eventually lead to a store-depleted influx of Ca^{2+} through calcium channels.[23,30] PAF, which is known to activate phospholipase C (PLC), could enhance aggregation through an increase in Ca^{2+} influx via receptor-operated Ca^{2+} channels, with such an effect at low concentrations of PAF (picomolar) being independent of PLC activation.[21]

Studies provided evidence that curcumin possibly inhibited Ca^{2+} influx in platelets, as demonstrated by the inhibition of platelet aggregation induced by calcium ionophore A-23187. Because both ADP- and collagen-mediated aggregation had distinct signaling pathways that did not involve activation of COX activity, higher concentrations of curcumin ($IC_{50} = 450$–600 mM) were required to block their effect. In conclusion, these results implied that curcumin not only inhibited COX activity as shown previously,[15] but also impaired Ca^{2+} ionophore-mediated platelet aggregation.

4. ANTICOAGULANT ACTIVITY

According to Pilgeram,[6] the origin of the two current concepts of atherogenesis was: one based on the formation of fibrous tissue and the other based on the accumulation of cholesterol. He concluded that the two mechanisms might act in synergy and that the probable atherogenic role of the fiber-forming blood protein fibrinogen might help in the conceptual integration of the above mechanisms. His findings showed an age-related increase in the plasma levels of fibrinogen, which was more striking in atherosclerotic subjects.

Additional evidence of the key pathogenetic role of fibrinogen and its byproducts was offered in the extensive review by Kaplan and Bini.[39] The authors suggested that the accumulation of fibrinogen in the intima might precede that of LDL and thus might be a more important risk factor than these lipoproteins for the development of atherosclerotic lesions in cerebral arteries.

According to Kaplan and Bini, the mechanisms by which the conversion of fibrinogen to fibrin and degradation of fibrin might occur in the vessel wall were not yet known, but monocytes might play a role, as they synthesized coagulation factors and fibrinolytic factors.[39] Further, it seems highly likely that fibrin formation might occur within the vessel wall and that fibrin might affect the function of the overlying endothelium and also interact with plaque components, especially monocytes, to aggravate the atherosclerotic process. In addition, Smith et al.[40] noted that fibrin degradation products might be chemotactic to monocyte-macrophages (and therefore stimulated smooth muscle cell proliferation), and Ernst and Resch,[41] on the basis of their meta-analysis of six prospective epidemiological studies, concluded that high plasma fibrinogen levels were associated with subsequent myocardial infarction or stroke.

In view of the above, it is interesting to note that the administration of an extract of the plant *Curcuma longa* to human subjects lowered their plasma fibrinogen level. Previous research had already shown that this product decreased the levels of lipid peroxides and oxidized lipoproteins.[42,43] The subjects chosen were in the age group 24–75 years and were in apparent good health and held managerial, scientific, or technical jobs. After obtaining their informed consent, blood was taken from the cubital vein and fibrinogen was determined using the method of Claus.[44] On the basis of the results obtained, eight subjects showing abnormally high values of plasma fibrinogen (i.e., over 350 mg/dLl) were chosen for the treatment with the curcumin product. This treatment consisted in the daily intake of the same dose used in our previous studies[42,43] [i.e., ingestion for 15 days of two tablets of hydroalcoholic extract of *Curcuma longa*,[45] containing approximately 10 mg of curcumin per tablet (A.S.A.C. Pharmaceutical International A.I.E., Alicante, Spain)]. On the15thday after the start of the treatment, blood was withdrawn from the cubital vein of the treated subjects for determination of the fibrinogen levels. Turmeric administration decreased the fibrinogen levels to 240–290 mg/dL, with no side effects such as nausea, diarrhea, or constipation observed throughout the course of the treatment.

Studies also showed that, at the doses and length of treatment indicated, turmeric had no apparent liver or kidney toxicity, as shown by the data on serum glutamate oxaloacetate transaminase (SGOT), serum glutamate pyruvate transaminase, alkaline phosphatase, and total bilirubin (direct and indirect).[42,43,46] Moreover, the treatment did not result in any significant change of the key parameters of blood coagulation. This apparent lack of toxicity was in agreement with data from the National Bureau of Standards of Hyderabad, according to which the daily adult intake of turmeric in diet (as a food additive) ranged between 0.1 and 3.8 g. In our opinion, this preliminary finding justified further work to confirm the fibrinogen-lowering effects of turmeric on a larger group of subjects. This work might have both fundamental interest (with regard to the role of pro-oxidant and antioxidant mechanisms in atherogenesis) and practical implications, as no proper drug is available for the safe and selective lowering of fibrinogen levels.

The cascade of reactions involved in the coagulation of blood was an auto-catalytic and self-limiting process converting zymogen to the active form in the presence of proteolytic enzyme thrombin.[11] The unique specificity of thrombin led us to search for highly potent and selective inhibitors for this enzyme. The anticoagulant action of curcumin was shown to prolong the clotting time, as observed by thrombin time (TT), prothrombin time (PT), and activated partial thromboplastin time (APTT) assays and this could be due to the presence of hydrophobic groups in curcumin moiety.[18]

5. CARDIOVASCULAR PROPERTIES OF CURCUMIN

Atherosclerosis is considered a chronic and progressive disease arising from the inflammatory processes and oxidative stress within the vessel wall.[47,48] Atherosclerosis is a multifactorial disease in which a major alteration of the vascular lipid metabolism is produced. It was observed that curcumin reduced plasma lipid peroxides, molecules that play an important role in the pathogenesis of the disease.[49] Curcumin also had different properties that contributed to combat this disease: It reduced the susceptibility of LDL to oxidation,[50] inhibited the proliferation of vascular smooth muscle cells,[51] had an antithrombotic effect, had a transient hypotensive effect, and inhibited platelet aggregation *in vivo* and *ex vivo*. Curcumin also provided an additional benefit by potentially reducing the risk of cardiovascular-related disease by inhibiting platelet aggregation and significantly decreasing the level of lipid peroxides.[18]

Observation of curcumin's mechanism of action showed that it blocked the formation of TXA_2, a promoter of platelet aggregation, thereby inhibiting abnormal blood clot formation. Curcumin also increased the level of prostacyclin, a natural inhibitor of platelet aggregation.[52] Although the molecular mechanisms of action of curcumin are not fully understood, in several animal models it was demonstrated to exert potent antioxidant, anti-inflammatory and antitumor properties.[53–55]

The question arises, whether curcumin could suppress the inflammatory component of atherogenesis. Indeed, recently, curcumin derivatives were demonstrated to reduce aortic fatty streak formation in cholesterol-fed rabbits.[5] In 1992, excellent animal models became available for experimental atherosclerosis research. At that time, the first line of gene-targeted animal models, namely apolipoprotein E (apoE) knockout mice was developed.[56] More recently, a apoE and LDLR–double knockout (apoE/LDLR-DKO) mice model that developed more severe hyperlipidemia and atherosclerosis than mice deficient for apoE alone was demonstrated.[57] In both strains of gene-targeted mice, lesion formation was greatly accelerated and feeding with an atherogenic Western diet increased the lesion size. Thus, apoE/LDLR–DKO mouse is currently considered one of the most relevant models for the study of the antiatherogenic potential of drugs.

Curcumin, given orally at a relatively low dose, was able to decrease formation of atherosclerotic changes in apoE/LDLR–DKO mice fed a Western diet. The antiatherosclerotic action of curcumin was relatively weak, as it decreased the lesion formation by 20%, as evidenced by the "en face" method.

Also, a more accurate cross-section analysis of aortic root showed almost a twofold decrease of Oil red O (ORO) staining. It should be noted that the action of curcumin was present in apoE/LDLR–DKO mice, despite feeding them by Western diet, which accelerated lesion formation, increased lesion size, and promoted development of advanced lesions at a significantly earlier age.[58] Taking for granted the poor curcumin bioavailability due to its rapid metabolism in the liver and intestinal wall[59] as well as the relatively low dose used (0.3 mg/mouse/day); the antiatherogenic action of curcumin seemed to be quite significant. In order to demonstrate mechanisms responsible for the antiatherosclerotic action of curcumin, which might interfere with atherogenesis at several critical points, many authors addressed NF-κB as an important therapeutic target for atherosclerosis.[60]

Recently, it was reported that PDTC, an NF-κB inhibitor, significantly decreased atherosclerosis in apoE/LDLR–DKO mice fed a Western diet.[61] It was demonstrated that oral supplementation of caffeic acid phenethyl ester (CAPE), a compound similar to curcumin, attenuated the atherosclerotic process in apoE knockout mice due to the inhibition of transcription factor NF-κB.[62] Interestingly, curcumin was demonstrated to inhibit the activity of NF-κB in stimulated endothelial cells.[63]

However, whether inhibition of NF-κB could be responsible for the antiatherogenic action of curcumin remains to be elucidated. Another plausible mechanism responsible for the antiatherogenic action of curcumin might depend on the induction of heme oxygenase-1 (HO-1), a potent antioxidant and vascular protective enzyme.[64,65] Induction of HO-1 was claimed to inhibit development of atherosclerosis in apoE-deficient mice.[66,67] Interestingly, it was shown that curcumin might induce HO-1 in endothelial cells *in vitro*.[68] Thus, the possible involvement of HO-1 induction in the antiatherogenic action of curcumin requires further investigation. Natural, polyphenolic compounds presented a wide spectrum of biological activities[69,70,70a] and might represent a promising group of antiatherosclerotic compounds.[71]

For example, flavonoids protected LDL from oxidation, enhanced endothelium-derived nitric oxide bioactivity,[72,73] inhibited endothelial activation,[74,75] and platelet aggregation.[76]

Curcumin might be able to prevent the oxidation of cholesterol in the body. Because oxidized cholesterol is what damages blood vessels and builds up in the plaques that can lead to heart attack or stroke, preventing the oxidation of new cholesterol might help to reduce the progression of atherosclerosis and diabetic heart disease. In addition, turmeric is a very good source of vitamin B_6, which is needed to keep homocysteine levels from getting too high. Homocysteine, an intermediate product of an important cellular process called methylation, is directly damaging to blood vessel walls. High levels of homocysteine are considered a significant risk factor for blood vessel damage, atherosclerotic plaque buildup, and heart disease; whereas a high intake of vitamin B_6 is associated with a reduced risk of heart disease.

6. ANTIOXIDANT EFFECT OF CURCUMIN

Many studies showed the potential of curcumin to prevent lipid peroxidation, a key process in the onset and progression of many diseases. The capacity of curcumin to stabilize membranes was demonstrated.[50] Venkatesan observed a protective effect of curcumin against the cardiotoxicity produced by adriamycin in rats, showing a reduction in the parameters that indicated lipid peroxidation.[77]

Oxidative stress played a major role in the pathogenesis of various diseases, including myocardial ischemia, cerebral ischemia–reperfusion injury, hemorrhage and shock, neuronal cell injury, hypoxia, and cancer. Curcumin exhibited strong antioxidant activity, comparable to vitamins C and E.[78] Curcumin, with its proven anti-inflammatory and antioxidant properties, was shown to have several therapeutic advantages. It was shown to be a potent scavenger of a variety of reactive oxygen species, including superoxide anion radicals, hydroxyl radicals,[79] and nitrogen dioxide radicals.[80,81] It was also demonstrated to inhibit lipid peroxidation in different animal models.[82,83] Curcumin protected oxidative cell injury of kidney cells (LLC-PK1) by inhibiting lipid degradation, lipid peroxidation, and cytolysis[84] and decreased ischemia-induced biochemical changes in heart in a cat model.[85]

Vascular endothelial cells treated with curcumin prevented oxidant-mediated injury by increased heme oxygenase production.[68] Curcumin pretreatment and cotreatment with isoprenaline (ISO) to protect rat myocardium against ISO-induced myocardial necrosis,[86,87] and the protective effect was attributed to its antioxidant properties by inhibiting free-radial generation.[18] It caused a decrease in the degree of degradation of the existing collagen matrix and collagen synthesis, 2 weeks after the second dose of ISO. These effects were attributed to free-radical scavenging properties and inhibition of lysosomal enzyme release by curcumin.[88]

Studies in our laboratory showed that pretreatment with curcumin resulted in significant restoration of the liver cytokines interleukin (IL)-1α, IL-1β, IL-2, IL-6, and IL-10 to normal levels that were increased by hemorrhage/resuscitation regimen in rats. In fact, IL-1β levels were lower than sham levels. Nuclear factor (NF)-κB and activator protein (AP)-1 were differentially activated at 2 and 24 h posthemorrhage and were inhibited by curcumin pretreatment. Serum aspartate transaminase estimates indicated decreased liver injury in curcumin-pretreated animals subjected to hemorrhage. These results suggested that protection by curcumin pretreatment against hemorrhage/resuscitation injury might have resulted from the inactivation of transcription factors involved and regulation of cytokines to beneficial levels.[89] Similarly, in chronically hypoxic rabbit hearts, Hsp70i translocated from the particulate to the cytosolic fraction and curcumin reversed this subcellular redistribution[90] through protein kinase pathways.[91]

Sreejayan and Rao claimed that the presence of phenolic groups in the structure of curcumin was fundamental in explaining its ability to eliminate oxygen-derived free radicals from the medium largely responsible for the peroxidation of cell lipids.[92] They were able to eliminate the hydroxyl radical,[79] superoxide radical,[92] singlet oxygen,[93] nitrogen dioxide,[80] and NO.[81] It was also demonstrated that curcumin inhibited the generation of the superoxide radical.[94]

Curcumin, at a relatively low concentration, exhibited remarkable anti-inflammatory and antioxidant effects.[4] Although the exact mechanism by which curcumin promoted these effects remains to be elucidated, the antioxidant properties of this yellow pigment appeared to underlie its pleiotropic biological activities. Oxidative stress and formation of reactive oxygen species (ROS) could set off a cascade of biochemical and molecular sequel such as the xanthine dehydrogenase/xanthine oxidase (XD/XO) conversion, leading to production of ROS.[95] Oxidative ischemic injury was suggested to be a central mechanism of the cellular damage affecting all organs and tissues after ischemia; however, the mechanisms that triggered and modulated this damage were not studied in detail.[96]

Oxidative injury was associated with the generation of ROS. It is well documented that XO is an important prerequisite factor in the process of $O_2^{\cdot-}$ generation in acute ischemic injury.[97] and this observation concurred with our finding wherein a significant rise in percentage of XO was noticed in rat heart after ischemic insult followed by a significant increase in $O_2^{\cdot-}$ generation.[18] This finding was also in agreement with that of Terada et al.,[98] who demonstrated increased $O_2^{\cdot-}$ production from endothelial cells exposed to hypoxia. In this study, the elevated level of %XO activity in the diseased group was shown to be effectively counteracted by the administration of curcumin. In addition, curcumin was noticed to significantly decrease $O_2^{\cdot-}$ burst, and the one proposed mechanism underlying this protective effect of curcumin could be through its antioxidant properties restoring endogenous glutathione (GSH) levels and thereby detoxifying free radicals.[83] An alternative mechanism for such protection could be through curcumin free-radical scavenging activity, particularly against $O_2^{\cdot-}$ radical, which would inhibit sulfhydryl (SH) oxidation, leading to inhibition of reversible XD/XO conversion.[99] Curcumin, as a free-radical scavenger, might also inhibit proteases, which are known to be activated by free radicals and this would eventually lead to inhibition of the irreversible proteolytic XD/XO conversion.[100] Thus, the protective role of curcumin could be related to inhibition of free-radical propagation with subsequent inhibition of XD/XO conversion and resultant decreased production of $O_2^{\cdot-}$ through its antioxidant and/or free-radical scavenging activity. This could be explained on the basis of the low bioavailability of curcumin as well as heart disposition to curcumin,[101] and dual treatment (posttreatment and pretreatment) could result in a concentration sufficient to elicit antioxidant activity.[99]

It is already known that lipids are the most susceptible macromolecules to oxidative stress[83] and our results showed that the level of lipid peroxides, measured in terms of malonaldehyde (MDA), significantly increased due to ischemic insult.[18] In this study, curcumin (pretreatment and posttreatment) was shown to significantly reduce the lipid peroxides level by scavenging free radicals and inhibiting the propagating chain reaction of lipid peroxides, and our finding was in consonance with an earlier report.[83] It was noticed earlier that the neutrophils, a major source of free radicals, characteristically invaded the myocardial tissue during ischemia.[4] The observations of this study showing that curcumin pretreatment and posttreatment reduced the myeloperoxidase (MPO) levels indicated that curcumin suppressed neutrophil infiltration into the injured myocardium.

Reduced GSH, considered the most prevalent and important intracellular non-protein thiol, had a crucial role as a free-radical scavenger and a decline in GSH could reflect oxidative stress.[102] In this study, GSH content was significantly reduced due to ischemic insult and this could be explained by the assimilation of GSH by the rapidly generating free radicals. Nevertheless, a number of studies reported that low myocardial GSH levels might not provide evidence for deleterious free-radical reactions during and following ischemia.[103] This observation could be explained on the basis of a Michael-type addition reaction of GSH (as a nucleophile) in the chromophore region of curcumin (as an electrophile), which would result in the depletion of cellular GSH defense.[104] However, curcumin antioxidant activity still imparted some degree of GSH salvation, as GSH levels were shown to increase significantly in the pretreatment and posttreatment groups.[18] ISO administration resulted in an increase in the level of free radicals, which, in turn, induced cellular damage, and this observation could be substantiated by the low levels of free-radical scavenging enzymes such as catalase, superoxide dismutase, glutathione peroxidase, and glutathione-*S*-transferase (GST) that formed the first line of cellular defense against the oxidation injury. The second line of defense consisting of ascorbic acid, α-tocopherol, and ceruloplasmin, which scavenged residual free radicals escaping decomposition, was also found to be affected. Indeed, curcumin (pretreatment and post-treatment) proved beneficial in restoring the levels of both enzymatic and nonenzymatic antioxidants and this finding was in accordance with that of Jovanovic et al.,[105] who showed regeneration of curcumin radicals by α-tocopherol and ascorbic acid.

The antioxidant activity of curcumin could be attributed to the phenolic and methoxy groups in conjunction with the 1,3-diketone-conjugated diene system, for scavenging of the oxygen radicals.[83] In addition, the curcumin primary metabolite tetrahydrocurcumin, a major antioxidant with diketone moiety, exhibited antioxidant activity by cleavage of the C–C bond at the active methylene carbon between the two carbonyls.[101] These antioxidant properties seemed to have a role in inhibiting $O_2^{\cdot-}$ generation directly or indirectly (via inhibiting XD/XO conversion). However, the impact of this effect on the generation of other damaging species such as hydroxyl radical ($OH^{\cdot}$) or peroxynitrite ($ONOO^{\cdot}$) is yet to be resolved. In addition to its inherent ability to attenuate the reactivity of oxygen free-radical species, curcumin was shown *in vivo* to enhance the activities of detoxifying enzymes such as GST.[107]

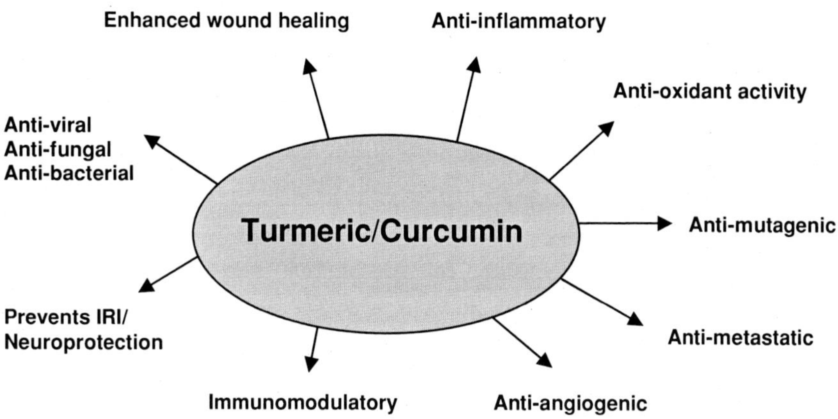

Figure 3. Schematic showing multiple biological activities of turmeric/curcumin. (From Ref. 106.)

7. CONCLUSIONS

In conclusion, curcumin exhibits a variety of beneficial effects and appears to have a significant potential in the treatment of multiple diseases that are a result of oxidative stress[106] (Figure 3). These protective effects of curcumin are attributed mainly to its antioxidant properties and should be further exploited to develop novel drugs. In most of these disease states, a long-term treatment will be necessary. Thus, oral administration of turmeric/curcumin with minimal acute or chronic toxicity will be of great value in combating these chronic illnesses.

REFERENCES

1. H. P. T. Ammon and M. A. Wahl, Pharmacology of *Curcuma longa*. *Planta Med* **57**, 1–7 (1991).
2. R. C. Srimal, Turmeric. A brief review of medicinal properties. *Fitoterapia*. **68**, 483–493 1997).
3. S. Toda, Antioxidative components isolated from rhizome of *Curcuma longa L. Chem Pharm Bull*33, 1725–1728 (1985).
4. Y. Abe, S. Hashimoto, and T. Horie, Curcumin inhibition of inflammatory cytokine production by human peripheral blood monocytes and alveolar macrophages. *Pharmacol Res* **39**, 41–47 (1999).
5. J. L. Quiles, M. D. Mesa, C. L. Ramirez-Tortosa , C. M. Aguilera , M. Battino, A. Gil, M. C. Ramirez-Tortosa, *Curcuma longa* extract supplementation reduces oxidative stress and attenuates aortic fatty streak development in rabbits. *Arterioscler Thromb Vasc Biol* **22**,1225-1231 (2002).
6. L. Pilgeram, Atherogenesis and fibrinogen: Historical perspective and current status. *Naturwissenschaften* **80**, 547–555 (1993).

7. K. Von Rokitansky, *Handbuch der Pathologischen Anatomie*, Vols. 1–3. Vienna: Braunmuller und Seidel, 1846.
8. C. Anitschow, In: E. V. Cowdry, ed. *Arteriosclerosis: A Survey of the Problem*. New York: McMillan, New York, 1933, pp.107–121.
9. R. O. Hynes, Integrins: A family of cell surface receptors. *Cell* **48**, 549–554 (1987).
10. S. S. Smyth, C. C. Joneckis, and L. V. Parise, Regulation of vascular integrins. *Blood* **81**, 2827–2843 (1993).
11. W. H. Frishman, B. Burns, B. Atac, N. Alturk B. Altajar, and K. Lerrick, Novel antiplatelet therapies for treatment of patients with ischemic heart disease: Inhibitors of the platelet glycoprotein IIb/IIIa integrin receptor. *Am Heart J* **130**, 877–892 (1995).
12. K. C. Srivastava and T. Mustafa, Spices: Antiplatelet activity and prostanoid metabolism. *Prostaglandins Leukotr Essent Fatty Acids* **8**, 255–266 (1989).
13. K. C. Srivastava and O. D. Tyagi, Effects of a garlic-derived principle (ajoene) on aggregation and arachidonic acid metabolism in human blood platelets. *Prostaglandins Leukotr Essent Fatty Acids* **49**, 587–595 (1993).
14. K. C. Srivastava, Extracts of two frequently consumed spices—cumin *(Cuminum cyminum)* and turmeric (*Curcuma longa*)—inhibit aggregation and alter eicosanoid biosynthesis in human blood platelets. *Prostaglandins Leukotr Essent Fatty Acids* **37**, 57–64 (1989).
15. K. C. Srivastava, A. Bordia, and S. K. Verma, Curcumin, a major component of food spice turmeric (*Curcuma longa*), inhibits aggregation and alters eicosanoid metabolism in human blood platelets. *Prostaglandins Leukotr Essent Fatty Acids* **52**, 223–227 (1995).
16. J. Lefkovits, E. F. Plow, and E. J. Topol, Platelet glycoprotein IIb/IIIa receptors in cardiovascular medicine. *N Engl J Med* **332**, 1553–1559 (1995).
17. M. T. Huang, T. Lysz, T. Ferraro, T. F. Abidi, J. D Laskin, and A. H. Conney, Inhibitory effects of curcumin on in vitro lipoxygenase and cyclooxygenase activities ischemia in rats. *Int J Biochem Cell Biol* **36**, 1967–1980 (1991).
18. P. Manikandan, M. Sumitra, S. Aishwarya, B. M. Manohar, B. Lokanadam, and R. Puvanakrishnan, Curcumin modulates free radical quenching in myocardial in mouse epidermis. *Cancer Res* **51**, 813–819 (2004).
19. S. Offermans , Kl. Laugwitz, K. Spicher, and G. Schultz, G proteins of the G12 family are activated via thromboxane A2 and thrombin receptors in human platelets. *Proc Natl Acad Sci USA* **91**, 504–508 (1994).
20. S. D. Shukla, C. C. Franklin, and M. G. Carter, Activation of phospholipase C in platelets by platelet activating factor and thrombin causes hydrolysis of a common pool of phosphatidylinositol 4,5- bisphosphate. *Biochim Biophys Acta* **929**, 134–141 (1987).
21. M. R. James-Kracke, R. B. Sexe, and S. D. Shukla, Picomolar platelet activating factor mobilizes Ca21 to change platelet shape without activating phospholipase C or protein kinase C; simultaneous measurements of intracellular free Ca21 concentration and aggregation. *J Pharmacol Exp Ther* **271**, 824–831 (1994).
22. W. Siess, Molecular mechanisms of platelet activation. *Physiol Rev* **69**, 58–178 (1989).
23. J. W. M. Heemskerk and O. Sage, Calcium signaling in platelets and other cells. *Platelets* **5**, 295–316 (1994).
24. D. E. Clapham, Calcium signaling. *Cell* **80**, 259–268 (1995).
25. M. Crabos, D. Fabbro, S. Stabel, and P. Erne, Effect of tumor promoting phorbol ester, thrombin platelets and regulation by calcium. *Biochem J* **288**, 891–896 (1992).

26. A. C. Newton, Protein kinase C: Structure, function and regulation. *J Biol Chem* **270**, 28,495–28,498 (1995).
27. T. M. Quinton and W. L. Dean, Multiple inositol 1,4,5-triphosphate receptor isoforms are present in platelets. *Biochem Biophys Res Commun* **224**, 740–746 (1996).
28. L. F. Brass, J. A. Hoxie, and D. R. Manning, Signaling through G proteins and G protein-coupled receptors during platelet activation. *Thromb Haemost* **70**, 217–223 (1993).
29. W. Siess, B. Grunberg, and K. Luber, Functional relationship between cyclic AMP-dependent protein phosphorylation and platelet inhibition. *Adv Exp Med Biol* **344**, 229–235 (1993).
30. S. M. O. Hourani and D. A. Hall, Receptors for ADP on human blood platelets. *Trends Pharmacol Sci* **15**, 103–108 (1994).
31. W. Chao and M. S. Olson, Platelet-activating factor: Receptors and signal transduction. *Biochem J* **292**, 617–629 (!993).
32. S. A. Saeed and B. H. Shah, Diversity of agonist-mediated signal transduction pathways in human platelets. *Adv Exp Med Biol* **407**, 531–535 (1997).
33. B. H. Shah , D. J. McEwan, and G. Milligan, Gonadotrophin releasing hormone receptor agonist-mediated down-regulation of Gqa/G11 a (pertussis toxin-insensitive) G proteins in aT3–1 gonadotroph cells reflects increased G protein turnover but not alterations in mRNA levels. *Proc Natl Acad Sci USA* **92**, 1886–1889 (1995).
34. J. Kawabe, G. Iwami, T. Ebina, S. Ohno, T. Katada, Y. Ueda, C. J. Homcy, and Y. Ishikawa, Differential activation of adenylyl cyclase by protein kinase C isoenzymes. *J Biol Chem* **269**, 16,554–16,558 (1994).
35. J. Y. Liu, S. J. Lin, and J. K. Lin, JInhibitory effects of curcumin on protein kinase C activity induced by 12-*O*-tetradecanoylphorbol-13-acetate in NIH 3T3 cells. *Carcinogenesis* **14**, 857–861 (1993).
36. J. K. Lin, Y. C. Chen, Y. T. Huang, and S. Y. Lin-Shiau, Suppression of protein kinase C and nuclear oncogene expression as possible molecular mechanisms of cancer chemoprevention by apigenin and curcumin. *J Cell Biochem Suppl* **28–29**, 39–48 (1997).
37. X. Wang, S. Yanagi, C. Yang, R. Inatome, and H. Yamamura, Tyrosine phosphorylation and SYK activation are involved in thrombin-induced aggregation of epinephrine-potentiated platelets. *J Biochem* **121**, 325–330 (1997).
38. Y. Banno, T. Asano, and Y. Nozawa, Stimulation by G protein bg subunits of phospholipase Cb isoforms in human platelets.*Thromb Haemost* **79**, 1008–1013 (1998).
39. K. L. Kaplan. and A. Bini, Thrombosis in atherogenesis. *Crit Rev Oncol Hematol* **9**, 305–318 (1989).
40. E. B. Smith, R. S. Slater, and J. A. Hunter, Quantitative studies on fibrinogen and low-density lipoprotein in human aortic intima. *Atherosclerosis* **55**, 171–178 (1973).
41. E. Ernst and K. L. ReschFibrinogen as a cardiovascular risk factor: A meta-analysis and review of the literature. *Ann Intern Med* **118**, 956–963 (1993).
42. A. Ramý'rez-Bosca', M. A Carrio'n Gutie'rrez, A. Soler, et al., Effects of the antioxidant turmeric on lipoprotein peroxides: Implications for the prevention of atherosclerosis. *Age* 20,165–168 (1997).
43. A. Ramý'rez-Bosca', A. Soler, M. A. Carrio'n-Gutie'rrez, A. Laborda Alvarez, and E. Quintanilla Almagro, Antioxidant curcuma extracts decrease the blood lipid peroxide levels of human subjects. *Age* **18**, 167–169 (1995).
44. A. Claus, Fibrinogens. *Acta Haemat* **7**, 237 (1957).

45. T. Masuda, J. Isobe, A. Jitoe, and N. Nakatani, Antioxidative curcuminoids from rhizomes of *Curcumaxantorrhyza*. *Phytochemistry* **31**, 3645–3647 (1992).
46. T. N. Bhavani Shankar, N. V. Shantha, H. P. Ramesh, I. A. S. Murthy, and V. S. Murthy, Toxicity studies on turmeric (*Curcuma longa*): Acute toxicity studies in rats, guinea pigs and monkeys. *Ind J Exp Biol* **18**, 73–75 (1980).
47. G. K. Hansson, Inflammation, atherosclerosis, and coronary artery disease. *N Engl J Med* **352**, 1685–1695 (2005).
48. A. Doria, Y. Sherer, P. L. Meroni and Y. Shoenfeld, Inflammation and accelerated atherosclerosis: Basic mechanisms. *Rheum Dis Clin North Am* **31**, 355–362 (2005).
49. J. Miquel, M. Martínez, A. Diez, E. De Juan, A. Solar, A. Ramírez-Boscá, J. Laborda, and M. Carriona, Effects of turmeric on blood and liver lipoperoxide levels of mice: Lack of toxicity. *Age* **18**, 171–174 (1995).
50. M. C. Ramírez-Tortosa , M. D. Mesa, M. C. Aguilera, J. L. Quiles, L. Baró, C. L. Ramírez-Tortosa, E. Martínez-Victoria, and A. Gil, Oral administration of a turmeric extract inhibits LDL oxidation and has hypocholesterolemic effect in rabbits with experimental atherosclerosis. *Atherosclerosis* **147**, 371–378 (1999).
51. H. W. Chen and H. C. Huang, Effect of curcumin on cell cycle progression and apoptosis in vascular smooth muscle cells. *Br J Pharmacol* **124**, 1029–1040 (1998).
52. K. C. Srivastava, Evidence for the mechanism by which garlic inhibits platelet aggregation. *Prostaglandins Leukotr Med* **22**, 313–321 (1986).
53. C. C. Araujo, and L. L. Leon,.Biological activities of *Curcuma longa* L. *Mem Inst Oswaldo Cruz* **96**,723–728 (2001).
54. B. B. Aggarwal, A. Kumar, and A. C. Bharti, Anticancer potential of curcumin: Preclinical and clinical studies. *Anticancer Res* **23**, 363–398 (2003).
55. A. Duvoix, R. Blasius, S. Delhalle, M. Schnekenburger , F. Morceau, E. Henry, et al., Chemopreventive and therapeutic effects of curcumin. *Cancer Lett* **223**, 181–190 (2005).
56. J. A. Piedrahita, S. H. Zhang, J. R. Hagaman, P. M. Oliver, and N. Maeda, Generation of mice carrying a mutant apolipoprotein E gene inactivated by gene targeting in embryonic stem cells. *Proc Natl Acad Sci USA* **89**, 4471–4475 (1992).
57. S. Ishibashi , J. Herz, N. Maeda, J. L. Goldstein, and M. S. Brown, The two-receptor model of lipoprotein clearance: Tests of the hypothesis in "knockout" mice lacking the low density lipoprotein receptor, apolipoprotein E, or both proteins. *Proc Natl Acad Sci USA* **91**, 4431–4435 (1994).
58. Y. Nakashima, A. S. Plump, E. W. Raines, J. L. Breslow, and R. Ross, ApoE-deficient mice develop lesions of all phases of atherosclerosis throughout the arterial tree. *Arterioscler Thromb* **14**, 133–140 (1994).
59. G. Shoba, D. Joy, T. Joseph, M. Majeed, R. Rajendran, and P. S. Srinivas, Influence of piperine on the pharmacokinetics of curcumin in animals and human volunteers. *Planta Med* **64**, 353–356 (1998).
60. C. Monaco and E. Paleolog, Nuclear factor kappaB: A potential therapeutic target in atherosclerosis and thrombosis. *Cardiovasc Res* **61**, 671–682 (2004).
61. J. Jawien, M. Gajda, L. Mateuszuk, R. Olszanecki, A. Jakubowski, A Szlachcic, et al., Inhibition of nuclear factor-kappaB attenuates artherosclerosis in apoE/LDLR-double knockout mice. *J Physiol Pharmacol* **56**, 483–489 (2005).
62. K. Hishikawa, T. Nakaki, and T. Fujita, Oral flavonoid supplementation attenuates atherosclerosis development in apolipoprotein E-deficient mice. *Arterioscler Thromb Vasc Biol* **25**, 442–446 (2005).

63. B. Gupta and B. Ghosh, *Curcuma longa* inhibits TNF-alpha induced expression of adhesion molecules on human umbilical vein endothelial cells. *Int J Immunopharmacol* **21**, 745–757 (1999).
64. K. A. Hoekstra, D. V. Godin, and K. M. Cheng, Protective role of heme oxygenase in the blood vessel wall during atherogenesis. *Biochem Cell Biol* 82, 351–359 (2004).
65. N. G. Abraham and A. Kappas, Heme oxygenase and the cardiovascular-renal system. *Free Radical Biol Med* **39**, 1–25 (2005).
66. S. H. Juan, T. S. Lee, K. W. Tseng, J. Y. Liou, S. K. Shyue, , K. K. Wu, et al., Adenovirus-mediated heme oxygenase-1 gene transfer inhibits the development of atherosclerosis in apolipoprotein E deficient mice.*Circulation*. **104**, 1519–1525 (2001).
67. S. F. Yet, M. D. Layne X. Liu, Y. H. Chen, B. Ith, N. E. Sibinga, et al., Absence of heme oxygenase-1e xacerbates atherosclerotic lesion formation and vascular remodeling. *FASEB J* **17**, 1759–1761 (2003).
68. R. Motterlini, R. Foresti, R. Bassi, and C. J. Green, Curcumin, an antioxidant and anti-inflammatory agent, induces heme oxygenase-1 and protects endothelial cells against oxidative stress. *Free Radical Biol Med* **28**, 1303–1312 (2000).
69. E. Middleton, Jr., C. Kandaswami, T. C. Theoharides, The effects of plant flavonoids on mammalian cells: Implications for inflammation, heart disease, and cancer. *Pharmacol Rev* **52**, 673–751 (2000).
70. S. Schmitt-Schillig , S. Schaffer, C. C. Weber , G. P. Eckert , and W. E. Muller, Flavonoids and the aging brain. *J Physiol Pharmacol* **56**(Suppl 1), 23–36 (2005).
70a. O. S. Zayachkivska, S. J. Konturek, D. Drozdowicz, P. C. Konturek , T. Brzozowski, and M. R. Ghegotsky, Gastroprotective effects of flavonoids in plant extracts. *J Physiol Pharmacol* **56**, 219–231 (2005).
71. C. Manach, A. Mazur, and A. Scalbert, Polyphenols and prevention of cardiovascular diseases. *Curr Opin Lipidol* **16**, 77–84 (2005).
72. M. Aviram and B. Fuhrman, wine flavonoids protect against LDL oxidation and atherosclerosis. *Ann NY Acad Sci* **957**, 146–161 (2002).
73. R. Olszanecki, A. Gebska, V. I. Kozlovski, and R. J. Gryglewski, Flavonoids and nitric oxide synthase. *J Physiol Pharmacol* **53**, 571–584 (2002).
74. S. J. Duffy and J. A. Vita, Effects of phenolics on vascular endothelial function. *Curr Opin Lipidol* **14**, 21–27 (2003).
75. M. Strzelecka, M. Bzowska, J. Koziel, B. Szuba, O. Dubiel, N. D. Riviera, et al., Anti-inflammatory effects of extracts from some traditional Mediterranean diet plants. *J Physiol Pharmacol* **56**, 139–156 (2005).
76. J. C. Ruf, Wine and polyphenols related to platelet aggregation and atherothrombosis. *Drugs Exp Clin Res* **25**, 125–131 (1999).
77. N. Venkatesan, Pulmonary protective effects of curcumin against paraquat toxicity. *Life Sci* **66**, 21–28 (2000).
78. S. Toda,, T. Miyase, H. Arichi, H. Tanizawa, and Y. Takino, Natural antioxidants. III. Antioxidative components isolated from rhizome of *Curcuma longa L. Chem Pharma Bull* **33**, 1725–1728 (1985).
79. A. C. Reddy and B. R. Lokesh, Studies on the inhibitory effects of curcumin and eugenol on the formation of reactive oxygen species and the oxidation of ferrous iron. *Mol Cell Biochem* **137**, 1– 8 (1994).
80. M. K. Unnikrishnan and M. N. Rao, Curcumin inhibits nitrogen dioxide induced oxidation of hemoglobin. *Mol Cell Biochem* **146**, 35–37 (1995).
81. N. Sreejayan and M. N. A. Rao, Nitric oxide scavenging by curcuminoids. *J Pharm Pharmacol*. **49**, 105–107 (1997).

82. A. C. Reddyand B. R. Lokesh, Studies on spice principles as antioxidants in the inhibition of lipid peroxidation of rat liver microsomes. *Mol Cell Biochem* **111** , 117–124 (1992).
83. N. Sreejayan and M. N. A. Rao, Curcuminoids as potent inhibitors of lipid peroxidation. *J Pharm Pharmacol* **46**, 1013–1016 (1994).
84. H. H. Cohly, A. Taylor, M. F. Angel, and A. K. Salahudeen, Effect of turmeric, turmeric and curcumin on H_2O_2-induced renal epithelial (LLCPK1) cell injury. *Free Radical Biol Med* **24**, 49–54 (1998).
85. M. Dikshit, L. Rastogi, R. Shukla, and R. C. Srimal,. Prevention of ischemia-induced biochemical changes by curcumin and quinidine in the cat heart. *Ind J Med Res* **101**, 31–35 (1995).
86. C. Nirmala and R. Puvanakrishnan, Effect of curcumin on certain lysosomal hydrolases in isoproterenol-induced myocardial infarction in rats. *Biochem Pharmacol* **51**, 47–51 (1996).
87. C. Nirmala and R. Puvanakrishnan, Protective role of curcumin against isoproterenol induced myocardial infarction in rats. *Mol Cell Biochem* **159**, 85–93 (1996).
88. C. Nirmala, S. Anand, and R. Puvanakrishnan, Curcumin treatment modulates collagen metabolism in isoproterenol induced myocardial necrosis in rats. *Mol Cell Biochem* **197**, 31–37 (1999).
89. J. P. Gaddipati, S. V. Sundar, J. Calemine, P. Seth, G. S. Sidhu, and R. K. Maheshwari, Differential regulation of cytokines and transcription factors in liver by curcumin following hemorrhage/resuscitation. *Shock* **19**, 150–156 (2003).
90. P. Rafiee, Y. Shi, K. A. Pritchard, H. Ogawa, A. L. Eis, R. A. Komorowski, C. M. Fitzpatrick, J. S. Tweddell, S. B. Litwin, K. Mussatto, R. D. Jaquiss, and J. E. Baker, Cellular redistribution of inducible Hsp70 protein in the human and rabbit heart in response to the stress of chronic hypoxia: Role of protein kinases. *J Biol Chem* **278**, 43,636–43,644 (2003).
91. P. Rafiee, Y. Shi, X., Kong, K. A. Pritchard, Jr., J. S. Tweddell, S. B. Litwin, K. Mussatto, R. D. Jaquiss, J. Su, and J. E. Baker, Activation of protein kinases in chronically hypoxic infant human and rabbit hearts: Role in cardioprotection. *Circulation* **106**, 239–245 (2002).
92. N. Sreejayan and M. N. A. Rao, Free radical scavenging activity of curcuminoids. *Arzneimittelforschung* **6**, 169–171 (1996).
93. C. V. Rao, A. Rivenson , B. Simi, and B. S. Reddy, Chemoprevention of colon carcinogenesis by dietary curcumin, a naturally occurring plant phenolic compound. *Cancer Res* **55**, 259–266 (1995).
94. A. J. Ruby, G. Kuttan, K. D. Babu, K. N. Rajasekharan, and R. Kuttan, Anti-tumor and antioxidant activity of natural curcuminoids. *Cancer Lett* **94**, 79–83 (1995).
95. J. M. McCord, Oxygen-derived free radicals in post-ischaemic tissue injury. *N Eng J Med* **312**, 159–163 (1985).
96. Y. Xia and J. L. Zweier, Substrate control of free radical generation from xanthine oxidase in the post-ischaemic heart. *J Biol Chem* **270**, 18,797–18,803 (1995).
97. W. F. Saavedra, N. Paolocci, M. E. St John, M. W. Skaf, G. C. Stewart, J. S. Xie, et al., Imbalance between xanthine oxidase and nitric oxide synthase signaling pathways underlies mechanoenergetic uncoupling in the failing heart. *Circ Res* **90**, 297–304 (2002).
98. L. S. Terada, D. M. Guidot, J. A. Leff, I. R. Willingham, M. E. Hanley, D. Piermattei, and J. E. Repine, Hypoxia injures endothelial cells by increasing endogenous xanthine oxidase activity. *Proc Natl Acad Sci USA* **89**, 3362–3366 (1992).

99. E. Kunchandy and M. N. A. Rao, Oxygen radical scavenging activity of curcumin. *Int J Pharm* **58**, 237–240 (1990).
100. T. Matsuyama, Free radical-mediated cerebral damage after hypoxia/ischemia and stroke. In: G. J. Ter Horst and J. Korf, eds. *Clinical Pharmacology of Cerebral Ischemia*. Totowa, NJ: Humana Press, 1997. pp. 153–184.
101. M. Pan, T. Huang, and J. Lin, Biotransformation of curcumin through reduction and glucuronidation in mice. *Drug Metab Dispos* **27**, 486–494 (1999).
102. J. A. Thomas, Oxidative stress, oxidant defense, and dietary constituents. In *Modern Nutrition in Health and Disease*, 8th ed. Lea & Febiger, Phil.; 1994, pp. 501–512.
103. M. Seif-El-Nasr and A. A. Abd-El-Fattah, Lipid peroxide, phospholipids, glutathione levels and superoxide dismutase activity in rat brain after ischemia: Effect of Ginkgo biloba extract. *Pharmacol Res* **32**, 273–278 (1995).
104. S. Mathews and M. N. A. Rao, Interaction of curcumin with glutathione. *Int J Pharm* **76**, 257–259 (1991).
105. 105. S. V. Jovanovic, C. W. Boone, S. Steenken, M. Trinoga, and R. B. Kaskey, How curcumin works preferentially with water-soluble antioxidants. *J Am Chem Soc* **123**, 3064–3068 (2001).
106. R. K. Maheshwari, A. K. Singh, J. Gaddipati, and R. C. Srimal, Multiple biological activities of curcumin: A short review. *Life Sci* **78**, 2081–2087 (2006).
107. K. I. Priyadarsini, Free radical reactions of curcumin in membrane models. *Free Radical Biol Med* **23**, 838–843 (1997).

PROTECTION FROM ACUTE AND CHRONIC LUNG DISEASES BY CURCUMIN

Narayanan Venkatesan, Durairaj Punithavathi, and Mary Babu

Abstract: The aim of this review has been to describe the current state of the therapeutic potential of curcumin in acute and chronic lung injuries. Occupational and environmental exposures to mineral dusts, airborne pollutants, cigarette smoke, chemotherapy, and radiotherapy injure the lungs, resulting in acute and chronic inflammatory lung diseases. Despite major advances in treating lung diseases, until now disease-modifying efficacy has not been demonstrated for any of the existing drugs. Current medical therapy offers only marginal benefit; therefore, there is an essential need to develop new drugs that might be of effective benefit in clinical settings. Over the years, there has been increasing evidence that curcumin, a phytochemical present in turmeric (*Curcuma longa*), has a wide spectrum of therapeutic properties and a remarkable range of protective effects in various diseases. Several experimental animal models have tested curcumin on lung fibrosis and these studies demonstrate that curcumin attenuates lung injury and fibrosis caused by radiation, chemotherapeutic drugs, and toxicants. The growing amount of data from pharmacological and animal studies also supports the notion that curcumin plays a protective role in chronic obstructive pulmonary disease, acute lung injury, acute respiratory distress syndrome, and allergic asthma, its therapeutic action being on the prevention or modulation of inflammation and oxidative stress. These findings give substance to the possibility of testing curcumin in patients with lung diseases.

1. INTRODUCTION

The lungs are responsible for the vital process termed "gas exchange," wherein the tiny air sac-like structures, alveoli, take oxygen into the lungs and drive out carbon dioxide. Thus, any substance that is breathed in affects the lungs, and many of these substances can be dangerous and intimidate the lungs' ability to function properly. Occupational and environmental exposures to mineral dusts, airborne pollutants, cigarette smoke, pharmacologic therapy with anticancer drugs, and radiotherapy injure the lungs in various fashions, resulting in acute and chronic inflammatory lung diseases, including lung fibrosis, allergic asthma, acute lung injury (ALI)/acute respiratory distress syndrome (ARDS), and chronic obstructive pulmonary disease (COPD)/emphysema. These types of lung disease pose a significant health risk to humans and are associated with high morbidity and mortality.

Although major advances exist in the field of treatment of lung diseases, the incidence of acute and chronic inflammatory lung diseases continue to rise, causing significant worries to the patients and clinicians alike. A variety of new medications have appeared for the treatment of acute and chronic lung injuries and new research on traditional therapies has been performed. However, until now disease-modifying efficacy has not been demonstrated for any of the existing drugs. Preclinical and clinical research shows that modern anti-inflammatory therapy is effective in that it gradually reduces inflammatory response, but with significant toxic side effects, which is no longer a sufficient end point for therapy. Therefore, there is an imperative need to develop new drugs that have broad-based anti-inflammatory properties plus a safety profile.

Over the past several years, there is a renewed interest in alternative inexpensive therapies without any apparent toxic effects. In this regard, curcumin, a natural phytochemical present in turmeric (*Curcuma longa*), is noted for its anti-inflammatory, antioxidant, chemopreventive, and chemotherapeutic properties. Curcumin has a wide spectrum of therapeutic properties and a remarkable range of protective effects in various diseases; therefore, it is pertinent to state what actions it does not have than to say what it does to various disease processes. The therapeutic utility of curcumin has undergone a swift development during the last decade and several biochemical, pharmacological, and clinical aspects have been previously reviewed.[1,2] The present chapter is an endeavor to update the rapidly expanding information on curcumin's action relevant to the various lung injuries. Becausee this chapter is concerned primarily with pulmonary protection by curcumin, the reader is referred to earlier reviews for detailed information on pulmonary fibrosis, COPD, ARDS, and allergic asthma. However, a brief summary of salient clinical and biological information about various types of lung injury is presented to provide the reader with material that has a bearing on the objective of treatment and the proper evaluation of therapeutic results.

2. CURCUMIN THERAPY IN ANIMAL MODELS AND CLINICAL DISEASE

The growing amount of data from pharmacological and animal studies supports the notion that curcumin plays a protective role not only in lung fibrosis but also in COPD, ALI/ARDS, and allergic asthma, its therapeutic action being on the prevention or repression of the inflammatory response and oxidative stress (Table 1). This section covers the diverse array of reported therapeutic effects of curcumin and its application in the treatment of acute and chronic lung diseases.

Table 1. *In vivo* effects of curcumin on various animal models of lung disease.

LUNG INJURY MODEL	SPECIES/THERAPY	OUTCOME (REF.)
Radiation-induced fibrosis	Rats/oral treatment	↓ Lung hydroxyproline and lung structural damage[3]
Paraquat-induced fibrosis	Rat/oral treatment	↓ Mortality, BAL proteins, enzymes, neutrophil accumulation, lung MPO activity, and lipid peroxidation ↑ Lung glutathione levels[4]
Cyclophos-phamide-induced fibrosis	Rat/oral treatment	↓ BAL biomarkers, influx of inflammatory cells, lipid peroxidation ↑ Lung antioxidant defense[5]
Bleomycin-induced fibrosis	Rat/oral treatment	Mortality, inflammatory cell influx, and total lung hydroxyproline[6,7]
Amiodarone-induced fibrosis	Rat/oral treatment	↓ Lung MPO, TGF-β1, hydroxyproline type I collagen, and c-Jun protein[8]
Nicotine-induced injury	Rat/oral treatment	↓ Biochemical marker enzymes and lipid peroxidation in BAL ↑ Antioxidant status[9]
Endotoxic shock	Mouse/oral treatment	↓ Endotoxin-induced lung ICAM-1 expression[10]
Allergen-induced airway hyper responsiveness	Guinea pigs/oral treatment	↓ Airway constriction and airway hyperreactivity[11]

3. CURCUMIN INHIBITION OF DRUG/CHEMICAL-INDUCED LUNG INJURY AND FIBROSIS

Several experimental animal models have tested curcumin on lung fibrosis and these studies demonstrate that curcumin attenuates lung injury and fibrosis caused by radiation, drugs and toxicants.

3.1. Pulmonary Fibrosis: Disease Mechanisms

Pulmonary fibrosis (PF) is a chronic inflammatory interstitial lung disease with poor prognosis. Although various factors, including environmental, infectious, immunologic, toxic, or pharmacologic, have been suggested as causes of the disease,[12] so far the etiology and pathogenesis have not been completely elucidated. Drug-induced PF is one of the major health problems encountered in clinical medicine. It is caused from the increasing use of anticancer drugs in chemotherapeutic interventions. The list of drugs causing pulmonary injury[13,14] is on the rise and has been extensively reviewed. Diagnosis of lung fibrosis is usually based on the clinical recognition of the characteristic signs that include shortness of breath, evident diffuse pulmonary infiltrates, and varying degrees of inflammation and fibrosis. The clinical course generally progresses to death, secondary to respiratory failure. Mean survival in patients has been estimated to be 3–6 years from onset of symptoms. Histologically, it is characterized by the accumulation of inflammatory cells in the airways, fibroblast and myofibroblast proliferation in lung parenchyma, and disproportionate extracellular matrix (ECM) deposition.[15] Infiltration of multiple cell types such as activated macrophages, neutrophils, and lymphocytes produce numerous soluble mediators, cytokines, chemokines, growth factors, and reactive oxygen species (ROS) that are implicated in the development of interstitial pulmonary fibrosis (IPF). These inflammatory mediators trigger the proliferation of capillary endothelial cells, fibroblasts, and smooth muscle cells that form the basis of the fibrotic scar (Figure 1).

Although inflammation has traditionally been considered the major factor of lung fibrosis, recent evidence indicates that dysregulation of the ECM metabolism plays a critical role in the pathogenesis of lung fibrosis, supporting studies on matrix production and matrix deposition.[16] In this regard, several antifibrotic compounds

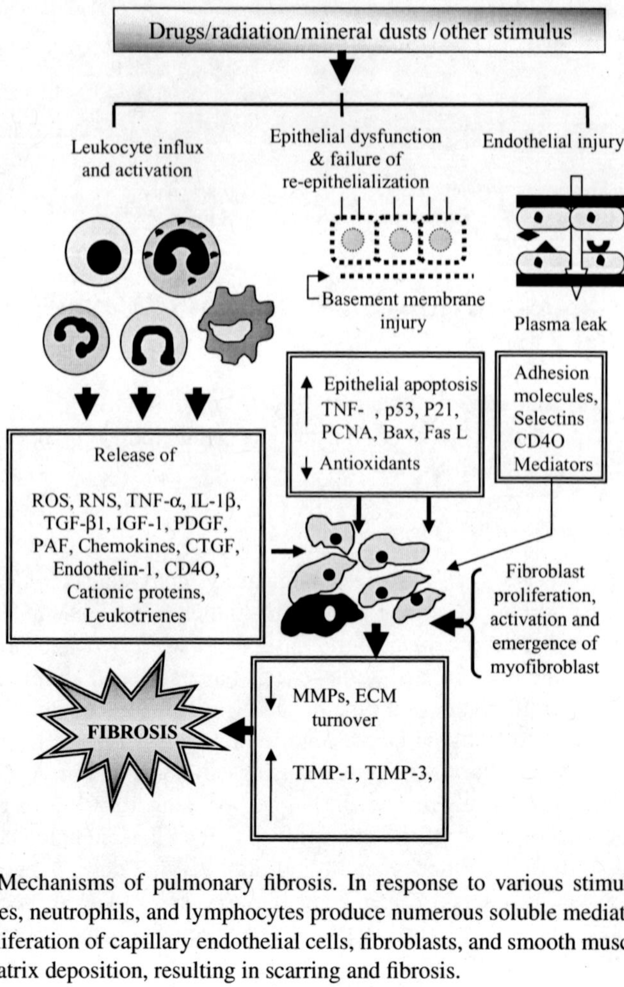

Figure 1. Mechanisms of pulmonary fibrosis. In response to various stimuli, activated macrophages, neutrophils, and lymphocytes produce numerous soluble mediators, triggering the proliferation of capillary endothelial cells, fibroblasts, and smooth muscle cells that promote matrix deposition, resulting in scarring and fibrosis.

such as penicillamine, colchicines, pirfenidone, and interferon-γ have been used in the treatment of PF.[17,18] Also, increasing evidence suggests that activation of the coagulation cascade, with the resultant generation of coagulation proteases, plays an important role in fibrotic lung disease that is associated with acute and chronic lung injury.[19] More recently, it has been proposed that bone-marrow-derived stem cells might play a pivotal role in the fibroproliferative response as well as in epithelial regeneration.[20] Current medical therapy (corticosteroids, azathioprine, cyclophosphamide) offers only marginal benefit; consequently, lung transplantation is considered for patients with PF who do not respond to medical therapy.[12,17,18] Given the poor response to on-hand anti-inflammatory therapy, other approaches to therapy are being followed. Therefore, molecules that suppress epithelial injury, proinflammatory cytokine action, and fibroblast proliferation and/or induce fibroblast apoptosis might be of effective benefit in clinical settings.

3.2. Curcumin Protects Against Radiation-Induced Lung Injury

Initial evidence concerning the protective action of curcumin on lung injury was given by Thresiamma and co-workers, who showed that if curcumin was administered orally to rats carrying radiation, the content of lung hydroxyproline was greatly reduced.[3] Their studies also indicated the inhibitory activity of curcumin against paraquat and cyclophosphamide-induced lung injury in mice.[21] Curcumin had multiple inhibitory actions on these lung injury models, such as lowering lipid peroxide formation and tissue cholesterol levels.

3.3. Curcumin Protects Against Cyclophosphamide-Induced Lung Injury

Using several *in vivo* animal models, we investigated the efficacy of curcumin in suppressing the development of lung injury and fibrosis in rats. First, we investigated the potential of curcumin against cyclophosphamide-induced lung injury in rats.[5] Lung injury was induced in rats by an intraperitoneal injection of of cyclophosphamide. Prior to cyclophosphamide injection, rats were orally treated with curcumin daily for 7 days. At various time intervals, biochemical markers in serum and lung lavage fluid were measured. The lavage cells and lung tissues were examined for lipid peroxidation and glutathione content. Antioxidant enzyme levels were analyzed in lung tissue. Biochemical analyses revealed increases in lavage fluid protein biomarkers and lipid peroxide levels, whereas a reduction in glutathione and ascorbic acid content was observed in cyclophosphamide-injected rats. Interestingly, curcumin decreased leukocytes and inflammatory mediators and increased lung antioxidant defenses in cyclophosphamide-injected rats. These experiments demonstrated that anti-inflammatory and antioxidant activity serves to spare the lung against damaging activities of cyclophosphamide.

3.4. Curcumin Protects Againt Paraquat-Induced Lung Injury

The interesting findings observed in curcumin-treated cyclophosphamide lungs prompted us to question the effect of curcumin on chemical-induced lung injury and fibrosis. To test this project, we injected rats with paraquat,[4] a broadly used nonselective herbicide that causes harsh lung injury and fibrosis in humans and animal models. As reported previously, the administration of paraquat had a marked effect, provoking inflammation and oxidative stress, as judged by increases in lung lavage fluid biomarkers and lipid peroxide formation accompanied by glutathione depletion. Remarkably, curcumin treatment abrogated the inflammatory response and inhibited the general toxicity and mortality that was induced by paraquat injection. Thus, results from these experiments confirmed that curcumin is indeed capable of protecting paraquat-induced lung injury.

3.5. Curcumin Protects Against Bleomycin-Induced Lung Fibrosis

Further experiments were performed to address the protection of curcumin in bleomycin (BLM)-induced lung injury) in rats.[6,7] First, to analyze the protective effect of curcumin on lung structure alterations in disease, we examined the histopathological changes associated with experimental fibrosis. We observed significant structural changes in fibrotic lungs (Figures 2C and 2E) compared with normal lungs (Figures 2A and 2B). In BLM controls, these changes included influx of inflammatory cells, particularly neutrophils, macrophages, and granulocytes, in the alveolar space and interstitium at 7 days after BLM instillation, as evaluated by hematoxylin–eosin staining (Figure 2C). Consistent with its protective effect, curcumin treatment to bleomycin rats showed a significant reduction in inflammation and presented little structural damage (Figures 2D and 2F). These findings indicated that curcumin is capable of preserving the lung architecture and organ dysfunction in fibrotic disease. Next, we evaluated the effect of curcumin on BLM-induced increases in inflammatory biomarkers. BLM administration caused increased leukocyte influx, biomarkers, and malondialdehyde and a reduction in glutathione content. Stimulated tumor necrosis factor (TNF)-α, superoxide anion, hydrogen peroxide, and nitric oxide release by alveolar macrophages from BLM rats were higher. Curcumin significantly suppressed leukocyte influx (Figure 3), reduced lavage fluid biomarkers and restored the antioxidant status in BLM rats. Also, curcumin treatment resulted in attenuation of BLM-induced macrophage release of TNF-α, superoxide anion, hydrogen peroxide, and nitric oxide. These results indicated that curcumin reduces the development of BLM-induced inflammatory and oxidant activity. Furthermore, experiments were conducted to evaluate the ability of curcumin to inhibit BLM-induced pulmonary fibrosis in rats. BLM-injected rats had a significant increase in the amounts of α-smooth muscle actin (α-SMA; an indicator of lung myofibroblast content) and lung hydroxyproline (an indicator of collagen deposition) content. Oral administration of curcumin (300 mg/kg, 10 days before and daily thereafter throughout the experimental time period) suppressed BLM-induced increases in lung α-SMA (Figure 4) and hydroxyproline (Figure 5). These results suggested that curcumin acts as a potent anti agent against BLM-induced pulmonary fibrosis in rats.

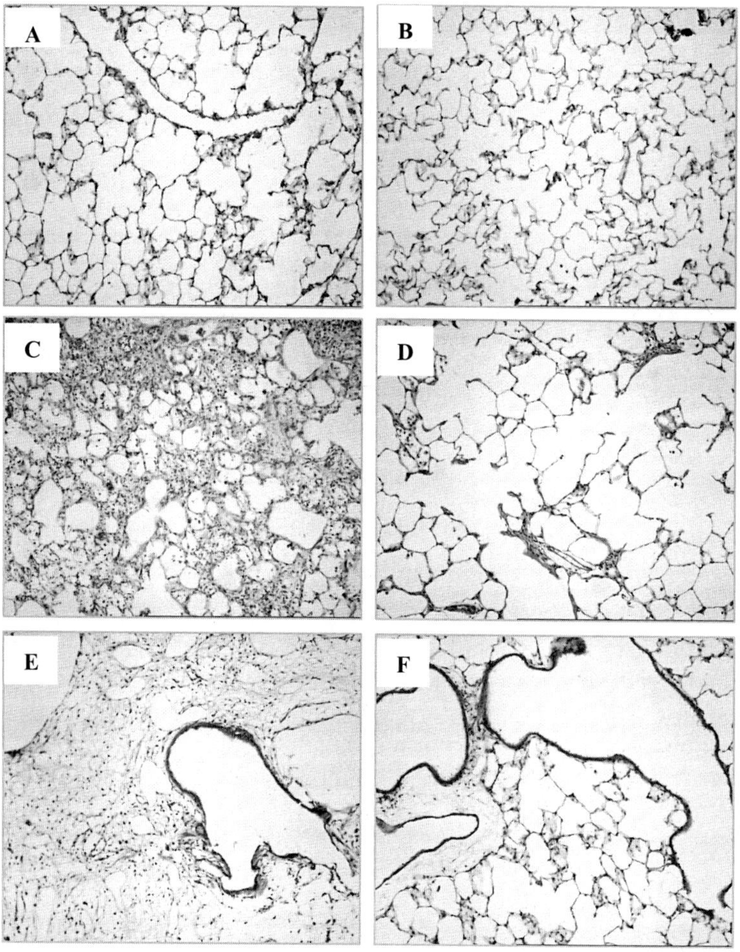

Figure 2. Effect of curcumin on BLM-induced inflammation and fibrosis in rats: (A) saline-treated lung; (B) curcumin-treated lung; (C) day 7 BLM-treated lung; (D) day 7 curcumin + BLM-treated lung; (E) day 28 BLM-treated lung; (F) day 28 curcumin + BLM-treated lung. Curcumin significantly attenuated BLM-induced accumulation of inflammatory cells and fibrotic changes in rat lungs.

3.6. Curcumin Protects Against Amiodarone-Induced Lung Fibrosis

Buoyed by the above observations, we attempted to investigate the antifibrotic effect of curcumin in the amiodarone (AMD), an antiarrhythmic drug, model of lung injury and fibrosis in rats.[8] As described earlier, obvious instances of protective responses against inflammation were seen in amiodarone-instilled rats treated with curcumin. For the AMD fibrosis model, we focused in particular

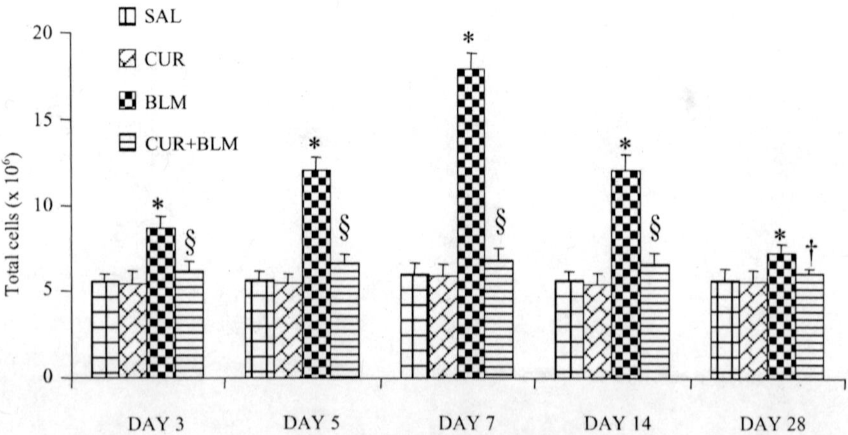

Figure 3. Protective effect of curcumin on BLM-induced changes in total cell counts in the BAL fluid of rats. BLM caused a significant increase in the total cell count in the BAL fluid, whereas curcumin treatment prevented these effects. SAL: saline; CUR: curcumin; BLM: bleomycin. *Significantly ($p < 0.001$) higher than all groups; [†,§]significantly ($p < 0.01$, $p < 0.005$, respectively) lower than in BLM-injected rats.

on two molecules: transforming growth factor (TGF)-β1, which controls ECM deposition in the fibrotic lungs, and c-Jun, a nuclear transcription factor involved in inflammatory and fibroproliferative processes. These two molecules constitute essential links between the inflammatory and fibroproliferative manifestations of lung injury,[22] and both are known to precipitate fibrotic tissue remodeling, which

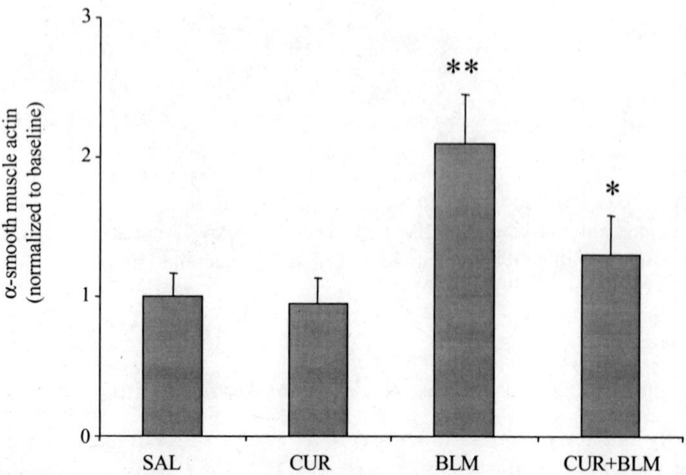

Figure 4. Effect of curcumin on BLM-induced changes in α-SMA expression in rat lung. Data are mean ± SD of six rats in each group. **Significantly ($p < 0.001$) higher than all groups; *significantly ($p < 0.005$) lower than BLM-injected rats. SAL: saline; CUR: curcumin; BLM: bleomycin.

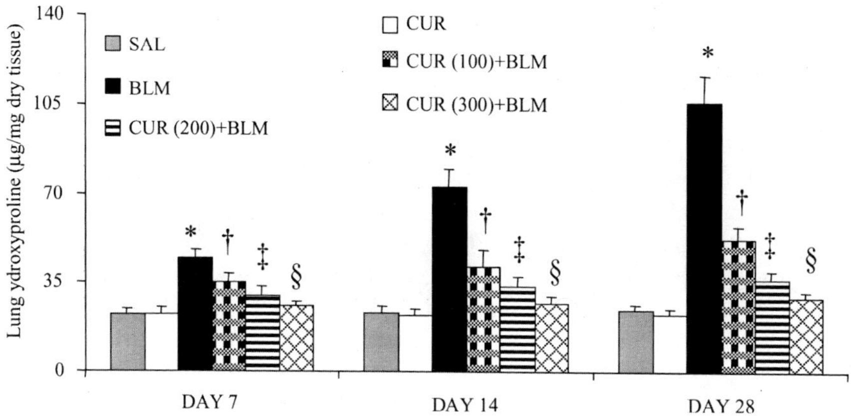

Figure 5. Dose-dependent protective effects of curcumin on BLM-induced changes in lung hydroxyproline content. Data are mean ± SD of six rats in each group. *Significantly ($p < 0.001$) higher than all groups; †,‡,§significantly ($p < 0.01$, $p < 0.005$, $p < 0.0001$, respectively) lower than inBLM-injected rats. SAL: saline; CUR: curcumin; BLM: bleomycin. CUR (100)–CUR (300): fibrotic rats treated with 100–300 mg/kg curcumin.

showed significant increases in the fibrotic lungs. In these instances, clear evidence of curcumin inhibition of TGF-β1 expression and c-Jun protein was observed in amiodarone rats. Results of this study suggested that treatment of rats with curcumin was able to provide a significant protection against AMD-induced lung injury and fibrosis. These findings were largely on par with the extent of protection offered by curcumin in BLM-induced inflammation and fibrosis.

3.7. Proapoptotic Effect of Curcumin on Scleroderma Lung Fibroblast

Scleroderma, also known as systemic sclerosis (SSc), is a multisystem disease characterized by pathological remodeling of connective tissues. Fibrosis in the skin and internal organs is a key feature of scleroderma.[23] A distinctive characteristic of fibroblasts explanted from lesional SSc tissues is the retention of an abnormal phenotype, characterized by enhanced ECM synthesis and the expression of the myofibroblast marker, α-SMA, during serial passage *in vitro*. Unfortunately, despite the major burden of fibrosis in scleroderma, there are no agents that have yet been shown to have an antifibrotic effect in this disease. The discovery by Tourkina and colleagues[24] that curcumin caused apoptosis in scleroderma lung fibroblasts (SLFs), but not in normal lung fibroblasts (NLFs), and that a signaling pathway involving protein kinase CPKCε and phase 2 detoxification enzymes offered protection against curcumin-induced apoptosis in NLFs and is defective in SLFs is a stimulating and fascinating development. It increases our conception of the overall protective effect of curcumin in tissue fibrosis in a significant manner.

In this regard, it is of great interest that curcumin did not induce apoptosis of NLFs, but it does in SLFs. Importantly, PKCε-overexpressing SLFs show suppressive activity toward apoptosis, whereas loss of function studies in NLFs makes it sensitive to apoptosis, which suggests that expression of PKCε is linked to suppressor functions. In this study, they provide direct evidence that PKCε signaling, which is highly expressed in NLFs but is deficient in SLFs, is required to resist apoptosis. Both NLFs and SLFs differ in the subcellular distribution of PKCε, and, interestingly, fibrotic lung tissue had reduced expression of PKCε. Taken together, these data strongly support the idea that curcumin might uniquely cause apoptosis of SLF cells. They also found that levels of the phase 2 detoxification enzymes heme oxygenase-1 (HO-1) and glutathione S-transferase P1 (GST P1) varied in NLF and SLF cells. In curcumin-treated SLF cells, in contrast to NLF cells, the levels of these phase 2 detoxification enzymes have been found to be decreased.

In essence, experimental studies clearly demonstrate that curcumin prevents lung fibrosis in various animal models by blocking leukocytes influx, by inhibiting the activation of inflammatory cells and subsequent release of proinflammatory and toxic mediators, and by attenuating excess accumulation of extracellular matrix proteins.

4. CURCUMIN INHIBITION OF ACUTE AND CHRONIC INFLAMMATORY LUNG INJURIES

Spurred by the encouraging findings in experimental fibrosis, several laboratories have recently begun to dissect the pulmonary protective effect of curcumin in other lung diseases. These studies complement the earlier findings, which reveal that curcumin not only provides protection against pulmonary fibrosis but also inhibits nicotine lung injury, endotoxin acute lung injury, and allergic asthma. Here, we will review the results of experimental and clinical studies evaluating the efficacy of curcumin in these lung diseases.

4.1. Curcumin Inhibition of Chronic Obstructive Lung Disease

In this section we have extracted the data from both *in vitro* and *in vivo* studies, which illustrate the effectiveness of curcumin in attenuating cigarette smoke- and nicotine-induced lung injuries.

4.1.1. Chronic Obstructive Lung Disease: Disease Mechanisms

Cigarette smoking (CS) is associated with inflammatory diseases of the lung, especially COPD, including increased airway reactivity, exacerbations of asthma, and an increased frequency of pulmonary infections.[25] COPD is currently the fourth killer disease in the United States, affecting over 18 million Americans, which also intimidates a worldwide epidemic. Although COPD occurs primarily in cigarette smokers, the precise relationship between cigarette smoking and COPD is multifarious because only 10–15% of active smokers develop clinical COPD.[26] Clinical and pathological characteristics of COPD have been reviewed in detail

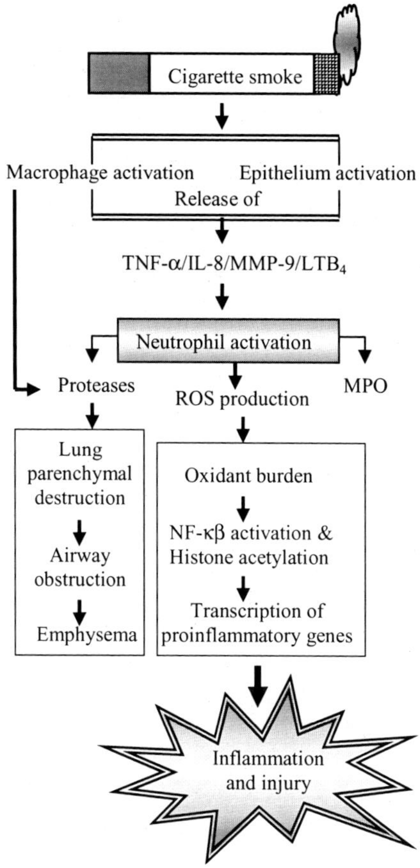

Figure 6. Pathogenesis of COPD. Following exposure to cigarette smoke (and other irritants), alveolar macrophages and epithelial cells in the respiratory tract are activated, releasing neutrophil chemotactic factors, including interleukin-8 and leukotriene B_4 (LTB_4). Also, these activated cells along with neutrophils release matrix-degrading proteases, resulting in the destruction of lung parenchyma and emphysema.

previously.[27] Generally, COPD is characterized by a massive infiltration of inflammatory cells, particularly neutrophils, macrophages, and T-lymphocytes, and increased production and release of cytokines, oxidants, and proteases in the lung. Numerous studies have reported increases in total cell recovery, macrophages, neutrophils, lymphocytes (often CD8+), and eosinophils in BAL fluids and/or lung biopsies from patients with COPD.[28,29]

A considerable number of studies suggest that chronic exposure to cigarette smoke results in the vast infiltration of inflammatory cells in the lung, and these cells release exaggerated amounts of elastolytic proteinases, destroying lung ECM components (Figure 6). Unsuccessful repair of alveoli and ECM macromolecules causes permanent enlargement of peripheral airspaces and loss of lung function.[27] CS has been reported to induce emphysema or emphysematous, inflammatory, and metaplasic changes in the lungs of experimental animals.[30] Cigarette smoking causes an increase in oxidative stress in the lower respiratory tract, either directly from its oxidative components,or indirectly from free radicals released from activated alveolar inflammatory cells.[31] Numerous studies of COPD patients have shown a decrease in lung and plasma levels of antioxidants, indicating a systemic oxidative stress, but inconsistent results have been attained by administering exogenous antioxidants, particularly vitamins such as vitamin E, to patients with COPD.[32] Although cigarette smoke contains a huge number of compounds, oxidants and free radicals, nicotine is the chief toxic constituent that has been implicated in the development of cardiorespiratory injuries, vascular disease, and lung cancer. It has been reported that nicotine triggers an inflammatory response in the airway epithelium leading to the generation of proinflammatory cytokines and chemokines, which injures lung epithelium, thereby resulting in increased permeability, and recruitment of neutrophils and macrophages to the airways. Currently, there is a need for new agents that will be more effective than the current therapeutic options for patients with COPD.[33]

Given the importance of oxidative stress in the pathogenesis of COPD, one rational approach would be therapeutic interventions targeted against oxidative processes to eventually relieve the burden of oxidative stress in COPD. Also, compounds that increase glutathione (GSH) homeostasis might have therapeutic potential in protecting the lung from increased oxidant burden.[31]

4.1.2. Curcumin Protects Against Nicotine-Induced Lung Injury

To explore the therapeutic importance of curcumin in smoking-related lung injury, Kalpana and colleagues treated rats with nicotine, a component of cigarette smoke primarily responsible for the toxic effects. In their studies, they provided preliminary evidence that indicates that curcumin exerts a protective effect against nicotine-injured rat lungs.[9,34] As reported in their studies, lung tissue from nicotine-treated rats exhibited increased levels of lipid peroxides coupled with a decrease in both enzymatic and nonenzymatic antioxidants. Additionally, nicotine-injected rats had higher amounts of protein biomarkers in bronchoalveolar lavage fluid. Interestingly, curcumin reduced the levels of nicotine-induced oxidant burden, restored the antioxidant levels and modulated the biomarkers in the lung lavage fluid.

Consistent with the results of the above study, data from our laboratory also indicate that curcumin offered protection against nicotine-induced lung toxicity in rats. Results from our findings show that oral administration of curcumin (200 mg/kg body weight, 1 week before nicotine injection and daily thereafter for 3 weeks following nicotine injection) attenuated nicotine-induced neutrophil infiltration and activation, as judged by BALF cellularity (Figure 7A) and myeloperoxidase activity (Figure 7B). When all of the effects of curcumin on the various biochemical parameters in nicotine-induced lung toxicity are examined, one does see some novel

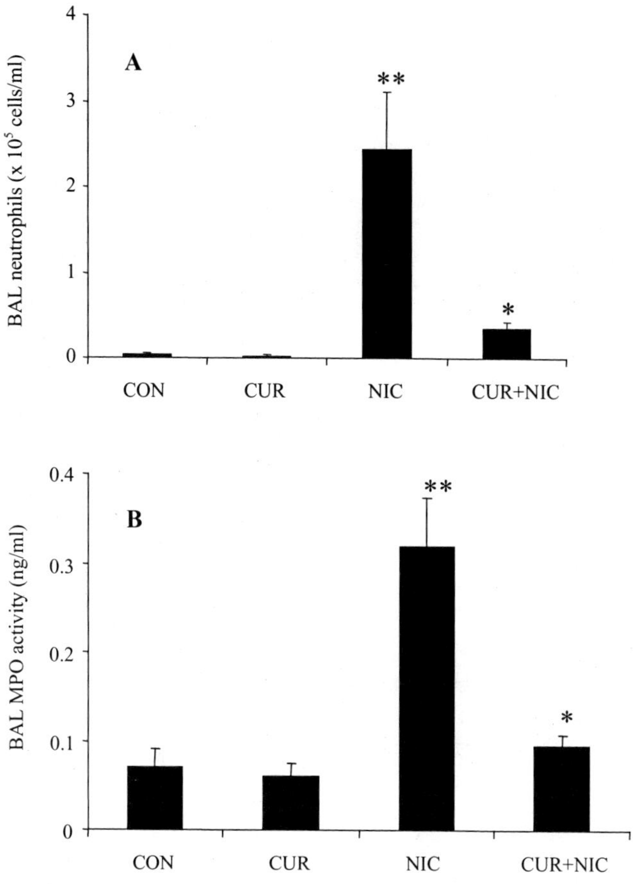

Figure 7. Protective effects of curcumin on nicotine-induced changes in neutrophil cell count and myeloperoxidase activity in BAL fluid. Nicotine injection resulted in a significant increase in neutrophil cell numbers accompanied by increased myeloperoxidase activity, whereas curcumin treatment prevented these effects. Values are mean ± SD of six rats in each group. **Significantly ($p < 0.001$) higher than all groups; *significantly ($p < 0.01$) lower than nicotine groups. CON: control; CUR: curcumin; NIC: nicotine.

findings. However, what is (are) the *in vivo* mechanism(s) of action of curcumin in this system? Further studies are in progress in our lab to explore mechanistic insights underlying curcumin's protection in this system. Perhaps one of the most relevant studies to appear regarding the molecular mechanisms whereby curcumin inhibit the deleterious effects of cigarette smoke was that described by Shishodia et al.[35] These authors investigated cigarette smoke-induced alterations in nuclear factor (NF)-κB, a nuclear transcription factor implicated in chemical carcinogenesis, tumorigenesis, and the regulation of proinflammatory gene expression. The authors discovered the ability of curcumin to inhibit CS-induced NF-κB activation in the human lung epithelial cell lines. These investigators reported that exposure of human lung epithelial cells to cigarette smoke caused relentless activation of NF-κB, and pretreatment with curcumin eliminated the CS-induced DNA binding of NF-κB, IκBα kinase activation, IκBα phosphorylation and degradation, p65 nuclear translocation, and CS-induced NF-κB-dependent reporter gene expression. Also, these authors showed that curcumin inhibition of NF-κB activation correlated with suppression of CS-induced NF-κB-dependent cyclin D1, cyclooxygenase-2, and matrix metalloproteinase-9 expression.

Collectively, these findings raise the possibility that dietary curcumin could modulate the development of smoking-related lung diseases. It will be worthwhile to study the effects of curcumin in animal models of emphysema/COPD to determine whether curcumin treatment could attenuate cigarette smoke-induced changes in lung function, morphology, and upregulation of gene expression of various proinflammatory mediators and activation of specific signaling cascades.

4.2. Curcumin Inhibition of Acute Lung Injury/Acute Respiratory Distress Syndrome

4.2.1. Acute Respiratory Distress Syndrome/Acute Lung Injury: Disease Mechanisms

The acute respiratory distress syndrome (ARDS), a clinical syndrome of noncardiogenic pulmonary edema, occurs due to sepsis, trauma, and aspiration and is associated with pulmonary infiltrates, stiff lungs, and severe hypoxemia, eventually resulting in respiratory failure. Numerous studies have improved our understanding of ARDS,[36–38] which is now documented as an acute lung injury as well as the pulmonary sign of a systemic disorder called the multiple organ dysfunction syndrome (MODS), the leading cause of morbidity and mortality in surgical intensive care units. Despite intense efforts to alleviate the development of ALI/ARDS, currently there is no successful treatment[39] and the mortality remains disappointingly high at 40–60%. Furthermore, although mechanical ventilation has been shown to have a bearing on mortality, ventilator-induced lung injury remains a significant problem because it has been shown to initiate or exacerbate the inflammatory response in the lung.[40] Several experimental models such as the administration of lipopolysaccharide (LPS), live bacteria, or zymosan-activated plasma to the animals and cecal ligation and puncture[41] are used to study the pathogenesis of sepsis or endotoxin shock. Although these models vary in clinical and pathological features, they all present a severe multiorgan failure, including acute lung injury.[38] The molecular mechanism of ARDS has been extensively studied and was reviewed recently.[36–38] In brief, inflammation plays a pivotal role in ARDS and studies have implicated several proinflammatory mediators, together with endotoxin-activated complement and cytokines in the pathogenesis of this disease. These mediators account for local effects at the site of inflammation (i.e., vasodilatation, increased vascular permeability, and migration of leukocytes into the affected area), for interstitial and alveolar accumulation of denatured protein and fluid, and for reduced levels of surfactant and surfactant proteins. Increased numbers of intravascular and extravascular polymorpho nuclear leukocytes (PMNs), platelets, and fibrin, together with endothelial and epithelial injury have been reported in lungs of patients dying of ALI secondary to sepsis.[36] Although multiple therapeutic interventions have been considered so far, none has shown promising effects on the survival of patients dying of ARDS. Unfortunately, clinical trials using corticosteroids, prostaglandins, nitric oxide, prostacyclin, surfactant, lisofylline, ketoconazole, *N*-acetylcysteine, and fish oil have been incapable of showing statistically significant improvement in patient mortality[39] Clinical and experimental studies have shown that oxidative stress leading to increased production of ROS might play a role in the pathological defects observed in ALI.[42,43] Consequently, numerous reports have shown a decrease in the levels of antioxidant status both in experimental models and patients with ALI.[43] These findings indicate that strategies aimed at limiting inflammation and increased oxidant production could be effective in both treating ARDS and examining the mechanisms that lead to its development (Figure 8).

4.2.2. Curcumin Protects Against Endotoxin-Induced Acute Lung Injury

Given that curcumin treatment attenuates both inflammation and oxidant burden in various lung injuries, it is sensible to hypothesize that curcumin treatment might also alter the course of disease development in experimental models of ALI. Indeed, there is some evidence in the literature to support the plausibility of this hypothesis. In a recent study of the possible therapeutic value of curcumin for the treatment of sepsis and systemic inflammatory response syndrome, Madan and Ghosh found that curcumin offered protection in high-dose endotoxin shock by improving survival and reducing the severity of endotoxin shock symptoms such as lethargy, diarrhea, and watery eyes following endotoxin challenge.[10] The authors reported that curcumin inhibited the transmigration and infiltration of neutrophils from blood vessels to the liver tissue and also suppressed the induced expression of ICAM-1 and VCAM-1 in the liver and lungs. Although these authors have studied the protective effects of curcumin in the context of liver injury, their data also indicated that curcumin treatment effectively inhibited the expression of ICAM-1 in lung tissues. Thus, curcumin inhibition of ICAM-1 expression could play a protective role in the epithelial–endothelial integrity by modulating the interactions between these surface molecules and thereby blocking the adhesion of neutrophils to the endothelial cell monolayers *in vivo*. In this regard, antibodies to ICAM-1, an important ligand for the β2 integrin family of leukocyte membrane glycoproteins, have been shown to inhibit both neutrophil sequestration and lung injury induced by complement activation.[44] In line with the above studies, we envisaged that endotoxin-induced ALI would be protected by curcumin. Consistent with this expectation, the data we have presented here indicate that curcumin might also protect endotoxin-induced lung injury in mice. Curcumin's protection against endotoxin lung injury was reflected in a reduction of the increase in lung wet weight/dry weight ratio (Figure 9A), indicating that curcumin inhibited endotoxin-induced lung edema. Also, our data demonstrate that curcumin is fully

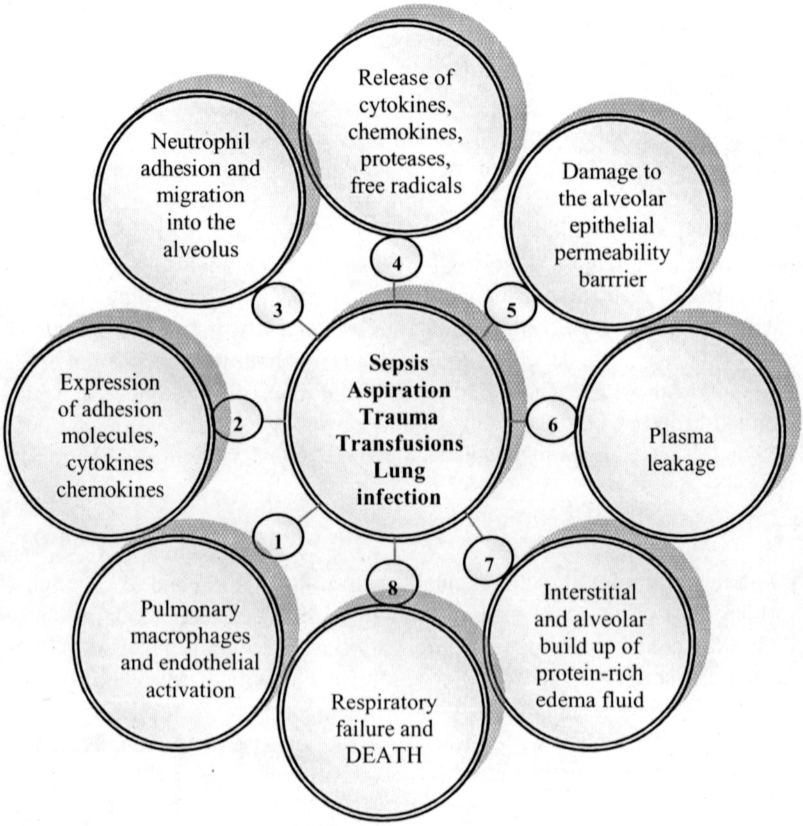

Figure 8. A schematic representation of cellular mechanisms involved in ALI. A variety of direct (aspiration, lung infection) and indirect (sepsis, trauma, multiple transfusions, endotoxin shock) stimuli cause an acute inflammatory response, followed by influx of neutrophils, resulting in injury to epithelial and endothelial cells. Subsequent necrosis of type I epithelial cells causes damage to epithelial barrier function of the lung leading to flooding of the air spaces with protein-rich edema fluid, resulting in diffuse alveolar damage.

capable of restricting the increase in neutrophil influx into the alveolar spaces and inhibiting increased BAL MPO activity (Figure 9B).

4.2.3. Proapoptotic Effect of Curcumin on Human Neutrophils

ALI/ARDS is characterized by a neutrophilic inflammation and considerable data implicate neutrophils as primary mediators of disease induction. Also, delayed apoptosis of neutrophils is associated with lung injury and organ failure under these conditions. Thus, agents that are able to block recruitment and activation of neutrophils and/or to promote apoptosis might provide protection against this disease. Hu et al. hypothesized that curcumin would exert a proapoptotic effect

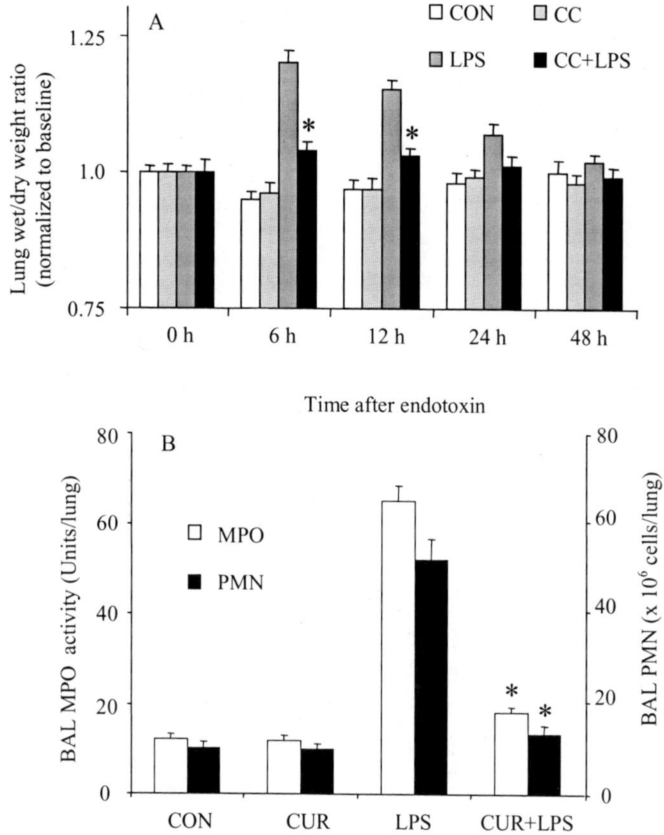

Figure 9. Protective effects of curcumin on endotoxin-induced changes in lung weight (A) and neutrophil cell count and myeloperoxidase activity (B) in mice. Endotoxin injection caused significant increases in the lung wet weight/dry weight ratio, increased neutrophil influx, and myeloperoxidase activity in BAL fluid, whereas curcumin treatment signficantly prevented these effects. Values are mean ± SD of six rats in each group. *Significantly ($p < 0.001$) lower than LPS groups. CON: control; CUR: curcumin; LPS: lipopolysaccharide, BAL: bronchoalveolar lavage; MPO: myeloperoxidase, PMN: polymorphonuclear leukocyte.

on neutrophils.[45] These authors investigated both spontaneous neutrophil apoptosis and apoptosis of neutrophils following transmigration across a human lung endothelium–epithelium bilayer by morphology and terminal dUTP nucleotide end-labeling analyses, respectively. In addition, myeloperoxidase activity and migration assays were carried out to determine the effect of curcumin on neutrophil function. To determine the mechanism by which curcumin acts, they studied the p38 mitogen-activated protein kinase pathway and caspase-3 activity. Their data indicate that curcumin increased constitutive neutrophil apoptosis and attenuated the transbilayer migration-induced delay in neutrophil apoptosis. Curcumin decreased neutrophil activation, as judged by a reduction in migration and myeloperoxidase release. Curcumin induced increases in p38 phosphorylation and caspase-3 activity, contributing to the proapoptotic effect of human neutrophil apoptosis by curcumin. These findings indicate that curcumin might be used as a potential agent to treat neutrophil-induced lung injury and sepsis.

4.2.4. Curcumin Protects Against Cytokine Production in Chronic Inflammation

Certainly, the most compelling data for the protective effect of curcumin in chronic inflammation comes from clinical studies of Literat et al.[46] These investigators demonstrated that curcumin inhibited the LPS-induced increased production of proinflammatory cytokines such as TNF-α, interleukin (IL)-1β, and IL-8 by the lung inflammatory cells isolated from preterm newborns at risk for development of chronic lung disease. Their findings also showed that curcumin inhibited the expression of IL-8 by adult peripheral blood mononuclear cells. Collectively, these data show that curcumin has the potential to be used as an alternative therapeutic agent in inflammatory lung diseases.

Future studies should address whether curcumin is capable of altering lung cellular function leading to decreased production of inflammatory mediators and inhibition of lung injury in experimental models of ALI/ARDS. Also, these studies should investigate whether there is an association between curcumin treatment and inhibition of lung function abnormalities.

4.3. Curcumin Inhibition of Allergic Asthma

4.3.1. Asthma: Disease Mechanisms

Allergic inflammatory diseases such as bronchial asthma is a chronic inflammatory disease of the airways characterized by airway eosinophilia, goblet cell hyperplasia with mucus hypersecretion and hyperresponsiveness to inhaled allergens and to nonspecific stimuli. Clinical features of asthma include intermittent wheezing, dyspnea, cough, and chest tightness. The effector mechanisms in this disease process are less well understood; however, a number of inflammatory and structural cell types, including eosinophils, macrophages, T-lymphocytes, mast cells, epithelial cells, fibroblasts, and smooth muscle cells, have been implicated (Figure 10). Excellent detailed reviews concerning the cellular and molecular pathways leading to allergic asthma has been published previously.[47,48] Eosinophil recruitment and activation in bronchial tissues play a major role in the pathogenesis of bronchial asthma. Numerous studies have shown that eosinophils are able of producing inflammatory mediators, releasing cytotoxic granule constituents, and generating toxic free radicals that might be involved in promoting inflammatory injury in asthma. In this regard, eosinophils have been identified as a key source of transforming growth factor (TGF)-β1, a profibrotic cytokine involved in fibrosis.[49,50] It has been also documented that the cytokine IL-5 plays indispensable roles in controlling eosinophil function by regulating the trafficking of eosinophils from

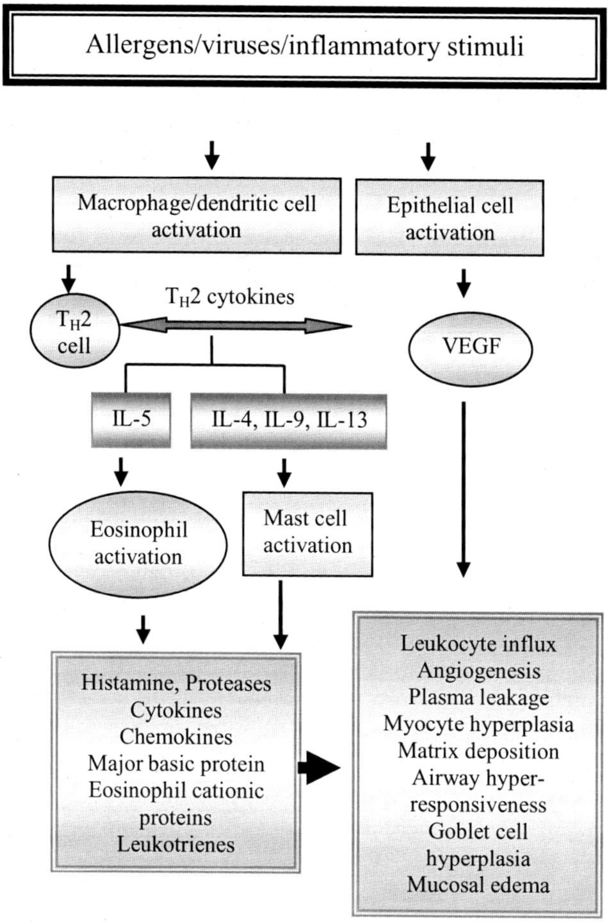

Figure 10. A schematic diagram depicting the mechanisms by which T$_H$2 cells, mast cells, and eosinophils contribute to major pathophysiology of asthma.

bone marrow to the lung. In most eosinophilic disorders, IL-5 contributes to eosinophil proliferation, differentiation, activation, and tissue survival of eosinophils. Indeed, a contribution of IL-5 to the pathological airway has been shown in experimental models of asthma.[51–53] Interestingly, IL-4 has also been implicated in tissue eosinophilia, as it promotes IL-5 production and homing of eosinophils within the tissues. Therefore, antagonizing the effect of these cytokines has been considered a potential new treatment strategy in asthma.[54–57]

4.3.2. Curcumin Inhibition of Cytokine Production by Lymphocytes From Bronchial Asthmatics

Along this line, Kobayashi and colleagues have shown in vitro evidence that curcumin inhibited Dermatophagoides farinea-induced cytokine and chemokine production by lymphocytes from bronchial asthmatics.[58] These investigators provide data indicating that curcumin dose-dependently inhibited IL-5, GM-CSF, and IL-4 production. Overall, their findings suggest that curcumin may have therapeutic potential in controlling allergic diseases by disrupting the production of cytokines that affect eosinophil function and IgE synthesis.

4.3.3. Curcumin Inhibition Of Allergen-induced Asthma

Evidence pointing to the anti-allergic/anti-asthmatic effect of curcumin comes from the studies of Ram et al., (2003) in a guinea pig model of airway hyperresponsiveness.[11] The authors sensitized guinea pigs with ovalbumin (OVA) that manifests particular characteristics of asthma such as allergen induced airway constriction and airway hyperreactivity to histamine. In this study, in order to assess the preventive effect of curcumin, guinea pigs were treated with curcumin during sensizitation and then airway constriction and airway hyperreactivity were determined by measuring specific airway conductance using a non-invasive technique, constant-volume body plethysmography. Their data indicated that curcumin treatment significantly inhibited OVA-induced airway constriction and airway hyperreactivity. Another set of their data indicates that curcumin was also effective in reversing the allergen-impaired airway constriction and hyperreactivity. These results indicate that curcumin exerted both preventive as well as therapeutic efficacy in recovering the impaired airways features in the OVA-sensitized guinea pigs. An apparent question inspired by the findings of Ram et al., (2003) relate to the biochemical and molecular mechanisms that underlie the antiasthmatic effect of curcumin in this experimental model.[11] However, their interesting findings add further support to the notion that curcumin could be a potential treatment for allergic disorders. This study has implications for both curcumin-related response to allergic disease and influence of curcumin on improvement of disease indicators and lung function in an experimental model of asthma.

5. MECHANISMS OF CURCUMIN PROTECTION

Curcumin has been shown for years to possess anti-inflammatory and antioxidant activities,[1,2] and the demonstration that curcumin protects against various lung diseases by participating in multiple steps of disease progression has considerable clinical applications (Tables 2 and 3, Figure 11).[59–64] It seems highly likely that the anti-inflammatory activity of curcumin results from inhibiting enzymes like cyclooxygenase-2, 5-lipoxygenase, and nitric oxide synthase. It also affects arachidonic acid metabolism and prostaglandin production. Numerous *in vitro* and *in vivo* studies have reported that curcumin blocks the activation of proinflammatory transcription factors like NF-κB and c-Jun/AP-1. Also, curcumin

Table 2. Molecular and cellular effects of curcumin on various cultured cells.

CELLS	SPECIES	MECHANISM (REF.)
Peripheral neutrophils	Human	↑ Constitutive neutrophil apoptosis, p38 MAPK phosphorylation and caspase-3 activity ↓ Transbilayer migration-induced delay in neutrophil apoptosis, myeloperoxidase release[45]
Alveolar epithelial cells	Human	↓ H_2O_2- and TNF-α-mediated NF-κB and AP-1 activation and IL-8 release ↑ GSH levels and glutamylcysteine ligase mRNA expression[65]
Airway smooth muscle cells	Human	↓ IL-1β-induced eotaxin and MCP-1 protein expression and IL-17-induced IL-8 production[66,67]
Alveolar macrophages	Rat	↓ LPS and PMA-induced TNF-α, superoxide anion, nitric oxide, hydrogen peroxide production in bleomycin and amiodarone-injured rat lungs[6–8]
Embryonic lung fibroblasts	Human	↓ Silica-induced increases in cyclin D1 and CDK4 expression[68]
Scleroderma lung fibroblast	Human	↑ Apoptosis[24]
Lung fibroblasts	Chinese Hamster	↓ *N*-Methyl-*N*′-nitro-*N*-nitrosoguanidine (MNNG)-induced DNA damage[69]
Lymphocytes from atopic asthmatics	Human	↓ House dust mites-induced lymphocyte proliferation, IL-2, IL-5, GM-CSF and IL-4 production[58]
Lung epithelial cells	Human	↓ Silicon carbide-mediated activation and translocation of NF-κB[70]
	Human	↓ Quartz and titanium dioxide-induced IL-8 production[71]

Table 3. Actions and primary mechanisms underlying pulmonary protection of curcumin.

MECHANISMS	ACTION
Anti-inflammation	COX-2, 5-LOX, and iNOS inhibition Suppression of prostaglandin production Inhibition of cytokine and chemokine expression Induction of heme oxygenase-1
Signal transduction and transcription factors	NF-κB and AP-1 inhibition (reduced nuclear translocation) p38 MAPK inhibition Inhibition of EGR-1 expression Inhibition of JNK1/2 phosphorylation
Antioxidant	Scavenging of reactive oxygen/nitrogen species Inhibition of oxidation of biomolecules Removal of hydrogen and lipid peroxides Increases GSH levels by upregulating glutamylcysteinyl ligase mRNA
Antifibrotic	Inhibition of ECM proteins production and α-SMA expression

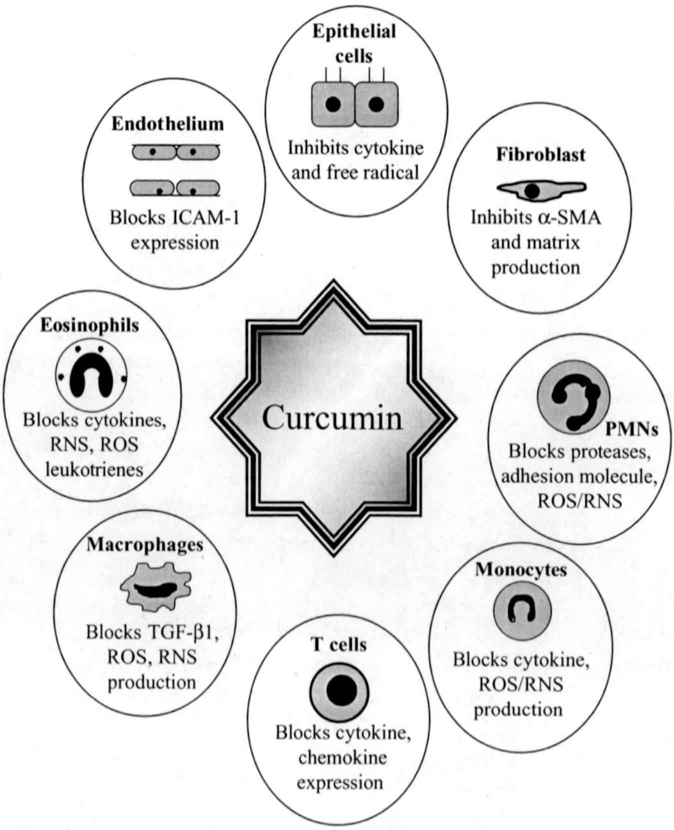

Figure 11. A schematic illustrating the protective effects of curcumin on lung inflammatory and structural cells. ROS: reactive oxygen species, RNS: reactive nitrogen species, ICAM-1: intercellular adhesion molecule-1, α-SMA: α-smooth muscle actin.

was shown to inhibit proinflammatory cytokines and chemokines production by blood monocytes and lung inflammatory cells and also to modulate the expression of adhesion molecules. Additionally, curcumin has the potential to induce heme oxygenase-1, an enzyme that has anti-inflammatory properties. It has been widely reported that neutrophil–endothelial interactions are a prerequisite for the progression of inflammatory response. Accordingly, curcumin inhibition of adhesion molecule expression might prevent inflammation by blocking infiltration of inflammatory cells into the respiratory tract and prove to be beneficial against lung injuries. The mechanism by which curcumin also protects lung appears to be attenuation of the increased oxidant burden in these lung injuries. A role for oxidative stress as important mediators of pathological events in acute and chronic lung diseases has been shown in numerous reports. Evidence available to date indicates the role of curcumin in protecting alveolar macrophages, epithelial cells, and endothelial cells against cytokine and toxic free radicals. Whatever the source of underlying toxicity in both *in vitro* and *in vivo* findings illustrate curcumin as a scavenger of oxygen radical and hydroxyl radical. Importantly, curcumin augments GSH (the most abundant intracellular antioxidant molecule) content in alveolar epithelial cells by upregulation of glutamylcysteine ligase.[65] Thus, curcumin inhibition of increased oxidative stress might modulate the inflammatory response, increase the cellular antioxidant capacity, and influence ECM remodeling in lung diseases.

6. CONCLUSION

In summary, substantial amounts of evidence provide adequate reasons to suggest that, in all probability, curcumin can be considered an alternative nonsteroidal anti-inflammatory drug for inflammatory lung diseases. Studies on experimental animal models clearly demonstrate that curcumin largely prevented the pulmonary dysfunction and attenuated organ damage. Moreover, these findings suggest that it is feasible to develop a curcumin-based therapeutic strategy for human lung diseases, particularly pulmonary fibrosis, through the application of the knowledge acquired from the animal studies using curcumin. Further, the potential use of curcumin in experimental models of other lung diseases such as COPD, ALI, and asthma is obvious. Interestingly, there are certain attributes common to these lung injuries, namely inflammation, oxidative stress, and tissue remodeling, which are important therapeutic targets for curcumin-mediated pulmonary protection. An added advantage of curcumin is that it is nontoxic and a natural product with an excellent safety profile. Unfortunately, curcumin has not yet been tested in human lung diseases. However, in order to make concrete recommendations for the evaluation of curcumin in patients with lung diseases, not only will more studies be warranted to test curcumin at various stages of disease progression, but also its mechanism(s) leading to significant protection still remains to be elucidated. Finally, as our knowledge of curcumin's mechanisms expands, a combination of other treatment approaches could be applied to increase curcumin's efficacy in human lung injuries.

REFERENCES

1. B. B. Aggarwal and S. Shishodia, Suppression of the nuclear factor-kappaB activation pathway by spice-derived phytochemicals, Reasoning for seasoning. *Ann NY Acad Sci* **1030**, 434 (2004).
2. S. Shishodia, S. Gautam, and B. B. Aggarwal, Curcumin: Getting back to the roots. *Ann NY Acad Sci* **1056**, 206 (2005).
3. K. C. Thresiamma, J. George, and R. Kuttan, Protective effect of curcumin, ellagic acid and bixin on radiation induced toxicity. *Indian J Exp Biol* **34**, 845 (1996).

4. N. Venkatesan, Pulmonary protective effects of curcumin against paraquat toxicity *Life Sci* **66**, PL21 (2000).
5. N. Venkatesan and G. Chandrakasan, Modulation of cyclophosphamide-induced early lung injury by curcumin, an anti-inflammatory antioxidant. *Mol Cell Biochem* **142**, 79 (1995).
6. N. Venkatesan, V. Punithavathi, and G. Chandrakasan, Curcumin protects bleomycin-induced lung injury in rats, *Life Sci* **61**, PL51 (1997).
7. D. Punithavathi, N. Venkatesan, and M. Babu, Curcumin inhibition of bleomycin-induced pulmonary fibrosis in rats. *Br J Pharmacol* **131**, 169 (2000).
8. D. Punithavathi, N. Venkatesan, and M. Babu, Protective effects of curcumin against amiodarone-induced pulmonary fibrosis in rats. *Br J Pharmacol* **139**, 1342 (2003).
9. C. Kalpana and V. P. Menon, Modulatory effects of curcumin on lipid peroxidation and antioxidant status during nicotine-induced toxicity. *Pol J Pharmacol* **56**, 581 (2004).
10. B. Madan and B. Ghosh, Diferuloylmethane inhibits neutrophil infiltration and improves survival of mice in high-dose endotoxin shock. *Shock* **19**, 91 (2003).
11. A. Ram, M. Das, and B. Ghosh, Curcumin attenuates allergen-induced airway hyperresponsiveness in sensitized guinea pigs. *Biol Pharm Bull* **26**, 1021 (2003).
12. T. J. Gross and G. W. Hunninghake, Idiopathic pulmonary fibrosis. *N Engl J Med* **345**, 517 (2001).
13. P. Camus, A. Fanton, P. Bonniaud, C. Camus, and P. Foucher, Interstitial lung disease induced by drugs and radiation. *Respiration*. **71**, 301 (2004).
14. J. A. Copper, Jr. Drug-induced lung disease. *Adv Intern Med* **42**, 231 (1977).
15. E. Crouch, Pathobiology of pulmonary fibrosis. *Am J Physiol* **259**, L159 (1990).
16. M. P. Keane, R. M. Strieter, and J. A. Belperio, Mechanisms and mediators of pulmonary fibrosis. *Crit Rev Immunol* 25, 429 (2005).
17. M. Gharaee-Kermani and S. H. Phan, Molecular mechanisms of and possible treatment strategies for idiopathic pulmonary fibrosis. *Curr Pharm Des* **11**, 3943 (2005).
18. D. Bouros and K. M. Antoniou, Current and future therapeutic approaches in idiopathic pulmonary fibrosis. *Eur Respir J* **26**, 693 (2005).
19. R. C. Chambers and G. J. Laurent, Coagulation cascade proteases and tissue fibrosis. *Biochem Soc Trans* **30**, 194 (2002).
20. N. Hashimoto, H. Jin, T. Liu, S. W. Chensue, and S. H. Phan, Bone marrow-derived progenitor cells in pulmonary fibrosis *J Clin Invest* **113**(2), 243–252 (2004).
21. K K. Soudamini, M. C. Unnikrishnan, K. B. Soni, and R. Kuttan, Inhibition of lipid peroxidation and cholesterol levels in mice by curcumin. *Indian J Physiol Pharmacol* **36**, 239 (1992).
22. W. H. Chung, B. M. Bennett, W. J. Racz, J. F. Brien, and T. E. Massey, Induction of c-jun and TGF-β1in Fischer 344 rats during amiodarone-induced pulmonary fibrosis, *Am J Physiol* **281**, L1180 (2001).
23. C. P. Denton and C. M. Black, Targeted therapy comes of age in scleroderma. *Trends Immunol* **26**, 596 (2005).
24. E. Tourkina, P. Gooz, J. C. Oates, A. Ludwicka-Bradley, R. M. Silver, and S. Hoffman, Curcumin-induced apoptosis in scleroderma lung fibroblasts, role of protein kinase cepsilon. *Am J Respir Cell Mol Biol* **31**, 28 (2004).
25. R. M. Senior and S. D. Shapiro, Chronic obstructive pulmonary disease: Epidemiology, pathophysiology, and pathogenesis. In: *Fishman's Pulmonary Diseases and Disorders. Volume 1*, A. P. Fishman et al., eds. New York: McGraw-Hill, 1998, pp. 659–681.
26. P. M. O'Byrne and D. S. Postma, The many faces of airway inflammation: Asthma and chronic obstructive pulmonary disease. *Am J Respir Crit Care Med* **15**, S41 (1999).

27. J. R. Spurzem and S. I. Rennard, Pathogenesis of COPD. *Semin Respir Crit Care Med* **26**, 142 (2005).
28. M. Saetta, Activated T-lymphocytes and macrophages in bronchial mucosa of subjects with chronic bronchitis. *Am Rev Respir Dis* **147**, 301 (1993).
29. M. Saetta, CD8+ve cells in the lungs of smokers with chronic obstructive pulmonary disease. *Am J Respir Crit Care Med* **160**, 711 (1999).
30. J. L. Wright and A. Churg, Animal models of cigarette smoke-induced COPD. *Chest* **122**, 301S (2002).
31. V. L. Kinnula, Focus on antioxidant enzymes and antioxidant strategies in smoking related airway diseases. *Thorax* **60**, 693 (2004).
32. W. Boots, G. R. Haenen, and A. Bast, Oxidant metabolism in chronic obstructive pulmonary disease *Eur Respir J* **46**(Suppl),14 (2003).
33. P. J. Barnes, COPD: Is there light at the end of the tunnel? *Curr Opin Pharmacol* **4**, 263 (2004).
34. C. Kalpana and V. P. Menon, Curcumin ameliorates oxidative stress during nicotine-induced lung toxicity in Wistar rats. *Ital J Biochem* **53**, 82 (2004).
35. S. Shishodia, P. Potdar, C. G. Gairola, and B. B. Aggarwal, Curcumin (diferuloylmethane) down-regulates cigarette smoke-induced NF-κB activation through inhibition of IκBα kinase in human lung epithelial cells: correlation with suppression of COX-2, MMP-9 and cyclin D1. *Carcinogenesis* **24**, 1269 (2003).
36. M. A. Matthay and G. A. Zimmerman, Acute lung injury and the acute respiratory distress syndrome: Four decades of inquiry into pathogenesis and rational management. *Am J Respir Cell Mol Biol* **33**, 319 (2005).
37. M. A. Schwarz, Acute lung injury: cellular mechanisms and derangements. *Paediatr Respir Rev* **2**, 3 (2001).
38. G. J. Bellingan, The pulmonary physician in critical care: The pathogenesis of ALI/ARDS. *Thorax* **57**, 540 (2002).
39. R. Jain and A. DalNogare, Pharmacological therapy for acute respiratory distress syndrome. *Mayo Clin Proc* **81**, 205 (2006).
40. N. R. MacIntyre, Current issues in mechanical ventilation for respiratory failure. *Chest* **128**, 561S (2005).
41. H. Freise, U. B. Bruckner, and H. U. Spiegel, Animal models of sepsis. *J Invest Surg* **14**, 195 (2003)
42. C. W. Chow, M. T. Herrera Abreu, T. Suzuki, and G. P. Downey, Oxidative stress and acute lung injury. *Am J Respir Cell Mol Biol* **29**, 427 (2003).
43. M. P. Fink, Role of reactive oxygen and nitrogen species in acute respiratory distress syndrome. *Curr Opin Crit Care* **8**, 6 (2002).
44. C. M. Doerschuk, W. M. Quinlan, N. A. Doyle, D. C. Bullard, D. Vestweber, M. L. Jones, F. Takei, P. A. Ward, and A. L. Beaudet, The role of P-selectin and ICAM-1 in acute lung injury as determined using blocking antibodies and mutant mice. *J Immunol* **157**, 4609 (1996).
45. M. Hu, Q. Du, I. Vancurova, X. Lin, E. J. Miller, H. H. Simms, and P. Wang, Proapoptotic effect of curcumin on human neutrophils: Activation of the p38 mitogen-activated protein kinase pathway. *Crit Care Med* **33**, 2571 (2005).
46. A. Literat, F. Su, M. Norwicki, M. Durand, R. Ramanathan, C.A. Jones, P. Minoo, and K. Y. Kwong, Regulation of pro-inflammatory cytokine expression by curcumin in hyaline membrane disease (HMD). *Life Sci* **70**, 253 (2001).
47. L. Cohn, J. A. Elias, and G. L. Chupp, Asthma: Mechanisms of disease persistence and progression. *Annu Rev Immunol* **22**, 789 (2004).

48. J. A. Elias, C. G. Lee, T. Zheng, B. Ma, R. J. Homer, and Z. Zhu, New insights into the pathogenesis of asthma. *J Clin Invest* **111**, 291 (2003).
49. B. Kay, S. Phipps, and D. S. Robinson, A role for eosinophils in airway remodelling in asthma. *Trends Immunol* **25**, 477 (2004).
50. G. M. Walsh, M. Al-Rabia, M. G. Blaylock, D. W. Sexton, C. J. Duncan, and A. Lawrie, Control of eosinophil toxicity in the lung. *Curr Drug Targets Inflamm Allergy* **4**, 481 (2005).
51. A. Sher, R. L. Coffman, S. Hieny, P. Scott, and A. W. Cheever, Interleukin-5 is required for the blood and tissue eosinophilia but not granuloma formation induced by infection with *Schistosoma mansoni. Proc Natl Acad Sci USA.* **87**, 61 (1990).
52. T. T. Kung, D. M. Stelts, J. A. Zurcher, G. K. Adams 3rd, R. W. Egan, W. Kreutner, A. S. Watnick, H. Jones, and R. W. Chapman, Involvement of IL-5 in a murine model of allergic pulmonary inflammation, prophylactic and therapeutic effect of an anti-IL-5 antibody. *Am J Respir Cell Mol Biol* **13**, 360 (1995).
53. M. J. Leckie, Anti-interleukin-5 monoclonal antibodies: Preclinical and clinical evidence in asthma models. *Am J Respir Med* **2**, 245 (2003).
54. G. M. Walsh, Novel therapies for asthma: Advances and problems. *Curr Pharm Des* **11**, 3027 (2005).
55. G. Caramori, K. Ito, and I. M. Adcock, Targeting Th2 cells in asthmatic airways. *Curr Drug Targets Inflamm Allergy* **3**(3), 243–255 (2004).
56. M. Ichinose and P. J. Barnes, Cytokine-directed therapy in asthma. *Curr Drug Targets Inflamm Allergy* **3**, 263 (2004).
57. J. E. Pease, Asthma, allergy and chemokines. *Curr Drug Targets* **7**, 3 (2006).
58. T. Kobayashi, S. Hashimoto, and T. Horie, Curcumin inhibition of *Dermatophagoides farinea*-induced interleukin-5 (IL-5) and granulocyte macrophage-colony stimulating factor (GM-CSF) production by lymphocytes from bronchial asthmatics. *Biochem Pharmacol* **54**, 819 (1997).
59. Y. Hu, J. Peng, D. Feng, L. Chu, X. Li, Z. Jin, Z. Lin, and Q. Zeng, Role of extracellular signal-regulated kinase, p38 kinase, and activator protein-1 in transforming growth factor-beta1-induced alpha smooth muscle actin expression in human fetal lung fibroblasts in vitro. *Lung* **84**, 33 (2006).
60. Y. Moon, W. C. Glasgow, and T. E. Eling, Curcumin suppresses interleukin-1β-mediated microsomal prostaglandin E synthase 1 by altering early growth response gene 1 and other signaling pathways. *J Pharmaco. Ex. The.* **315**, 788 (2005).
61. C. T. Shun, S. K. Lin, C. Y. Hong, S. H. Kok, Y. H. Juan, C. C. Wang, M. C. Hsu, C. M. Liu, C.-C., Chemokine ligand 2 gene expression in nasal polyp fibroblasts, possible implication in the pathogenesis of nasal polyposis. *Ann Otol Rhinol Laryngol* **114**, 879 (2005).
62. J. Hong, J., M. Bose, J. Ju, J. H. Ryu, X. Chen, S. Sang, M. J. Lee, and C. S. Yang, Modulation of arachidonic acid metabolism by curcumin and related beta-diketone derivatives: Effects on cytosolic phospholipase A(2), cyclooxygenases and 5-lipoxygenase. *Carcinogenesis* **25**, 1671 (2004).
63. Y. Abe, S. Hashimoto, and T. Horie, Curcumin inhibition of inflammatory cytokine production by human peripheral blood monocytes and alveolar macrophages. *Pharmacol Res* **39**, 41 (1999).
64. T. Yokoyama, H. Oono, A. Miyamoto, S. Ishiguro, and A. Nishio, Magnesium-deficient medium enhances NO production in alveolar macrophages isolated from rats. *Life Sci.* **72**, 1247 (2003).

65. S. K. Biswas, D. McClure, L. A. Jimenez, I. L. Megson, and I. Rahman, Curcumin induces glutathione biosynthesis and inhibits NF-κB activation and interleukin-8 release in alveolar epithelial cells: Mechanism of free radical scavenging activity. *Antioxid Redox Signal.* **7**, 32 (2005).
66. W. A. Wuyts, B. M. Vanaudenaerde, L. J. Dupont, M. G. Demedts, and G. M. Verleden, Involvement of p38 MAPK, JNK, p42/p44 ERK and NF-κB in IL-1β-induced chemokine release in human airway smooth muscle cells. *Respir Med* **97**, 811 (2003).
67. W. A. Wuyts, B. M. Vanaudenaerde, L. J. Dupont, D. E. Van Raemdonck, M. G. Demedts, and G. M. Verleden, Interleukin-17-induced interleukin-8 release in human airway smooth muscle cells: Role for mitogen-activated kinases and nuclear factor-kappaB. *J Heart Lung Transplant* **24**, 875 (2005).
68. F. Shen, X. Fan, B. Liu, X. Jia, H. Du, B. You, M. Ye, C. Huang, and X. Shi, Overexpression of cyclin D1-CDK4 in silica-induced transformed cells is due to activation of ERKs, JNKs/AP-1 pathway. *Toxicol Lett* **160**, 185 (2006).
69. S. Chakraborty, M. Roy, and R. K. Bhattacharya, Prevention and repair of DNA damage by selected phytochemicals as measured by single cell gel electrophoresis. *J Environ Pathol Toxicol Oncol* **23**, 215 (2004).
70. D. M. Brown, D. M., P. H. Beswick, and K. Donaldson, Induction of nuclear translocation of NF-kappaB in epithelial cells by respirable mineral fibres. *J Pathol* **189**, 258 (1999).
71. R. P. Schins, A. McAlinden, W. MacNee, L. A. Jimenez, J. A. Ross, K. Guy, S. P. Faux, and K. Donaldson, Persistent depletion of I kappa B alpha and interleukin-8 expression in human pulmonary epithelial cells exposed to quartz particles. *Toxicol Appl Pharmacol* **167**, 107 (2000).

NEPHROPROTECTIVE AND HEPATOPROTECTIVE EFFECTS OF CURCUMINOIDS

Toshihiko Osawa

Abstract: Curcumin (U1) has a wide spectrum of therapeutic effects such as antitumor and anti-inflammatory effects, including antibacterial, antiviral, antifungal, and antispasmodic activities. By comparison of the structure–activity relationship, tetrahydrocurcumin (THU1), one of the major metabolites, showed the highest antioxidative activity in both *in vitro* and *in vivo* systems.

U1 has been reported to have the nephroprotective effect to improve creatinine and urea clearance and also protected the chronic renal allograft nephropathy. These beneficial effects have been explained by the protection of oxidative stress and the induction of antioxidative enzymes. The protective effect of THU1 against ferric nitrilotriacetate (Fe-NTA)-induced oxidative renal damage using male ddY mice was greater than that of U1, by monitoring not only radical scavenging activity measured by ESR, and TBARS, 4-HNE-modified protein and 8-OHdG formation but also induction of anioxidative enzymes and detoxification enzymes. THU1 was also expected to improve redox regulation through glutathione and suppress the oxidative stress in diabetic nephropathy and neuropathy.

Earlier studies reported that U1 reduced the iron-induced hepatic damage, aflatoxin- and benzo[*a*]pyrene- induced mutagenicity and hepatocarcinogenecity and also the formation of the DNA adduct by inhibiting cytochrome P450 in the liver. The hepatoprotective role of U1 has been examined using carbone terachloride-induced liver damage in rats and alcoholic liver disease model rats, but not examined using THU1. Our recent data suggests that THU1 is a more promising hepatprotective agent because of its strong induction activity of antioxidant and phase 2-metabolizing enzymes in liver compared to kidney, although more detaied examinations are required.

1. INTRODUCTION

The rhizome of turmeric (*Curcuma Longa* Linn.) has been widely used as a yellow coloring agent and spice in many foods, and it has also been used in indigenous medicine.[1] for the treatment of pain and inflammation. Curcumin is the main component of turmeric, and two minor components are also present as

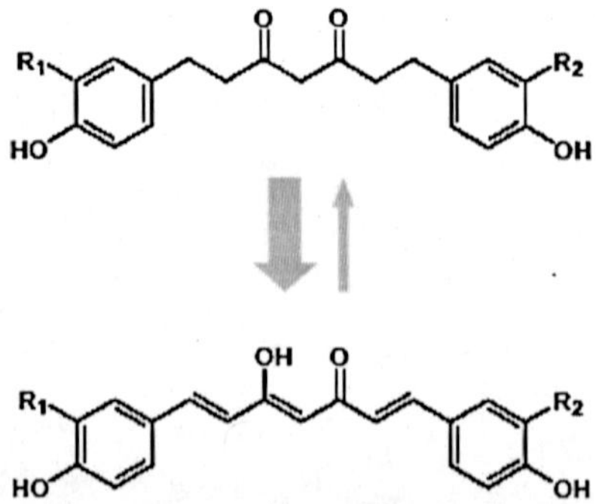

	R_1	R_2
curcumin (U1)	OCH_3	OCH_3
demethoxycurcumin (U2)	H	OCH_3
bisdemethoxycurcumin (U3)	H	H

Figure 1. Structure of curcuminoids present in turmeric.

the curcuminoids; about 70–76% curcumin (U1) is present along with about 16% demethoxycurcumin (U2) and 8% bisdemethoxycurcumin (U3) (Figure 1). Extensive scientific research on U1 have demonstrated a wide spectrum of therapeutic effects such as antitumor and anti-inflammatory effects, including antibacterial, antiviral, antifungal, and antispasmodic activities.[2]

Many studies indicated that compounds that possess antioxidant or anti-inflammatory effect inhibit 12-*O*-tetradecanoyl-phorbol-13-acetate (TPA)-induced tumor promotion in mouse skin. Conney and his colleagues reported that the topical application of U1 inhibited TPA-induced tumor promotions as well as benzo(*a*)pyrene (BP)- and 7,12-dimethyl-benz(*a*)anthracene (DMBA)- induced tumor initiation on mouse skin and these series of reports were the first studies to demonstrate an inhibitory effect of U1 on carcinogenesis.[3] Our research group recently reported that curcuminoids possess antioxidant activity[4] and inhibited the microsome-mediated mutagenicity of BP and DMBA, and it was also reported that U1 acts as a strong inhibitor of tumor promotion; this effect can be explained to roughly parallel the relative antioxidant activity.[5] Extensive studies of the inhibitory effects of dietary U1 on colon, duodenal, stomach, esophageal, and oral carcinogenesis[3] have been carried out using cell lines and it was shown that curcumin suppresses both cyclooxygenase-2 and inducible nitric oxide synthase as potential targets.[6,7] U1 has also been shown to inhibit the activation of the transcription factors activator protein (AP)-1 and nuclear factor (NF)-κB in human leukemia cell lines.[8,9] It was reported that topical U1 application to mouse skin inhibited TPA-mediated activation of ERK and p38 mitogen-activated protein kinase and subsequent activation of NF-κB.[10] These data indicated that U1 might act primarily as an antiinflammatory agent, but further studies with *in vivo* biomarkers are needed.

Although the orally administered U1 has been reported to have poor bioavailability and only low blood levels were observed,[11] it was reported that dietary curcumin inhibited chemically induced skin[12] and liver carcinogenesis.[13] Recently, we reported that oral administration of U1 also inhibited the initiation of radiation-induced mammary and pituitary tumors[14] as well as diethylstilbesterol-induced tumor promotion in the mammary glands of rats initiated with radiation.[15] During this study, we found that U1 was converted to tetrahydrocurcumin (THU1), one of the major metabolites, and THU1 was observed in the serum.[15] To date, not many studies have been carried out on absorption, metabolism, and biological activities other than anti-inflammatory effects. In this chapter, we focus on absorption, metabolism, *in vitro* and *in vivo* biological activities such as antioxidant activity, and nephroprotective and hepatoprotective effects.

2. ABSORPTION AND METABOLISM OF CURCUMINOIDS

Several studies on the absorption and metabolism of U1 have been reported. In an early study that elucidated the metabolic disposition of U1 in rats, U1 labeled with deuterium and tritium was prepared. Oral and intraperitoneal doses of [^{3}H]U1 led to fecal excretion of most of the radioactivity. The major biliary metabolites were glucuronides of THU1 and hexahydrocurcumin.[16] We also made an extensive study on the chemical analyses of the metabolic pathway of U1 after scavenging lipid hydroperoxide using *in vitro* systems[4] and proposed that dihydroferulic acid must be a candidate as the final product during the antioxidative process of THU1 (Figure 2).

A recent study investigated the pharmacokinetic properties of U1 in mice and further clarified the nature of the metabolites of U1.[17] U1 (0.1 g/kg) was administered intraperitoneally to mice and about 2.25 μg/mL were found in the plasma in the first 15 min. Treatment of the plasma with β-glucuronidase was reported to give the free form of THU1 and U1, respectively. The chemical structures of these metabolites, determined by mass spectrometry, suggested that U1 was first biotransformed to dihydrocurcumin and then to THU1 and that these compounds subsequently were converted to monoglucuronide conjugates. The stability of U1 and THU1 at physiological pH was also reported, and it was found that U1 and THU1 were stable at different pHs; in particular, THU1 was very stable in 0.1 M phosphate buffers of various pH values. These results suggest that U1-glucuronoside, dihydrocurcumin-glucuronoside, THU1-glucuronoside, and THU1 are major metabolites of U1 *in vivo*.[17]

In the course of our investigation of the inhibitory effect of diethylstilbesterol-induced tumor promotion in the mammary glands of rats initiated with radiation after the oral administration of U1, we found that THU1 was observed in the serum.[15] As shown in Table 1, nearly 20 times the amount ofTHU1 was found in the serum even after administration of U1.

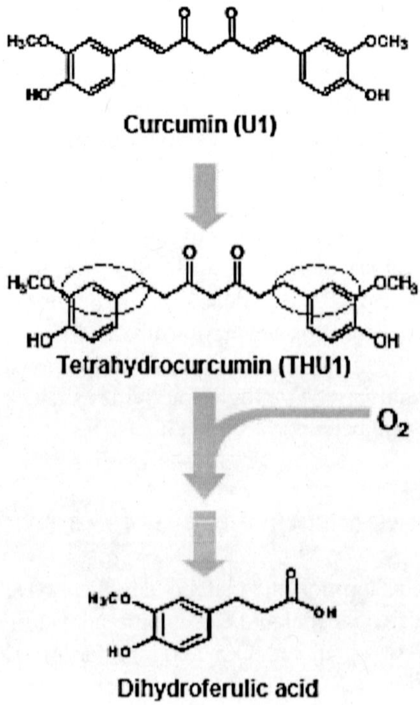

Figure 2. Metabolic pathway of curcumin in the biological system.

We studied extensively the chemopreventive effects of THU1 feeding[18] and observed strong chemopreventive effects of carotenoids such as fucoxanthin, lycopene, and lutein as well as U1 and THU1 on the development of putative preneoplastic aberrant crypt foci in colons of mice initiated with the tumor promoter 1,2-dimethylhydrazine dichloride (DMH).

Table 1. Concentration of U1 and THU1 in the serum after administration of 0.2% of U1 in rats.

SUBSTANCE ASSAYED	CONTROL DIETS[a] ($n = 5$)	CURCUMIN DIET[a] ($n = 5$)
Thiobarbituric acid reactive substances (TBARS) (nmol Malondialdehyde (MDA)/mL)	10.8 ± 0.4^b	$9.3 + 0.5^b$
U1 (ng/mL)	ND[c]	$6.0 + 2.0$
THU1 (ng/mL)	ND[c]	$112.0 + 27.0$

[a]Mean + SE.
[b]Significant difference between control and curcumin diet group, $p < 0.05$.
[c]Not detected.
Source: Ref. 15.

Of the compounds tested, dietary fucoxanthin (0.01% in drinking water), lutein (0.05% in the diet), and THU1 (0.5% in the diet) significantly reduced the number of aberrant crypt foci (ACF) when administered weeks 5 to 12 of the study. A significant inhibition of ACF development in the colons of mice treated with fucoxanthin, lutein, or THU1 when given in the postinitiation phase (tumors were initiated using DMH) was observed. The influence of the proliferation of colonic crypt epithelial cells was also assessed in terms of 5-bromo-2′-deoxyuridine (BrdU) incorporation. BrdU labeling indexes (LI) in mice treated with lutein and 0.5% THU1 was significantly decreased in both the upper-half and lower-half compartments of colonic crypts compared to the controls.

The dose-dependent decreases of BrdU LI observed for lycopene and THU1 indicate that larger doses might be more effective for the inhibition of ACF development. This study demonstrated that THU1 is more active than the parent compound, U1, in terms of the inhibition of ACF development and cell proliferation. This observation, combined with the fact that THU1, which has both phenolic and β-diketone moieties in the same molecule, is a stronger antioxidant,[4,19] suggests that THU1 might be particularly suitable for application as a chemopreventive agent.

3. PREVENTION OF OXIDATIVE STRESS BY CURCUMINOIDS

Excess production of reactive oxygen species (ROS) such as hydroxy radical (·OH) can easily initiate the lipid peroxidation in the cell membranes to form the lipid peroxides. Lipid peroxidation is known to be a free chain reaction, which takes place both *in vivo* and *in vitro*, and forms lipid hydroperoxides and secondary products. These lipid peroxidation products are highly reactive and have been shown to interact with many biological components such as proteins, amino acids, amines, phospholipids, and DNA. Much attention has been focused on the importance of the protective defense systems in living cells against damage caused by ROS and free radicals. Several endogenous antioxidants such as vitamins C and E, carotenoids, uric acid, bilirubin, carnosine, and ubiquinols have been found to play an important role in nonenzymatic protection. In addition to these endogenous defense systems, there is increasing interest in the protective biochemical functions of dietary antioxidants, which are candidates for the prevention of aging-related diseases such as cancer, atherosclerosis, and diabetes mellitus. Many studies have been carried out to establish the ability of U1 to scavenge the hydroxyl radical,[20] superoxide radical,[21] singlet oxygen,[22] nitrogen dioxide,[23] and nitrogen monoxide.[24] It has also been demonstrated that U1 inhibits the generation of the superoxide radical.[25]

From this background, we have been involved in measuring the *in vitro* antioxidant activity of curcuminoids. After spectroscopic analyses of these curcuminoids, the antioxidative activity was evaluated in four *in vitro* antioxidative assay systems.[19] We have evaluated the comparative antioxidant activity of curcuminoids and tetrahydrocurcumin *in vitro* using linoleic acid as the substrate in an

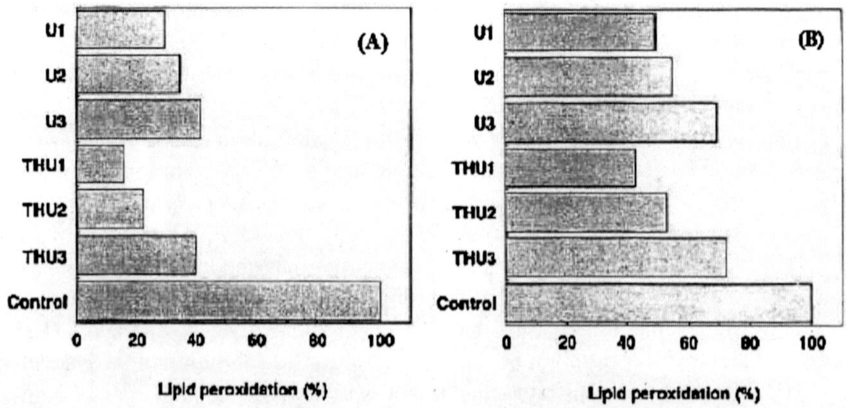

Figure 3. Antioxidative activities of curcuminoids in the rabbit erythrocyte membrane ghost system (A) and in the rat liver microsome system determined by the TBA method (B).

ethanol/water system. The antioxidant activity of curcuminoids in the linoleic acid auto-oxidation model were determined by the thiocyanate method and the 2-thiobarbituric acid (TBA) method, and it was found that THU1 had the strongest antioxidant activity among all of the curcuminoids. We also measured the antioxidative activity using the rabbit erythrocyte membrane ghost (Figure 3A) and rat liver microsome (Figure 3B) determining the 2-thiobarbituric acid-reactive substances (TBARS) formation after inducing lipid peroxidation by *tert*-butylhydroperoxide (*t*-BuOOH), and it was found that THU1 had the strongest antioxidant activity in all of theantioxidant assay systems.

The results demonstrated that the reducing forms of curcuminoids, tetrahydrocurcuminoids (THU1, THU2, and THU3; Figure 4) showed a greater inhibitory effect than the original curcuminoids (U1, U2, and U3), and THU1 showed the strongest antioxidant activity. We concluded that the tetrahydro- form of curcuminoids must scavenge free radicals, such as the *tert*-butoxyl radical and peroxyl radical, efficiently. We attempted to explain the mechanism of the antioxidant action of the tetrahydro- form of curcuminoids, especially, THU1, on the basis of the

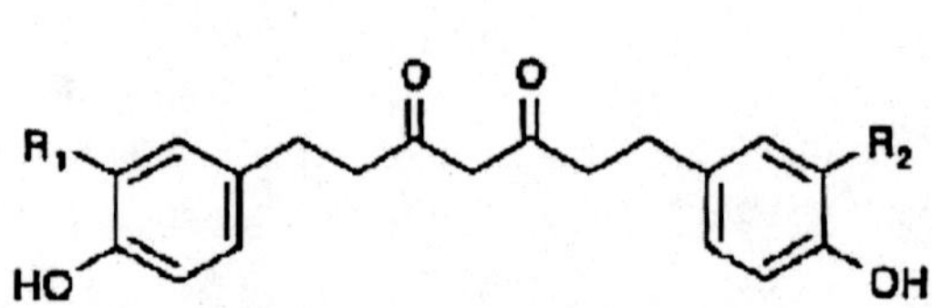

THU1 : R1 = R2 = OCH_3
THU2 : R1 = OCH_3, R2 = H
THU3 : R1 = R2 = H

Figure 4. Structures of the tetrahydro- form of curcuminoids.

molecular structure.[19] We concluded that the β-diketone moiety of the tetrahydro-form of curcuminoids must exhibit antioxidant activity by cleavage of the C—C bond at the active methylene carbon between two carbonyls in the β-diketone moiety. Because THU1 is one of the major metabolites of U1, we propose that this compound might exhibit the observed physiological and pharmacological properties *in vivo* by means of the β-diketone moiety as well as phenolic hydroxyl groups.[4]

More recently, we examined the protective role of THU1 on oxidative stress in cholesterol-fed rabbits,[26] in order to evaluate the antioxidative activity of THU1 *in vivo*. We fed rabbits with diets containing 1% cholesterol with or without 0.5% THU1 and examined their effects on oxidative stress together with the inhibitory effect on atherosclerosis. At the age of 13 weeks, a significantly low amount of TBARS was found in the kidney and liver of in the THU1-treated rabbits. More interestingly, N^{ε}-(hexanoyl)lysine (HEL) in the liver was significantly inhibited in the THU1-treated group compared to the control after 13 weeks, suggesting that the early peroxidation product, HEL, might be a useful biomarker of peroxidation *in vivo*. Other oxidation products, 4-hydroxy-2-nonenal (4-HNE) and dityrosine in the liver, also had a tendency to be inhibited by the THU1 treatment (Table 2).

The levels of THU1 were higher in the liver than the serum, explaining why the HEL concentration is lower in liver, but not in serum, of the THU1-treated group compared with the control. There was also a tendency for the formation of HEL and 4-HNE in kidney to be inhibited in the THU1-treated group.[27] At an early stage of lipid peroxidation, the lipid hydroperoxide is formed

Table 2. HEL, dityrosine and 4-HNE in liver and kidney.

			CONTROL	THU1	P VALUE
HEL					
	Liver	7 weeks	0.268 ± 0.029	0.247 ± 0.030	0.44
		13 weeks	0.343 ± 0.060	0.274 ± 0.039	0.04**
	Kidney	7 weeks	0.300 ± 0.082	0.254± 0.075	0.50
		13 weeks	0.333 ± 0.057	0.289 ± 0.045	0.19
4-HNE					
	Liver	7 weeks	0.330 ± 0.108	0.302 ± 0.084	0.73
		13 weeks	0.438 ± 0.085	0.368 ± 0.036	0.09*
	Kidney	7 weeks	0.310 ± 0.063	0.266 ± 0.096	0.12
		13 weeks	0.375 ± 0.041	0.300 ± 0.072	0.07*
Dityrosine					
	Liver	7 weeks	0.282 ± 0.060	0.245 ± 0.033	0.30
		13 weeks	0.332 ± 0.062	0.283 ± 0.028	0.11
	Kidney	7 weeks	0.302 ± 0.060	0.283 ± 0.020	0.12
		13 weeks	0.334 ± 0.062	0.284 ± 0.056	0.20

Data are expressed as the optical density at 490 nm and the mean ± SD.
$n = 3$ at 7 weeks, $n = 6$ at 13 weeks.
* $p < 0.1$.
** $p < 0.05$

and then decomposed into several aldehydes, including 4-HNE,[28] malondialdehyde (MDA),[29] acrolein,[30] and crotonaldehyde.[31] These reactive aldehydes can easily react with proteins, nucleic acids, and amino-phospholipids, accompanied by stable and unstable adduct formation modification.[32–34] During the course of these studies, the formation of HEL derived from the reaction between the peroxidized lipid and a lysine moiety has been reported.[35] We have identified hexanoylethanolamine in oxidized LDL and erythrocytes.[36] Whereas the mechanism for the formation of the amide-type linkage including the precursor derived from lipid peroxidation is almost unknown, the HEL adduct is considered to be one of the earlier and stable markers for lipid peroxidation-derived protein modification compared to aldehyde-derived protein adducts.[34] The immunochemical evidence of HEL presence in human atherosclerotic plaque,[34] exercised rat tissues,[37,38] lipopolysaccharides-induced liver injury in D-galactosamine-sensitized mice,[39] hepatic ischemia–reperfusion injury in rats,[40] and kidneys in diabetic Akita mice[41] have already been reported. Recently, the chemical identification of HEL *in vivo* has been accomplished by application of the liquid chromatography–mass spectrometry (LC/MS) technique.[42] The oxidation of tyrosine generates a tyrosyl radical, and dityrosine is then formed by the reaction of two tyrosyl radicals. The antidityrosine antibody was developed and it demonstrated that dityrosine can be detected in lipofuscin from the aged human brain[33] and in atherosclerotic lesions in the aorta of Apo-E-deficient mice.[43] These results suggest that dityrosine might also become a useful marker for the estimation of protein cross-linking under oxidative stress.

4. NEPHROPROTECTIVE EFFECTS OF CURCUMINOIDS

Oxidative stress can promote the formation of a variety of vasoactive mediators that can affect renal function directly by causing renal vasoconstriction or decreasing the glomerular capillary ultrafiltration coefficient, thus reducing the glomerular filtration rate.[44] U1 significantly and dose-dependently improved creatinine and urea clearance and decreased the elevated levels of serum creatinine and BUN. Earlier studies have also shown that U1 pretreatment decreases an ischemia–reperfusion-induced rise in serum creatinine levels in the kidney.[45]

Chronic renal allograft nephropathy is associated with both immune and ischemic injury, which might act synergistically to promote an inflammatory response. Nephrotoxicity and hypertension are the major adverse effects that often limit cyclosporine (CsA) treatment following solid-organ transplantation and autoimmune diseases.[46] The functional changes caused by CsA are dose dependent and are usually reversible after short-term CsA treatment.[47] Recently, Jones and Shoskes (2000) examined the possible beneficial effect of U1 in preventing the acute renal failure and related oxidative stress caused by chronic administration of CsA in rats.[48] U1 was administered concurrently with CsA (20 mg/kg/day s.c.)

for 21 days. Oxidative stress in kidney tissue homogenates was estimated using TBARS, reduced glutathione (GSH) content, superoxide dismutase (SOD), and catalase (CAT). Nitrite levels were estimated in serum and tissue homogenates.

Our research group reported the protective effects of U1 and THU1 against ferric nitrilotriacetate (Fe-NTA)-induced oxidative renal damage using male ddY mice.[27] Single Fe-NTA treatment (5 mg Fe/kg body weight intraperitoneally) transiently causes oxidative stress, as shown by the accumulation of lipid peroxidation products and 8-hydroxy-2′-deoxyguanosine (8-OHdG) in the kidney. Mice were fed with a diet containing 0.5 g/100 g U1 or THU1 for 4 weeks. THU1 significantly inhibited TBARS and 4-HNE-modified proteins and 8-OHdG formation in the kidney; U1 inhibited only 4-HNE-modified protein formation. To elucidate the mechanisms of protection by U1 and THU1, the pharmacokinetics and radical-scavenging capacities of U1 and THU1 were investigated by high-performance liquid chromatography (HPLC) and electron spin resonance spin trapping with 5,5-dimethyl-1-pyrroline-*N*-oxide, respectively. Induction of antioxidant enzymes was also investigated. The amounts of THU1 and its conjugates (as sulfates and glucuronides) in the liver and serum were larger in the THU1 group than in the U1 group. The amounts of U1 and its conjugates were small, even in the U1 group. These results suggest that THU1 is more easily absorbed from the gastrointestinal tract than U1. Furthermore, THU1 induced antioxidant enzymes, such as glutathione peroxidase (GPx), glutathione-*S*-transferase (GST), and NADPH:quinone reductase, as well as or better than U1 and scavenged Fe-NTA-induced free radicals *in vitro* better than U1 (Figure 5). From these results, we suggest that U1 is converted to THU1 *in vivo* and that THU1 is a more promising nephroprotective agent.

Protective effects of U1 and THU1 have also been examined on the development of diabetic cataract in 25% galactose-fed SD rats. Both of orally administrated U1 and THU1 (0.2%) prevented cataractogenesis in galactosemic rats effectively, but THU1 showed a greater inhibitory effect than U1. THU1 also showed strong preventive activity using a xylose-induced cataract in cultured monkey lenses (Ueno et al., unpublished data). By the detailed examination of the protective mechanisms of THU1, it was found that THU1 not only scavenges ROS formed during hyperglycemia but also induces antioxidative enzymes (GPx and SOD) significantly. THU1 exhibited no effects on polyol metabolism, but THU1 showed a significant increase of glutathione concentration in the cultured rat lens (Figure 6).

We have already examined the induction mechanisms of GST at the mRNA level, and we now focus on redox regulation by glutathione (γ-glutamylcysteinyl glycine, GSH). GSH is thought to be an important factor in cellular function and defense against oxidative stress, and we found that dietary GSH suppresses oxidative stress *in vivo* in the prevention of diabetic complications such as diabetic nephropathy and neuropathy.[49] Recently, much attention has been focused on the role of oxidative stress as the cause of various forms of tissue damage in patients with diabetes. The aim of our study is to examine the involvement of oxidative stress in the progression of kidney dysfunction and neuropathy in diabetes and to evaluate the potential usefulness of GSH in diabetes. In the present study, we

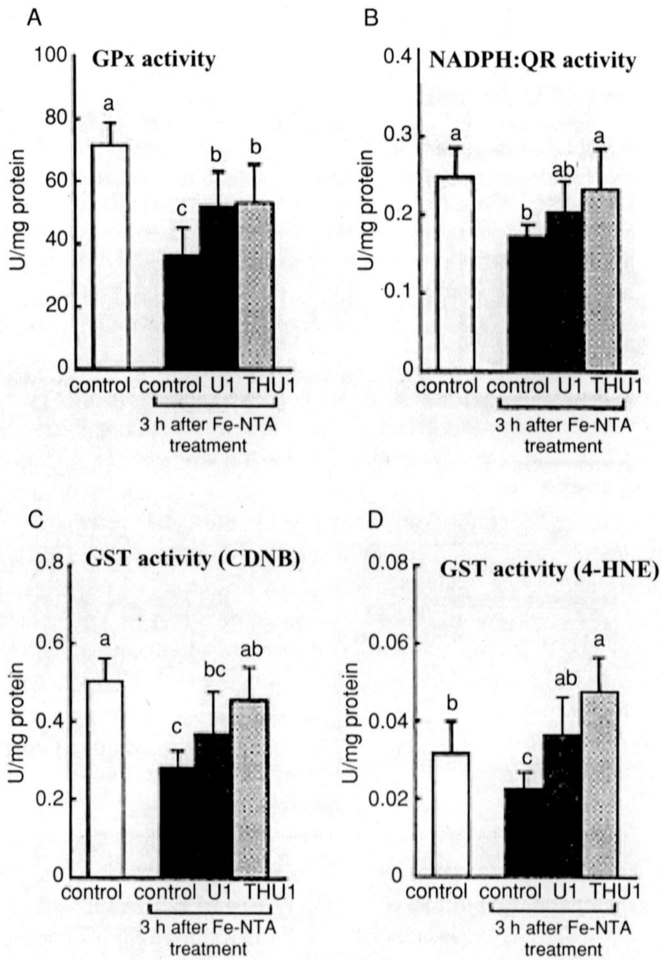

Figure 5. Effect of dietary U1 and THU1 on GPx (A), NADPH:QR (B), and GST (C and D) activities in the kidney of mice treated with Fe-NTA. (C) GST activity toward CDNB and (D) GST activity toward (4-HNE) ($p < 0.05$)

examined the effect that treatment of streptozotocin-induced diabetic rats with GSH has on the renal and neural functions. Diabetic rats were treated with 1% GSH as a diet supplement. The administration of GSH significantly suppressed the diabetes-induced increase in urinary 8-hydroxy-2'-deoxyguanosine, which is one of the markers of oxidative stress. The administration of GSH in diabetic rats significantly prevents the diabetes-induced increase in albumin and creatinine in the urine. The diabetes-induced increase in the tail flick reaction time to thermal stimuli was improved by treatment with GSH. In conclusion, our observations indicate that GSH treatment can exert beneficial effects in diabetes, with preservation of *in vivo*

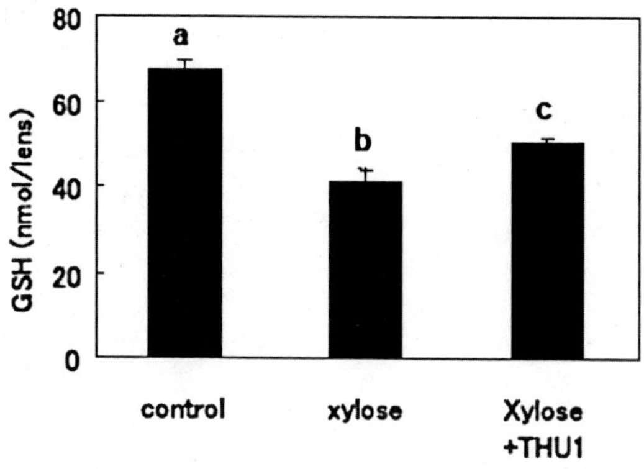

Figure 6. Decrease of glutathione (GSH) in the cultured rat lens with xylose and increase with supplementation with THU1 ($p < 0.05$).

renal and neural function. This finding suggests a potential usefulness of GSH for treating diabetes and provides further support for the implication of oxidative stress in diabetic nephropathy and neuropathy.

In conclusion, it was found that dietary antioxidants such as curcuminoids possess the direct scavenging activity of ROS as the protective mechanism and also induction of antioxidative enzymes, including detoxification enzymes. Furthermore, curcuminoids also have interesting protective role through redox regulation by GSH. Thus, a sufficient supply of dietary antioxidants might prevent or delay diabetes complications, including renal and neural dysfunctions, by providing protection against oxidative stress.

5. HEPATOPROTECTIVE EFFECTS OF CURCUMINOIDS

Earlier studies reported that turmeric and U1 protect the liver against several toxicants both *in vitro* and *in vivo*. Reddy and Lokesh found that oral administration of U1 (30 mg/kg body weight) for 10 days lowered the liver and serum lipid peroxide levels, serum alanine aminotransferase (ALAT), aspartate aminotransferase (ASAT), and lactate dehydrogenase (LDH), enhanced by i.p. injection of iron in rats.[20] This study indicates that U1 reduces the iron-induced hepatic damage by lowering lipid peroxidation.

Soni et al. explored the liver-protective properties of several food additives in inhibiting mutagenesis induced by aflatoxin (AF) B1.[50] In a comparative study which examined the protective effects of various food additives on aflatoxin-induced mutagenicity and hepatocarcinogenicity, U1 was found to offer the best protection. Mutagenesis induced by AFB1 (0.5 mg/plate) in *Salmonella* tester strains TA 98 and TA 100 was 80% inhibited by U1 at concentrations of 2 mg/plate. U1 was found to inhibit the formation of the covalent adduct between AFB1 and DNA, as catalyzed by microsomes or a reconstituted microsomal monooxygenase system in a dose-dependent manner.[51] They reported that a strong affinity of U1 toward cytochromes was further substantiated from the observation that U1-pretreated cytochrome P450 had reduced the ability to catalyze AFB1–DNA adduct formation in the reconstituted system, and they suggested that U1 might inhibit chemical carcinogenesis by modulating cytochrome P450 function.

It was also reported that U1 strongly inhibited the formation of [3H]benzo[*a*]pyrene-derived DNA adduct *in vitro* employing mouse liver S9, compard to other curcuminoids such as U2 and U3.[52] Investigation on the inhibitory effect of curcumin showed a dose-dependent decrease in cytochrome P450 and aryl hydrocarbon hydroxylase (AHH) activity resulting in relatively larger amounts of unmetabolized B(a)P in the presence of curcumin. Thapliyal et al., (2001) reported that curcumin strongly inhibits cytochrome 4501A1/1A2 in the liver.[53] These are isoenzymes involved in the bioactivation of several toxins, including benzo[*a*]pyrene.

Recently, Park et al reported on the protective effects of U1 on acute or subacute carbon-tetrachloride-induced liver damage in rats.[54] Acute hepatotoxicity was induced by intraperitoneal injection of carbon tetrachloride (CCl_4) and subacute hepatotoxicity was induced by oral administration of CCl_4. One hundred-milligrams per kilogram of U1 treatment of rats with acute liver injury by $CCl_4$was found to lower serum ALAT and alkaline phosphatase. U1 was also found to reduce the liver hydroxyproline content and malonaldehyde levels in rats with subacute liver injury, and it was concluded that U1 improved both acute and subacute liver injury induced by CCl_4. More recently, Sugiyama et al. investigated the effects of U1 on hepatic cytochrome P450 (CYP) activity in rats with or without injection of CCl_4, and they speculated that dietary intake of U1 might protect against CCl_4-induced hepatic CYP inactivation via its antioxidant properties, without inducing hepatic CYPs.[55]

Recently, Nanii et al. reported that U1 inhibited alcoholic liver disease (ALD) model rats effectively. Induction of NF-κB-mediated gene expression has been implicated in the pathogenesis of ALD.[56] They determined whether treatment with U1 would prevent experimental ALD and elucidated the underlying mechanism. Four groups of rats (six rats/group) were treated by intragastric infusion for 4 weeks. One group received fish oil plus ethanol (FE); a second group received fish oil plus dextrose (FD). The third and fourth groups received FE or FD supplemented with U1, and liver samples were analyzed for histopathology, lipid peroxidation, NF-κB binding, transforming growth factor (TNF), interleukin (IL)-12, monocyte chemotactic protein-1, macrophage inflammatory protein-2, cyclooxygenase-2 (COX-2), inducible nitric oxide synthase (iNOS), and nitrotyrosine. Rats fed FE developed fatty liver, necrosis, and inflammation, which was accompanied by activation of NF-κB and the induction of cytokines, chemokines, COX-2, iNOS, and nitrotyrosine formation. Treatment with U1 prevented both the pathological and biochemical changes induced by alcohol. Because endotoxin and the Kupffer cell are implicated in the pathogenesis of ALD, they investigated whether curcumin suppressed the stimulatory effects of endotoxin in isolated Kupffer cells. U1 blocked endotoxin-mediated activation of NF-κB and suppressed the expression of cytokines, chemokines, COX-2, and iNOS in Kupffer cells. Thus, U1 prevents experimental ALD, in part by suppressing induction of NF-κB-dependent genes.

Previous studies have shown that U1 causes an increase in GST activity in rodent liver, which might contribute to its anticancer and anti-inflammatory activities. Piper et al. studied the effect of U1 on hepatic nonprotein sulfhydryls and GSH-linked enzymes in male Sprague–Dawley rats, and they found that GST activity toward 4-HNE increased in a saturable, dose-dependent manner.[57] Western blot analyses of liver cytosols revealed that U1 caused a dose dependent induction of rGST 8-8, an isozyme known to display the highest activity toward 4-HNE, a highly toxic product of lipid peroxidation. Iqbal et al. found that dietary supplementation of U1 to male ddY mice significantly increased the activities of GST and quinone reductase to 1.7 and 1.8 times in the liver and 1.1 and 1.3 times in the kidney.[58] In general, the increase in the activities of antioxidant and phase 2-metabolizing enzymes was more pronounced in the liver than in the kidney.

We have also found that THU1 induces antioxidative enzymes such as GPx, GST, and NADPH:quinone reductase, as well as or better than U1, and scavenged Fe-NTA-induced free radicals *in vitro* better than U1.[27] Recently, we also found that THU1 induced phase 2 enzymes in the liver better than U1 (Ueno et al., unpublished data), and it was suggested that U1 is converted to THU1 *in vivo* and that THU1 is a more promising chemopreventive agent.

REFERENCES

1. K. M. Nadkarni, *Curcuma longa*. In: K. M. Narkarni, ed. *India Materia Medica*. Bombay, Popular Prakashan Publishing, 1976, pp. 414–416.
2. K. Kohli, J. Ali, M. J. Ansari, and Z. Raheman, Curcumin: A natural anti-inflmmatory agent. *Ind J Pharmacol* **37,** 141–147 (2005).
3. A. H. Conney, Enzyme induction and dietary chemicals as approaches to cancer chemoprevention: The Seventh DeWitt S. Goodman lecture. *Cancer Res* **63**, 7005–7031 (2003).
4. Y. Sugiyama, S. Kawakishi, and T. Osawa, Involvement of the β-diketone moiety in the antioxidative mechanism of tetrahydrocurcumin. *Biochem Pharmacol* **52**, 519–525 (1996).
5. Y. Nakamura, Y. Ohto, A. Murakami, T. Osawa, and H. Ohigashi, Inhibitory effects of curcumin and tetrahydro- curcuminoids on the tumor promoter-induced reactive oxygen species generation in leukocytes *in vitro* and *in vivo*. *Jpn J Cancer Res* **89**, 361–370 (1998).
6. J. P. Gaddipati, S. V. Sundar, J. Calemine, P. Seth, G. S. Sidhu, and R. K. Maheshwari, Differential regulation of cytokines and transcription factors in liver by curcumin following hemorrhage/resuscitation. *Shock* **19**,150–156 (2003).
7. C. D. Huang, O. Tliba, R. A. Panettieri, Jr., and Y. Amrani Bradykinin induces interleukin-6 production in human airway smooth muscle cells: Modulation by Th2 cytokines and dexamethasone. *Am J Respir Cell Mol Biol* **28,** 330–338 (2003).

8. C. Natarajan and J. J. Bright, Curcumin inhibits experimental allergic encephalomyelitis by blocking IL-12 signaling through Janus kinase-STAT pathway in T lymphocytes. *J Immunol* **168,** 6506–6513 (2002).
9. U. R. Pendurthi, J. T. Williums, and L. V. Rao, Inhibition of tissue factor gene activation in cultured endothelial cells by curcumin. Suppression of activation of transcription factors Egr-1, AP-1 and NF kappa B. *Arterioscler Thromb Vasc Biol* **17,** 3406–3413 (1997).
10. A. Bierhaus, Y. Zhang, P. Quehenberger, T. Luther, M. Haase and M. Muller, The dietary pigment curcumin reduces endothelial tissue factor gene expression by inhibiting binding of AP-1 to the DNA and activation of NF-kappa B. *Thromb Haemost* **77,** 772–782 (1997).
11. S. W. Perkins, R. D. Verschoyle, K. Hill, L. Parveen, M. D. Threadgill, R. A. Sharma, RM. L. Williams, W. P. Steward, and A. J. Gescher, Chemopreventive efficacy and pharmacokinetics of U1 in the min/+ mouse, a model of familial adenomatous polyposis. *Cancer Epidemiol Biomark Prev* **11**, 535–540 (2002).
12. P. Limtrakul, S. Lipigorngoson, O. Namwong, A. Apisariyakul, and F. W. Dunn, Inhibitory effect of dietary curcumin on skin carcinogenesis in mice.*Cancer Lett* **116**, 197–203 (1997).
13. S. W. E. Chuang, M. L. Kuo, C. H. Hsu, C.R. Chen, J. K. Lin, G. M. Lai, C. Y. Hsieh, and A. L. Cheng, Curcumin-containing diet inhibits diethylnitrosamine- induced murine hepatocarcinogenesis. *Carcinogenesis* **21,** 331–335 (2000).
14. H. Inano, M. Onoda, N. Inafuku, M. Kubota, Y. Kamada, and T. Osawa, Potent preventive action of curcumin on radiation-induced initiation of mammary tumorigenesis in rats. *Carcinogenesis* **21**, 1836–1841 (2000).
15. H. Inano, M. Onoda, N. Inafuku, M. Kubota, Y. Kamada, T. Osawa, H. Kobayashi, and K.Wakabayashi, Chemoprevention by curcumin during the promotion stage of tumorigenesis of mammary gland in rats irradiated with gamma-rays. *Carcinogenesis* **20**, 1011–1018 (1999).
16. G. M. Holder, J. L. Plummer, and A. J. Ryan, The metabolism and excretion of curcumin in the rat.*Xenobiotica* **8**, 761–768 (1978).
17. M.-H. Pan, T.-M. Huang, and J.-K. Lin, Biotransformation of curcumin through reduction and glucuronidation in mice. *Drug Metab Dispos*, **27,** 486–494 (1999).
18. J.-M. Kim, D.-J. Arakis, Kim, C.-B. Park, N. Takasuka, H. Baba-Toriyama, T. Ota, Z. Nir, F. Khachik, N. Shimizu, Y. Tanaka, and T. Osawa, Chemopreventive effects of carotenoids and curcumins on mouse colon carcinogenesis after 1,2-dimethylhydrazine Initiation. *Carcinogenesis* **19**, 81–85 (1998).
19. T. Osawa, Y. Sugiyama, M. Inayoshi, and S. Kawakishi, Antioxidative activity of tetrahydrocurcuminoids. *Biosci Biotech Biochem* **59,** 1609–1612 (1995).
20. A. C. Reddy and B. R. Lokesh, Studies on the inhibitory effects of curcumin and eugenol on the formation of reactive oxygen species and the oxidation of ferrous iron.. *Mol Cell Biochem* **137,** 1–8 (1994).
21. N. Sreejayan and M. N. Rao, Free radical scavenging activity of curcuminoids. *Arzneimittelforschung* **46,** 169–171 (1996).
22. C. V. Rao, A. Rivenson, B. Simi, and B. S. Reddy, Chemoprevention of colon carcinogenesis by dietary curcumin, a naturally occurring plant phenolic compound. *Cancer Res* **55,** 259–266 (1995).
23. M. K. Unnikrishnan and M. N. Rao, Curcumin inhibits nitrogen dioxide induced oxidation of hemoglobin. *Mol Cell Biochem* **146,** 35–37 (1995).

24. N. Sreejayan and M. N. Rao, Nitric oxide scavenging by curcuminoids. *J Pharm Pharmacol* **49,** 105–107 (1997).
25. A. J.Ruby, G. Kuttan, K. D. Babu, K. N. Rajasekharan, nd R. Kuttan, Anti-tumour and antioxidant activity of natural curcuminoids. *Cancer Lett* **94,** 79–83 (1995).
26. M. Naito, X. Wu, H. Nomura, M. Kodama, Y. Kato, and T. Osawa, The protective effects of tetrahydrocurcumin on oxidative stress in cholesterol-fed rabbits. *J Atehroscler Thromb* **9**, 243–250 (2002).
27. K. Okada, C. Wanpoengfrakul, T. Tanaka, S. Toyokuni, K. Uchida, and T. Osawa, Curcumin and especially tetrahydricurcumin ameliorate oxidative stress-induced renal injury in mice. *J Nutr* **131**, 2090–2095 (2001).
28. K. Itakura, T. Osawa, and K. Uchida, Structure of a fluorescent compound from 4-hydroxy-2-nonenal and N^{ε}-hippuryllysine: A Model for fluorophores derived from protein modifications by lipid peroxidation. *J Org Chem* **63**, 185–187 (1998).
29. S. Yamada, S. Kumazawa, J. Ishii, T. Nakagawa, K. Itakura, N. Shibata, M. Kobayashi, K. Suzuki, T. Osawa, and K. Uchida, Lipofuscin-like fluorescent pigments derived from malondialdehyde. *J Lipid Res* **42**, 1187–1196 (2001).
30. K. Uchida, M. Kanematsu, K. Sakai, T. Matsuda, N. Hattori, Y. Mizuno, D. Suzuki, T. Miyata, N. Noguchi, E. Niki, and T. Osawa, , Protein-Bound acrolein: Potential markers for oxidative stress. *Proc Natl Acad Sci USA* **95**, 4882–4887 (1998).
31. K. Ichihashi, T. Osawa, S. Toyokuni, and K. Uchida, KEndogenous formation of protein adducts with carcinogenic aldehydes. Implication for oxidative stress. *J Biol Chem* **276**, 23,903–23,913 (2001).
32. Y. Kato, Y. Makino, and T.Osawa, Characterization of a specific polyclonal antibody against 13-hydroperoxyoctadecadienoic acid-modified protein. Formation of lipid hydroperoxide-modified apo B-100 in oxidized LDL.*J Lipid Res* **38**, 1334–1346 (1997).
33. Y. Kato, W. Maruyama, M. Naoi, Y. Hashizume, and T. Osawa, Immunohistochemical detection of dityrosine in lipofuscin pigments in the aged human brain. *FEBS Lett* **439**, 231–234 (1998).
34. Y. Kato, Y. Mori, Y. Morimitsu, S. Hiroi, T. Ishikawa, and T. Osawa, Formation of N^{ε}-(Hexanonyl)lysine in protein exposed to lipid hydroperoxide: A plausible marker for lipid hydroperoxide-derived protein modification. *J Biol Chem* **274**, 20,406–20,414 (1999).
35. H. Esterbauer, R. J. Schaur, and H. Zdlner, Chemistry and biochemistry of 4-hydroxynonenal, malonaldehyde and related aldehydes. *Free Radical. Biol Med* **11**, 81–128 (1991).
36. K. Tsuji, Y. Kawai, Y. Kato, and T. Osawa, Formation of N^{ε}-(hexanoyl)ethanolamine, a novel phosphatidylethanolamine adduct, during the oxidation of erythrocyte membrane and low-density lipoprotein.*Biochem. Biophys Res Commun.* **306**, 706–711 (2003).
37. Y. Kato, X. Wu, M. Naito, H. Nomura, N. Kitamoto, and T. Osawa, Preparation of a monoclonal antibody to N^{ε}-(hexanonyl)lysine: application to the evaluation of protective effects of flavonoid supplementation against exercise-induced oxidative stress in rat skeletal muscle.*Biochem Biophys Res Commun* **274**, 389–393 (2000).
38. K. Minato, Y. Miyake, S. Fukumoto, K. Yamamoto, Y. Shimomura, and T. Osawa, Lemon flavonoid, eriocitrin, suppresses exercise-induced oxidative damage in rat liver. *Life Sci* **72**, 1609–1616 (2003).
39. N. Osakabe, A. Yasuda, M. Natsume, C. Sanbongi, Y. Kato, T. Osawa, and T. Yoshikawa, Rosmarinic acid, a major polyphenolic component of *Perilla Frutescens*, reduces lipopolysaccharide (LPS)-induced liver injury in D-galactosamine (D-GalN)-sensitized mice. *Free Radical Biol Med* **33**, 798–806 (2002).

40. T. Tsuda, F. Horio, Y. Kato, and T. Osawa, Cyanidin 3-*O*-β-D-glucoside attenuates the hepatic ischemia–reperfusion injury through a decrease in the neutrophil chemoattractant production in rats. *J Nutr Sci Vitaminol* **48**, 134–141 (2002).
41. Y. Ueno, F. Horio, K. Uchida, M. Naito, M. Nomura, Y. Kato, T. Tsuda, S. Toyokuni, and T. Osawa, Increase in oxidative stress in kidneys of diabetic Akita mice. *Biosci Biotechno. Bioche.* **66**, 869–872 (2002).
42. Y. Kato, A. Yoshida, M. Naito, Y. Kawai, K. Tsuji, M. Kitamura, N. Kitamoto, and T. Osawa, Identification and Quantification of N^{ε}-(hexanoyl)lysine in human urine by liquid chromatography/tandem mass spectrometry. *Free Radical. Biol Med* **37**, 1864–1874 (2004).
43. Y. Kato, X. Wu, M. Naito, H. Nomura, N. Kitamoto, and T. Osawa, Immunochemical detection of protein dityrosine in atherosclerotic lesion of apo-E-deficient mice using a novel monoclonal antibody. *Biochem Biophys Res Commun* **275**, 11–15 (2000).
44. E. C. Garcia-Cohen, J. Marin, L. D. Diez-Picazo, A. B. Baena, M. Salaices, and M. A. Rodriguez-Martinez, Oxidative stress induced by tert-butyl hydroperoxide causes vasoconstriction in the aorta from hypertensive and aged rats: role of cyclooxygenase-2 isoform. *J Pharmacol Exp Ther* **293,** 75–81 (2000).
45. D. A. Shoskes, Effect of bioflavonoids quercetin and curcumin on ischemic renal injury: A new class of renoprotective agents. *Transplantation* **66**(2)**,** 147–152 (1998).
46. J. Mason, Pharmacology of cyclosporine (sandimmune). VII. Pathophysiology and toxicology of cyclosporine in humans and animals. *Pharmacol Rev* **41,** 423–434 (1990).
47. G. Remuzzi and N. Perico, Cyclosporine-induced renal dysfunction in experimental animals and humans. *Kidney Int* **52**(Suppl), S70–S74 (1995).
48. E. A. Jones and D. A. Shoskes, The effect of mycophenolate mofetil and polyphenolic bioflavonoids on renal ischemia reperfusion injury and repair. *J Urol*, **163**, 999–1004 (2000).
49. Y. Ueno, M. Kizaki, R. Nakagiri, T. Kamiya, H. Sumi, and T. Osawa, Dietary glutathion protects rats from diabetic nephropathy and neuropathy. *J. Nutr* **132**, 897–900 (2002).
50. K. B. Soni, M. Lahiri, P. Chackradeo, S. V. Bhide, and R. Kuttan, Protective effect of food additives on aflatoxin-induced mutagenicity and hepatocarcinogenicity. *Cancer Lett* **115,** 129–133 (1997).
51. P. F. Firozi, V. S. Aboobaker, and R. K. Bhattacharya, Action of curcumin on the cytochrome P450-system catalyzing the activation of aflatoxin B1. *Chem-Biol Interact* **100,** 41–51 (1996).
52. S. S. Deshpande and G. B. Maru, Effects of curcumin on the formation of benzo[a]pyrene derived DNA adducts *in vitro. Cancer Lett* **96**, 71–80 (1995).
53. R. Thapliyal, S. S. Deshpande, and G. B. Maru, Effects of turmeric on the activities of benzo(a)pyrene-induced cytochrome P-450 isozymes. *J Environ Pathol Toxicol Oncol* **20,** 59–63 (2001).
54. E. J. Park, C. H. Jeon, G. Ko, J. Kim, and D. H. Sohn, Protective effect of curcumin in rat liver injury induced by carbon tetrachloride. *J Pharm Pharmacol* **52**, 437–440 (2000).
55. T. Sugiyama, J. Nagata, A.Yamagishi, K. Endoh, M. Saito, K. Yamada, S. Yamada, and K. Umegaki, Selective protection of curcumin against carbon tetrachloride-induced inactivation of hepatic cytochrome P450 isozymes in rats. *Life Sci* **78,** 2188–2193 (2006).
56. A. A. Nanji, K. Jokelainen, G. L. Tipoe, A. Rahemtulla, P. Thomas, and A. J. Dannenberg, Curcumin prevents alcohol-induced liver disease in rats by inhibiting the

expression of NF-kappa B-dependent genes. *Am J Physiol: Gastrointest Liver Physiol* **284(2),** G321–G327 (2003).

57. J. T. Piper, S. S. Singhal, M. S. Salameh, R. T. Torman, Y. C. Awasthi, and S. Awasthi, Mechanisms of anticarcinogenic properties of curcumin: The effect of curcumin on glutathione linked detoxification enzymes in rat liver. *Intl J Biochem Cell Biol* **30,** 445–456 (1998).
58. M. Iqbal, S. D. Sharma, Y. Okazaki, M. Fujisawa, and S. Okada, Dietary supplementation of curcumin enhances antioxidant and phase II metabolizing enzymes in ddY male mice: Possible role in protection against chemical carcinogenesis and toxicity. *Pharmacol Toxicol* **92(1),** 33–38 (2003).

CURCUMIN AND AUTOIMMUNE DISEASE

John J. Bright

Abstract: The immune system has evolved to protect the host from microbial infection; nevertheless, a breakdown in the immune system often results in infection, cancer, and autoimmune diseases. Multiple sclerosis, rheumatoid arthritis, type 1 diabetes, inflammatory bowel disease, myocarditis, thyroiditis, uveitis, systemic lupus erythromatosis, and myasthenia gravis are organ-specific autoimmune diseases that afflict more than 5% of the population worldwide. Although the etiology is not known and a cure is still wanting, the use of herbal and dietary supplements is on the rise in patients with autoimmune diseases, mainly because they are effective, inexpensive, and relatively safe. Curcumin is a polyphenolic compound isolated from the rhizome of the plant *Curcuma longa* that has traditionally been used for pain and wound-healing. Recent studies have shown that curcumin ameliorates multiple sclerosis, rheumatoid arthritis, psoriasis, and inflammatory bowel disease in human or animal models. Curcumin inhibits these autoimmune diseases by regulating inflammatory cytokines such as IL-1β, IL-6, IL-12, TNF-α and IFN-γ and associated JAK-STAT, AP-1, and NF-κB signaling pathways in immune cells. Although the beneficial effects of nutraceuticals are traditionally achieved through dietary consumption at low levels for long periods of time, the use of purified active compounds such as curcumin at higher doses for therapeutic purposes needs extreme caution. A precise understanding of effective dose, safe regiment, and mechanism of action is required for the use of curcumin in the treatment of human autoimmune diseases.

1. INTRODUCTION

The immune system has evolved to discriminate self from non-self antigens, thereby protecting the host from microbial infection and cancer.[1] Nevertheless, a breakdown in the fundamental immune response often results in the development of chronic infectious diseases, malignant tumors, and organ-specific autoimmune diseases. Although the etiology of autoimmune disease is not known, it is generally believed to be mediated by autoimmune cells that are influenced by genetic, environmental, and behavioral factors. Although the induction of an immune response involves the orchestrated interaction of phagocytes and lymphocytes, autoimmune diseases are characterized by deregulated immune responses. Traditionally, these diseases are categorized as either cell mediated, with a particular role for Th1 cells, or humoral, with autoantibodies playing a key role in disease manifestation,[2,3] It is now realized that cytokines, chemokines, adhesion molecules, and other components of inflammatory responses also mediate tissue damage in autoimmune diseases. Despite recent improvements in patient care, a cure for autoimmune disease is still wanting. Human diets of plant origin, containing many hundreds of biologically active compounds called nutraceuticals, appear to play a role in the regulation of immune diseases and maintenance of health. In view of their ability to alleviate pain and inflammation with fewer side effects, the use of herbal medicine and dietary supplements is on the rise in patients with autoimmune diseases. In some cases, the disease process is partially understood, where the elements of protection can be related to a single compound or group of compounds in the diet. These bioactive components are featured with antioxidant, anti-inflammatory, and anticancer properties. Curcumin is a polyphenolic compound isolated from the rhizome of the plant *Curcuma longa,* which has traditionally been used in the treatment of inflammation and cancer. Recent studies have demonstrated promise in the use of curcumin for the treatment of autoimmune diseases.[5] A precise understanding of the effect and mechanism of action of curcumin will help develop new strategies to use it in the treatment of autoimmune diseases.

2. AUTOIMMUNE DISEASES

An immune response is initiated when phagocytic cells, such as macrophage, microglia and dendritic cells, and endocytose foreign antigens, degrade to peptides and present to $CD4^+$ T-lymphocytes in conjunction with major histocompatibility complex (MHC) or human leukocyte antigen (HLA) antigens.[6] The activated antigen-presenting cells (APCs) migrate to the regional lymph node and spleen and encounter naive or memory lymphocytes, leading to differentiation of Th1/Th2 cells, maturation of antigen-reactive B-cells, and migration to inflammatory sites. Whereas immunological memory is the basis for beneficial effects of vaccines, the autoimmune memory B- and T-cells mediate autoimmune diseases. The immune system is highly evolved to react only to foreign antigens but to maintain tolerance to self antigens.[7–9] Despite the highly evolved immune system, the autoimmune cells escape immune tolerance and induce organ-specific autoimmune diseases such as multiple sclerosis, rheumatoid arthritis, type 1 diabetes, inflammatory bowel disease, myocarditis, thyroiditis, uveitis, systemic lupus erythromatosis, and myasthenia gravis, all of which are major health problems throughout the world. The autoimmune disease usually begins in young adulthood and affects women three times more frequently than men.[11] Although the etiology is not known, it is generally believed that genetic, environmental, and behavioral factors influence the pathogenesis of autoimmune diseases. In humans, the most potent genetic contribution to autoimmunity is from alleles of the MHC class II (HLA-DR) locus.[10] Although microbial infection and autoantigens can trigger organ-specific autoimmune diseases, cytokines, chemokines, and signaling molecules in the target organs determine the final outcome of the disease. Thus, the identification of drugs

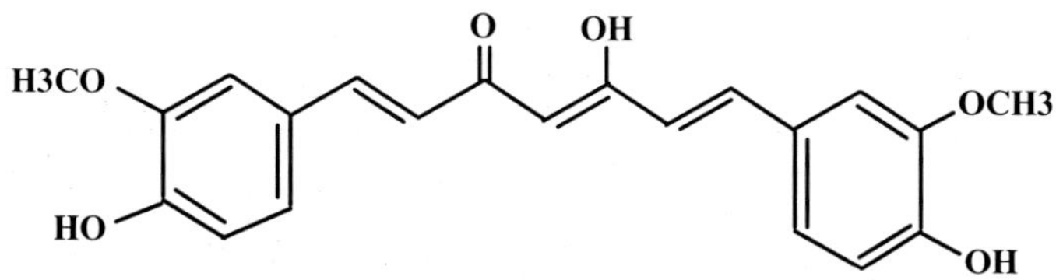

Figure 1. Chemical structure of curcumin: Curcumin [1,7-bis(4-hydroxy-3-methoxy-phenyl)-1,6-heptadiene-3,5-dione] belongs to curcuminoids and its chemical structure is similar to diarylheptanoids.[30,31] The anti-inflammatory activity of curcumin is associated with the hydroxyl and phenol groups. It also has a b-dicarbonylic system with conjugated double bonds and the diene ketone system provides lipophylicity and better penetration.

that regulate autoimmune responses is critical in the treatment of organ-specific autoimmune diseases.

3. CURCUMIN IN THE TREATMENT OF AUTOIMMUNE DISEASES

Traditional medicines have used edible and medicinal plants to treat human diseases in different parts of the world.[12] There is a large variety of phytochemicals that can be extracted and purified from these edible and medicinal plants for the treatment of human diseases.[13]Curcumin (diferuloylmethane) (Figure 1) is a naturally occurring yellow pigment isolated from the rhizomes of the plant *Curcuma longa* (Linn) (turmeric) and is commonly used as a coloring and flavoring agent in food products. Traditional medicine in India and China uses curcumin to treat sprain and swelling caused by injury.[14] The medicinal value of curcumin has been well recognized, as it has profound anti-inflammatory and antitumor activities. *In vivo* treatment with curcumin induces complete protection in chronic and acute models of inflammation.[15–20]Curcumin inhibits reactive oxygen-generating enzymes such as lipoxygenase (LOX), cyclooxygenase (COX), xanthine dehydrogenase, and inducible nitric oxide synthase (iNOS) associated with inflammation.[21,22] Curcumin inhibits lipopolysaccharide (LPS) and interferon (IFN)-γ-induced nitric oxide production in macrophages.[23–25] and protein kinase C (PKC) activation and c-jun expression in fibroblasts.[26] It also inhibits skin inflammation and associated c-Fos and c-Jun expression and hydrogen peroxide formation.[27] Curcumin inhibits COX and LOX activities associated with inflammation *in vivo* and *in vitro*.[28,29] In view of its anti-inflammatory property, we and others have examined the use of curcumin in the treatment of autoimmune diseases (Table 1).

4. CURCUMIN AND MULTIPLE SCLEROSIS

Multiple sclerosis (MS) is an inflammatory demyelinating disease of the central nervous system (CNS) that afflicts millions of people worldwide.[44] About 30% of

Table 1. Effect of curcumin in autoimmune diseases.

DISEASE	EFFECT OF CURCUMIN	REFERENCE
MS	No report on human MS	—
	Ameliorates EAE model of MS in mice	(32)
Type I diabetes	Inhibits glucose in human diabetic patients	(3, 34)
	Protects pancreatic β-cell death	(35)
	No report in animal models of type 1 diabetes	—
RA	Alleviates RA in humans	(36)
	Inhibits RA in animal models	(37)
Psoriasis	Inhibits psoriasis in humans	(38)
	Inhibits psoriasis in animal model	(39)
IBD	Inhibits IBD in humans	(40)
	Inhibits IBD in animal models	(41–43)
Myocarditis	No report in animal or human myocarditis	—
SLE	No report in animal or human SLE	—
Myasthenia gravis	No report	—

MS patients develop clinical paralysis and become wheelchair-bound for the rest of their lives.[45] Although the destruction of the oligodendrocyte myelin sheath in the CNS is the pathological hallmark of MS,[46] axonal degeneration contributes to irreversible long-term disability.[47] Activation of immune cells, secretion of inflammatory cytokines, and differentiation of encephalitogenic Th1 cells are key processes associated with the pathogenesis of MS.[48,49] Immunosuppressive agents have been commonly used to treat MS, but there is no medical treatment available that can cure MS. Experimental allergic encephalomyelitis (EAE) is an autoimmune disease of the CNS. EAE can be induced in susceptible rodents and primates by immunization with whole-brain homogenate or purified neural antigens such as myelin basic protein (MBP), myelin oligodendrocyte glycoprotein (MOG), and proteolipid protein (PLP). The clinical and pathological features of EAE show close similarity to human MS; therefore, EAE has been commonly used as a model system to study the mechanism of MS pathogenesis and to test the efficacy of potential therapeutic agents for the treatment of MS.[4,5,50–53] To test the use of curcumin in the treatment of MS, we examined the protective effect of curcumin on EAE in SJL/J mice. We found that *in vivo* treatment (i.p.) with 50 or 100 μg curcumin every other day decreased the clinical and pathological severity of EAE in SJL/J mice.[5] Curcumin also induced a dose-dependent decrease in neural antigen-induced T-cell proliferation, Th1 differentiation, and IFN-γ production. These results suggest the use of curcumin in the treatment of MS.[5] However, there is no study so far examining the effect of curcumin in human MS. Although daily intake of low doses of curcumin as a dietary beverage might reduce the incidence and severity of autoimmune inflammation, controlled systematic studies in human patients are required before this can be used to treat MS and other human autoimmune diseases.

5. CURCUMIN AND TYPE 1 DIABETES

Type 1 diabetes is an autoimmune disease of the pancreas in which the pancreatic β-cell destruction leads to insulin deficiency in 5–10% of diabetes cases worldwide.[53,54] Patients with type 1 diabetes are also susceptible to other autoimmune conditions, including Hashimoto's thyroiditis, Graves' disease, Addison's disease, coeliac disease, myasthenia gravis, and vitiligo.[55,56] In susceptible individuals, the autoimmune B-cells produce antibodies to β-cell antigens such as insulin (IAA) and glutamic acid decarboxylase (GADA/GAA), and the protein tyrosine phosphatase IA2 (IA-2AA) and autoimmune T-cells mediate inflammation within the islets. Continuing destruction of β-cells leads to progressive loss of insulin reserve and insulin deficiency, resulting in the development of diabetes.[53] Although nutritional planning and insulin pumps help patients manage their disease,[57] a cure for diabetes is still wanting. Earlier studies have shown that dietary curcumin inhibits blood sugar levels in diabetic patients and its animal models.[33,34] Curcumin treatment also inhibits diabetes associated complications such as renal lesion, wound-healing, and cataracts in human patients and animal models.[58–61] The islet β-cells are susceptible to damage caused by oxygen free radicals, and curcumin protects pancreatic β-cells against reactive oxygen species (ROS)-mediated damage by enhancing antioxidants and reduces hyperglycemia in chemically induced diabetes.[35] Although there is no systematic study examining the effect of curcumin on human or animal models of type 1 diabetes, these results suggest the potential use of curcumin in the treatment of type 1 diabetes.

6. CURCUMIN AND RHEUMATOID ARTHRITIS

Rheumatoid arthritis (RA) is a chronic systemic inflammatory condition that affects approximately 0.8% of the population worldwide. RA is characterized by synovitis within diarthrodial joints, and angiogenesis is an important early event. Activated macrophages and dendritic cells are important sources of key proinflammatory cytokines, such as tumor necrosis factor (TNF)-α and interleukin (IL)-1, that promote the accumulation of inflammatory cells and the synthesis of cytokines, chemokines, matrix metalloproteinases (MMPs), COX-2, and other inflammatory mediators. $CD4^+$ T-cells with a Th1 phenotype appear to play a key role in orchestrating the immune response, and B-cells contribute to the ongoing inflammation by activating T-cells and producing potentially pathogenic autoantibodies in RA.[62,63] The primary goals of therapy for RA are relief of pain, reduction of inflammation, preservation of functional status, prevention of complications, and resolution of the pathogenic process. Historically, RA has been managed with nonsteroidal anti-inflammatory drugs (NSAIDs). Howeer, now there are earlier treatments with disease-modifying antirheumatic drugs, including methotrexate, hydroxychloroquine (HCQ), sulfasalazine, and leflunomide.[64] With a better understanding of the immunopathogenesis of the disease, researchers are now investigating other immunomodulatory approaches. Earlier studies have shown that curcumin has antirheumatic activity in humans.[36] Many recent studies have shown that curcumin inhibits RA in association with inhibition of inflammatory cytokines and matrix metalloproteinase by blocking signaling pathways, including mitogen-activated protein kinases (MAPKs), activator protein (AP)-1 and nuclear factor (NF)-κB transcription factors in articular chondrocytes.[65–67] Curcumin also protects human chondrocytes from IL-1β-induced inhibition of collagen type II and β1-integrin expression and activation of caspase-3. Curcumin synergistically potentiates the growth-inhibitory and proapoptotic effects of celecoxib in osteoarthritis synovial adherent cells.[68] Recent studies have also shown that curcumin inhibits disease in an animal model of RA.[37] These reports suggest that curcumin is useful in the treatment of human RA.[69]

7. CURCUMIN AND PSORIASIS

Psoriasis is an autoimmune inflammatory disease of the skin and joints in which intralesional T-lymphocytes trigger primed basal stem keratinocytes to proliferate and perpetuate the disease process.[70–72] Although the self-antigens have not been identified, drugs that regulate complex interactions among susceptibility genes, immunologic effector mechanisms, and environmental triggers that elicit the disease process in skin will prove to be useful in the treatment of psoriasis. Interestingly, recent studies have shown the antipsoriatic actions of curcumin. *In vitro* treatment with curcumin results in a significant decrease in the proliferation of keratinocytes in culture. Topical administration of curcumin inhibits the symptoms of psoriasis in a mouse model, suggesting its use in the treatment of psoriasis.[39] Curcumin-induced suppression of phosphorylase kinase activity correlates with the resolution of human psoriasis as assessed by clinical, histological, and immunohistochemical parameters.[38] Curcumin also inhibited keratinocyte transferrin receptor expression, severity of parakeratosis, and density of epidermal $CD8^+$ T-cells in psoriasis.[38,67,73] These studies suggest that curcumin is useful in the treatment of psoriasis.

8. CURCUMIN AND INFLAMMATORY BOWEL DISEASE

Inflammatory bowel disease (IBD) characterized by Crohn's disease and ulcerative colitis is a common health problem worldwide.[74] Several cytokines, including TNF-α and IL-1β, have been shown to be upregulated in IBD and amplify and perpetuate tissue damage. Furthermore, chemokines are upregulated, thus providing a continuous signal for the influx of leukocytes.[75] Animals with knockouts of inflammatory factors such as IL-2, IL-10, and T-cell receptor serve as models of bowel inflammation.[76]Management of IBD involves the use of immunosuppressives, such as corticosteroids and monoclonal antibodies against TNF-α, which demonstrate clinical efficacy.[77,78] However, these agents are expensive and not without side effects, thus warranting the need for alternative drugs that might be equally or more effective and inexpensive. Interestingly, recent studies have shown the beneficial effects of curcumin in the murine models of IBD.[41–43] The anti-inflammatory effects of curcumin involve a reduction in myeloperoxidase activity, a reduction in the number of infiltrating neutrophils, as well as a reduced expression of IL-1β.[40] A recent study has also shown the effect of curcumin in reducing the clinical symptoms of IBD in human patients, but precisely how curcumin ameliorates IBD is not clear. However, these encouraging studies suggest that curcumin might prove to be an inexpensive, well-tolerated, and effective therapy for the treatment of IBD in human.

9. CURCUMIN AND MYOCARDITIS

Myocarditis is an inflammatory disease of the myocardium.[79] This disease might be idiopathic, infectious, or autoimmune and might heal or lead to dilated cardiomyopathy (DCM), the most common cause of heart failure.[80,81] Thus, in a patient subset, myocarditis and DCM are thought to represent the acute and chronic stages of an organ-specific autoimmune disease of the myocardium. The cardiac autoantibodies are predictive markers of progression to DCM.[82,83] Cellular as well as humoral autoimmune responses are critically associated with the pathogenesis and progression of myocarditis and cardiomyopathy. Animal models greatly advanced our knowledge of the pathogenesis of myocarditis and inflammatory cardiomyopathy. In susceptible mice, for example, infection with enteroviruses results in a biphasic myocarditis, with an early acute stage 5–8 days after inoculation, followed by a chronic stage of low-grade inflammation.[84] Interestingly, T-cells from mice with enteroviral myocarditis transfer the disease into syngeneic severe combined immunodeficiency (SCID) recipients lacking B- and T-cells, which suggests a crucial role for autoreactive T-cells in disease pathogenesis.[85,86] Furthermore, immunization of susceptible mice with α-myosin-derived peptides (MyHCα) results in $CD4^+$ T-cell-mediated experimental autoimmune myocarditis.[87] Cytokines play critical roles in accentuating or regulating autoimmunity; hence, cytokines represent new therapeutic targets in the treatment and prevention of autoimmunity-mediated myocarditis and cardiomyopathy. Curcumin is a potent inhibitor of inflammation and autoimmune diseases[88] and recent studies have shown that curcumin inhibits myocardial inflammation associated with ischemia.[89,90] However, there is no study thus far examining the effect of curcumin in human or animal models of autoimmune myocarditis.

10. CURCUMIN AND SYSTEMIC LUPUS ERYTHROMATOSUS

Systemic lupus erythematosus (SLE) is an autoimmune disease characterized by the presence of multiple autoantibodies that react with various components of the cell nucleus.[91] Specific autoantibodies might correlate with particular organ involvement and prognosis in SLE and related autoimmune conditions. Many autoantibodies, such as SS-A and SS-B and anti-Smith, double-stranded DNA (dsDNA) have been shown to be pathogenic in SLE. A higher titer of dsDNA and its deposition along the glomeruli is associated with active glomerulonephritis, and antiphospholipid antibodies are associated with a hypercoagulable state in SLE.[92,93] Malar rash, discoid rash, photosensitivity, oral ulcers, arthritis, serositis, renal disorder, neuropsychiatric disorder, hematologic disorder, immunologic disorder, and antinuclear antibody are the symptoms of SLE. Therapy for SLE depends on the particular organ system involved. Patients with minor manifestations can often be controlled with low-dose steroids, but moderate and severe disease might require higher doses of steroids or other immunosuppressive agents. A great deal of research is currently ongoing to assess the efficacy of targeting B-cells with specific inhibitors.[94,95] Although curcumin has been known to inhibit B-cell activation in inflammation, there is no study examining the effect of curcumin in human or animal models of SLE.

11. CURCUMIN AND MYASTHENIA GRAVIS

Myesthenia gravis (MG) is an autoantibody-mediated neuromuscular disease.[96] Weakness and fatigability of voluntary muscles characterize both MG and experimental autoimmune MG (EAMG).[96–98] Although the autoantibodies produced by B-cells cause the symptoms of MG, there is ample evidence that T-cells have a key role in the etiopathology of the disease in humans and animals.[99,100] Peptides representing different sequences of the human acetylcholine receptor (AchR) α-subunit or its peptides p195-212 and p259-271 are able to stimulate the peripheral blood lymphocytes (PBL) of patients with MG.[96,97] EAMG can be induced in mice and rats by immunization with AChR in complete Freund's adjuvant.[99,100] The cytokines IFN-γ and IL-12 upregulate and IFN-α downregulates the pathogenesis of EAMG.[101,102] However, the Th2 cytokine IL-4 fails to play a significant role in the development of antibody-mediated EAMG. Antigen-specific tolerance and downregulation of pathogenic cytokines could achieve effective therapy of EAMG and probably MG. Although curcumin has been shown to inhibit inflammatory cytokines, there is no study examining the use of curcumin in the treatment of myasthenia gravis.

12. CURCUMIN REGULATION OF AUTOIMMUNE CELLS

Antigen-presenting cells such as macrophage, microglia, and dendritic cells play a critical role in mediating innate immunity and pathogenesis of autoimmune diseases.[103] Toll-like receptors (TLRs) are a family of cell-surface receptors expressed on APCs that are key components of the innate immune response.[104] When microbial ligands or autoantigens engage a TLR, a cascade of signaling through the NF-κB pathway will be initiated, leading to secretion of inflammatory cytokines, including IL-12, NO, IL-1β, and TNF-α, and surface receptors. CD40

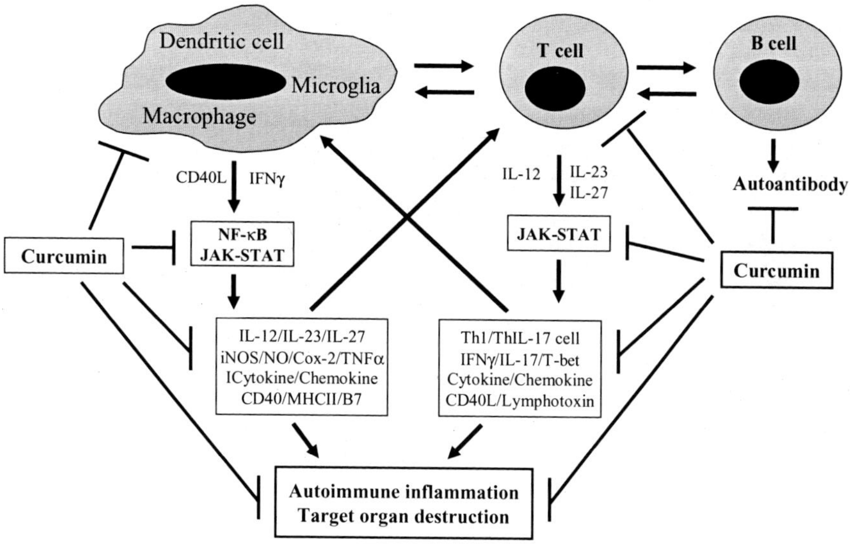

Figure 2. Curcumin regulates autoimmune cells. Upon interaction with autoimmune T-cells, the professional APCs secrete proinflammatory cytokines, which, in turn, induce the differentiation of autoimmune T- and B-cells and secretion of autoantibodies and inflammatory cytokines, resulting in target-organ destruction. Curcumin targets activation and differentiation of APC, T- and B-cells, secretion of inflam-matory cytokines, and target-organ destruction in autoimmune diseases.

is another receptor that activates APCs upon interaction with CD40L expression on Th1 cells, resulting in the secretion of inflammatory cytokines.[105] There is emerging evidence that TLRs and CD40 are involved in the pathophysiology of autoimmune diseases.[104,105] Recent studies have shown that curcumin inhibits lipopolysaccharide (LPS)-induced expression of TLR2 in mouse macrophages[106] and TLR4-dependent chemokine MIP-2 expression in kidney cells.[107]Further investigations are required to define the molecular mechanisms in the regulation of innate immune responses by curcumin in autoimmune diseases.

Among the many lymphocytes, the CD4$^+$ Th1 cells play critical roles in mediating autoimmune diseases, including MS, RA, type 1 diabetes, SLE, myocarditis, thyroiditis, and uveitis.[108] Systematic studies in human and animal models demonstrate that inflammatory cytokines such as IFN-γ and lymphotoxin produced by Th1 cells determine the final outcome of these autoimmune diseases. Alternatively, CD4$^+$ type 2 helper T-cells (Th2) represent an anti-inflammatory population of lymphocytes that produce large amounts of immunoregulatory cytokines (e.g., IL-4 and IL-5).[2] Earlier studies have shown that curcumin inhibits neural antigen-induced Th1 differentiation in an EAE model of multiple sclerosis.[5] Curcumin also inhibits IL-12-induced Th1 differentiation in culture,[5,109] suggesting that the regulation of Th1 differentiation is a mechanism by which curcumin inhibits Th1-cell-mediated autoimmune diseases (Figure 2).

B-Lymphocytes also play a key role in mediating autoimmune responses by producing antibodies, acting as APCs, providing support to other mononuclear cells, and contributing directly to inflammatory pathways. It has long been recognized that auto-antibodies produced by B-cells play a critical role in the pathogenesis of autoimmune diseases such as myasthenia gravis and SLE.[3,110] The autoantibodies present in the cerebrospinal fluid of patients with MS recognize myelin antigens, suggesting the importance of B-cells in the pathogenesis of Th1-cell-mediated autoimmune diseases.[111] These studies provide a rationale for targeting B-cells as a potential therapeutic strategy in autoimmune disorders.[112] Further investigations are required to examine the effect of curcumin on B-cell activation and antibody production in autoimmune diseases.

There are several lymphocyte subsets that can regulate autoimmune diseases. The CD4+ Th2 cells that secrete IL-4 are known to inhibit Th1 responses and Th1-cell-mediated autoimmune diseases. $CD4^+CD25^+$ regulatory T-cells secrete IL-10 and transforming growth factor (TGF)-β and suppress $CD4^+$ and $CD8^+$ T-cell responses in autoimmune diseases.[113,114] NKT cells that express TCR and NKR recognize glycolipids presented in the context of CD1d, resulting in the secretion of IFN-γ and IL-4, and they regulate Th1/Th2 differentiation in autoimmune diseases. Although there is no study examining the effect of curcumin on Treg or NKT cells in autoimmune diseases,[115,116] understanding the effect of curcumin on these regulatory cells is important for determining its mechanism of action in autoimmune diseases.

13. CURCUMIN REGULATES INFLAMMATORY CYTOKINES IN AUTOIMMUNE DISEASES

The organ-specific autoimmune diseases are characterized by the presence of many inflammatory cytokines in the target organs. (See Table 2.) Among the many proinflammatory cytokines, TNF-α, IL-1β, and IL-12 play critical roles in the pathogenesis of autoimmune diseases, whereas anti-inflammatory cytokines such as TGF-β, IFN-α, IL-10, and IL-4 confer recovery. Thus, the amelioration of autoimmune diseases by curcumin might be associated with the inhibition of

Table 2. Curcumin regulates inflammatory cytokines.

AUTOIMMUNE DISEASE	INFLAMMATORY CYTOKINES	REFERENCE
MS	No report	—
EAE	Inhibits IL-12 and IFNγ	(5)
Type I diabetes	No report	—
RA	Inhibits TNFα, IL-1, IL-6, NO, IL-12, IFNγ	(117, 118, 120)
Psoriasis	No report	—
IBD	Inhibits IL-1β, TNFα, enhances IL-10	(24, 121–124)
Myocarditis	No report	—
SLE	No report	—
Myasthenia gravis	No report	—

proinflammatory cytokines or upregulation of anti-inflammatory cytokines. TNF-α is a proinflammatory cytokine produced by macrophages, polymorphonuclear cells, mast cells, NK cells, activated T-cells, and endothelial cells.[117] When stimulated with TNF-α, the target cells produce cytokines, chemokines, iNOS, COX-2, and adhesion molecules.[118] Overproduction of TNF-α is associated with septic shock, MS, psoriasis, RA, and IBD. The importance of TNF-α in inflammation has been demonstrated by the efficacy of TNF-αα-targeting agents in the treatment of autoimmune diseases.[119,120]

Earlier studies have shown that curcumin inhibits TNF-α production and TNF-α-induced responses in immune cells.[24,121–124] IL-β is another proinflammatory cytokine that also plays a key role in mediating cartilage degradation in osteoarthritis (OA) and RA. At the cellular level, IL-β activates matrix-degrading enzymes, downregulates the expression of matrix components, and induces chondrocyte apoptosis.[125] Curcumin inhibits IL-1β secretion from macrophage cells and IL-1β-induced responses in immune cells.[126] Curcumin also inhibits IL-β-induced degenerative changes and caspase-3 activation in human chondrocytes. These studies suggest that curcumin targets proinflammatory cytokines in autoimmune diseases (Table 2).

14. CURCUMIN REGULATION OF IL-12 FAMILY CYTOKINES IN AUTOIMMUNE DISEASES

Interleukin-12, IL-23 and IL-27 are three IL-12 family cytokines produced by macrophage, microglia, and dendritic cells that play critical roles in the pathogenesis of autoimmune diseases. (See Figure 3.) The biologically active IL-12 is a 70-kDa heterodimeric protein composed of covalently linked p35 and p40 subunits.[127] IL-12 induces T-cell proliferation, Th1 differentiation, and pathogenesis of autoimmune diseases. The increased expression of IL-12 in the target and lymphoid organs has been shown in many autoimmune diseases. Interestingly, treatment with neutralizing anti-IL-12 antibodies or agents that inhibit IL-12 production was sufficient to inhibit the pathogenesis of autoimmune diseases.[128–130]

Interleukin-23 is another IL-12 family heterodimeric cytokine composed of a common IL-12 p40 subunit and an IL-23 p19 subunit specific to IL-23.[131] Like IL-12, IL-23 is also secreted by macrophage, microglia, and dendritic cells that induce the differentiation of ThIL-17 from memory T-cells and pathogenesis of autoimmune disease.[132,133] Targeted disruption of IL-23 p19 was effective in preventing the pathogenesis of EAE and suggested that IL-23 plays a critical role in the pathogenesis of EAE.[134] Recent studies have also shown the importance of T-bet transcription factor in the differentiation of Th1 cells.[135] IFN-γ and IL-27 are potent inducers of T-bet in naive T-cells,[136] and targeted disruption or siRNA inhibition of T-bet is sufficient to prevent the pathogenesis of EAE[137,138] IL-27 is a heterodimeric cytokine composed of EBI3 and IL-27 p28 that induces the proliferation of naive $CD4^+$ T-cells.[139] Recent studies have also shown the inhibition of EAE by neutralizing antibodies to IL-27 in mice.[140] Thus, a thorough

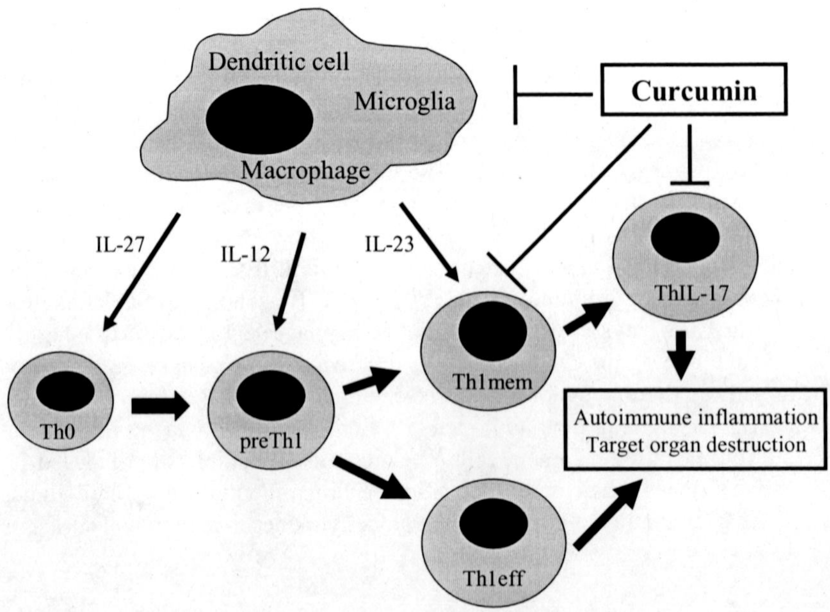

Figure 3. Curcumin inhibits IL-12 in autoimmune diseases. Activation of macrophage, microglia, and dendritic cells through TLR/CD40 leads to secretion of IL-12, IL-23, and IL-27 and differentiation of Th1 and ThIL-17 cells, which mediate autoimmune inflammation and organ destruction. Curcumin inhibits IL-12 family cytokines and prevents organ-specific autoimmune diseases.

understanding of the effect of curcumin on the activation of IL-12/IFN-γ, IL-23/IL-17, and IL-27/T-bet axes is essential to understand its mechanism of action in autoimmune diseases (Figure 3). Because Th2 cytokines such as IL-4, IFN-β, and IL-10 inhibit the activation of macrophage and microglial cells, expression of inflammatory cytokines, and differentiation of Th1 cells, it is important to test the effect of curcumin on these cytokines in autoimmune diseases as well.[108,141]

15. CURCUMIN REGULATION OF NF-κB PATHWAY IN AUTOIMMUNE DISEASES

The IL-12 family cytokines are produced by macrophage, microglia and dendritic cells in response to autoantigens, TLR ligands and CD40 ligands.[127] In earlier studies, we and others have shown that autoimmune cells secrete IL-12 in response to antigens and that this response was inhibited by treatment with curcumin in culture.[5,110] Curcumin also inhibits LPS and CD40L-induced secretion of IL-12 from macrophage, microglia and dendritic cells.[5,110]The induction of IL-12/IL-23 gene expression involves activation of the NF-κB signaling pathway in APCs.[142] NF-κB is a heterodimeric transcription factor composed of p50 and p65 subunits from the Rel family of proteins. It is sequestered in the cytoplasm as an inactive complex when associated with its inhibitor, IκB.

Upon stimulation with specific inducers, IκB becomes phosphorylated and degraded through proteosome-mediated pathways. The activated NF-κB then translocates into the nucleus and binds to specific 10-bp response elements of the IL-12/-23/-27 gene.[142,145] Activation of NF-κB is a complex process involving the successive action of proximal NF-κB-inducing kinase (NIK) and the IκB kinases, IKKα, IKKβ, and IKKγ.[146] The expression of the IL-12 p40 subunit is controlled by proximal cis-acting elements (NF-κB half-site) interacting with NF-κB family members.[147] Inhibitors of IL-12 gene expression, including retinoids, acetyl salicylic acid, and 1,25 dihydroxyvitamin D3, block NF-κB activation and binding within the IL-12p40 promoter.[148,149] Earlier studies have also shown that curcumin inhibits the NF-κB pathway leading to IL-12 gene expression in phagocytic cells,[43,143,] suggesting that the blockade of the NF-κB pathway is a mechanism by which curcumin regulates IL-12 production in autoimmune diseases (Figure 4, Table 3).

16. CURCUMIN REGULATION OF JAK–STAT SIGNALING PATHWAY IN AUTOIMMUNE DISEASES

The antigen-induced proliferation of autoimmune T-cells is a two-step process in which signaling through the T-cell receptor (signal 1) drives T-cells from the resting G0 phase to the activated G1 phase of the cell cycle, whereas signaling through the IL-2 or IL-12 receptor (second signal) is required for T-cells to transit from the G1 phase to the S/G2/M phase of the cell cycle (proliferation). IL-12 is a potent inducer of G1 to S/G2/M phase transition and differentiation of Th1 cells that are critical in the pathogenesis of EAE. IL-12 signals through IL-12 receptor β1 and β2, members of the gp130 cytokine receptor super-family, expressed primarily on activated NK cells and T-cells. Coexpression of IL-12Rβ1 and Il-12Rβ2 leads to the formation of high-affinity IL-12 receptors.[127] Signaling through its receptor, IL-12 induces tyrosine phosphorylation and activation of JAK2, TYK2, STAT3, and STAT4 in T-cells and NK cells.[150,151] Activation of the JAK–STAT pathway leads to the transcription of IL-12 response genes associated with proliferation, Th1 differentiation, and IFN-γ production. IL-23 receptor is composed of a common IL-12Rβ1 and a specific IL-23 receptor subunit.[152] Signaling through its receptor, IL-23 induces the activation of JAK2, TYK2, STAT1, STAT3, STAT4, and STAT5 in T-cells.[152] Activation of the JAK–STAT pathway leads to transcription of IL-23 response genes, including IL-17, which are associated with the proliferation of memory T-cells,[153] whereas IL-27 and IFN-γ activate a specific JAK–STAT pathway in T-cells, resulting in the induction of T-bet in naive T-cells.[154] Modulation of cytokine signaling by targeting protein tyrosine kinases or transcription factors has been considered a novel strategy for the treatment of autoimmune diseases.[155,156] We have shown earlier that the blockade

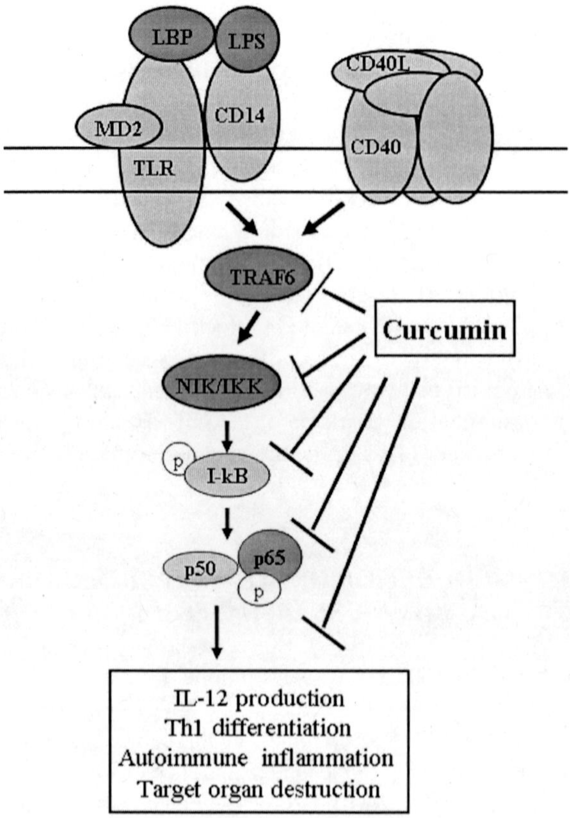

Figure 4. Curcumin targets the NF-κB pathway in autoimmune disease. The activation of the NF-κB pathway through TLR/CD40 involving NIK/IKK-mediated phosphorylation of IκB leads to expression of IL-12 and other inflammatory cytokines and pathogenesis of autoimmune diseases. Curcumin ameliorates autoimmune diseases by targeting the NF-κB signaling pathway, leading to the secretion of IL-12 and other inflammatory cytokines.

Table 3. Curcumin targets inflammatory signaling molecules in autoimmune diseases.

AUTOIMMUNE DISEASE	INFLAMMATORY CYTOKINES	REFERENCE
MS	No report	—
EAE	Inhibits IL-12-induced JAK- STAT pathway	(5)
Type I diabetes	No report	—
RA	Inhibits AP-1, ERK, p38, JNK, MAPK, NF-κB	(43,143, 144)
IBD	Inhibits NF-κB, p38 MAPK	(43)
Psoriasis	Inhibits AP-1, Lipoxigenase, Phosphorylase kinase	(38, 39)
Myocarditis	No report	—
SLE	No report	—
Myasthenia gravis	No report	—

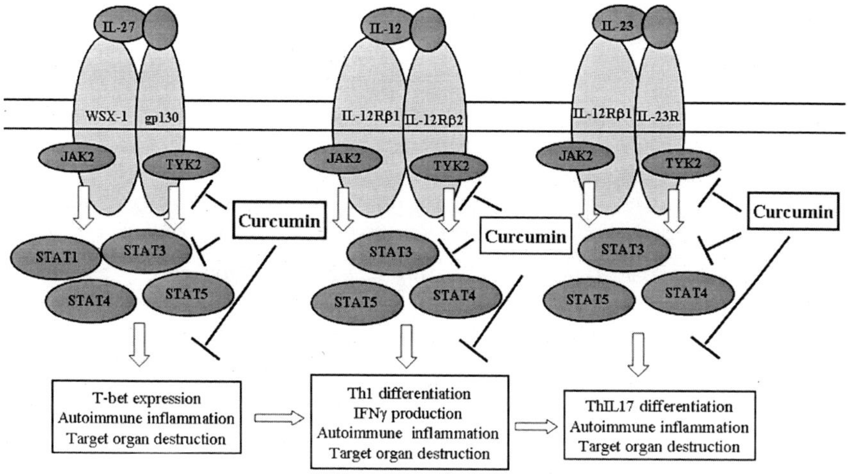

Figure 5. Curcumin regulates IL-12 signaling in T-cells. The IL-12 family cytokines, IL-12, IL-23, and IL-27, signals through the JAK–STAT pathway, leading to induction of IFN-γ/T-bet/IL-17 genes associated with Th1 and ThIL-17 differentiation, autoimmune inflammation, and target-organ destruction. Curcumin targets IL-12 signaling through the JAK–STAT pathway, leading to Th1 differentiation in autoimmune diseases.

of IL-12 signaling through the JAK–STAT pathway by treatment with a JAK-2 inhibitor, tyrphostin AG490, peroxisome proliferator-activated receptor-γ (PPARγ) ligands, Quercetin, and vitamin D inhibit Th1 differentiation and pathogenesis of EAE.[5,51,157–159] We have also shown recently that curcumin inhibits IL-12-induced tyrosine phosphorylation of JAK2, TYK2, STAT3, and STAT4 in T-cells, differentiation of Th1 cells, and pathogenesis of EAE.[5,157] These findings suggest that IL-12 signaling through the JAK–STAT pathway is a molecular target in the regulation of autoimmune diseases by curcumin (Figure 5, Table 3).

17. CURCUMIN REGULATION OF MAPK AND AKT PATHWAYS IN AUTOIMMUNE DISEASES

MAPK and AKT and other immune signaling pathways also play critical roles in the pathogenesis of autoimmune diseases. It was interesting to note that curcumin attenuates inflammatory activity in association with the inhibition of MAPKs in an experimental model of IBD. More recent work has demonstrated that the use of p38 MAPK inhibitors can be effective for human autoimmune diseases. Thus, it was anticipated that the inflammatory response might entail activation of p38 MAPK in the infiltrating immune cells.[160] The exact role played by activated p38 MAPKs at these sites remains unclear, but it might involve secretion of chemokines, neuropeptides and other trophic factors. Similarly, activation of the PI3K–AKT signaling pathway has been linked to macrophage activation and inflammation.[161] Thus, the inhibition of MAPK and AKT might be other molecular targets in the amelioration of autoimmune diseases by curcumin (Table 3).[162]

18. CURCUMIN REGULATES CHEMOKINES IN AUTOIMMUNE DISEASES

Chemokines are small heparin-binding proteins that promote the movement of circulating leukocytes to sites of inflammation and injury throughout the body and play crucial roles in mediating adaptive immune responses and pathogenesis of a variety of autoimmune diseases. There are approximately 50 human chemokines grouped into 4 families on the basis of differences in structure and function.[163] The largest family consists of CC chemokines, which attract mononuclear cells to sites of chronic inflammation. The most studied CC chemokine, monocyte chemoattractant protein 1 (MCP-1), is a potent agonist for monocytes, dendritic cells, memory T-cells, and basophils. Other CC chemokines include macrophage inflammatory protein (MIP)-1α (CCL3), MIP-1β (CCL4), and RANTES (CCL5). IL-8 (CXCL8) is the prototypic CXC chemokine that attracts polymorphonuclear leukocytes to sites of acute inflammation. CXCL8 also activates monocytes and might direct the recruitment of these cells to vascular lesions. Chemokines affect cells by activating surface receptors that are seven-transmembrane-domain G-protein-coupled receptors. The binding of the chemokine to the receptor activates signaling cascades that culminate in the rearrangement, change of shape, and cell movement of actin. Chemokine receptors are important drug targets that regulate inflammation and antoimmunity.[164,163]There are reports showing the inhibition of chemokine and chemokine receptors by curcumin in immune cells,[166] suggesting this is a molecular target in the regulation of autoimmune diseases by curcumin.

19. THERAPEUTIC POTENTIAL AND FUTURE PROSPECTS OF CURCUMIN IN AUTOIMMUNE DISEASES

The ability of herbal medicines and dietary supplements to alleviate pain and clinical symptoms has led to an increased use of complementary and alternative medicine in patients with autoimmune diseases. Although many more new compounds are being isolated and tested for their anti-inflammatory and anticancer properties, these nutraceuticals are of considerable interest because they are effective, inexpensive, and relatively safe. The protective effect of curcumin in the treatment of autoimmune disease has been proven, but its use in the treatment of many human autoimmune diseases is yet to be determined. With the available information, it is difficult to predict the type of dietary modifications containing curcumin that can better reduce the risk, incidence, or severity of autoimmune diseases. Although it is generally believed that the nutraceuticals induce minor side effects during dietary consumption at low levels, one needs to be extremely cautious about the use of purified active compounds such as curcumin at higher doses for therapeutic purposes. Much work needs to be done to determine the effective dose, safe regimens, and molecular mechanisms of action before these nutraceuticals can be used for the treatment of human autoimmune diseases. We believe that curcumin ameliorates autoimmune diseases by inhibiting proinflammatory responses or by enhancing anti-inflammatory responses in the target organs. Interestingly, the research on nutraceuticals has now taken a new dimension that will unravel many unanswered questions on the use of curcumin in the treatment of organ-specific autoimmune diseases.

20. CONCLUSION

Autoimmune diseases such as multiple sclerosis, rheumatoid arthritis, type 1 diabetes, inflammatory bowel disease, myocarditis, thyroiditis, uveitis, systemic lupus erythromatosis, and myasthenia gravis are common health problems affecting more than 5% of the population worldwide. Although the etiologies are not known, these autoimmune diseases manifest following deregulated activation of self-reactive immune cells influenced by genetic, environmental, and behavioral factors. Curcumin is a polyphenolic compound isolated from the rhizome of the plant *Curcuma longa* that has traditionally been used in the treatment of inflammation and cancer. Recent studies have shown promise in the use of curcumin to treat autoimmune diseases.

REFERENCES

1. C. A. Janeway, Jr., The immune system evolved to discriminate infectious nonself from noninfectious self. *Immunol Today* **13**, 11 (1992).
2. I. J. Crane and J. V. Forrester, Th1 and Th2 lymphocytes in autoimmune disease. *Crit Rev Immunol* **25**, 75 (2005).
3. T. Tsubata, B cell abnormality and autoimmune disorders. *Autoimmunity* **38**, 331 (2005).
4. J. J. Bright, C. Du, M. Coon, S. Sriram, and S. J. Klaus, Prevention of experimental allergic encephalomyelitis via inhibition of IL-12 signaling and IL-12-mediated Th1 differentiation: An effect of the novel anti-inflammatory drug lisofylline. *J Immunol* **161**, 7015 (1998).
5. C. Natarajan and J. J. Bright, Curcumin inhibits experimental allergic encephalomyelitis by blocking IL-12 signaling through Janus kinase-STAT pathway in T lymphocytes. *J Immunol* **168**, 6506 (2002).
6. P. Friedl, A. T. den Boer, and M. Gunzer, Tuning immune responses: Diversity and adaptation of the immunological synapse. *Nat Rev Immunol* **5**, 532 (2005).
7. M. Kronenberg, Self-tolerance and autoimmunity. *Cell* **65**, 537 (1991).
8. S. Anderton, C. Burkhart, B. Metzler, and D. Wraith, Mechanisms of central and peripheral T-cell tolerance: Lessons from experimental models of multiple sclerosis. *Immunol Rev* **169**, 123 (1999).

9. A. W. Goldrath and S. M. Hedrick, Central tolerance matters.[comment]. *Immunity* **23**, 113 (2005).
10. E. Thorsby and B. A. Lie, HLA associated genetic predisposition to autoimmune diseases: Genes involved and possible mechanisms. *Transplant Immunol* **14**, 175 (2005).
11. S. G. Sukkar and E. Rossi, Oxidative stress and nutritional prevention in autoimmune rheumatic diseases. *Autoimmun Rev* **3**, 199 (2004).
12. S. M. Rates, Plants as source of drugs. *Toxicon* **39**, 603 (2001).
13. M. M. Chan, C. T. Ho, and H. I. Huang, Effects of three dietary phytochemicals from tea, rosemary and turmeric on inflammation-induced nitrite production. *Cancer Lett* **96**, 23 (1995).
14. Y. Surh, Molecular mechanisms of chemopreventive effects of selected dietary and medicinal phenolic substances. *Mutat Res.* **428**, 305 (1999).
15. R. B. Arora, V. Kapoor, N. Basu, and A. P. Jain, Anti-inflammatory studies on Curcuma longa (turmeric). *Indian J Med Res.* **59,** 1289 (1971).
16. D. Chandra and S. S. Gupta, Anti-inflammatory and anti-arthritic activity of volatile oil of Curcuma longa (Haldi). *Indian J Med Res* **60**, 138 (1972).
17. N. Ghatak and N. Basu, Sodium curcuminate as an effective anti-inflammatory agent. *Indian J Exp Biol* **10**, 235 (1972).
18. A. Mukhopadhyay, N. Basu, N. Ghatak, and P. K. Gujral, Anti-inflammatory and irritant activities of curcumin analogues in rats. *Agents Actions* **12**, 508 (1982).
19. R. C. Srimal and B. N. Dhawan, Pharmacology of diferuloyl methane (curcumin), a non-steroidal anti-inflammatory agent. *J Pharm Pharmacol* **25**, 447 (1973).
20. H. P. Ammon, H. Safayhi, T. Mack, and J. Sabieraj, Mechanism of antiinflammatory actions of curcumine and boswellic acids. *J Ethanopharmacol* **38**, 113 (1993).
21. A. C. Reddy and B. R. Lokesh, Studies on spice principles as antioxidants in the inhibition of lipid peroxidation of rat liver microsomes, *Mol Cell Biochem.* **111**, 117 (1992).
22. M. N.Sreejayan Rao, Curcuminoids as potent inhibitors of lipid peroxidation. *J Pharm Pharmacol* **46**, 1013 (1994).
23. I. Brouet and H. Ohshima, Curcumin, an anti-tumour promoter and anti-inflammatory agent, inhibits induction of nitric oxide synthase in activated macrophages. *Biochem Biophys Res Commun* **206**, 533 (1995).
24. M. M. Chan, H. I. Huang, M. R. Fenton, and D. Fong, In vivo inhibition of nitric oxide synthase gene expression by curcumin, a cancer preventive natural product with anti-inflammatory properties. *Biochem Pharmacol* **55**, 1955 (1998).
25. F. Zhang, N. K. Altorki, J. R. Mestre, K. Subbaramaiah, and A. J. Dannenberg, Curcumin inhibits cyclooxygenase-2 transcription in bile acid- and phorbol ester-treated human gastrointestinal epithelial cells. *Carcinogenesis* **20**, 445 (1999).
26. J. Y. Liu, S. J. Lin, and J. K. Lin, Inhibitory effects of curcumin on protein kinase C activity induced by 12-*O*-tetradecanoyl-phorbol-13-acetate in NIH 3T3 cells. *Carcinogenesis* **14,** 857 (1993).
27. S. S. Kakar and D. Roy, Curcumin inhibits TPA induced expression of c-fos, c-jun and c-myc proto-oncogenes messenger RNAs in mouse skin. *Cancer Lett* **87**, 85 (1994).
28. A. H. Conney, T. Lysz, T. Ferraro, T. F. Abidi, P. S. Manchand, J. D. Laskin, and M. T. Huang, Inhibitory effect of curcumin and some related dietary compounds on tumor promotion and arachidonic acid metabolism in mouse skin. *Adv Enzyme Regul* **31**, 385 (1991).
29. Y. J. Surh, K. S. Chun, H. H. Cha, S. S. Han, Y. S. Keum, K. K. Park, and S. S. Lee, Molecular mechanisms underlying chemopreventive activities of anti-inflammatory

phytochemicals: Down-regulation of COX-2 and iNOS through suppression of NF-kappa B activation. *Mutat Res* **480–481**, 243 (2001).

30. P. Claeson, U. Pongprayoon, T. Sematong, P. Tuchinada, V. Reutrakul, P. Soontornsaratune, an W. C. Taylor, Non-phenolic linear diarylheptanoids from *Curcuma xanthorrhiza*. A novel type of topical anti-inflammatory agents: structure–activity relationship. *Planta Med* **62**, 236 (1996).
31. P. Venkatesan and M. N. Rao, Structure–activity relationships for the inhibition of lipid peroxidation and the scavenging of free radicals by synthetic symmetrical curcumin analogues. *J Pharm Pharmacol* **52**, 1123 (2000).
32. C. Natarajan and J. J. Bright, Peroxisome proliferator-activated receptor-gamma agonists inhibit experimental allergic encephalomyelitis by blocking IL-12 production, IL-12 signaling and Th1 differentiation. *Genes Immun* **3**, 59 (2002).
33. M. Srinivasan, Effect of curcumin on blood sugar as seen in a diabetic subject. *Indian J Med Sci* **26**, 269 (1972).
34. P. S. Babu and K. Srinivasan, Influence of dietary curcumin and cholesterol on the progression of experimentally induced diabetes in albino rat. *Mol Cell Biochem* **152**, 13 (1995).
35. A. Srivivasan, V. P. Menon, V. Periaswamy, and K. N. Rajasekaran, Protection of pancreatic beta-cell by the potential antioxidant bis-*o*-hydroxycinnamoyl methane, analogue of natural curcuminoid in experimental diabetes. *J Pharm Pharm Sci* **6**, 327 (2003).
36. S. D. Deodhar, R. Sethi, and R. C. Srimal, Preliminary study on antirheumatic activity of curcumin (diferuloyl methane). *Indian J Med Res* **71**, 632 (1980).
37. J. L. Funk, J. N. Oyarzo, J. B. Frye, G. Chen, R. C. Lantz, S. D. Jolad, A. M. Solyom, and B. N. Timmermann, Turmeric extracts containing curcuminoids prevent experimental rheumatoid arthritis. *J Nat Prod* **69**, 351 (2006).
38. M. C. Heng, M. K. Song, J. Harker, and M. K. Heng, Drug-induced suppression of phosphorylase kinase activity correlates with resolution of psoriasis as assessed by clinical, histological and immunohistochemical parameters. *Br J Dermatol* **143**, 937 (2000).
39. B. Bosman, Testing of lipoxygenase inhibitors, cyclooxygenase inhibitors, drugs with immunomodulating properties and some reference antipsoriatic drugs in the modified mouse tail test, an animal model of psoriasis, *Skin Pharmacol* **7**, 324 (1994).
40. P. R. Holt, S. Katz, and R. Kirshoff, Curcumin therapy in inflammatory bowel disease, a pilot study. *Dig Dis Sci* **50**, 2191 (2005).
41. K. Sugimoto, H. Hanai, K. Tozawa, T. Aoshi, M. Uchijima, T. Nagata, and Y. Koide, Curcumin prevents and ameliorates trinitrobenzene sulfonic acid-induced colitis in mice. *Gastroenterology* **123**, 1912 (2002).
42. B. Salh, K. Assi, V. Templeman, K. Parhar, D. Owen, A. Gomez-Munoz, and K. Jacobson, Curcumin attenuates DNB-induced murine colitis. *Am J Physiol Gastrointest Liver Physiol* **285**, G235 (2003).
43. Y. T. Jian, G. F. Mai, J. D. Wang, Y. L. Zhang, R. C. Luo, and Y. X. Fang, Preventive and therapeutic effects of NF-kappaB inhibitor curcumin in rats colitis induced by trinitrobenzene sulfonic acid. *World J Gastroenterol* **11**, 1747 (2005).
44. G. Dean, How many people in the world have multiple sclerosis? *Neuroepidemiology* **13**, 1 (1994).
45. S. Donoghue and C. Greenlees, Drugs in development for the treatment of multiple sclerosis, antigen non-specific therapies: An update. *Expert Opin Investig Drugs* **9**, 167 (2000).

46. J. W. Prineas, R. O. Barnard, T. Revesz, E. E. Kwon, L. Sharer, and E. S. Cho, Multiple sclerosis. Pathology of recurrent lesions. *Brain* **116**, 681 (1993).
47. B. D. Trapp, J. Peterson, R. M. Ransohoff, R. Rudick, S. Mork, and L. Bo, Axonal transection in the lesions of multiple sclerosis. [see comment]. *N Engl J Med* **338,** 278 (1998).
48. C. S. Raine, Multiple sclerosis: Immunopathologic mechanisms in the progression and resolution of inflammatory demyelination. *Res Publ Assoc Res Nerv Ment Dis* **68**, 37 (1990).
49. E. M. Frohman, M. K. Racke, and C. S. Raine, Multiple sclerosis: The plaque and its pathogenesis. *N Engl J Med* **354**, 942 (2006).
50. J. J. Bright, B. F. Musuro, C. Du, and S. Sriram, Expression of IL-12 in CNS and lymphoid organs of mice with experimental allergic encephalitis. *J Neuroimmunol* **82**, 22 (1998).
51. J. J. Bright, C. Du, and S. Sriram, Tyrphostin B42 inhibits IL-12-induced tyrosine phosphorylation and activation of Janus kinase-2 and prevents experimental allergic encephalomyelitis. *J Immunol* **162**, 6255 (1999).
52. G. Muthian, H. P. Raikwar, C. Johnson, J. Rajasingh, A. Kalgutkar, L. J. Marnett, and J. J. Bright, COX-2 inhibitors modulate IL-12 signaling through JAK–STAT pathway leading to Th1 response in experimental allergic encephalomyelitis. *J Clin Immunol* **26,** 73 (2006).
53. D. Devendra, E. Liu, and G. S. Eisenbarth, Type 1 diabetes: Recent developments. *Br Med J* **328**, 750 (2004).
54. G. S. Eisenbarth, Type I diabetes mellitus. A chronic autoimmune disease. *N Engl J Med* **314**, 1360 (1986).
55. J. M. Barker, J. Yu, L. Yu, J. Wang, D. Miao, F. Bao, E. Hoffenberg, J. C. Nelson, P. A. Gottlieb, M. Rewers, and G. S. Eisenbarth, Autoantibody "subspecificity" in type 1 diabetes: Risk for organ-specific autoimmunity clusters in distinct groups. *Diabetes Care* **28**, 850 (2005).
56. O. Kordonouri, R. Hartmann, D. Deiss, M. Wilms, and A. Gruters-Kieslich, Natural course of autoimmune thyroiditis in type 1 diabetes: Association with gender, age, diabetes duration, and puberty. *Arch Dis Child* **90**, 411 (2005).
57. M. J. Franz, J. P. Bantle, C. A. Beebe, J. D. Brunzell, J.-L. Chiasson, A. Garg, L. A. Holzmeister, B. Hoogwerf, E. Mayer-Davis, A. D. Mooradian, J. Q. Purnell, M. Wheeler, American Diabetes, Association, Nutrition principles and recommendations in diabetes. *Diabetes Care* **27**, S36 (2004).
58. G. S. Sidhu, H. Mani, J. P. Gaddipati, A. K. Singh, P. Seth, K. K. Banaudha, G. K. Patnaik, and R. K. Maheshwari, Curcumin enhances wound healing in streptozotocin induced diabetic rats and genetically diabetic mice. *Wound Repair Regen* **7**, 362 (1999).
59. P. Suresh Babu and K. Srinivasan, Amelioration of renal lesions associated with diabetes by dietary curcumin in streptozotocin diabetic rats. *Mol Cell Biochem* **181**, 87 (1998).
60. P. A. Kumar, P. Suryanarayana, P. Y. Reddy, and G. B. Reddy, Modulation of alpha-crystallin chaperone activity in diabetic rat lens by curcumin. *Mol Vis* **11**, 561 (2005).
61. P. Suryanarayana, M. Saraswat, T. Mrudula, T.P. Krishna, K. Krishnaswamy, and G. B. Reddy, Curcumin and turmeric delay streptozotocin-induced diabetic cataract in rats. *Invest Ophthalmol Vis Sci* **46**, 2092 (2005)
62. G. S. Firestein, Immunologic mechanisms in the pathogenesis of rheumatoid arthritis. *J Clin Rheumatol* **11**, S39 (2005).

63. C. J. Edwards and C. Cooper, Early environmental factors and rheumatoid arthritis. *Clin Exp Immunol* **143**, 1 (2006).
64. E. M. Ruderman, Current and future pharmaceutical therapy for rheumatoid arthritis. *Curr Pharm Des* **11**, 671 (2005).
65. A. Liacini, J. Sylvester, W. Q. Li, and M. Zafarullah, Inhibition of interleukin-1-stimulated MAP kinases, activating protein-1 (AP-1) and nuclear factor kappa B (NF-kappa B) transcription factors down-regulates matrix metalloproteinase gene expression in articular chondrocytes. *Matrix Biol* **21**, 251 (2002).
66. A. Liacini, J. Sylvester, W. Q. Li, W. Huang, F. Dehnade, M. Ahmad, an M. Zafarullah, Induction of matrix metalloproteinase-13 gene expression by TNF-alpha is mediated by MAP kinases, AP-1, and NF-kappaB transcription factors in articular chondrocytes. *Exp Cell Res* **288**, 208 (2003).
67. B. B. Aggarwal and S. Shishodia, Suppression of the nuclear factor-kappaB activation pathway by spice-derived phytochemicals: Reasoning for seasoning. *Ann NY Acad Sci* **1030**, 434 (2004).
68. M. Shakibaei, G. Schulze-Tanzil, T. John, and A. Mobasheri, Curcumin protects human chondrocytes from IL-l1beta-induced inhibition of collagen type II and beta1-integrin expression and activation of caspase-3: An immunomorphological study, *Ann Anat* **187**, 487 (2005).
69. S. Shishodia, G. Sethi, and B. B. Aggarwal, Curcumin: Getting back to the roots. *Ann NY Acad Sci* **1056**, 206 (2005).
70. A. M. Bowcock, The genetics of psoriasis and autoimmunity. *Annu Rev Genomics Hum Genet* **6**, 93 (2005).
71. S. Chow, C. Rizzo, L. Ravitskiy, and A. A. Sinha, The role of T cells in cutaneous autoimmune disease. *Autoimmunity* **38**, 303 (2005).
72. J. G. Krueger and A. Bowcock, Psoriasis pathophysiology: Current concepts of pathogenesis. *Ann Rheum Dis* **64**, 30 (2005).
73. J. Miquel, A. Bernd, J. M. Sempere, J. Diaz-Alperi, and A. Ramirez, The curcuma antioxidants: Pharmacological effects and prospects for future clinical use. A review. *Arch Gerontol Geriatr* **34**, 37 (2002).
74. D. Shi, J. Das, and G. Das, Inflammatory bowel disease requires the interplay between innate and adaptive immune signals. *Cell Res* **16**, 70 (2006).
75. E. Ricart, R. Panaccione, E. V. Loftus, Jr., W. J. Tremaine, W. S. Harmsen, A. R. Zinsmeister, and W. J. Sandborn, Autoimmune disorders and extraintestinal manifestations in first-degree familial and sporadic inflammatory bowel disease: A case-control study. *Inflamm Bowel Dis* **10**, 207 (2004).
76. F. R. Byrne and J. L. Viney, Mouse models of inflammatory bowel disease. *Curr Opin Drug Discov Devel* **9**, 207 (2006).
77. S. Ardizzone and G. Bianchi Porro, Biologic therapy for inflammatory bowel disease. *Drugs* **65**, 2253 (2005).
78. E. Domenech, Inflammatory bowel disease: Current therapeutic options. *Digestion* **73**, 67 (2006).
79. A. M. Feldman and D. McNamara, Myocarditis.[see comment]. *N Engl J Med* **343**, 1388 (2000).
80. G. W. Dec, Jr., I. F. Palacios, J. T. Fallon, H. T. Aretz, J. Mills, D. C. Lee, and R. A. Johnson, Active myocarditis in the spectrum of acute dilated cardiomyopathies. Clinical features, histologic correlates, and clinical outcome. *N Engl J Med* **312**, 885 (1985).

81. R. E. McCarthy 3rd, J. P. Boehmer, R. H. Hruban, G. M. Hutchins, E. K. Kasper, J. M. Hare, and K. L. Baughman, Long-term outcome of fulminant myocarditis as compared with acute (nonfulminant) myocarditis.[see comment]. *N Engl J Med* **342**, 690.
82. B. Lauer, M. Schannwell, U. Kuhl, B. E. Strauer, and H. P. Schultheiss, Antimyosin autoantibodies are associated with deterioration of systolic and diastolic left ventricular function in patients with chronic myocarditis. *J Am Coll Cardiol* **35**, 11 (2000).
83. A. L. Caforio, N. J. Mahon, F. Tona, and W. J. McKenna, Circulating cardiac autoantibodies in dilated cardiomyopathy and myocarditis, pathogenetic and clinical significance. *Eur J Heart Fail* **4**, 411 (2002).
84. D. Fairweather, Z. Kaya, G. R. Shellam, C. M. Lawson, and N. R. Rose, From infection to autoimmunity. *J Autoimmun* **16**, 175 (2001).
85. Y. Furukawa, K. Kobuke, and A. Matsumori, Role of cytokines in autoimmune myocarditis and cardiomyopathy. *Autoimmunity* **34**, 165 (2001).
86. M. Afanasyeva, D. Georgakopoulos, and N. R. Rose, Autoimmune myocarditis: Cellular mediators of cardiac dysfunction. *Autoimmun Rev* **3**, 476 (2004).
87. W. Liu, W.-M. Li, C. Gao, and N.-L. Sun, Effects of atorvastatin on the Th1/Th2 polarization of ongoing experimental autoimmune myocarditis in Lewis rats. *J Autoimmun* **25**, 258 (2005).
88. C. Nirmala and R. Puvanakrishnan, Protective role of curcumin against isoproterenol induced myocardial infarction in rats. *Mol Cell Biochem* **159**, 85 (1996).
89. C. H. Yeh, T. P. Chen, Y. C. Wu, Y. M. Lin, and P. Jing Lin, Inhibition of NFkappaB activation with curcumin attenuates plasma inflammatory cytokines surge and cardiomyocytic apoptosis following cardiac ischemia/reperfusion. *J Surg Res* **125**, 109 (2005).
90. C. H. Yeh, Y. M. Lin, Y. C. Wu, and P. J. Lin, Inhibition of NF-kappa B activation can attenuate ischemia/reperfusion-induced contractility impairment via decreasing cardiomyocytic proinflammatory gene up-regulation and matrix metalloproteinase expression. *J Cardiovasc Pharmaco.* **45**, 301 (2005).
91. R. R. Singh, SLE: Translating lessons from model systems to human disease. *Trends Immunol* **26**, 572 (2005).
92. R. Lyons, S. Narain, C. Nichols, M. Satoh, and W. H. Reeves, Effective use of autoantibody tests in the diagnosis of systemic autoimmune disease. *Ann NY Acad Sci* **1050**, 217 (2005).
93. G. Nagy, A. Koncz, A. and A. Perl, T- and B-cell abnormalities in systemic lupus erythematosus. *Crit Rev Immunol* **25**, 123 (2005).
94. J. A. Croker and R. P. Kimberly, SLE: Challenges and candidates in human disease. *Trends Immunol* **26** , 580 (2005).
95. S. G. O'Neill and L. Schrieber, Immunotherapy of systemic lupus erythematosus. *Autoimmun Rev* **4**, 395 (2005).
96. D. B. Drachman, Myasthenia gravis. *N Engl J Med* **330**, 1797 (1994).
97. J. Lindstrom, D. Shelton, and Y. Fujii, Myasthenia gravis. *Adv Immunol* **42**, 233 (1988).
98. K. Shigemoto, S. Kubo, N. Maruyama, N. Hato, H. Yamada, C. Jie, N. Kobayashi, K. Mominoki, Y. Abe, N. Ueda, and S. Matsuda, Induction of myasthenia by immunization against muscle-specific kinase. *J Clin Invest.* **116**, 1016 (2006).
99. D. Asthana, Y. Fujii, G. E. Huston, and J. Lindstrom, Regulation of antibody production by helper T cell clones in experimental autoimmune myasthenia gravis is mediated by IL-4 and antigen-specific T cell factors. *Clin Immunol Immunopathol* **67**, 240 (1993).

100. G. X. Zhang, B. G. Xiao, M. Bakhiet, P. van der Meide, H. Wigzell, H. Link, and T. Olsson, Both CD4+ and CD8+ T cells are essential to induce experimental autoimmune myasthenia gravis. *J Exp Med* **184**, 349 (1996).
101. L. Moiola, F. Galbiati, G. Martino, S. Amadio, E. Brambilla, G. Comi, A. Vincent, L. M. Grimaldi, and L. Adorini, IL-12 is involved in the induction of experimental autoimmune myasthenia gravis, an antibody-mediated disease. *Eur J Immunol* **28**, 2487 (1998).
102. S. Sitaraman, D. W. Metzger, R. J. Belloto, . A. J. Infante, and K. A. Wall, Interleukin-12 enhances clinical experimental autoimmune myasthenia gravis in susceptible but not resistant mice. *J Neuroimmunol* **107**, 73 (2000).
103. H. Tlaskalova-Hogenova, L. Tuckova, R. Stepankova, T. Hudcovic, L. Palova-Jelinkova, H. Kozakova, P. Rossmann, D. Sanchez, J. Cinova, T. Hrncir, M. Kverka, L. Frolova, H. Uhlig, F. Powrie, and P. Bland, Involvement of innate immunity in the development of inflammatory and autoimmune diseases. *Ann NY Acad Sci* **1051**, 787 (2005).
104. D. N. Cook, D. S. Pisetsky, and D. A. Schwartz, Toll-like receptors in the pathogenesis of human disease. *Nat Immunol* **5**, 975 92004).
105. G. Cheng and S. P. Schoenberger, CD40 signaling and autoimmunity. *Curr Dir Autoimmun* **5**, 51 (2002).
106. 106. H. S. Youn, S. I. Saitoh, K. Miyake, and D. H. Hwang, Inhibition of homodimerization of Toll-like receptor 4 by curcumin. *Biochem Pharmacol* **72**, 62 (2006).
107. S. Kato, Y. Yuzawa, N. Tsuboi, S.Maruyama, Y. Morita, T. Matsuguchi, and S. Matsuo, Endotoxin-induced chemokine expression in murine peritoneal mesothelial cells: The role of toll-like receptor 4. *J Am Soc Nephrol* **15**, 1289 (2004).
108. R. S. Liblau, S. M. Singer, and H. O. McDevitt, Th1 and Th2 CD4+ T cells in the pathogenesis of organ-specific autoimmune diseases.[see comment]. *Immunol Today* **16**, 34 (1995).
109. B. Y. Kang, S. W. Chung, W. Chung, S. Im, S. Y. Hwang, and T. S. Kim, Inhibition of interleukin-12 production in lipopolysaccharide-activated macrophages by curcumin. *Eur J Pharmacol* **384**, 191 (1999).
110. B. Y. Kang, Y. J. Song, K. M. Kim, Y. K. Choe, S. Hwang, and T. S. Kim, Curcumin inhibits Th1 cytokine profile in CD4+ T cells by suppressing interleukin-12 production in macrophages. *Br J Pharmacol* **128**, 380 (1999).
111. M. Sospedra and R. Martin, Immunology of multiple sclerosis. *Annu Rev Immunol* **23**, 683 (2005).
112. J. C. W. Edwards and G. Cambridge, B-cell targeting in rheumatoid arthritis and other autoimmune diseases. *Nat Rev Immunol* **6**, 394 (2006).
113. S. Hori, T. Takahashi, and S. Sakaguchi, Control of autoimmunity by naturally arising regulatory CD4+ T cells. *Adv Immunol* **81**, 331 (2003).
114. L. A. Stephens, D. Gray, and S. M. Anderton, CD4+CD25+ regulatory T cells limit the risk of autoimmune disease arising from T cell receptor crossreactivity. *Proc Natl Acad Sci USA* **102**, 17,418 (2005).
115. K. J. Hammond and D. I. Godfrey, NKT cells: Potential targets for autoimmune disease therapy? *Tissue Antigens* **59**, 353 (2002).
116. S. Sharif, G. A. Arreaza, P. Zucker, Q. S. Mi, and T. L. Delovitch, Regulation of autoimmune disease by natural killer T cells. *J Mol Med* **80**, 290 (2002).
117. R. M. Strieter, S. L. Kunkel, and R. C. Bone, Role of tumor necrosis factor-alpha in disease states and inflammation. *Crit Care Med* **21**, S447 (1993).

118. H. Korner and J. D. Sedgwick, Tumour necrosis factor and lymphotoxin: Molecular aspects and role in tissue-specific autoimmunity. *Immunol Cell Biol* **74**, 465 (1996).
119. F. Atzeni, M. Turiel, F. Capsoni, A. Doria, P. Meroni, and P. Sarzi-Puttini, Autoimmunity and anti-TNF-{alpha} agents. *Ann NY Acad Sci* **1051**, 559 (2005).
120. K. Hosaka, J. Ryu, S. Saitoh, T. Ishii, K. Kuroda, and K. Shimizu, The combined effects of anti-TNFalpha antibody and IL-1 receptor antagonist in human rheumatoid arthritis synovial membrane. *Cytokine* **32**, 263 (2005).
121. V. S. Yadav, K. P. Mishra, D. P. Singh, S. Mehrotra, and V. K. Singh, Immunomodulatory effects of curcumin. *Immunopharmacol Immunotoxicol* **27**, 485 (2005).
122. B. Gupta and B. Ghosh, Curcuma longa inhibits TNF-alpha induced expression of adhesion molecules on human umbilical vein endothelial cells. *Int J Innumopharmacol* **21**, 745 (1999).
123. S. M. Plummer, K. A.Holloway, M. M. Manson, R. J. Munks, A. Kaptein, S. Farrow, and L. Howells, Inhibition of cyclo-oxygenase 2 expression in colon cells by the chemopreventive agent curcumin involves inhibition of NF-kappaB activation via the NIK/IKK signalling complex. *Oncogene* **18**, 6013 (1999).
124. Y. R. Chen and T. H. Tan, Inhibition of the c-Jun N-terminal kinase (JNK) signaling pathway by curcumin. *Oncogene* **17**, 173 (1998).
125. Y. Iwakura, Roles of IL-1 in the development of rheumatoid arthritis:Consideration from mouse models. *Cytokine Growth Factor Rev* **13**, 341 (2002).
126. N. Jurrmann, R. Brigelius-Flohe, and G. F. Bol, Curcumin blocks interleukin-1 (IL-1) signaling by inhibiting the recruitment of the IL-1 receptor-associated kinase IRAK in murine thymoma EL-4 cells. *J Nutr* **135**, 1859 (2005).
127. G. Trinchieri, S. Pflanz, and R. A. Kastelein, The IL-12 family of heterodimeric cytokines: New players in the regulation of T cell responses.[comment]. *Immunity* **19**, 641 (2003).
128. K. E. Balashov, D. R. Smith, S. J. Khoury, D. A. Hafler, and H. L. Weiner, Increased interleukin 12 production in progressive multiple sclerosis: Induction by activated CD4+ T cells via CD40 ligand. *Proc Natl Acad Sci USA* **94**, 599 (1997).
129. J. J. Bright, M. Rodriguez, and S. Sriram, Differential influence of interleukin-12 in the pathogenesis of autoimmune and virus-induced central nervous system demyelination. *J Virol* **73**, 1637 (1999).
130. 130. J. P. Leonard, K. E. Waldburger, and S. J. Goldman, Prevention of experimental autoimmune encephalomyelitis by antibodies against interleukin 12. *J Exp Med* **181**, 381 (1995).
131. B. Oppmann, R. Lesley, B. Blom, J. C. Timans, . Xu, B. Hunte, F. Vega, N. Yu, J. Wang, K. Singh, F. Zonin, E. Vaisberg, T. Churakova, M. Liu, D. Gorman, J. Wagner, S. Zurawski, Y. Liu, J. S. Abrams, K. W. Moore, D. Rennick, R. de Waal-Malefyt, C. Hannum, J. F. Bazan, and R. A. Kastelein, Novel p19 protein engages IL-12p40 to form a cytokine, IL-23, with biological activities similar as well as distinct from IL-12. *Immunity* **13**, 715 (2000).
132. X. T. Ma, X. j. Zhang, B. Zhang, Y. Q. Geng, Y. M. Lin, G. Li, and K. F.Wu, Expression and regulation of interleukin-23 subunits in human peripheral blood mononuclear cells and hematopoietic cell lines in response to various inducers. *Cell Biol Int* **28**, 689 (2004).
133. A. Wada, Y. Tada, O. Shimozato, Y. Takiguchi, K. Tatsumi, T. Kuriyama, and M. Tagawa, Expression of CD40 ligand in CD40-positive murine tumors activates transcription of the interleukin-23 subunit genes and produces antitumor responses. *Anticancer Res* **24**, 2713 (2004).

134. D. J. Cua, J. Sherlock, Y. Chen, C. A. Murphy, B. Joyce, B. Seymour, L. Lucian, W. To, S. Kwan, T. Churakova, S. Zurawski, M. Wiekowski, S. A. Lira, D. Gorman, R.A. Kastelein, and J. D. Sedgwick, Interleukin-23 rather than interleukin-12 is the critical cytokine for autoimmune inflammation of the brain.[see comment]. *Nature* **421**, 744 (2003).
135. S. J. Szabo, S. T. Kim, G. L. Costa, X. Zhang, C. G. Fathman, and L. H. Glimcher, A novel transcription factor, T-bet, directs Th1 lineage commitment. *Cell* **100**, 655 (2000).
136. A. Takeda, S. Hamano, A. Yamanaka, T. Hanada, T. Ishibashi, T. W. Mak, A. Yoshimura, and H. Yoshida, Cutting edge: Role of IL-27/WSX-1 signaling for induction of T-bet through activation of STAT1 during initial Th1 commitment. *J Immunol* **170**, 4886 (2003).
137. E. Bettelli, B. Sullivan, S. J. Szabo, R. A. Sobel, L. H. Glimcher, and V. K. Kuchroo, Loss of T-bet, but not STAT1, prevents the development of experimental autoimmune encephalomyelitis. *J Exp Med* **200**, 79 (2004).
138. A. E. Lovett-Racke, A. E. Rocchini, J. Choy, S. C. Northrop, R. Z. Hussain, R. B. Ratts, D. Sikder, and M. K. Racke, Silencing T-bet defines a critical role in the differentiation of autoreactive T lymphocytes. *Immunity* **21**, 719 (2004).
139. S. Pflanz, J. C. Timans, J. Cheung, R. Rosales, H. Kanzler, J. Gilbert, L. Hibbert, T. Churakova, M. Travis, E. Vaisberg, W. M. Blumenschein, J. D. Mattson, J. L. Wagner, W. To, S. Zurawski, T. K. McClanahan, D. M. Gorman, J. F. Bazan, R. de Waal Malefyt, D. Rennick, and R. A. Kastelein, IL-27, a heterodimeric cytokine composed of EBI3 and p28 protein, induces proliferation of naive CD4(+) T cells. *Immunity* **16**, 779 (2002).
140. R. Goldberg, Y. Zohar, G. Wildbaum, Y. Geron, G. Maor, and N. Karin, Suppression of ongoing experimental autoimmune encephalomyelitis by neutralizing the function of the p28 subunit of IL-27. *J Immunol* **173**, 6465 (2004).
141. B. M. Segal, B. k. Dwyer, and E. M. Shevach, An interleukin (IL)-10/IL-12 immunoregulatory circuit controls susceptibility to autoimmune disease. *J Exp Med* **187**, 537 (1998).
142. S. Ghosh, M. J. May, and E. B. Kopp, NF-kappa B and Rel proteins: Evolutionarily conserved mediators of immune responses. *Annu Rev Immunol* **16**, 225 (1998).
143. G. Y. Kim, K. H. Kim, S. H. Lee, M. S. Yoon, H. J. Lee, D. O. Moon, C. M. Lee, S. C. Ahn, Y. C. Park, and Y. M. Park, Curcumin inhibits immunostimulatory function of dendritic cells: MAPKs and translocation of NF-kappa B as potential targets. *J Immunol* **174**, 8116 (2005).
144. G. Kang, P. J. Kong, Y. J. Yuh, S. Y. Lim, S. V. Yim, W. Chun, and S. S. Kim, Curcumin suppresses lipopolysaccharide-induced cyclooxygenase-2 expression by inhibiting activator protein 1 and nuclear factor kappab bindings in BV2 microglial cells. *J Pharmacol Sci* **94**, 325 (2004).
145. W. C. Sha, Regulation of immune responses by NF-kappa B/Rel transcription factor. *J Exp Med* **187**, 143 (1998).
146. J. D. Woronicz, X. Gao, Z. Cao, M. Rothe, and D. V. Goeddel, IkappaB kinase-beta: NF-kappaB activation and complex formation with IkappaB kinase-alpha and NIK. *Science* **278** , 866 (1997).
147. T. L. Murphy, M. G. Cleveland, P. Kulesza, J. Magram, and K. M. Murphy, Regulation of interleukin 12 p40 expression through an NF-kappa B half-site. *Mol Cell Biol* **15**, 5258 (1995).

148. D. D'Ambrosio, M. Cippitelli, M. G. Cocciolo, D. Mazzeo, P. Di Lucia, R. Lang, F. Sinigaglia, and P. Panina-Bordignon, Inhibition of IL-12 production by 1,25-dihydroxyvitamin D3. Involvement of NF-kappaB downregulation in transcriptional repression of the p40 gene. *J Clin Invest* **101**, 252 (1998).
149. D. Mazzeo, P. Panina-Bordignon, H. Recalde, F. Sinigaglia, and D. D'Ambrosio, Decreased IL-12 production and Th1 cell development by acetyl salicylic acid-mediated inhibition of NF-kappaB. *Eur J Immunol* **28**, 3205 (1998).
150. C. M. Bacon, E. F. Petricoin 3rd, J. R. Ortaldo, R. C. Rees, A. C. Larner, J. A. Johnston, and J. J. O'Shea, Interleukin 12 induces tyrosine phosphorylation and activation of STAT4 in human lymphocytes. *Proc Natl Acad Sci USA* **92**, 7307 (1995).
151. N. G. Jacobson, S. J. Szabo, R. M. Weber-Nordt, Z. Zhong, R. D. Schreiber, J. E. Darnell, Jr., and K. M. Murphy, Interleukin 12 signaling in T helper type 1 (Th1) cells involves tyrosine phosphorylation of signal transducer and activator of transcription (Stat)3 and Stat4. *J Exp Med* **181**, 1755 (1995).
152. C. Parham, M. Chirica, J. Timans, E. Vaisberg, M. Travis, J. Cheung, S. Pflanz, R. Zhang, K. P. Singh, F. Vega, W. To, J. Wagner, A. M. O'Farrell, T. McClanahan, S. Zurawski, C. Hannum, D. Gorman, D. M. Rennick, R. A. Kastelein, R. de Waal Malefyt, and K. W. Moore, A receptor for the heterodimeric cytokine IL-23 is composed of IL-12Rbeta1 and a novel cytokine receptor subunit, IL-23R. *J Immunol* **168**, 5699 (2002).
153. S. Aggarwal, N. Ghilardi, M. H. Xie, F. J. de Sauvage, and A. L. Gurney, Interleukin-23 promotes a distinct CD4 T cell activation state characterized by the production of interleukin-17. *J Biol Chem* **278**, 1910 (2003).
154. S. Kamiya, T. Owaki, N. Morishima, F. Fukai, J. Mizuguchi, and T. Yoshimoto, An indispensable role for STAT1 in IL-27-induced T-bet expression but not proliferation of naive CD4+ T cells. *J Immunol* **173**, 3871 (2004).
155. J. J. O'Shea, H. Park, M. Pesu, D. Borie, and P. Changelian, New strategies for immunosuppression, interfering with cytokines by targeting the Jak/Stat pathway. *Curr Opin Rheumatol* **17**, 305 (2005).
156. H. M. Seidel, P. Lamb, and J. Rosen, Pharmaceutical intervention in the JAK/STAT signaling pathway. *Oncogene* **19**, 2645 (2000).
157. J. J. Bright, Targeting autoimmune diseases through nutraceuticals. *Nutrition* **20**, 39 (2004).
158. G. Muthian and J. J. Bright, Quercetin, a flavonoid phytoestrogen, ameliorates experimental allergic encephalomyelitis by blocking IL-12 signaling through JAK-STAT pathway in T lymphocyte. *J Clin Immunol* **24**, 542 (2004).
159. G. Muthian, H. P. Raikwar, J. Rajasingh, and J. J. Bright, 1,25 Dihydroxyvitamin-D3 modulates JAK-STAT pathway in IL-12/IFNgamma axis leading to Th1 response in experimental allergic encephalomyelitis. *J Neurosci Res* **83**, 1299 (2006).
160. L. Neff, M. Zeisel, J. Sibilia, M. Scholler-Guinard, J. P. Klein, and D. Wachsmann, NF-kappaB and the MAP kinases/AP-1 pathways are both involved in interleukin-6 and interleukin-8 expression in fibroblast-like synoviocytes stimulated by protein I/II, a modulin from oral streptococci. *Cell Microbiol* **3**, 703 (2001).
161. R. K. Patel and C. Mohan, PI3K/AKT signaling and systemic autoimmunity. *Immunol Res* **31**, 47 (2005).
162. J. W. Cho, K. Park, G. R. Kweon, B. C. Jang, W. K. Baek, M. H. Suh, C. W. Kim, K. S. Lee, and S. I. Suh, Curcumin inhibits the expression of COX-2 in UVB-irradiated human keratinocytes (HaCaT) by inhibiting activation of AP-1, p38 MAP kinase and JNK as potential upstream targets. *Exp Mol Med* **37** , 186 (2005).

163. S. L. Kunkel and N. Godessart, Chemokines in autoimmunity: From pathology to therapeutics. *Autoimmun Rev* **1**, 313 (2002).
164. S. Arimilli, W. Ferlin, N. Solvason, S. Deshpande, M. Howard, and S. Mocci, Chemokines in autoimmune diseases. *Immunol Rev* **177**, 43 (2000).
165. X. Chen, J. J. Oppenheim, and O. M. Howard, Chemokines and chemokine receptors as novel therapeutic targets in rheumatoid arthritis (RA): Inhibitory effects of traditional Chinese medicinal components. *Cell Mol Immunol* **1**, 336 (2004).
166. H. Hidaka, T. Ishiko, T. Furuhashi, H. Kamohara, S. Suzuki, M. Miyazaki, O. Ikeda, S. Mita, T. Setoguchi, and M. Ogawa, Curcumin inhibits interleukin 8 production and enhances interleukin 8 receptor expression on the cell surface: Impact on human pancreatic carcinoma cell growth by autocrine regulation. *Cancer* **95**, 1206 (2002).

PHARMACOKINETICS AND PHARMACODYNAMICS OF CURCUMIN

Ricky A. Sharma, William P. Steward, and Andreas J. Gescher

Abstract: *Curcuma* spp. contain turmerin, essential oils, and curcuminoids, including curcumin. Curcumin [1,7-bis-(4-hydroxy-3-methoxyphenyl)-1,6-heptadiene-3,5-dione] is regarded as the most biologically active constituent of the spice turmeric and it comprises 2–8% of most turmeric preparations. Preclinical data from animal models and phase I clinical studies performed with human volunteers and patients with cancer have demonstrated low systemic bioavailability following oral dosing. Efficient first-pass metabolism and some degree of intestinal metabolism, particularly glucuronidation and sulfation of curcumin, might explain its poor systemic availability when administered via the oral route. A daily oral dose of 3.6 g of curcumin is compatible with detectable levels of the parent compound in colorectal tissue from patients with cancer. The levels demonstrated might be sufficient to exert pharmacological activity. There appears to be negligible distribution of the parent drug to hepatic tissue or other tissues beyond the gastrointestinal tract. Curcumin possesses wide-ranging anti-inflammatory and anticancer properties. Many of these biological activities can be attributed to its potent antioxidant capacity at neutral and acidic pH, its inhibition of cell signaling pathways at multiple levels, its diverse effects on cellular enzymes, and its effects on cell adhesion and angiogenesis. In particular, curcumin's ability to alter gene transcription and induce apoptosis in preclinical models advocates its potential utility in cancer chemoprevention and chemotherapy. With regard to considerable public and scientific interest in the use of phytochemicals derived from dietary components to combat or prevent human diseases, curcumin is currently a leading agent.

1. INTRODUCTION

There has been considerable public and scientific interest in the use of phytochemicals derived from dietary components to combat or prevent human diseases, especially the two commonest killers in the developed world: cardiovascular disease and cancer. The dried, ground rhizome of the perennial herb *Curcuma longa* Linn. has been used in Asian medicine since the second millenium BC.[1] Its utility is referred to in the ancient Hindu scripture, the Ayurveda. In addition to its aromatic, stimulant, and coloring properties in the diet, turmeric is mixed with other natural compounds such as slaked lime and has been used topically as a treatment for wounds, inflammation, and tumors. In contrast to the high dietary consumption (up to 1.5 g curcumin per person per day) in certain Southeast Asian communities, smaller quantities of turmeric tend to be used for medicinal purposes.[2] The appeal of turmeric as a coloring, food preservative, and flavoring is global; according to the Food and Agriculture Organization of the United Nations, over 2400 metric tons of turmeric are imported annually into the United States for consumer use.

Curcuma spp. contain turmerin (a water-soluble peptide), essential oils (such as turmerones, atlantones, and zingiberene), and curcuminoids, including curcumin [1,7-bis-(4-hydroxy-3-methoxyphenyl)-1,6-heptadiene-3,5-dione]. Curcumin, which was first chemically characterized in 1910, is regarded as the most biologically active constituent of turmeric and comprises 2–8% of most turmeric preparations.[3a,4] Curcumin has been the subject of hundreds of published articles over the past three decades, studying its antioxidant, anti-inflammatory, cancer chemopreventive, and potentially chemotherapeutic properties. The pharmacology and putative anticancer properties of curcumin have been the subject of several review articles,[5–7] publication of which predates a number of clinical studies of curcumin that have been completed and reported within the last 2 years. The purpose of this chapter is to appraise the current level of knowledge of the fate of curcumin in mammalian organisms and its pharmacodynamic properties, with particular reference to its potential use in the chemoprevention of human cancer.

2. PHARMACOKINETICS AND METABOLISM

2.1. Pharmacokinetics and Metabolism in Rodents

The absorption, distribution, metabolism, and excretion of curcumin in rodents have been described in at least 10 studies over the past three decades. Collectively, these studies support the notion that curcumin undergoes a rapid and efficient metabolism that severely curtails the availability of parent compound in the biophase. In an early study in rats, a dietary dose (1 g/kg) administered to rats resulted in about 75% of species related to curcumin being detected in feces, whereas negligible amounts appeared in the urine.[8] In another early article, absorption of oral curcumin was 60%; urinary agent-derived species were characterized as glucuronide and sulfate conjugates.[9] When curcumin bioavailability was investigated using a ^{3}H-radiolabeled agent, the vast majority of the oral dose was excreted in the feces and one-third was excreted unchanged.[10] Intravenous and intraperitoneal administration of curcumin in rats resulted in large quantities of curcumin and metabolites in bile.[11,12] The metabolites were characterized mainly as glucuronides of tetrahydrocurcumin and hexahydrocurcumin. After intravenous dosing, more than

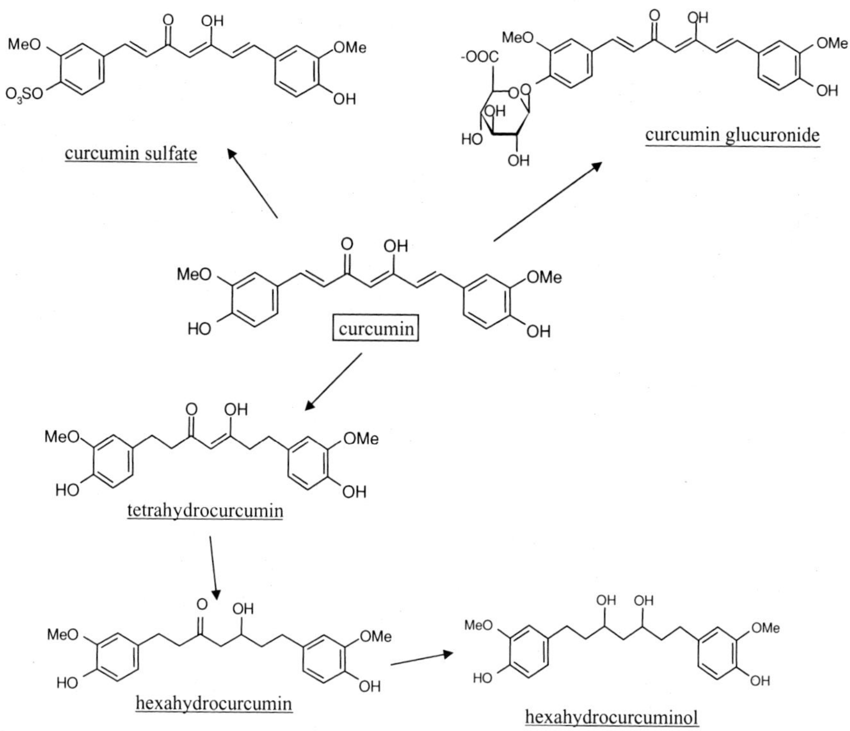

Figure 1. Major metabolites of curcumin detected in rodents and humans.

50% of the dose was excreted in the bile within 5 h. This finding was interpreted as evidence in support of the hypothesis that curcumin undergoes biotransformation during absorption in the intestinal tract and enterohepatic recirculation.[11] More recently, curcumin (0.1 g/kg) administered intraperitoneally to the mouse was found to undergo metabolic reduction to dihydrocurcumin and tetrahydrocurcumin, which, in turn, were converted to monoglucuronide conjugates.[13] High-pressure liquid chromatography (HPLC) analysis of plasma from rats that had received oral curcumin demonstrated substantial levels of curcumin glucuronide and curcumin sulfate, small quantities of hexahydrocurcumin, hexahydrocurcuminol, and hexahydrocurcumin glucuronide and negligible amounts of curcumin (see Figure 1).[14] In suspensions of isolated human hepatocytes, or in microsomes derived from liver or gut tissues of rats and humans, curcumin was rapidly reduced to metabolites, as shown in Figure 1.[15] In a separate study in rats, a high dose of curcumin mixed into the diet (2%, equating to approximately 1.2 g/kg) for 14 days yielded low-nanomolar levels of the parent compound in the plasma, with concentrations in the liver and colon mucosa ranging from 0.1 to 1.8 nmol/g tissue.[16] It is also conceivable that other constituents of the dietary matrix might alter the bioavailability of curcumin. When oral curcumin (2 g/kg) is coadministered to rats with l-piperoylpiperidine (piperine), a constituent of the fruit of the pepper vine (piper nigrum) that induces glucuronyl transferase enzymes, the systemic bioavailability of curcumin might be increased by as much as 154%.[17]

2.2. Clinical Pharmacokinetics and Metabolism

In comparison to the preclinical work summarized above, comprehensive pharmacokinetic data in humans do not exist. Healthy volunteers who ingested 2 g pure curcumin powder after fasting showed less than 10 ng/mL curcumin in their plasma 1 h postdose.[17] In the same study, coingestion of curcumin with 20 mg of piperine appeared to increase the bioavailability of curcumin by 2000%. In a study of oral curcumin, patients with preinvasive malignant or high-risk premalignant conditions of the bladder, skin, cervix, stomach, or oral mucosa received 0.5–8 g curcumin by mouth daily for 3 months.[18] Plasma curcumin concentrations were found to peak 1–2 h after intake and gradually declined within 12 hours. The 8-g/day dose resulted in a peak serum concentration of 1.75 ± 0.80 μM. When curcumin in micronized form was administered orally with orange juice at doses of 50–200 mg to 18 healthy volunteers, curcumin was not found in the plasma at or above the limit of quantitation (approximately 0.63 ng/mL).[19]

In a clinical phase I dose-escalation study using a standardized oral *Curcuma* extract comprised mainly of curcumin, doses up to 180 mg of curcumin per day were administered to patients with advanced colorectal cancer for up to 4 months without toxicity or detectable systemic bioavailability.[20] In a follow-up study in 15 patients with advanced colorectal cancer refractory to standard chemotherapy, curcumin in the form of "Curcuminoids C3" (Sabinsa Corp., 90% curcumin) was consumed orally for up to 4 months at doses of curcumin between 0.45 and 3.6 g daily.[21] Oral consumption of 3.6 g of curcumin per day resulted in levels of drug and glucuronide/sulfate conjugates in plasma near the limit of detection (5 pmol/mL). Curcumin and its conjugates were also detected in 24-h urine collections. In the six patients who had consumed 3.6 g curcumin, urinary levels (in μM) varied between 0.1 and 1.3 for curcumin, between 0.019 and 0.045 for curcumin sulfate, and between 0.21 and 0.51 for curcumin glucuronide. The presence of curcumin and its conjugates in the urine of patients taking 3.6 g of curcumin daily suggests that urinary curcumin/curcumin metabolites might serve as measures of compliance with treatment.

Exploratory studies have also been performed in patients undergoing operations for colorectal cancer who consented to have tissues analysed for research purposes.[22,23] Twelve patients with confirmed colorectal cancer received oral curcumin at 0.45, 1.8, or 3.6 g/day for 7 days prior to surgery. Levels of agent-derived species were determined in blood and colorectal tissue obtained at the time of surgical resection. The mean concentrations of curcumin in normal and malignant colorectal tissue of patients who had ingested 3.6 g curcumin daily were 12.7 and 7.7 nmol/g tissue, respectively.[22] Curcumin sulfate and curcumin glucuronide were also found in the intestinal tissue taken from these patients; trace levels of curcumin were detected in peripheral blood samples. Compatible with the preclinical data presented earlier, these preliminary results in humans suggest that a daily dose of 3.6 g curcumin achieves measurable levels in colorectal tissue with negligible distribution of the parent drug outside of the gut. When 12 patients with liver metastases from colorectal cancer received oral curcumin (0–3.6 g) daily for 7 days prior to hepatic surgery, curcumin was not found in liver tissue resected 6–7 h after the last dose of curcumin, whereas trace levels of products of its metabolic reduction were detected.[23] Levels of curcumin and glucuronide and sulfate conjugates in the low-nanomolar range were found in blood samples taken 1 h after the last dose. The results of this pilot study suggest that doses of oral curcumin required to produce hepatic levels sufficient to exert pharmacological activity are probably not feasible in humans.

To summarize the data from pilot and phase I clinical studies performed with curcumin, it appears that low systemic bioavailability following oral dosing is consistent with the findings in preclinical models presented earlier. Efficient first-pass and some degree of intestinal metabolism of curcumin, particularly glucuronidation and sulfation, might explain its poor systemic availability when administered via the oral route. A daily oral dose of 3.6 g of curcumin results in detectable levels in colorectal tissue, which might be sufficient to exert pharmacological activity, with negligible distribution of the parent drug in hepatic tissue or other tissues beyond the gastrointestinal tract.

3. PHARMACODYNAMICS AND SAFETY

3.1. Pharmacodynamics in Preclinical Models

Curcumin has been shown to exert a fascinating array of pharmacological effects in cells *in vitro* at physiologically attainable and supraphysiological concentrations (see Chapter 10). Table 1 lists some of these activities, many of which are highly relevant to the cancer chemopreventive activity and pharmacodynamic properties of curcumin. In the following subsections, evidence is reviewed pertaining to the ability of curcumin to interfere with carcinogenesis, drug- and carcinogen-metabolizing enzymes, and biological oxidative processes in rodents *in vivo*.

3.1.1. Inhibition of Carcinogenesis

Following oral administration, curcumin prevented cancer in the colon, skin, stomach, liver, lung, duodenum, soft palate, and breasts of rodents.[24,25] In particular, the effects of dietary curcumin (0.05–2.0%) on colorectal carcinogenesis have been demonstrated in both carcinogen-induced and genetic rodent models. Curcumin inhibited carcinogenic initiation, as reflected by decreased levels of adducts induced by benzo[*a*]pyrene or by aflatoxin B_1.[26,27] In intestinal cancer induced in mice by azoxymethane, oral curcumin (2000 ppm) for 14 weeks produced a significant increase in the apoptotic histological index when compared to controls.[28] In the azoxymethane-induced rat colon cancer model, dietary curcumin (0.8%)

Table 1. Molecular targets of curcumin in cells grown *in vitro* relevant to the chemoprevention of cancer.

TARGETS	EFFECTS	REFERENCES
Protein kinases (MAPK, JNK, PKA, PKC, src tyrosine kinase, IκBα kinase, growth factor receptor protein tyrosine kinases)	Inhibition, downregulation	67
Antiapoptotic proteins	Induction of cytochrome-*c* release, Bid cleavrage, activation of caspase-3 and caspase-9, downregulation of Bcl-2 and BclX2	68–71
Proinflammatory proteins	Downregulation of COX-2, 5-LOX, and iNOS	
Cytokines/growth factors	Downregulation of TNF, IL-6, IL-8, IL-12, and fibroblast growth factor-2	72–74
Transcription factors	Suppression of NF-κB, STAT3, Egr-1, AP-1, and PPAR-γ, activation of β-catenin	59, 60, 75
Oxidant systems	Upregulation of heme oxygenase, glutathione transferases, downregulation of xanthine oxidase, scavenging reactive oxygen/superoxide	76, 77
Phase I and Phase II drug/carcinogen-metabolizing enzymes	Inhibition of CYP1A1	78, 79
Metalloproteinases	MMP-9 expression	80

Abbreviations: AP, activator protein; JNK, c-Jun NH_2-terminal kinases; PK, protein kinase; COX, cyclooxygenase; LOX, lipooxygenase; NOS, nitric oxide synthase; NF, nuclear factor; TNF, tumor necrosis factor; IL, interleukin; PPAR, peroxisome proliferators-activated factor; CYP, cytochromes p450; MMP, matrix metalloproteinase; STAT, signal transducer and activator of transcription; EGR, early growth response; CYP, cytochromes p450.

halved the number of aberrant crypt foci compared with control.[29] Genetic models, such as the multiple intestinal neoplasia (e.g. Apc^{Min}) mouse, permit the study of the inhibition of the promotion phase of carcinogenesis. Curcumin interfered with adenoma formation in the Apc^{Min} mouse, which harbors an adenomatous polyposis coli (*APC*) gene mutation and is a model of the human disease familial adenomatous polyposis.[30] When administered in the diet at 0.1% and 0.2 % for the animals' lifetimes, a significant decrease in adenoma number was observed compared to control animals.[3,31] The decrease after the latter dose was accompanied by downregulation of the expression of the enzyme cyclooxygenase-2 (COX-2) and attenuation of tissue oxidative status, as reflected by levels of the oxidative DNA adduct pyrimido-[1,2α]purin-10(3H)-one-2′-deoxyguanosine (M_1dG).[32] The dietary dose of 0.2%, which equates to approximately 300 mg/kg per day, furnished only trace levels of curcumin and metabolites in the plasma, but concentrations of curcumin in the 100-nmol/g tissue range in the gastrointestinal mucosa.[31] This result might provide a tentative "target concentration," although reliable strategies to extrapolate these levels to the equivalent levels in human gastrointestinal mucosa do not currently exist.

Topical application of curcumin (3 or 10 μmol curcumin, 5 min prior to the application of carcinogen) has been shown to inhibit chemical carcinogenesis of the skin.[33] In this series of studies, tumor initiation was induced by benzo[*a*]pyrene or 7,12-dimethylbenz[*a*]antracene (DMBA) and tumor promotion was induced by 12-*O*-tetradecanoylphorbol-13-acetate. Potential mechanisms of these effects were considered to involve inhibition of arachidonic acid-induced inflammation, inhibition of hydrogen peroxide formation, and inhibition of ornithine decarboxylase activity/transcription, which is a rate-limiting step in polyamine biosynthesis.[33] Topical application of curcumin (10 mmol) three times weekly to the buccal pouch of Syrian golden hamsters has also demonstrated inhibition of DMBA-induced oral carcinogenesis.[34] In this early example of "combinatorial chemoprevention," the effect of topical curcumin appeared to be enhanced by the concomitant consumption of green tea (6 mg tea solids/mL) for 18 weeks.[34] Subsequent studies combining curcumin with other chemopreventive agents have also shown augmented growth inhibitory effects. It should also be noted that studies have also been performed that have demonstrated no attenuation of chemically induced carcinogenesis by curcumin. For example, dietary curcumin (500 ppm) did not affect prostate carcinogenesis in rats exposed to 3,2′-dimethyl-4-aminobiphenol (DMAB)- or 2-amino-1-methylimidazo[4,5-*b*]pyridine (PhIP).[35] Although there are numerous reports in the published literature to suggest that curcumin augments the cytotoxicity of anticancer drugs such as paclitaxel in cells *in vitro*,[36,37] observations that confirm this notion *in vivo* are lacking at present. In a human breast cancer xenograft model, nude mice bearing the human-derived MDA-MB-435 breast tumor received dietary curcumin (2%) after excision of the primary tumor: Curcumin consumption decreased the load of breast cancer metastases and concomitantly suppressed the expression of nuclear factor (NF)-κB, COX-2, and matrix metalloproteinase-9 (MMP-9) in the lung metastases that did form.[36] Interestingly, certain rodent studies have suggested a potential for curcumin to confound unwanted detrimental effects of cytotoxic anticancer drugs. For example, curcumin administered to rats by gavage (100 or 200 mg/kg daily for 7 days) ameliorated chromosomal mutations induced by cyclophosphamide in the bone marrow.[38] The diversity of the biological actions of curcumin in mammalian species was recently emphasized by a noteworthy study demonstrating its beneficial effects in mice homozygous for a complete knockout of a gene linked with cystic fibrosis.[39]

3.1.2. Effects on Carcinogen-Metabolizing Enzymes and Oxidant Systems

Phase II drug- and carcinogen metabolizing enzymes such as glutathione-*S*-transferases (GST) render xenobiotics more water soluble, thus facilitating their excretion. Because induction of phase II enzymes stimulates carcinogen excretion, it is generally thought to confer a protective effect. Epoxide hydrolase and various hepatic GST isoenzymes were significantly increased upon curcumin feeding in mice[40]; in other studies, total GST activity was induced by dietary curcumin in both mice and rats.[41–44] A structure–activity study of the potency of curcumin analogues suggested that their ability to induce phase II enzymes might be linked to the presence of the phenolic hydroxy and β-diketone groups.[44] Although induction of GST activity might be desirable in the prevention of the early stages of carcinogenesis, in patients with advanced malignancy GST isozymes might be aberrantly overexpressed and linked with resistance to chemotherapy.[45] Paradoxically, curcumin also appears capable of inhibiting GST isoenzyme expression. An example is provided by a study of GSTP1 expression in leukemia cells, in which an association was observed between the level of inhibition by curcumin and the induction of apoptosis.[46]

Nitric oxide (NO) is a short-lived, lipophilic molecule generated from L-arginine by various NADPH-dependent enzymes called NO synthases (NOS).[47] NO is involved physiologically in vasorelaxation, neurotransmission, inhibition of platelet aggregation, immune defense, and intracellular signaling. NO has an unpaired electron and is therefore a free-radical species; its bioactivity is related to the production of many reactive intermediates and many of these reactive nitrogen species are capable of damaging DNA or hindering DNA repair.[48,49] Peak inducible NOS (iNOS) activity might relate to the transition of colonic adenomas to carcinomas.[50] Upregulation of COX-2 via NF-κB or activator protein (AP)-1 pathways, or increasing intracellular concentrations of reduced glutathione, appears to confer resistance to NO-induced apoptosis in malignant cells *in vitro*.[51,52] *Ex vivo* studies have suggested that the inducibility of macrophage NOS activity is inhibited by 1–20 μM concentrations of curcumin.[53] In mice that received curcumin at a dose of as little as 92 ng/g body weight in aqueous alkaline solution with the drinking water, hepatic lipopolysaccharide-induced *iNOS* gene expression was significantly inhibited.[54] Because inhibition of iNOS activity might represent a mechanism of intervention during carcinogenesis, the activity of curcumin at such low concentrations has considerable implications for cancer chemoprevention. Nevertheless, this effect needs to be reproduced in other experiments.

3.2. Clinical Pharmacodynamics

3.2.1. Dose–Effect Relationships

Substantial data supporting a dose–response relationship for any biomarker of the pharmacological efficacy of curcumin in humans are currently lacking. Nevertheless, several observations in volunteers and patients suggest that curcumin might possess systemic biological activity at low oral doses. A single oral dose of 20 mg of curcumin appeared to induce contraction of the gallbladder as adjudged by ultrasound scanning in human volunteers, compared to an amylum placebo.[55] Two potential surrogate biomarkers of the efficacy of curcumin were evaluated in the blood of patients with advanced colorectal cancer who received up to 180 mg of curcumin per day for up to 4 months.[56] In three patients on 36 mg of curcumin daily, lymphocytic activity of GST was decreased with time to reach 41% of control (untreated) on day 29 of treatment. This decline was not observed at the higher dose levels and was not reproduced in a subsequent study of higher doses in the patients with the same disease.[21] Similarly, consumption of curcumin did not affect blood leukocyte levels of the oxidative DNA adduct, pyrimido-[1,2α]purin-10(3H)-one-2′-deoxyguanosine (M_1dG), although interesting observations were made regarding GST isoenzyme genotypes and baseline leukocytic M_1dG adduct levels.

In contrast to leukocyte M_1dG and GST, the inducibility of prostaglandin (PG) E_2 production in whole blood *ex vivo* might represent a surrogate biomarker for assessing the pharmacological activity of curcumin at a systemic level. COX-2 is an important target for chemoprevention, and its pharmacological modulation mght hold implications for cancer treatment. At least part of the effect of curcumin on inducible PGE_2 production in human blood can be attributed to inhibition of *COX2* transcription, which might be due to the inhibition of the NF-κB-activating enzymes IKK-α/β.[57,58] The effect of curcumin described in an *ex vivo* assay developed using blood from healthy volunteers[56] was associated with a daily oral dose which furnished plasma levels in the 10^{-8} M range in patients with advanced colorectal cancer.[16] This concentration of curcumin is less than a hundredth of that shown *in vitro* to elicit an effect in blood or colon cells.[14,57] Blood was taken immediately predose or 1 h postdose on days 1, 2, 8, and 29 of treatment with 3.6 g of curcumin daily.[21] Following the addition of acetylsalicylic acid (200 μM) to eliminate COX-1 activity, whole blood was incubated for 24 h in the presence of lipopolysaccharide (LPS, 10 μg/mL).[56] In the same trial, oral administration of curcumin did not impact on nonstimulated PGE_2 levels in leukocytes, nor did doses of 0.45–1.8 g daily alter LPS-induced PGE_2. In contrast, consumption of 3.6 g of curcumin daily affected LPS-induced PGE_2 levels.[21] When values obtained immediately predose or 1 h postdose on days 1, 2, 8, and 29 were pooled for the six patients consuming this dose, PGE_2 levels observed postdose were 46% lower than those measured immediately predosing. The difference reached significance on days 1 and 29 of treatment but not on day 2 or day 8; this discrepancy could not be explained scientifically by the study investigators.[21] Although these results tentatively suggest that consumption of 3.6 g of curcumin daily was linked with inhibition of PGE_2 induction in blood taken postdose compared to blood taken predose, overall time-dependent trends were not identified and dose–response was not demonstrated for this biomarker. Although the *ex vivo* assay described using human blood is limited in its clinical application by the high interindividual and high intraindividual variability,[56] the results suggest the feasibility and potential utility of measurement of PGE_2 levels in target tissue as a biomarker to reflect potential anticancer activity of curcumin. It should also be noted that curcumin sulfate and products of metabolic reduction of curcumin also inhibit PGE_2 production in colon cells grown *in vitro*, although their inhibitory potency appeared lower than that of parent curcumin in colon cancer cells.[14] Similarly, the activity of curcumin metabolites, relative to the parent compound, on levels of enzymes involved in glucose and lipid metabolism has been demonstrated in other preclinical models.[59]

In parallel with studies in which potential changes in blood taken from patients with advanced colorectal cancer were analyzed, exploratory clinical investigations have also been performed in patients undergoing operations for resectable colorectal cancer in whom colorectal and hepatic tissues have been analyzed to study potential pharmacodynamic effects.[22,23] Twelve patients with confirmed colorectal carcinoma received oral curcumin at 0.45, 1.8, or 3.6 g/day for 7 days prior to surgery. Whereas ingestion of 3.6 g of curcumin daily for 1 week affected M_1dG levels in patients' colorectal tissue, it did not decrease COX-2 protein expression in this tissue.[22] Interestingly, M_1dG adduct levels were 2.5-fold higher in malignant colorectal tissue than normal colorectal mucosa. Whereas administration of curcumin did not affect M_1dG levels in normal colorectal mucosa, it caused a 58% decrease in adduct levels in malignant colorectal tissue. The effect was only observed at the highest dose level; it requires replication in a larger study before definite conclusions can be made. As in the above-presented results, a dose–response relationship was not established. A similar study of hepatic tissue with the same oral dosing regime suggested that the levels of curcumin attained in normal and malignant liver tissues were insufficient to exert biological activity.[23]

3.2.2. Anti-inflammation

Data from preclinical models would suggest that suppression by curcumin of the inflammatory response might involve inhibition of the induction of COX-2 and iNOS and the production of cytokines such as interferon-γ, at least in part due to its suppression of the Janus kinase (JAK)–STAT signaling cascade via its effect on the Src homology 2 domain-containing protein tyrosine phosphatases (SHP)-2.[59] Compatible with these immunological effects, data from an experiment involving chemical-induced inflammatory bowel disease in mice suggested that curcumin might be of value in the treatment of this disease.[60]

A number of studies have addressed the effect of oral curcumin on inflammatory diseases in humans. Curcumin at 400 mg three times daily for 5 days caused a significant anti-inflammatory effect measured objectively and subjectively in postoperative patients.[61] In a double blind study, 300 mg curcumin was administered four times daily to 18 patients with rheumatoid arthritis for 2 weeks.[62] The authors reported a significant improvement in their inflammatory symptoms without apparent toxicity.[62]Ten patients with inflammatory bowel disease received pure curcumin at doses between 0.55 and 1.65 g daily for up to 2 months; all patients showed encouraging clinical improvement.[63]

One research team has studied the effects of oral curcumin on ophthalmological conditions. In their first study, 375 mg of curcumin was administered three times daily to patients with chronic anterior uveitis for 12 weeks, resulting in a suggestion of improvement in the condition.[64] In their second study, the same dose of curcumin was administered to eight patients with idiopathic inflammatory orbital pseudotumors for 6–22 months.[65] Complete responses were observed in half of the patients up to 2 years of follow-up. Although histopathological details were not presented in this report, inflammatory orbital pseudotumor is currently regarded as low-grade non-Hodgkin's lymphoma in most cases; hence, this result hints at potential anticancer activity.

3.2.3. Anticarcinogenesis

Induction of apoptosis by curcumin in cancer cells is mediated via a variety of mechanisms (see Chapter 11). Such findings suggest that curcumin might have some potential for chemotherapeutic activity in the treatment of cancer. In this regard, there are published anecdotes of the activity of curcumin as a topical treatment for cancer. One example is the use of turmeric as a treatment for oral cancers and leukoplakia, in which the authors report a significant reduction in the size of the lesions in 10% of the 62 patients treated.[66] Unfortunately, there was no control group, no assessment of anti-inflammatory activity and no chemical analysis of the preparation applied in this study.

In a pilot trial performed in the United Kingdom, low doses (36–180 mg) of curcumin were administered daily for up to 4 months to patients with progressive advanced colorectal cancer, refractory to standard chemotherapies.[20] Five out of 15 patients treated in this study experienced radiologically stable disease for 3 months or longer. In one patient, venous levels of the tumor marker carcinoembryonic antigen (CEA) decreased during treatment. In a subsequent study in patients with progressive advanced colorectal cancer, doses of 0.45–3.6 g of curcumin were administered daily. Radiologically stable disease was observed in 2 out of 15 patients for up to 4 months of treatment.[21]

Considering both of these studies together, although there were perhaps hints of cytostatic activity using macroscopic measures in these patients, the variable natural history of colorectal cancer renders the results difficult to interpret.

In a study performed in Taiwan, the potential anticancer activity of 1–8 g of curcumin (500 mg of curcumin per capsule, 99% pure) daily for 3 months was studied in patients with preinvasive malignant or high-risk premalignant conditions of the bladder, skin, cervix, stomach, or oral mucosa.[18] Doses above 8 g per day were not tolerated by patients because of the excessive number of capsules that had to be consumed daily. Histological improvement was noted in one of two patients with presumed bladder carcinoma *in situ*, two of seven patients with oral leukoplakia, one of six patients with stomach metaplasia, one of four patients with cervical intraepithelial neoplasia (CIN), and two of six patients with Bowen's disease of the skin. Conversely, in one of four patients with CIN and one of seven patients with oral leukoplakia, the treatment failed to prevent the development of invasive malignancy during the 3-month study period. The small numbers of patients with each condition, the variable natural history of premalignant lesions, and the lack of blinding of the interpreting pathologists make it difficult to draw definite conclusions. Nevertheless, the histological representations of treatment effects presented in this report reemphasise the biological activity that curcumin might possess in a range of human tissues.

3.3. Safety

Although the experience of using curcumin in the diet for many centuries inspires confidence in its safety, one cannot assume a priori that diet-derived agents are innocuous when administered as pharmaceutical formulations at doses that generally exceed those consumed in the dietary matrix. Anecdotal reports suggest that dietary consumption of turmeric up to 1.5 g per person per day, equating to a probable maximum of 150 mg of curcumin daily, are not associated with adverse effects in humans.[2] Studies of curcumin in animals have confirmed a lack of significant toxicity. In an early investigation, doses up to 5 g/kg were administered orally to Sprague–Dawley rats, resulting in no demonstrable toxicity.[8] Systematic preclinical safety studies orchestrated by the US National Cancer Institute (NCI) did not discover any adverse effects in rats, dogs, or monkeys at doses of up to 3.5 g/kg administered up to 3 months in duration.[24] One early report suggested a potentially ulcerogenic effect of dietary curcumin in the stomach of the albino rat,[67] but this finding has not been replicated in subsequent rodent studies. In more recent preclinical investigations of dietary curcumin, toxicity has not been observed at 2% of the diet in rats[16] (approximately 1.2 g/kg) or at 0.2% of the diet in mice[31] (approximately 300 mg/kg).

Only a few clinical studies of oral curcumin and curcuminoids have reported discernible adverse effects. Unfortunately, many investigators have not stated in their reports which methods or scales have been used to assess potential toxicity. Administration of 1.2–2.1 g of oral curcumin daily to patients with rheumatoid arthritis for 2–6 weeks did not result in any adverse effects.[62] Similarly, 10 patients with inflammatory bowel disease received pure curcumin daily at between 0.55 and 1.65 g for up to 2 months without clinical manifestations of toxicity.[63] No significant adverse events were observed in a study of up to 8 g of oral curcumin daily for 3 months in patients with preinvasive malignant or high-risk premalignant conditions.[18] In patients with advanced colorectal cancer, curcumin was well tolerated at all dose levels up to 3.6 g daily for up to 4 months.[21] Two types of gastrointestinal adverse event were reported by patients in this study, which were probably related to curcumin consumption: Two patients (on 0.45 and 3.6 g curcumin daily) developed diarrhea (NCI grades 1 and 2) 1 month and 4 months into treatment, respectively. One patient, who was receiving 0.9 g of curcumin daily, experienced nausea (NCI toxicity grade 2), which resolved spontaneously despite continuation of treatment. Two abnormalities were detected in blood tests, both of which might have been related either to curcumin treatment or malignant disease progression: Increases in serum alkaline phosphatase levels were observed in four patients and in serum lactate dehydrogenase in three patients.

Overall, the comprehensive preclinical data and the phase I clinical data currently available inspire cautious confidence that curcumin possesses a safety spectrum suitable for a chemopreventive agent. In the planning of future studies using curcumin in any patient group, special attention must to be paid to the systematic documentation of potential toxicities that ultimately might influence the feasibility of administering curcumin to healthy individuals over prolonged periods in the setting of clinical chemoprevention.

4. CONCLUSIONS

Curcumin possesses anti-inflammatory and anticarcinogenic properties in rodent models and there are suggestions of related pharmacodynamic effects in a few clinical pilot studies. Many of these activities can be attributed to its potent antioxidant capacity, its inhibition of cell signaling pathways at multiple levels, and its diverse effects on cellular enzymes, angiogenesis, and cell adhesion. In particular, the ability of curcumin to affect gene transcription and induce apoptosis in malignant cells advocates its potential utility in cancer chemoprevention and perhaps chemotherapy. Phase I clinical data have confirmed that the low systemic bioavailability of curcumin following oral dosing limits the tissues that the parent compound can reach at efficacious concentrations to exert beneficial effects. Nevertheless, the attainment of physiologically active levels of curcumin in the gastrointestinal tract, particularly the colon and rectum, has been demonstrated in animals and humans. In view of the pharmacological properties of curcumin presented here, its phase II clinical evaluation in individuals at risk of developing cancers of the gastrointestinal tract appears opportune.

REFERENCES

1. B. Brouk, *Plants Consumed by Man*. New York: Academic Press, 1975, p. 331.
2. D. Eigner and D. Sholz, Ferula *asa–foetida* and curcuma longa in traditional medical treatment and diet in Nepal. *J Ethnopharmacol* **67**, 1–6 (1999).
3. J. Milobedzka, V. Kostanecki, and V. Lampe, Structure. *Chem Ber* **43**, 2163 (1910).
4. D. D. Heath, F. Khwaja, and C. L. Rock, Curcumin content of turmeric and curry powders. *FASEB J* **18**, A125 (2004).
5. H. P. Ammon and M. A. Wahl, Pharmacology of curcuma longa. *Planta Med* **57**, 1–7 (1991).
6. B. B. Aggarwal, A. Kumar, and A. C. Bharti, Anticancer potential of curcumin: Preclinical and clinical studies. *Anticancer Res* **23**, 363–398 (2003).
7. B. Joe, M. Vijaykumar, and B. R. Lokesh, Biological properties of curcumin: Cellular and molecular mechanisms of action. *Crit Rev Food Sci Nutr* **44**, 97–111 (2004).
8. B. Wahlstrom and G. Blennow, A study on the fate of curcumin in the rat. *Acta Pharmacol Toxicol* **43**, 86–92 (1978).
9. V. Ravindranath and N. Chandrasekhara, Absorption and tissue distribution of curcumin in rats. *Toxicology* **16**, 259–265 (1980).
10. V. Ravindranath and N. Chandrasekhara, In vitro studies on the intestinal absorption of curcumin in rats. *Toxicology* **20**, 251–257 (1981).
11. V. Ravindranath and N. Chandrasekhara, Metabolism of curcumin: Studies with [3H] curcumin. *Toxicology* **22**, 337–344 (1981).

12. G. M. Holder, J. L. Plummer, and A. J. Ryan, The metabolism and excretion of curcumin [1,7-bis-(4-hydroxy-3-methoxyphenyl)-1,6-heptadiene-3,5-dione] in the rat. *Xenobiotica* **8**, 761–768 (1978).
13. M. H. Pan, T. M. Huang, and J. K. Lin, Biotransformation of curcumin through reduction and glucuronidation in mice. *Drug Metab Dispos* **27**, 486–494 (1999).
14. C. Ireson, S. Orr, D. J. L. Jones, et al., Characterization of metabolites of the chemopreventive agent curcumin in humans and rat hepatocytes and in the rat in vivo, and evaluation of their ability to inhibit phorbol ester-induced prostaglandin E_2 production. *Cancer Res* **61**, 1058–1064 (2001).
15. C R. Ireson, D. J. L. Jones, S. Orr, M. W. H. Coughtrie, D. Boocock, M. L. Williams, P. B. Farmer, W. P. Steward, and A. J. Gescher, Metabolism of the cancer chemopreventive agent curcumin in human and rat intestine. *Cancer Epidemiol Biomarkers Prev* **11**, 97–104 (2002).
16. R. A. Sharma, C. R. Ireson, R. D. Verschoyle, et al., Effects of dietary curcumin on glutathione S-transferase and malondialdehyde-DNA adducts in rat liver and colon mucosa: Relationship with drug levels. *Clin Cancer Res* **7**, 1452–1458 (2001).
17. G. Shoba, D. Joy, T. Joseph, M. Majeed, R. Rajendran, and P. S. S. R. Srinivas, Influence of piperine on the pharmacokinetics of curcumin in animals and human volunteers. *Planta Med* **64**, 353–356 (1998).
18. A. L. Cheng, C, H. Hsu, J. K. Lin, et al., Phase I clinical trial of curcumin, a chemopreventive agent, in patients with high-risk or pre-malignant lesions. *Anticancer Res* **21**, 2895–2900 (2001).
19. M. T. Ruffin, D. P. Normolle, D. D. Heath, J. M. Bailey, S. I. Murray, M. E. Boggs, J. A. Crowell, C. L. Rock, and D. E. Brenner, Dose escalation of curcumin in healthy adults. *Cancer Epidemiol Biomarkers Prev* **12**(Pt 2 Suppl S), 1324S–1324S (2003).
20. R. A. Sharma, H. R. McLelland, K. A. Hill, et al., Pharmacodynamic and pharmacokinetic study of oral *Curcuma* extract in patients with colorectal cancer. *Clin Cancer Res* **7**, 1894–1900 (2001).
21. R. A. Sharma, S. A. Euden, S. L. Platton, et al., Phase I clinical trial of oral curcumin: Biomarkers of systemic activity and compliance. *Clin Cancer Res* **10**, 6847–6854 (2004).
22. G. Garcea, D. P. Berry, D. J. L. Jones, et al., Consumption of the putative chemopreventive agent curcumin by cancer patients: Assessment of curcumin levels in the colorectum and their pharmacodynamic consequences. *Cancer Epidemiol Biomarkers Prev* **14**, 120–125 (2005).
23. G. Garcea, D. J. L. Jones, R. Singh, et al., Detection of curcumin and its metabolites in hepatic tissue and portal blood of patients following oral administration. *Br J Cancer* **90**, 1011–1015 (2004).
24. NCI, DCPC, Clinical development plan: Curcumin. *J Cell Biochem* **26S**, 72–85 (1996).
25. C. V. Rao, A. Rivenson, B. Simi, and B. S. Reddy, Chemoprevention of colon carcinogenesis by dietary curcumin, a naturally occurring plant phenolic compound. *Cancer Res* **55**, 259–266 (1995).
26. K. B. Soni, A. Rajan, and R. Kuttan, Reversal of aflatoxin induced liver damage by turmeric and curcumin. *Cancer Lett* **66**, 115–121 (1992).
27. T. Kawamori, R. Lubet, V. E. Steele, et al., Chemopreventative effect of curcumin, a naturally occurring anti-inflammatory agent, during the promotion/progression stages of colon cancer. *Cancer Res* **59**, 597–601 (1999).

28. H. S. Samaha, G. J. Kelloff, V. Steele, C. V. Rao, and B. S. Reddy, Modulation of apoptosis by sulindac, curcumin, phenylethyl-3-methylcaffeate, and 6-phenylhexyl isothiocyanate, Apoptotic index as a biomarker in colon cancer chemoprevention and promotion. *Cancer Res* **57**, 1301–1305 (1997).
29. S. R. Volate, D. M. Davenport, S. J. Muga, and M. Wargovich, Modulation of aberrant crypt foci and apoptosis by dietary herbal supplements (quercetin, curcumin, silymarin, ginseng and rutin). *Carcinogenesis* **26**, 1450–1456 (2005).
30. C. Luongo, A. R. Moser, S. Gledhill, and W. F Dove, Loss of *Apc*(+) in intestinal adenomas from Min mice. *Cancer Res* **54**, 5947–5952 (1994).
31. S. Perkins, R. D. Verschoyle, K. A. Hill, I. Parveen, M. D. Threadgill, R. A. Sharma, M. L. Williams, W. P. Steward, and A. J. Gescher, Chemopreventive efficacy and pharmacokinetics of curcumin in the Min/+ mouse, a model of familial adenomatous polyposis. *Cancer Epidemiol Biomarkers Prev* **11**, 535–540 (2002).
32. R. G. Tunstall, R. A. Sharma, S. Perkins, S. Sale, R. Singh, P. B. Farmer, W. P. Steward, and A. J. Gescher, Cyclooxygenase-2 expression and oxidative DNA adducts in murine intestinal adenomas: Modification by dietary curcumin and implications for clinical trials. *Eur J Cancer* **42**, 415–421 (2006).
33. A. H. Conney, Enzyme induction and dietary chemicals as approaches to cancer chemoprevention: The seventh DeWitt S. Goodman Lecture. *Cancer Res* **63**, 7005–7031 (2003).
34. N. Li, X. Chen, J. Liao, et al. Inhibtion of 7,12-dimethylbenz[a]anthracene (DMBA)-induced oral carcinogenesis in hamsters by tea and curcumin. *Carcinogenesis* **23**, 1307–1313 (2002).
35. K. Imaida, S. Tamano, K. Kato, Y. Ikeda, M. Asamoto, S. Takahashi, Z. Nir, M. Murakoshi, H. Nishino, and T. Shirai, Lack of chemopreventive effects of lycopene and curcumin on experimental rat prostate carcinogenesis. *Carcinogenesis* **22**, 467–472 (2001).
36. B. B. Aggarwal, S. Shishodia, Y. Takada, S. Banerjee, R. A. Newman, C. E. Bueso-Ramos, and J. E. Price, Curcumin suppresses the paclitaxel-induced nuclear factor-kappa B pathway in breast cancer cells and inhibits lung metastasis of human breast cancer in nude mice.*Clin Cancer Res* **11**, 7490–7498 (2005).
37. S. V. Bava, V. T. Puliappadamba, A. Deepti, A. Nair, D. Karunagaran, and R. J. Anto, Sensitization of taxol-induced apoptosis by curcumin involves down-regulation of nuclear factor-kappa B and the serine/threonine kinase Akt and is independent of tubulin polymerization. *J Biol Chem* **280**, 6301–6308 (2005)
38. Y. Shukla, A. Arora, and P. Taneja, Antimutagenic potential of curcumin on chromosomal aberrations in Wistar rats. *Mutat Res: Genet Toxicol Environ Mutag* **515**, 197–202 (2002).
39. M. E. Egan, M. Pearson, S. A. Weiner, et al., Curcumin, a major constituent of turmeric, corrects cystic fibrosis defects. *Science* **304**, 600–602 (2004).
40. S. V. Singh, X. Hu, S. K. Srivastava, et al., Mechanism of inhibition of benzo[*a*]pyrene-induced forestomach cancer in mice by dietary curcumin. *Carcinogenesis* **19**, 1357–1360 (1998).
41. M. Susan and M.A. Rao, Induction of glutathione S-transferase activity by curcumin in mice. *Drug Res* **42**, 962–964 (1992).
42. J. T. Piper, S. S. Singhal, M. Salameh, R. T. Torman, Y. C. Awasthi, and S. Awasthi, Mechanisms of anticarcinogenic properties of curcumin: The effect of curcumin on glutathione linked detoxification enzymes in rat liver. *Int J Biochem Cell Biol* **30**, 445–456 (1998).

43. W. A. Nijhoff, G. M. Groen, and W. H. M. Peters, Induction of rat hepatic and intestinal glutathione S-transferases and glutathione by dietary naturally occurring anticarcinogens. *Int J Oncol* **3**, 1131–1139 (1993).
44. A. T. Dinkova-Kostova and P. Talalay, Relation of structure of curcumin analogs to their potencies as inducers of Phase 2 detoxification enzymes. *Carcinogenesis* **20**, 911–914 (1999).
45. J. D. Hayes and D. J. Pulford, The glutathione S-transferases supergene family. *Crit Rev Biochem Mol Biol* **30**, 445–600 (1995).
46. A. Duvoix, S. Delhalle, R. Blasius, et al., Effect of chemopreventive agents on glutathione S-transferase P1-1 gene expression mechanisms via activating protein 1 and nuclear factor kappaB inhibition. *Biochem Pharmacol* **68**, 1101–1111 (2004).
47. P. K. Lala and C. Chakraborty, Role of nitric oxide in carcinogenesis and tumour progression. *Lancet Oncol* **2**, 149–156 (2001).
48. T. deRojas-Walker, S. Tamir, H. Ji, J. S. Wishnok, and S. R. Tannenbaum, Nitric oxide induces oxidative damage in addition to deaminiation in macrophage DNA. *Chem Res Toxicol* **8**, 473–477 (1995).
49. M. Graziewicz, D. A. Wink, and F. Laval, Nitric oxide inhibits DNA ligase activity: Potential mechanisms for NO-mediated DNA damage. *Carcinogenesis* **17**, 2501–2505 (1996).
50. S. Ambs, W. P. Bennett, W. G. Merrium, et al., Relationship between p53 mutations and inducible nitric oxide synthase expression in human colorectal cancer. *J Natl Cancer Inst* **91**, 86–88; reply 1510–1511 (1999).
51. A. Von Knethen and B. Brune B, Cyclooxygenase-2: An essential regulator of NO-mediated apoptosis. *FASEB J* **11**, 887–895 (1997).
52. A. Von Knethen, D. Callsen, and B. Brune, NF-κB and AP-1 activation by nitric oxide attenuated apoptotic death in RAW 264.7 macrophages. *Mol Biol Cell* **10**, 361–370 (1999).
53. I. Brouet and H. Ohshima, Curcumin, an anti-tumour promoter and anti-inflammatory agent, inhibits induction of nitric oxide synthase in activated macrophages. *Biochem Biophys Res Commun* **206**, 533–540 (1995).
54. M. M. Chan, H. I. Huang, M. R. Fenton,and D. Fong, *In vivo* inhibition of nitric oxide synthase gene expression by curcumin, a cancer preventive natural product with anti-inflammatory properties. *Biochem Pharmacol* **55**, 1955–1962 (1998).
55. A. Raysid and A. Lelo, The effect of curcumin and placebo on human gall-bladder function: An ultrasound study. *Aliment Pharmacol Ther* **13**, 245–249 (1999).
56. S. M. Plummer, K. A. Hill, M. F. W. Festing, W. P. Steward, A. J. Gescher, and R. A. Sharma, Clinical development of leukocyte cyclooxygenase 2 activity as a systemic biomarker for cancer chemopreventive agents. *Cancer Epidemiol Biomarkers Prev* **10**, 1295–1299 (2001).
57. S. M. Plummer, K. A. Holloway, M. M. Manson, et al., Inhibition of cyclo-oxygenase 2 expression in colon cells by the chemopreventive agent curcumin involves inhibition of NF-κB activation via the NIK/IKK signalling complex. *Oncogene* **18**, 6013–6020 (1999).
58. F. Zhang, N. K. Altorki, J. R. Mestre, et al., Curcumin inhibits cyclooxygenase-2 transcription in bile acid- and phorbol ester-treated human gastrointestinal epithelial cells. *Carcinogenesis* **20**, 445–451 (1999).
59. A. C. Bharti, N. Donato, and B. B. Aggarwal, Curcumin (diferuloylmethane) inhibits constitutive and IL-6-inducible STAT3 phosphorylation in human multiple myeloma cells. *J Immunol* **171**, 3863–3871 (2003).

60. K. Sugimoto, H. Hanai, K. Tozawa, et al., Curcumin prevents and ameliorates trinitrobenzene sulfonic acid-induced colitis in mice. *Gastroenterology* **123**, 1912–1922 (2002).
61. R. R. Satoskar, S. J. Shah, and S. G. Shenoy, Evaluation of anti-inflammatory property of curcumin (diferuloyl methane) in patients with post-operative inflammation. *Int J Clin Pharmacol Ther Toxicol* **24**, 651–654 (1986).
62. S. D. Deodhar, R. Sethi, and R. C. Srimal, Preliminary study on antirheumatic activity of curcumin (diferuloyl methane). *Indian J Med Res* **71**, 632–634 (1980).
63. P. R. Holt, S. Katz, and R. Kirshoff, Curcumin therapy in inflammatory bowel disease: A pilot study. *Dig Dis Sci* **11**, 2191–2193 (2005).
64. B. Lal, A. K. Kapoor,O. P. Asthana, et al., Efficacy of curcumin in the managememtn of chronic anterior uveitis. *Phytother Res* **13**, 318–322 (1999).
65. B. Lal, A. K. Kapoor, P. K. Agrawal, et al., Role of curcumin in idiopathic inflammatory orbital pseudotumours. *Phytother Res* **14**, 443–447 (2000).
66. R. Kuttan, P. C. Sudheeran, and C. D. Josph, Turmeric and curcumin as topical agents in cancer therapy. *Tumori* **73**, 29–31 (1987).
67. W. H.Chan and H. Wu, Anti-apoptotic effects of curcurnin on photosensitized human epidermal carcinoma A431 cells. *J Cell Biochem* **92**, 200–212 (2004).
68. M. C. Jiang, H. F. Yang Yen, J. J. Y. Yen, et al., Curcumin induces apoptosis in immortalized NIH 3T3 and malignant cancer cell lines. *Nutr Cancer* **26**, 111–120 (1996).
69. C. H. Yan, M. S. Jamaluddin, B. Aggarwal, J. Myers, and D.D. Boyd, Gene expression profiling identifies activating transcription factor 3 as a novel contributor to the proapoptotic effect of curcumin. *Mol Cancer Therapeut.* **4**, 233–241 (2005).
70. M. L. Kuo, T. S. Huang, and J. K. Lin, Curcumin, an antioxidant and anti-tumor promoter, induces apoptosis in human leukemia cells. *Biochim Biophys Acta* **1317**, 95–100 (1996).
71. G. P. Collett and F. C. Campbell, Curcumin induces c-jun N-terminal kinase-dependent apoptosis in HCT116 human colon cancer cells. *Carcinogenesis* **25**, 2183–2189 (2004).
72. M. M. Y. Chan, Inhibition of tumor necrosis factor by curcumin, a phytochemical. *Biochem Pharmacol* **49**, 1551–1556 (1995).
73. Y. Abe, S. Hashimoto, and T. Horie, Curcumin inhibition of inflammatory cytokine production by human peripheral blood monocytes and alveolar macrophages. *Pharmacol Res* **39**, 41–47 (1999).
74. B. Gupta and B. Ghosh, Curcuma longa inhibits TNF-alpha induced expression of adhesion molecules on human umbilical vein endothelial cells. *Int J Immnopharmacol* **21**, 745–757 (1999).
75. Y. Moon, W. C. Glasgow, and T. E. Eling, , Curcumin suppresses interleukin 1 beta-mediated microsomal prostaglandin E synthase 1 by altering early growth response gene 1 and other signaling pathways. *J Pharmacol Exp Ther* **315**, 788–795 (2005).
76. E. Balogun, M. Hoque, P. F. Gong, E. Killeen, C. J. Green, R. Foresti, J. Alam, and R. Motterlini, , Curcurmin activates the haem oxygenase-1 gene via regulation of Nrf2 and the antioxidant-responsive element. *Biochem J* **371**, 887–895 (2003).
77. C. K. Andreadi, L. M. Howells, P. A. Atherfold, and M. M. Manson, Involvement of Nrf2, p38, B-Raf, and nuclear factor-kappa B, but not phosphatidylinositol 3-kinase, in induction of hemeoxygenase-1 by dietary polyphenols. *Mol Pharmacol* **69**, 1033–1040 (2006).

78. P. F. Firozi, V. S. Aboobaker, and R. K. Bhattacharya, Action of curcumin on the cytochrome P450-system catalyzing the activation of aflatoxin B1. *Chem-Biol Interact* **100**, 41–51 (1996).
79. H. P. Ciolino, P. J. Daschner, T. T. Wang, and G. C. Yeh, Effect of curcumin on the aryl hydrocarbon receptor and cytochrome P450 1A1 in MCF-7 human breast carcinoma cells. *Biochem Pharmacol* **56**, 197–206 (1998).
80. M. S. Woo, S. H. Jung, S. Y. Kim, J. W. Hyun, K. H. Ko, W. K. Kim, and H. S. Kim, Curcumin suppresses phorbol ester-induced matrix metalloproteinase-9 expression by inhibiting the PKC to MAPK signaling pathways in human astroglioma cells. *Biochem Biophys Res Commun* **335**, 1017–1025 (2005).

CLINICAL STUDIES WITH CURCUMIN

Chih-Hung Hsu and Ann-Lii Cheng

Abstract: Curcumin has long been expected to be a therapeutic or preventive agent for several major human diseases because of its antioxidative, anti-inflammatory, and anticancerous effects. In phase I clinical studies, curcumin with doses up to 3600–8000 mg daily for 4 months did not result in discernible toxicities except mild nausea and diarrhea. The pharmacokinetic studies of curcumin indicated in general a low bioavailability of curcumin following oral application. Nevertheless, the pharmacologically active concentration of curcumin could be achieved in colorectal tissue in patients taking curcumin orally and might also be achievable in tissues such as skin and oral mucosa, which are directly exposed to the drugs applied locally or topically. The effect of curcumin was studied in patients with rheumatoid arthritis, inflammatory eye diseases, inflammatory bowel disease, chronic pancreatitis, psoriasis, hyperlipidemia, and cancers. Although the preliminary results did support the efficacy of curcumin in these diseases, the data to date are all preliminary and not conclusive. It is imperative that well-designed clinical trials, supported by better formulations of curcumin or novel routes of administration, be conducted in the near future.

1. INTRODUCTION

The medicinal use of the turmeric plant has been described in ancient Indian and Chinese thousands of years ago. In India, turmeric has traditionally been used primarily for arthritis and muscular disorders, whereas in China, it has been used as a topic analgesic and for conditions ranging from flatulence, colic, hepatitis, to chest pain.[1] Over the past decades, numerous preclinical studies have suggested that curcumin has beneficial effects in human diseases such as arthrosclerosis, hyperlipidemia, thromboembolism, myocardial infarction, diabetes, rheumatoid arthritis, human immunodeficiency virus infection, and cancers.[2–5] The chapter will review the results of clinical studies of curcumin in published journals.

2. PHASE I CLINICAL STUDIES OF CURCUMIN

2.1. Safety of Curcumin in Humans

Although curcumin is presumed to be safe by the long history of human use in India and Asia, it was not until 2001 when the first phase I study of curcumin in humans reported by Cheng and colleagues.[6] Curcumin, manufactured by chemical methods, was formulated as a 500-mg tablet with 99.3% purity. The study enrolled patients with premalignant lesions or high-risk individuals, including those with oral leukoplakia, intestinal metaplasia of the stomach, uterine cervical intraepithelial neoplasia (CIN), Bowen's disease of the skin, and recently resected bladder cancer. The study subjects took curcumin at doses of 500, 1000, 2000, 4000, and 8000 mg daily for 3 months. The dose was escalated to the next level when less than one-third of three to six patients at a certain dose level experienced no toxicity greater than grade 1 during the 3-month treatment period. The results revealed no treatment-related toxicity up to 8000 mg/day for 3 months. However, further escalation of curcumin was prohibited due to the bulky volume of curcumin tablets.

Sharma and colleagues conducted two phase I studies of curcumin with different formulations in patients with colorectal cancer who were refractory to conventional chemotherapeutic agents.[7,8] In the first study, curcumin was formulated as a "*Curcuma* extract," which was a 220-mg soft gelatin capsule containing 20 mg curcuminoids (18 mg of curcumin and 2 mg of demethoxycurcumin) suspended in 200 mg of essential oils derived from *Curcuma* spp.[7] The constitutents of *Curcuma* essential oil mixtures were tumoerone, atlantone, and zingiberene. Three patients were enrolled for each dose level, which was escalated from 440 mg *Curcuma* extract (equivalent to 36 mg curcumin) to 2200 mg *Curcuma* extract (equivalent to 180 mg curcumin) per day for up to 4 months. The treatment was well tolerated in 15 patients, and there was no dose-limiting toxicity. One patient on 1320 mg of *Curcuma* extract daily experienced grade 1 nausea; and two patients (one on 880 mg and the other on 2200 mg of Curcuma extract daily) developed grade 1 to grade 2 diarrhea. In the subsequent phase I study, Sharma and colleagues evaluated another formulation, which was a 500-mg curcuminoid capsule containing 450 mg curcumin, 40 mg demethoxycurcumin, and 10 mg bisdemethoxycurcumin.[8] The dose levels of curcumin were 450, 900, 1800, or 3600 mg per day for up to 4 months. A total of 15 patients with refractory colorectal cancer were enrolled. Again, the drug was well tolerated, except three patients experienced minor gastrointestinal adverse events, including grade 1 diarrhea and grade 2 nausea. In addition, a minor rise of serum alkaline phosphatase level and serum lactate dehydrogenase (compatible with grade 1–2 toxicity) was observed in four and three patients, respectively.

Lao and colleagues studied the safety of curcumin in healthy volunteers.[9] Healthy volunteers were enrolled for a single oral administration of curcumin with doses escalating from 500 to 12,000 mg. The curcumin was provided in a capsule form containing a standardized powder extract from Alleppey finger turmeric (C3 Complex™, Sabinsa Corp.). The curcumin capsule was composed of a minimum 95% concentration of three curcuminoids: curcumin (75%), bisdemethoxycurcumin (2%), and demethoxycurcumin (23%). Safety was assessed for 72 h following the curcumin administration. Among 24 enrolled patients, 7 developed adverse effects, including diarrhea, headache, rash, and yellowish stool.

All of the toxicities were grade 1 and not dose-related. The maximal tolerated dose of curcumin was not reached because doses more than 12,000 mg were unacceptable to patients due to the bulky volume of the tablets.

In conclusion, these phase I clinical studies confirmed the safety of curcumin in humans for a period up to 4 months of continuous treatment. In patients with premalignant lesions or patients with advanced colorectal cancers treated with curcumin with doses as high as 3600–8000 mg daily for up to 4 months, there was almost no discernible toxicity related to curcumin, except for a few gastrointestinal side effects, including nausea and diarrhea, which were usually very mild and easily manageable. In healthy volunteers, the single oral application of curcumin with doses up to 12,000 mg did not result in significant adverse effects. The maximal tolerated dose, a traditional end point for anticancer chemotherapy, of curcumin has not been reached in these studies.

2.2. Bioavailability of Curcumin

In the phase I study conducted by Cheng and colleagues, the serum concentration of curcumin was determined by high-performance liquid chromatography (HPLC) in selected patients after ingestion of the first dose of curcumin.[6] The average C_{max} (maximum serum concentration) of curcumin in patients receiving 4000, 6000, and 8000 mg/day was 0.51 ± 0.11, 0.64 ± 0.06, and 1.77 ± 1.87 μM, respectively, peaking at 1–2 h following oral ingestion. No curcumin could be detected in the urine.

Two phase I studies conducted by Sharma and colleagues also evaluated the pharmacokinetics of curcumin in advanced colorectal cancer patients.[7,8] In the first study using *Curcuma* extract with the doses of curcumin ranging from 36 to 180 mg daily, neither curcumin nor its metabolites were detected in blood or urine, but curcumin was recovered from feces.[7] In the subsequent phase I study using a 500-mg curcuminoid capsule (450 mg curcumin, 40 mg demethoxycurcumin, and 10 mg bisdemethoxycurcumin per capsule), measurement of curcumin and its metabolites (curcumin glucuronide and curcumin sulfate) from plasma, urine, and feces was analyzed by HPLC and mass spectrometry.[8] Curcumin and its metabolites were detected in plasma and urine only in patients receiving 3600 mg daily, but not in those who received 1800 mg or less. Plasma curcumin level was detectable only in three of six patients receiving 3600 mg daily, with a mean of 11.1 ± 0.6 nM. However, the presence of curcumin metabolites was detected in the plasma of all six patients consuming 3600 mg daily, with a mean of 15.8 ± 0.9 and 8.9 ± 0.7 nM for curcumin glucuronide and curcumin sulfate, respectively. Interestingly, high levels of curcumin and its metabolites were detected in the urine of all patients consuming 3600 mg daily, ranging between 0.1 and 1 μM for curcumin, between 210 and 510 nM for curcumin glucuronide, and between 19 and 45 nM for curcumin sulfate. The findings indicate that a certain amount of curcumin is absorbed, metabolized, and excreted in urine in patients taking curcumin at doses of 3600 mg per day.

Two additional studies were conducted to understand the distribution of curcumin and its metabolites in intestinal tissues and liver.[10,11] Patients with hepatic metastases of colorectal cancer were enrolled. Patients were treated with curcuminoid capsules (450 mg curcumin, 40 mg demethoxycurcumin, and 10 mg bisdemethoxycurcumin per capsule) with three different dose levels: 450, 1800, and 3600 mg of curcumin daily for 7 days prior to surgery. Curcumin and its metabolites in portal circulation, liver tissues, and colorectal tissues were determined. Curcumin was detected in normal mucosa and in malignant colorectal tissues in patients receiving 1800 or 3600 mg of curcumin daily.[11] The tissue concentrations were 19.6–12.7 and 6.7–7.7 nmol/g for benign and malignant colorectal tissues, respectively. Curcumin sulfate and curcumin glucuronide were identified in the intestinal tissues, too. Curcumin and its metabolites were not found in bile or normal and malignant liver tissues in any patient.[10] However, trace amount of hexahydrocurcumin and hexahydrocurcuminol was detected in the normal liver tissue from one patient receiving 3600 mg of curcumin daily.

In the study conducted in healthy volunteers by Lao and colleagues, sera taken prior to and 1, 2, and 4 h after application of curcumin were analyzed by HPLC.[9] Curcumin was not detectable in all samples except that of two subjects who were at 10,000 and 12,000 mg dose levels, respectively. The serum curcumin levels of these two subjects were 29.7–30.4 ng/mL (equivalent to ~0.1 μM) at 1 h after taking curcumin and 50.5–51.2 ng/mL (equivalent to ~0.15 μM) at 4 h, respectively.

In conclusion, data from the aforementioned phase I studies, albeit using different formulations of curcumin, indicate a low bioavailability of curcumin following oral application. However, oral intake with curcumin at doses as high as 3600–12,000 mg could result in detectable levels of curcumin and its metabolites in plasma and urine, indicating that an active absorption and metabolism of curcumin does occur. Importantly, the pharmacologically active levels of curcumin could be achieved in colorectal tissue in patients taking curcumin orally. Although relevant data are lacking, it is very likely that the biologically active concentrations of curcumin can be achieved in tissues that are directly exposed to the drugs applied locally or topically; examples of these, in addition to the gastrointestinal tract, include oral lozenges for leukoplakia or oropharyngeal lesions, vaginal suppository tablets for uterine cervix lesions, and topical applications for skin lesions.

3. PHASE II HUMAN CLINICAL STUDIES OF CURCUMIN

A number of clinical studies, most of which were single-arm phase II design, have suggested that curcumin might be beneficial in diseases such as chronic inflammation, malignancies, and premalignant lesions. Unfortunately, most of these studies employed only small numbers of patients, and none of these observations has been verified by other groups of investigators. Further, although some of these trials did include a control group as a part of the study design, a large-scale randomized control study of curcumin has never been conducted.

3.1. Anti-inflammatory Effect of Curcumin

3.1.1. Oral Application of Curcumin for Inflammatory Diseases

Deodhar and colleagues randomized 18 patients with rheumatoid arthritis to receive either curcumin (1200 mg/day) or phenylbutazone (300 mg/day) for 2 weeks.[12] Patients receiving curcumin had improvement of rheumatoid symptoms as much as that of patients receiving phenylbutazone. The administration of curcumin caused no discernible side effects. However, the long-term effect of curcumin in rheumatoid arthritis has not been reported.

Lal and colleagues studied the effect of curcumin in two inflammatory eye diseases.[13,14] In one study, 53 patients with chronic anterior uveitis were enrolled for the treatment of curcumin, 375 mg three times daily, for 12 weeks.[13] The curcumin preparation, isolated from rhizomes of *Curcuma longa* with more than 95% purity, was formulated in a gelatin capsule. Only 32 patients who had completed 12 weeks of curcumin treatment were reported. Uveitis-related symptoms and signs, such as pain, redness, lacrimation, poor vision, circumciliary congestion, keratic precipitates, flare, and vitreous turbidity, improved in 86–100% of cases. However, during the subsequent 3 years of follow-up, 47% of patients experienced relapse of diseases, and around 20% of patients lost their vision. In the other study, eight patients of idiopathic inflammatory orbital pseudotumor, previously treated by corticosteroids, were treated with curcumin, 375 mg three times a day.[14] One patient had histological diagnosis, and the other patients were diagnosed solely by clinical and imaging studies. Three patients dropped out after 2–8 weeks of treatment. Among the remaining five patients who had been treated for 6–22 months, four showed complete regression. The other patient had complete normalization of the imaging study, but still suffered from a partial limitation of ocular movement. Taken together, curcumin in such a dose and schedule was well tolerated and appeared effective in patients with chronic anterior uveitis or inflammatory orbital pseudotumor.

Holt and colleagues reported the use of curcumin in the treatment of inflammatory bowel disease.[15] Five patients with ulcerative proctitis were treated with curcumin, 550 mg twice daily for 1 month, followed by 550 mg three times daily for another month. All patients improved. Another five patients with Crohn's disease were treated with curcumin, 360 mg three times daily for 1 month, followed by 360 mg four times daily for another 2 months. Four of five patients with Crohn's disease improved, as evidenced by improvement of the surrogate end points including Crohn's Disease Activity Index (CDAI) and erythrocyte sedimentation rate (ESR).

Durgaprasad and colleagues reported the efficacy of curcumin in patients with nonalcoholic chronic pancreatitis of the tropics.[16] A total of 20 patients were randomized to receive either placebo or curcumin; the latter was a mixture of

500 mg of pure extract of curcumin and 5 mg of piperine, three times daily for 6 weeks. Only 15 patients (75%) returned for evaluation following 6 weeks of treatment. There was no improvement in pain as assessed by the visual analog score. Nevertheless, in the curcumin-treated group, there was a significant reduction in the serum erythrocyte malonyldialdehyde level and an increase in the serum glutathione level, suggesting a reversion of excessive lipid peroxidation.

3.1.2. Topical Use of Curcumin for Inflammatory Diseases

Psoriasis is a skin disease characterized by hyperproliferation and inflammation. Heng and colleagues reported the topical use of curcumin in the treatment of active plaque lesions.[17] The topical curcumin was prepared in an alcoholic gel containing 1% curcumin. In the first cohort of 10 patients treated by curcumin, 5 had more than a 90% resolution of psoriasis after 2–6 weeks of treatment, and the remaining 5 patients showed a 50–85% resolution after 3–8 weeks. In the second cohort of six patients, two untreated plaques of similar disease activity were selected—one treated with curcumin gel twice daily and the other with the alcoholic gel base. After 3–4 weeks, six of six curcumin-treated plaques improved by 25–70%, compared with no improvement (two of six) or worsening (four of six) of the vehicle-treated plaques.

3.2. Anticancer Effect of Curcumin

3.2.1. Effect of Curcumin for Advanced Cancers

Kuttan and colleagues reported the use of turmeric as a topical treatment for oral cancers and leukoplakia. Of 62 patients enrolled, 10% showed a reduction in lesion sizes.[18]

Although there is no formally reported phase II trial for curcumin in the treatment of human cancers, some anticancer activity was noticed in the aforementioned phase I studies. Two phase I studies conducted by Sharma and colleagues evaluated the effect of curcumin in patients with advanced colorectal cancer.[7,8] In the first study, *Curcuma* extract with the doses of curcumin containing 36–180 mg curcumin were administered daily for 4 months.[7] Five out of 15 patients had radiologically stable disease for more than 3 months. Another patient had a significant reduction of the carcinoembryonic antigen (CEA). In the subsequent phase I study using a 500-mg curcuminoid capsule (450 mg curcumin, 40 mg demethoxycurcumin, and 10 mg bisdemethoxycurcumin per capsule), 2 out of 15 patients had radiologically stable disease for up to 4 months of treatment.[8] Although these results might suggest a cytostatic effect of curcumin in some of the advanced colorectal cancers, confirmatory studies are needed.

3.2.2. Effect of Curcumin for Precancerous Lesions (Chemopreventive Effect)

In the phase I study conducted by Cheng and colleagues, patients with premalignant lesions were enrolled; and indicative lesions before and after curcumin treatment were biopsied and compared histologically.[6] Histological improvement was seen in one of two patients with recently -resected bladder cancer, two of seven patients with oral leukoplakia, one of six patients with intestinal metaplasia of the stomach, one of four patients with uterine CIN, and two of six patients with Bowen's disease. On the other hand, one patient with CIN and another patient with oral leukoplakia progressed to frank malignancies during the study period. These data were suggestive for a chemopreventive effect of curcumin, but large-scale prospective studies are needed for confirmation.

4. MISCELLANEOUS ACTIVITY OF CURCUMIN

Curcumin has also been evaluated as an antiviral agent for human immunodeficiency virus (HIV). However, in a 40-patient cohort, there was no evidence that curcumin reduced viral load or increased CD4 counts.[19]

The effect of curcumin in reducing the serum levels of cholesterol and lipid peroxides was studied in 10 healthy volunteers, who received 500 mg of curcumin per day for 7 days.[20] A significant decrease in the level of serum lipid peroxides (33%), an increase in high-density lipoprotein cholesterol (29%), and a decrease in total serum cholesterol (11.63%) were reported.

5. FUTURE CHALLENGE FOR THE CLINICAL DEVELOPMENT OF CURCUMIN

In the past few decades, a number of phase I and phase II studies of curcumin for different disease entities have been conducted. However, the true value of curcumin as a therapeutic agent for human diseases remains elusive. To advance our knowledge in the clinical application of curcumin, several issues related to clinical trials should be addressed.

5.1. Formulation Issues

The compositions and preparations of curcumin used in previous clinical studies were heterogenous, a situation that makes the comparison of the results from different studies difficult. It is therefore mandatory to use pure curcumin, manufactured by a standardized procedure, for future clinical studies.

A relatively poor oral bioavailability of curcumin has been noticed in all pharmacokinetic studies. This low oral bioavailability of curcumin would inevitably limit the application of this agent. Development of new formulations or analogues of curcumin with better bioavailability, as well as novel routes of administration, will be critical for future development of curcumin.

5.2. Trial Design

Much of the enthusiasm of curcumin has been focused on the treatment of inflammatory and neoplastic diseases. Unfortunately, the natural courses of these two human diseases are quite variable and make the results of most single-arm studies

Table 1. Examples of ongoing clinical trials with curcumin.

DISEASE	INSTITUTION	STUDY DESIGN	STARTING DATE
Multiple myeloma	M.D. Anderson Cancer Center, USA	Pilot study with two arms: curcumin vs. curcumin + bioperine	Nov. 2004
Pancreatic cancer	M.D. Anderson Cancer Center, USA	Phase II single-arm: curcumin	Nov. 2004
Pancreatic cancer	Rambam Medical Center, Israel	Phase II single-arm: curcumin + gemcitabine	July 2004
Familial adenomatous polyposis	Johns Hopkins University, USA	Phase II : curcumin	Nov. 2005
Sporadic adenomatous polyps, recently resected	University of Pennsylvania, USA	Phase II, placebo-controlled: curcuminoids	July 2005
Aberrant crypt foci in colon	University of Medicine and Dentistry New Jersey, USA	Phase II: curcuminoids vs. sulindac	Apr. 2004
Chronic psoriasis vulgaris	University of Pennsylvania, USA	Phase II, single-arm: curcuminoids	Oct. 2005
Alzheimer's disease	University of Pennsylvania, USA	Phase II, placebo-controlled: curcumin C3 complex	July 2003
Alzheimer's disease	Chinese University of Hong Kong, Hong Kong	Phase I/II, placebo-controlled: curcumin + ginkgo extract	Oct. 2004

not quite interpretable. Therefore, it is imperative that future clinical studies should try to include a control arm for comparison. Further, inclusion of surrogate end points such as relevant biomarkers, might also help in defining the efficacy of curcumin.

5.3. On-going Clinical Trials of Curcumin

Currently, curcumin is investigated as a therapeutic agent for various types of human cancer. Curcumin is also being studied in other diseases that are related to oxidative stress and chronic inflammation, such as Alzheimer's disease.[21] Table 1 is a selected list of ongoing clinical trials (for details, please refer to www.ClinicalTrials.gov).

6. CONCLUSIONS

As demonstrated in numerous preclinical and early-phase clinical studies, curcumin is safe and is expected to be effective in human diseases such as chronic inflammation and cancer. However, the results of clinical trials to date are all preliminary and not conclusive. It is imperative that well-designed clinical trials, supported by better formulations of curcumin or novel routes of administration, be conducted in the near future.

REFERENCES

1. K. L. Grant and C. D. Schneider, Tumeric. *Am J Health-Syst Pharm* **57**, 1121–1122 (2000).
2. J. K. Lin and S. Y. Lin-Shiau, Mechanisms of cancer chemoprevention by curcumin. *Proc Natl Sci Counc Repub China B.* **25**(2), 59–66 (2001).
3. B. B. Aggarwal, A. Kumar, and A. C. Bharti, Anticancer potential of curcumin: Preclinical and clinical studies. *Anticancer Res* **23**, 363–398.
4. B. Joe, M. Vijaykumar, and B. R. Lokesh, Biological properties of curcumin-cellular and molecular mechanisms of action. *Crit Rev Food Sci Nutr* **44**(2), 97–111 (2004).
5. R. A. Sharma, A. J. Gescher, and W. P. Steward, Curcumin: The story so far. *Eur J Cancer* **41**, 1995–1968 (2004).
6. Cheng, A. L., Hsu, C. H., Lin, J. K., Hsu, M. M., Ho, Y. F., Shen, T. S., Ko, J. Y., Lin, J. T., Lin, B. R., Wu, M. S., Yu, H. S., Jee, S. H., Chen, G. S., Chen, T. M., Chen, C. A., Lai, M. K., Pu, Y. S., Pan, M. H., Wang, Y. J., Tsai, C. C., and Hsieh, C. Y., 2001, Phase I clinical trial of curcumin, a chemopreventive agent, in patients with high–risk or pre–malignant lesions. *Anticancer Res* **21(4B)**, 2895–2900.
7. R. A. Sharma, H. R. McLelland, K. A. Hill, C. R. Ireson, S. A. Euden, M. M. Manson, M. Pirmohamed, L. J. Marnett, A. J. Gescher, and W. P. Steward, Pharmacodynamic and pharmacokinetic study of oral Curcuma extract in patients with colorectal cancer. *Clin Cancer Res* **7**(7), 1894–1900 (2001).
8. R. A. Sharma, S. A. Euden, S. L. Platton, D. N. Cooke, A. Shafayat, H. R. Hewitt, T. H. Marczylo, B. Morgan, D. Hemingway, S. M. Plummer, M. Pirmohamed, A. J. Gescher, and W. P. Steward, Phase I clinical trial of oral curcumin: Biomarkers of systemic activity and compliance. *Clin Cancer Res* **10**(20), 6847–6854 (2004).
9. C. D. Lao, M. T. Ruffin 4th, D. Normolle, D. D. Heath, S. I. Murray, J. M. Bailey, M. E. BoggsJ. Crowell, C. L. Rock, and D. E. Brenner, Dose escalation of a curcuminoid formulation. *BMC Complement Altern Med* **6**, 10–13 (2006).
10. G. Garcea, D. J. Jones, R. Singh, A. R. Dennison, P. B. Farmer, R. A. Sharma, W. P. Steward, A. J. Gescher, and D. P. Berry, Detection of curcumin and its metabolites in hepatic tissue and portal blood of patients following oral administration. *Br J Cancer* **90**(5), 1011–1015 (2004).
11. G. Garcea, D. P. Berry, D. J. Jones, R. Singh, A. R. Dennison, P. B. Farmer, R. A. Sharma, W. P. Steward, and A. J. Gescher, Consumption of the putative chemopreventive agent curcumin by cancer patients: Assessment of curcumin levels in the colorectum and their pharmacodynamic consequences. *Cancer Epidemiol Biomarkers Prev* **14**(1), 20–125 (2005).
12. S. D. Deodhar, R. Sethi, and R. C. Srimal, Preliminary study on antirheumatic activity of curcumin (diferuloyl methane). *Indian J Med Res* **71**, 632–634.
13. B. Lal, A. K. Kapoor, O. P. Asthana, P. K. Agrawal, R. Prasad, P. Kumar, and R. C. Srimal, Efficacy of curcumin in the management of chronic anterior uveitis. *Phytother Res* **13**(4), 318–322 (1999).

14. B. Lal, A. K. Kapoor, P. K. Agrawal, O. P. Asthana, and R. C. Srimal, Role of curcumin in idiopathic inflammatory orbital pseudotumours. *Phytother Res* **14**(6), 443–447 (2000).
15. P. R. Holt, S. Katz, and R. Kirshoff, Curcumin therapy in inflammatory bowel disease: A pilot study. *Dig Dis Sci* 50(11), 2191–2193 (2005).
16. S. Durgaprasad, C. G. Pai, Vasanthkumar, J. F. Alvres, and S. Namitha, A pilot study of the antioxidant effect of curcumin in tropical pancreatitis. *Indian J Med Res* **122**(4), 315–318 (2005).
17. M. C. Heng, M. K. Song, J. Harker, and M. K. Heng, Drug-induced suppression of phosphorylase kinase activity correlates with resolution of psoriasis as assessed by clinical, histological and immunohistochemical parameters. *Br J Dermatol* **143**(5), 937–949 (2000).
18. R. Kuttan, P. C. Sudheeran, and C. D. Josph, Turmeric and curcumin as topical agents in cancer therapy. *Tumori* **73**(1), 29–31 (1987).
19. J. S. James, Curcumin: Clinical trial finds no antiviral effect. *AIDS Treat News* **242**, 1–2 (1996).
20. K. B. Soni and R. Kuttan, Effect of oral curcumin administration on serum peroxides and cholesterol levels in human volunteers. *Indian J Physiol Pharmacol* **36**(4), 273–275 (1992).
21. J. M. Ringman, S. A. Frautschy, G. M. Cole, D. L. Masterman, and J. L. Cummings, A potential role of the curry spice curcumin in Alzheimer's disease. *Curr Alzheimer Res* **2**(2), 131–136 (2005).

INDEX

Printed in the USA